GEOTECHNICAL INSTRUMENTATION FOR MONITORING FIELD PERFORMANCE

GEOTECHNICAL INSTRUMENTATION FOR MONITORING FIELD PERFORMANCE

John Dunnicliff
Geotechnical Instrumentation Consultant
Lexington, Massachusetts

With the assistance of:

Gordon E. Green
Geotechnical Engineer
Seattle, Washington

A Wiley-Interscience Publication
John Wiley & Sons, Inc.
New York / Chichester / Brisbane / Toronto / Singapore

This text is printed on acid-free paper

Library of Congress Cataloging in Publication Data:

Dunnicliff, John.
Geotechnical instrumentation for monitoring field performance/
John Dunnicliff with the assistance of Gordon E. Green.
"A Wiley-Interscience publication."
ISBN 0-471-00546-0 (pbk)
1. Engineering geology—Instruments. 21. Soil—Testing.
I. Green, Gordon E. II. Title.
TA705.D86 1988
624.1'51'028—dc19 87-27709

10

To the manufacturers of
geotechnical instruments,
without whom there would be no
geotechnical instrumentation
for monitoring field performance

FOREWORD

Every geotechnical design is to some extent hypothetical, and every construction job involving earth or rock runs the risk of encountering surprises. These circumstances are the inevitable result of working with materials created by nature, often before the advent of human beings, by processes seldom resulting in uniform conditions. The inability of exploratory procedures to detect in advance all the possibly significant properties and conditions of natural materials requires the designer to make assumptions that may be at variance with reality and the constructor to choose equipment and construction procedures without full knowledge of what might be encountered.

Field observations, including quantitative measurements obtained by field instrumentation, provide the means by which the geotechnical engineer, in spite of these inherent limitations, can design a project to be safe and efficient, and the constructor can execute the work with safety and economy. Thus, field instrumentation is vital to the practice of geotechnics, in contrast to the practice of most other branches of engineering in which people have greater control over the materials with which they deal. For this reason geotechnical engineers, unlike their colleagues in other fields, must have more than casual knowledge of instrumentation: to them it is a working tool, not merely one of the components of research.

Notwithstanding its vital role, instrumentation is not an end in itself. It cannot guarantee good design or trouble-free construction. The wrong instruments in the wrong places provide information that may at best be confusing and at worst divert attention from telltale signs of trouble. Too much instrumentation is wasteful and may disillusion those who pay the bills, while too little, arising from a desire to save money, can be more than false economy: it can even be dangerous.

Every instrument installed on a project should be selected and placed to assist in answering a specific question. Following this simple rule is the key to successful field instrumentation. Unfortunately, it is easier to install instruments, collect the readings, and then wonder if there are any questions to which the results might provide an answer. Instrumentation is currently in vogue. Some design agencies and many regulatory bodies mandate instrumentation whether the results might be useful or not. It is a widely held dogma, for instance, that every earth dam should be instrumented, in the hope that some unsuspected defect will reveal itself in the observations and give warning of an impending failure. Part of the criticism directed at Teton Dam following its failure was paucity of instrumentation. Yet, it is extremely doubtful that any instrumental observations could have given timely warning of the particular failure that occurred. Instruments cannot cure defective designs, nor can they indicate signs of impending deterioration or failure unless, fortuitously, they happen to be of the right type and in the right place.

The engineer should bring the best knowledge and judgment to bear on every geotechnical problem that arises and should analyze the quality of the information on which a design is based. The engineer should judge not only the way the design will function if the information is essentially correct, but how the gaps or shortcomings might influence the performance of the project. Then, and only then, can specific items be identified that will reveal whether the project is performing in accordance with design assumptions or, if not, in what significant way the performance differs. Then the critical questions can be framed—the answers to which will fill the gaps or correct the errors in the original design assumptions—and the engineer can determine what instruments, at what locations, can answer those questions.

Of course, not all instruments are installed to monitor the safety of a structure or construction operation or to confirm design assumptions. Some are used to determine initial or background conditions. Observations of groundwater prior to construction, or in situ stresses in rock masses, or of elevations of structures before the start of adjacent construction are examples. Certain types of construction, such as the installation of tiebacks, are inherently dependent on instrumentation. Furthermore, advancements in the state of our knowledge require large-scale or full-scale observations of an extent and complexity far beyond the requirements of the practicing engineer. Yet, in all these applications, it is equally true that every instrument should be selected and located to assist in answering a specific question.

Instrumentation needs to be kept in perspective. It is one part of the broader activities of observation and surveillance. Trained people, using the best of all instruments, the human eye, can often provide all the information necessary and are always an essential part of the field observations on any project. Even when instruments are used because the necessary quantities are too small to be observed by eye, or the events are taking place out of the reach of a human being, the findings must be related to other activities. Without good records of the progress of excavation and of details of the excavation and bracing procedures, for example, the results of measurements of deformations or earth pressure associated with a braced cut become almost meaningless. Among the most valuable uses of instrumentation are empirical correlations between construction procedures and deformations or pressures, correlations that can be used immediately to improve the procedures so as to reduce the movements or pressures. Thus, highly sophisticated, fully automated installations for obtaining and presenting data, sometimes held in high favor by those intrigued with gadgetry, may fail to serve a useful purpose because the simple visual observations of what may be affecting the readings are overlooked.

Not only is instrumentation not an end in itself, but neither is sophistication or automation. The two prime requirements are sensitivity sufficient to provide the necessary information and reliability to ensure that dependable data can be obtained throughout the period when the observations are needed. Usually, the most dependable devices are the simplest.

If one can with sufficient accuracy make a direct visual observation with a graduated scale, then a micrometer should not be used. If one can use a micrometer, a mechanical strain gage should not be used. If one can use a mechanical strain gage, an electrical one should not be used. Mechanical instruments are to be preferred to electrical devices and simple electrical devices depending on simple circuits are to be preferred to more complex electronic equipment. That is, where a choice exists, the simpler equipment is likely to have the best chance for success.

Nevertheless, simple instruments are sometimes inappropriate and more complex ones must be used. An open standpipe may be the simplest device for observing a piezometric level, but the point at which the pore pressure needs to be measured may be located where direct access is impossible. A more sophisticated arrangement is then necessary. If one wishes to determine the state of stress in a mass of rock, there is no choice but to install sophisticated equipment to make measurements at a considerable distance from the position of the observer, and the strains likely to be observed will be too small to detect by any mechanical device. Thus, nothing but a sophisticated system will serve.

Not all sophisticated systems are equally reliable. Equipment that has an excellent record of performance can be rendered unreliable if a single essential but apparently minor requirement is overlooked during the installation. The best of instruction manuals cannot provide for every field condition that may affect the results. Therefore, even slavish attention to instructions cannot guarantee success. The installer must have a background in the fundamentals of geotechnics as well as knowledge of the intricacies of the device being installed. Sometimes the installer must consciously depart from the installation manual.

The installer must also want desperately to do the job well and must often work under difficult and unpleasant conditions, trying to do precision work while surrounded by workers whose teamwork or operation of equipment is being interrupted, or working the graveyard shift in an attempt to reduce such interruptions. Dedication of this sort is the price of success, and it is rarely found at the price tendered by the lowest bidder. Moreover, the installer can hardly be motivated to be dedicated to the task of installing instruments of inferior quality that are likely to fail prematurely or to produce questionable data. Rugged, reliable instruments are not necessarily expensive, but lowest cost of the hardware is rarely a valid reason for its choice. No arrangement for a program of instrumentation is a candidate for success if it sets cost above quality of instruments or fee above experience and dedication of the installer.

Instruments are discontinuities, nonrepresentative objects introduced into soil/rock structure systems. Their presence or the flows or displacements required to generate an observation alter the very quantities they are intended to measure. The alteration may be significant or negligible; its extent depends on the nature of the phenomenon being observed, on the design of the instrument, and on the operations required for installation. The engineer who embarks on a program of field instrumentation needs to understand the fundamental physics and mechanics involved and how the various available instruments will perform under the conditions to which they will be subjected. In addition, the engineer needs to know whether corrections can be made by calibration or by theoretical calculations, or whether under the circumstances no valid result is possible. Perhaps the classic examples of the latter eventuality were the attempts in the early 1920s to measure earth pressures against braced cuts by observing deflections of the wales between struts, a procedure made futile because of the then unsuspected phenomenon of arching. The same phenomenon affects to greater or lesser degree the results of all earth-pressure cells.

Finally, there must be enough instruments not only to allow for the inevitable losses resulting from malfunction and damage by construction activities, but also to provide a meaningful picture of the scatter in results inherent in geotechnics as a consequence of variations in geology and in construction procedures. For example, in the early days of construction of the Chicago subways, when the loads in the struts for bracing open cuts were measured, it was observed that struts at the same elevation in the same cut carried widely different loads. To some extent the difference was the result of slight differences in soil properties. To a much greater extent, however, the differences were associated with construction procedures. When a strut was placed as soon as possible, it always carried a load substantially greater than the load in another strut at the same elevation but not placed until excavation had advanced considerably beyond its future location. If the strut loads had been measured on only a few struts, for example, along a single vertical line, the influence of the construction procedure on strut loads could not have been detected, and entirely erroneous conclusions might have been drawn about the magnitude of the earth pressures resisted by the bracing. Since the variability of the quantities measured by geotechnical instrumentation depends not only on the kind of measurement itself—whether it be pore pressure, displacement, or load in a structural member—but also on the geology and on details of the construction procedures, the design of a system of measurements requires mature judgment based on experience and understanding of the geotechnical problems at hand.

Use of field instrumentation therefore requires a thorough grounding in geotechnical principles, a detailed conception of the variations that may be expected in the natural or artificial deposit in which the observations are to be made, a realistic notion of the construction procedures likely to be followed, a thorough knowledge of the capabilities and shortcomings of the instruments themselves, and an appreciation of the practical problems of installation. It also requires a clear perception of the way in which the results of the observations will be obtained, recorded, digested, and used on the particular project for which the design is being prepared. Small wonder that the need exists for a book dealing comprehensively with this subject.

RALPH B. PECK

PREFACE

This is intended to be a practical book for use by practitioners. There is information for all those who plan or implement geotechnical instrumentation programs: owners, project managers, geotechnical engineers, geologists, instrument manufacturers, specialty geotechnical contractors, civil engineers, and technicians. The book should also be helpful to students and faculty members during graduate courses in geotechnical engineering.

A practical book about geotechnical instrumentation must go beyond a mere summary of the technical literature and manufacturers' brochures: it must hold the hands of readers and guide them along the way. This need has created two difficulties for me.

First, my own practical experience is that of one person and does not arm me to write a comprehensive guide on my own. I have tried to fill this gap by drawing on the experience and opinions of many colleagues, who are identified elsewhere.

Second, it is certain that, soon after publication of this book, I will alter some of my opinions as my experience increases. I am well aware that the subject of geotechnical instrumentation is a contentious subject, made so by strongly held views among practitioners and by vested commercial interests. The guidelines in this book are an attempt to convey the "best ways" as I see them today. You, the reader, will have your own experience and your own best ways, which may differ from mine. I therefore have a plea: when you see possibilities for improving the content of this book, send me reasonable evidence. My address is in the Directory of the American Society of Civil Engineers. Not only will I learn from you, but I will try to disseminate the improvements, perhaps ultimately in a second edition of this book.

Length restrictions have strongly influenced the contents. There is no attempt to describe every instrument, either currently available or described in published papers. Some available instruments are not well suited to their intended purpose, and the literature abounds with descriptions of prototype gadgets that have found little real use in practice. In selecting information to be included, I have been guided by the title of the book, and thus there is nothing on geotechnical instrumentation for in situ measurement of soil and rock properties. Detailed case histories have been excluded in favor of guidelines directed toward the problem-oriented reader. However, summaries of selected case histories are included in Part 5.

Finally, a few words about the organization of the book and how it may be used. The book is divided into seven parts, each with a self-explanatory title. Readers looking for an overview may start with the Foreword and Chapter 1, then scan through Chapter 26, *The Key to Success*. In my view, the greatest shortcoming in the state-of-the-practice is inadequate planning of monitoring programs, and therefore problem-oriented readers should give their first concentrated attention to Chapter 4, *Systematic Approach to Planning Monitoring Programs Using Geotechnical Instrumentation*. The various steps in this chapter lead readers to each of the chapters in Parts 2, 3, and 4: Chapter 4 is therefore the hub of the book. The chapters in Part 5, *Examples of Instrumentation Applications*, are intended as supplementary chapters to open the minds of readers to the possible role of geotechnical instrumentation on various types of construction projects and to guide them toward implementation. They are not intended as exhaustive summaries, state-of-the-art papers, or "cookbooks." If a reader uses this book by (1) turning to the chapter in Part 5 that discusses his or her type of project, (2) noting the types of instruments suggested in Part 5, (3) noting the sketched layouts in Part 5, (4) studying Part 3, *Monitoring Methods*, for details of the instruments, and (5) proceeding with a monitoring program, that reader is misusing the book. Turn back to Chapter 4!

John Dunnicliff

ACKNOWLEDGMENTS

I would not have been able to write this book without the help of many people.

First, Gordon E. Green, Associate, Golder Associates, Inc., Seattle, WA (formerly with Shannon & Wilson, Inc., Seattle, WA), has made an extraordinary commitment of time and enthusiasm to this book. He has given many hours to guide me with content and format, to provide a second opinion on numerous issues of judgment, and to make detailed reviews of the chapters in Parts 2, 3, and 4. For this untiring dedication, I can only express my great gratitude. No author could hope for more.

Second, some colleagues have coauthored or assisted with sections or chapters. When I felt that my own experience of a subject has been too limited, I asked one or more engineers to help me with the text. Their names and affiliations are given in footnotes on appropriate pages. To these colleagues I say, thank you—for enduring my persistent questioning and for helping me to convert my shaky drafts into texts fit for a book. Your roles have been crucial.

Third, some colleagues have helped me by providing facts or opinions on a multitude of subjects. These include the following:

Arild Andresen
William R. Beloff
Douglas J. Belshaw
Bradford P. Boisen
Jean-Louis Bordes
Jean Boucher
Michael Bozozuk
Ed Brylawski
Georgi A. Buckley
Thomas G. Bumala
Roy W. Carlson
Pierre Carrier
David J. Clements
J. Barrie Cooke
Christopher B. H. Cragg
J. Clive P. Dalton
Richard R. Davidson
Brian J. Dawes
George B. Deardorff
Elmo B. DiBiagio
Walter Dieden
Herbert J. Dix
James Dorsey
Edward J. Drelich
Charles N. Easton
Alex I. Feldman
Bengt H. Fellenius
Rainer Glötzl
Charles W. Hancock, Jr.
Leo D. Handfelt
Richard K. Harris
David G. F. Hedley
Anwar Hirany
Robert D. Holtz
Gary R. Holzhausen
Robert G. Horvath
Bob Joy
Kunsoo Kim
Peter Lang
Pierre LeFrancois
Jeffrey M. Lingham
G. Stuart Littlejohn
Thomas K. Liu
C. Leroy McAnear
Verne C. McGuffey
John B. McRae
P. Erik Mikkelsen
Anthony Minnitti
Dewayne L. Misterek
Ian Mitchell
John G. Morrison
Michael W. O'Neill
Walter Nold
Ralph B. Peck
Arthur D. M. Penman
Edward C. Pritchett
Red Robinson
Arthur Ross
Birger Schmidt
Ernest T. Selig
Dale Shoup
Tony Simmonds
Patrick D. K. Smith

Terry Stevens
Kalman Szalay
Duncan Tharp
Petur Thordarson
Arno Thut
Bengt-Arne Torstensson
Peter R. Vaughan
James Warner
Robert C. Weeks
William A. Weiler, Jr.
Clark Welden
James R. Wheeler
Stanley D. Wilson
Anwar E. Z. Wissa
Stephen P. Wnuk
John R. Wolosick

To these colleagues—thank you for passing on your experience and for guiding me as I tried to resolve so many uncertainties.

Fourth, I thank Judy Grande and Sarah Matthews, whose word processing skills and responsiveness to deadlines have been outstanding.

Last, but never least, my children Christopher, Jonathan, and Tanya. Their tolerance and understanding during the 5 years of writing have made it all possible.

CONTENTS

PART 1 INTRODUCTION

Chapter 1 Geotechnical Instrumentation: An Overview / 3

1.1 What Is Geotechnical Instrumentation?, 3
1.2 Why Do We Need to Monitor Field Performance?, 3
1.3 What Capabilities Must the People Have?, 5
1.4 What Capabilities Must the Instruments Have?, 5
1.5 Where Have We Been?, 5
1.6 Where Are We Now?, 5
1.7 Where Are We Going?, 10
1.8 The Key to Success, 12

Chapter 2 Behavior of Soil and Rock / 13

2.1 Behavior of Soil, 13
2.2 Behavior of Rock, 23

PART 2 PLANNING MONITORING PROGRAMS

Chapter 3 Benefits of Using Geotechnical Instrumentation / 33

3.1 Benefits During Design, 33
3.2 Benefits During Construction, 34
3.3 Benefits After Construction Is Complete, 36
3.4 General Considerations, 36

Chapter 4 Systematic Approach to Planning Monitoring Programs Using Geotechnical Instrumentation / 37

4.1 Define the Project Conditions, 37
4.2 Predict Mechanisms that Control Behavior, 38
4.3 Define the Geotechnical Questions that Need to Be Answered, 38
4.4 Define the Purpose of the Instrumentation, 38
4.5 Select the Parameters to Be Monitored, 38
4.6 Predict Magnitudes of Change, 38
4.7 Devise Remedial Action, 39
4.8 Assign Tasks for Design, Construction, and Operation Phases, 39
4.9 Select Instruments, 40
4.10 Select Instrument Locations, 42

4.11 Plan Recording of Factors that May Influence Measured Data, 43
4.12 Establish Procedures for Ensuring Reading Correctness, 43
4.13 List the Specific Purpose of Each Instrument, 43
4.14 Prepare Budget, 43
4.15 Write Instrument Procurement Specifications, 44
4.16 Plan Installation, 44
4.17 Plan Regular Calibration and Maintenance, 44
4.18 Plan Data Collection, Processing, Presentation, Interpretation, Reporting, and Implementation, 44
4.19 Write Contractual Arrangements for Field Instrumentation Services, 44
4.20 Update Budget, 44

Chapter 5 Specifications for Procurement of Instruments / 45

5.1 Task Assignment for Procurement, 45
5.2 Specifying Method, 46
5.3 Basis for Determining Price, 47
5.4 Content of Specifications for Procurement of Instruments, 49

Chapter 6 Contractual Arrangements for Field Instrumentation Services / 57

6.1 Goals of Contractual Arrangements, 57
6.2 Definition of Terms, 57
6.3 Contractual Arrangements for Instrument Installation, 59
6.4 Contractual Arrangements for Regular Calibration and Maintenance, 61
6.5 Contractual Arrangements for Data Collection, Processing, Presentation, Interpretation, and Reporting, 61
6.6 Content of Specifications for Field Instrumentation Services, 63

PART 3 MONITORING METHODS

Chapter 7 Measurement Uncertainty / 75

7.1 Conformance, 75
7.2 Accuracy, 75
7.3 Precision, 75
7.4 Resolution, 76
7.5 Sensitivity, 76
7.6 Linearity, 76
7.7 Hysteresis, 76
7.8 Noise, 77
7.9 Error, 77

Chapter 8 Instrumentation Transducers and Data Acquisition Systems / 79

8.1 Mechanical Instruments, 79
8.2 Hydraulic Instruments, 80
8.3 Pneumatic Instruments, 87
8.4 Electrical Instruments, 92

Chapter 9 Measurement of Groundwater Pressure / 117

9.1 Instrument Categories and Applications, 117
9.2 Observation Wells, 118
9.3 Open Standpipe Piezometers, 118
9.4 Twin-Tube Hydraulic Piezometers, 123
9.5 Pneumatic Piezometers, 126
9.6 Vibrating Wire Piezometers, 127
9.7 Electrical Resistance Piezometers, 128
9.8 Miscellaneous Single-Point Piezometers, 130
9.9 Multipoint Piezometers, 136
9.10 Hydrodynamic Time Lag, 139
9.11 Types of Filter, 141
9.12 Recommended Instruments for Measuring Groundwater Pressure in Saturated Soil and Rock, 141
9.13 Recommended Instruments for Measuring Pore Water Pressure in Unsaturated Soil, 144
9.14 Saturation of Filters, 146
9.15 Installation of Piezometers in Fill, 148
9.16 Installation of Piezometers by the Push-in Method, 148
9.17 Installation of Piezometers in Boreholes in Soil, 150
9.18 Installation of Piezometers in Boreholes in Rock, 163

Chapter 10 Measurement of Total Stress in Soil / 165

10.1 Instrument Categories and Applications, 165
10.2 Embedment Earth Pressure Cells, 165
10.3 Contact Earth Pressure Cells, 177

Chapter 11 Measurement of Stress Change in Rock / 185

11.1 Applications, 185
11.2 Instrument Categories, 185
11.3 Soft Inclusion Gages, 186
11.4 Rigid Inclusion Gages, 191
11.5 Recommended Procedures for Measurement of Stress Change in Rock, 195

Chapter 12 Measurement of Deformation / 199

12.1 Instrument Categories, 199
12.2 Surveying Methods, 199
12.3 Surface Extensometers, 209
12.4 Tiltmeters, 216
12.5 Probe Extensometers, 219
12.6 Fixed Embankment Extensometers, 233
12.7 Fixed Borehole Extensometers, 237
12.8 Inclinometers, 250
12.9 Transverse Deformation Gages, 268
12.10 Liquid Level Gages, 275
12.11 Miscellaneous Deformation Gages, 292

Chapter 13 Measurement of Load and Strain in Structural Members / 297

13.1 Instrument Categories and Applications, 297
13.2 Load Cells, 297

13.3 Surface-Mounted Strain Gages, 306
13.4 Embedment Strain Gages, 320
13.5 Determination of Existing Stress, 326
13.6 Concrete Stress Cells, 327

Chapter 14 Measurement of Temperature / 331
14.1 Applications, 331
14.2 Mercury Thermometer, 332
14.3 Bimetal Thermometer, 332
14.4 Thermistor, 333
14.5 Thermocouple, 333
14.6 Resistance Temperature Device (RTD), 334
14.7 Frost Gages, 335
14.8 Other Transducers for Measurement of Temperature, 336
14.9 Comparison Among Transducers for Remote Measurements, 336
14.10 Installation of Transducers for Measurement of Temperature, 338

PART 4 GENERAL GUIDELINES ON THE EXECUTION OF MONITORING PROGRAMS

Chapter 15 A Recipe for Reliability of Performance Monitoring / 341
15.1 Instrument Ingredients in a Recipe for Reliability, 341
15.2 People Ingredients in a Recipe for Reliability, 342

Chapter 16 Calibration and Maintenance of Instruments / 343
16.1 Instrument Calibration, 343
16.2 Instrument Maintenance, 345

Chapter 17 Installation of Instruments / 347
17.1 Contractual Arrangements for Installing Instruments, 347
17.2 Locations of Instruments, 347
17.3 Detailed Installation Procedures, 348
17.4 Installation at the Ground Surface, 348
17.5 Installation in Boreholes, 348
17.6 Installation in Fill, 358
17.7 Installation in Underground Excavations, 363
17.8 Protection from Damage, 363
17.9 Acceptance Tests, 364
17.10 Installation Records, 364
17.11 Installation Schedule, 364
17.12 Coordination of Installation Plans, 365
17.13 Field Work, 365
17.14 Installation Report, 366

Chapter 18 Collection, Processing, Presentation, Interpretation, and Reporting of Instrumentation Data / 367
18.1 Collection of Instrumentation Data, 367
18.2 Processing and Presentation of Instrumentation Data, 374
18.3 Interpretation of Instrumentation Data, 382
18.4 Reporting of Conclusions, 384

PART 5 EXAMPLES OF INSTRUMENTATION APPLICATIONS

Chapter 19 Braced Excavations / 389

19.1 General Role of Instrumentation, 389
19.2 Principal Geotechnical Questions, 390
19.3 Overview of Routine and Special Applications, 400
19.4 Selected Case Histories, 400

Chapter 20 Embankments on Soft Ground / 407

20.1 General Role of Instrumentation, 407
20.2 Principal Geotechnical Questions, 407
20.3 Overview of Routine and Special Applications, 410
20.4 Selected Case Histories, 410

Chapter 21 Embankment Dams / 417

21.1 General Role of Instrumentation, 417
21.2 Principal Geotechnical Questions, 418
21.3 Long-Term Performance Monitoring of Embankment Dams, 423
21.4 General Guidelines on the Execution of Monitoring Programs for Embankment Dams, 432
21.5 Selected Case Histories, 435

Chapter 22 Excavated and Natural Slopes / 443

22.1 General Role of Instrumentation, 443
22.2 Principal Geotechnical Questions, 443
22.3 Overview of Routine and Special Applications, 448
22.4 Selected Case Histories, 448

Chapter 23 Underground Excavations / 453

23.1 General Role of Instrumentation, 453
23.2 Principal Geotechnical Questions, 455
23.3 Overview of Routine and Special Applications, 461
23.4 Selected Case Histories, 461

Chapter 24 Driven Piles / 467

24.1 General Role of Instrumentation, 467
24.2 Principal Geotechnical Questions, 467
24.3 Overview of Routine and Special Applications, 479
24.4 Selected Case Histories, 479

Chapter 25 Drilled Shafts / 483

25.1 General Role of Instrumentation, 483
25.2 Principal Geotechnical Questions, 483
25.3 Overview of Routine and Special Applications, 489
25.4 Selected Case Histories, 489

PART 6 THE KEY TO SUCCESS

Chapter 26 The Key to Success: The Chain with 25 Links / 493

PART 7 APPENDIXES

A. Checklist for Planning Steps / 501
B. Checklist for Content of Specifications for Procurement of Instruments / 505
C. Checklist for Content of Specifications for Field Instrumentation Services / 507
D. Commercially Available Geotechnical Instruments / 511
E. Details of Twin-Tube Hydraulic Piezometer System / 519
F. Dimensions of Drill Rods, Flush-Joint Casing, Diamond Coring Bits, Hollow-Stem Augers, and U.S. Pipe / 527
G. Example of Installation Procedure, with Materials and Equipment List / 533
H. Conversion Factors / 539
References / 541
Index / 563

GEOTECHNICAL INSTRUMENTATION FOR MONITORING FIELD PERFORMANCE

Every instrument on a project should be
selected and placed to assist with answering
a specific question: if there is no question,
there should be no instrumentation.

Part 1

Introduction

Part 1 is intended to serve as a general introduction. Chapter 1 sets the stage for the book, describing the role of geotechnical instrumentation and giving a historical perspective and a look into the future. It is hoped that Chapter 1 will motivate the reader toward a deeper study of the subject. Chapter 2 presents an overview of key aspects of soil and rock behavior, targeted for the practitioners who become involved with geotechnical instrumentation programs and who do not have formal training in soil or rock mechanics.

CHAPTER 1

GEOTECHNICAL INSTRUMENTATION: AN OVERVIEW

1.1. WHAT IS GEOTECHNICAL INSTRUMENTATION?

The engineering practice of geotechnical instrumentation involves a marriage between the capabilities of measuring instruments and the capabilities of people.

There are two general categories of measuring instruments. The first category is used for in situ determination of soil or rock properties, for example, strength, compressibility, and permeability, normally during the design phase of a project. Examples are shown in Figure 1.1. The second category is used for monitoring performance, normally during the construction or operation phase of a project, and may involve measurement of groundwater pressure, total stress, deformation, load, or strain. Examples are shown in Figure 1.2. This book is concerned only with the second category.

During the past few decades, manufacturers of geotechnical instrumentation have developed a large assortment of valuable and versatile products for the monitoring of geotechnically related parameters. Those unfamiliar with instrumentation might believe that obtaining needed information entails nothing more than pulling an instrument from a shelf, installing it, and taking readings. Although successful utilization may at first appear simple and straightforward, considerable engineering and planning are required to obtain the desired end results. The use of geotechnical instrumentation is not merely the selection of instruments but a comprehensive step-by-step engineering process beginning with a definition of the objective and ending with implementation of the data. Each step is critical to the success or failure of the entire program, and the engineering process involves combining the capabilities of instruments and people.

1.2. WHY DO WE NEED TO MONITOR FIELD PERFORMANCE?

The term *geotechnical construction* can be used for construction requiring consideration of the engineering properties of soil or rock. In the design of a surface facility, the ability of the ground to support the construction must be considered. In the design of a subsurface facility, consideration must also be given to the ability of the ground to support itself or be supported by other means. In both cases, the engineering properties of the soil or rock are the factors of interest. The designer of geotechnical construction works with a wide variety of naturally occurring heterogeneous materials, which may be altered to make them more suitable, but exact numerical values of their engineering

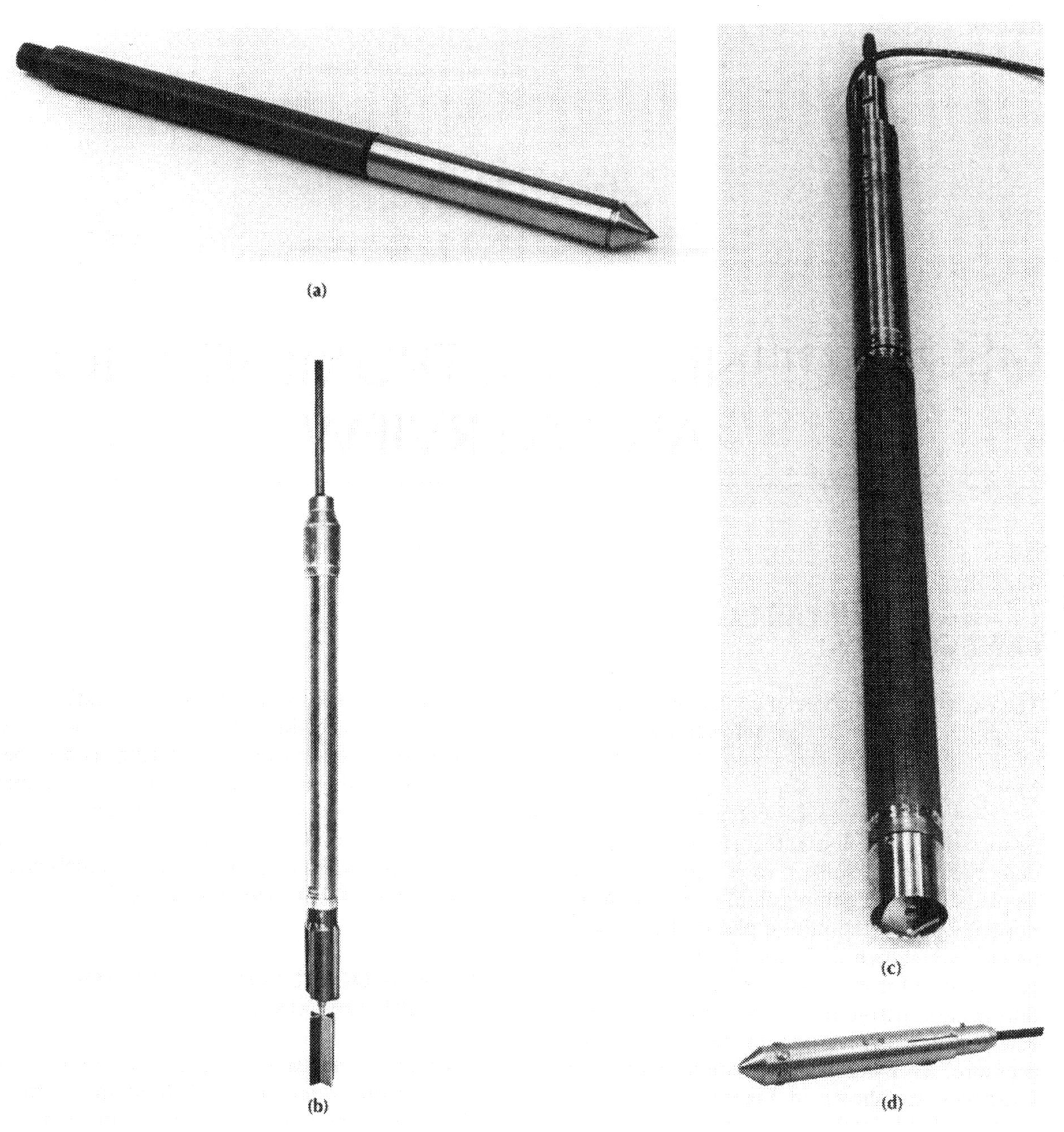

Figure 1.1. Examples of measuring instruments for in situ determination of soil or rock properties: (a) Piezocone: combined static cone and pore pressure probe (courtesy of Geotechniques International, Inc., Middleton, MA); (b) vane shear equipment (courtesy of Geonor A/S, Oslo, Norway); (c) self-boring pressuremeter (courtesy of Cambridge Insitu, Cambridge, England); and (d) borehole deformation gage (courtesy of Geokon, Inc., Lebanon, NH).

properties cannot be assigned. Laboratory or field tests may be performed on selected samples to obtain values for engineering properties, but these tests will only provide a range of possible values.

The significance of these statements about geotechnical construction can be demonstrated by comparison with steel construction. A designer of a steel structure works with manufactured materials. The materials are specified, their manufacture is controlled, and fairly exact numerical values of engineering properties are available for design. An accurate analysis can be made and design plans and specifications prepared. Then, provided construction is in accordance with those plans, the structure will perform as designed. There will generally be no need to monitor field performance. Similar remarks apply to reinforced concrete. In contrast, the design of geotechnical construction will be based on judgment in selecting the most probable values within the ranges of possible values for engineering properties. As construction progresses and geotechnical conditions are observed or behavior monitored, the design judgments can be evaluated and, if necessary, updated. Thus, engineering observations during geotechnical construction are often an integral part of the design process, and geotechnical instrumentation is a tool to assist with these observations.

1.3. WHAT CAPABILITIES MUST THE PEOPLE HAVE?

Basic capabilities required for instrumentation personnel are reliability and patience, perseverance, a background in the fundamentals of geotechnical engineering, mechanical and electrical ability, attention to detail, and a high degree of motivation.

1.4. WHAT CAPABILITIES MUST THE INSTRUMENTS HAVE?

Reliability is the overriding desirable capability for instruments. Inherent in reliability is maximum simplicity, and in general the order of decreasing simplicity and reliability is optical, mechanical, hydraulic, pneumatic, electrical. Also inherent in reliability is maximum quality. Lowest cost of an instrument is rarely a valid reason for its choice, and unless high quality can be specified adequately, instrument procurement on a low-bid basis will remain a stumbling block to good field performance.

1.5. WHERE HAVE WE BEEN?

Figures 1.3–1.15 show examples of past uses of geotechnical instrumentation.

The birth of geotechnical instrumentation, as a tool to assist with field observations, occurred in the 1930s and 1940s. During the first 50 years of its life, a general trend can be observed. In the early years, simple mechanical and hydraulic instruments predominated, and most instrumentation programs were in the hands of diligent engineers who had a clear sense of purpose and the motivation to make the programs succeed. There were successes and failures, but the marriage between instruments and people was generally sound. In more recent years, as technology has advanced and the role of geotechnical instrumentation has become more secure, more complex devices with electrical and pneumatic transducers have become commonplace. Some of these devices have performed well, while others have not. At the same time, the technology has attracted an increasingly large proportion of the geotechnical engineering profession, and an increasing number of instrumentation programs have been in the hands of people with incomplete motivation and sense of purpose. There have continued to be successes and failures but, in contrast to the early years, a significant number of the failures can be attributed to an unsound marriage between instruments and people.

1.6. WHERE ARE WE NOW?

The state of the art of instrument design is now far ahead of the state of the practice by users, and many more imperfections in current instrumentation programs result from user-caused people problems rather than from manufacturer-caused instrument problems. As users we are fortunate in having access to such a wide variety of good instruments. It is **our** responsibility to develop an adequate level of understanding of the instruments that we select and to maximize the quality of our own work if we are to take full advantage of instrumentation technology. The greatest shortcoming in the state of the practice is failure to plan monitoring programs in a

Figure 1.2. Examples of measuring instruments for monitoring field performance: (a) twin-tube hydraulic piezometer (courtesy of Geotechnical Instruments (U.K.) Ltd., Leamington Spa, England); (b) vibrating wire piezometer (courtesy of Telemac, Asnières, France); (c) vibrating wire stressmeter (courtesy of Geokon, Inc., Lebanon, NH); (d) load cell (courtesy of Proceq SA, Zürich, Switzerland); (e) embedment earth pressure cell (courtesy of Thor International, Inc., Seattle, WA); (f) surface-mounted vibrating wire strain gage (courtesy of Irad Gage, a Division of Klein Associates, Inc., Salem, NH); (g) multipoint fixed borehole extensometer (courtesy of Soil Instruments Ltd., Uckfield, England); and (h) inclinometer (courtesy of Slope Indicator Company, Seattle, WA).

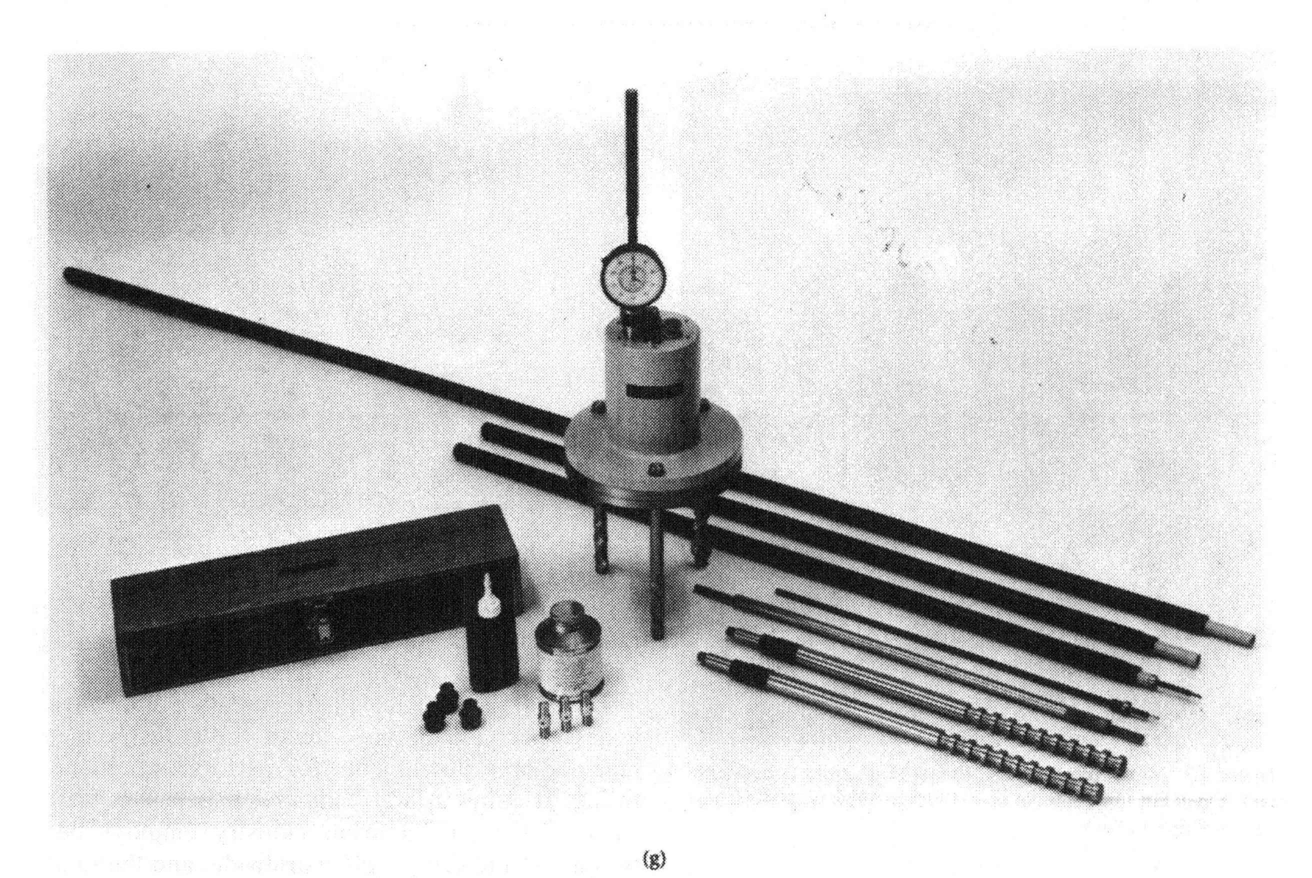

(g)

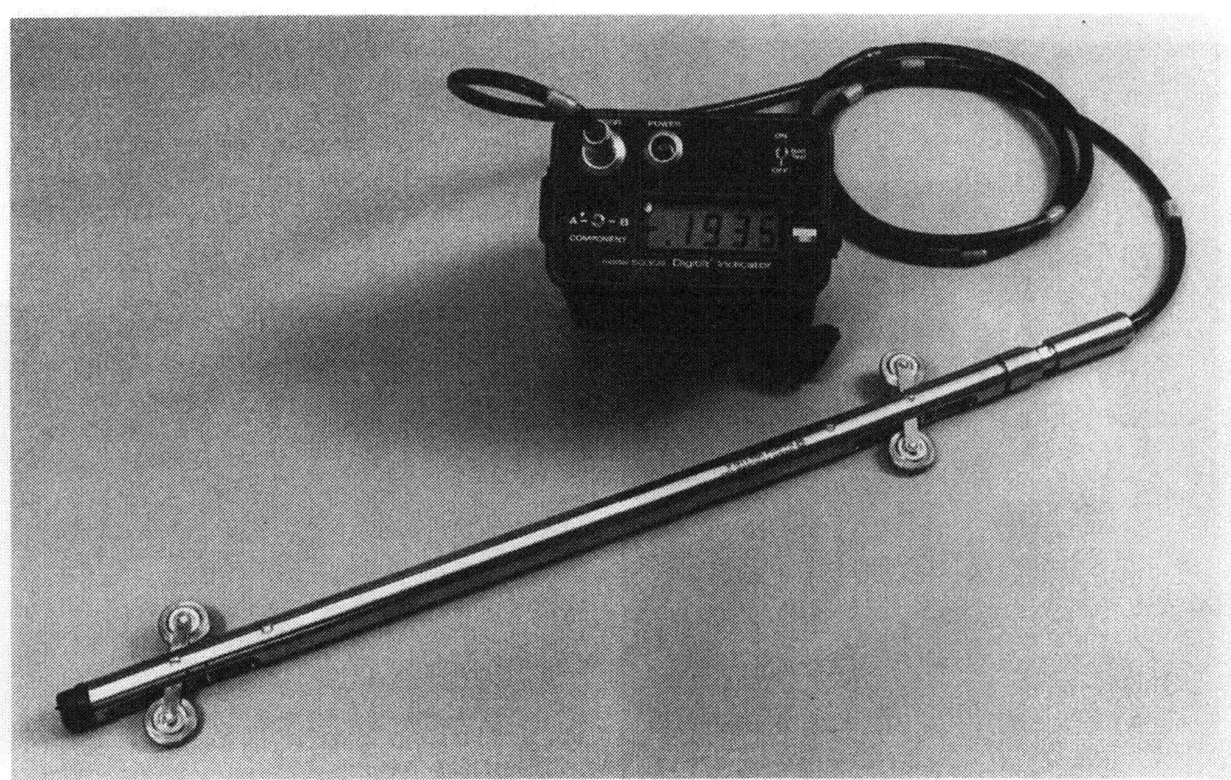

(h)

Figure 1.3. Measuring load in a timber strut, using a hydraulic jack. Open cut for station in clay. Chicago Subway, 1940 (courtesy of Ralph B. Peck).

Figure 1.4. Determination of load in a steel strut, using a mechanical strain gage. Open cut for station in clay. Chicago Subway, 1948 (courtesy of Ralph B. Peck).

Figure 1.5. Installing twin-tube hydraulic piezometers. Usk Dam, England, 1952 (courtesy of Arthur D. M. Penman).

rational and systematic manner, and therefore planning procedures are emphasized in this book.

Users of geotechnical instrumentation often have a misconception of the size of the industry that manufactures instruments for performance monitoring. It is **not** a large industry: it is in fact very small. The manufacturing industry employs between 300 and 400 people worldwide, and the total annual volume of sales is about 30 million U.S. dollars. This misconception sometimes leads to unreasonable expectations on the part of users: we cannot expect the manufacturers to make large expenditures on research, development, and testing of special instruments, unless justified by the size of the market. If the market is small, special funding is needed.

Figure 1.6. Installing fixed embankment extensometer with vibrating wire transducer. Balderhead Dam, England, 1963 (courtesy of Arthur D. M. Penman).

Figure 1.7. Manometer panels for twin-tube hydraulic piezometers. Plover Cove Main Dam, Hong Kong, 1965 (after Dunnicliff, 1968). Reprinted by permission of Institution of Civil Engineers, London.

Figure 1.8. Installing fixed embankment extensometers in embankment dam. Ludington Pumped Storage Project, Ludington, MI, 1972.

This book includes chapters that describe available methods for monitoring various geotechnically related parameters. The following is a summary rating of our current ability to obtain reliable measurements of these parameters, in order of increasing reliability:

- Total stress in soil and stress change in rock
- Groundwater pressure
- Load and strain in structural members
- Deformation
- Temperature

Figure 1.9. Bonded resistance strain gages on segmented steel liner for soft ground tunnel. Port Richmond Water Pollution Control Project, Staten Island, NY, 1974.

Figure 1.10. Installing multipoint fixed borehole extensometer above alignment of rock tunnel. East 63rd Street Subway, New York, NY, 1976.

Figure 1.11. Pneumatic piezometer and earth pressure cell on opposite faces of precast concrete pile, prior to concreting. Keehi Interchange, Honolulu, HI, 1977.

Reliability is strongly influenced by the extent to which measurements are dependent on local characteristics of the zone in which the instruments are installed. Most measurements of pressure, stress, load, strain, and temperature are influenced by conditions within a very small zone and are therefore dependent on local characteristics of that zone. They are often essentially *point* measurements, subject to any variability in geologic or other characteristics, and may therefore not represent conditions on a larger scale. When this is the case, a large number of measurement points may be required before confidence can be placed in the data. On the other hand, many deformation measuring devices respond to movements within a large and representative zone. Data provided by a single instrument can therefore be meaningful, and deformation measurements are generally the most reliable and least ambiguous.

1.7. WHERE ARE WE GOING?

As we look ahead, there is no reason to believe in a decreasing role for geotechnical instrumentation. Geotechnical design and construction will always be subject to uncertainties, and instrumentation will continue to be an important item in our tool box. However, several current trends can be identified, each of which will continue in the future and change the state of the practice.

First, there is the advent of automatic data acquisition systems and computerized data processing and presentation procedures. Clearly, these systems and procedures have many advantages, yet we must remain aware of their limitations. No automatic system can replace engineering judgment. When automatic data acquisition systems are used,

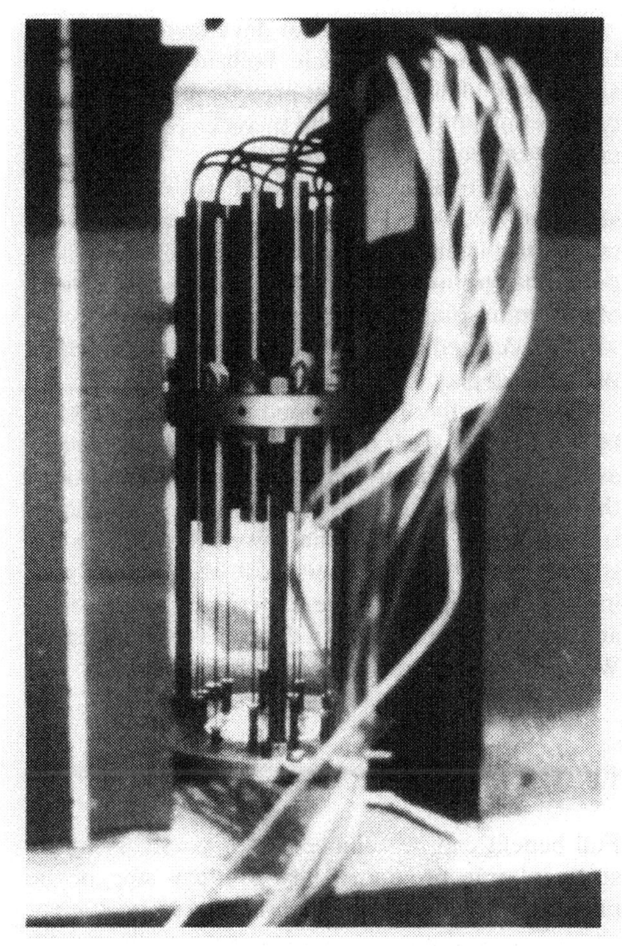

Figure 1.12. Electrical transducers for monitoring movement of multiple telltales during load test of precast concrete pile. Keehi Interchange, Honolulu, HI, 1977.

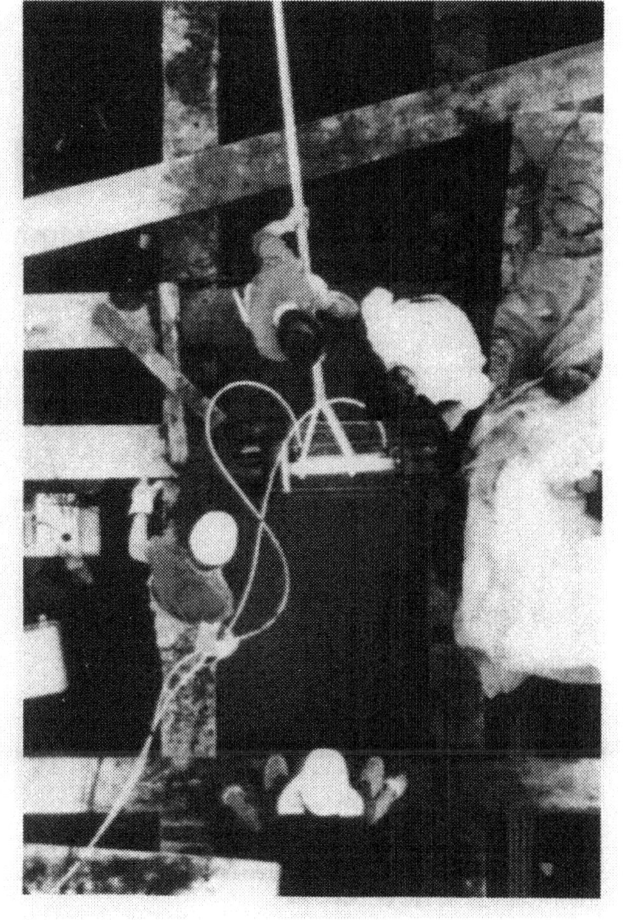

Figure 1.13. Installing gage with induction coil transducers, for monitoring convergence of slurry trench test panel in soft clay, Alewife Station, Cambridge, MA, 1978.

there is a real possibility that visual observations will not be made, that other factors influencing measured data will not be recorded, and that information will therefore not be available for relating measured effects to their likely causes. When computerized data processing and presentation procedures are used, there is a real possibility that engineering judgment will be given second place and that correlations between causes and effects will not be made. We should take all possible advantage of this exciting new technology but should never forget that judgment plays an important and often overriding role in the practice of geotechnical engineering.

Second, increasing labor costs in many countries have sharply reduced the availability of competent personnel. This trend of course encourages the use of automatic systems and procedures, yet reduces

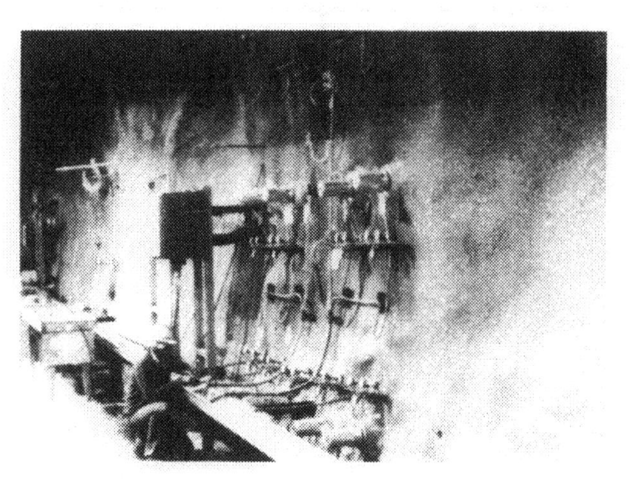

Figure 1.14. Heads of multipoint fixed borehole extensometers, installed to monitor rock movements during full-scale heater test for studies relating to disposal of high-level nuclear waste. Basalt Waste Isolation Program, Hanford, WA, 1980 (courtesy of Department of Energy).

Figure 1.15. Reading inclinometer during construction of underwater test fill. Chek Lap Kok Replacement Airport, Hong Kong, 1981 (courtesy of Leo D. Handfelt).

the number of personnel available for exercising engineering judgment.

Third, the use of design tools such as finite and boundary element modeling techniques leads to a need for field verification. Geotechnical instrumentation is likely to play an increasing role in providing a check on these advanced analytic predictions.

Fourth, there is a trend to develop and improve transducers and to include built-in features that create redundancy and that provide direct output in engineering units. The trend includes provisions for calibrations and zero checks.

Fifth, there is a trend toward use of new construction methods. Examples of innovations in the recent past include earth reinforcement, lateral support, and ground modification. These innovations often require field verification before they become widely accepted, and geotechnical instrumentation will always play a role.

The above five trends, good or bad, are inevitable. There is a sixth trend, which the author views as wholly bad: the procurement of instruments and the awarding of field instrumentation service contracts on the basis of the lowest bid. If an instrumentation program sets cost above quality of instruments, or fee above experience, dedication, and motivation of people, it deserves to be a failure. We must work hard to reverse this trend.

1.8. THE KEY TO SUCCESS

Full benefit can be achieved from geotechnical instrumentation programs only if every step in the planning and execution process is taken with great care. The analogy can be drawn to a chain with many potential weak links: this chain breaks down with greater facility and frequency than in most other geotechnical engineering endeavors. The links in the chain are defined in Chapter 26: their strength depends both on the capabilities of measuring instruments and the capabilities of people. The success of performance monitoring will be maximized by maximizing the strength of each link.

CHAPTER 2

BEHAVIOR OF SOIL AND ROCK

Many practitioners who become involved with geotechnical instrumentation programs do not have formal training in soil or rock mechanics. The purpose of this chapter is to present a brief and simple overview of the key aspects of soil and rock behavior that relate to the use of geotechnical instrumentation. For a thorough treatment of soil behavior readers are referred to Holtz and Kovacs (1981) and Terzaghi and Peck (1967). McCarthy (1977) presents similar material oriented for students at technical colleges. Rock behavior is well described by Blyth and DeFreitas (1974) and Franklin (1988).

2.1. BEHAVIOR OF SOIL

2.1.1. Constituents of Soil

Soil is composed of solid particles with intervening spaces. As shown in Figure 2.1, the particles are referred to as the *mineral skeleton* and the spaces as *pore spaces, pores,* or *voids*. The pore spaces are usually filled with air and/or water. A soil in which the pore spaces are completely filled with water is called a *saturated soil*. If any gas is present in the pore spaces, the soil is called an *unsaturated soil*. The term *partially saturated* is sometimes used, but because *it's either saturated or it isn't,* this is not a satisfactory term.

2.1.2. Basic Types of Soil

Soil can be categorized into two broad groups: *cohesionless soil* and *cohesive soil*. Cohesionless soils include sand and gravel, which consist of fragments of rocks or minerals that have not been altered by chemical decomposition. Inorganic silt is a fine-grained soil with little or no plasticity and can generally be classified as cohesionless. Organic silt is a fine-grained soil with an admixture of organic particles and behaves as a plastic cohesive soil. Clay is a cohesive soil consisting of microscopic and submicroscopic particles derived from the chemical decomposition of rock constituents.

2.1.3. Stress and Pressure

Stress and *pressure* are defined as force per unit area, with typical units of pounds per square inch (lb/in.^{2}) or pascals (Pa). Strictly, pressure is a general term meaning force per unit area, and stress is the force per unit area that exists *within* a mass. However, in geotechnical engineering, the terms

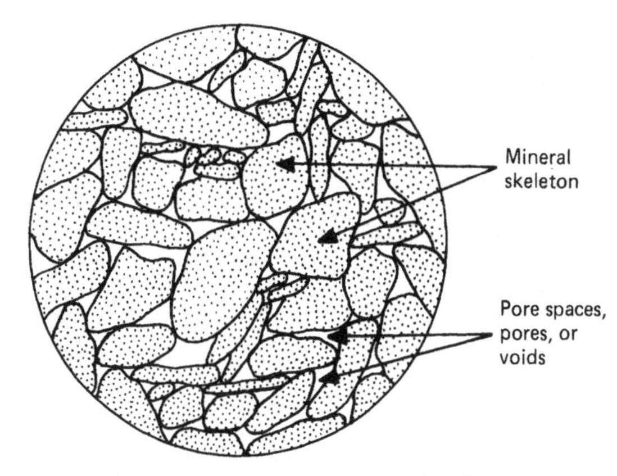

Figure 2.1. Constituents of soil.

are applied more loosely; for example, as will be discussed later, *earth pressure* and *soil stress* are used as synonyms.

2.1.4. Pore Water Pressure

When the pore spaces are filled with water (saturated soil), the pressure in the water is called the *pore water pressure* (Figure 2.2). It acts in all directions with equal intensity.

Figure 2.2. Pore water pressure caught in 10^{-6} fill (after Partially Integrated, 1962).

2.1.5. Total and Effective Stresses

Total stress is the total force transmitted across a given area, divided by that area. Thus, if a 2-foot square piece of wood is placed on the ground surface and a person weighing 200 pounds stands on the wood, the total stress in the ground immediately below the wood is increased by 50 lb/ft^2.

Effective stress can be explained by use of an analogy. Figure 2.3a shows saturated soil placed in a cylindrical container with a cross-sectional area, in plan, of 1 square inch. Figure 2.3b shows an analogy in which the resistance of the mineral skeleton to compression is represented by a spring, and the porous piston is replaced by an impermeable piston with a valved orifice. The orifice represents the resistance to the flow of water through the soil. The piston is assumed to be weightless, and the water is incompressible. Initially, the valve is closed. The spring has a stiffness of 10 lb/in., meaning that a force of 10 pounds is required to produce an axial

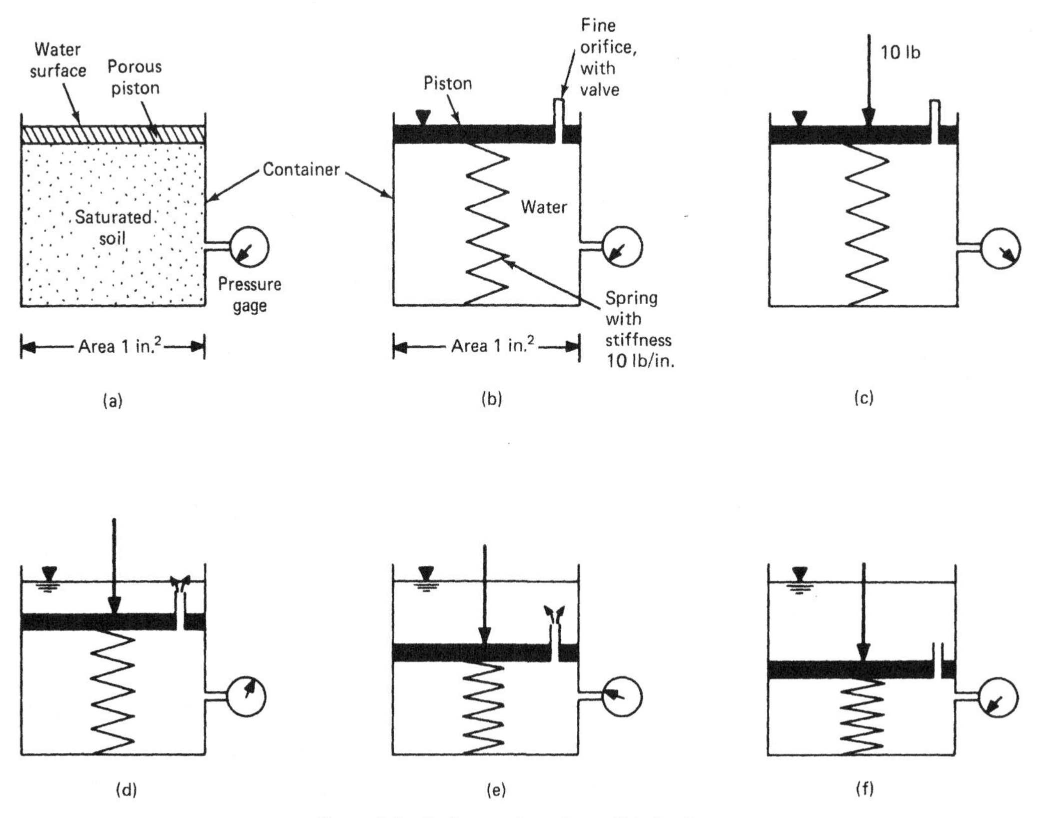

Figure 2.3. Spring analogy for soil behavior.

Table 2.1. Sharing of Applied Force

Figure 2.3	Condition	Valve Position	Force on Piston (lb)	Force Carried by Spring (lb)	Force Carried by Water (lb)
(b)	Initial	Closed	0	0	0
(c)	10 lb force applied	Closed	10	0	10
(d)	Piston descended 0.4 in.	Open	10	4	6
(e)	Piston descended 0.8 in.	Open	10	8	2
(f)	Piston descended 1.0 in.	Open	10	10	0

deflection of 1 inch. In Figure 2.3c, a 10-pound force has been applied to the piston. The water is not free to escape; therefore, the spring cannot compress and cannot carry the newly applied force. The water must therefore carry **all** the force, and the pressure gage will show an increase immediately as the force is applied. If the valve is now opened, water will pass through the orifice and the piston will descend. Figures 2.3d and 2.3e show intermediate steps, and Figure 2.3f shows the condition when the piston has descended 1 inch and there is no further flow of water. Because the spring has now been compressed 1 inch, it must be carrying a force of 10 pounds. The spring is now carrying **all** the force, and the pressure gage has returned to the same reading as in Figure 2.3b. Table 2.1 summarizes the steps and shows the sharing of applied force between the spring and water. It can be seen from the table that the sum of the forces carried by the spring and the water is always equal to the force on the piston.

Effective stress is defined as the force acting between the points of the mineral skeleton **per total area.** Because a cross-sectional area of 1 square inch has been chosen in the above analogy, all the forces in Table 2.1 are numerically equal to stresses in lb/in.^{2} if a real soil is considered. By thinking now in terms of stresses, it can be seen that the force on the piston represents the total stress, the force carried by the spring represents the effective stress, and the force carried by the water represents the pore water pressure. The following relationship always applies:

total stress = effective stress + pore water pressure.

This is Terzaghi's *principle of effective stress*. The following symbols are normally used:

Total stress, σ
Effective stress, σ'
Pore water pressure, u

Thus,

$$\sigma = \sigma' + u.$$

Forces and stresses are plotted in Figure 2.4a. It can be seen from the figure that the rates of pressure change decrease as time increases: this is consistent with the observation that the flow of water through the orifice in the piston decreases as the water pressure in the container decreases.

2.1.6. Consolidation

The process of gradual squeezing out of water, with the accompanying transfer of total stress to effective stress and decrease in pore water pressure, is called *consolidation*. Figure 2.4b shows the volume change that occurs during consolidation. The amount by which the pore water pressure exceeds the equilibrium pore water pressure is called the *excess pore water pressure*, and the gradual decrease of this pressure is often referred to as *dissipation* of pore water pressure.

As a practical example of consolidation, consider

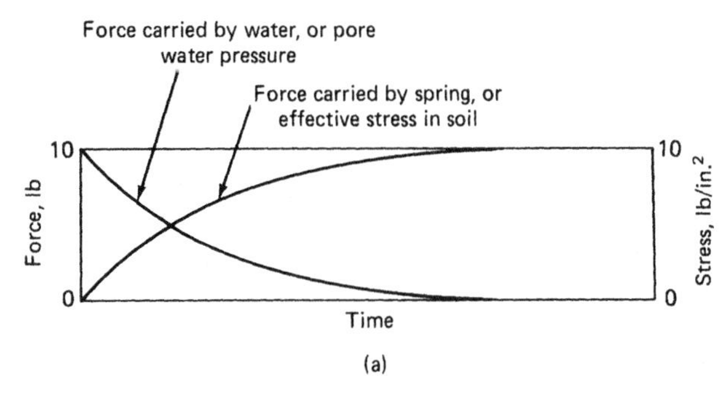

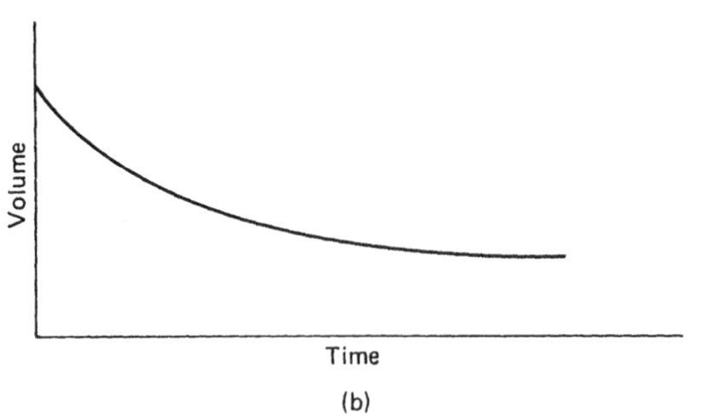

Figure 2.4. (a) Sharing of applied force and stress. (b) Volume change.

a layer of fill for a highway embankment, placed on a clayey foundation soil. As the fill is placed, pore water pressure in the foundation immediately increases and then starts to dissipate, resulting in *settlement*. The rate of settlement depends primarily on the *permeability* of the foundation soil. Permeability is a measure of the rate at which water can move through the soil. Cohesive soils have lower permeability than cohesionless soils, and therefore consolidation and settlement of cohesive soils occur more slowly.

2.1.7. Shear Strength

It has been shown above that effective stresses increase as consolidation progresses. Because an increase of effective stress means that the grains within the mineral skeleton are pressing more tightly together, it becomes increasingly harder to cause sliding between the grains. As an analogy, a brick can be placed on a concrete floor and pushed sideways to cause it to slide. If a second brick is now placed on top of the first brick, it takes more sideways force to cause sliding. It is therefore evident that the ability of a soil to resist sliding is related to the effective stress: the larger the effective stress in a particular soil, the greater is its *shear strength*. The shear strength is a measure of the resistance to sliding between grains that are trying to move laterally past each other.

It can now be seen that the gain in shear strength during the consolidation process can be monitored by measuring pore water pressure.

2.1.8. Normally Consolidated and Overconsolidated Soil

A *normally consolidated* soil is one that has never been subjected to an effective stress greater than the existing overburden pressure. Examples include ocean and lake-bed clays. An *overconsolidated* soil is one that has been subjected to an effective stress greater than the existing overburden pressure. Examples include clays such as London clay, where thousands of feet of overburden have been eroded.

2.1.9. Difference Between Pore Water Pressure and Groundwater Level

The *groundwater level* is defined as the upper surface of a body of groundwater at which the pressure is atmospheric.

Figure 2.5 shows three perforated pipes installed in a soil within which there is no flow of groundwater; therefore, groundwater pressure increases uniformly with depth. When such equilibrium conditions exist, the level of water within the pipe will rise to the groundwater level, independent of the location of the perforations.

Now consider what happens when a layer of fill is placed above the sand shown in Figure 2.5. Fig-

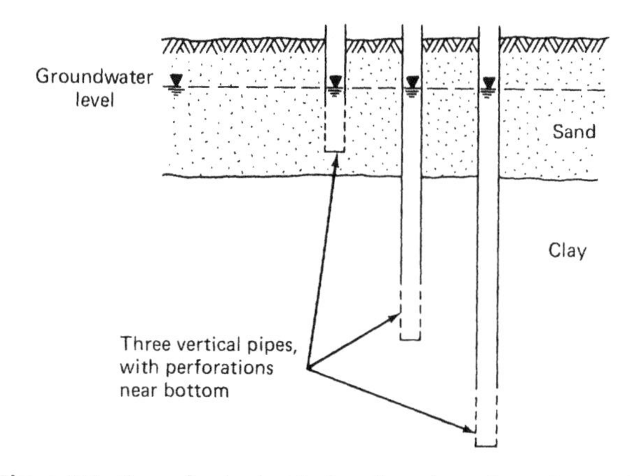

Figure 2.5. Groundwater level when there is no flow of groundwater.

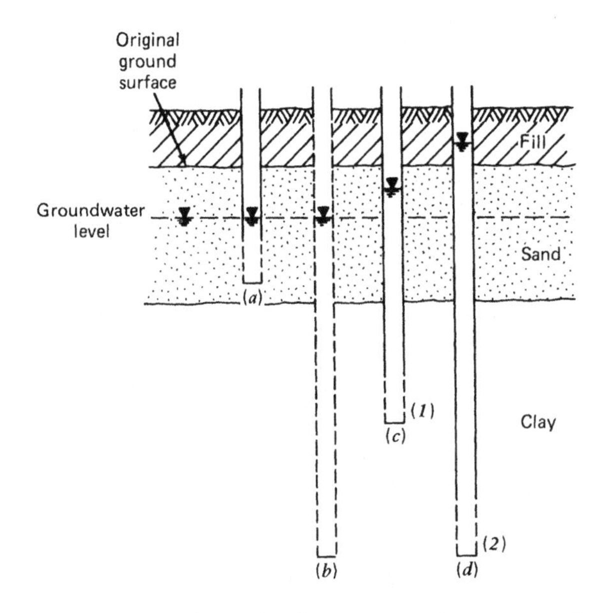

Figure 2.6. Groundwater level and pore water pressure when there is flow of groundwater.

ure 2.6 shows the condition soon after fill placement, when consolidation is not yet complete; therefore, excess pore water pressures exist in the clay and the groundwater is no longer in equilibrium. The four perforated pipes in Figure 2.6 are installed such that soil is in intimate contact with the outsides of the pipes. Pipe (*b*) is perforated throughout its length; the remaining pipes are perforated only near the bottom. Because of the high permeability of sand, excess pore water pressures in the sand dissipate almost immediately and do not exist there. As in Figure 2.5, pipe (*a*) indicates the groundwater level. Pipes (*c*) and (*d*) indicate the pore water pressures in the clay at locations (*1*) and (*2*). The water level in pipe (*c*) is shown lower than in pipe (*d*) because more dissipation of pore water pressure has occurred at level (*1*) than at level (*2*): the drainage path for excess pore water pressure is shorter, and therefore the rate of dissipation is greater. In the case illustrated, pipe (*b*) is likely to indicate the groundwater level because the permeability of the sand is substantially greater than that of the clay: excess pore water pressures in the clay will cause an upward flow of water from the clay to the sand, via the pipe.

In the more general case of a perforated pipe installed through two or more strata, either with perforations throughout or surrounded with sand throughout, an undesirable vertical connection between strata is created, and the water level in the pipe will usually be misleading. This situation is discussed further in Sections 9.1 and 9.12.2, and pipe (*b*) should not be used in practice.

Pipe (*b*) is called an *observation well*. Pipes (*a*), (*c*), and (*d*) indicate pore water pressure and its dissipation within the sand or clay and are called *piezometers*. Details of both types of instrument are given in Chapter 9. As a general rule, piezometers are sealed within the soil so that they respond only to changes of pore water pressure at a local zone, whereas observation wells are **not** sealed within the soil, so that they respond to changes of groundwater pressure throughout their length.

2.1.10. Positive and Negative Pore Water Pressures

All references to pore water pressures in earlier parts of this chapter have been to pressures that are above atmospheric pressure. These are called *positive* pore water pressures. As shown in Figures 2.4a and 2.6, pore water pressure can be increased by applying a compressive force to the soil. The example of placing fill for a highway embankment has been given. Pore water pressure can also increase when a shear force is applied to a soil in which the mineral skeleton is in a loosely packed state. When the array shown in Figure 2.7a is sheared, it decreases in volume. When the pore spaces are filled with water, and water is prevented from leaving, pore water pressure will increase. As a practical example, consider a foundation failure of an embankment on soft ground, with a foundation of loose alluvial material. The material beneath the toe is subjected to lateral shear forces as the embankment is constructed. These shear forces cause deformation, increased pore water pressure, and decreased strength, and therefore increase the tendency toward failure.

Pore water pressure can also be *negative*, defined as pore water pressure that is less than atmospheric pressure. Negative pore water pressure can sometimes be caused by removing a compressive force that has been applied to a soil. For example, when an excavation is made in clay, the soil below the base of the excavation is unloaded, causing an initial decrease of pore water pressure, which may become negative. Pore water pressure can also decrease when a shear force is applied to a soil in which the mineral skeleton is in a densely packed state. When the array shown in Figure 2.7b is sheared, it increases in volume. If the pore spaces

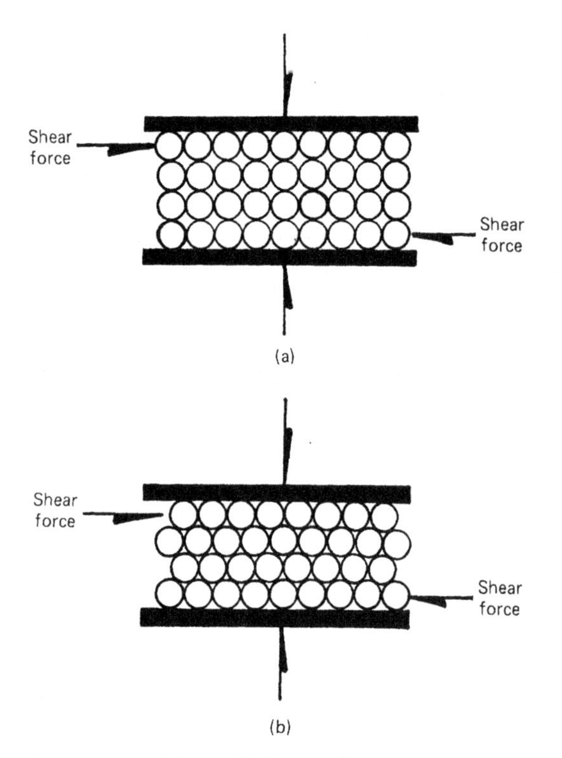

Figure 2.7. (a) Positive and (b) negative pore water pressure caused by application of shear force.

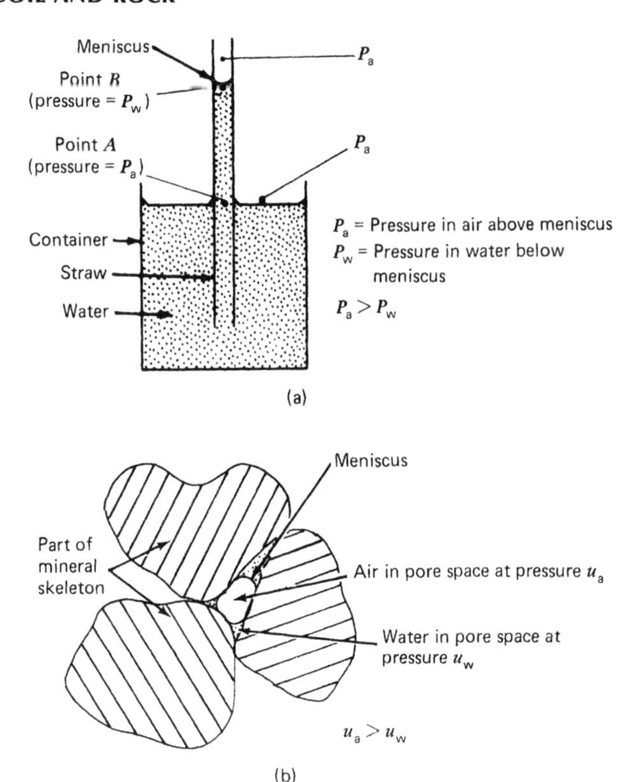

Figure 2.8. Pore gas and pore water pressure: (a) straw in container of water and (b) element of unsaturated soil.

are filled with water, and additional water is prevented from entering, pore water pressure decreases. As a practical example, consider the excavation of a slope in overconsolidated clay. Pore water pressure decreases as a result of unloading, but significant additional decrease can be caused by the development of lateral shear forces. These shear forces cause deformation, a temporary decrease in pore water pressure, and a temporary increase in strength.

2.1.11. Pore Gas Pressure

In an *unsaturated* soil, both gas and water are present in the pore spaces, and the pressure in the gas is called the *pore gas pressure*. As with pore water pressure, pore gas pressure act in all directions with equal intensity. Examples of unsaturated soil include compacted fills for embankment dams and organic soil deposits in which gas is generated as organic material decomposes. When the gas is air, the term *pore air pressure* may be used.

The pore gas pressure is always greater than the pore water pressure. Consider the analogy of a straw placed in a container of water, as shown in Figure 2.8a. The water level in the straw rises to a level higher than in the container and is "held up" by surface tension forces between the straw and water at the meniscus. The pressure in the air is atmospheric pressure P_a; therefore, the pressure at point A must also be P_a: if it were not so, there would be a flow of water to create equality of pressures at the same level. The pressure at point A must be greater than that at point B (P_w), because the pressure in water increases as depth below the surface increases. P_a is therefore greater than P_w, and the pressure at the air side of the meniscus is greater than the pressure at the water side. It is well known that water in a smaller diameter straw rises to a greater height: the pressure difference across the meniscus is therefore greater, and the radius of curvature of the meniscus is smaller.

Now consider a meniscus between air and water in a pore space within soil, as shown in Figure 2.8b. The same rule applies as in the analogy: the pore air pressure u_a is greater than the pore water pressure u_w. The smaller the pore space, the smaller the radius of curvature of the meniscus, and therefore the difference between pore air pressure and pore water pressure is greater.

2.1.12. Other Terms Relating to Behavior of Soil

A *perched groundwater level* is above and not hydraulically connected to the more general groundwater level. For example, groundwater may be trapped above a clay layer at shallow depth, whereas the general groundwater level is in a lower layer of sand.

An *aquifer* is a pervious soil or rock stratum that contains water. An *artesian aquifer* is confined between two relatively impervious layers and capable of carrying groundwater under pressure.

The *piezometric elevation,* or *piezometric level,* is the elevation to which water will rise in a pipe sealed within the soil, as shown for pipes (c) and (d) in Figure 2.6.

2.1.13. Primary Mechanisms that Control Behavior of Soil*

The primary mechanisms that control the engineering behavior of soil may be categorized as hydraulic, stress-deformation, and strength mechanisms.

When a soil is subjected to excess pore water pressure, the water flows through the pore spaces in the soil. These may be fairly large, as are the spaces in openwork gravel, or microscopic, as in the spaces between the finest clay particles. As it flows, water reacts against the particles, causing friction and a resistance to flow. The amount of friction depends on both the velocity of flow and the size of the soil particles. In a very fine-grained soil, there is a larger area of contact between the water and soil particles than in a coarse-grained soil, so that there is more friction. The permeability of the soil is governed by the amount of friction and resistance to flow. Just as the soil acts on the water to retard the flow, so also does the water act on the soil. Water exerts a tractive force on the soil, in the direction of flow, owing to the friction. If the soil is saturated, water acts also with a buoyant force on the soil, whether or not the water is moving.

Stress-deformation characteristics of cohesive soil are governed by the time required for water to flow through the pore spaces in the soil and for consequent volume changes to occur. For cohesionless soils, most of the volume change is caused by rearrangement of the relative positions of grains as shear deformation occurs.

Shear strength is governed by the nature, size and shape of the soil grains, packing density and effective stresses within the soil. Shear failure occurs when the stresses increase beyond those that can be sustained by the soil and the strength is exceeded.

Footing Foundations

Figure 2.9 depicts a load applied to a footing foundation located on a soil mass. In the upper figure, the load is substantially less than that required to cause failure. The footing transfers the stress to the soil, causing the soil to compress in the vertical direction, and the footing settles. In the lower figure, a larger load is applied to the footing, stressing the soil and causing a rotation that tends to push the adjacent block of soil out of the ground, as if it were a rigid body. This movement is opposed by the shearing resistance of the soil along the potential failure surface. In a cohesionless soil, this shearing resistance is frictional, and the magnitude of friction depends directly on the normal stresses against the potential failure surface. In a cohesive soil, this shearing resistance is often purely cohesive, independent of the normal stresses during the shearing process but dependent on the history of previous effective stresses.

Deep Foundations

Figure 2.10 shows a loaded pile or drilled shaft, embedded in soil. The pile moves downward in response to the load and, as it moves with respect to the soil, it mobilizes the skin friction or shearing stress, which resists the sliding along the soil/pile interface. This accounts for part of the support that the pile receives from the soil. The rest of the support is from point bearing, the same mechanism that supports a footing on soil.

When soil settles with respect to a pile or drilled shaft, the pile is loaded by frictional forces at the interface, causing an increase in stress within the pile and/or increased settlement. The loading is referred to as *downdrag* or *negative skin friction*. Downdrag loading may be caused by several events. First, when piles are driven through fill that overlies clay, the loading from the fill may cause continuing primary consolidation or secondary time effects. Second, settlements are caused

*Written with the assistance of Norbert O. Schmidt, Professor of Civil Engineering, University of Missouri–Rolla, Rolla, MO.

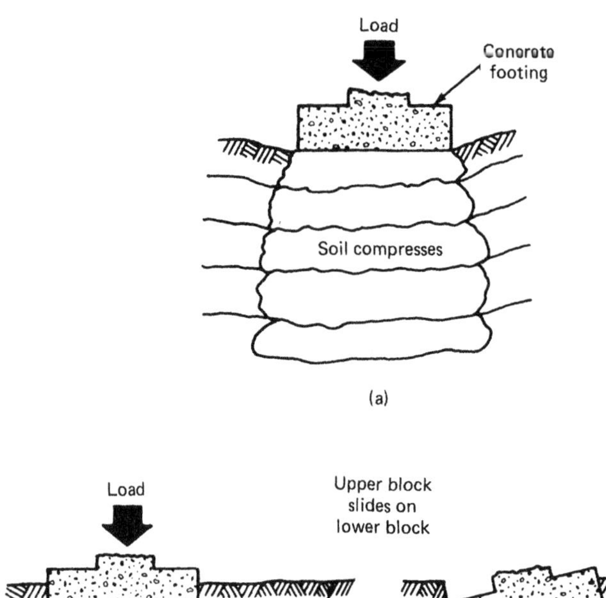

Figure 2.9. Behavior of a footing foundation: (a) settlement of footing and (b) bearing failure.

when ground surface loading is increased after piles are driven, for example, by placing an approach fill alongside a pile-supported bridge abutment. Third, dewatering, site grading, and/or vibration during pile driving may cause settlement and downdrag loading.

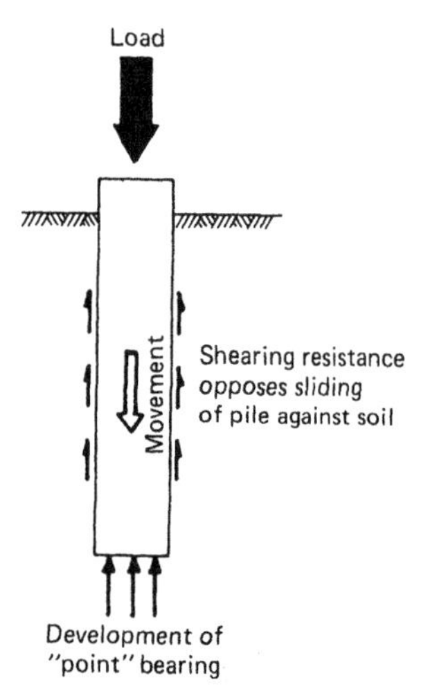

Figure 2.10. Behavior of a driven pile or drilled shaft.

Excavated and Natural Slopes

Excavated and natural slopes often involve layered sediments, which may be parallel to the original surface of the slope. One of these layers is usually weaker than the others, so the potential surface of sliding is noncircular and tends to follow the weakest layer as depicted in Figure 2.11. When the soil is relatively homogeneous, the potential surface of sliding may be more circular. The stability of slopes in soil is controlled by the ratio between available shearing resistance along a potential surface of sliding and the shear stress on that surface. Any increase in pore water pressure along the potential surface of sliding decreases the shearing resistance and the factor of safety against sliding.

Retaining Walls

Soil is only partially capable of supporting itself. Sand requires a retaining structure if it is to stand vertically, but clay may support itself vertically for limited heights for some period of time. Figure 2.12 shows a retaining wall with forces against it. The earth pressure shown pushing the wall towards the

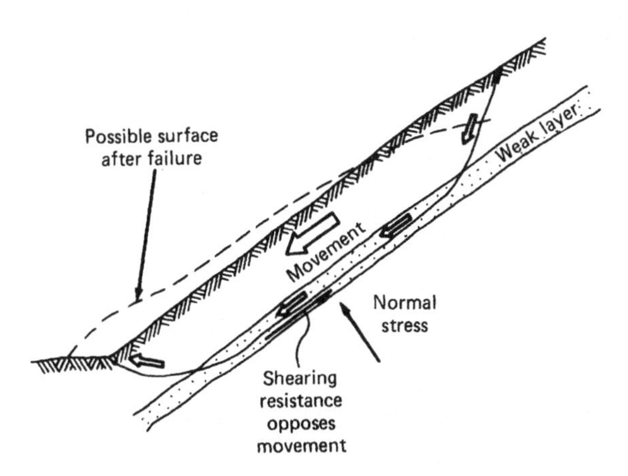

Figure 2.11. Failure of a slope in layered soil.

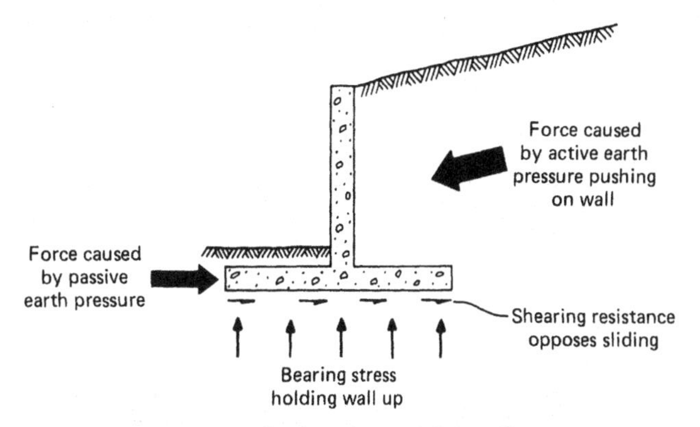

Figure 2.12. Behavior of a retaining wall.

left is that portion of the pressure that remains after the soil tries to support itself. It is known as the *active earth pressure* and is reduced to its minimum value when the wall slides and tilts slightly to the left. The sliding also mobilizes the shearing resistance between the soil and the base of the wall. In front of the wall on the left, the *passive earth pressure* is mobilized as the soil resists sliding.

Braced Excavations

When compared with retaining walls, braced excavations differ in their mobilization of earth resistance because significant sliding or tilting toward the excavation is not permitted. The wall of a braced excavation may be braced against the opposite wall with *cross-lot bracing,* braced against the bottom or the adjacent wall with *rakers,* or held in place by *tieback anchors* that pull against the soil outside the excavation. Stresses against a braced wall are greater than against a retaining wall because the passive pressure is greater. However, if bracing is properly installed, deformation of the soil is small and buildings on adjacent sites may therefore experience only small settlements and little distress. When a deep braced excavation is made in soft soil, soil tends to move into the bottom of the excavation: this is called *bottom heave,* and special construction techniques may be required to control deformation.

Embankments

Embankments of relatively homogeneous soils overlying hard ground tend to fail by rotation along an almost circular arc. This is depicted in Figure 2.13. As with the bearing failure of a footing foundation, shown in Figure 2.9, the soil behaves as if it were a rigid body. Rotation is resisted by mobilization of shearing resistance along the arc, as described for footing foundations.

Embankments on Soft Ground

The behavior of embankments on soft ground tends to be dominated by the properties of the soft ground. A potential circular failure surface may develop, with a large portion of the surface in the weak foundation material as shown in Figure 2.14a. However, the loading of the embankment may cause settlement and lateral bulging of the foundation, as shown in Figure 2.14b, long before the rotational failure occurs. The lateral bulging of the soft ground transfers horizontal tension to the embankment, which may experience tension cracking, since it is less deformable than the soft foundation.

Embankment Dams

Embankment dams may experience distress or failure in a variety of ways.

Overtopping may result from incorrect estima-

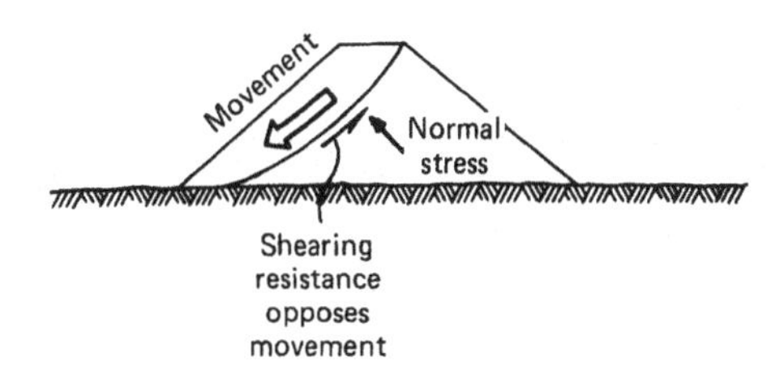

Figure 2.13. Behavior of an embankment of relatively homogeneous soil.

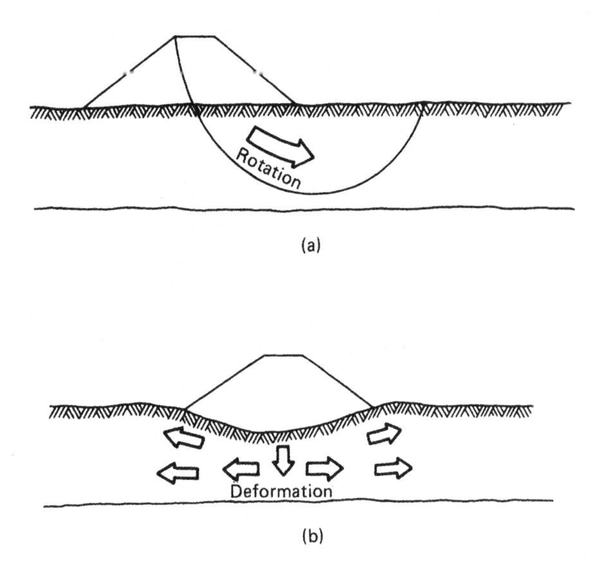

Figure 2.14. Behavior of an embankment on soft ground: (a) rotational slide along arc and (b) settlement and lateral bulging of soft foundation.

tion of storm water volume or duration. It may also be the result of slope failure on the reservoir rim, causing a large volume of material to slide into the reservoir and induce a wave, as occurred at Viaont Dam in Italy.

Internal erosion, or *piping,* can cause distress to an embankment dam. Erosion occurs where seepage water is flowing at a velocity sufficient to carry soil particles along with the water. If the soil has some measure of cohesion, a *pipe* or small tunnel may form at the downstream exit of the seepage path as the soil is eroded. As the pipe lengthens inside the embankment, there is less resistance to flow, the flow in the pipe increases, and piping accelerates. Silts and fine sands are most prone to piping, because they erode easily, and when moist are sufficiently cohesive to form the walls of a pipe without collapsing. In a well-designed embankment dam, piping is prevented by sizing downstream material such that the pore spaces of that material are just smaller than the larger sizes of the mineral skeleton in the adjacent upstream material, so that migration of particles is blocked. Such a downstream material is known as a *filter*.

When the slopes of a dam are too steep, the dam may fail as described above in the discussion on embankments. Many river sediments consist of *soft ground,* and a dam constructed over these materials may behave as shown in Figure 2.14.

If a dam is built on loose granular material, especially silty sand, an earthquake can cause the foundation to liquify and flow, as with the near failure of the San Fernando Dam in California.

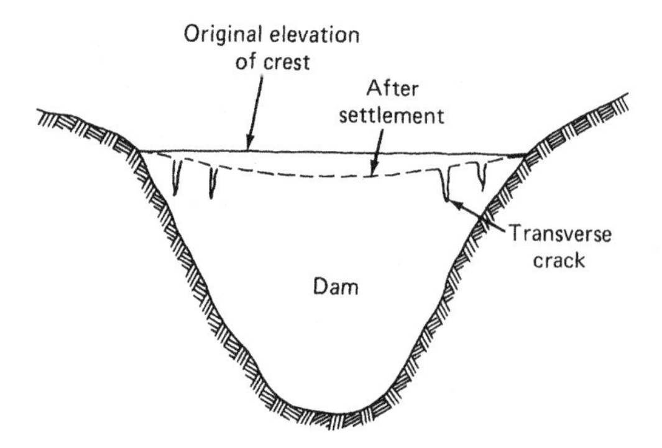

Figure 2.15. Transverse cracking of an embankment dam.

Even if the design of the dam is adequate, the weight of the embankment dam on the underlying soil or rock must be considered. Heavily loaded soil under the dam may settle, and there will be downward and lateral movements of the base of the dam. Moreover, even well-compacted fill material will experience settlements when loaded with overlying material, and poor compaction procedures will result in greater settlements. If the crest of the dam is initially level, with time it will settle, and the center of the dam will settle the most. If the abutments are steep, the settlements may put the crest of the dam in tension, as shown in Figure 2.15, possibly causing cracks transverse to the axis of the dam.

Soft Ground Tunnels

When a tunnel is excavated in soft ground, it may be excavated to a diameter slightly larger than its lining so that the lining may be placed, and the soil may be unsupported for a short time period. Stresses are relieved and soil tends to move inwards toward the tunnel cross section as well as into the face. Although the space between the soil and lining is usually small and may be grouted, the ground surface may settle, with greatest settlements occurring directly over the tunnel as shown in Figure 2.16.

If the groundwater level was previously above the invert of the tunnel, the tunnel may drain the soil. In granular soils, the groundwater may have to be predrained to the level of the invert, so that the tunnel can be constructed. The reduction in groundwater level will increase effective stresses in the soil

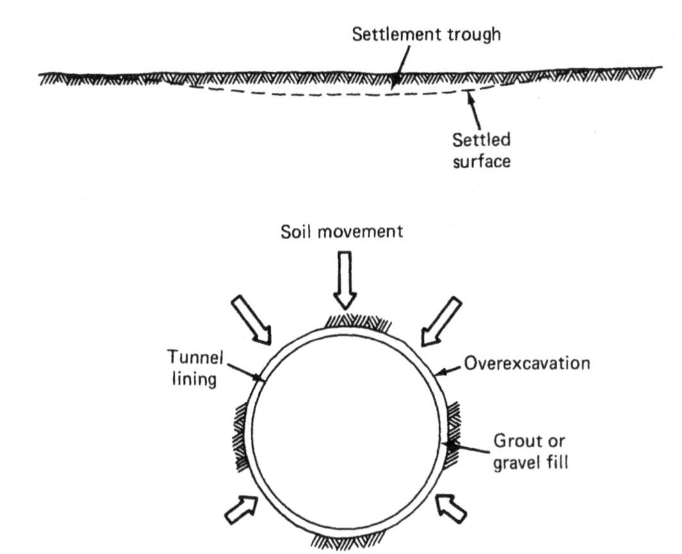

Figure 2.16. Behavior of a soft ground tunnel. Note that soil also tends to move toward the face of the tunnel.

profile below the original water level, and in cohesive soils this will cause the initiation of consolidation settlements. In cohesionless soils, the increased stresses will tend to cause immediate strains in the soil and therefore settlement.

2.2. BEHAVIOR OF ROCK*

2.2.1. Geologists' View of Rock

Rock is by definition a material older than about 1 million years; a material formed before the *Pleistocene* ice age. The oldest rocks, which are more than 2 billion years old, have been welded by geologic processes and so are usually stronger than the more recent rock formations.

Rock types are classified geologically according to their *genesis,* the way in which they were formed. *Igneous rocks,* such as granites and basalts, are those formed by the cooling and solidification of a hot molten magma originating in the earth's core. Although in our life span they are strong and durable, over geologic time periods they dissolve in the groundwater or break down to form soils through heating and cooling, freezing and thawing, and chemical attack. *Sedimentary rocks,* such as sandstones, limestones, and shales, are those formed when soils are welded or cemented together by overburden pressure or by intergranular cements deposited from groundwater solutions. Other sedimentary rocks, termed *evaporites,* such as rock salt, potash, anhydrite, and gypsum, are formed as crystalline residues when a saline lake dries up. The third main category, *metamorphic rocks,* such as slates, marbles, and schists, are those formed from either igneous or sedimentary types as a result of extreme geologic heating or pressure, sufficient to cause recrystallization and the formation of new minerals.

The microtexture of most rocks is characteristically granular, as shown in Figure 2.17. Igneous rocks are composed of grains that are angular and interlocked, with microcracks along the grain boundaries that give a very low level of intergranular porosity. Sedimentary rocks, in contrast, usually have rounded grains with not much interlocking. The intergranular porosity of sedimentary rocks is generally greater than igneous rocks, but depends on the degree to which the pore spaces have been filled by finer-grained fragments or cement. Metamorphic rocks tend to have elongated grains, oriented subparallel to each other as a result of recrystallization, which makes them *fissile,* meaning easily split along the alignment of the grains.

Rocks are composed of minerals. Simple rock types are formed from just one mineral variety, for example, marble is composed of interlocking crystals of calcite only and quartzite from interlocking quartz. Most rocks, however, are *polymineralic,* meaning composed of several mineral types. Granite, for example, is composed of quartz, feldspar, and mica.

Geologic names are assigned to rock materials according to their mineral composition and also their grain size. For example, rocks composed mainly of calcite fragments are called limestones, and rocks composed mainly of quartz fragments are called siltstones, sandstones, or conglomerates, depending on whether the typical grain size is small, medium, or large.

On the larger scale, different *primary structures* have become solidified into rock formations by the very different processes of sedimentary and igneous rock formation. Sedimentary rocks are characteristically *bedded,* being composed of *beds* (layers) of materials such as sandstone and shale, separated by *bedding planes* that run continuously through the

*This section has been coauthored by John A. Franklin, Consultant and Research Professor, University of Waterloo, Ontario, Canada.

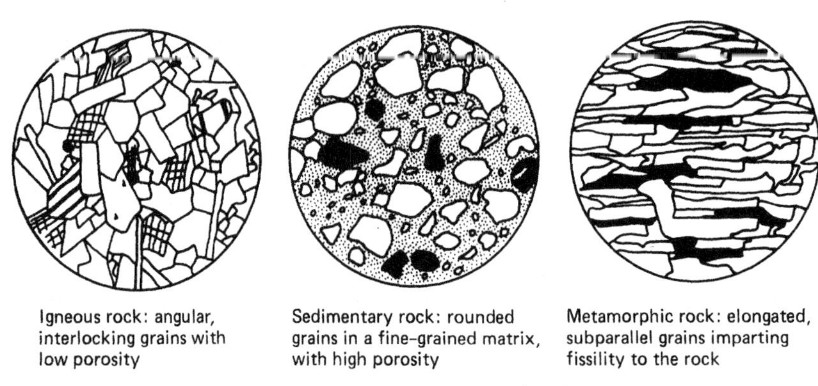

Igneous rock: angular, interlocking grains with low porosity

Sedimentary rock: rounded grains in a fine-grained matrix, with high porosity

Metamorphic rock: elongated, subparallel grains imparting fissility to the rock

Figure 2.17. Microtexture of rocks.

deposit and form planes of weakness. *Flow banding* is a characteristic of igneous volcanic rocks such as basalt, in which layers of lava are often separated by volcanic froth (pumice stone) and windborne or water-deposited layers of ash or bentonite clay. *Tension joints* (cracks) have been formed by contraction of igneous rocks during cooling and of sedimentary rocks during the accumulation of overlying material.

Secondary structures have been generated by the subsequent action of metamorphism, heat, and pressure, caused mainly by drifting of rigid continental blocks on the more fluid mantle and core of the earth. Slowly but continuously the continents are being thrust one beneath the other, in a process of *continental drift* that is creating mountain chains and belts of volcanic and earthquake activity along the lines of impact. This has resulted in the folding and sometimes complete overturning of beds that were originally horizontal. The same forces acting on more brittle rocks have created not only further tension joints but also *faults,* defined as surfaces of sliding and shearing. *Fault zones* are common, consisting of several subparallel faults or shears (small faults) with interposed *fault breccia* (broken rock) and *fault gouge* (finely crushed rock and clay).

2.2.2. Engineers' View of Rock

The many hundreds of rock names assigned by geologists often mean little in the context of an engineering project. The engineer is concerned not so much with the history of the rock or its precise mineral composition as with its potential behavior in an excavation or foundation. When classifying rocks for engineering purposes, the engineer must consider properties that often go unreported by the geologist yet are critical to the mechanical character and behavior of the rock mass.

On the small scale, the engineer has to consider how mineral composition and microtexture affect properties such as rock strength and durability. Pores are by far the weakest rock-forming component, and porous rocks are much weaker and more deformable than dense ones. *Porosity* (the ratio of pore volume to total volume) can exceed 50% in some sandstones and highly weathered granites. *Hard rocks (competent rocks)* are usually formed from hard and resistant minerals. *Soft rocks* such as shales are usually composed of clay and similar minerals that are themselves weak and deformable. These rocks often break down when subjected to wetting and drying because their minerals tend to attract a boundary layer of water. Limestones composed of calcite and salt rocks composed of saline minerals, although quite strong and brittle when found in near-surface engineering construction, deform and flow quite readily when subjected to the higher temperature and pressures found in deep underground mines.

Mineral composition has an important influence on how rocks deform in response to stress. Those like granite that are composed mainly of hard minerals are usually *elastic,* because when moderate levels of stress are applied and then removed the rocks return to their original size and shape. If the small deformations that occur are in direct proportion to the magnitude of the applied stresses, the rock is described as *linearly elastic.* At higher stresses and after very little deformation, they fail suddenly in a *brittle* manner. Rocks like potash that contain soft minerals are often *inelastic* and may *creep* (continue to distort) when a high level of stress is maintained. This is termed *viscous* or *vis-*

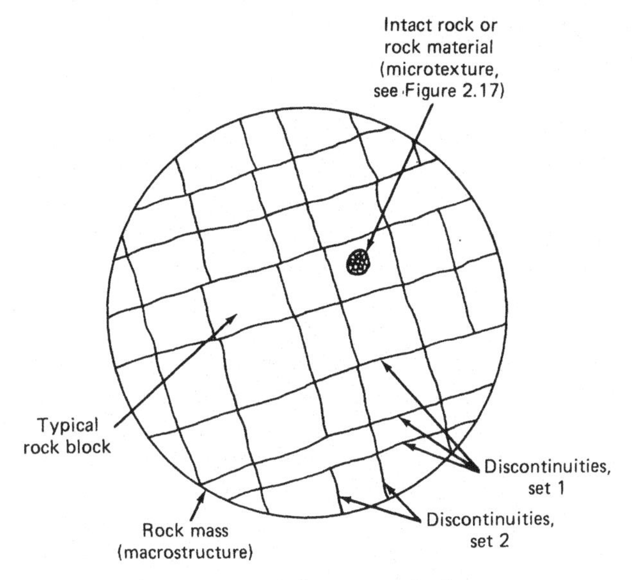

Figure 2.18. Rock mass terminology.

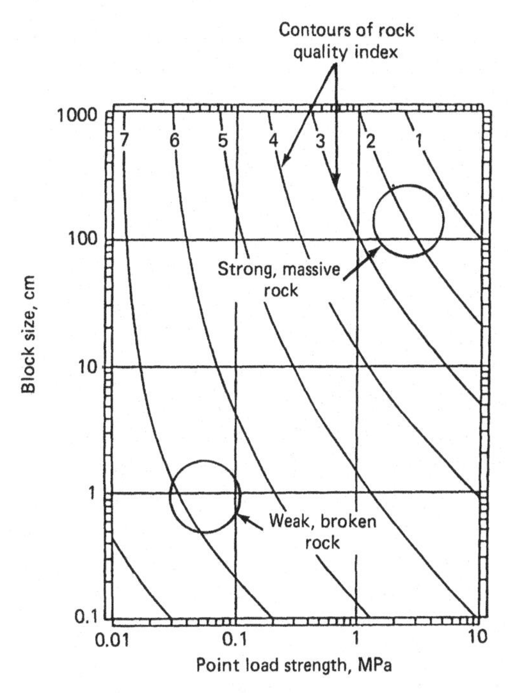

Figure 2.19. Size–strength rock mass classification diagram (after Franklin, 1986).

coelastic behavior. Their failure is *ductile* (gradual and accompanied by large deformations).

Rocks with subparallel platy minerals, as noted earlier, tend to be fissile and are also *anisotropic* (with strengths that vary according to the direction of loading). Slates are characterized by *slaty cleavage* so that they can be split readily into thin plates, and schists and gneisses by *foliation surfaces* along which platy mica minerals are abundant. In general, such rocks are much weaker than those with randomly oriented, equidimensional, and interlocking grains.

Even more important in an engineering context than minerals and microtexture are the features of rock on a larger scale. Rock engineers distinguish between *intact rock* and *rock mass*. The properties of intact rock, otherwise termed *rock material,* are those that can be measured by testing a small specimen of solid rock in the laboratory. The same properties measured on the scale of the rock mass are affected by large-scale structural features and can only be measured directly by large-scale in situ tests. Figure 2.18 shows the difference between intact rock and rock mass. The rock mass is nearly always much weaker, more deformable, and more permeable than the intact material it contains. This is because of the presence of *discontinuities,* a global term used for the various types of fracture and weakness planes discussed earlier: joints, faults, shears, bedding planes, and surfaces of cleavage and foliation. Joints are the most common, and the terms *jointing* and *jointed* (e.g., widely or closely jointed) are often used to apply to discontinuities irrespective of their origin. Nearly all aspects of rock behavior, both at the surface and underground, are controlled much more by the characteristics of such discontinuities than by those of the intact rock blocks within the mass.

Figure 2.19 shows a multipurpose *size–strength* rock mass classification that takes into account the two features of the rock mass that are usually the most important in engineering applications: the *size* of rock blocks and their *strength,* plotted respectively along the vertical and horizontal axes. Zones in the rock mass that plot toward the lower left of the diagram are broken and weak and easy to excavate but difficult to support in excavations. Those toward the upper right are massive and strong and require blasting for excavation but are often self-supporting. The diagram is contoured to give a *rock quality index,* which is a combined measure of block size and strength.

Note must be taken of other discontinuity characteristics that affect rock mass character and behavior, such as the orientation (dip magnitude and direction) of joint sets. In a rock slope, for example, horizontal jointing generally has little effect on stability, whereas joints that dip steeply into the excavation may well be a source of sliding failures. Other characteristics of discontinuities such as roughness, continuity, and the presence or absence of a clay infilling determine the shear strength of the discontinuity surfaces. Rough joints with no infilling are strong compared with smooth, slickensided, and clay-filled fault planes.

2.2.3. Water in Rock

Seepage through rock masses occurs almost exclusively along discontinuities, with little or no flow through the intact rock blocks unless these are extremely porous. Discontinuity characteristics govern water pressures and flow rates through the rock mass, just as they govern mechanical characteristics such as rock mass strength and deformability. Flow and pressures are governed by the *aperture* (openness or tightness) of the discontinuities and by their spacing and continuity.

Joint water pressures within a rock mass can have a most important influence on stability. When there is a need to monitor these pressures, measurements must be made within zones of open jointing. The principle of effective stress discussed for soil in Section 2.1.5 applies equally to the rock mass, except that the pore water and mineral skeleton system are replaced by the joint water and rock block system.

In strong and durable rock masses, high joint water pressures are the main problem related to groundwater, and these can usually be relieved effectively by drainage. In weaker and less durable rock types such as shales, further problems may result from swelling, and rocks containing clay minerals are particularly prone to breakdown. Closely jointed or less durable rock types are also susceptible to internal erosion within the jointing or to external erosion by processes of raveling and slaking. Vertical rock cliffs are often undercut by erosion, and groundwater can be a hazard also in tunnels and underground chambers. Clayey fault gouge acts as a barrier to high-pressure water. Tunnels penetrating such faults can encounter sudden and catastrophic inflows of water and broken rock.

In other respects, the groundwater regime in rocks is similar to that found in soils, so that terms such as groundwater level, perched groundwater level, aquifer, artesian aquifer, and piezometric elevation and level, defined in Section 2.1, apply with the same definitions in a rock engineering context.

2.2.4. Stress in Rock

In contrast to soils, which are usually deformable, rocks are rigid and permit the transfer and storage of high stresses. Stability of a rock mass therefore depends on three principal factors: the characteristic of the intact rock and particularly of the discontinuities, the characteristics of the groundwater regime, and the magnitudes of ground stress in relation to rock strength and brittleness.

The stresses acting vertically at depth can be estimated, as for soils, from the weight of overlying materials. The magnitudes of horizontal stresses cannot be estimated in this way and can be much greater than those acting vertically. At some locations, even close to the surface, horizontal stresses can be ten or more times greater than vertical stresses. High horizontal stresses close to the surface can result in buckling and heaving of shallow excavations and in squeezing and cracking of tunnel linings and buried concrete pipes. When high stresses surround an underground excavation, they can be a stabilizing influence, helping to hold in place the rock arch above a tunnel, a mine stope, or an underground powerhouse cavern. However, if excessive they can cause extensive damage in the form of squeezing or rockbursting.

Rock engineers distinguish between *virgin stresses* (also called *in situ stresses*) created by geologic processes and *induced stresses* caused by excavation. We can speculate that the virgin stresses were first established by continental drift as discussed in Section 2.2.1. They have since been modified by the deposition and erosion of geologic materials. In the past, many locations on the surface of the earth were overlain by several miles of soil, rock, or ice. Geologic weathering and erosion over many millions of years have removed this overburden. However, the horizontal stresses existing in rocks that were once deeply buried may not be entirely relieved by erosion. Rigid rock types, in particular, often contain large virgin horizontal stresses that reflect their geologic history and that have important influences on present-day engineering construction.

When an excavation is made in rock, the stresses originally carried by the excavated material are transferred to the rock that remains in place. Stress levels around tunnels are amplified to several times their virgin values. The rocks may become overstressed and may suffer from squeezing or explosive bursting, depending on whether the materials are comparatively ductile or brittle.

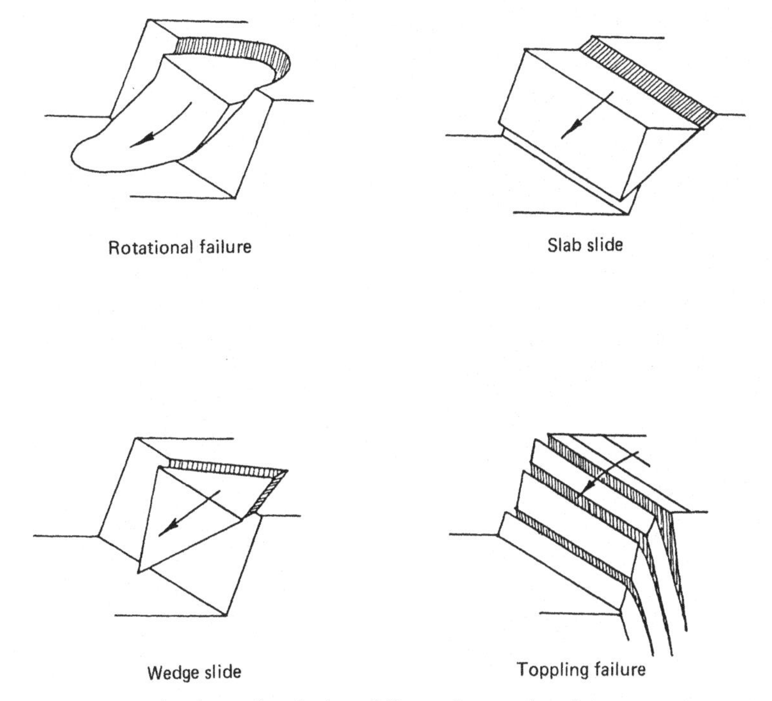

Figure 2.20. Primary mechanisms of rock slope failure (after Hoek and Bray, 1974). By permission of the Institution of Mining and Metallurgy.

2.2.5. Primary Mechanisms that Control Behavior of Rock

The important differences between rocks and soils, and between their respective groundwater and ground stress regimes, result in different patterns of behavior. Whereas soft rocks behave similarly to soils, the stronger, more competent and massive rock types behave quite differently. Examples are given below.

Foundations on Rock

Foundations on rock are rarely subject to *generalized* or *rotational failures* such as outlined in Section 2.1.13 for footing foundations on soil. Generalized failure is a real possibility only in the case of foundations near the crest of a rock cut, which can be considered as a special type of slope stability problem. Slopes in rock are discussed in the next subsection.

Differential settlement, however, remains a potential problem. Settlement magnitudes are again governed by the presence and characteristics of discontinuities. They depend on whether the discontinuities are closely or widely spaced, open or tight, filled or unfilled. Foundation loading may cause open discontinuities to close, unless they have been filled with a suitable grout.

Slopes in Rock

Slopes in rock fail predominantly by sliding along preexisting discontinuities, particularly along those bedding planes or faults that cut extensively through the rock and appear in the face of the slope. The *rotational failures* that are characteristic of soil happen only occasionally in rock because they require rupturing of intact rock blocks, which are much stronger than discontinuities. Rotational failures do occur, however, in very closely jointed and weak rock masses that behave like soils. Groundwater pressures that develop within discontinuities and within tension cracks behind the crest of the slope are often the cause of instability. Depending on the number of discontinuity sets and their relative orientations, failure of a rock slope may occur either as *slab sliding, wedge sliding,* or *toppling* (Figure 2.20). *Undercutting* and *raveling* of steep rock faces may also result from toe erosion, particularly if the rock is closely jointed or of low durability.

Underground Openings in Rock

Yield and failure of rock around underground openings occur by a variety of mechanisms, depending on the stress levels, groundwater conditions and rock characteristics, the shape and size of the open-

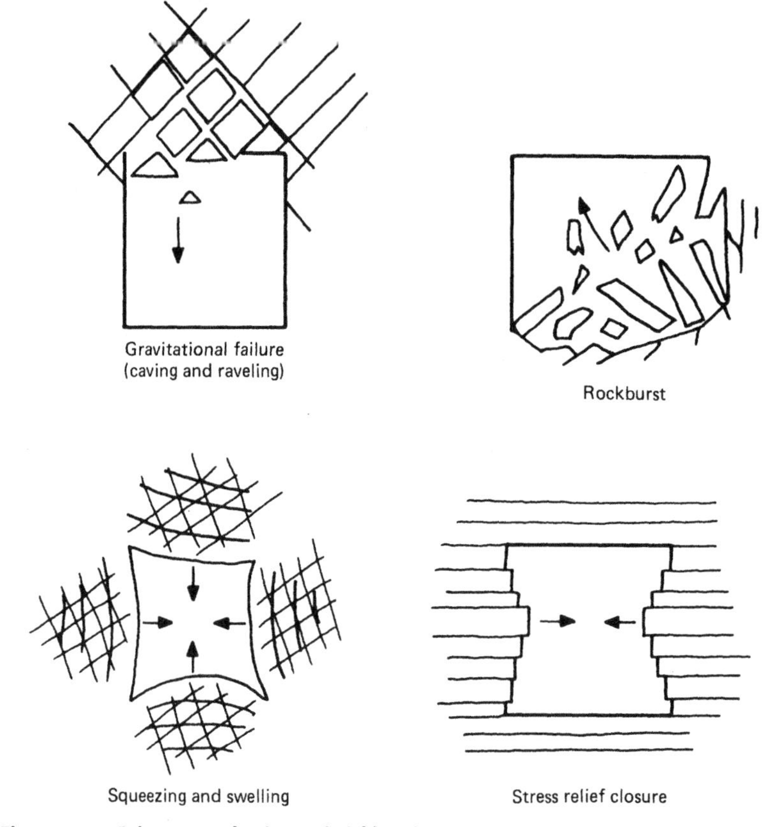

Figure 2.21. Primary mechanisms of yield and failure of underground openings in rock.

ing, the methods used for excavation (e.g., blasting or boring machine), and the type and quality of support installed. Each of these mechanisms, shown in Figure 2.21, is characteristically different and calls for its own individual approach to monitoring and design of instrumentation systems.

Gravitational failures (caving and raveling mechanisms) occur when rock blocks fall from the roof of an underground opening without appreciable breakage of intact rock material. Such failures are governed entirely by discontinuities. They start with the fall of small keystone blocks and progress upward until they either reach the ground surface or are arrested by *bulking* of broken rock within the opening (broken rock occupies a greater volume than undisturbed rock). Uncontrolled caving in a Quebec mine in 1980 migrated upward until it reached the saturated overburden soils, which liquified and flowed into the workings. Miners were killed, and the affected levels of the mine had to be closed.

Rockbursting is the characteristic mode of failure in brittle rock materials where the strength of the intact rock is exceeded by the magnitudes of excavation-induced stresses. Classic examples occur in South African mines at depths approaching 2–3 miles (3–5 km) where even the high-strength rocks are incapable of sustaining the stresses that develop.

Squeezing behavior is characteristic of rock masses that are overstressed but relatively weak or closely jointed. The rock mass yields by shearing along weak discontinuity surfaces or by crushing or distortion of the rock blocks. Tunnel closures in excess of 3 ft (1 m) have been recorded, for example, by convergence monitoring in the Austrian Alps.

Swelling behavior is encountered when the minerals of the rock are themselves of a swelling variety. Clay minerals are the most common in this category. Some shales contain montmorillonite and similar clay minerals that adsorb water, expand, and eventually generate high pressures on tunnel liners and support systems. Swelling problems are also experienced in anhydrite rock types that convert to gypsum when wetted.

Stress relief closure occurs most commonly in openings excavated in horizontally bedded sedimentary rocks. The beds expand as a result of stress relief and slide inward, exerting very high pressures on liners and supports. Some sliding occurs immediately when the rock is blasted, but continuing creep along the bedding planes results in the development of squeeze over much longer time periods, sometimes of years or decades. Hydroelectric turbine pits in Niagara have suffered from stress relief closure. The inward movement in one pit has been monitored monthly since 1905, and total movement has exceeded 4 in. (100 mm), sufficient to cause severe misalignment problems and shattering of a cast iron support strut.

Part 2

Planning Monitoring Programs

Part 2 is addressed primarily to people who make important decisions during the design phase of civil engineering projects: owners, project managers, project engineers, and project geologists.

Chapter 3 enumerates the benefits of using geotechnical instrumentation. Probably the greatest shortcoming in the state of the practice of geotechnical instrumentation is inadequate planning of monitoring programs, and this is the subject of Chapter 4. Chapter 4 is the hub of the book and sets out a logical step-by-step planning process that is vital to success. Chapters 5 and 6 describe the various contractual arrangements that are available for procuring instruments and field instrumentation services. Recommendations are made for contractual arrangements that maximize the quality of instrumentation data.

CHAPTER 3

BENEFITS OF USING GEOTECHNICAL INSTRUMENTATION

The role of geotechnical instrumentation is described in Chapter 1. In the foreword to this book, Ralph Peck states: "every instrument installed on a project should be selected and placed to assist in answering a specific question. Following this simple rule is the key to successful field instrumentation." Unfortunately, many practitioners select and place instrumentation when there is no specific question, perhaps because *everybody is doing it*. Benefits of using geotechnical instrumentation are described in this chapter, and examples of instrumentation applications for various project types are given in Part 5.

3.1. BENEFITS DURING DESIGN

Instrumentation is used to provide input to the initial design of a facility or for the design of remedial treatment.

3.1.1. Definition of Initial Site Conditions

Instrumentation often plays a role in defining site conditions during the design phase of a project. For example, groundwater pressures and fluctuations must often be determined for design purposes, requiring use of piezometers. Knowledge of in situ stress and deformability conditions is sometimes required to permit rational design of a tunnel lining or as input to predictions of movements around a large excavation. Preconstruction conditions surveys are often made to define initial ground elevations and the condition of any structures that may be influenced by future construction.

3.1.2. Proof Testing

Because of uncertainties inherent in a design, specifications for geotechnical construction may require that the contractor conduct one or more proof tests to verify adequacy of design. Ideally, proof tests are performed as part of the design phase so that construction specifications can reflect test results, but time constraints or contractual restrictions often make this impossible.

A proof test will always include observations, which may include instrumentation. For example, specifications for pile-supported foundations usually call for one or more load tests before production pile driving, requiring the use of deformation gages and a load cell. Similar instrumentation is required for proof testing of tiebacks. When ground conditions differ from past experience and construction methods are uncertain, the designer can choose between an ultraconservative design, or an economical design based on results of a full-scale proof test. There are many examples in geotechnical engineering practice of full-scale tests designed to answer specific questions, including test embankments, test excavations, evaluation of vertical drain performance, and evaluation of effectiveness of bitumen coating in reducing downdrag load on driven piles.

3.1.3. Fact-Finding in Crisis Situations

If a crisis situation occurs, its characteristics must be defined so that remedial measures can be planned and put into practice. Instrumentation often plays a role in defining these characteristics. For example, measurements of water table position and fluctuation, together with failure plane depth, are needed to define the nature of a landslide.

3.2. BENEFITS DURING CONSTRUCTION

Instrumentation is used during construction to ensure safety, minimize construction costs, control construction procedures or schedules, provide legal protection, provide data for measurement of quantities, enhance public relations, and advance the state of the art.

Inherent in the use of instrumentation for construction reasons is the absolute necessity for deciding, in advance, a positive means for solving any problem that may be disclosed by the results of the observations (Peck, 1973). If the observations should demonstrate that remedial action is needed, that action must be based on appropriate, previously anticipated plans.

3.2.1. Safety

Safety is an essential consideration in all construction projects. Instrumentation programs can provide the needed safeguards, by indicating behavior with respect to threshold limits and by providing a forewarning of any adverse effects of construction. For example, there is often a need to monitor the effect of construction on adjacent structures, such as the measurement of deformation in and around an excavation as a means of ensuring safety of the lateral support. Use of instrumentation for safety monitoring is routine during excavations for buildings, subway tunnels, and highways in urban areas.

3.2.2. Observational Method

Construction is becoming increasingly expensive, nullifying the justification for overconservative designs, and construction costs can be reduced by the use of the *observational method*. Peck (1969a) lists the following ingredients for complete application of the observational method:

(a) Exploration sufficient to establish at least the general nature, pattern and properties of the deposits, but not necessarily in detail.
(b) Assessment of the most probable conditions and the most unfavourable conceivable deviations from these conditions. In this assessment geology often plays a major role.
(c) Establishment of the design based on a working hypothesis of behaviour anticipated under the most probable conditions.
(d) Selection of quantities to be observed as construction proceeds and calculation of their anticipated values on the basis of the working hypothesis.
(e) Calculation of values of the same quantities under the most unfavourable conditions compatible with the available data concerning the subsurface conditions.
(f) Selection in advance of a course of action or modification of design for every foreseeable significant deviation of the observational findings from those predicted on the basis of the working hypothesis.
(g) Measurement of quantities to be observed and evaluation of actual conditions.
(h) Modification of design to suit actual conditions.

Peck cites several examples of the method, two of which will illustrate the application. First, rather than accept the owner's conservative design for support of a braced excavation, the contractor opted to work with a lower factor of safety and ensure adequacy of support by measuring load in every strut. The contractor had additional struts available for immediate insertion if needed and achieved overall economy while providing positive assurance that no strut would become overloaded. Second, Peck describes use of the observational method during construction of a soft ground tunnel. It was feared that tunneling beneath a large building would cause settlement and resultant building damage. Protective work was designed—not included in the tunnel contract but available if the need should develop. The contractor started work several thousand feet from the building and advanced toward it, making extensive deformation measurements to judge the effect of construction on surface facilities. Well before reaching the critical building it became evident that the planned protective work would not be required.

3.2.3. Construction Control

Uncertainties in engineering properties or behavior during the design phase often affect construction procedures or schedules. The designer may therefore specify a program to monitor actual behavior during construction so that procedures or schedules can be modified in accordance with actual behavior. This use of instrumentation is normally referred to as *construction control,* even though it also plays a role in ensuring safety and reducing construction costs. For example, in the construction of an embankment on a soft clay deposit using staged construction procedures, instrumentation will normally be used to determine when the clay can support the next stage of fill. Similarly, instrumentation can be used to monitor contractor performance, thereby ensuring that contract requirements are met. For example, during excavation alongside an existing building, the designer may require that movements are limited to a specified amount to avoid damage to the building.

3.2.4. Providing Legal Protection

Where construction may affect neighboring property, instrumentation is useful in determining if there is a relationship between construction and changing conditions of that property. For example, if an open cut is to be excavated in a city, the designer or a building owner may use instrumentation to provide a bank of data concerning performance of adjacent structures during excavation, for possible use in the event of litigation. Instrumentation to provide legal protection will be used to a greater extent if the construction procedure is relatively new, or if there is a possible direct link between the construction procedure and damage to the property off the right of way, such as dewatering, or if nearby structures are particularly sensitive to ground movements or vibrations.

3.2.5. Measurement of Fill Quantities

When embankments are constructed on soft foundations and payment for fill quantities is based on measurements to the actual bottom of the fill, there is a need to determine the final elevation of the bottom of the fill. Settlement gages can be used for this purpose.

3.2.6. Enhancing Public Relations

Plans for an instrumentation program, indicating that the construction will be watched carefully, can give reassurance to the public and thus can expedite approval of the project. In situations where community or political obstacles threaten delays to approval, it may be appropriate for an owner to specify extensive instrumentation—more than is needed for technical reasons—to reassure the public that safety will be enhanced and that adverse problems will be minimized. Although such an approach will result in higher-than-normal instrumentation costs, it may in fact create an overall cost saving because the effects of inflation and other costs of delays will be reduced.

3.2.7. Advancing the State of the Art

Many advances in geotechnical engineering have resulted from field measurements. Often these measurements have been made for one of the project-specific reasons described above, and the general advance of knowledge has been a by-product. However, several notable practical research tests have been made to check and extend existing theories for soil and rock behavior and thus provide a basis for extending the state of the art for design of geotechnical construction. These research-oriented investigations, which usually require much more extensive field instrumentation than is required for other purposes, include the following:

- Measurement of stress and deformation in embankment dams to advance the state of the art of dam design.
- Measurement of total stress on culverts beneath highway embankments.
- Full-scale tests on individual piles and pile groups to determine load transfer relationships.
- Measurement of tunnel support load and ground deformation as input to improved support design procedures.
- Measurement of the effectiveness of various types of vertical drains.
- Measurement of rock behavior at elevated temperatures during studies relating to the disposal of high-level nuclear waste.

3.3. BENEFITS AFTER CONSTRUCTION IS COMPLETE

Engineers have an obligation to build safe structures, particularly if loss of life would result from lack of safety. Performance monitoring over the life of a structure, using observations and instrumentation, may be the expedient way to ensure long-term safety. Examples are given below:

- Long-term measurements of leakage, pore water pressure, and deformation are made frequently during the operating life of embankment dams.
- If permanent tiebacks are used for support of an excavation, surface and subsurface ground deformations may be measured, and perhaps also load in representative tiebacks.
- Where rockbolts have been used to stabilize a natural or an excavated slope in rock, fixed borehole extensometers may be installed to provide a means of indicating long-term performance of the rockbolts.
- If drainage arrangements have been provided to increase the stability of a slope or retaining wall, piezometers may be installed to check on long-term performance.

3.4. GENERAL CONSIDERATIONS

When the need for instrumentation is properly and correctly established, and when the program is properly planned, cost savings may be a direct result, as indicated by previous examples. However, instrumentation does not have to reduce costs to be justified. In some cases, instrumentation has been valuable in proving that the design is correct. In other cases, instrumentation might show that the design is inadequate, which may result in increased construction costs. However, the value of added safety and the avoidance of failure (and saving the cost of repairs) will make the instrumentation program cost effective.

CHAPTER 4

SYSTEMATIC APPROACH TO PLANNING MONITORING PROGRAMS USING GEOTECHNICAL INSTRUMENTATION

Planning a monitoring program using geotechnical instrumentation is similar to other engineering design efforts. A typical engineering design effort begins with a definition of an objective and proceeds through a series of logical steps to preparation of plans and specifications. Similarly, the task of planning a monitoring program should be a logical and comprehensive engineering process that begins with defining the objective and ends with planning how the measurement data will be implemented.

Unfortunately, there is a tendency among some engineers and geologists to proceed in an illogical manner, often first selecting an instrument, making measurements, and then wondering what to do with the measurement data. Franklin (1977) indicates that a monitoring program is a chain with many potential weak links and breaks down with greater facility and frequency than most other tasks in geotechnical engineering.

Systematic planning requires special effort and dedication on the part of responsible personnel. The planning effort should be undertaken by personnel with specialist expertise in applications of geotechnical instrumentation. Recognizing that instrumentation is merely a tool, rather than an end in itself, these personnel should be capable of working in a *team-player* capacity with the project design team.

Planning should proceed through the steps listed below. The steps are summarized in checklist form in Appendix A. All steps should, if possible, be completed before instrumentation work commences in the field.

4.1. DEFINE THE PROJECT CONDITIONS

If the engineer or geologist responsible for planning a monitoring program is familiar with the project, this step will usually be unnecessary. However, if the program is planned by others, a special effort must be made to become familiar with project conditions. These include project type and layout, subsurface stratigraphy and engineering properties of subsurface materials, groundwater conditions, status of nearby structures or other facilities, environmental conditions, and planned construction method. If the monitoring program has been instigated to assist in finding facts during a crisis situation (Chapter 3), all available knowledge of the situation should also be assimilated.

4.2. PREDICT MECHANISMS THAT CONTROL BEHAVIOR

Prior to developing a program of instrumentation, one or more working hypotheses must be developed for mechanisms that are likely to control behavior. The hypotheses must be based on a comprehensive knowledge of project conditions, as described above. Various mechanisms that control the behavior of soil and rock are described in Chapter 2.

4.3. DEFINE THE GEOTECHNICAL QUESTIONS THAT NEED TO BE ANSWERED

Every instrument on a project should be selected and placed to assist in answering a specific question: if there is no question, there should be no instrumentation. Before addressing measurement methods themselves, a listing should be made of geotechnical questions that are likely to arise during the design, construction, or operation phases. Various potential geotechnical questions are posed in Part 5 of this book.

4.4. DEFINE THE PURPOSE OF THE INSTRUMENTATION

Instrumentation should not be used unless there is a valid reason that can be defended. Benefits of using instrumentation are discussed in Chapter 3. When using this chapter to assist with planning a monitoring program, if engineers or geologists are unable to define a clear purpose for the program, they should cancel the program and proceed no further through this chapter. Peck (1984) states, "The legitimate uses of instrumentation are so many, and the questions that instruments and observation can answer so vital, that we should not risk discrediting their value by using them improperly or unnecessarily."

4.5. SELECT THE PARAMETERS TO BE MONITORED

Parameters include pore water pressure, joint water pressure, total stress, deformation, load and strain in structural members, and temperature. The question *which parameters are most significant?* should be answered.

Variations in parameters can result both from *causes* and *effects*. For example, the primary parameter of interest in a slope stability problem is usually deformation, which can be considered as the *effect* of the problem, but the *cause* is frequently groundwater conditions. By monitoring both cause and effect, a relationship between the two can often be developed, and action can be taken to remedy any undesirable effect by removing the cause.

Most measurements of pressure, stress, load, strain, and temperature are influenced by conditions within a very small zone and are therefore dependent on local characteristics of that zone. They are often essentially *point* measurements, subject to any variability in geologic or other characteristics, and may therefore not represent conditions on a larger scale. When this is the case, a large number of measurement points may be required before confidence can be placed in the data. On the other hand, many deformation measuring devices respond to movements within a large and representative zone. Data provided by a single instrument can therefore be meaningful, and deformation measurements are generally the most reliable and least ambiguous.

4.6. PREDICT MAGNITUDES OF CHANGE

Predictions are necessary so that required instrument ranges and required instrument sensitivities or accuracies can be selected.

An estimate of the maximum possible value, or the maximum value of interest, leads to a selection of instrument range. This estimate often requires substantial engineering judgment, but on occasion it can be made with a straightforward calculation, as is the case with maximum pore water pressure in a clay foundation beneath the centerline of an embankment.

An estimate of the minimum value of interest leads to a selection of instrument sensitivity or accuracy. There is a tendency to seek unnecessarily high accuracy, when in fact high accuracy should often be sacrificed for high reliability if the two are in conflict. High accuracy often goes hand in hand with delicacy and fragility. In some instances, high accuracy may be necessary where small changes in the measured variable have significant meaning, or where only a short time is available for defining trends, for example, when establishing the rate of slide movement from inclinometer data. Parametric studies with the aid of a computer can often be carried out to assist in establishing range, accuracy, and sensitivity.

If measurements are for construction control or safety purposes, a predetermination should be made of numerical values that indicate the need for remedial action. These numerical values will often be in terms of rate of measured change, rather than absolute magnitude. In his ingredients for the observational method, defined in Chapter 3, Peck (1969a) includes the following:

- Selection of quantities to be observed . . . and calculation of their anticipated values on the basis of the working hypothesis.
- Calculation of values of the same quantities under the most unfavourable conditions.

The first of the above steps allows any abnormalities to be recognized. The first and second steps allow the determination of *hazard warning levels*. Hazard warning levels may be based on clearly defined performance criteria—for example, where an acceptable differential settlement has been established for a structural foundation—or may be based on substantial engineering judgment, requiring a general assessment of ground behavior modes and mechanisms of potential problems or failures. When in doubt, several hazard warning levels should be established. As an example, Table 4.1 shows a hypothetical example of hazard warning levels and contingency actions for slope monitoring at an open pit mine.

The concept of green, yellow, and red hazard warning levels is also useful. Green indicates that all is well, yellow indicates the need for cautionary measures including an increase in monitoring frequency, and red indicates the need for timely remedial action. Fellenius et al. (1982) present a case history that illustrates this concept.

Table 4.1. Example of Hazard Warning Levels

Warning Level	Criterion	Action
1	Movement greater than 10 mm at any one survey station	Report to mine management
2	Movement greater than 15 mm at two adjacent stations; **or** velocity exceeding 15 mm per month at any one station	Verbal report and site meeting followed by written report and recommendations
3	Movement greater than 15 mm plus acceleration at any one station	Immediate site inspection by consulting engineer, site meeting and probable remedial measures (according to contingency plans)

Source: Franklin (1977). Reprinted with permission.

4.7. DEVISE REMEDIAL ACTION

Inherent in the use of instrumentation for construction purposes is the absolute necessity for deciding, in advance, a positive means for solving any problem that may be disclosed by the results of the observations (Peck, 1973). If the observations should demonstrate that remedial action is needed, that action must be based on appropriate, previously anticipated plans.

As described above, several hazard warning levels may be identified, each requiring a different plan. Planning should ensure that required labor and materials will be available so that remedial action can proceed with minimum and acceptable delay and so that personnel responsible for interpretation of instrumentation data will have contractual authority to initiate remedial action. An open communication channel should be maintained between design and construction personnel, so that remedial action can be discussed at any time. A special effort will often be required to keep this channel open, both because the two groups sometimes tend to avoid communication and because the contract for design personnel may have been terminated. Arrangements should be made to determine how all parties will be forewarned of the planned remedial actions.

4.8. ASSIGN TASKS FOR DESIGN, CONSTRUCTION, AND OPERATION PHASES

When assigning tasks for monitoring, the party with the greatest vested interest in the data should be given direct line responsibility for producing it accurately. The various tasks involved in accomplishing a monitoring program, together with alternative choices of the parties available for performing

Table 4.2. Chart Used for Task Assignment

Task	Responsible Party			
	Owner	Design Consultant	Instrumentation Specialist	Construction Contractor
Plan monitoring program				
Procure instruments and make factory calibrations				
Install instruments				
Maintain and calibrate instruments on regular schedule				
Establish and update data collection schedule				
Collect data				
Process and present data				
Interpret and report data				
Decide on implementation of results				

them, are listed in Table 4.2. It is useful to complete this chart during the planning stage by indicating the responsible party for each task.

Several of the tasks involve the participation of more than one party. In cases where the owner is also the designer, there will be no design consultant. Instrumentation specialists may be employees of the owner or the design consultant or may be consultants with special expertise in geotechnical instrumentation. All tasks assigned to instrumentation specialists should be under the supervision of one individual.

Chapters 5 and 6 include guidance on assigning tasks for instrument procurement and field services, and Section 6.6.5 indicates required qualifications of instrumentation specialists.

If construction contractors have economic or professional incentive to contribute toward good data, they should be assigned major responsibilities. If the instrumentation program has been instigated by the contractor, clearly the contractor will have responsibility for all tasks. However, if the instrumentation program has been instigated by the owner, as is usually the case, the construction contractor will often regard it as an interference with normal construction work and the contractor's participation should be minimized. The contractor will usually be responsible for providing support services during installation and access during the data collection phase. Instrument selection and procurement, factory calibration, installation, regular calibration and maintenance, and data collection, processing, and presentation should preferably be under the direct control of the owner or instrumentation specialist selected by the owner. When any of these tasks are performed by the construction contractor, data quality is often in doubt. Data interpretation and reporting should be the direct responsibility of the owner, the design consultant, or instrumentation specialist selected by the owner. Table 4.3 gives an example of task assignments for an owner-instigated monitoring program for which the contractor's cooperation is not assured. It is emphasized that Table 4.3 should not be used as a "cookbook": it is merely an example, and the needs of each project should be considered individually.

While completing Table 4.2, it may become evident that personnel are not available for all tasks, leading either to assignment of additional personnel or to a change in direction of the monitoring program. For example, if personnel available for data collection are insufficient, it may be appropriate to turn toward use of automatic data acquisition systems: this decision will affect instrument selection.

Task assignment should include planning of liaison and reporting channels. Assignments should clearly indicate who has overall responsibility and contractual authority for implementing the results of the observations.

4.9. SELECT INSTRUMENTS

The preceding eight steps should be completed before instruments are selected. Instruments are de-

Table 4.3. Example of Task Assignment for Owner-Instigated Monitoring Program

Task	Owner	Design Consultant	Instrumentation Specialist	Construction Contractor
Plan monitoring program	•	•	•	
Procure instruments and make factory calibrations		•	•	
Install instruments			•	•
Maintain and calibrate instruments on regular schedule			•	•
Establish and update data collection schedule		•	•	
Collect data			•	•
Process and present data			•	
Interpret and report data		•	•	
Decide on implementation of results	•	•		

(Responsible Party: Owner, Design Consultant, Instrumentation Specialist, Construction Contractor)

scribed in Part 3, and commercial sources are listed in Appendix D.

When selecting instruments, the overriding desirable feature is **reliability**. A recipe for reliability is given in Chapter 15. Inherent in reliability is maximum simplicity, and in general transducers can be placed in the following order of decreasing simplicity and reliability:

- Optical
- Mechanical
- Hydraulic
- Pneumatic
- Electrical

Lowest cost of an instrument should never be allowed to dominate the selection, and the least expensive instrument is not likely to result in minimum total cost. In evaluating the economics of alternative instruments, the overall cost of procuring, calibration, installation, maintenance, monitoring, and data processing should be compared.

The state of the art in hardware design is far ahead of the state of the art in user technology. It is the responsibility of users to develop an adequate level of understanding of the instruments that they select, and users will often benefit from discussing their application with geotechnical engineers or geologists on the manufacturer's staff before selecting instruments. They should discuss as much as possible about the application and seek out any limitations of the proposed instruments.

Instruments should have a good past performance record and should always have maximum durability in the installed environment. The environment for geotechnical instruments is harsh, and unfortunately some instruments are not sufficiently well designed for reliable operation in such an environment. Table 4.4 is a list of some of the main features of the instrument environment. Transducer, readout unit, and the communication system between the transducer and readout unit should be considered separately because different criteria may apply to each.

Table 4.4. Instrument Environments

1. Large deformations—often shearing deformations
2. High pressures—both solids and fluids
3. Corrosive—chemical (groundwater, grouts, concrete additives, bacteria) and electrolytic (electrolysis of dissimilar materials, stray electrical currents)
4. Temperature extremes—subfreezing to 100°F+ in the sun (temperature can be higher in certain instances, such as nuclear waste storage)
5. Shock—blasting, construction activities, rough handling during transportation to and from site
6. Vandalism, destruction by construction equipment, fly rock
7. Dust, dirt, mud, rain, chemical precipitates
8. High humidity, flowing or standing water
9. Erratic power supplies (electrical instruments)
10. Loss of accessibility to instruments when covered by rock, soil, shotcrete, and other supports

Source: After Cording et al. (1975).

With certain instruments, if a reading can be obtained, that reading is necessarily correct, while other instruments have a feature whereby calibration can be verified after installation; clearly, either feature is very desirable.

Instrument selection should recognize any limitations in skill or quantity of available personnel, identified while completing Table 4.2, and should consider both construction and long-term needs and conditions. Criteria for the two phases may be different and may entail selection of two different monitoring methods.

Other goals for instrument selection include good conformance (Chapter 7), minimum interference to construction, and minimum access difficulties while installing and reading.

The need for an automatic data acquisition system should be determined, and readouts should be selected in recognition of planned frequency and duration of readings. Unnecessary sophistication and automation should be avoided.

Action should be planned in the event any part of the system malfunctions, and the need for spare parts and standby readout units should be identified. Lead time for delivery and time available for instrument installation may affect instrument selection.

The final question is: *Will the selected instrument achieve the objective?* If an unproven instrument is selected, all parties should recognize the experimental nature of the instrument, and maximum backup should be provided, as described in Section 4.12.

4.10. SELECT INSTRUMENT LOCATIONS

The selection of instrument locations should reflect predicted behavior and should be compatible with the method of analysis that will later be used when interpreting the data. Finite element analyses are often helpful in identifying critical locations and preferred instrument orientations. A practical approach to selecting instrument locations entails three steps.

First, zones of particular concern are identified, such as structurally weak zones, most heavily loaded zones, or zones where highest pore water pressures are anticipated, and appropriate instrumentation is located. If there are no such zones, or if instruments are also to be located elsewhere, a second step is taken. A selection is made of zones, normally cross sections, where predicted behavior is considered representative of behavior as a whole. When considering which zones are representative, variations in both geology and construction procedures should be considered. These cross sections are then regarded as *primary instrumented sections,* and instruments are located to provide comprehensive performance data. There should usually be at least two such primary instrumented sections. Third, because the selection of representative zones may be incorrect, instrumentation should be installed at a number of *secondary instrumented sections,* to serve as indices of comparative behavior. Instruments at these secondary sections should be as simple as possible and should also be installed at the primary sections so that comparisons can be made. For example, instrumentation of a tieback wall might entail selection of two or three primary cross sections for installation of optical survey points, inclinometers, and load cells. Optical survey points would also be installed at a large number of secondary sections and used for monitoring both horizontal and vertical deformation of the wall. If in fact the behavior at a secondary section appears to be significantly different from the behavior at the primary sections, additional instrumentation may be installed at the secondary section as construction progresses.

When selecting locations, survivability of instruments should be considered, and additional quantities should be selected to replace instruments that may become inoperative. For example, Abramson and Green (1985) report on a survey of users, conducted to establish the required number of strain gages and load cells to compensate for losses occurring after installation. The survey indicates an average survivability rate for load cells of 75%, while for strain gages it is 60%.

Locations should generally be selected so that data can be obtained as early as possible during the construction process. Because of the inherent variability of soil and rock, it is usually unwise to rely on a single instrument as an indicator of performance.

Wherever possible, locations should be arranged to provide cross-checks between instrument types. For example, if both subsurface settlement and pore water pressure are to be measured in a clay subject to consolidation, piezometers should be located at mid-depth between settlement points. If both vertical inclinometers and horizontal fixed embankment extensometers are installed near each other and at the same cross section in the ground, an extensometer anchor should be installed near the inclinometer casing. However, care should be taken to avoid creating nonconformance or zones of weakness by excessive concentration of instruments in clusters.

Although instrument locations will usually be shown on the contract plans, flexibility should be maintained so that locations can be changed as new information becomes available during construction; thus, flexible installation specifications are required. Installation specifications are discussed in Chapter 6.

4.11. PLAN RECORDING OF FACTORS THAT MAY INFLUENCE MEASURED DATA

Measurements by themselves are rarely sufficient to provide useful conclusions. The use of instrumentation normally involves relating measurements to causes, and therefore complete records and diaries must be maintained of all factors that might cause changes in the measured parameters. As discussed in Section 4.5, a decision may have been made to monitor various causal parameters, and these should always include construction details and progress. Visual observations of expected and unusual behavior should also be recorded. Records should be kept of geology and other subsurface conditions and of environmental factors that may, in themselves, affect monitored data, for example, temperature, rainfall, snow, sun, and shade.

Details of each instrument installation should be recorded on *installation record sheets,* because local or unusual conditions often influence measured variables. Installation record sheets are discussed further in Chapter 17.

4.12. ESTABLISH PROCEDURES FOR ENSURING READING CORRECTNESS

Personnel responsible for instrumentation must be able to answer the question: *Is the instrument functioning correctly?* The ability to answer depends on availability of good evidence, for which planning is required. The answer can sometimes be provided by visual observations. For example, visual observations of tunnel lining behavior are essential when questioning correctness of apparently large lining strains or ground deformations during tunneling.

In critical situations, duplicate instruments can be used. A backup system is often useful and will often provide an answer to the question even when its accuracy is significantly less than that of the primary system. For example, optical survey can often be used to examine correctness of apparent movements at surface-mounted heads of instruments installed for monitoring subsurface deformation. Convergence measurements across an excavation can sometimes be used in a similar way when fixed borehole extensometers or inclinometers have been installed in adjacent ground.

Data correctness can also be evaluated by examining consistency. For example, in a consolidation situation, dissipation of pore water pressure should be consistent with measured settlement, and increase of pore water pressure should be consistent with added loading. Repeatability can also give a clue to data correctness, and it is often worthwhile to take many readings over a short time span to disclose whether or not lack of normal repeatability indicates suspect data.

Certain instruments have features that allow an in-place check to be made, and these checks should be made on a regular basis. For example, permeability tests can be made in twin-tube hydraulic piezometers to examine correct functioning. Some instruments have dual transducers. Some fixed borehole extensometers can be checked for free-sliding of the wire or rod by moving the instrument head outward and measuring elongation of the wire or rod.

4.13. LIST THE SPECIFIC PURPOSE OF EACH INSTRUMENT

At this point in the planning, it is useful to question whether all planned instruments are justified. Each planned instrument should be numbered and its purpose listed. If no viable specific purpose can be found for a planned instrument, it should be deleted.

4.14. PREPARE BUDGET

Even though the planning task is not complete, a budget should be prepared at this stage for all tasks listed in Table 4.2, to ensure that sufficient funds are indeed available. A frequent error in budget preparation is to underestimate the duration of the project and the real data collection and processing costs. If insufficient funds are available, the instrumentation program may have to be curtailed or more funds sought from the owner on a timely basis. Clearly, an application for more funds must be supported by reasons that can be defended.

4.15. WRITE INSTRUMENT PROCUREMENT SPECIFICATIONS

Attempts by users to design and manufacture instruments generally have not been successful, although joint efforts by user and manufacturer are sometimes undertaken. Instruments should therefore be purchased from established manufacturers, for which procurement specifications are usually needed. Alternative approaches are described in Chapter 5. At this time, the requirements for factory calibration should be determined and acceptance tests planned to ensure correct functioning when instruments are first received by the user. Responsibility for performing acceptance tests should be assigned. Guidelines for factory calibrations and acceptance tests are given in Chapter 16.

4.16. PLAN INSTALLATION

Installation procedures should be planned well in advance of scheduled installation dates, following the guidelines given in Chapter 17.

Written step-by-step procedures should be prepared, making use of the manufacturer's instruction manual and the designer's knowledge of specific site geotechnical conditions. The written procedures should include a detailed listing of required materials and tools, and installation record sheets should be prepared, for documenting factors that may influence measured data. The fact that the owner's personnel will install the instruments does not eliminate the need for written procedures. An example of an instrument installation procedure and an installation record sheet is included in Appendix G.

Staff training should be planned. Installation plans should be coordinated with the construction contractor and arrangements made for access and for protection of installed instruments from damage. An installation schedule should be prepared, consistent with the construction schedule.

4.17. PLAN REGULAR CALIBRATION AND MAINTENANCE

Regular calibration and maintenance should be planned, following the guidelines given in Chapter 16.

4.18. PLAN DATA COLLECTION, PROCESSING, PRESENTATION, INTERPRETATION, REPORTING, AND IMPLEMENTATION

Written procedures for data collection, processing, presentation, and interpretation should be prepared, following the guidelines given in Chapter 18.

The effort required for these tasks should not be underestimated. Many consulting engineering firms have files filled with large quantities of partially processed and undigested data because sufficient time or funds were not available for these tasks. The computer is a substantial aid but is no panacea.

Staff training should be planned. At this stage in the planning a verification should be made to ensure that remedial actions have been planned, that personnel responsible for interpretation of instrumentation data have contractual authority to initiate remedial action, that communication channels between design and construction personnel are open, and that arrangements have been made to forewarn all parties of the planned remedial actions.

4.19. WRITE CONTRACTUAL ARRANGEMENTS FOR FIELD INSTRUMENTATION SERVICES

Field services include instrument installation, regular calibration and maintenance, and data collection, processing, presentation, interpretation, and reporting. Contractual arrangements for the selection of personnel to provide these services may govern success or failure of a monitoring program. Alternative approaches are described in Chapter 6.

4.20. UPDATE BUDGET

Planning is now complete, and the budget for all tasks listed in Table 4.2 should be updated in light of all planning steps.

CHAPTER 5

SPECIFICATIONS FOR PROCUREMENT OF INSTRUMENTS*

Watson (1964) defines a specification as follows:

> A statement containing a minute description or enumeration of particulars, as of the terms of a contract or the details of construction not shown in an architect's drawings. A specification is definite, determinate, distinctly and plainly set forth, and stated in full and explicit terms. The reader will note from this definition that specifications are expected to be all-encompassing and exact. We may simply say that specifications should be clear, concise, complete and correct.

The last sentence of this definition should be emphasized. If specifications are **clear, concise, complete, and correct,** there is less chance of misunderstanding, delay, and conflict.

Procurement of other than the most simple geotechnical instruments should not be considered as a routine construction procurement item because, if valid measurements are to be made, extreme attention must be paid to quality and details. However, simple devices such as settlement platforms may be procured as routine construction items.

When planning the monitoring program in accordance with guidelines given in Chapter 4, an assignment of tasks will have been made, including responsibility for instrument procurement and factory calibration. Justification is given in this chapter for the procurement task assignment recommended in Chapter 4.

After the procurement task has been assigned, the next steps in preparation of procurement specifications are selection of specifying method and basis for determining price. In making these selections, the specifier should remember that the primary needs are high quality and reliability. The final step is to write the detailed specifications. These steps are described in turn.

5.1. TASK ASSIGNMENT FOR PROCUREMENT

Instruments can be procured by the construction contractor, by the owner, or by the design consultant. Alternatively, an assigned subcontract approach can be used, as described in Chapter 6, with instrument suppliers acting as assigned subcontractors. Advantages and limitations of the four options are given in Table 5.1. Procurement by the construction contractor is the least desirable option. Selection among the other three options depends on factors specific to each project.

In its manual on instrumentation for concrete structures, the U.S. Army Corps of Engineers (Corps of Engineers, 1980) acknowledges the need

*Written with the assistance of Robert E. Vansant, Black & Veatch, Kansas City, MO.

Table 5.1. Advantages and Limitations of Task Assignments for Procurement[a]

Procurement by	Advantages	Limitations
Construction contractor	Contractor's liability is clear	If *or equal* provision is required, specification covering all salient points is needed to guard against supply of an undesirable substitution Contractor will generally buy lowest-cost instruments, with risk of low quality and invalid measurement data
Owner	Minimum cost (because no markup) Owner has direct control over substitutions, inspection during manufacture, acceptability on receipt, and warranty service Can select between competitive bid method or (if permitted by owner's regulations) negotiation with one or more suppliers Flexible to accommodate changes	Contractor has no liability for nonperformance Owner may purchase lowest price instruments rather than highest quality
Design consultant	Design consultant has direct control over substitutions, inspection during manufacture, acceptability on receipt, and warranty service Can select between competitive bid method or (if permitted by owner's regulations) negotiation with one or more suppliers Flexible to accommodate changes	Contractor has no liability for nonperformance Cost is included in design fee
Instrument suppliers acting as assigned subcontractors	Owner or design consultant has direct control over substitutions, inspection during manufacture, and acceptability on receipt Can select between competitive bid method or (if permitted by owner's regulations) negotiation with one or more suppliers Flexible to accommodate changes	Contractor has no liability for nonperformance

[a] Procurement should preferably be under the direct control of the owner or design consultant; thus, the first option is the least desirable. Selection among the other three options depends on factors specific to each project.

to avoid procurement by the construction contractor:

> The general policy of the Chief of Engineers is to perform all civil works by contract unless it is in the best interest of the United States to accomplish the work by Government forces. However, the specialized nature of instrumentation facilities and the care required in the preparation, calibration, and placement of test apparatus demands that these features of work be retained under close operational control of the Corps of Engineers. In view of these conditions, direct procurement by the Government of embedded meters, cable, tubing, . . . , indicating or recording equipment, and similar items not normally encountered in construction work . . . is recommended.

5.2. SPECIFYING METHOD

There are two general categories of specifications: *descriptive specifications,* in which the details of the required instruments are specified, and *performance specifications,* in which the end result is specified. Descriptive specifications can be divided further into specifications with or without brand name and model number. A definition of each is given next, followed by recommendations.

5.2.1. Descriptive Specification, with Brand Name and Model Number

The specifier selects the method for making a measurement, including one or more preferred instrument models. The selection may be based on the specifier's previous experience, on the experience of others, or on the reputation of a particular manufacturer. When brand names are used in public work contracts, there should be at least two and preferably at least three sources to make sure that there is effective competition. If only one model is considered suitable, the specifier may write a sole source procurement specification, excluding use of a substitution. On public work where there is only a single source, a rigorous sole source procurement justification will generally be required, and it will usually be preferable for the owner to purchase the instruments and furnish them to the construction contractor. Such an approach facilitates the justification of sole source, allows more effective price negotiation, and precludes the sole source manufacturer from "packaging" that instrument with other instruments required for the project.

The owner may require the specifier to add an *or equal* provision to the brand name and model number, in which case great care must be taken to define what is acceptable as a substitution. *Or equal* specifications are discussed in Section 5.4.6.

5.2.2. Descriptive Specification, without Brand Name and Model Number

The specifier selects the method for making a measurement and writes a generic specification to define the required characteristics of the system and components. For example, a specification may call for pneumatic piezometers, with certain size, mechanical, pneumatic, and other requirements. A descriptive specification may be so limiting that the requirements can be satisfied only by one commercial model.

5.2.3. Performance Specification

In this method the end result is specified. For example, the specification might state that water pressure measurement devices shall have a certain range and accuracy and must operate for a certain period in the geotechnical environment described in the specifications. The specification should state how the manufacturer's proposed instrument will be evaluated.

5.2.4. Advantages and Limitations of Specifying Methods

Advantages and limitations of the three methods are given in Table 5.2. The second method requires a thorough generic specification, generally not within the capability of geotechnical engineer specifiers, and therefore it is not often used. The third method assumes that the manufacturer will understand instrument design criteria dictated by the specific geotechnical environment and thus is applicable only if the manufacturer's staff includes geotechnical engineers and if the geotechnical environment is clearly specified. Although the second or third method may on occasion be the method of choice, the first method is the most common and is satisfactory provided that a specification covering all salient features is written to guard against supply of an undesirable substitution.

5.3. BASIS FOR DETERMINING PRICE

If instruments are procured directly by the owner or design consultant, price can be determined either by bidding or by negotiation. If instruments are procured through the construction contract, bid items can be included, or the assigned subcontract approach (Chapter 6) can be used to allow the owner or design consultant to retain control over selection of the equipment and basis for determining price.

Advantages and limitations of negotiation and bidding are given in Table 5.3.

If instruments are procured by the owner or design consultant, negotiation is generally preferable, but public agencies may be required to take bids. However, it was a surprise to the author to read (Perlman, 1985):

> Despite the fact that the general policy of the Government is to acquire goods and services on a competitive basis, most Government procurement dollars are spent on contracts awarded *noncompetitively* (i.e., sole source contracts). Of the $146.9 billion spent by the Federal Government in fiscal year (FY) 1982 for goods, $79.2 billion (54%) was expended noncompetitively.
>
> Questions regarding the use (or overuse) of sole source procurements concern every member of the procurement community. From the Government's standpoint, if a contract is improperly awarded on a sole source basis, the Government may pay higher prices than if competition had been sought. On the other hand, *failure* to permit a sole source award in *appropriate* circumstances may unnecessarily (a) increase the price paid for the item, (b) place a greater administrative burden on Government personnel, since competitive procurements require ap-

Table 5.2. Advantages and Limitations of Various Specifying Methods for Procurement of Instruments[a]

Method	Advantages	Limitations
Descriptive specification, with brand name and model number	Most direct way of defining preferred instrument	If *or equal* provision is required, specification covering all salient features is needed to guard against supply of an undesirable substitution Owner may have to accept responsibility if specified model number does not perform adequately
Descriptive specification, without brand name and model number	No bias toward particular model(s) Can specify general material requirements	Specifier must have necessary knowledge to write specification Specified requirements may limit innovation on part of manufacturer
Performance specification	Allows maximum innovation on part of manufacturer Manufacturer has contractual commitment to furnish a device that performs adequately Permits maximum competition No bias toward particular model(s)	Assumes manufacturer will understand design criteria dictated by the specific geotechnical environment; thus, applicable only if manufacturer's staff includes geotechnical engineers Can be difficult to evaluate whether manufacturer's proposed instrument will perform as needed

[a] The first method is the most common and is satisfactory provided that a specification covering all salient features is written to guard against supply of an undesirable substitution.

> proximately twice as much work as sole source awards, and (c) cause the Government user to receive a product that merely meets the minimum requirements of a solicitation, instead of the one that is *best* for the job.

It is therefore apparent that there is ample precedent for procuring geotechnical instruments on a sole source basis, in order to avoid *receiving a product that merely meets the minimum requirements of a solicitation, instead of the one that is best for the job.*

If instruments are procured through the construction contract, the assigned subcontract approach is preferable, with price negotiation by the owner or design consultant.

Two discussions of a paper by Sherard (1981) reinforce quality concerns inherent in the bid procedure. Green (1982) states:

Table 5.3. Advantages and Limitations of Procedures for Determining Price[a]

Procedure	Advantages	Limitations
Bid	Procurement may be least costly	Requires preparation of comprehensive bid specification Lowest cost runs risk of low quality and invalid measurement data Not flexible to accommodate changes Adequate lead time required
Negotiation	Owner or design consultant has direct control over quality and price Can use preferred sources Flexible to accommodate changes	Often resisted by public agency procurement community

[a] Negotiation is generally preferable. If instruments are procured through the construction contract, negotiation procedure entails use of the assigned subcontract approach.

> On the subject of cost, the writer is in entire agreement with the author [Sherard] that instrumentation hardware cost is so small that it should not be the deciding factor. However, this is not the case in practice. In far too many instances, instrumentation hardware is put out to bid on the basis of a . . . make and model number, or approved equal. Despite specification writers' good intentions, it is possible for manufacturers to use lower quality components, for example, the leads, and the commonly used low bid process encourages this. Until high quality can be adequately specified, instrument procurement on a low bid basis will remain a stumbling block to good field performance.

Mikkelsen (1982) presents a manufacturer's perspective:

> Sales often go to the lowest bidder on materials. Such practices are often unwise since a customer generally gets what he pays for. Also, it is unwise because, as the author [Sherard] also points out, material costs are only a small part of the instrument program cost. This practice also promotes use of marginal and inferior materials. A *manufacturer's dilemma* is created because there is little incentive to make product improvements and use higher quality materials that increase the product cost. A well-designed, high quality product, thoroughly tested, is a total loss to the manufacturer unless it is sold. Unless more informed buyers come forth and a change in the practices of *low bid* procurement occurs, desirable advances in field instrumentation will be slow and unsatisfying.

In his closure, Sherard (1982) agrees:

> The writer agrees completely with Mikkelsen's comments under the heading **Cost,** and similar opinions of Green. Clearly, there is an important problem in the industry created by common *low-bid* procurement and specification of a certain instrument, *or equal.* To combat this problem, the writer suggests that designers should consider to select the instrument (or two) believed to be most suitable, and specify it with no substitution allowed. This practice would allow instrument manufacturers to market products of consistently high quality. The common *or equal* clause, combined with competitive bidding, leads inevitably to excessive emphasis on economy, with the result that high-quality instruments cannot compete. This keeps the quality of the average instrument on the market just above the "acceptable" level, a highly undesirable situation.

5.4. CONTENT OF SPECIFICATIONS FOR PROCUREMENT OF INSTRUMENTS

Depending on the selection of specifying method and basis for determining price, the procurement specification should address some or all of the requirements discussed in this section. Greatest detail is required when the low-bid procedure with an *or equal* provision is used. Some requirements apply to the entire instrumentation system, usually consisting of a transducer, readout unit, and a communication system, and some apply only to one or more of these components.

The specification will usually be written with three main headings. The first heading, **Part 1. General,** will include wording applicable to all instruments, as described in Sections 5.4.1–5.4.10. The second heading, **Part 2. Instrument Details,** will include wording under separate subheadings for each instrument, as described in Sections 5.4.11–5.4.17. The third heading, **Part 3. Measurement and Payment,** will include content described in Section 5.4.18.

The procurement specification is sometimes combined with a specification for field services, following the recommendations given in Chapter 6.

Technical information for the specification should be provided by geotechnical engineers working on the project design, and these same engineers may draft the technical specification, following guidelines given in this chapter. However, the final specification should be reviewed by an experienced specification writer. A review by an instrumentation specialist may also be necessary. If instruments are to be procured as part of a construction contract, the specification should also be reviewed by personnel familiar with the General Conditions and other sections of the technical specifications, to ensure consistency.

Before enumerating the items that should be considered when writing specifications for procurement of instruments, the author will reiterate a key point. The quotations in Section 5.3 will illustrate problems inherent in the low-bid procedure. **Unless you must, don't use the low-bid procedure! The author regrets very much the need to write such a lengthy section as Section 5.4: the length is necessary only because the low-bid procedure continues to be used.** Its use creates lengthy specifications that appear out of proportion to other parts of the specifications, and the author believes that its use is not in the best interests of either manufacturers or users.

5.4.1. Division of Responsibilities

As described in Chapter 4, the division of responsibilities among owner, design consultant, instrumentation specialist, and construction contractor will have been determined during the planning phase. Section 5.1 indicates preferred responsibility for instrument procurement. The procurement specification should contain a clear statement of the work included and should indicate who will be responsible for factory calibration and quality assurance, for review of proposed instruments, and for performing acceptance tests to ensure correct functioning when instruments are first received by the user.

5.4.2. Submittals

The specification will usually include a summary listing of required submittals to the owner or design consultant. Submittals are described in later sections and may include experience lists, requests for review of proposed instruments, calibration certificates, quality assurance checklists, warranties, instruction manuals, shipping documents, and instrument samples. The schedule for all submittals should be specified.

5.4.3. Operating Environment

Instruments are normally required to operate in adverse environments. The specification should provide a general description of the operating environment, including soil or rock type, and all relevant environmental factors listed in Table 4.4. Recognizing that many instrument failures occur as the ground deforms around them, potential deformations should be evaluated and defined, with particular attention given to deformations of buried tubes, cables, and pipes.

5.4.4. Experience

Manufacturers may be required to submit an experience list indicating records of prior instrument use. However, requirements for extensive experience should not be imposed if they may exclude competent new suppliers. Experience lists should include names of firms who have used the instrument, project names, and names and telephone numbers of responsible individuals. When the procurement entails innovation, manufacturers may be required to cite examples of previous successful innovations.

5.4.5. General Material Requirements

Descriptive specifications should include appropriate mechanical, hydraulic, pneumatic, and electrical requirements. The transducer, readout unit, and communication system should each be covered separately. Recommendations for selection of transducers and components are given in Chapter 8.

To assist the manufacturer in selecting materials, desired longevity should be stated, but it is difficult to hold manufacturers to longevity specifications. If an instrument malfunctions after installation, it will generally be unclear whether the problem has been caused by a manufacturing defect or an installation error. When a brand name *or equal* specification is used, good practice dictates that the specification identify all salient characteristics that are considered important for the evaluation of substitutions. Table 5.4 indicates some of the factors to be considered when specifying general material requirements and acceptability of substitutions.

5.4.6. Review of Proposed Instruments

For all cases except sole source procurement, the specification should include requirements for review of proposed instruments, including review of substitutions if brand names and model numbers have been specified.

If instruments are procured by the construction contractor, using the low-bid procedure, the following wording, under a heading **Or Equal** in the General Conditions or in the procurement specification, will minimize the possibility of an unacceptable substitution. Similar wording is applicable when instruments are procured by the owner or design consultant and the low-bid procedure is used.

> The term *or equal* shall be understood to indicate that the *equal* product is the same or better than the product named in the Specifications in function, performance, reliability, quality, and general configuration.
>
> Except for Owner-selected materials and materials where no substitution is clearly specified, whenever any material is indicated or specified by proprietary name, by name of manufacturer, or by catalog number, such specifications shall be deemed to be used for the purpose of establishing a standard of quality and facilitating the description of the material desired. This procedure is not to be con-

Table 5.4. General Material Requirements

Instrument Type	Requirement
All (mechanical, hydraulic, pneumatic, and electrical)	Transducer type Materials specification Environmental factors in Table 4.4, with particular attention to (1) surviving ground deformations, (2) waterproofing needs, (3) chemical corrosion, (4) electrolytic corrosion (dissimilar metals can be separated by inert materials or by passivation methods), (5) temperature sensitivity, and (6) packaging for outdoor operation and rough handling Permanent labeling of all controls and connectors Dust covers on all connectors, anchored to the connector Humidity indicator in readout unit
Hydraulic	Constituents of liquid Dissolved gas content in liquid Liquid freezing point Tubing material Tubing valves and fittings Pressure indicator
Pneumatic	Container for gas Constituents of gas Moisture content in gas Filter for gas Tubing material Tubing valves and fittings Pressure indicator Gas flow rate Gas flow controller Gas flow indicator
Electrical	Battery characteristics Auxillary power supply Fuse Lightning protection Cable Connectors Cable splices Indicator Electrical safety Power consumption

strued as eliminating from competition other materials of equal quality by other manufacturers where fully suitable in design, and shall be deemed to be followed by the words *or equal*. The Contractor may, in such cases, submit complete comparative data to the Engineer for consideration of another material which shall be substantially equal in every respect to that so indicated or specified. Substitute materials shall not be ordered, delivered to the site, or used in the work unless approved by the Engineer in writing. The Engineer will be the sole judge of the substituted article or material.

Requests for review of equivalency will not be accepted from anyone except the Contractor and such request will not be considered until after the bids have been opened. All requests from the Contractor for consideration of a substitution shall clearly state any deviation from the Contract, and differences in cost.

5.4.7. Factory Calibration and Quality Assurance

Instruments should be calibrated, inspected, and tested before shipment to the user, and the procurement specification should indicate required factory calibrations.

Calibrations should be made while the variable parameter is both increasing and decreasing for at least two cycles, to document hysteresis, throughout the maximum range expected to occur in the field. Readings should be taken at about ten equal increments, and the manufacturer should be required to supply a calibration curve with data points clearly indicated, and a tabulation of the data. Each instrument should be marked with a unique identification number. In general, the readout unit that is to be supplied to the user should be used during factory calibrations, and calibrations should be made at several different temperatures to determine the effect of temperature change on the instruments. Calibration specifications should include the following wording:

> Certification shall be provided with each shipment to indicate that the manufacturer's test equipment is calibrated and maintained in accordance with the test equipment manufacturer's calibration requirements, and that all calibrations are traceable to . . .

In the United States, traceability should be to the National Bureau of Standards. However, the practicability of such a requirement should be verified with manufacturers whose products are being considered, since calibration of some commonly used instruments such as inclinometer probes is generally not traceable to standard agencies. Manufacturers, of course, may make an additional charge for calibrations requiring more effort than their standard procedure.

Manufacturers will generally apply their own quality control procedures, but these may not always be adequate, and it is often worthwhile to specify preshipment inspections:

> A final quality assurance inspection shall be made of each instrument prior to shipment. During the inspection, a checklist shall be completed to indicate each inspection and test detail. A completed copy of the checklist shall be supplied with each instrument.

5.4.8. Warranty

Manufacturers will generally warrant their products against defects in material and workmanship for a period ranging from 3 months to 1 year after delivery, and such a requirement should be included in the procurement specification. Standard warranties of the manufacturers whose products are being considered should be studied before specifying warranty terms and duration. However, unless indisputable evidence is provided, most manufacturers will not accept responsibility for malfunction of embedded components after installation, and therefore every effort should be made to specify instruments with proven longevity.

5.4.9. Instruction Manual

The procurement specification should indicate instruction manual requirements. Instruction manuals of the various instrument manufacturers vary greatly in content and quality. Although a comprehensive manual is valuable to the user, it may be impracticable to require a custom-made manual because of time or cost. The requirements should be realistic in light of manufacturers' existing manuals, available time, and the cost of preparing a custom manual.

A comprehensive instruction manual need not and should not be complicated: if it is, it will rarely be read. However, it should contain at least the following:

Purpose of Instrument

- Parameter measured.
- Applications.

Theory of Operation

- Basic measuring principle of instrument with appropriate illustrations, schematics, and circuit diagrams for each component.
- Limitations of system.
- Factors that affect measurement uncertainty.
- Specification sheet.

Calibration Procedure

- Step-by-step acceptance test procedure to ensure correct functioning when instruments are first received by the user. The manual should emphasize the importance of making acceptance tests immediately on receipt.
- Procedure for any possible regular in-place calibration checks on embedded components, including equipment required and recommended frequency.
- Procedure for regular calibration of readout instruments, including equipment required and recommended frequency. If possible, these procedures should be in such a form that check calibrations can be performed by the user or by local calibration houses, thereby avoiding the necessity of returning instruments to the factory for calibration.

Installation Procedure

- Step-by-step procedure for installation, with illustrations of the system and components, showing their correct juxtaposition when installed. Alternative procedures may be required if installation methods are dependent on the application or on field conditions. The procedures should include a listing of materials, tools, and spares required during assembly and installation and any borehole requirements, indicating maximum and minimum borehole diameters and restrictions on wall roughness. Procedures should emphasize the most critical steps and should indicate pitfalls to be avoided. (The user should recognize that, although the manufacturer can suggest installation procedures, those procedures should be planned to suit the specific site geotechnical conditions, well in advance of scheduled installation dates. Planning guidelines are given in Chapter 4.)
- Statement of all factors that should be recorded during installation for later use during data

evaluation. (These factors, as discussed in Chapter 17, will be included on the installation record sheet.)

Maintenance Procedure

- Regular maintenance procedures, with illustrations, including any appropriate disassembly instructions.
- Cleaning, drying, and lubricating instructions.
- Battery service and charging instructions, if required.
- Frequency of required maintenance.
- Recommended spare parts list, including consumables.
- List of part numbers and manufacturers of standard parts to facilitate local procurement by user.
- Troubleshooting guide, with appropriate illustrations, including a list of failure indications and probable cause and corrective action requirements for each listed failure.
- Names, addresses, and telephone numbers of instrument service representatives.

Data Collection Procedure

- Step-by-step procedure for equipment set-up and turn-on.
- Functional explanation of each connector and control.
- Cautions pertaining to personnel and equipment.
- Statement of procedure for obtaining initial reading, with appropriate illustrations.
- Statement of procedure for obtaining readings subsequent to initial reading.
- Statement of construction or environmental factors that might cause changes in the measured data. (Users should recognize that, although the manufacturer can suggest these factors, specific procedures should be planned for recording factors that may influence measured data, well in advance of instrument installation. Planning guidelines are given in Chapter 4.)
- List of equipment and tools required during reading.
- Field data sheet.
- Sample completed field data sheet.

Data Processing, Presentation, and Interpretation Procedures

- Data calculation sheet.
- Step-by-step calculation procedure, including an instruction manual for any computer program supplied by the manufacturer.
- Sample data calculation.
- Alternative methods of plotting data.
- Sample data plots.
- Notes on data interpretation.

5.4.10. Shipment and Delivery

Delivery dates should be specified, or suppliers should be required to state the time for delivery. If lead time is short and installations are to be made at several different times, a schedule of partial shipments and dates should be prepared. The method of shipment can be specified, or manufacturers can be given freedom to ship at their own option to meet the specified delivery dates. Specifications should indicate whether freight charges should be prepaid.

Unless specified otherwise, the manufacturer's responsibility for the instruments ends when they have been delivered to the carrier for shipment. Therefore, the specification should indicate requirements for insurance against loss and damage in transit. Standard transit insurance often requires notice of damage within a relatively short period after delivery; thus, the user should make a careful inspection before this time has expired.

5.4.11. Instrument Operating Principles

This is the first section of the specification under the heading, **Part 2. Instrument Details**, requiring wording under separate subheadings for each instrument.

A general description of each system and its components should be given. The operating principle should be stated, and general characteristics of the transducer, readout unit, and communication system should be defined.

5.4.12. Component Specifications

The general approach to specification of components will depend on the selection of specifying method. With this in mind, the general criteria given in Section 4.9 for selecting instruments should be reviewed. Component specifications should formalize decisions made while systematically planning the monitoring program: instrument range, acceptable uncertainty, conformance requirements, and method of verifying reading correctness. If nonstandard instruments or components are required, detail design may be necessary at this stage.

Manufacturers' price lists are often very helpful when identifying and specifying components that make up an instrumentation system. The product literature may not convey clearly the need for each component, but component listings in price lists will usually do so, and a study of these lists will help to avoid omission of an essential component.

5.4.13. Compatibility with Other Instruments

A newly procured instrument may be used in conjunction with previously procured instruments. For example, a new inclinometer probe may be required for use in existing inclinometer casing, or additional strain gages may be required for use with a previously purchased readout unit. In these cases, care must be taken to specify the features required for mechanical, hydraulic, pneumatic, and/or electrical compatibility.

5.4.14. Physical Size Limitations

Physical size limitations of each component should be defined. Components may have to fit in a certain diameter or length of borehole, or terminal size may be limited by terminal enclosure dimensions. Physical size and weight of readout instruments are often limited to permit easy carrying.

5.4.15. Submittal of Samples

The specification may require submittal of a sample instrument or instruments for testing, in which case the characteristics to be tested and a submittal schedule should be indicated.

5.4.16. Installation Tools and Materials

As described in Chapter 4, written installation procedures should include a listing of required tools and materials. The tools and materials should either be included in the instrument procurement specification or be procured separately.

5.4.17. Spare Parts

Spare parts may be required either to replace components damaged during installation or to replace accessible malfunctioning components during operation. A listing of appropriate spare parts should be made and included in the procurement specification.

Provision should be made for spare or standby readout units, for use in case of malfunction. Options are procurement of a spare unit, arrangement with the manufacturer for rental of a spare unit if needed, and a standby equipment lease. A standby equipment lease entails making arrangements with the manufacturer for a spare unit to be held in readiness at the factory, to be called on if needed. The manufacturer will generally charge either a lump sum or a monthly amount for holding the equipment. If the spare is in fact called on, payment will either be covered by the lease, or an additional payment will be made. However, the standby equipment lease procedure is not available from all instrument manufacturers. Some prefer to rent spare items and allow some or all of the rental cost to apply to a later purchase. A readout unit may also be available at short notice on a temporary basis from another user whose current day-to-day needs are low.

5.4.18. Measurement and Payment

When instruments are procured as part of a construction contract, a measurement and payment specification section is required. Instrument procurement should be measured and paid for on a unit price basis rather than a lump sum, to allow for changes in quantities as work proceeds. The unit price schedule should contain sufficient items so that major component prices are defined. For example, when requiring piezometers with attached tubing, either the length of tubing should be specified for each piezometer, or a separate unit price item should be provided for tubing. When requiring multipoint fixed borehole extensometers, either the distances between each anchor and the head should be specified, or a separate unit price item should be provided for heads, anchors, and connecting linkage.

A special effort should be made to ensure that a unit price pay item is included for every specified requirement. If payment for a specified requirement is deemed to be included in the price bid for a particular item (e.g., if payment for factory calibration is deemed to be included in the unit price for a diaphragm piezometer), this should be stated clearly.

If instruments are to be procured by use of the assigned subcontract approach, clear wording is required to define quantities, payment, and markup. The following wording is suitable:

> Payment for procurement of instruments will be made under allowance item no. ______. The Owner's Representative will determine instrument descriptions, sources, quantities, and prices and will provide this information to the Contractor. The Contractor shall place orders with assigned instrument suppliers within three working days after receipt of instructions from the Owner's Representative. The Contractor shall submit copies of suppliers' invoices and will be reimbursed in accordance with those invoices, plus the specified (or bid) markup. If the actual total cost of instruments is more or less than the allowance, the contract price will be adjusted to the actual approved amount.

Payment for procurement of instruments is often combined with payment for installation by using several *furnish and install* unit price items in the bid. Measurement and payment recommendations for this case are given in Chapter 6.

5.4.19. Checklist for Specification Content

Appendix B contains a checklist for content of specifications for procurement of instruments. Use of the checklist will help ensure completeness. However, each individual specification may require only some of this content.

CHAPTER 6

CONTRACTUAL ARRANGEMENTS FOR FIELD INSTRUMENTATION SERVICES*

Field instrumentation services include installation, regular calibration and maintenance, and data collection, processing, presentation, and interpretation. Selection of personnel responsible for field services may govern the success or failure of a monitoring program. Even if the program has been planned in a complete and systematic way and appropriate instruments have been procured, measured data may not be reliable unless field personnel perform high-quality work. Geotechnical instrumentation field work should not be considered a routine item of construction work, because successful measurements require extreme dedication to detail throughout all phases of the work. Specifications should be clear, concise, complete, and correct (Watson, 1964).

When planning the monitoring program in accordance with guidelines given in Chapter 4, an assignment of tasks will have been made. Contractual arrangements for each field task are discussed in this chapter, and justifications are given for task assignments recommended in Chapter 4.

*Written with the assistance of Robert E. Vansant, Black & Veatch, Kansas City, MO.

6.1. GOALS OF CONTRACTUAL ARRANGEMENTS

The primary goals of contractual arrangements for field instrumentation services are (1) to ensure high-quality work at acceptable cost to the owner, (2) to create a cooperative working relationship between specialty intsrumentation personnel and the construction contractor, and (3) to permit flexibility to accommodate changes during progress of the work. Flexibility is needed because unforeseen factors are often revealed during construction, requiring a change of instrument types and/or locations.

6.2. DEFINITION OF TERMS

Definitions of terms used in the discussion on contractual arrangements are given below.

6.2.1. Support Work

Tasks that are within the capability of the average construction contractor are considered *support work*.

6.2.2. Biddable Support Work

Support work of a production nature, for example, production drilling, protection from damage, and furnishing and installing simple instruments such as settlement platforms, is considered *biddable support work.* Support work that can be bid by the hour, for example, drill rig and crew to assist during instrument installation, is also considered biddable support work.

6.2.3. Nonbiddable Support Work

Support work that will be controlled by the owner's instrumentation schedules or procedures, for example, access for reading; support work that is not defined at the time of bid, for example, revised instrumentation owing to geologic conditions revealed during construction; and contractor's assistance that will be needed on an *as required* basis, for example, assistance of tradespeople, are all considered *nonbiddable support work.*

6.2.4. Specialist Work

All instrumentation tasks outside the capability of the average construction contractor are considered *specialist work.*

6.2.5. Force Account and Contingency Allowance

Force account is a method of payment to the construction contractor for work additional to the specified work, whereby reimbursement is on a time and materials basis. The cost of force account work may be included in the total bid price by adding a *contingency allowance* in the contract. A line item in the bid schedule is identified as a contingency allowance, and the owner's cost estimate is entered in the amount column. Because the work is not specified, a single contingency allowance can be used for many items of work. The cost of force account work is sometimes not included in the total bid price, in which case there must be a separate contingency fund.

The contingency allowance procedure can also be used to pay for nonbiddable support work, in which case the item description in the bid schedule will be *Instrumentation support work* and the specification will indicate the work included and the method of establishing payment rates, which are normally the same as force account rates.

Payment rates for labor are normally based on wage rates actually paid by the contractor, plus direct payroll overhead such as social security, unemployment, and pension funds. Payment rates for plant and equipment are normally based on rental rates accepted by the industry. Payment rates for materials are based on actual cost, supported by receipts. The construction contactor is normally allowed a specified or agreed markup on all payment rates. If force account work is performed by a subcontractor, both the construction contractor and subcontractor are normally allowed a markup.

6.2.6. Assigned Subcontract

For an *assigned subcontract,* the owner negotiates the purchase and assigns the contract to the construction contractor for administration. Payment is made on the basis of actual work done, and the cost is included in the total bid price. A line item in the bid schedule is designated as an *allowance item* and appropriate wording entered in the description column to define the work category. The owner's cost estimate is included in the bid schedule. After contract award, the construction contractor is instructed to enter into a subcontract with the assigned subcontractor, and payment is made to the subcontractor via the construction contractor under the allowance item. The contractor's monthly payment requests to the owner are supported by including copies of subcontractor invoices.

There are two circumstances in which assigned subcontracts are expedient in geotechnical instrumentation specifications:

1. *For Procurement of Instruments.* When instruments are procured through the construction contractor (Chapter 5), the assigned subcontractor approach can be used to allow the owner or design consultant to retain control over selection of equipment and basis for determining price. The item description will be *Furnish instruments.* The owner specifies that, after contract award, the owner's representative will determine instrument descriptions, sources, quantities, and prices and will provide this information to the contractor. The contractor is then required to place orders, within a specified time period, and the instrument suppliers become assigned subcontractors.
2. *For Specialist Field Work.* When specialist field work is to be performed by specialist personnel selected by the owner, the cost can be included in the total bid price by use of the assigned subcontractor approach. The item

description will be *Provide services of specialty field instrumentation personnel.* The owner specifies that, after contract award, the contractor will be instructed to enter into a subcontract with an organization selected by the owner and agreeable to the contractor, and the organization becomes an assigned subcontractor.

Payment to the construction contractor can be at actual cost without markup, but this inhibits the contractor's willingness to cooperate and is not recommended. A markup for overhead and profit can be specified, in which case the cost estimate entered by the owner in the bid schedule will be the marked up amount. Alternatively, bidders can be invited to bid a markup, in which case bidders will enter the percent markup in the bid schedule and extend the marked up estimate to the amount column.

The owner's estimate should not be regarded as a *not to exceed* figure, and the contract price should be increased by change order if needed.

6.3. CONTRACTUAL ARRANGEMENTS FOR INSTRUMENT INSTALLATION

Five basic contractual arrangements for installing instruments are described below. Recommendations are given in Section 6.3.6.

6.3.1. Method 1. Specialist Installation Work by Owner's Personnel

The owner's personnel perform all *specialist work,* sometimes using their own equipment such as drill rigs, and *support work* is performed by the construction contractor. *Biddable support work* is bid, and *nonbiddable support work* is paid for at *force account* rates under a *contingency allowance.*

6.3.2. Method 2. Bid Items in Construction Contract, without Prequalification

All instrument installation work is included in the construction contract, usually specified as *furnish and install,* and bid on a unit price or lump sum basis. No qualification requirements for specialist personnel are included, and no distinction is made between *specialist* and *support work.* If the work is specified as *furnish and install,* the specification should include sections on procurement of instruments, as described in Chapter 5.

6.3.3. Method 3. Bid Items in Construction Contract, with Prequalification

Contractual arrangements are similar to the preceding method, and all instrument installation work is bid. However, the specification includes a requirement that all installation work shall be performed under the direct supervision of suitably qualified personnel. Qualification requirements are discussed in Section 6.6.5.

6.3.4. Method 4. Instrumentation Specialist Selected by and Contracting with Owner

Specialist work is performed by an instrumentation specialist under contract to the owner, the design consultant, or the construction manager. The owner or design consultant determines which tasks are within the capability of the average construction contractor and designates these as *support work.* The remainder of the instrument installation tasks are designated as *specialist work,* and the owner or design consultant selects personnel to perform *specialist work,* in accordance with appropriate parts of the following recommendations given in the proposed new text of *ASCE Manual No. 45* (ASCE, 1987). In this extract, *consulting engineers* refers to instrumentation specialists:

1. By invitation or by public notice, state the general nature of the project and services required and request statements of qualifications and experience from consulting engineers who appear to be capable of meeting the project requirements. . . .
2. Evaluate the statements of qualifications received. Select at least three consulting engineers that appear to be best qualified for the specific project. . . .
3. Write a letter to each consulting engineer, describing the proposed project in as much detail as possible, including a scope of work and outline of services required, and asking for a proposal describing in detail the consulting engineer's plan for managing and performing the required work, the personnel to be assigned, the proposed work schedule, experience in similar work, and other appropriate information such as office location in which work is to be performed, financial standing, present work-

load, and references. Each selected firm should have an opportunity to visit the site, review all pertinent data, and obtain clarification of any items as required.

4. On receipt of proposals, invite the firms to meet individually with the selection committee for interviews and discussions of the desired end results of the project and the engineering services required. During each interview, the selection committee should review the qualifications and experience of each firm, its capability to complete the work within the time allotted, and the specific key personnel to be assigned to the project; and should be satisfied that each firm understands the scope and requirements of the project.
5. Check carefully with recent clients of each firm to determine the quality of their performance. This check should not be limited only to references specified by the consulting engineer.
6. List the firms in the order of preference, again taking into account their reputation, experience, financial standing, size, personnel available, quality of references, workload, location, and any other factors peculiar to the project being considered.
7. Invite the consulting engineer considered to be best qualified to appear for a second interview to discuss the project and agree on a detailed scope of work and work products, the schedule, and negotiate fair compensation for the work and services.
8. The compensation proposed by the consulting engineer should be evaluated on the basis of the client's previous experience, taking account of the range of charges reported by other users of engineering services, . . . giving consideration to the project's special characteristics and the scope of services agreed upon. Fair and reasonable compensation to the consulting engineer is vital to the success of the project to enable the expertise of the consulting engineer to be fully utilized.
9. If satisfactory agreement is not reached with the first consulting engineer, the negotiations should be terminated and the consulting engineer notified in writing to that effect. An interview should then follow with the second firm; and failing an accord, the third should be called in for negotiations. Such a procedure will usually result in development of a satisfactory contract. All such negotiations should be on a strictly confidential basis, and in no case should the compensation discussed with one consulting engineer be disclosed to another.
10. When agreement has been reached on scope of work, schedule and compensation the client and selected consulting engineer should formalize their agreement in a written contract. . . .

Qualification guidelines are given in Section 6.6.5. If the owner has had satisfactory experience with one or more instrumentation specialists in the past, it may not be necessary to follow all the above steps. In fact, as discussed in Chapter 5 for procurement of instruments, a sole source approach may be used, both by private and public agencies. Perlman (1985) indicates that in 1982 more than half of procurement dollars spent by the U.S. government to acquire goods and services were spent on sole source contracts.

The contract will be for specialist work only, and payment provisions will normally be on a time and materials basis. Contracts and charges for engineering services are also discussed in *ASCE Manual No. 45* (ASCE, 1987).

Support work is performed by the construction contractor. As for Method 1, *biddable support work* is bid, and *nonbiddable support work* is paid for at *force account* rates under a *contingency allowance*.

6.3.5. Method 5. Instrumentation Specialist Selected by Owner and Contractor, and Contracting with Contractor as an Assigned Subcontractor

Specialist work is performed by an instrumentation specialist under contract to the construction contractor. As with Method 4, the owner or design consultant determines which tasks are *support work* and which are *specialist work*. After the construction contract is awarded, the owner and contractor mutually select an instrumentation specialist in the following manner. The owner supplies a list of suitable instrumentation specialists to the contractor. From this list the contractor selects and submits the names of three specialists. The owner finally selects an instrumentation specialist from this short list, following the qualification guidelines given in Section 6.6.5 and the selection procedures described in *ASCE Manual No. 45* (ASCE, 1987) and Section 6.3.4. The owner negotiates a time and materials payment method with the selected instrumentation specialist, who then becomes an *assigned subcontractor*.

Support work is performed by the construction contractor and paid for in the same way as Method 4.

6.3.6. Recommended Contractual Arrangements for Installing Instruments

The major advantages and limitations of the five arrangements are given in Table 6.1.

Method 2 can be used for simple installations that can be considered as normal construction items, such as settlement platforms, but should never be used for more complex installations. Method 3 should be used only when owner regulations require use of the low-bid method because even with prequalification requirements, construction contractors will generally shop for the lowest price, with risk of subcontractor cost cutting and possible invalid measurement data. Use of Method 3 entails a very comprehensive specification addressing all issues discussed in Section 6.6.

Methods 1, 4, and 5 are all satisfactory, the selection depending on factors specific to each project.

Methods 1 and 4 sometimes raise concerns about adversary relationships between owner's and construction contractor's personnel, but in practice cooperation can normally be achieved by following the suggestions given in Section 6.6.7.

Use of Method 1 is supported by a major U.S. public agency. In its manual on instrumentation for concrete structures, the U.S. Army Corps of Engineers (Corps of Engineers, 1980) states:

> The general policy of the Chief of Engineers is to perform all civil works by contract unless it is in the best interest of the United States to accomplish the work by Government forces. However, the specialized nature of instrumentation facilities and the care required in the preparation, calibration, and placement of test apparatus demands that these features of work be retained under close operational control of the Corps of Engineers. In view of these conditions . . . , utilization of Government personnel to accomplish certain phases of the fabrication and installation work, such as embedment of instruments and splicing of cables, is recommended.

Method 5 sometimes raises concerns about professionalism on the part of the instrumentation specialist, who has negotiated with the owner but contracted with the construction contractor. However, in practice, again by following the suggestions in Section 6.6.7, this concern is generally unfounded. Guertin and Flanagan (1979) used Method 5 for a major tunnel project in Rochester, New York, and comment that the day-to-day operations during the monitoring program proceeded smoothly with a minimum of conflict between the instrumentation specialist, design consultant, construction contractor, and owner. A spirit of common purpose existed during the project that, the writers believe, would not have developed as readily under a different contractual arrangement. Where space limitations are severe (and they are particularly severe during tunnel construction), the construction contractor and instrumentation specialist **must** have a cooperative working relationship so that instrument installation schedules and access needs can be coordinated with construction needs. In these cases Method 5 is most suitable.

6.4. CONTRACTUAL ARRANGEMENTS FOR REGULAR CALIBRATION AND MAINTENANCE

Contractual arrangements for regular calibration and maintenance should preferably follow Method 1, 4, or 5. If Method 3 is used, procedures, schedules, and personnel qualifications should be defined. Use of Method 2 entails a similar specification, but without personnel qualifications. However, Method 2 should be used only for very simple calibration and maintenance.

6.5. CONTRACTUAL ARRANGEMENTS FOR DATA COLLECTION, PROCESSING, PRESENTATION, INTERPRETATION, AND REPORTING

Contractual arrangements for data collection, processing, and presentation should preferably follow Method 1, 4, or 5. If Method 3 is used, the pay item unit should be *data set* or *man-hour* with clear definition of procedures and personnel qualifications. Use of Method 2 entails a similar specification, but without personnel qualifications. However, Method 2 should be used only for very simple data collection, processing, and presentation.

Data interpretation and reporting must be the direct responsibility of the owner or the owner's representative; therefore, Methods 2 and 3 are not suitable.

Table 6.1. Advantages and Limitations of Various Contractual Arrangements for Instrumentation Installation[a]

Method	Advantages	Limitations
1. Specialist installation work by owner's personnel	Owner has direct control over cost and quality Flexible to accommodate changes	Potential problems with contractor cooperation if instrumentation work interferes with other work, but see Section 6.6.7 Owner must plan for workload well in advance Assumes owner has necessary in-house skills Cannot always be financed by construction funds
2. Bid items in construction contract, without prequalification	Installation costs will usually be low Financed by construction funds	Generally, contractor will shop for lowest price subcontractor, with risk of lowest quality and invalid measurement data Requires strong and experienced supervision by owner's representative Not flexible to accommodate changes
3. Bid items in construction contract, with prequalification	Installation costs will usually be low Excludes inexperienced instrumentation subcontractors Financed by construction funds	Generally, contractor will shop for lowest price "qualified" subcontractor, with risk of subcontractor having inadequate price, cutting corners, and thus invalid measurement data Often difficult to substantiate desire to reject questionably qualified subcontractor Usually requires strong and experienced supervision by owner's representative Not flexible to accommodate changes
4. Instrumentation specialist selected by and contracting with owner	Owner has direct control over cost and quality Flexible to accommodate changes Instrumentation specialist can, if retained early enough, assist with design of monitoring program	Potential problems with contractor cooperation if instrumentation work interferes with other work, but see Section 6.6.7 Cannot always be financed by construction funds Requires some effort by owner to select specialist
5. Instrumentation specialist selected by owner and contractor, and contracting with contractor as an assigned subcontractor	Owner has direct control over cost and, via the owner's representative, quality Facilitates cooperation and scheduling with contractor Flexible to accommodate changes Financed by construction funds	Selection is made after award of construction contract: instrumentation specialist therefore cannot assist with design of monitoring program Assumes "professionalism" on part of instrumentation specialist, who has negotiated with the owner but contracted with the construction contractor, but see Section 6.6.7 Requires some effort by owner to select specialist Not permitted under some public agency regulations

[a] Method 2 should be used only for simple installations that can be considered as normal construction items. Methods 1, 4, and 5 are more likely to result in valid measurement data than Method 3.

6.6. CONTENT OF SPECIFICATIONS FOR FIELD INSTRUMENTATION SERVICES

Depending on selection among the five methods for instrument installation, regular calibration and maintenance, and data collection, processing, presentation, and interpretation, the specification for field instrumentation services should address some or all of the requirements discussed in this section. It is not intended that all specifications should address all these requirements but, if Method 2 or 3 has been selected for any task, extra care must be taken to write a comprehensive specification and all relevant requirements should be specified.

The specification for field instrumentation services will sometimes be combined with a specification for procurement of instruments, following the recommendations given in Chapter 5.

The specification will often be written with four main headings. The first heading, **Part 1. General,** will include wording relating to the overall approach, as described in Sections 6.6.1–6.6.7. The second heading, **Part 2. Products,** will include wording relating to materials, including procured instruments, as described in Section 6.6.8 and part of Section 6.6.9. The third heading, **Part 3. Execution,** will include detailed wording related to the field work, as described in Sections 6.6.9–6.6.18. The fourth heading, **Part 4. Measurement and Payment,** will include content as described in Section 6.6.19.

Technical information for the specification should be provided by geotechnical engineers working on the project design, and these same engineers may draft the technical specification, following guidelines given in this chapter. However, the final specification should be reviewed by an experienced specification writer. A review by an instrumentation specialist may also be necessary. The specification should also be reviewed by personnel familiar with the General Conditions and other sections of the technical specifications, to ensure consistency.

Before enumerating the items that should be considered when writing specifications for field instrumentation services, the author will reiterate a key point. When the low-bid procedure is used, even when prequalification requirements are included, construction contractors will generally shop for the lowest price, with risk of subcontractor cost cutting and possible invalid measurement data. An engineer who has a serious need for valid data cannot afford to take this risk. **Unless you must, don't use the low-bid procedure! The author regrets very much the need to write such a lengthy section as Section 6.6.: the length is necessary only because the low-bid procedure continues to be used.** Its use creates lengthy specifications that appear out of proportion to other parts of the specifications, and the author believes that its use is not in the best interests of the owner.

6.6.1. Purpose of Instrumentation Program

Many construction contractors view instrumentation as an interference with their normal construction work. However, if contractors are aware of a clear purpose for an instrumentation program, their willingness to cooperate is likely to be increased. Therefore, specifications should include a brief statement of purpose, of the parameters monitored by each instrument, and of how the data will be used. More detailed statements on use of data should be included under the heading **Implementation of Data** (see Section 6.6.15).

6.6.2. Division of Responsibilities

As described in Chapter 4, the division of responsibilities among owner, design consultant, instrumentation specialist, and construction contractor will have been determined during the planning phase. Specifications should include a clear statement of the work items for which the construction contractor is responsible and the work items to be undertaken by others.

6.6.3. Specification Method

The specification should indicate which of the five contractual arrangements has been selected for each field task. Table 6.2 summarizes the recommendations made earlier in this chapter. For Methods 1, 4, and 5, a clear definition should also be made of which tasks are *specialist work* and which are *support work.*

If a contingency allowance or an assigned subcontract procedure is being used, a brief definition should be given, based on appropriate parts of Sections 6.2.5 and 6.2.6, and the relevant work items should be listed. A more detailed specification should be included under the heading **Measurement and Payment** (see Section 6.6.19).

Table 6.2. Summary of Recommendations for Specifying Field Instrumentation Services

	Method				
Task	1	2	3	4	5
Installation, regular calibration and maintenance, and data collection, processing, and presentation	Satisfactory	Only very simple instrumentation	Less satisfactory than Methods 1, 4, and 5	Satisfactory	Satisfactory
Data interpretation and reporting	Satisfactory	Unsatisfactory	Unsatisfactory	Satisfactory	Satisfactory

6.6.4. Related Work Specified Elsewhere

A listing should be made of related work that is specified elsewhere, for example, dewatering, support of excavation, and other work items that are affected by implementation of instrumentation data. Section numbers of related parts of the specification should be given.

6.6.5. Qualifications of Specialist Field Instrumentation Personnel

Successful operation of an instrumentation program requires a special effort. Reliability and patience, perseverance, a background in the fundamentals of geotechnical engineering, mechanical and electrical ability, attention to detail, and a high degree of motivation are the basic requirements for qualities needed in instrumentation personnel.

If the safety of the project depends on the measurements, then the requirements for reliable personnel are doubly important. Personnel must also be flexible and ingenious to adapt to constantly changing situations. In selecting personnel to perform or supervise field work, consulting firms with special expertise in geotechnical instrumentation should be considered. Geotechnical consulting firms are listed in various professional magazines, in the directory of the American Consulting Engineers Council,* and in the membership list of the Association of Soil and Foundation Engineers.† Some instrumentation manufacturers do not maintain a staff of experienced geotechnical engineers and field technicians; therefore, caution should be exercised in specifying that *installation of the instruments shall be supervised by a representative of the manufacturer*. If instrumentation field work requires combining a knowledge of an instrument with a knowledge of appropriate geotechnical conditions, it will usually be necessary to involve a **geotechnical engineer** having comprehensive experience with the selected instrument.

Qualification wording should address requirements for the firm responsible for field instrumentation services and also for the individuals who will be directly performing the field work. The specification will usually require the demonstration of certain specific previous experience. All specialist field instrumentation personnel should be subject to the owner's approval.

When specification Method 3 is used, if instrumentation work is the primary part of the construction contract, qualifications will usually be required with the bid, and attention will normally be drawn to this requirement both in the invitation to bid and at the prebid conference when one is held. If instrumentation work is not the primary part of the construction contract, qualifications will be required within a certain period after award. For example, the following wording may be appropriate for an instrumentation program of substantial size, but which is not the primary part of the construction contract:

> Installation of piezometers and inclinometers shall be under the full-time supervision of a geotechnical engineering consulting firm with previous experience in installing similar instruments. The two senior individuals who will actually be performing the supervision shall be a qualified Geotechnical Instrumentation Engineer and a qualified Field Supervisor. The Geotechnical Instrumentation Engineer

* 1015 15th Street, N.W., Suite 802, Washington, DC 20005.
† 8811 Colesville Road, Suite G106, Silver Spring, MD 20910.

shall be a registered professional engineer with a minimum of a Master of Science degree in geotechnical engineering. The Geotechnical Instrumentation Engineer shall have at least 4 years experience in geotechnical engineering and geotechnical instrumentation and shall have direct field experience in the installation of geotechnical instruments including piezometers and inclinometers. The Geotechnical Instrumentation Engineer shall prepare detailed written procedures for installation of each type of instrument, shall personally supervise the installation of at least the first two instruments of each type, shall review and sign all submittals, shall instruct the installation crews regarding installation procedures, and shall be available for consultation with the crews at all times. All geotechnical instruments shall be installed under the full-time field supervision of the Field Supervisor, who shall have a minimum of a Bachelor of Science degree in civil engineering or geology, and at least 2 years field experience in the drilling and logging of soil and rock borings. The firm, the Geotechnical Instrumentation Engineer, and the Field Supervisor shall be subject to the approval of the Authority. The firm's name and individuals' names shall be submitted to the Authority for approval no later than 120 days prior to installing the first piezometer or inclinometer. The submission shall include a statement of past similar experience of the firm and the individuals.

6.6.6. Submittals During Construction

If details of any work items are to be determined by the construction contractor or subcontractor, drawings and procedures should be submitted to the owner for approval, a specified time period prior to commencing that work in the field. Sufficient time should be allowed for suppliers' lead times and for one or more resubmittals in the event that the first submittal is not approved. Such a requirement creates a contractual commitment to plan the work well in advance and gives the owner an opportunity for timely review. Submittals may include detailed step-by-step installation procedures with a listing of materials and equipment, logs of installation activities, and procedures for any other tasks assigned to the construction contractor (Table 4.2).

6.6.7. Cooperation Between Construction Contractor and Owner's Specialist Field Instrumentation Personnel

When Method 1 or 4 is used, a cooperative working relationship between the construction contractor and the owner's field personnel is essential. The specification should convey that extreme attention to detail is necessary and should include a general requirement for the contractor's cooperation. However, if the contractor has no economic incentive to cooperate, it can be difficult to enforce such requirements, and the owner's field personnel should make a special effort to establish a cooperative relationship. The best way of establishing such a relationship is for instrumentation personnel to initiate thorough communication with all levels of the contractor's personnel several weeks or months before anything has to be accomplished physically in the field. The instrumentation personnel should meet with the contractor's engineers and supervisors to explain what will be done, why it must be done, and what will be required of the contractor. They should prepare sketches to forewarn the contractor of the impact on normal construction work, provide lists of materials that will be required for support work, and be willing to adjust their own plans to create minimum interference to the contractor's work. In short, mutual respect can usually be established by thorough communication, dispelling the adversary relationship that sometimes arises between architect-engineers and construction contractors. Having established this respect, instrumentation personnel should maintain the contractor's respect by performing top-quality work, be responsive to the effects of the program on the contractor, and work **with** the contractor to minimize any adverse effects.

When Method 5 is used, the specification should also include a general requirement for the contractor's cooperation. Concerns related to professionalism on the part of the instrumentation specialist, who has negotiated with the owner but contracted with the construction contractor, can usually be nullified by establishing mutual respect as suggested above.

6.6.8. Procurement of Instruments

This is the first section of the specification under the heading, **Part 2. Products.**

If instruments are to be procured under a separate contract and provided to the construction contractor for installation, the specification for field instrumentation services should include a full listing of all instruments, spare parts, tools, and associated materials that will be provided. The construction contractor's understanding and therefore cooperation will be increased by including a brief description of instrument operating principles and physical sizes. Information should be included on factory calibration and quality assurance, checking and testing by the owner on first receipt (acceptance tests), warranties, shipment and delivery arrangements, schedules and insurances, and procedures for handing over to and checking by the construction contractor. These last procedures must ensure that the contractor accepts responsibility for the condition of the instruments at the time of handing over.

If instrument procurement is included in the same specification as field instrumentation services, the Chapter 5 guidelines should be followed. In this case a special effort should be made to include in the plans and specifications the required quantities of instruments, communication systems, readout units, and terminals. All too often this information is unclear, and a tabular listing would resolve the difficulty.

The specification will normally require the construction contractor to provide clean, dry, secure storage space for instruments, including requirements for size, windows, doors, benches, furniture, tools, power outlets, lighting, heating, air conditioning, locks, completion date, and ownership at the end of the construction period.

6.6.9. Support Work

Typical support work items are listed in Table 6.3. The first two items relate to products and will be included under the heading **Part 2. Products.** The remaining items relate to execution and will be included under the heading **Part 3. Execution.**

Each biddable support work item should be specified in sufficient detail, either in this section or in later sections, so that bidders can understand requirements and bid an appropriate amount. Use of the term *as directed by the Owner's Representative* is strongly discouraged: if the owner is unable to specify a work item, how can bidders be expected to prepare a bid for the item? Support work specified as *incidental* (i.e., when no separate pay item is provided) should also be specified in detail. Nonbiddable support work that is to be paid for under a contingency allowance should be described but need not be specified in detail.

If the owner will be responsible for specialist work (specification Method 1 or 4), scheduling of support work should be under the control of the owner, and the following wording is appropriate:

> The Contractor's instrument-related support work shall be scheduled with the Owner. These work items shall not be conducted or paid for unless the Owner's Representative is present.

6.6.10. Locations of Instruments

The exact location of instruments should usually be determined in the field, when geologic details and construction procedures are defined more closely than during the design phase. The specification should state that the owner's representative will indicate the final locations, orientations, depths, and number of instruments in the field and will confer with the contractor as to the suitability of all locations. The approximate locations of instruments, communication systems, and terminals should be shown on the contract plans, but a note should be added to indicate that they are approximate and subject to change. The contractor should be required to perform a record survey of all instrument locations to define the vertical and lateral positions of the exposed parts, and the required measurement accuracy should be specified.

6.6.11. Installation of Instruments

The five basic contractual arrangements for installing instruments are described in Section 6.3. Installation will normally require both specialist and support work.

If Method 2 or 3 has been selected, comprehensive step-by-step installation procedures should be specified, and details should be included on the contract plans. The procedures will be abbreviated versions of the procedures described in Chapter 17, retaining key items for enforcement by the owner's representative. The wording *installation procedures shall be in accordance with the manufacturer's recommendations* is applicable **only** if the manufacturer's procedures are first reviewed, in order to ensure that they are complete and applicable

Table 6.3. Typical Support Work Items

Item	Usual Category: Nonbiddable	Usual Category: Biddable	Usual Bid Unit
Take delivery of instruments procured under a separate contract		•	Lump sum
Furnish simple components and instruments such as settlement platforms		•	Each
Check instruments procured under a separate contract		•	Lump sum (usually combined with the first item in this table)
Install simple instruments such as settlement platforms		•	Each (usually combined with the second item in this table)
Provide storage space for instruments		•	Month or lump sum
Obtain permits for drilling	•		None: incidental work item
Production drilling		•	Linear foot
Drilling or grouting as directed by the owner		•	Hour
Drill rig and crew to assist during instrument installation		•	Hour
Standby of drill rig and crew		•	Hour
Excavation of trenches, backfilling and compaction for instrument tubes		•	Linear foot or cubic yard
Tradespeople to assist owner on *as needed* basis, e.g., cutting, welding, fabrication services	•		None: force account or contingency allowance
Use of air, water, and electrical power supplies	•		None: incidental work item
Hoisting and transportation equipment to move instruments	•		None: incidental work item, force account, or contingency allowance
Protection from damage	•		None: incidental work item
Optical survey work to lay out instruments and to record location	•		None: incidental work item
Access for installation or reading by owner	•		None: incidental work item, force account, or contingency allowance
Make initial instrument readings jointly with the owner's representative	•		None: incidental work item
Optical survey work to read instruments[a]		•	Each, data set, hour, or lump sum
Provide information to owner on construction events and progress that may influence measured data	•		None: incidental work item
Revised instrumentation due to geologic conditions revealed during construction	•		None: force account or contingency allowance
Remove or salvage instruments or restore surface or subsurface	•		None: incidental work item

[a] But data are usually more reliable if optical survey measurements are made as specialist work.

to the specific site geotechnical conditions. The contractor should be required to submit detailed step-by-step installation procedures to the owner for review, a specified time period prior to commencing field installation work. The contractor should also be required to submit logs of installation activities for inclusion on installation record sheets. A listing of items to be recorded is given in Section 17.10. Installation schedules should be specified, including a deadline for instruments to be operational, normally in relation to a construction activity. If instruments are to be installed as geologic or construction conditions indicate, the contractor should be required to complete installations within a certain fixed time after being notified by the owner's representative. Any work restrictions, such as disposal of surplus grout, should also be specified.

If Method 1, 4, or 5 has been selected, installation procedures should also be specified, but in less detail.

If instrument installation requires drilling in soil or rock, drilling requirements will usually be specified in a separate section. However, when using site investigation drilling specifications as a basis for wording, care should be taken to tailor the wording, recognizing that the primary purpose is instrument installation rather than sampling and definition of stratigraphy. The separate drilling section will usually include requirements for drilling equipment, soil boring and sampling methods, rock drilling and coring methods, drilling fluid, drilling records, sample containers, sample storage and submittal. The following wording is often applicable in the general drilling section:

> Whenever withdrawing drill casing during instrument installation, care shall be taken to minimize the increments and rate of casing withdrawal so that collapse of the borehole does not occur and to ensure that backfill material does not build up inside the casing such that the instrument is lifted as the casing is withdrawn. The casing shall be withdrawn without rotation. The casing may be omitted only where permitted in accordance with specifications herein describing installation of each instrument, and where it can be shown to the satisfaction of the Engineer that instrument installation without the casing will not cause collapse of the borehole or in any way adversely affect instrument installation.

The sections describing installation of each instrument will usually include maximum and minimum borehole or casing diameter, tolerance on borehole alignment, and sampling frequency. Specifications for soil borings will include acceptability of drilling without casing and acceptability of using hollow-stem augers or drilling mud. Where grout is to be used, the mix for each instrument should be specified.

If practicable, acceptance tests should be specified to verify that installations have been completed satisfactorily. For example, falling and rising head tests can be made in open standpipe and twin-tube hydraulic piezometers, groove tracking tests and spiral survey measurements can be made in inclinometer casings, and rod sliding tests can be made in fixed borehole extensometers equipped with a disconnect between rod and anchor. Repeatability should be examined by taking several instrument readings.

6.6.12. Regular Calibration and Maintenance

Recommendations for specifying regular calibration and maintenance are given in Section 6.4.

Use of Method 3 requires specifying personnel qualifications, as well as procedures and schedules following the guidelines in Chapter 16. Method 2 requires a similar specification, but without personnel qualifications. If Method 5 has been selected, a similar specification is required, but in less detail, since detailed requirements will be addressed in the subcontract. No specification is necessary if Method 1 or 4 is used.

6.6.13. Data Collection, Processing, and Presentation

Recommendations for specifying data collection, processing, and presentation are given in Section 6.5.

Use of Method 3 requires specifying personnel qualifications and procedures following the guidelines in Chapter 18, including the format and schedule for submission of data to the owner's representative. If the construction contractor is responsible for any data collection, for example, optical survey work, the required accuracy should be specified. Reading frequency should preferably be as directed by the owner's representative—thus necessitating a *data set* or *man-hour* pay item unit. If reading frequency must be specified, the start, termination, and frequency should be defined. For example, *daily* may or may not include nonworking days and will normally terminate in relation to construction activity. However, such specifications do not allow flexibility to collect data when it is most beneficial to do so and should therefore be avoided wherever possible.

Use of Method 2 requires a specification similar to Method 3, but without personnel qualifications. Method 5 also requires a specification similar to Method 3, but in less detail, since detailed requirements will be addressed in the subcontract. No specification for regular data collection, processing, and presentation is necessary if Method 1 or 4 is used.

To avoid the possibility of disputed measurements of change, the contractor and owner's representative should jointly make initial readings and agree on appropriate values. The method and schedule should be specified for each instrument.

6.6.14. Availability of Data

If the owner is responsible for reading instruments, the data should be made available to the construction contractor. The specification should state whether raw or interpreted data will be made available and on what schedule. Care should be taken to ensure that construction contractors are not relieved of their responsibility to ensure safety. The following wording is usually appropriate:

> Raw data will be made available to the Contrator within one working day of reading. The Contractor may observe the readings at any time or take supplementary readings at no additional cost to the Owner. The Owner's Representative will interpret the data and make interpretations available to the Contractor as soon as practicable. The Contractor is expected to make his own interpretations of the data for his own purposes. Furthermore, the Contractor shall install, monitor, and interpret data from additional instrumentation that the Contractor deems necessary to ensure the safety of the work. Raw data collected by the Contractor, whether from specified instruments or from additional instruments of the same type, shall be taken in the same manner as adopted by the Engineer. All raw data collected by the Contractor shall be made available to the Engineer within one working day of reading. The Owner is not responsible for guaranteeing the safety of the work based on the instrumentation data. The Contractor shall not disclose instrumentation data to third parties and shall not publish data without prior approval of the Owner.

6.6.15. Implementation of Data

Data interpretation and reporting must be the direct responsibility of the owner or the owner's representative; therefore, Method 1, 4, or 5 must be used.

If instrumentation is used to provide benefits during construction (Chapter 3), the specification should include clear statements defining how the data will be used and what action will be taken. If observations are to be used to control construction schedules or quality of work, these requirements should also be clearly specified. If remedial action has been devised to solve problems that may be disclosed by observations (Chapter 4), the construction contractor should be forewarned of the actions. If hazard warning levels have been assigned, the action for each level should be indicated. Contractual responsibility for initiating remedial action and the method of forewarning all parties should be stated.

Great care should be taken to ensure that this section of the specifications is consistent with all other sections of the specifications.

6.6.16. Delay to Construction

If Method 1 or 4 is used for any task and if the owner's instrumentation work is likely to delay construction progress, the specification should provide for appropriate payment. The following wording can be used:

> The Owner will schedule ______ to minimize interference with the Contractor's operations. Delays of up to ______ hours per ______ for a total of ______ hours should be anticipated. No additional payment will be made to the Contractor for such delays.

Delay estimates should be conservative, because field instrumentation work often takes longer than expected. Alternatively, delay compensation can be paid in accordance with actual delay. An hourly bid item can be included for *delay to construction*, or an hourly rate can be stipulated in the contract. However, this is not applicable where a delay can upset an entire construction cycle, as is often the case with tunnel construction, and in such cases the problem should be avoided by using Method 5.

6.6.17. Damage to Instrumentation

Installed instruments are usually prone to damage by construction activities. The specification should include requirements for protection and should indicate the responsibility of the construction contractor in the event of damage.

If the contractor installs and damages instruments, repair or replacement should be required at no additional cost to the owner, and timeliness should be specified. The owner's representative should be the sole judge of whether repair or replacement is required. If data are vital, a work stoppage within a specified distance of the damaged instruments may be appropriate, until instruments are operational. If repair or replacement is not possible, payment for that installation can be negated. Increased incentive to avoid damage can be created by imposing an additional financial penalty. However, the owner will normally lose needed data and therefore should make every effort during the planning phase to select measurement methods with minimum damage potential. If the construction contractor damages instruments installed by the owner, some of the same specification provisions can be included, and the owner's costs can be deducted from payments to the contractor. However, development of mutual respect and cooperation (Section 6.6.7) is usually more productive than penalty clauses.

In general, the owner's representative must exercise good judgment in deciding whether damage was avoidable by the contractor or was due to conditions beyond the contractor's control.

6.6.18. Disposition of Instruments

The specification should define who will own portable or recoverable components at the end of the construction period. If not required for long-term monitoring, instruments such as inclinometer casing, borehole extensometers, and piezometers are usually abandoned in place, and the specification should require the construction contractor to cut off and dispose of protruding parts below the finished surface, grout the open casing or tubing and restore the surface. Sometimes instruments such as load cells and vibrating wire strain gages can be salvaged, overhauled, recalibrated, and reused. Portable instruments and readout units will often be reusable. If the instruments are to be used for long-term measurements, the specification should require them to be overhauled to first class condition and to have permanent protection and access provisions.

6.6.19. Measurement and Payment

Measurement and payment should be specified, in accordance with recommendations given earlier in this chapter. The work should be measured and paid for on a unit price basis rather than a lump sum basis, to allow for changes in quantities as work proceeds. Great care should be taken to ensure that, for every work item specified for the construction contractor, there is either a pay item or a statement that the item will be considered incidental for payment purposes.

There should be no ambiguity in the method of measurement: for example, if an instrument is to be installed near the end of a borehole, and measurement of drilling and installation is by the linear foot, the specification should indicate whether measurement will be made to the specified instrument location or to the actual end of the borehole.

When payment for procurement of instruments is combined with payment for installation, by using several *furnish and install* unit price items in the bid, special care must be taken to satisfy the requirements of Section 5.4.18 for definition of major component prices. For example, a single expensive readout unit should be included in a separate bid item, so that a change in the quantity of installed instruments can be accommodated under the *furnish and install* item.

Each pay item should be clarified by using a clear statement to define the work included. For example, the introductory part of the measurement and payment section might include the following wording:

> The contract prices for furnishing and installing each instrument shall be full compensation for all materials, labor, tools, and equipment necessary for submittals, furnishing all materials for installation, drilling, sampling, withdrawing casing, installation, acceptance test-

ing, protection, replacement or repair of any instruments damaged as a result of the Contractor's operation and/or vandalism, initial readings, restoration of surface and subsurface, providing access to the Engineer for reading, and all other specified items of work for which no separate bid item is provided, all to the satisfaction of the Engineer.

For a bid item per linear foot for *Furnish and Install Piezometers,* the specification might state:

> Measurement will be based on the total number of linear feet of drilling. Payment will be based on the unit price per linear foot for furnish and install piezometers as stated in the Bid.

If a *data-set* unit is used for any instrument readings, the unit must be defined.

If nonbiddable support work is to be paid for under a contingency allowance, the specification should indicate which are the applicable work items and how payment rates for labor, plant, equipment, and materials will be determined, including provisions for a markup.

If an assigned subcontract procedure has been used for specialist work, the specification should indicate which are the applicable work items and how payment will be made, including provisions for a markup. The following wording is suitable:

> Payment for specialist geotechnical instrumentation work will be made under allowance item no. ______. After contract award the Owner's Representative will select a specialist organization, acceptable to the Contractor, to perform specialist work, and will instruct the Contractor to enter into a subcontract with that organization, which will then become an assigned subcontractor. The Contractor shall submit copies of the subcontractor's invoices and will be reimbursed in accordance with those invoices plus the specified (or bid) markup. Measurement will be in accordance with the units included in the subcontract. If the actual total cost of the work is more or less than the allowance, the contract price will be adjusted to the actual approved amount.

6.6.20. Checklist for Specification Content

Appendix C contains a checklist for content of specifications for field instrumentation services. Use of the checklist will help ensure completeness. However, each individual specification may require only some of this content.

Part 3

Monitoring Methods

Part 3 is addressed to people who are involved in the details of instrumentation use. The primary emphasis is on a presentation of various monitoring methods, including comparative information and recommendations for selection among the methods.

The first two chapters are applicable to all monitoring methods. Chapter 7 defines the meaning of terms associated with measurement uncertainty, examines the various types of error that can affect a measurement, and suggests how they may be eliminated. Chapter 8 provides a description of various mechanical, hydraulic, pneumatic, and electrical transducers and communication and data acquisition systems that are used with geotechnical instrumentation.

Chapters 9–14 describe instruments for monitoring six parameters: groundwater pressure, total stress in soil, stress change in rock, deformation, load and strain in structural members, and temperature. Each chapter includes a categorization of available instruments and a brief indication of applications. Monitoring methods are described and compared, and guidelines are given for selection and use. In many cases the comparisons are summarized by listing advantages and limitations in tabular form. The comparative data in each of these tables are with respect to other entries in the same table and are generally qualitative rather than quantitative. Significant emphasis is given in Chapters 9–14 to installation of instruments and to appropriate aspects of calibration and data collection and processing.

CHAPTER 7

MEASUREMENT UNCERTAINTY*

Instrumentation is used for making measurements, and every measurement involves error and uncertainty. The purpose of this chapter is to define the meaning of terms associated with uncertainty, to examine the various types of error that can affect a measurement, and to suggest how they may be minimized.

7.1. CONFORMANCE

Ideally, the presence of a measuring instrument should not alter the value of the parameter being measured. If in fact the instrument alters the value, it is said to have poor *conformance*. For example, fixed borehole extensometers and any surrounding grout should be sufficiently deformable so that deformation of the soil or rock is not inhibited, and embedment earth pressure cells should ideally have the same deformability characteristics as the soil in which they are placed. In addition, the act of drilling a borehole or compacting fill around an instrument should not result in a significantly different condition within the geologic material. Thus, piezometers should not create drainage paths that would reduce the measured pore water pressure below the value elsewhere. Conformance is a desirable ingredient of high accuracy.

*Written with the assistance of Ralph S. Carson, Professor of Electrical Engineering, University of Missouri–Rolla, MO, and J. Barrie Sellers, President, Geokon, Inc., Lebanon, NH.

7.2. ACCURACY

Accuracy is the closeness of approach of a measurement to the true value of the quantity measured. Accuracy is synonymous with *degree of correctness*. Accuracy of an instrument is evaluated during calibration, when the true value is the value indicated by an instrument whose accuracy is verified and traceable to an accepted standard. In the United States, traceability should be, wherever possible, to the National Bureau of Standards. It is customary to express accuracy as a ± number. An accuracy of ±1 mm means that the measured value is within 1 mm of the true value, and an accuracy of ±1% means that the measured value is within 1% of the true value. However, an accuracy of ±1% full scale (FS) indicates a lesser accuracy, because the percentage applies to the full scale of the indicator rather than the measured value.

When selecting an instrument with appropriate accuracy, the entire system must be considered, including accuracy of each component and each source of error. Errors are discussed in Section 7.9.

7.3. PRECISION

Precision is the closeness of approach of each of a number of similar measurements to the arithmetic mean. Precision is synonymous with *reproducibility* and *repeatability*. It is customary to express precision as a ± number. The number of significant figures quoted for a measurement reflects the preci-

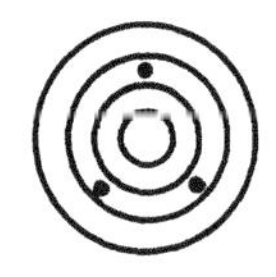

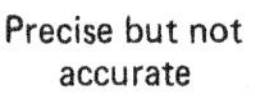

Figure 7.1. Accuracy and precision.

sion of the measurement; thus, ± 1.00 indicates a higher precision than ± 1.0. Conversely, a recorded measurement should reflect the precision of the instrument used. The fact that an instrument yields a measurement to three significant figures is no assurance that it is accurate to ± 0.1, and it is pointless to attempt to obtain a reading to three significant figures from an instrument known to be accurate to only $\pm 10\%$.

The difference between accuracy and precision is illustrated in Figure 7.1. The bull's eye represents the true value. In the first case the measurements are precise but not accurate, as would occur when using a survey tape with a bad kink or a pressure gage with a zero shift. Such errors are *systematic*. In the second case the measurements lack precision but, if sufficient readings are taken, the average will be accurate. Such errors are *random*. In the third case the measurements are both precise and accurate.

7.4. RESOLUTION

Resolution is the smallest division on the instrument readout scale. In some cases it may be possible and convenient to interpolate between divisions on the readout scale, but interpolation is subjective and does not increase the resolution of the instrument. The resolution for a digital display is one digit of change in the last digit.

7.5. SENSITIVITY

Sensitivity refers to the amount of output response that an instrument or transducer produces when an input quantity is applied to it. For example, the sensitivity of a linear variable differential transformer (LVDT) used to monitor displacement across a rock joint might be given as 1000 millivolts per inch (mV/in.). When the same amount of input is applied to several instruments or transducers, the one whose sensitivity is the highest produces the most output. High sensitivity does not imply high accuracy or precision.

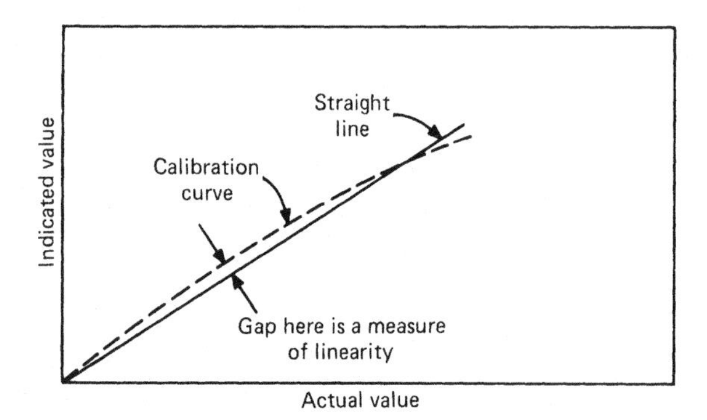

Figure 7.2. Linearity.

7.6. LINEARITY

An instrument is said to be linear when its indicated measured values are directly proportional to the quantity being measured. As illustrated in Figure 7.2, the graphic relationship between indicated and actual value is often slightly curved, owing to instrument limitations. If a straight line is drawn on this plot such that the widest gap between straight line and curve is a minimum, the gap is a measure of *linearity*. Thus, a linearity of 1% FS means that the maximum error incurred by assuming a linear calibration factor will be 1% of the full-scale reading.

7.7. HYSTERESIS

When the quantity being measured is subjected to cyclic change, the indicated measured value sometimes depends on whether the measurement is increasing or decreasing. If the two relationships are plotted as shown in Figure 7.3, the separation between the two curves is a measure of *hysteresis*. Hysteresis is commonly caused by friction or backlash and occurs, for example, in hydraulic jacks as

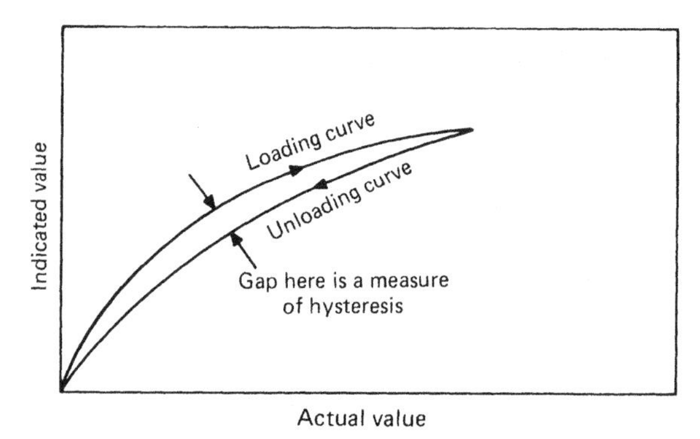

Figure 7.3. Hysteresis.

load is increased and decreased while proof testing tieback anchors. Instruments with large hysteresis are not suitable for measurement of rapidly changing parameters.

7.8. NOISE

Noise is a term used to cover random measurement variations caused by external factors, creating lack of precision and accuracy. Excessive noise in a system may mask small real changes. Radio frequency (RF) interference from high-voltage sources and TV or radio transmitters is an example of an external factor that creates noise.

7.9. ERROR

Error is the deviation between the measured value and the true value; thus, it is mathematically equal to accuracy. Errors arise from many causes, described below and summarized in Table 7.1.

Gross errors are caused by carelessness, fatigue, and inexperience. They include misreading, misrecording, computational errors, failure to operate the readout instrument correctly, incorrect installation, improper electrical connections, and wrong switch positions. Gross errors are avoidable and can be minimized by taking duplicate readings, using more than one observer, checking present readings against previous readings, and insisting on proper care and training.

Systematic errors are caused by improper calibration, by alteration of calibration with time such that readings are consistently high or low, and also by hysteresis and nonlinearity. The first data set in Figure 7.1 shows systematic error. These errors can be minimized by periodic recalibration and by checking readings against "standards" and "no-load" gages kept in the laboratory or in the field. If readings on standards or no-load gages change, cor-

Table 7.1. Causes and Remedies of Measurement Error

Type of Error	Causes	Remedies
Gross error	Inexperience Misreading Misrecording Computational error	Care Training Duplicate readings Dual observers Checking against previous readings
Systematic error	Improper calibration Loss of calibration Hysteresis Nonlinearity	Use of correct calibration Recalibration Use of standards Use of consistent reading procedures
Conformance error	Inappropriate installation details Instrument design limitations	Select appropriate instrument Modify installation procedure Improve instrument design
Environmental error	Weather Temperature Vibration Corrosion	Record environmental changes and apply corrections Make correct choice of instrument materials
Observational error	Variation between observers	Training Use of automatic data acquisition systems
Sampling error	Variability in the measured parameter Incorrect sampling techniques	Install a sufficient number of instruments at representative locations
Random error	Noise Friction Environmental effects	Correct choice of instrument Temporary elimination of noise Multiple readings Statistical analysis
Murphy's law	If something can go wrong, it will	None—any attempt to remedy the situation will only make things worse

rection factors can be applied. Errors caused by hysteresis can sometimes be minimized by operating an indicator such that a measurement is always decreasing or always increasing. Linearity errors can be minimized by using the correct calibration.

Conformance errors are caused by poor selection of installation procedures and by limitations in the instrument design. They can be avoided by use of correct installation procedures and by ensuring that the instrument design is appropriate for the application.

Environmental errors arise because of the influence of heat, humidity, vibration, shock waves, moisture, pressure, corrosion, and so on. Two approaches are possible for minimizing environmental errors: first, by measuring the extent of the influence, such as temperature, and applying suitable correction factors; and second, by choosing an instrument that is not adversely affected by the environment. Environmental conditions should always be recorded when data are collected, because they may correlate with real changes in measured quantities.

Observational errors arise when different observers use different observation techniques. They are minimized by regular training sessions and periodic refresher courses on readout procedures. Use of automatic data acquisition systems will prevent observational errors.

Sampling errors are common when making measurements of geotechnical parameters, because of inherent variability in the geologic materials. Correct measurements made at one location may not be representative of overall behavior. Sampling errors are minimized by installing a sufficient number of instruments at representative locations.

Even when errors are recognized and remedied, readings will still show variation due to *random error*. The second data set in Figure 7.1 shows random error. These errors are caused by noise, internal friction, hysteresis, and environmental effects. They can be minimized by correct choice of instrument and by multiple readings and can be treated mathematically using statistical analyses, so that measurements are presented as average values with a standard deviation and confidence limits. However, average values should be used with caution, because extreme values may in fact be true and may indicate a critical situation. As an example, a person standing with one foot in boiling water and one foot in ice water might, on the basis of the average water temperature, be said to be comfortable.

CHAPTER 8

INSTRUMENTATION TRANSDUCERS AND DATA ACQUISITION SYSTEMS

Most geotechnical instruments consist of a transducer, a data acquisition system, and a communication system between the two. A transducer is a device that converts a physical change into a corresponding output signal. Data acquisition systems range from simple portable readout units to complex automatic systems. This chapter provides a description of various mechanical, hydraulic, pneumatic, and electrical transducers and communication and data acquisition systems that are used with geotechnical instrumentation.

Many geotechnical instruments are not required to survive longer than the construction period. However, many are required to survive as long as practicable, for example, for monitoring long-term performance of embankment dams. Factors that influence the selection of transducer and communication and data acquisition systems, when they are required to survive as long as practicable, are discussed in this chapter.

8.1. MECHANICAL INSTRUMENTS

The two mechanical devices used most frequently in geotechnical instrumentation are *dial indicators* and *micrometers*. Although dial indicators are more common, both can be used in mechanical crack gages, convergence gages, mechanical tiltmeters, fixed borehole extensometers, mechanical strain gages, and mechanical load cells.

8.1.1. Dial Indicator

A dial indicator is used to convert the linear movement of a spring-loaded plunger to larger and more visible movement of a pointer that rotates above a dial. Mechanical parts include a rack and pinion and gear train.

Accuracies are generally ±0.001 in. (±0.025 mm) or ±0.0001 in. (±0.0025 mm), with typical ranges up to 2 in. (50 mm). Special long-range dial indicators can be obtained, with ranges up to 12 in. (300 mm). Dial indicators are somewhat fragile and susceptible to the effect of dirt entering the gear train but, because they are mass produced for many industrial purposes, they are inexpensive and can often be considered expendable. Sealed and waterproof versions are available, with rubber bellows around the plunger where it exits from the indicator body. Others are oil-filled, also with rubber bellows, and are preferable for long-term applications and for use in adverse environments.

8.1.2. Micrometer

Rotation of a finely threaded plunger causes the plunger to travel in or out of a housing. Longitudinal movement of the plunger is measured, using a scale on the housing, to indicate the number of revolutions of the plunger. Fractional revolutions are determined using graduations marked around the plunger and a vernier on the housing.

Accuracies are limited to about ±0.001 in. (±0.025 mm), because the reading is sensitive to the force applied to the plunger as it is rotated to touch the reference point. Good quality micrometers incorporate a simple clutch to minimize this effect. By changing the length of the plunger, the range of a micrometer can readily be extended to 6 in. (150 mm). Verniers can be difficult to read, and micrometers are now available with digital counters that greatly ease the reading task. Micrometers are more robust than dial indicators and thus are often favored for portable use when they are not built into a protective housing, for example, when used to read fixed borehole extensometers.

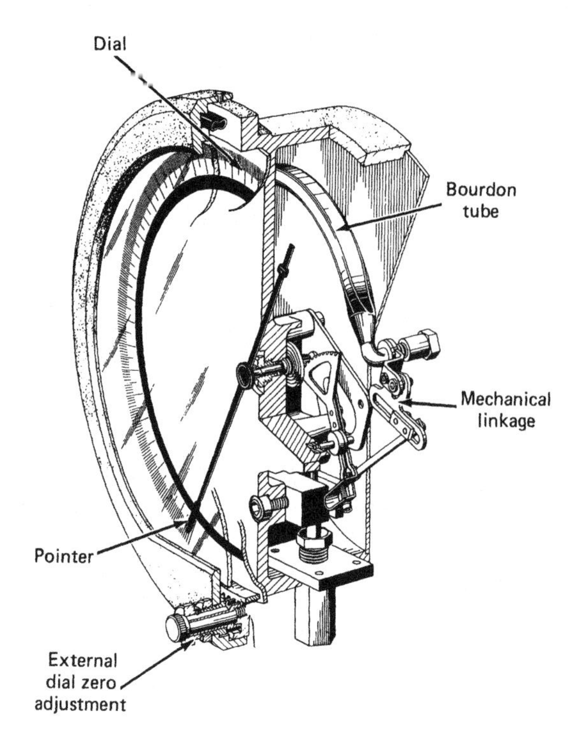

Figure 8.1. Bourdon tube pressure gage (courtesy of Dresser Industries, Newtown, CT).

8.2. HYDRAULIC INSTRUMENTS

The two hydraulic devices used most frequently in geotechnical instrumentation to measure liquid pressure are *Bourdon tube pressure gages* and *manometers*. Bourdon tube pressure gages are more common and are used with hydraulic piezometers, hydraulic load cells, borehole pressure cells, and in some readout units for pneumatic transducers. Manometers are sometimes used with twin-tube hydraulic piezometers and liquid level settlement gages.

8.2.1. Bourdon Tube Pressure Gage

A Bourdon tube is made by flattening a metal tube and coiling it into a C-shaped configuration. When the tube is pressurized internally, the flattened cross section expands, causing the tube to straighten. The uncoiling motion is transmitted via a mechanical linkage to a pointer that rotates over a circular scale, as shown in Figure 8.1.

Typical accuracy is ±0.5% full scale (FS) for commonly used gages with 4 or 6 in. (100 or 150 mm) dial sizes, but ±0.1% FS "test quality" gages are available. Caution should be used in relying on the accuracy of ±0.5% FS gages, because when checked with a dead-weight tester these gages sometimes do not meet specification. The author recommends use of ±0.25% FS gages when ±0.5% FS accuracy is required. It should be noted that accuracy is usually expressed as percentage of full-scale reading, not as percentage of gage reading. Thus, in selecting an appropriate range for a Bourdon gage, an unnecessarily high range should be avoided.

For long-term permanent installations, Bourdon gages must be of high quality, without risk of galvanic corrosion. The primary long-term application is in terminal enclosures for twin-tube hydraulic piezometers, and recommendations for gage selection are given in Appendix E.

8.2.2. Manometer

A manometer is formed by a liquid-filled U-tube. A pressure on one side of the U-tube is balanced by an equal pressure on the other side.

Figure 8.2 shows pressure readings using an open-ended tube, a Bourdon tube pressure gage, and a mercury manometer. Manometers have been used extensively for long-term monitoring of pore water pressure in embankment dams with twin-tube hydraulic piezometers where adequate headroom is available in the terminal enclosure. They are also useful for measuring very small positive or negative pressures and are easily calibrated. Manometers

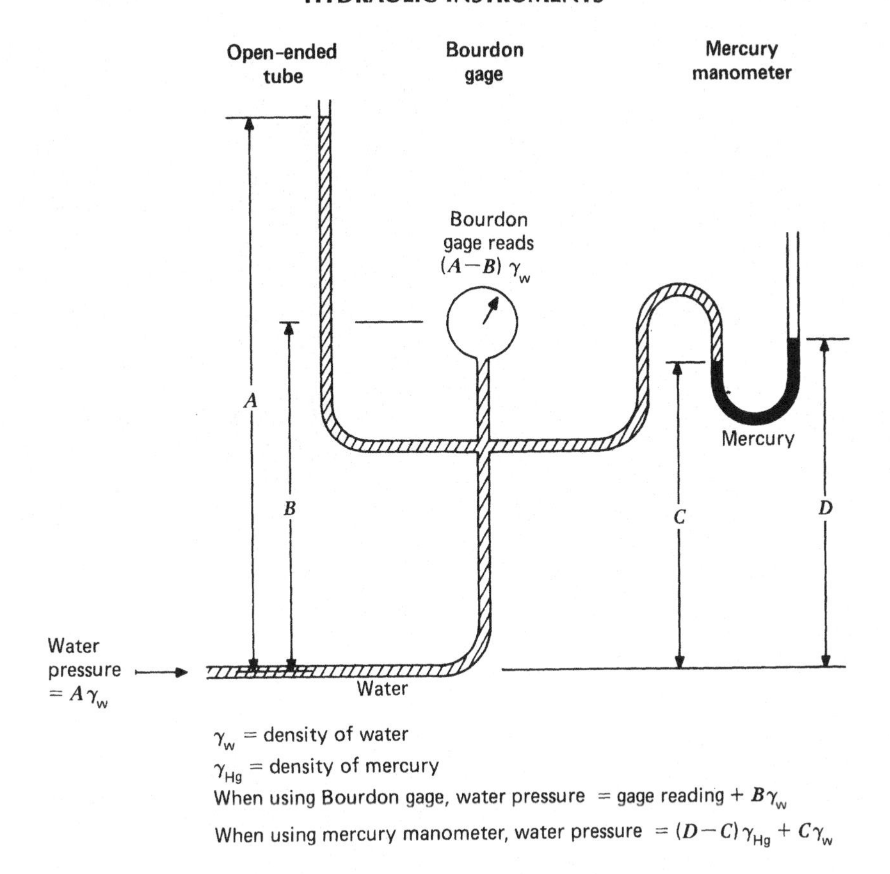

Figure 8.2. Schematic of hydraulic measuring devices.

have greater longevity than most Bourdon tube pressure gages.

8.2.3. Liquid-Filled Tubes for Hydraulic Instruments

Twin-tube hydraulic piezometers and liquid level settlement gages incorporate liquid-filled tubes as an essential part of the pressure measuring system. In these systems the pressure difference between the liquid surfaces at opposite ends of a tube is assumed to bear a known relationship to the elevation difference, via an assumed or measured density of liquid. The three primary causes of error that have long plagued this measurement method appear to be discontinuity in the liquid, liquid density changes caused by temperature variation, and surface tension effects. Each is discussed in turn, after which recommendations are given for tubing material and diameter, tubing fittings, liquid, and routing of liquid-filled tubes.

Discontinuity in Liquid

The source of greatest potential error is discontinuity of liquid in supposedly liquid-filled tubes. A 2 in. (50 mm) long gas bubble in the vertical segment of a tube will cause a reading error equivalent to 2 in. (50 mm) of liquid head. A single large gas bubble in the horizontal segment of a tube is of no consequence, but in practice it may migrate to a nonhorizontal part of the tube and cause an error.

Penman (1978) reports on findings by Jamin that a small tube can effectively be blocked if it contains a sufficient number of gas/water interfaces. At each gas/water meniscus, surface tension causes the gas bubbles to be at a slightly higher pressure than the water, as in unsaturated soil (Figure 2.8). When pressure is applied at one end of the tube to cause flow, the gas/water menisci yield to the attempted direction of flow, thereby opposing the flow. To achieve flow, the applied pressure must be increased sufficiently. As an example of this effect, about 1000 gas bubbles in an otherwise liquid-filled tube (or 1000 water drops in an otherwise gas-filled tube) of 0.11 in. (3 mm) inside diameter will resist an applied pressure of 15 lb/in.^{2} (100 kPa). Dissolved gas is also a major source of corrosion to metallic components. There is therefore ample evidence that all possible sources of gas should be eliminated from liquid-filled tubes. Gas can enter the system and form bubbles from three sources: dissolved gas in the liquid, diffusion through the tubing itself, and through the ends of the tubing.

When water is exposed to the atmosphere for an adequate length of time, it becomes saturated with air gases and thereafter acts somewhat as a sponge. Nitrogen and oxygen are absorbed when temperature decreases and are expelled when the temperature rises. When fully saturated with air, at room temperature and sea level, water contains approximately 10 parts per million of dissolved oxygen (ppm DO) by weight, equivalent to approximately 2% of air gases (oxygen and nitrogen) by volume. DO terminology is generally accepted as a method of specifying dissolved gas levels in water. Unless water is specifically indicated as de-aired (the word "de-aired" is used in this book to describe a liquid from which dissolved gases have been removed), it should be assumed to be saturated with gas. Commercially available bottled distilled and drinking waters are generally saturated with gas.

Many soil mechanics laboratories have "home-made" de-airing systems for use in preparing water for permeability and triaxial testing. These systems generally operate either by boiling water under a vacuum or by spraying water through a column to which a vacuum is applied and usually reduce the dissolved air gas content to between 2 and 5 ppm DO (0.4 and 1% air). The effectiveness of de-airing systems can be tested with a DO test kit, but measurements will be misleading if the liquid was initially saturated with gases other than oxygen. If a DO kit is to be used, air should be bubbled through the liquid for about half an hour before de-airing, to ensure initial saturation with air rather than with other gases.

For liquids in twin-tube hydraulic piezometers and liquid level settlement gages, dissolved gas content should preferably be reduced to less than 1 ppm DO (0.2% air); otherwise, frequent flushing may be required to ensure continuity of liquid. Two types of equipment are commercially available for preparing suitable liquid.

First, gas and electric boilers are supplied by several manufacturers of geotechnical instrumentation (e.g., Figure 8.3). The boiler is normally operated for about 4 hours and the liquid allowed to cool overnight. Dyer (1986) made tests to determine the effectiveness of this method on water that had initially been saturated by bubbling air for 2 hours, with a result of 0.5 ppm DO (0.1% air). The method may not be suitable for liquids other than water if their properties are affected by boiling.

Figure 8.3. Boilers for preparation of de-aired water (courtesy of Geotechnical Instruments (U.K.) Ltd., Leamington Spa, England).

Second, the *Nold DeAerator*™ can be used. The apparatus, shown in Figure 8.4, consists of a sealed tank, electric motor, impeller, and electric vacuum pump or water-powered aspirator for obtaining the necessary vacuum, which is applied to the space above the liquid. The phenomena of cavitation and nucleation generate an ultrahigh vacuum that vaporizes dissolved gases and volatile liquids. Centrifugal force directs the released vapors outward, where they bubble up to the partially evacuated space above the liquid surface. From there the gases are withdrawn through the aspirator or vacuum pump to the atmosphere. Six liters of water can be reduced from air saturation to less than 0.5 ppm DO (0.1% air) in 4 minutes running time without use of heat and to less than 0.05 ppm DO (0.01% air) in 15 minutes. This apparatus is suitable for de-airing antifreeze liquids as well as water. An automatic version is available for continuous preparation of de-aired liquid.

When transferring de-aired liquid from either a boiler or the DeAerator to the instrument tubes or to other vessels, agitation must be avoided to pre-

Figure 8.4. Nold DeAerator™ (courtesy of Walter Nold Company, Inc., Natick, MA).

vent reintroduction of air, and careless decanting can easily increase the DO content by 1 ppm or more. Transfer of de-aired liquid from one vessel to another is best done by evacuating the receiving vessel. De-aired liquid can be stored and transported either in a steel boiler (Figure 8.3) or in glass (**not** plastic) bottles with rubber stoppers; the container must be full. A convenient *Pressurized De-Aired Water Containment System* is available from the manufacturer of the DeAerator for transporting de-aired liquid, at up to 70 lb/in.^{2} (480 kPa), to instrument terminals in the field without reabsorbing air. The system is described in Appendix E.

Reabsorption of air must be minimized after the tubes are filled with liquid. This may be accomplished by appropriate selection of tubing and by inhibiting air entry at tubing ends. Recommendations for selection of tubing are given later in this section. Inhibition of gas entry at tubing ends depends on the type of instrument. When twin-tube hydraulic piezometers are installed in unsaturated fills, they must incorporate high air entry filters (Section 9.11). The ends of twin-tube hydraulic piezometer tubes will be connected in the terminal enclosure either to Bourdon gages or manometers, and gas-tight tubing fittings and valves should be used. Liquid level settlement gages are usually sealed at one end but open to the atmosphere at the readout end, and arrangements must be made to close this end when readings are not being taken, to avoid reabsorbing air.

Liquid Density Changes Caused by Temperature Variation

When the temperature of a liquid is increased, it expands and its density is reduced. Therefore, a temperature change in a nonhorizontal part of tubing will cause a reading error if the density variation is not taken into account. The error is unlikely to be significant for twin-tube hydraulic piezometer readings, but it is of great concern for liquid level settlement gages, which depend on the accurate measurement of small pressure changes. One solution is to measure temperature at various points along the tubing and apply a correction, but this is normally impracticable. Thus, a goal is selection of a liquid with the least coefficient of thermal expansion, but this may be in conflict with the need for a low freezing point, desirable density, and chemical compatibility with tubing.

As can be seen from Figure 8.5, of the aqueous solutions typically used in liquid level settlement gages, water has the least coefficient of thermal expansion and therefore introduces the least thermal error. The coefficient of thermal expansion increases with increasing ethylene glycol content; thus, unnecessarily high concentrations should be avoided. As an example of the error magnitude, if the temperature of a 50% ethylene glycol and water mix rises from −10°C to +20°C over a 4 ft (1.2 m) vertical height of liquid, an error of about 0.7 in. (18 mm) will be introduced into a settlement measurement. The error increases with increase in height of the liquid column affected by the temperature change. Attention is drawn to the unusual freezing point curve, shown in Figure 8.5, for high concentrations of ethylene glycol.

Surface Tension Effects

A decrease in surface tension will decrease the time required for equilibrium in a liquid-filled tube and

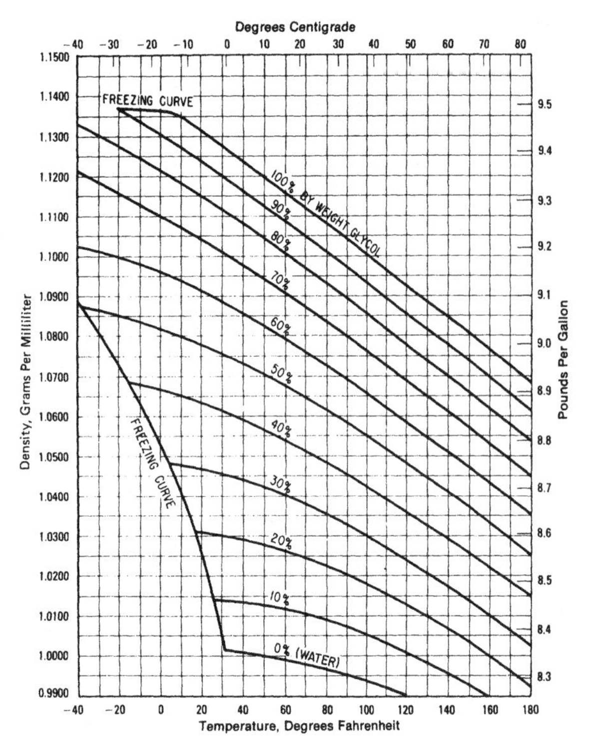

Figure 8.5. Densities of aqueous solutions of ethylene glycol at various temperatures (after Dow, 1981).

thus will tend to increase measurement accuracy. Water has a higher surface tension than typically used antifreeze mixes, but surface tension can be reduced to an acceptable level by adding a wetting agent, as discussed later in this section.

Recommendations for Tubing Material

Certain plastics are permeable to liquids but impermeable to gas; others exhibit the converse.

Tests reported by Penman (1960) indicate that polyethylene is permeable to air and that nylon absorbs water, thereby causing loss of water by evaporation. When filled with de-aired liquid under pressure and exposed to the atmosphere or to gas in unsaturated soil, the permeability of tubing to both liquid and gas must be minimized.

Unplasticized nylon 11 with a polyethylene (polythene) sheath is generally accepted as the tubing of choice for use with twin-tube hydraulic piezometer systems when making long-term measurements in embankment dams and is a good choice in most other cases. There is more than one type of nylon 11, and a specification for the appropriate type is given in Appendix E. Unplasticized nylon 11 with a polyurethane sheath also appears to be a good choice. Polyethylene with a polyvinyl choloride (PVC) sheath appears to be satisfactory from the standpoint of minimizing permeability to liquid and gas, but there is evidence that when compression fittings are attached to polyethylene tubing, the polyethylene creeps at the compression sleeve and long-term sealing is uncertain. The inner tube should therefore be unplasticized nylon 11. Nylon 66 (trade name *type H*) is not a suitable choice because of its water absorption characteristics. Saran tubing deteriorates with time and with exposure to sunlight and should never be used.

Because nylon tends to absorb water and also to leach gas until it becomes saturated with water, one or two flushes are usually required before the liquid-filled tubes can be fully commissioned.

Recommendations for Tubing Diameter

All tubing diameters referred to in this section are **inside** diameters, because the **inside** diameter impacts on the behavior of fluid within the tubing. The reader is cautioned about the possible confusion: industry standards use the **outside** diameter when referring to tubing sizes. The inside diameter of liquid-filled tubes depends on the application, but guidelines on upper and lower limits can be given.

When using water and aqueous solutions, 0.25 in. (6 mm) appears to be the largest **inside** diameter from which gas bubbles can readily be displaced during flushing. A lower limit results both from the need to avoid the Jamin effect described earlier in this section and the need to reach equilibrium within an acceptable time. Penman et al. (1975) indicate that, when using the overflow type of liquid level gage (Section 12.10) in dams, a 0.1 in. (2.5 mm) **inside** diameter water-filled tube could require more than half an hour to reach acceptable equilibrium. Poiseuille's equation for laminar flow in pipes shows that the time required for a given degree of equalization varies inversely as the fourth power of the pipe diameter. By using a tube of 0.25 in. (6 mm) **inside** diameter, the 99.9% equilibrium time was reduced to under 1 minute.

For liquid level settlement gages with aqueous solutions, 0.25 in. (6 mm) **inside** diameter appears to be a good choice for the liquid-filled tubes. This diameter is strongly recommended if any movement of the liquid column occurs at the time of reading, caused, for example, by deformation of a bladder or diaphragm. If no such movement occurs, 0.17 in.

(4.3 mm) **inside** diameter appears to be suitable. When using mercury in liquid level settlement gages, an **inside** diameter between 0.07 and 0.2 in. (2–5 mm) appears to be satisfactory. However, a general caution about use of mercury is appropriate. In the United States, mercury is considered to be a hazardous material, and environmental restrictions prevent its use in many applications. In addition, if mercury remains in plastic tubing in the long term, there is evidence that it can leach gas from the tubing and create discontinuities in the mercury. Also, oxidation can cause contamination and blockage of the tubing.

For twin-tube hydraulic piezometers, the need for rapid equilibrium is less critical, and an **inside** diameter between 0.1 and 0.2 in. (2.5–5 mm) appears to be the best choice.

Recommendations for Tubing Fittings

Compression fittings should be of the type that ensures positive alignment of the compression sleeve, by contact with either a shoulder in the body of the fitting or the nut. The type with a single *olive* compression sleeve that is aligned only by the tubing itself should not be used. Brass compression fittings are suitable for short-term measurements, for example, measurements made during a typical construction period, but nylon fittings are too weak and are not a good choice. For long-term measurements the possibility of corrosion should be recognized. Brass is subject to corrosion by dissolved gases in the liquid within the tubing and in any water surrounding the fittings and, although there is no specific evidence that excessive corrosion of inaccessible fittings has occurred, it appears worthwhile to use stainless steel fittings at all inaccessible locations when long-term measurements are required. Stainless steel should be type 316: this is a nonmagnetic stainless steel that is corrosion-resistant, and compression fittings of this type are readily available. When sheathed tubing is used, compression fittings should fit over the inner tubing, not the sheath. The sheath must therefore be pared back, using a cutting tool made for the purpose, and the protruding nylon tube inspected carefully to ensure that it has not been cut.

Recommendations for Liquid

Despite extensive research and trial use of a variety of liquids, no single liquid appears to be suitable for all purposes.

For twin-tube hydraulic piezometers, antifreeze solutions should not be used since they tend to plug piezometer filters and appear to have generated osmotic pressures. Tubes should be installed below the frost line and terminal enclosure temperatures should be maintained above freezing. Tap or well water may be polluted and should not be used. In the author's view the preferred liquid is filtered distilled water, produced by steam distillation. The water should be de-aired to reduce the dissolved gas content to less than 1 ppm DO. A wetting agent should be added and, for long-term applications, a bacterial inhibitor should also be added to minimize organic growth—two practices are in common use. First, quaternary ammonium compound (QAC)* is used (e.g., Houlsby, 1982; USBR, 1974) both as a wetting agent and bacterial inhibitor. Some users have argued that the chloride ions in QAC are corrosive to brass and copper, but this does not appear to be a realistic concern when the recommended concentration is used. The more likely cause of any observed corrosion is dissolved gases in the liquid. Second, copper sulfate† is used as a bacterial inhibitor and Aerosol OT‡ as a wetting agent (e.g., Corps of Engineers, 1971). Either practice appears to be satisfactory, with a preference for QAC, as this appears to be more effective as a bacterial inhibitor than copper sulfate. Houlsby (1982) reports that tubes for twin-tube hydraulic piezometers developed constrictions over a period of 10 or 15 years of regular flushing with uninhibited water. When QAC was added to the water, tubes were freed from constrictions, and regular usage of QAC was then able to keep tubes clean and free of organically caused bubbles.

For liquid level settlement gages not subjected to freezing temperatures, thermal errors are minimized by using de-aired distilled water as described above for twin-tube hydraulic piezometers. For short-term (construction period) applications, a wetting agent should be added. For long-term applications, QAC should be added, and inaccessible metal components should be type 316 stainless steel. Accessible components for long-term applications should, wherever possible, be selected for

*For example, Hyamine 1622 or 3500, available from Lonza AG. See Appendix D. Use a 60 parts per million solution.

†Use two approximately 0.25 in. (6 mm) size crystals per 10 U.S. gallons (40 liters) of liquid.

‡Available from American Cyanamid Company. See Appendix D. Use a 50 parts per million solution.

their resistance to corrosion (e.g., plastics or type 316 stainless steel), but if brass components are used they can be replaced if corrosion occurs. Recognition of deterioration assumes vigilance on the part of reading personnel, and it is better practice to design for a maintenance-free system. If tubes are subjected to freezing, the minimum required antifreeze should be added prior to de-airing. Antifreeze mixtures given in Figure 8.5 are suitable, but the commercial types that include a plugging antileak additive should not be used. A wetting agent is not necessary with ethylene glycol mixes. Specific gravity must be known in most cases (Section 12.10). If in doubt about specific gravity and its variation with temperature and dissolved gas content, the user should make laboratory determinations using a volumetric flask and an accurate balance having about ±0.1% accuracy. Some manufacturers of liquid level settlement gages recommend use of methanol/ethylene glycol/water mixes to create a specific gravity of 1.0 so that standard pressure indicator scales graduated in water head can be used. However, distillation occurs while de-airing this mixture, thus changing the specific gravity, and methanol can damage the plastic components of the de-airing system and liquid level gage. It is preferable to avoid use of methanol and to apply a calibration constant to correct for specific gravity.

There is some evidence that aqueous solutions of technical grade ethylene glycol may not remain uniformly mixed after filling tubes in liquid-filled settlement gages. However, it appears that if duplicate liquid-filled tubes are used and the liquid freely connects between both ends of the tube so that flow can occur throughout, the molecules of the mixture will remain in motion and specific gravity will remain uniform.

Some users have added a coloring agent (e.g., ink or food coloring) to liquid in full-profile liquid level settlement gages to facilitate regular inspection for gas bubbles. However, the coloring agent eventually stains the tubing and makes it harder to see bubbles. The author knows of no suitable coloring agent and recommends against its use.

Experience has shown that, despite all precautions discussed above, liquid in supposedly liquid-filled tubes is likely to become discontinuous after a period of months or perhaps years, requiring periodic flushing with fresh de-aired liquid. For twin-tube hydraulic piezometers with appropriate tubing and high air entry filters, flushing frequency is likely to be once every few years. Liquid level settlement gages require more frequent flushing, because highly accurate measurements of small pressure changes are required. For these gages, the author recommends developing a history of the need to flush, by periodic flushing, noting resulting reading change and adopting a job-specific standard procedure based on the data. Liquid-filled tubes in full-profile liquid level settlement gages should be inspected for continuity of liquid prior to each use and flushed when there is evidence of discontinuity.

Routing of Liquid-Filled Tubes

De-aired liquid can withstand significant subatmospheric pressure without breaking its continuity. When the dissolved gas content has been reduced to below 1 ppm DO (0.2% air), a practical limit of subatmospheric pressure appears to be about 20 ft (6 m), and liquid-filled tubes should be routed so that no part is above the internal pressure head level by more than this amount. Even within this limit there is a tendency for the liquid to become discontinuous with time, and if possible tubes should be routed so that a positive pressure exists at all times.

Closed Hydraulic Systems

All closed hydraulic systems unfortunately act as excellent thermometers, because changes in temperature cause the enclosed liquid to expand and contract, which in turn causes the internal pressure to rise and fall. For example, liquid-filled earth pressure cells are sensitive to temperature and may require temperature measurements and appropriate corrections. Liquid level settlement gages generally have valves on the open ends of liquid-filled tubes that are closed on completion of readings to avoid re-airation of the liquid, and thus they become closed systems. If any part of the gage could be damaged by significant rise in internal pressure owing to temperature or pressure surges during handling and transportation, precautions must be taken, for example, by including a cylinder with a spring-loaded piston in the liquid-filled tube.

8.2.4. Data Acquisition Systems for Hydraulic Instruments

Use of Bourdon tube pressure gages and manometers requires visual access to the readout location. Where remote reading is required, an electrical pressure transducer can replace the pressure gage or manometer.

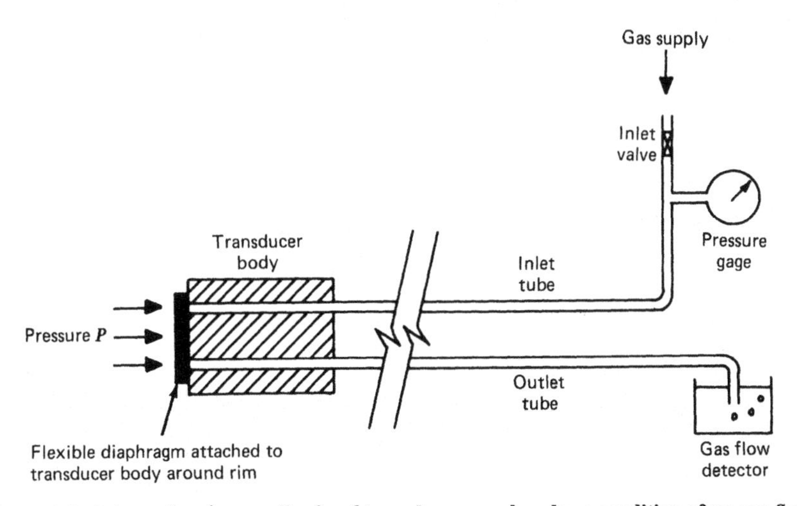

Figure 8.6. Schematic of *normally closed* transducer, read under a condition of no gas flow.

Automatic data acquisition systems for double-fluid full-profile settlement gages and twin-tube hydraulic piezometers are described in Chapter 12 and Appendix E respectively.

8.3. PNEUMATIC INSTRUMENTS*

Pneumatic transducers and data acquisition systems are used for pneumatic piezometers, earth pressure cells, load cells, and liquid level settlement gages.

8.3.1. Basic Types of Pneumatic Transducer

There are two basic types of pneumatic transducer, depending on whether a pneumatic circuit is normally closed or normally open when external pressure is applied to the transducer. Most modern transducers are of the first type.

8.3.2. Normally Closed Transducers

The *normally closed* type of transducer is also called *venting type*. The type can be subdivided according to whether the reading is made under a condition of no gas flow or as gas is flowing.

Normally Closed Transducers, Read Under a Condition of No Gas Flow

Figure 8.6 shows the basic arrangement. The pressure P is the pressure of interest. An increasing gas pressure is applied to the inlet tube and, while the gas pressure is less than P, it merely builds up in the inlet tube. When the gas pressure exceeds P, the diaphragm deflects, allowing gas to circulate behind the diaphragm into the outlet tube, and flow is recognized using a gas flow detector. The gas supply is then shut off at the inlet valve, and any pressure in the tubes greater than P bleeds away, such that the diaphragm returns to its original position when the pressure in the inlet tube equals P. This pressure is read on a Bourdon tube or electrical pressure gage.

This type of transducer has a variety of diaphragm types, depending on the manufacturer, including flat, hat-shaped, or convoluted (rolling) diaphragms of synthetic rubber or metal.

Normally Closed Transducers, Read as Gas is Flowing

The basic arrangement is the same as in Figure 8.6, but with a gas flow controller and sometimes also a gas flow meter in the inlet tube, and the outlet tube vents directly to the atmosphere. A schematic of the arrangement is shown in Figure 8.7. Gas pressure is increased, under a constant very small flow, causing a rise in pressure gage reading. When the gas pressure exceeds P, the diaphragm moves outward, allowing gas to circulate through the outlet tube, such that the maximum indicated pressure gage reading is equal to P.

As will be seen later, the readings of all types of pneumatic transducer are affected by the rate of gas flow. The version shown in Figure 8.8 incorporates a third tube, such that no gas flow occurs in the tube

*Written with the assistance of the manufacturers listed in Table 8.1.

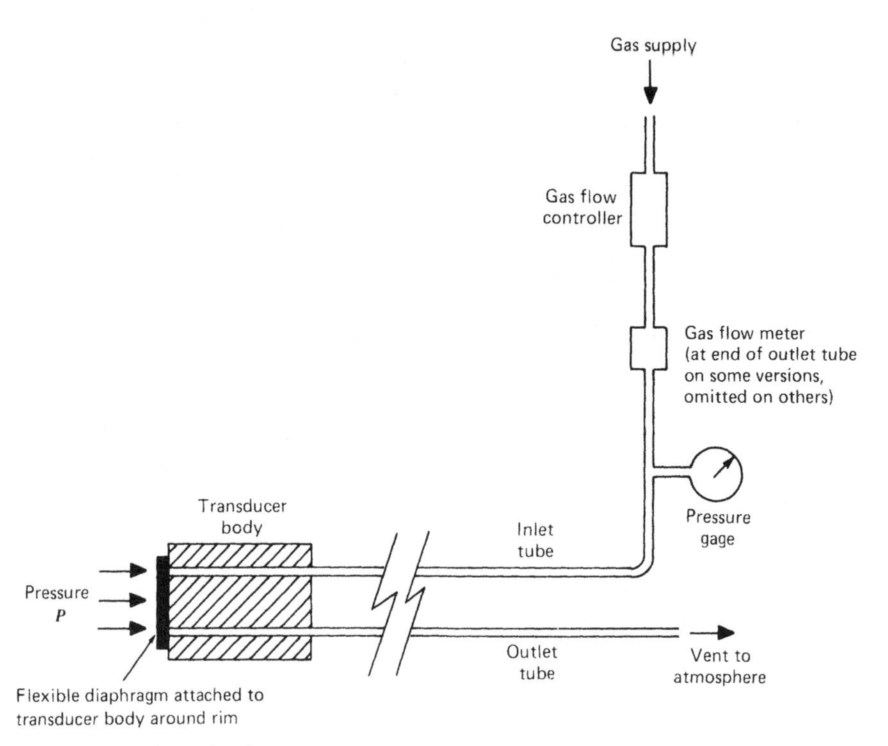

Figure 8.7. Schematic of *normally closed* transducer, with two tubes, read as gas is flowing.

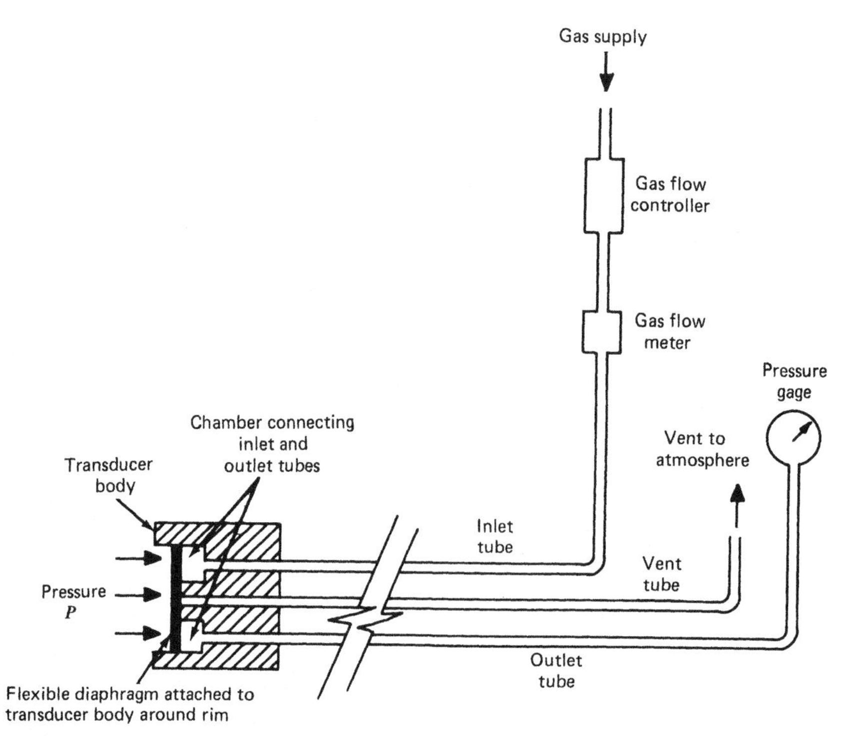

Figure 8.8. Schematic of *normally closed* transducer, with three tubes, read as gas is flowing.

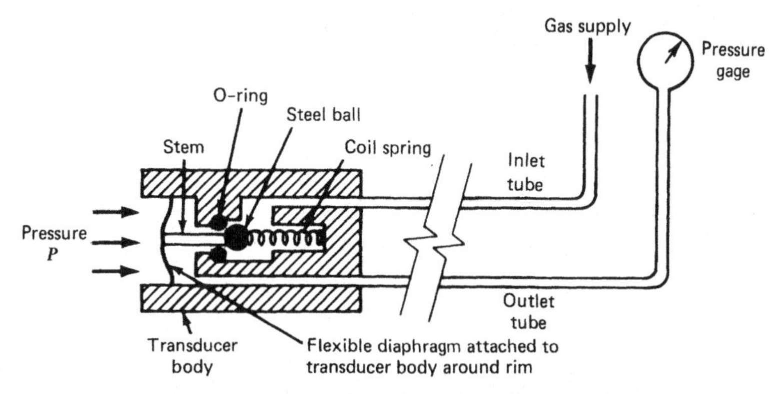

Figure 8.9. Schematic of *normally open* transducer.

to which the pressure gage is attached, thereby reducing errors caused by any variations in the rate of gas flow.

8.3.3. Normally Open Transducers

The *normally open* type of transducer is also called *check valve type*. The version shown in Figure 8.9 has a steel ball/coil spring/O-ring arrangement that serves as a check valve. Other versions have a different type of check valve. The diaphragm may be of synthetic rubber or may be metal bellows. The check valve remains open until the gas pressure first exceeds P, at which point the diaphragm moves outward, the check valve closes, and pressurized gas is locked in the outlet tube. The pressure gage is presumed to indicate P but, as discussed later, the reading is affected by the rate of gas flow. Reading errors caused by an excessive rate of gas flow can be minimized by including a flow controller and flow meter in the inlet tube, so that gas flows at a slow constant rate and gas pressure is reasonably uniform throughout the inlet and outlet tubes.

There is a potential problem with these transducers if water accidentally enters the tubes. If the transducer is at a lower elevation than the pressure gage, and if the pressure P is less than the head of water in the tubes, the check valve will close, and the water cannot be purged through the tubes. Water can sometimes be removed by connecting a vacuum pump, fitted to a water trap, but this is possible only if the head of water in the tube is less than about 25 ft (8 m).

8.3.4. Features of Commercially Available Systems

Table 8.1 indicates commercially available systems. Readers are cautioned that, although the text has been submitted to the listed manufacturers for review, new developments and product line changes are likely to occur on a regular basis; thus, readers are advised to contact manufacturers for a current status. Manufacturers' addresses are given in Appendix D.

Reading Mode

The reading mode, shown in Table 8.1, indicates whether the reading is made under a condition of no gas flow or as gas is flowing.

Sensitivity to Diaphragm Displacement

If a pneumatic transducer is to provide an accurate indication of the actual pressure P, it is important that diaphragm displacement should not cause a change of P at the moment when the transducer is being read. For example, when used in a piezometer installed in clayey soils, diaphragm displacement may in fact cause a change of P because outward diaphragm displacement causes a decrease in the volume of pore water on the outer side of the diaphragm. This volume decrease causes a pressure increase. All pneumatic transducers are sensitive to diaphragm displacement.

For the transducer shown in Figure 8.6, a rapid rise of gas pressure in the inlet tube "hammers" the diaphragm, and this shock causes an immediate pressure increase on the outer side of the diaphragm. When used in a piezometer, water is forced into the ground rapidly, owing to the high pressure differential existing across the diaphragm at that time. However, during diaphragm resetting as gas pressure falls, differential water/gas pressures are smaller and there is little driving force to encourage the diaphragm to return to its original position. Gas

Table 8.1. Commercially Available Pneumatic Transducers and Data Acquisition Systems

Basic Type of Transducer	Brand[a]	Figure Number	Normal Reading Mode
Normally closed (venting type)	Apparatus Specialties Co.	8.6	No flow
	Earl B. Hall, Inc.	8.7[b]	Flow
	Geotechnical Instruments (U.K.) Ltd.	8.6	No flow
	Geosistemas	8.6	No flow
	Glötzl GmbH	8.7	Flow
	Slope Indicator Co. (Model #514177, 514178)	8.7[c]	Flow
	Slope Indicator Co. (Model #51481)	8.8[d]	Flow
	Soil Instruments Ltd.	8.6	No flow
	Thor International, Inc.	8.7	Flow
Normally open (check valve type)	Slope Indicator Co. (Model #51401, 51471)	8.9[d]	No flow
	Terra Technology Corp.	8.9[e]	No flow

[a] Manufacturers' addresses are given in Appendix D.
[b] Gas flow meter is on outlet tube.
[c] Also available with three tubes as in Figure 8.8.
[d] No longer manufactured; included here for completeness.
[e] Also available, when used in a piezometer, with one or two additional tubes connecting to the space outside the diaphragm. Strongly **not** recommended. See Chapter 9.

pressure may therefore fall to a value significantly lower than the water pressure before the diaphragm reseats to seal the pressure in the inlet tube, and the resulting reading will be too low. The error described above is minimized by using a low-volume displacement diaphragm and by very careful control of gas pressure. Some commercially available systems have a reading sequence that is manually instigated but automatically conducted, thus permitting complete control of gas pressure and removing the variable quality of human input from the critical stages. Geotechnical Instruments (U.K.) Ltd. incorporates a "softening" arrangement on the outer side of the diaphragm, consisting of an air-filled rubber tube, such that sensitivity to diaphragm displacement is minimized.

The transducers shown in Figures 8.7 and 8.8 are sensitive to diaphragm displacement for the reasons given above. Errors are minimized by using a diaphragm with minimum volumetric displacement, and commercial versions with displacements of 0.01 cm^3 (0.0006 $in.^3$) or less are preferred. The effect of diaphragm displacement will be reduced if the operator waits until the reading stabilizes under a constant gas flow rate.

The transducer shown in Figure 8.9 is sensitive to diaphragm displacement because the volumetric displacement is large, and readings are made after the diaphragm has moved to a new position.

Sensitivity to Gas Flow Rate and Length of Tubing

Pressure loss (loss of velocity head) occurs as gas flows through tubing. For a given pressure at one end of a tube, the pressure at the other end will be affected by a change in the flow rate. The magnitude of pressure loss is also affected by the length of tubing. All pneumatic transducers are sensitive to pressure loss.

The above discussion on sensitivity to diaphragm displacement, for the transducer shown in Figure 8.6, concluded that the error is minimized by very careful control of gas pressure. This careful control also minimizes errors caused by sensitivity to gas flow rate and length of tubing. It is important to recognize that this type of transducer **must** be totally free from gas leaks in the inlet tube and at the diaphragm closure arrangement: this can sometimes be difficult to achieve when quick-connect fittings are used between inlet tube and readout unit, because the slightest dirt or damage to the O-ring seal in the fitting may create a leak.

The transducers shown in Figures 8.7 and 8.8 are read as gas is flowing, therefore they are sensitive to the gas flow rate and length of tubing. Addition of the third tube reduces errors caused by pressure loss, because no gas is flowing in the tube to which the pressure gage is attached. However, the addition does not eliminate these errors, because pressure is lost as the gas flow passes from the inlet to the vent tube. Tests by Slope Indicator Company (Mikkelsen, 1986a) to compare their two- and three-tube systems show that the sensitivity to gas flow rate and length of tubing is approximately halved by adding the third tube. For these transducers, a constant gas flow rate is therefore essential. The need is particularly critical for liquid level settlement gages, when required repeatability of liquid head measurement is usually a fraction of an inch. The purpose of the gas flow controller is to maintain a constant gas flow. Some manufacturers assume that the flow remains constant once the controller is set, and others provide a gas flow meter in the circuit so that the flow rate can be monitored. Recent experience has shown that gas flow controllers provided in some commercially available readout units do not maintain a gas flow that is adequately constant for use with liquid level settlement gages and that the position of the indicator ball in some gas flow meters is affected by static electrical charge on the ball. The charge builds up with time as gas passes by, the ball falls even when the gas flow remains constant, and the operator is likely to increase the flow to compensate. Improved flow controllers and flow meters are now available for use with liquid level settlement gages. For this application the flow meter should be provided with a bypass valve so that gas flows through it only momentarily when the operator turns the valve to check the flow.

The transducer shown in Figure 8.9 is sensitive to gas flow rate and length of tubing. The check valve will close when the gas pressure against the diaphragm first exceeds P. If significant pressure loss is occurring along the outlet tube as the check valve closes, subsequent pressure equalization along the outlet tube will cause the pressure gage to read less than P.

8.3.5. Readout Units

Gas pressure is measured by using either a Bourdon tube pressure gage or an electrical pressure transducer with a digital indicator. When an electrical pressure transducer is used, a keyboard selection can usually be made of the desired pressure measurement unit (e.g., lb/in.^{2}, kg/cm^2, or ft head). Indicators can be connected to a large gas tank and continuously energized, and the systems can be programmed to scan multiple transducers and display measurements sequentially on the digital display and on a printed paper tape. These systems can be provided with outputs for connection to a variety of automatic data acquisition and processing equipment.

8.3.6. Recommendations for System Selection

Normally closed transducers have largely superseded the *normally open* type, and use of the latter type is not recommended. They are more sensitive than *normally closed* transducers to diaphragm displacement, gas flow rate, and length of tubing, and water that may enter the tubing accidentally cannot always be removed.

All the commercially available *normally closed* transducers are suitable choices. When selecting a system from among those listed in Table 8.1, users should consider the general guidelines given in Section 4.9. Their own familiarity and previous experience with one or more brands will weigh heavily in the selection. The overriding need is very close adherence to the instrumentation manufacturer's recommendations for operation of the selected system.

8.3.7. Recommendations for Various System Components

Several general recommendations can be made for system components.

Components Within a Transducer

Transducer materials should not be subject to corrosion or electrolytic breakdown. Some stainless steels are not corrosion-free under geotechnical field conditions and some plastics are inadequately stable. The diaphragm should be resilient so that a permanent distortion will not be created if the tubes are accidentally overpressured, and nitrile rubber is preferable to steel if overpressurization is possible. Preferably, a positive stop should be provided to protect against overpressurization.

Gas

Most manufacturers of pneumatic instruments agree that a dry gas (carbon dioxide or dry nitrogen) should be used rather than air to avoid condensation of water in the tubing. Carbon dioxide can readily be obtained from companies that service fire extinguishers, and dry nitrogen is available from welding supply companies (*water pumped,* **not** *oil* *pumped* nitrogen). Portable indicators contain a small rechargeable gas cylinder, and an auxiliary larger cylinder can be used where tubes are long or where a large number of transducers are to be read.

Economy can be achieved by using a tank of compressed air and a desiccant canister. Use of a hand air pump and no desiccant runs the risk of water entering the tubing and, if the water is not blown out by flowing air, large reading errors may result from the summation of surface tension forces at gas/water menisci (Section 8.2.3).

Tubing

Section 8.2.3 describes requirements for liquid-filled tubes. Gas-filled tubes are not subject to concerns about discontinuity of liquid, temperature variation, surface tension, freezing, bacterial growth, or routing limitations, but nevertheless certain precautions are necessary. Materials for gas-filled tubes should be impermeable to water so that, when submerged below water or embedded in soil, moisture does not migrate through the walls of the tube. The following tubings appear to have appropriate properties: unplasticized nylon 11 with a polyethylene (polythene) sheath, unplasticized nylon 11 with a polyurethane sheath, and polyamide. Polyethylene with a polyvinylcholoride (PVC) or polyethylene sheath is not a good choice: it is less robust than the above three types; nylon creates a better seal than polyethylene at the entry point to the transducer; and there is evidence that when compression fittings are attached to polyethylene tubing, the polyethylene creeps at the compression sleeve and long-term sealing is uncertain. Saran tubing deteriorates with time and with exposure to sunlight and should never be used.

A 3/16 in. (5 mm) diameter tubing is suitable for all applications. A 1/8 in. (3 mm) tubing is sometimes used when lengths are less than about 250 ft (75 m), and 1/4 in. (6 mm) tubing is also used on occasion. It should be noted that these tubing dimensions are the **outside** diameters of the individual tubes, **not** the sheath.

The maximum length of tubing depends on the length of time considered tolerable for reading. A rule of thumb for reading time is 0.5–1 minute per 100 ft (30 m) of tubing, the longer time being necessary when requiring maximum accuracy. The tubing can be brought up to pressure with an initial high gas flow, but these reading times are required after such initial pressurization. Tubings of 3000 ft (900 m) length are in use, but 2000 ft (600 m) is considered to be a reasonable maximum length.

Tubing Fittings

Recommendations for tubing fittings for liquid-filled tubes are given in Section 8.2.3. The same recommendations apply to fittings for gas-filled tubes.

8.4. ELECTRICAL INSTRUMENTS*

Electrical transducers and data acquisition systems are used in numerous geotechnical instruments. Comprehensive references describing electrical transducers include Arthur (1970), Cerni and Foster (1965), Considine (1971), Herceg (1972), Lion (1959), Norton (1969), Perry and Lissner (1962), Prensky (1963), Spitzer and Howarth (1972), Stein (1964), and Wolf (1973).

Electrical transducers and associated manual data acquisition systems are described in turn below. Automatic data acquisition systems, power supplies, and communication systems are described in Sections 8.4.15 and 8.4.16. General guidelines on the use of electrical transducers and data acquisition systems, including selection of electrical cable, are given in Section 8.4.17.

8.4.1. Basic Types of Electrical Resistance Strain Gage

Electrical resistance strain gages are used in many geotechnical instruments. An electrical resistance strain gage is a conductor with the basic property that resistance changes in direct proportion to change in length. The relationship between resistance change ΔR and length change ΔL is given by the gage factor (GF), where

$$\frac{\Delta R}{R} = \frac{\Delta L}{L} \times \text{GF}.$$

The gage factor for bonded foil and bonded wire gages is usually close to 2. Semiconductor strain

*Sections 8.4.1–8.4.14 have been written with the assistance of Ralph S. Carson, Professor of Electrical Engineering, University of Missouri–Rolla, Rolla, MO, and J. Barrie Sellers, President, Geokon, Inc., Lebanon, NH. Leon J. Weymouth, Principal Engineer, Teledyne Engineering Services, Waltham, MA, assisted with writing Sections 8.4.1–8.4.4 on electrical resistance strain gages.

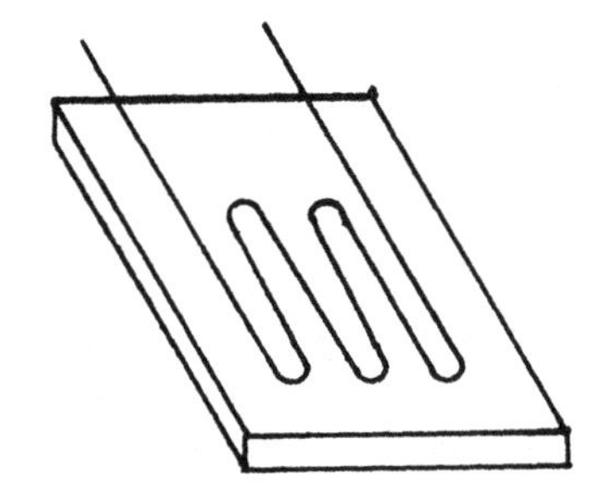

Figure 8.10. Schematic of uniaxial bonded wire resistance strain gage.

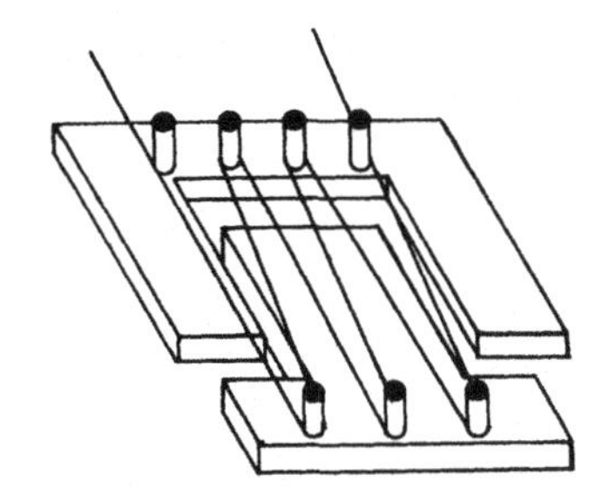

Figure 8.11. Schematic of unbonded wire resistance strain gage.

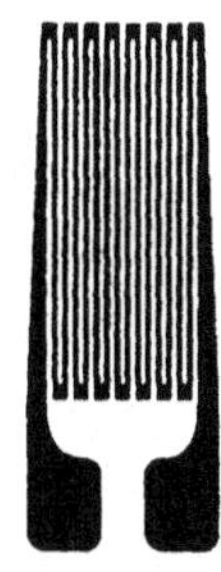

Figure 8.12. Uniaxial bonded foil resistance strain gage.

gages have a much larger gage factor, between 50 and 200.

The five basic types of electrical resistance strain gage—bonded wire, unbonded wire, bonded foil, semiconductor, and weldable—are described in turn below. Guidelines on the use of resistance strain gages for measurement of strain in structural members are given in Chapter 13.

Bonded Wire Resistance Strain Gage

A bonded wire strain gage is fabricated with fine copper–nickel or nickel–chromium wire, looped back and forth, and bonded to a thin elastic mounting of paper or plastic, which in turn is bonded to the structure being monitored. Figure 8.10 shows a uniaxial gage. They are commonly used where long gages are needed for testing rock or concrete specimens, enabling strains to be measured over a representative sample, but for short gage lengths the bonded foil gage is usually preferred.

Unbonded Wire Resistance Strain Gage

In the unbonded wire strain gage (Figure 8.11), a fine wire is looped around two sets of electrically insulated posts that are attached to the structure being monitored. The device is less robust than the bonded gage and was developed at a time when bonding techniques were unreliable. With the advent of improved bonding cements, unbonded wire strain gages have become less common. However, the Carlson unbonded wire strain gage transducer, referred to as an *elastic wire strainmeter,* is frequently used in embedment strain gages and concrete stress cells and has proven reliability and longevity. The Carlson transducer contains two similar coils of highly elastic carbon-steel wire each looped around two posts, arranged such that one increases in length and electrical resistance when strain occurs while the other decreases. The ratio of the two resistances is independent of temperature and therefore the change in resistance ratio is a measure of strain. The total resistance is independent of strain since the resistance change in one coil is equal and opposite to the resistance change in the other coil, and therefore total resistance is a measure of temperature. Wiring diagrams are given by the Corps of Engineers (1980).

Bonded Foil Resistance Strain Gage

A bonded foil strain gage is composed of a thin foil of resistance alloy, such as constantan or nichrome, bonded to a thin plastic film, which in turn is bonded to the structure being monitored. The foil is usually photofabricated and etched or die-cut during manufacture to produce grid patterns of varying designs. A uniaxial gage is shown in Figure 8.12. Temperature compensation is provided by matching the temperature characteristics of the foil to those of the material being studied and, if necessary, by adding additional resistors in the Wheatstone bridge circuit. Most strain gages have a nominal resistance of 120 or 350 ohms. The higher resistance varieties are generally favored for transducer applications because they permit the use of higher voltage inputs, with correspondingly higher outputs, thus minimizing effects of extraneous resistance changes.

Semiconductor Resistance Strain Gage

This gage uses highly doped semiconductor crystals of silicon or germanium. The doping is necessary to give the crystals specific gage factors and thermal coefficients. When strain is applied to the crystal, it undergoes a change in resistance proportional to the strain. Gage factors of 50–200 make this type of gage much more sensitive than other types. Its main

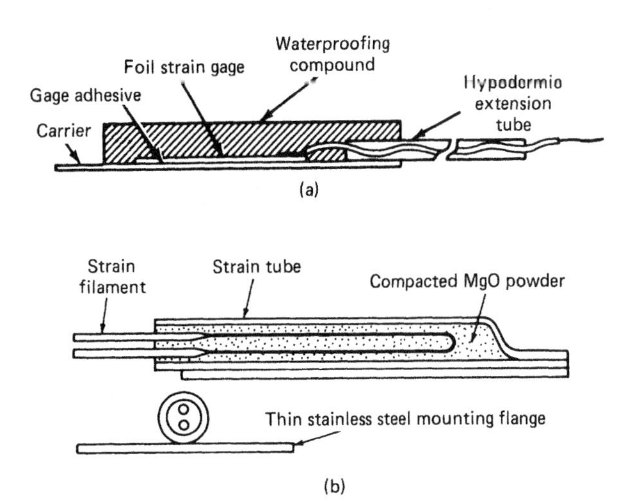

Figure 8.13. Weldable strain gages: (a) bonded foil transducer (courtesy of HITEC Products, Inc., Ayer, MA) and (b) strain filament encased in small tube (courtesy of Eaton Corporation, Los Angeles, CA).

disadvantage is the relatively complicated techniques required for correction of errors induced by temperature change. The effect can be minimized by choosing the correct gage type and by using temperature matched gages, but the gage is not suitable for environments with large temperature gradients or rapid fluctuations. The output of semiconductor gages is nonlinear, and their use is normally limited to monitoring strains of 100 microstrain or less.

Weldable Resistance Strain Gage

During manufacture of the weldable gage, a resistance element is permanently attached to a thin stainless steel mounting flange. The resistance element may be a conventional bonded foil gage (Figure 8.13a; Wnuk, 1981) or a strain filament encased in a small tube (Figure 8.13b). The mounting flange is later welded to the steel structure being monitored, using a small portable capacitive discharge spot welder. Weldable gages with integral leads are more suitable for field attachment than bonded gages.

Longevity of Resistance Strain Gages

Longevity of resistance strain gages is dependent primarily on methods of gage installation, sealing, and protection, rather than on inherent properties of the gages themselves. Bonded wire, unbonded wire, bonded foil, semiconductor, and weldable resistance strain gages can all be used for long-term performance measurements, provided that personnel responsible for gage selection, installation, sealing, and protection are specialists with wide experience in field application of resistance strain gages. Strain gages installed by inexperienced users normally have a short life. Recommendations for gage installation, sealing, and protection are given in Chapter 13.

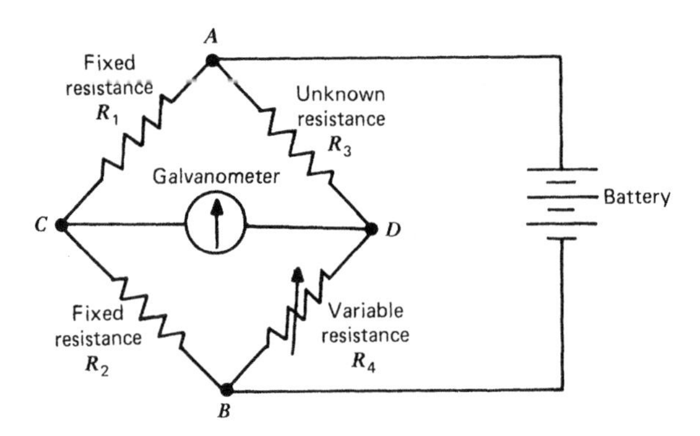

Figure 8.14. Wheatstone bridge circuit.

8.4.2. Wheatstone Bridge Circuits for Use with Electrical Resistance Strain Gages

Output from electrical resistance strain gages is normally measured using a Wheatstone bridge circuit, shown in Figure 8.14. The circuit has four "arms" formed by resistances, R_1, R_2, R_3, and R_4. A voltage is applied between A and B and the resistance R_4 is altered until no current flows between C and D. At this point the needle of the galvanometer is not deflected and the bridge is balanced. Under these conditions,

$$\frac{R_1}{R_2} = \frac{R_3}{R_4},$$

and since R_1, R_2, and R_4 are known, R_3 can be calculated. Bridge networks are described below, and advantages and limitations are summarized in Table 8.2.

Quarter Bridge Networks

The simplest network is the quarter bridge with two lead wires, shown in Figure 8.15a. However, the resistance of the lead wires in series with the active strain gage desensitizes the bridge and any change in lead wire resistance with temperature change cannot be differentiated from strain. The effect can be substantial if temperature changes and/or lead wire lengths are large but can be eliminated by using

Table 8.2. Wheatstone Bridge Networks

Type	Advantages	Limitations	Bridge Factor[a]	Usage
Quarter, two-wire system; Figure 8.15a	Least expensive Easiest to use	Sensitive to temperature change at gage and leads Nonlinear at high strain levels	1.0	Use only in laboratory environment at constant temperature Never use in geotechnical field applications
Quarter, three-wire system; Figure 8.15b	Eliminates error caused by temperature changes at leads	Sensitive to temperature change at gage Nonlinear at high strain levels	1.0	Most popular network for stress/strain analysis on structures
Half, with dummy gage; Figure 8.16a	No temperature effects	Dummy must be unstressed and bonded to same material and at same temperature as active gage	1.0	Long-term tests on structures where temperature variations are great and increased accuracy is required
Half, with both gages active, at 90° to each other (Poisson effect); Figure 8.16b	No temperature effects	Not suitable for biaxial stress fields	1.3	May be used on long columns or tendons subject to uniaxial loads
Half, with both gages fully active, equal tensile and compressive strains; Figure 8.16c	No temperature effects	Bridge network not always possible to achieve	2.0	Measurement of bending of beams Cantilever type transducers Torsional transducers
Full, with dummy gages; Figure 8.17a	No temperature effects	Most expensive	2.0	Rare
Full, with all gages active, two at 90° to other two (Poisson effect); Figure 8.17b	No temperature effects	Most expensive Not suitable for biaxial stress fields	2.6	Tension links or compression columns, for example, load cells
Full, with all gages fully active, two tensile, two compressive, equal strains; Figure 8.17c	No temperature effects Maximum output	Most expensive	4.0	Bending beams

[a]Output relative to quarter bridge network. See Figures 8.15–8.17. Assumes $\Delta r = 0$, $\mu = 0.3$, ΔR is small compared with R, uniaxial stress field, and all gages at same temperature.

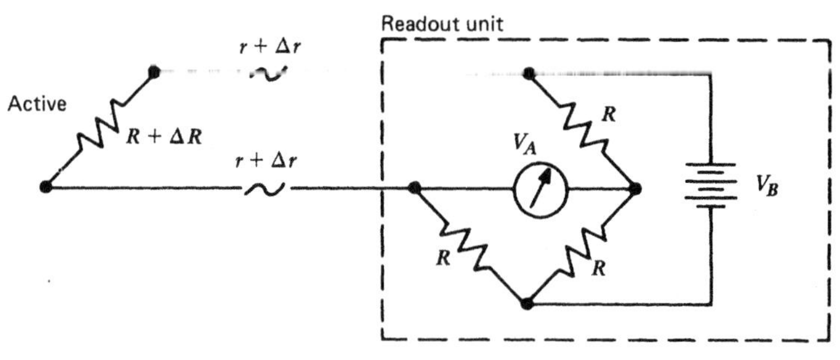

$R + \Delta R$ = gage resistance plus gage resistance change
$r + \Delta r$ = lead wire resistance plus lead wire resistance change
R = resistance of bridge completion resistors

$$\frac{V_A}{V_B} = \frac{\Delta R + 2\Delta r}{4[(R + 2r) + \frac{1}{2}(\Delta R + 2\Delta r)]}$$

(a)

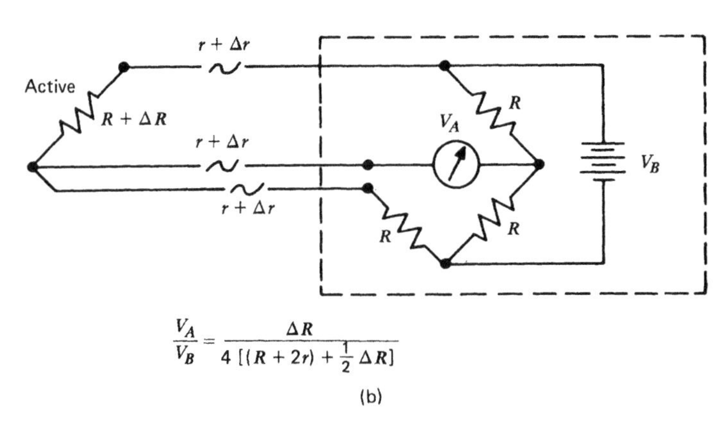

$$\frac{V_A}{V_B} = \frac{\Delta R}{4\,[(R + 2r) + \frac{1}{2}\Delta R]}$$

(b)

Figure 8.15. Wheatstone bridge quarter bridge networks: (a) two-wire system and (b) three-wire system.

a quarter bridge with three lead wires. As shown in Figure 8.15b, a third wire is connected to one end of the active strain gage and wired into the bridge network. This cancels the thermally induced error caused by lead wire resistance changes. However, the circuit remains sensitive to temperature change at the gage.

In practice, lead wire resistance is minimized and gage resistance is maximized, often by using 350 ohm gages instead of the originally standard 120 ohm type. Thus, because r may be small compared with R, the equations in Figure 8.15 can be rewritten:

$$\text{(a)} \quad \frac{V_A}{V_B} = \frac{\Delta R + 2\Delta r}{4\,[R + \frac{1}{2}(\Delta R + 2\Delta r)]}$$

and

$$\text{(b)} \quad \frac{V_A}{V_B} = \frac{\Delta R}{4\,(R + \frac{1}{2}\Delta R)}.$$

Note that because Δr may be significant with respect to ΔR, the term Δr remains in the equation for a two-wire system, indicating the effect of changing lead wire resistance caused by temperature change.

Half Bridge Networks

Temperature effects that plague a quarter bridge network can almost be eliminated by using a half bridge circuit. A second strain gage is connected to the bridge at the measuring location rather than in the readout unit, either as an active gage or a "dummy" gage. Two arrangements are possible for the active gage; thus, there are three possible configurations, as shown in Figure 8.16. Lead wire

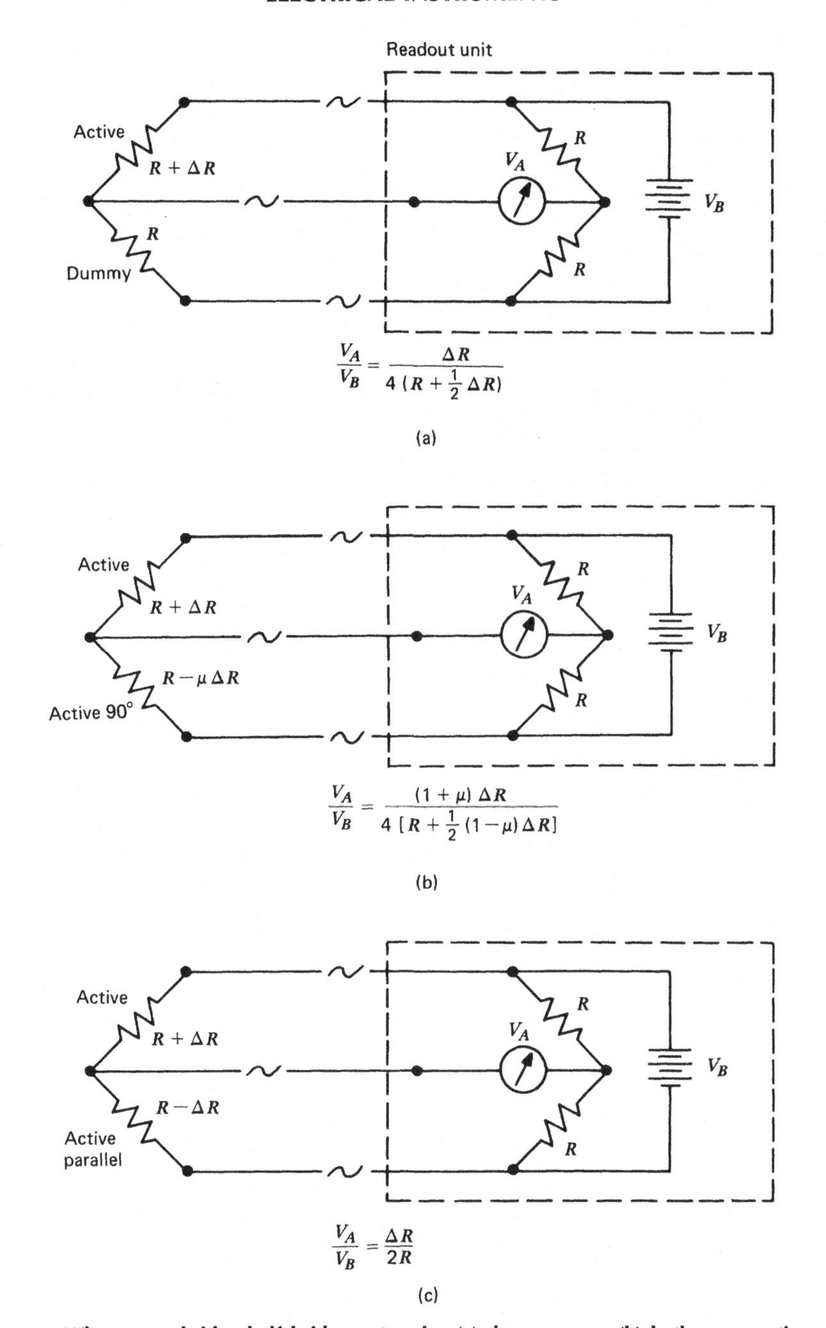

Figure 8.16. Wheatstone bridge half bridge networks: (a) dummy gage, (b) both gages active, at 90 degrees to each other (Poisson effect), and (c) both gages fully active, equal tensile and compressive strains.

resistance r and resistance change Δr have been omitted from the figure because Δr cancels in the bridge circuit and, if r is small compared with R, lead wire effects become negligible.

In the first configuration, Figure 8.16a, the dummy gage, also termed *inactive* or *compensating* gage, is mounted on a piece of the same material as the active gage but isolated from the stress field so that it undergoes no stress caused by temperature change. It is placed adjacent to the active gage so that both experience the same temperature change or gradient, thereby providing temperature compensation.

When the second gage is an active gage, it must be located on the structure in a position that will cause it to be in tension if the first gage is in compression, or vice versa. This can often be arranged by aligning the second gage at 90 degrees adjacent to the first gage, as in Figure 8.16b, thereby measuring the Poisson effect. This increases the output of the network by a factor of 1.3. If the second gage is positioned so that it experiences an equal and opposite strain to that at the first gage, as in Figure 8.16c, the output is doubled. Strain gages positioned on opposite sides of the neutral axis of a flexing beam fulfill this condition. Note that two gages experiencing the same strain in both magnitude and sign would cancel each other out if connected in a half bridge network. Such an arrangement allows use of a half bridge to isolate for bending measurements, because axial load has no effect on the measurements.

Full Bridge Networks

A full bridge network, shown in Figure 8.17, provides a maximum output and almost full compensation for temperature change at the gage and along the leads; thus, this is the optimum arrangement. Three networks are possible, corresponding to the half bridge networks. As for the half bridge networks, lead wire resistance r and resistance change Δr have been omitted from the figures.

Lead Wire Effects

The problem of changing lead wire temperature has been discussed previously and can be solved by avoiding use of two-wire quarter bridge networks. However, two additional errors remain, caused by *thermocouple* and *desensitization* effects. Thermocouple effects are discussed in Section 8.4.4 and can be overcome by use of an AC excitation voltage. Desensitization error can become significant if long runs of cable are used. Since the parasitic lead resistance appears as a larger gage resistance, with no equivalent change in resistance due to strain, the effective gage factor is reduced. This can be calculated by

$$\mathrm{GF_d} = \frac{R_g}{R_g + r}\,\mathrm{GF_n},$$

where $\mathrm{GF_d}$ = desensitized gage factor,
R_g = basic gage resistance,
r = cable resistance per lead,
$\mathrm{GF_n}$ = natural gage factor as reported by the manufacturer.

8.4.3. Use of a Circuit Tester to Check Integrity of Electrical Resistance Strain Gages and Cables

A circuit tester should be used after strain gage installation, and on a regular schedule thereafter, to check integrity of electrical resistance strain gages and cables by measuring gage and insulation resistance. The circuit tester should apply 2 volts or less for gage resistance testing and 15 volts or less for insulation testing.* The following tests should be made (the term *ground* refers to the structural member to which gages are attached):

- *Leakage, Bridge to Ground.* A connection is made between ground and any one (or all) of the bridge lead wires. Resistance should be a minimum of 1000 megohms, and preferably 10–20 gigohms.
- *Leakage, Shield to Ground.* A connection is made between shield and ground (shield isolated from structural member to which gage is attached). Resistance should be a minimum of 5 megohms.
- *Leakage, Bridge to Shield.* A connection is made between the shield and any one (or all) of the bridge lead wires. Resistance should be a minimum of 1000 megohms.
- *Gage Resistance.* If an individual gage can be isolated from the bridge, a connection is made between the pair of lead wires attached directly to that gage. If the gages are connected as a Wheatstone bridge and cannot be isolated, the readout unit should be disconnected and the bridge resistance measured across the input corners and output corners. The individual gage resistance, input bridge resistance, and output bridge resistance should not vary by more than ±2% from the nominal 120 or 350 ohm resistance plus any lead wire resistance in series. Lead wire resistance is determined from standard tables. However, commercial strain gage transducers may include auxiliary resistances in the bridge circuit for balancing, thermal correction, or calibration, and in these cases the transducer manufacturer should be contacted for gage resistance testing specifications.

*For example, Model 1300 gage installation tester, Measurements Group, Inc. (see Appendix D).

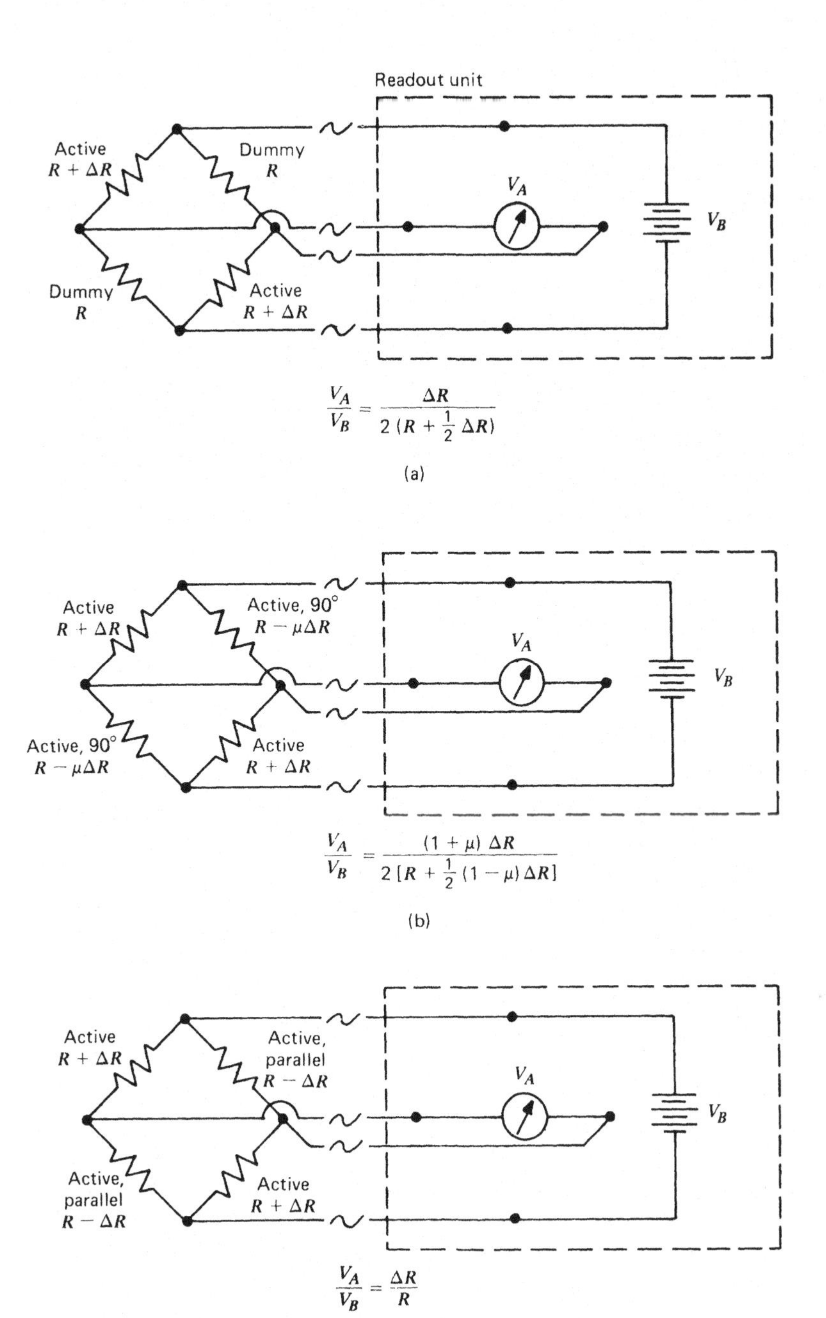

Figure 8.17. Wheatstone bridge full bridge networks: (a) dummy gages, (b) all gages active, two at 90 degrees to the other two (Poisson effect), and (c) all gages fully active, two in compression, two in tension, equal strains.

Low bridge resistance to ground or shield usually indicates failure of the waterproofing or other environmental protection over the gage or bridge, or breakdown of the insulation cover over the shield and individual lead conductors. It is sometimes possible to locate the damaged area, dry out the moisture causing the partial grounding condition, and repair the waterproofing or insulation. Soap bubble or helium leak detection techniques can sometimes be used to locate the damaged area. The soap bubble technique entails covering the gages and connections with liquid soap and applying light air pressure to the ends of the lead wires, but this procedure is very crude. A much more effective procedure is to apply a vacuum to the ends of the lead wires, connect the vacuum to a mass spectrometer, and spray helium around the gages and connections, using a fine nozzle. In this way, the exact location of minute leaks can be detected rapidly.

Low shield to ground resistance usually indicates moisture in the cable. Excessively high or low gage or bridge resistances usually indicate damage to one or more gages during installation, and gage performance and bridge balance will be affected adversely.

8.4.4. Manual Data Acquisition Systems for Electrical Resistance Strain Gages

A portable strain indicator is used for manual data acquisition. The indicator uses a second set of resistors to null the output voltage of the gage network and provides a measurement of the resistance required to null the circuit. Most portable indicators provide the excitation voltage to the gage. For geotechnical field applications, an AC excitation voltage is recommended and is available in most commercial strain indicators.* When a DC power supply is used, all conductors in the lead wire must be of the same material, otherwise a thermocouple junction is formed, and a signal error may be introduced if a thermal gradient is present between junctions of dissimilar materials.

Particular attention should be paid to keeping portable strain indicators clean and dry, because small current leakage to ground can influence readings. Also, some portable indicators may be influenced by temperature changes, and batteries used with these instruments have a much shorter life when subjected to below freezing temperatures.

*For example, Model P-350A or P-3500, Measurements Group, Inc., or Model HW1-D, Strainsert Company (see Appendix D).

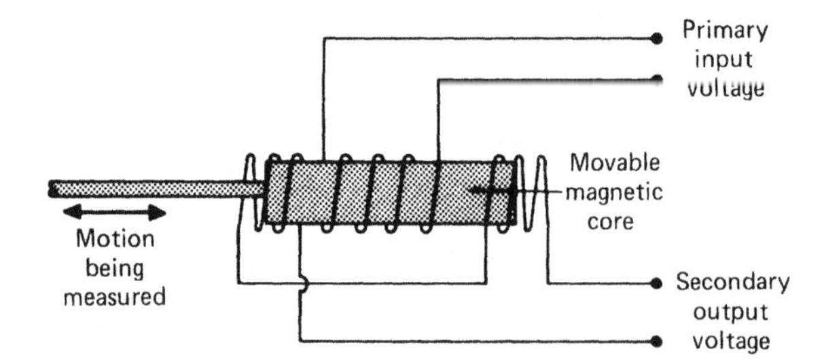

Figure 8.18. Schematic of linear variable differential transformer (LVDT).

8.4.5. Linear Variable Differential Transformer (LVDT)

LVDTs are used in fixed borehole extensometers and in other instruments for measurement of deformation.

An LVDT (Figure 8.18) consists of a movable magnetic core passing through one primary and two secondary coils. An AC voltage, called the *excitation voltage,* is applied to the primary coil, thereby inducing an AC voltage in each secondary coil, with a magnitude that depends on the proximity between the magnetic core and each secondary coil. The secondary voltages are connected in series opposition, so that the net output of the LVDT is the difference between these two voltages. When the core is at its midposition, the net output voltage is zero. When the core moves off center, the net output voltage increases linearly in magnitude with a polarity depending on the direction of core displacement.

A manual data acquisition system for an LVDT includes a means of generating the excitation voltage, a demodulator/amplifier, and a meter display. The meter display can be either analog or digital.

Since the core of an LVDT does not contact the coils, friction is avoided. There is no hysteresis and LVDTs are particularly suitable for measuring dynamic motions and very small displacements. Many types of LVDT have excellent resistance to humidity and corrosion and good long-term stability, and they can be protected within oil-filled housings to maximize longevity. However, the transmission of alternating currents through long lead wires introduces unwanted cable effects, which can seriously degrade the output signal.

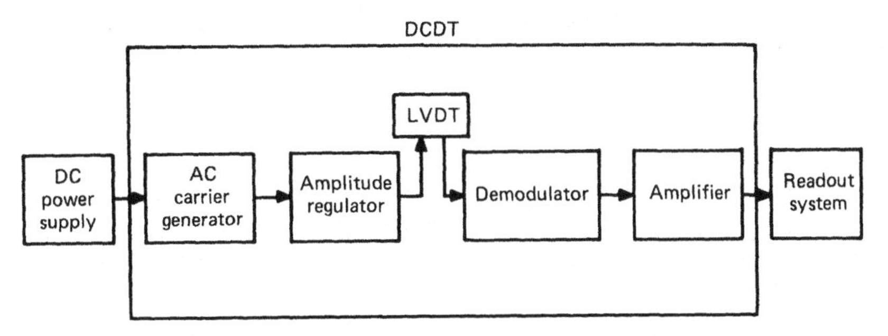

Figure 8.19. Schematic of direct current differential transformer (DCDT).

8.4.6. Direct Current Differential Transformer (DCDT)

DCDTs have similar applications to LVDTs and are usually preferred for geotechnical use. The need for special means of exciting the LVDT primary coil and modifying its secondary output voltage is provided for by miniaturizing the electrical circuitry and placing components within the transducer housing as shown in Figure 8.19. The device is now called a DC LVDT, or DCDT. The signal cable carries only DC voltages and unwanted cable effects are avoided.

DCDT manual data acquisition systems require only a stable DC power supply and a DC voltmeter. Most DCDTs have excellent resistance to humidity and corrosion effects and good long-term stability. They can also be protected within oil-filled housings.

8.4.7. Potentiometer

Linear potentiometers are an alternative to LVDTs and DCDTs for remote measurement of linear deformation. Rotary potentiometers are used for measurement of rotational deformation and where linear deformation can readily be converted to rotational deformation.

A potentiometer (*pot*) is a device with a movable slider, usually called a wiper, that makes electrical contact along a fixed resistance strip. As shown in Figure 8.20, a regulated DC voltage is applied to the two ends of the resistance strip and the voltage between *B* and *C* is measured as the output signal. The voltage varies between the voltage at *A* and the voltage at *B* as the wiper moves from *A* to *B*. When the device is used for measurement of linear deformation and the relationship between wiper position and output signal is proportional, the device is called a linear potentiometer. By reading voltages between *B* and *C*, and between *A* and *C*, a checking feature is created because the sum should equal the input voltage. An alternative method of readout is to measure the resistance between the wiper and one end of the resistance strip and to relate the measured resistance to the total resistance of the strip. In this way, the position of the wiper can be measured using an ohmmeter or digital multimeter. In a third readout method, the potentiometer forms two arms of a Wheatstone bridge circuit, and measurement of wiper position is made by using a portable indicator that contains a balancing potentiometer.

Potentiometers are not suitable for measuring rapidly varying motions. However, the readout is simple and can be arranged to give a high output voltage, which is not readily degraded by long lead wire effects or electrical noise. The resistance strip and wiper must be sealed to prevent moisture intrusion and, provided there are no leaks whatsoever, potentiometers can successfully be used for long-term measurements. In adverse environments, it may be worthwhile to seal potentiometers in oil-filled housings: oil-filled rotary potentiometers are available commercially, but the author is not aware of commercially available oil-filled linear potentiometers. However, because of the difficulty in en-

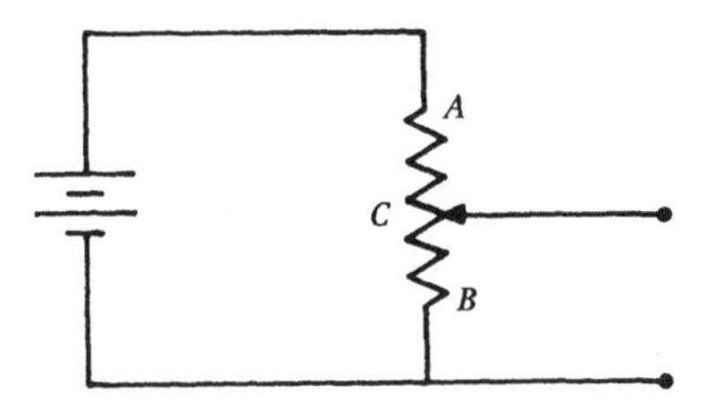

Figure 8.20. Schematic of linear potentiometer.

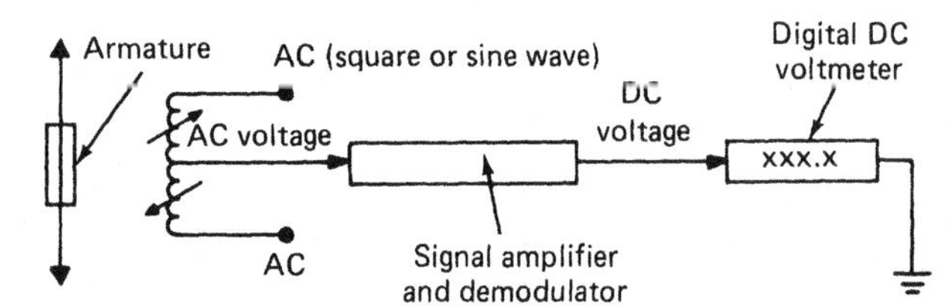

Figure 8.21. Schematic of variable reluctance transducer.

suring a long-term perfect seal on a linear potentiometer, LVDTs and DCDTs are now often preferred for long-term applications.

8.4.8. Variable Reluctance Transducer

Variable reluctance transducers (VRTs) are used in electrical crack gages and fixed embankment extensometers to measure linear deformation.

A VRT (Figure 8.21) consists of a center-tapped coil, positioned around a movable armature. The armature is magnetically permeable. An AC excitation voltage is applied to the ends of the coil, and an output voltage is sensed on the third wire. Movement of the armature away from the center, or null position, causes an imbalance in magnetic flux density between the two coil sections, which in turn causes the voltage in one section of the coil to increase, and the other to decrease. The signal amplifier and demodulator, shown in Figure 8.21, is normally positioned near the transducer. DC signals can be displayed on analog or digital meters or transferred to conventional data loggers and other voltage sensing devices. VRTs can also be read with portable strain indicators that are used with electrical resistance strain gages.

The coil is normally enclosed, both outside and inside, by stainless steel tubing sealed at the ends, and waterproofing of the lead wires is maintained to the signal amplifier and demodulator. Some leakage of moisture into the lead wires can be tolerated, and unless very low electrical resistance paths are created by free moisture between conductors, little degrading of the signal results. Water in the armature cavity is of no consequence. VRTs are therefore particularly suitable for installations below water, and they have been used very successfully in crack gages installed to monitor movement across perimeter joints on the upstream faces of concrete face rockfill dams. However, they should not be considered as the only transducers that are suitable for this purpose, since an LVDT or a DCDT could be

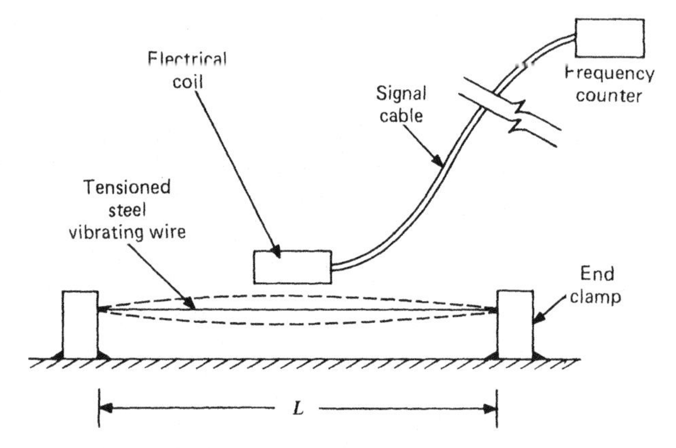

Figure 8.22. Schematic of surface-mounted vibrating wire strain gage.

sealed in a similar way to perform the same function.

8.4.9. Vibrating Wire Transducer

Vibrating wire transducers (Dreyer, 1977; Thomas, 1966) are used in pressure sensors for piezometers, earth pressure cells, and liquid level settlement gages, in numerous deformation gages, in load cells, and directly as surface and embedment strain gages.

Operating Principle

A length of steel wire is clamped at its ends and tensioned so that it is free to vibrate at its natural frequency. As with a piano string, the frequency of vibration varies with the wire tension, and thus with small relative movements between the two end clamps. The wire can therefore be used as a strain gage by plucking the wire, measuring natural frequency, and relating frequency change to strain. The wire is plucked magnetically by an electrical coil attached near the wire at its midpoint, and either this same coil or a second coil is used to measure the frequency of vibration. Figure 8.22 shows a vibrating wire transducer arranged for measuring surface strain.

Reading Methods

There are two methods of wire plucking and reading, the *pluck and read* method and the *continuous excitation* (or *autoresonant*) method. The latter allows measurement of low-frequency dynamic strains.

The pluck and read method entails application of one or more voltage pulses to the coil, thereby creating a magnetic attraction that causes the wire to vibrate. The coil then becomes a listening device because wire vibrations cause an alternating voltage to be induced in the plucking coil, of frequency identical to the natural frequency of the vibrating wire. The voltage signal is transmitted along the signal cable to a frequency counter, which is used to measure the time for a predetermined number of vibration cycles.

The continuous excitation or autoresonant method entails a similar procedure to initiate vibration, but a second coil is used to detect the frequency. The signal is fed back to the driving coil so that it applies a continuously pulsing voltage with frequency identical to the natural frequency of the vibrating wire. As the wire frequency changes, so does the driving frequency. The wire frequency can be determined, as above, by measuring the time for a predetermined number of vibration cycles. Alternatively, the pulsing voltage can be converted directly and continuously, by a frequency to voltage converter, into a voltage that is proportional to pulse frequency and can be displayed on a digital voltmeter or recorded on magnetic tape or a strip chart recorder. In this way, the transducer can be used for measuring dynamic strains, provided that the cycle frequency of the changing strain is less than about 100 hertz (Hz). It can also be used for control and alarm systems. However, when the continuous excitation method is used, readings can become unstable under the influence of radio frequency or electromagnetic interference, and the cable should be shielded as described in Section 8.4.17.

For both methods, frequency may be displayed directly on the frequency counter, requiring use of a calibration curve or table for calculating strain from frequency change, or the readout unit may include a linearizing circuit such that strain is displayed directly.

Relationship Between Frequency and Strain

The equation for frequency of a vibrating wire in terms of wire stress (Hawkes and Bailey, 1973) is

$$f = \frac{1}{2L}\sqrt{\frac{\sigma g}{\rho}},$$

where f = natural frequency (sec^{-1}),
L = length of vibrating wire (in.),
σ = stress in the wire (lb/in.^2),
ρ = density of the wire material (lb/in.^3),
g = acceleration due to gravity (in./sec^2).

In terms of wire strain,

$$f = \frac{1}{2L}\sqrt{\frac{Eg\varepsilon}{\rho}},$$

where E = modulus of elasticity of the wire (lb/in.^3),
ε = strain in the wire.

Thus,

$$\varepsilon = \frac{4L^2 f^2 \rho}{Eg} = Kf^2,$$

where

$$K = \frac{4L^2\rho}{Eg}.$$

Because the transducer is always installed with the wire under an initial tension, both the initial frequency f_0 and the new frequency f enter into the calibration relationship:

$$\varepsilon = K(f^2 - f_0^2).$$

Primary Advantage of Frequency Signal

The output signal contains the required information in the form of a frequency rather than magnitude of a resistance or voltage, and therefore undesirable effects involving signal cable resistance, contact resistance, leakage to ground, or length of signal cable are negligible. Very long cable lengths are acceptable. The stability of a frequency signal can be demonstrated dramatically by inserting bare conductors in water and observing no change in measured frequency: therein is a major reason for favoring a frequency rather than resistance or voltage signal when making measurements in a field environment.

Sources of Error

Historically, the major disadvantages of vibrating wire transducers have been wire corrosion, creep of the vibrating wire under permanent tension, and slippage at the wire clamping points, all of which usually result in a reduction in vibrating frequency.

Corrosion can be minimized by selection of materials that are not subject to galvanic corrosion and by drying and hermetically sealing the cavity around the wire. Attempts to minimize corrosion in vibrating wire pressure transducers by circulating dry nitrogen through them continuously, while successful in the short term, have proved difficult to maintain, for example, because of changes in operating personnel.

If the vibrating wire creeps under permanent tension, or if slippage occurs at the wire clamping points, a frequency reduction occurs that is unrelated to strain, and a zero drift has occurred. Although various investigators indicate that zero drift is minor (e.g., Bordes and Debreuille, 1985; Browne and McCurrich, 1967; DiBiagio, 1986; Londe, 1982; Thomas, 1966), others report that zero drift can be significant (e.g., Bozozuk, 1984; Jaworski, 1973; Kleiner and Logani, 1982; O'Rourke and Cording, 1975; Szalay and Marino, 1981).

Tharp (1986) reports on measurements of zero readings made 17 months after the original calibrations, prior to installation of 16 vibrating wire piezometers manufactured by Irad Gage. Data are shown in Table 8.3. The accuracy quoted by the manufacturer is ±0.5% full scale. During subsequent discussions, a representative of the manufacturer (LeFrancois, 1986) indicated his personal experience that evaluations of data should not be made on the basis of readings from a single instrument:

> Advanced manufacturing techniques greatly minimize drift, but no manufacturer can ever guarantee that every unit built will not drift beyond the accuracy specified over an indefinite period of time. The remedy is, whenever the piezometer is retrievable, to perform periodic verification of the zero and calibration. Otherwise the accuracy is achieved by an average from a group of piezometers.

As can be seen from Table 8.3, 5 of the 16 piezometers show a change in zero reading greater than the quoted accuracy, but the average change (−0.27 lb/in.^{2}) is within the quoted accuracy (±0.5 lb/in.^{2}).

The author is aware of two unpublished cases in which vibrating wire transducers have apparently experienced significant zero drift. They both involve measurements with piezometers, supplied by two different manufacturers, in two embankment dams. During first filling of each reservoir, one vibrating wire piezometer indicated a high piezometric level that caused concern, and filling was stopped. The piezometer reading continued to rise and, when the indicated piezometric level rose above pool level, the measurements were discounted and filling continued. It may be noted that

Table 8.3. Changes in Zero Reading of Vibrating Wire Piezometers During 17 Month Period

Instrument Number	Change in Zero Reading	
	lb/sq.in.[a]	% full scale[b]
10–603	−2.58	2.58
10–606	+0.65	0.65
10–607	−0.16	0.16
10–655	+0.09	0.09
10–663	+0.23	0.23
10–666	−0.33	0.33
10–667	−1.12	1.12
10–668	−1.15	1.15
10–669	−0.36	0.36
10–670	+0.53	0.53
10–671	−0.19	0.19
10–685	−0.06	0.06
10–686	+0.09	0.09
10–689	+0.08	0.08
10–697	−0.49	0.49
10–699	+0.49	0.49

[a] The negative sign indicates an increase in frequency of vibration.
[b] Range of piezometers: 100 lb/in.^{2}
Source: Tharp (1986).

an excessively high piezometer reading is consistent with wire slippage or creep. The fact that several recent versions of vibrating wire piezometers are provided with an in-place check feature whereby zero drift can be checked at any time during the life of the instrument (Chapter 9) appears to support the contention that the potential for zero drift should not be ignored.

Zero drift can be minimized by stress relieving the vibrating wire, clamps, and transducer body after they have been assembled, either by using high temperature or by load cycling, and most manufacturers insist that this is essential for long-term stability. On the other hand, some users report excellent results when this precaution was not taken. Wire attachment at the clamping points should not weaken the wire and squeezed capillary tube clamps or swaged pins appear to be the preferred methods. Bordes and Debreuille (1985) recommend that wire tension should be within the range 9–13% of yield strength, and comment:

> Experience shows it is desirable for the instruments to undergo a period of aging after manufacture, to provide time for strain hardening of the components and attachment points and for the relaxation of internal stresses that are inevitably set up in the

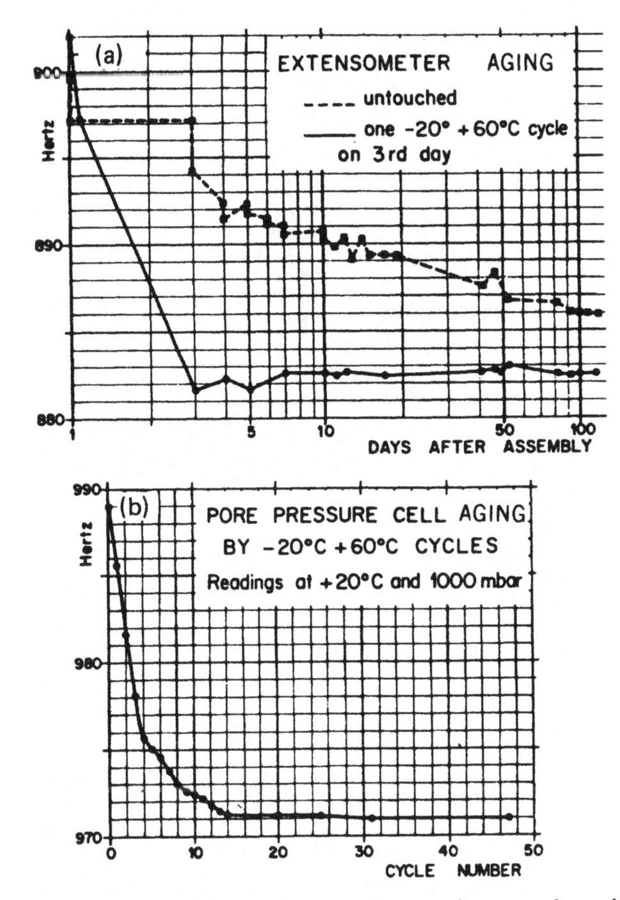

Figure 8.23. Response of vibrating wire transducers to thermal cycling (after Bordes and Debreuille, 1985).

> manufacturing process. The aging process can be accelerated by high temperature or a shaking table, although the two methods are not equally effective. [High] temperature accelerates the aging process.

Figure 8.23 illustrates the reactions of two Telemac vibrating wire transducers to thermal cycling. The upper figure shows readings of two embedment strain gages (the writers use the term *extensometer* for an embedment strain gage). The data indicate that readings stabilized immediately after a single thermal cycle (variations less than 1 hertz result from temperature and atmospheric pressure changes). The lower figure shows readings of a piezometer subjected to thermal cycles, indicating changes up to the 14th cycle. The writers make a distinction between the relatively simple strain gage arrangement, in which the only factors involved are changes in the wire itself and in its clamping arrangements, and the more complex transducer geometry of a wire attached to a diaphragm. In the second case, changes can occur at the point of attachment to the diaphragm and in the diaphragm itself, requiring more thermal cycles before stability is reached.

If a vibrating wire transducer is designed such that the electrical coil can be attached **after** the vibrating wire, clamps, and transducer body are assembled (e.g., the transducers manufactured by Geokon), the body can be stress relieved rapidly under high temperature, and no further aging is required.

In summary, on the issue of zero drift, four measures appear to be highly desirable: minimizing the potential for corrosion, use of appropriate wire attachment procedures at the clamping points, adherence to maximum tension limits, and appropriate aging by thermal and/or strain cycling. These measures are not adopted by all manufacturers. The author therefore does not support the viewpoint that **all** vibrating wire transducers are suitable for long-term applications. This opinion should not be taken as a vote in favor of alternative electrical transducers: when compared with other electrical transducers, a high-quality vibrating wire transducer is often the transducer of choice for long-term applications. However, the author favors the inherently more simple and reliable optical, mechanical, hydraulic, or pneumatic transducers for long-term measurements wherever they can provide the required data. There is a need for further documentation of long-term stability characteristics of vibrating wire transducers, both in the laboratory and in the field, and users are encouraged to examine the potential for zero drift and report their findings to the profession.

Applications when Transducers Are Subjected to Vibration

If transducers will be subjected to vibration, the user should ensure that zero drift will not be caused. O'Rourke and Cording (1975) indicate that impact in the vicinity of surface-mounted strain gages during laboratory testing caused large zero drift, but Elson and Reddaway (1980) report insignificant zero drift when gages were embedded in driven piles. The writers conclude:

> Generally, the first few blows of the hammer (up to 100) produced a slight permanent extension in the strain gauges (typically 10 to 20 microstrain). Thereafter, driving the pile had little effect on the readings. It is not clear whether the permanent ex-

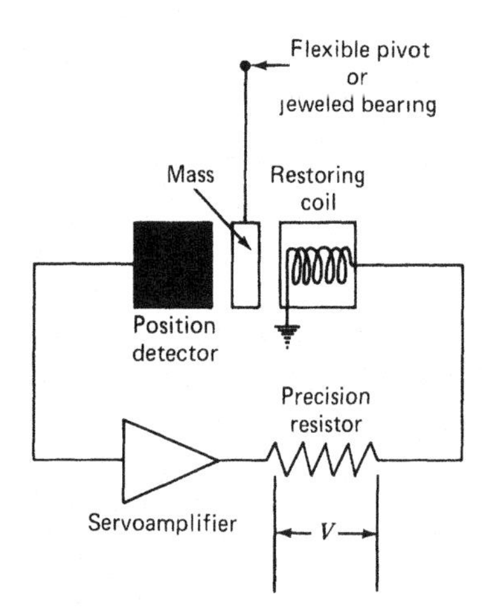

Figure 8.24. Schematic of force balance accelerometer. Voltage *V* is proportional to the force required to hold the mass in the null position.

Figure 8.25. Magnet/reed switch (after Burland et al., 1972). Reprinted by permission of Institution of Civil Engineers, London.

tension recorded is a real or an apparent value, although it would be reasonable to suppose that the recorded extension reflected the development of microcracks in the pile.

Strain measurement on driven steel piles is best accomplished with low-mass weldable vibrating wire gages, with additional epoxy encapsulation for extra mechanical strength during driving. When transducers will be subjected to vibration, the user is advised to contact manufacturers whose transducers are being considered for names of others who have used their transducers in similar applications. If in doubt, tests should be conducted to check transducer performance under conditions similar to those anticipated in the field.

8.4.10. Force Balance Accelerometer

Force balance accelerometers are used as tilt sensors in tiltmeters, inclinometers, and in-place inclinometers.

The device consists of a mass suspended in the magnetic field of a position detector (Figure 8.24). When the mass is subjected to a gravity force along its sensitive axis, it tries to move, and the motion induces a current change in the position detector. This current change is fed back through a servoamplifier to a restoring coil, which imparts an electromagnetic force to the mass that is equal and opposite to the initiating gravity force. The mass is thus held in balance and does not move. The current through the restoring coil is measured by the voltage across a precision resistor. This voltage is directly proportional to the input force.

The simplest and least expensive manual data acquisition system is a portable digital voltmeter. Automatic data acquisition systems for use with inclinometers are described in Chapter 12.

Force balance accelerometers have a good track record when used in portable tiltmeters and inclinometers where the transducer can be reversed to eliminate errors caused by zero shift. However, they are not yet proved for long-term measurements requiring permanent embedment of the transducer, although recent design improvements have increased their longevity.

8.4.11. Magnet/Reed Switch

The magnet/reed switch system (Burland et al., 1972) is used in probe extensometers.

It is an on/off position detector, arranged to indicate when the reed switch is in a certain position with respect to a ring magnet, as shown in Figure 8.25. The switch contacts are normally open and one of the reeds must be magnetically susceptible. When the switch enters a sufficiently strong magnetic field, the reed contacts snap closed and remain closed as long as they stay in the magnetic field. The closed contacts actuate a buzzer or indicator light in a portable readout unit.

The repeatability of closure depends on the radial position of the reed switch within the ring magnet and also on the orientation of the reed about its own axis, because it is difficult to manufacture ring magnets that are polarized uniformly. If the reed switch remains within the middle third of a 1.25 in. (32 mm) inside diameter ring magnet, the repeatability will be within about ±0.01 in. (±0.25 mm). However, tests have shown (Burland et al., 1972) that with very precise guidance repeatability can be improved to about ±0.001 in. (±0.025 mm). Moreover, no detectable drift has been observed over a period of 12 months in the laboratory. The system is simple, reliable, precise, inexpensive, and well suited for long-term performance measurements. The only hazard known to the author is the possibility of permanent loss of magnetism if magnets are allowed in prolonged contact with each other prior to installation.

8.4.12. Induction Coil Transducers

Induction coil transducers are used in probe extensometers, fixed embankment extensometers, fixed borehole extensometers, and crack gages.

If an electrical coil is powered by an AC source, a magnetic field is created around the coil. This coil is referred to as the *primary coil*. If a second coil is within the influence of the magnetic field, a voltage is induced in this *secondary coil*. The principle is termed *inductive coupling*. The inductive coupling principle is used in three transducers for geotechnical applications. In all cases the two coils are separated both electrically and mechanically. The difference is in output and configuration.

Soil Strain Gage

In the first transducer, the *soil strain gage* (Section 12.6.5), the magnitude of voltage induced in the secondary coil is a function of spacing between the two coils. Thus, by maintaining constant input voltage and measuring induced voltage, coil separation can be determined. The readout unit contains a Wheatstone bridge circuit to null out the voltage from the secondary coil. Coil spacing is calculated from the amount of adjustment needed to balance the bridge.

Transducer with Current-Displacement Induction Coil

The second transducer is used in probe extensometers (Section 12.5.6). The secondary coil is a single steel wire ring, with no external electrical connection. When the primary coil is placed inside the ring, a voltage is induced in the ring, which in turn alters the current in the primary coil because its inductance changes. The current in the primary coil is a maximum when the primary coil is centered inside the ring; thus, by measuring primary coil current the transducer can be used as a proximity sensor. The readout unit contains an ammeter for current indication. In some versions of the probe extensometer, the secondary coil is a steel plate with a central hole instead of a single steel wire ring.

Transducer with Frequency-Displacement Induction Coil

The third transducer functions as a linear displacement gage with frequency output and is used in crack gages (Section 12.3.2), probe extensometers (Section 12.5.6), fixed embankment extensometers (Section 12.6.4), and fixed borehole extensometers (Section 12.7.4). The secondary coil is a nonferrous ring, with no external electrical connection. The primary coil includes a coil-capacitance resonant frequency circuit. As the primary coil moves relative to the secondary coil, the mutual inductance between the two changes. As the mutual inductance changes so does the resonant frequency of the circuit. Frequency can be measured and displayed automatically by digital electronics or by a manually operated frequency-sensitive bridge circuit, and displacement is determined from a frequency/displacement calibration.

Induction coil transducers have excellent long-term stability. For long-term applications any embedded steel components should be protected from corrosion.

8.4.13. Magnetostrictive Transducer

The magnetostrictive transducer (Hawkes, 1978) is used in fixed borehole extensometers, probe extensometers, and tube convergence gages.

The transducer is also referred to as a *sonic probe*. It consists of at least two permanent magnets attached to points between which displacement is to be measured. A nickel–iron alloy tube is placed alongside the magnets and a copper wire is threaded inside the tube, as shown in Figure 8.26. The nickel–iron alloy experiences a physical distortion when subjected to a change in magnetization, thus the term *magnetostrictive*. An instantaneous electrical pulse is applied to the copper wire, thereby inducing an instantaneous magnetic field along its length. The magnetic field interacts with the magnetic field at each of the permanent magnets and causes a strain pulse to be initiated in the nickel–iron tube. These strain pulses travel along the tube

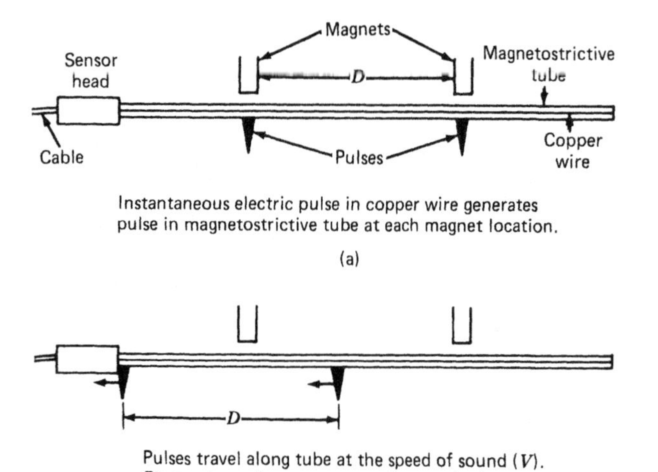

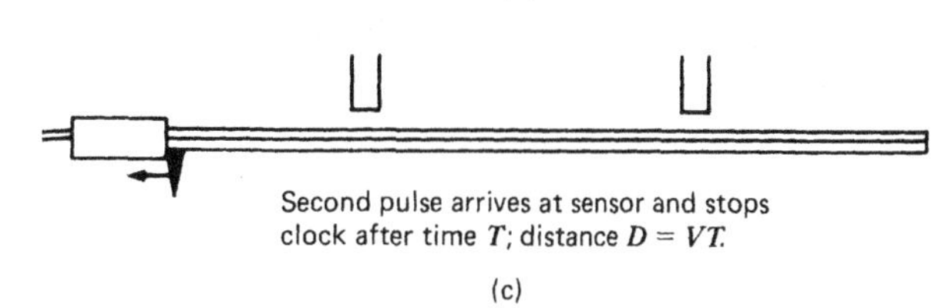

Figure 8.26. Schematic of magnetostrictive transducer.

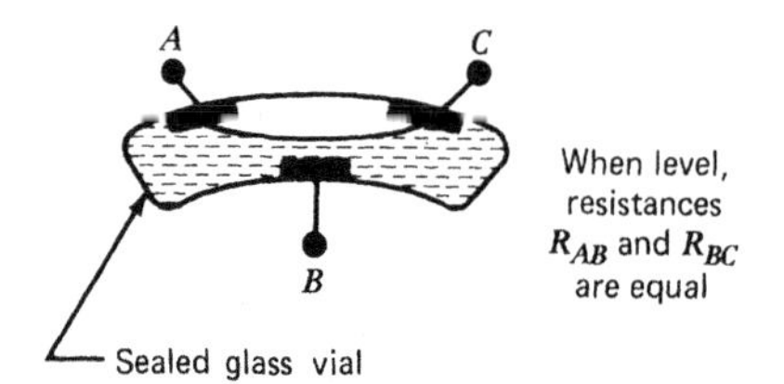

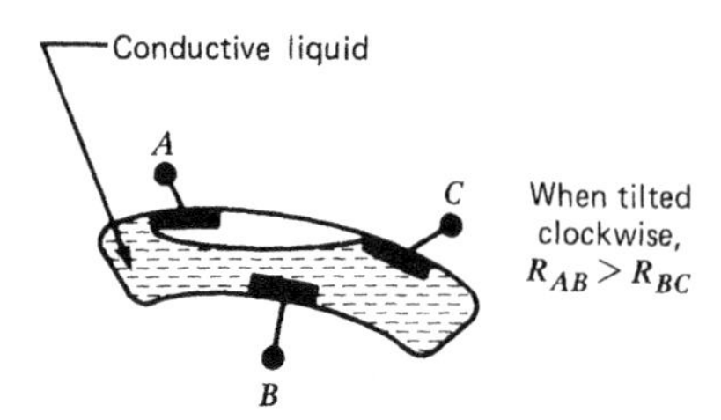

Figure 8.27. Schematic of electrolytic level.

at the speed of sound and are detected at the readout end of the tube. The first pulse, from the nearest magnet, arrives at the sensor head and starts a quartz crystal clock. The second pulse, from the next nearest magnet, stops the clock. Given the time between pulses and the velocity of pulse travel, the distance between adjacent magnets is determined readily.

Transducer accuracy is ±0.001 in. (±0.025 mm) and the range can be as much as several feet. The transducer is noncontacting and has low hysteresis and high stability. Temperature effects are minimal, because the time measurement depends only on the pulse travel velocity and the distance between magnets. The nickel–iron alloy tube can expand or contract without affecting the time measurement. The time measurement is affected by change in the pulse travel velocity with temperature, but materials have been selected such that the error is negligible. The magnetostrictive transducer was not used in geotechnical applications until 1977; therefore, its longevity has not yet been proved. However, performance to date has been good and there appear to be no basic reasons for questioning longevity if access remains available to the sensor head for any necessary maintenance.

8.4.14. Electrolytic Level

Electrolytic levels are used in tiltmeters and inclinometers.

An electrolytic level (Figure 8.27) consists of a sealed glass vial similar to the vial on a conventional builder's level, partly filled with a conductive liquid. Output resistance can be read with a portable strain indicator, in which case the two resistances form two arms of a Wheatstone bridge circuit, or the device can be read in the same way as a linear potentiometer. Criteria for the design of accurate electrolytic levels are discussed in Section 12.4.

8.4.15. Automatic Data Acquisition Systems for Electrical Transducers*

Advantages and limitations of automatic data acquisition systems, when contrasted with manual systems, are discussed in Chapter 18. This section is confined to a description of the equipment and a discussion of the applicability of automatic data acquisition systems for monitoring various transducers.

Description of Equipment

An automatic data acquisition system (ADAS) may consist of simple components or may be a complex computer system. All ADASs have certain common characteristics. First, the systems are programmed

*Written with the assistance of David A. Roberts, Senior Engineer, Shannon & Wilson, Inc., Seattle, WA.

to collect data automatically on a predetermined schedule, without human intervention. Second, the systems are designed to accommodate more than one transducer. Third, some type of signal conditioning is performed. Fourth, the data are either recorded or retransmitted to other equipment for recording.

The heart of an ADAS is either a data logger, data controller, or small computer. With the advancement of electronics, the differences between these three devices have blurred. The data logger is an instrument designed to record information from various transducers, usually at high speed. The data controller not only collects data but also uses a program to evaluate the data and to control various equipment such as an industrial process assembly line. The small mini, micro, or personal computer will generally be incorporated into the ADAS through the use of an interface device and will not have been designed for this particular application. Computers usually allow more sophisticated immediate data processing than the other two devices, but the most frequently used ADAS for geotechnical instrumentation applications is the data logger.

Data loggers can be separated into two categories: *dedicated loggers* and *flexible loggers*. A dedicated logger is a device designed for connection to one or two types of transducer. Data loggers manufactured by geotechnical instrumentation manufacturers are generally dedicated data loggers, designed for extended field use with that manufacturer's transducers. Flexible data loggers are applicable for use with a large variety of transducer types and a large number of transducers and may be capable of handling digital data. Flexible loggers generally have more sophisticated data handling, storage, and processing capability than dedicated loggers.

Figure 8.28 shows the basic configuration of an ADAS. The power supply and signal conditioning convert the output of analog transducers into a signal that can be measured and converted to a number by an analog-to-digital converter. In most ADASs, the signals from the transducers are conditioned to produce a DC voltage or a binary decimal signal. The electronics controlling the ADAS performs various functions such as controlling measurement frequency, scaling and displaying the data, converting to engineering units, averaging data, checking for alarm limits, and controlling external equipment such as alarms and annunciators. The controlling

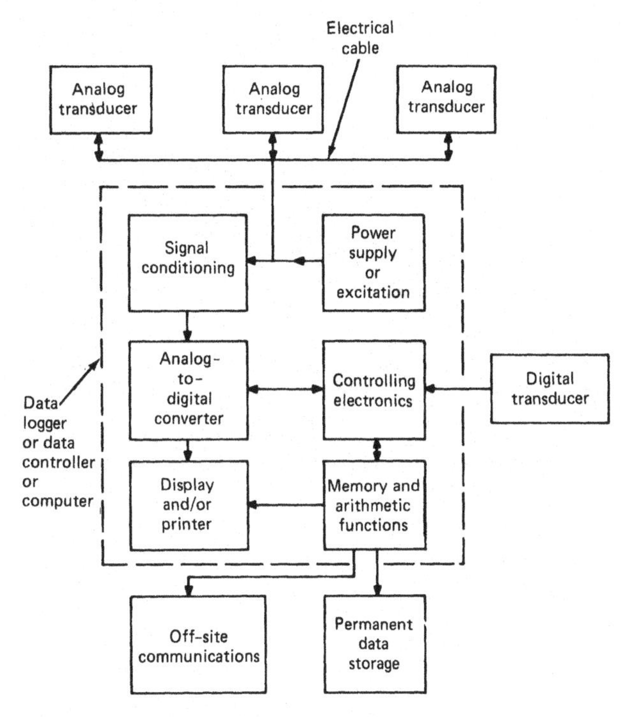

Figure 8.28. Generalized block diagram of automatic data acquisition system (ADAS).

electronics have some memory but data must then be stored or printed out for further analysis. The controlling electronics usually have the capability to store the data on a variety of media including paper, magnetic tape, and in some cases magnetic disk. Some ADASs also can be polled to collect data or to transmit data over the telephone to a separate processing center.

Recently, several manufacturers of geotechnical instrumentation have begun offering data loggers for field use, which are claimed to withstand extremes of temperature and humidity and yet are flexible enough to handle more than one or two types of transducer. The continuing improvement in low-power electronics permits these units to be programmed for unattended, battery operation from a month to up to a year. Figure 8.29 shows a data logger manufactured by Terrascience Systems Ltd., which is designed for installation within a protective enclosure at the collar of a borehole. A larger data logger is shown in Figure 8.30.

Slope Indicator Company has recently developed a line of "smart" vibrating wire transducers (Figure 8.31), the IDA™ (*intelligent data acquisition*) system, which includes additional electronics integral to the transducer package. The main pur-

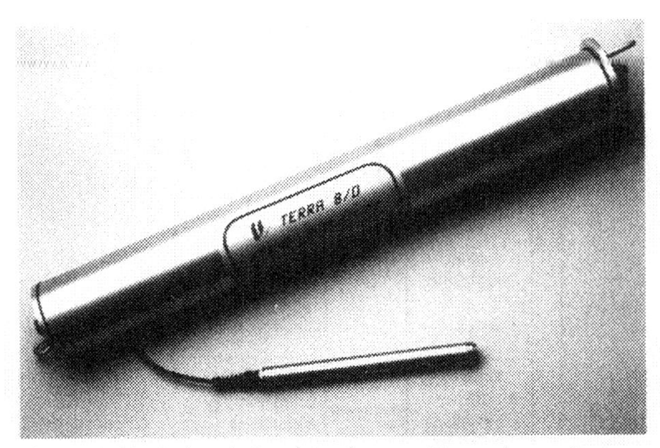

Figure 8.29. In-borehole data logger. Terrascience Systems Model TERRA 8/D™ (courtesy of Terrascience Systems Ltd., Vancouver, BC, Canada).

Figure 8.30. Computer-based data logger for 1000 instruments, Waste Isolation Pilot Plant, Carlsbad, NM (courtesy of Soil & Rock Instrumentation Division, Goldberg-Zoino & Associates, Inc., Newton, MA).

poses of these "smart" transducers are to simplify on-site connection of large numbers of transducers, simplify data acquisition, and reduce cabling costs on large projects. This line of transducers is new and therefore no information is yet available on field reliability.

Applicability of ADASs for Monitoring Various Transducers

The primary considerations for transducer selection should be reliability and required accuracy. If an ADAS is required, these primary considerations

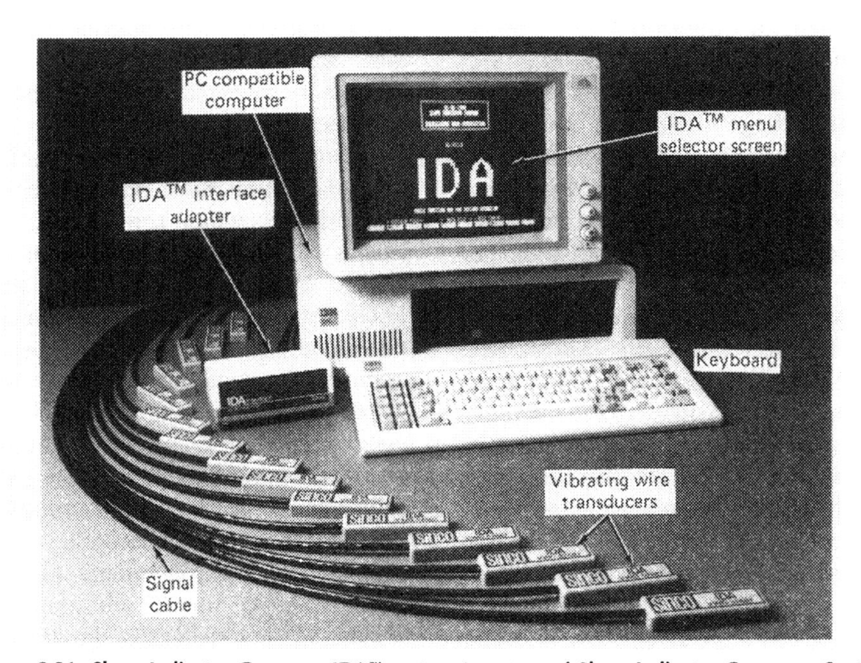

Figure 8.31. Slope Indicator Company IDA™ system (courtesy of Slope Indicator Company, Seattle, WA).

should override considerations of easy interfacing, and significant effort may be needed to design, build, and test interfacing between the preferred transducer and existing ADAS equipment.

A single model of data logger, data controller, or computer is not likely to be compatible with all transducers that are normally used in geotechnical instrumentation. If the transducers are supplied by a single manufacturer, and the same manufacturer sells a compatible dedicated data logger, this is likely to be the first choice. The following discussion is generally limited to flexible data loggers, data controllers, and computers. Applicability of ADASs for monitoring various transducers can be grouped into four categories, discussed in turn below.

Use of an ADAS is straightforward with some transducers. Included in this category are transducers that produce a full-scale DC output voltage of 1 volt or higher because these require minimum interfacing and signal conditioning. Examples are LVDTs, DCDTs, potentiometers, force balance accelerometers, electrolytic levels, high-output capacitance pressure transducers, and high-output electrical resistance strain gage networks. High-output electrical resistance strain gage networks are a relatively new development in which amplifying and signal conditioning electronics are built into the transducer, producing a full-scale DC output voltage of approximately 5 volts. Thermocouples, thermistors, and resistance temperature devices (RTDs), described in Chapter 14 and used for temperature measurement, require only minor interfacing. However, as discussed in Chapter 14, a different method is used for each of the three transducers.

The second category of transducers requires more complex signal conditioning. Examples include all five of the low-output electrical resistance strain gages described in Section 8.4.1, the type of signal conditioning depending on the bridge network. Some ADASs can provide this conditioning in the form of plug-in modules. Alternatively, standalone electronic interface equipment can be purchased to produce a high-level output voltage suitable for most commercially available ADASs. Other transducers in this second category include vibrating wire, magnetostrictive, and induction coil transducers of both the soil strain gage type and the type with frequency output, all of which require either special excitation or signal conditioning when used with flexible data loggers, data controllers, or computers. Vibrating wire transducers require an ADAS that is able to convert frequency to a digital signal. Soil strain gages require an adequate excitation voltage and an ADAS that is sufficiently sensitive to resolve small voltage changes resulting from changes in coil separation. Magnetostrictive transducers rely on the measurement of pulse travel time; therefore, an interface is required for connection to an ADAS, and at the present time there is no flexible data logger, data controller, or computer that could be connected directly to these transducers. For all transducers in this second category, a dedicated data logger is the best choice, preferably supplied by the manufacturer of the transducer.

The third category consists of electrical transducers that require manual operation, normally by traversing a sensing probe through a pipe, and thus there is little reason to use an ADAS. Examples are magnet/reed switch and induction coil transducers used with probe extensometers.

Fourth, several non electrical transducers can be read with ADASs, using arrangements described elsewhere in this book. These include pneumatic transducers (Section 8.3), plumb lines and double-fluid full-profile settlement gages (Chapter 12), and twin-tube hydraulic piezometers (Appendix E).

Publications describing geotechnical applications for automatic data acquisition systems include Bailey (1980), Brough and Patrick (1982), Carpenter (1984b), Carpentier and Verdonck (1986), Chedsey and Dorey (1983), DiBiagio (1979), DiBiagio et al. (1981), Green and Roberts (1983a, 1983b), ICOLD (1982), Lytle (1982), Murray (1986), Steenfelt (1983), USCOLD (1988) and Weeks and Starzewski (1986).

8.4.16. Power Supplies and Communication Systems for Electrical Transducers*

Electrical transducers generally require an electric power supply, either to a portable readout unit, an ADAS, or in some cases directly to the transducer. Options are mains power, batteries, portable generators, and solar, wind, and water power. Electrical instruments also require a communication linkage between the transducer and data acquisition system and in some cases between the data acquisition system and a remote facility. Options are hard-wiring, telephone lines, and radio transmission.

*Written with the assistance of David A. Roberts, Senior Engineer, Shannon & Wilson, Inc., Seattle, WA.

Overview of Electrical Power Supplies

The primary sources of electrical power are mains power and battery power. Mains power in the United States is 110 volts or 220 volts AC, 60 hertz single phase, and is usually available only near inhabited areas. Some other countries have 240 volts AC, 50 hertz single phase as a standard. Battery power is used in most portable instruments, and batteries may be disposable or rechargeable. Alternative power supplies, used when batteries cannot provide sufficient power economically and when the site is too remote for mains power, include diesel or gasoline generators, thermoelectric generators, and solar, wind, and water power. A good overview of some electrical power supply systems is given by Ball (1987).

Mains Power

Mains power is preferred for automatic data acquisition systems. Many components of an ADAS require 110 or 220 volts AC, and control of environmental conditions such as temperature and humidity is simplified when mains power is used. The quality and reliability of mains power varies considerably, and conditioning may be necessary where quality is poor. Low voltage, commonly called *brownout,* can cause erratic behavior and can damage electronic equipment. Overvoltage caused by lightning strikes can destroy unprotected equipment, and high-voltage transients can cause errors in data collected by ADASs. Undervoltage, overvoltage, and transients can be cured by a combination of lightning protection equipment (Section 8.4.17), circuit breakers, and power line conditioners. The problem of power outages can be addressed by use of uninterruptable power supplies: these are essentially switching devices that use rechargeable batteries to supply power during outages of mains power.

Disposable Batteries

Disposable batteries are used in portable readout units that have low power consumption and also in areas where rechargeable batteries would be difficult to service. Common types are carbon–zinc, mercury, and alkaline cells. Alternatively, air-depolarized cells can be used in remote areas where other power supplies are inadequate and where high capacity is required. These cells provide only 1.2 volts each but can be connected in series to provide the required voltage. They are available with capacities of up to 2000 ampere-hours and will operate at temperatures as low as −40°C. Their main disadvantages are high price, large bulk, and the caustic property of the electrolyte.

Rechargeable Batteries

Rechargeable batteries are of two general types, *nicad* and *lead–acid.* Nicad (nickel–cadmium) batteries are normally used where power requirements are small, physical space is limited, and repeated deep discharge cycles are anticipated. Lead-acid batteries are available with a wide variety of voltages and power capacities, but they are vulnerable to damage if fully discharged for a long period. Automobile batteries are lead–acid type. Low-capacity sealed lead–acid batteries that use a gelled electrolyte are now used widely in portable readout units. Shoup and Dutro (1985) describe various types of disposable and rechargeable batteries.

Diesel and Gasoline Generators

In remote areas where high-capacity power is required, diesel or gasoline electric generators are commonly used. Diesel units are generally preferred because maintenance is minimized and stored fuel is safer, but use of either type entails maintenance and refueling costs. Most diesel generators in the United States are designed to produce 110 or 220 volts AC, but convertors can be added to reduce the power output to a convenient DC voltage such as 12 volts.

Solar Power

Solar power is used with increasing frequency, but several variables must be considered before selecting this approach. Northern areas such as Canada and Alaska receive much smaller amounts of solar radiation than other parts of North America and have a wide variation in summer/winter radiation. When solar power is required in northern areas, either the system must be sized on winter needs, entailing significant overcapacity in summer, or additional batteries are required for winter use. In more southerly latitudes, similar needs can arise if a south-facing location is not available or if the solar panels are affected by shade. For high-capacity solar power, large arrays of panels must be used, and specialized batteries are required to smooth the variations in power supply.

Thermoelectric Generators

In remote locations that are unsuitable for solar power and where power requirements are moderate, thermoelectric generators are sometimes used in preference to diesel or gasoline generators or batteries. These generators are powered by natural gas, propane, or butane, with either a flame burner or catalytic burner. Capacities range from 10 to 90 watts, and various output voltages can be selected. These units are relatively small, weigh between 30 and 200 lb (14–90 kg), have very low maintenance needs, but of course require periodic refueling.

Wind and Water Power

Wind- and water-powered generators can also be used for power supply in remote areas. However, both methods are relatively expensive, require frequent maintenance, and are rarely used in geotechnical applications.

Ranges of Power Output for Various Electrical Power Supplies

Table 8.4 presents a range of power output available from various electrical power supplies. The values cited in this table are approximate, and the ranges are those that are economically feasible for instrumentation purposes.

Table 8.4. Typical Range of Power Output for Various Electrical Power Supplies

	Power Output	
Power Supply	Minimum (watts)	Maximum (watts)
Diesel generator	1000	50,000
Solar panel[a]	4	70
Thermoelectric generator	7	90
Rechargeable battery[b]	0.4	4
Nonrechargeable battery[c]	4×10^{-4}	3

[a] For a location in central United States. Normally more than one panel is required to obtain maximum power output.
[b] One month recharge cycle.
[c] Yearly replacement.
Note: All calculations based on a 12 volt DC system.

Communication Systems

Electrical transducers are generally *hard-wired* to the data acquisition system, using electrical cable. Cable may be buried in the ground or concrete, or attached to the surface of a structure, or suspended in air. Guidelines for selection of cable are given in Section 8.4.17.

When an automatic data acquisition system is used, data can be collected from the system by visiting the site. Alternatively, a communication system can link the on-site system to an off-site master collection station as shown in Figure 8.32, so that data can be collected and processed without visiting the site. Communication between the on-site junction boxes and collection station will often be via hard-wiring, or occasionally radio transmission may be used when rugged terrain, roads, or other obstacles prevent the use of hard-wiring. Radio transmission between remote on-site stations and the on-site collection station is usually performed with low-power line-of-sight transmitters.

Communication between on-site and off-site stations will usually be via telephone lines or radio. When available, telephone lines provide the most convenient and least expensive link. Telephone lines are rented from local telephone companies and can be either dedicated or standard voice grade. Dedicated lines can handle higher rates of transmission but are more expensive and may not be available. Standard dial-up voice grade lines are the more common method and require a modem at each end of the line. Transmission rates for dial-up lines vary from 10 to 120 characters per second and depend both on the sophistication of the transmitting and receiving equipment used and on the quality of the telephone lines over which the signal is sent. Data transmission over telephone lines can be either analog or digital but the digital method, using an ASCII protocol, is generally the preferred method. Data transmission from the on-site station to an off-site station can be initiated by any of four methods: manually from the on-site station, timed automatically from the on-site station, polled manually from the off-site station, or polled automatically from the off-site station.

Where telephone lines are not available, and the site is too remote for installation of a private line to the nearest telephone line, radio transmission can be used to link the on-site and off-site stations, but radio transmission over more than a few miles requires use of a relay system. Relay systems can be simple single channel relays, radio networks, or satellites. It is likely that, in the future, satellite

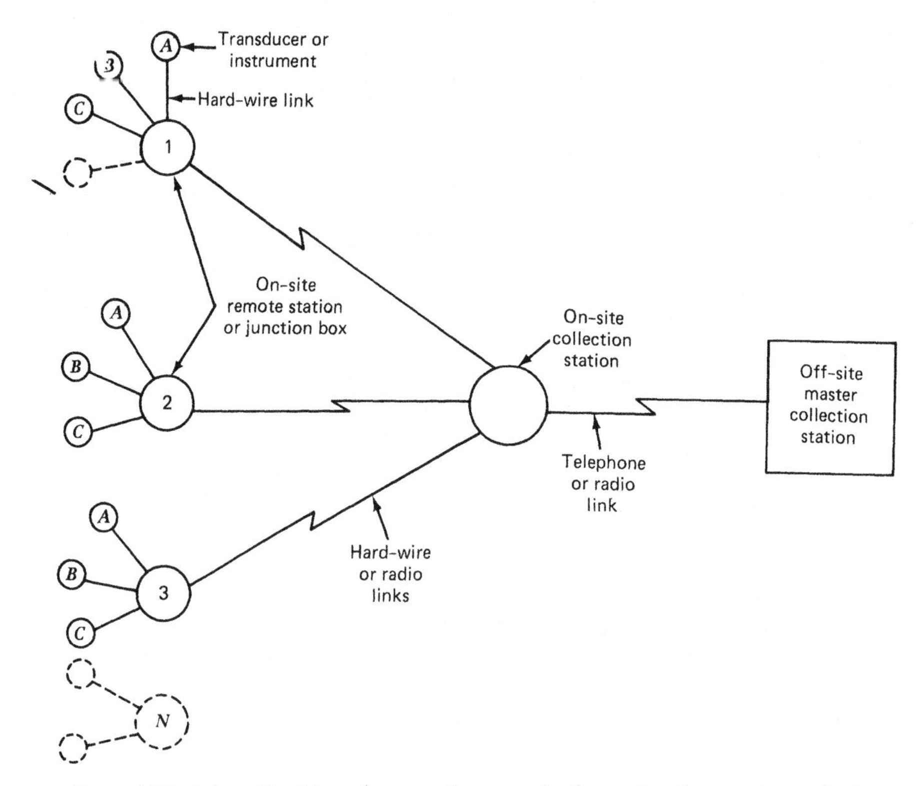

Figure 8.32. Schematic of transducers and communication system for remote monitoring.

links will play an increasing role in communication of data. As for telephone lines, radio transmission of data can be either analog or digital, but digital transmission usually has less errors. Radio frequencies, power, and protocols for transmitting data vary greatly among different systems and are usually regulated by government agencies. Baur (1987) presents a good overview of the role of radio transmission systems.

An innovative system is described by Teal (1986) for communicating data from inaccessible transducers to an on-site data collection station. The system uses a technique akin to that used in geophysical resistivity surveys. An alternating electrical potential field is established between a pair of buried electrodes and a digitally encoded signal is transmitted. This signal is received at the monitoring station by a second pair of electrodes, thereby avoiding hard-wiring. Arrangements are being made to install a prototype system in England.

Whenever data are transmitted without use of hard-wiring, there is a possibility of transmission error. The data should be transmitted more than once and any errors identified by comparing transmissions.

8.4.17. General Guidelines on Use of Electrical Transducers and Data Acquisition Systems*

This section includes only brief general guidelines. Most of the issues discussed require knowledge beyond that of the typical user of instrumentation. Unless users have sound experience with field electronics, they are encouraged to seek the detailed guidance of others with appropriate experience.

Selection of Cable

Manufacturers of geotechnical instruments normally standardize the type of cable for each instrument: to do otherwise would create difficulties in supply to the end user. However, the standard type may not be suitable for all applications, and the user should participate in selection of cable. The following items must all be considered:

*Written with the assistance of Howard B. Dutro, Vice President, Slope Indicator Company, Seattle, WA, and Charles T. McNeillie, Senior Instrumentation Technician, Soil & Rock Instrumentation Division, Goldberg-Zoino & Associates, Inc., Newton, MA.

- Length of run
- Frequency and magnitude of signal
- Environment (e.g., underground, aerial, indoor, marine sunlight)
- Temperature of environment
- Altitude
- Longevity requirements
- Susceptibility to damage (e.g., need for conduit or armoring, likelihood of axial or shear strains such as when embedded in fill)
- Proximity to sources of electrical noise

Electrical noise is a term used to cover random measurement variations caused by external factors, and excessive noise in a system may mask small real changes. It is therefore important that electrical signals should be as free as possible from noise. Primary causes of noise are radio frequency interference (RFI) and electromagnetic interference (EMI) from such sources as power lines, electrical generators and motors, commercial TV, radio or radar stations, defective fluorescent signs, electronic navigation systems, nearby thermostats and other switch closures, welding, and dirty terminals of power line transformers. Problems created by some of these sources can be corrected, for example by replacing defective fluorescent signs and cleaning terminals of power line transformers. The best protection against noise is the use of shielding.

Individually shielded twisted pairs, bundled inside an overall shield, are recommended for all applications. For some applications this may indeed be "overkill," but it is inexpensive insurance against the many unknowns associated with a field environment. As a rule of thumb, the shield should be grounded with respect to the measuring circuit at **one end** only, as close as possible to the chassis ground for a portable instrument and as close as possible to the earth ground on a mains instrument.

The primary sources of guidance on selection of cable are the manufacturers of cable, and several manufacturers who have applications engineers are listed in Appendix D. The end user should make the final decision, after discussions with manufacturers of the instrument and the cable. From a delivery and cost standpoint, it is always cost effective to use a standard stocked cable.

Cable Connectors

The most suitable types of connector for field instrumentation are the *Mil. Spec.* (*MS*) bayonet lock type. MS connectors are available with from two to several hundred connecting pins, which are normally gold plated to provide long-term stable connections. The crimp versions are preferable to the soldered versions because, although they are more expensive, they provide the most reliable connection and are less dependent on variations in quality of work when connecting cables to connectors. The back shells of the connectors are environmentally sealed, and when properly mated by turning 90°, the male/female connection is also sealed. However, the connectors are **not** waterproof.

It is advisable to discourage water from wicking along the cable from a cut in the outer insulation and entering the connector. This is best done by sealing the back shell around the individual conductors where they fan out. The two-part epoxy sealers used in cable splice kits (see next subsection) are suitable, and alternatively room temperature curing silicone rubber (e.g. *RTV, room temperature vulcanize*) can be used. *Noncorrosive* RTV should be used*—the type available from electronic supply houses, not the type available from hardware stores. RTV should be applied and allowed to cure in layers not exceeding 0.25 in. (6 mm) thick.

It must be emphasized that the connection method described above will not create **waterproof** connections. Sealing materials will generally not bond to cable insulation materials such as Teflon™ and polyethylene. In extreme cases, such as underwater applications, it is necessary to use special marine connectors, but these are very expensive.

Connectors should be kept scrupulously clean and dry. A non-contact cleaner—a high pressure inert gas contained in a spray can—is best for cleaning.† Whenever connectors are not mated, they should be sealed by dust caps and each attached to the connector by a chain so that it will not be lost.

Cable Splices

Cable splices should be reserved for repair work and should not be planned into a system unless there is no alternative. Commercial splice kits should be used, available from the manufacturers listed in Appendix D. The kits generally consist of crimp splices and a two-part epoxy sealer with a forming mold. It is essential to match the crimp splice size with the cable conductor size, and to use a rachet-type crimping tool that can be calibrated to

*For example, Dow Corning Type 3140 or 3141.
†For example, Chemtronics 70 p.s.i. cleaner.

the proper crimp size. Great attention to detail must be paid when making a splice, and the instructions provided with the kits assume significant knowledge. The guidance of a good electronics technician, such as a person who repairs TVs, should be sought.

"Homemade" splice kits, consisting of crimp splices, mastic, RTV, and electrical tape should not be used. They generally develop leakage to moisture over the long term and deteriorate more rapidly than the commercial kits. Unless made indoors by skilled personnel, soldered splices will generally be inferior to a high-quality crimp splice.

Lightning Protection

Lightning can damage electrical instruments in three ways: power surges caused by a direct strike, induced transients caused by a nearby strike, and electromagnetic pulses induced by the magnetic field of a strike. *Transient protection devices* should be incorporated into the system whenever there is a possibility of serious damage by lightning. Commercial sources are listed in Appendix D. Both the power input and the transducer should be protected. Good sources of information on lightning protection include Baker (1978, 1980), Burkitt (1980), DiBiagio and Myrvoll (1985), and General Electric (1976).

Protection from Water

If it can find a way to the inside of an electrical system, water will find a way. Methods of preventing water from entering cables, connectors, and splices have been described earlier in this section. *Water-blocked* cable is available to prevent migration of water along the conductors in case of damage to a cable.

Readout units should be kept dry, and a humidity indicator can be incorporated to indicate when there is a need to dry out the unit. The indicator contains a chemical that changes color when the humidity exceeds a predetermined level.

Field Checks with Volt–Ohm–Milliammeter

Section 8.4.3 describes the use of a circuit tester to check the integrity of electrical resistance strain gages and cables. The integrity of cables connected to other types of instrument can be checked with a volt–ohm–milliammeter (VOM). The VOM should be the type used for testing radios and TVs, not the type used for power line measurements. Specific procedures appropriate for checking each instrument should be obtained from the manufacturer of the instrument.

Charging and Maintenance of Batteries

Rechargeable batteries are of two general types, *nickel-cadmium* (*nicad*) and *lead-acid*. Recharging procedures for the two types are quite different. Nicad batteries must be discharged completely at maximum intervals of 3 months, whereas lead–acid batteries must never be discharged completely and should be fully charged at maximum intervals of 3 months when not being used. Shoup and Dutro (1985) give detailed guidelines on charging and maintenance of both types.

Operating Spares

Components of an instrumentation system can malfunction as a result of mechanical damage, water damage, deterioration, defects, or wear and tear. Chapter 5 includes a recommendation for procurement of appropriate spare parts at the time of initial procurement, together with suggestions for obtaining spare or standby readout units. The stock of operating spares should be reviewed on a regular basis during service life.

Handling and Transporting Transducers and Data Acquisition Systems

Instruments should be handled with great care. As an illustration of this need, when transporting instruments and personnel in a pickup truck, the instruments should be on the seat and the personnel in the back, not the other way around. A classic and recurrent blunder is to carry an instrument by its signal cable: this should never be permitted.

Regular Calibration and Maintenance of Readout Units

General guidelines for calibration and maintenance of readout units are given in Chapter 16.

Use of Electrical Transducers and Data Acquisition Systems in Cold Conditions

Guidelines for use of electrical instruments in cold conditions are given by Atkins (1981).

CHAPTER 9

MEASUREMENT OF GROUNDWATER PRESSURE

9.1. INSTRUMENT CATEGORIES AND APPLICATIONS

Definitions of the terms *groundwater level, pore water pressure,* and *joint water pressure* are given in Chapter 2. In this book the term *piezometer* is used to indicate a device that is sealed within the ground so that it responds only to groundwater pressure around itself and not to groundwater pressures at other elevations. Piezometers are used to monitor pore water pressure and joint water pressure. The term *pore pressure cell* is sometimes used as a synonym for *piezometer*. An *observation well* is a device that has no subsurface seals, and it creates a vertical connection between strata.

Applications for piezometers fall into two general categories. First, for monitoring the pattern of water flow and second, to provide an index of soil or rock mass strength. Examples in the first category include monitoring subsurface water flow during large-scale pumping tests to determine permeability in situ, monitoring the long-term seepage pattern in embankment dams and slopes, and monitoring uplift pressures below concrete dams. In the second category, monitoring of pore or joint water pressure allows an estimate of effective stress to be made and thus an assessment of strength. Examples include assessing the strength along a potential failure plane behind a cut slope in soil or rock, and monitoring of pore water pressure to control staged construction over soft clay foundations.

Applications for observation wells are very limited. In current practice they are frequently installed in boreholes during the site investigation phase of a project, ostensibly to define initial groundwater pressures and seasonal fluctuations. However, because observation wells create a vertical connection between strata, their **only** application is in continuously permeable ground in which groundwater pressure increases uniformly with depth. This condition can rarely be assumed. The author believes that many practitioners do not appreciate the need to make pore water pressure measurements at many different depths, rather than measuring at one or two points and assuming a straight line pressure–depth relation. Detailed pore water pressure data provided by recently developed multipoint piezometers with movable probes (Section 9.9) have shown that pressure–depth profiles can be quite irregular and very different from that which one would infer from measurements at one or two points. Perhaps the major reason for the frequent and continued use of observation wells is that they can be installed by drillers without the participation of geotechnical personnel: this is certainly not the case for installation of piezometers. Observation wells are therefore inexpensive, but in the view of the author are often misleading and should be used only when the groundwater regime is well known. If the groundwater regime is well known, measurements may be unnecessary: therefore, a practitioner must have a strong argument in favor of an observation well before selecting this option.

Additional applications for measurement of groundwater pressure are given in Part 5.

Piezometers can be grouped into those that have a diaphragm between the transducer and the pore or joint water and those that do not. Instruments in the first group are piezometers with pneumatic, vibrating wire, and electrical resistance strain gage transducers. Instruments in the second group are open standpipe and twin-tube hydraulic piezometers.

In general, piezometers used for measuring pore water pressure in soil are no different from piezometers used for measuring joint water pressure in a rock mass: the difference is in the installation arrangements. These differences are discussed in Section 9.18.

Various instruments for measuring groundwater pressure are described and compared below. Other descriptions and comparisons are given by Bozozuk (1960), Cording et al. (1975), Corps of Engineers (1971), Hanna (1985), USBR (1974), and Wilson and Mikkelsen (1978).

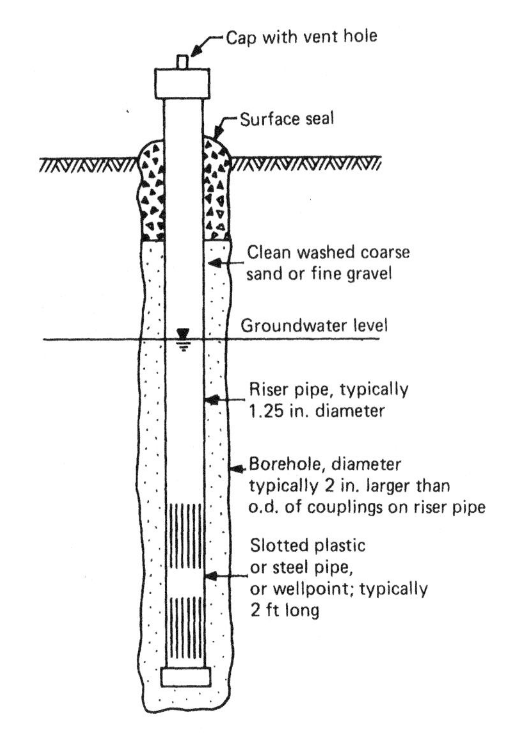

Figure 9.1. Schematic of observation well.

9.2. OBSERVATION WELLS

As shown in Figure 9.1, an observation well consists of a perforated section of pipe attached to a riser pipe, installed in a sand-filled borehole. The surface seal, with cement mortar or other material, is needed to prevent surface runoff from entering the borehole, and a vent is required in the pipe cap so that water is free to flow through the wellpoint. The elevation of the water surface in the observation well is determined by sounding with one of the probes described in Section 9.3.2 for measurements within open standpipe piezometers.

As discussed in Section 9.1, observation wells create an undesirable vertical connection between strata and should rarely be used. The water level within the observation well is likely to correspond to the head in the most permeable zone and will usually be misleading. At sites where a contaminant exists in one aquifer, installation of an observation well also leads to contamination of other aquifers.

The term *observation well* should not be confused with *monitoring well,* which is a system for sampling and monitoring water quality in a particular aquifer, requiring an arrangement similar to an open standpipe piezometer.

9.3. OPEN STANDPIPE PIEZOMETERS

9.3.1. Description

An open standpipe piezometer requires sealing off a porous filter element so that the instrument responds only to groundwater pressure around the filter element and not to groundwater pressures at other elevations. Piezometers can be installed in fill, sealed in boreholes (Figure 9.2), or pushed or driven into place.

The components are identical in principle to components of an observation well, with the addition of seals. The water surface in the standpipe stabilizes at the piezometric elevation and is determined by sounding with a probe. Care must be taken to prevent rainwater runoff from entering open standpipes, and an appropriate stopcock cover can be used, ensuring that venting of the standpipe is not obstructed.

The open standpipe piezometer is also referred to as a *Casagrande piezometer,* after publication of measurement methods for monitoring pore water

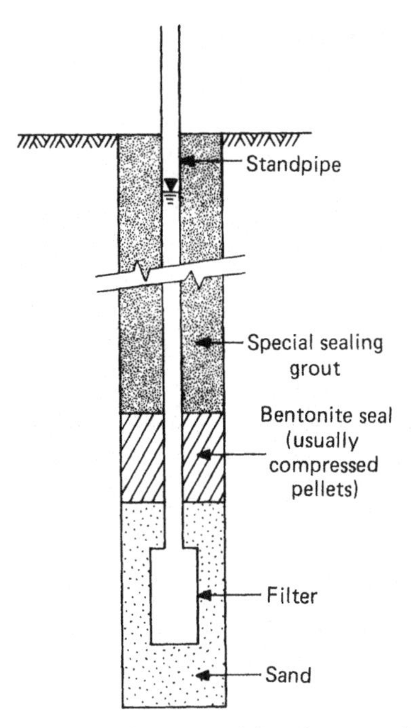

Figure 9.2. Schematic of open standpipe piezometer installed in a borehole.

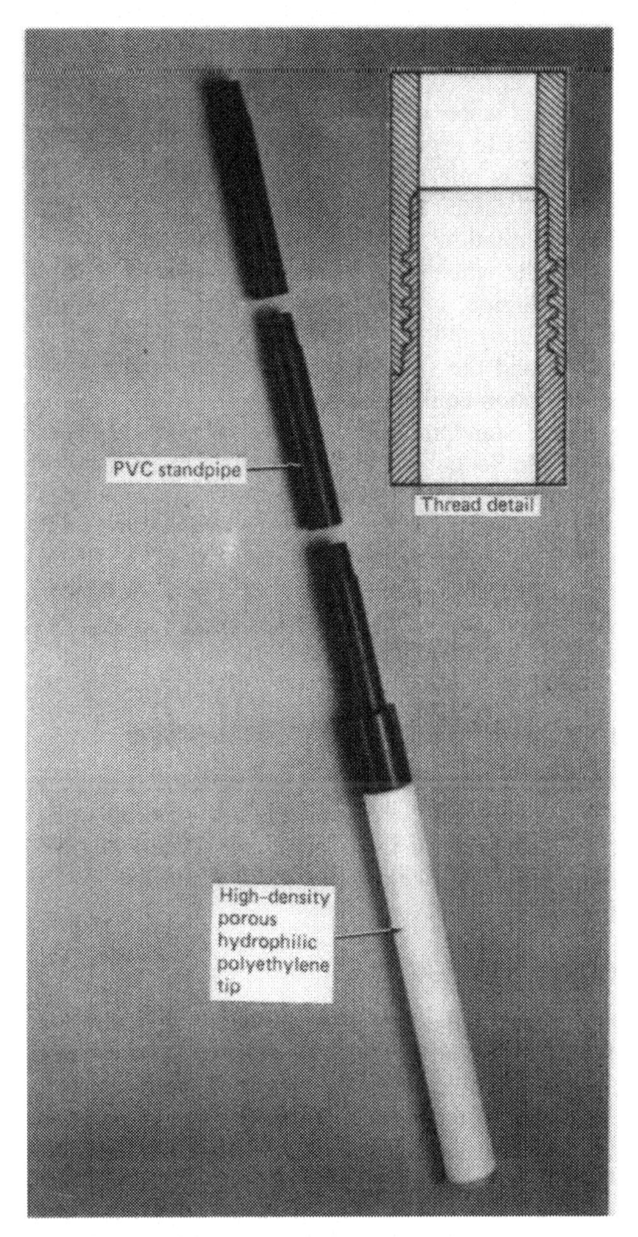

Figure 9.3. Open standpipe piezometer with porous polyethylene filter and self-sealing threaded PVC standpipe (courtesy of Piezometer Research & Development, Bridgeport, CT).

pressure during construction of Logan Airport in Boston (Casagrande, 1949, 1958). The Casagrande version consisted of a cylindrical porous ceramic tube, connected with a rubber bushing to 0.375 in. (10 mm) inside diameter saran plastic tubing. Today, high-density porous hydrophilic polyethylene (Figure 9.3) usually replaces the brittle ceramic, and PVC or ABS plastic pipe or polyethylene tubing replaces the saran tubing, which becomes brittle with age and exposure to sunlight. Standpipe inside diameter typically ranges from 0.2 to 3 in. (5–76 mm).

Flush-coupled Schedule 80 PVC or ABS pipe is a good choice for standpipes, with either cemented or threaded couplings that allow easy passage of the reading device and the sounding hammer (Section 9.17.8). When cemented couplings are used, one end of each length of pipe is machined as a male, the other as a female, and the coupling connected using solvent cement. A primer should be used to etch the surfaces and ensure proper adhesion of the cement. When threaded couplings are used, they should be of the self-sealing type shown in Figure 9.3. These special threads are undersized at the female end and create a watertight connection without any sealer, merely by tightening with vice grips or small pipe wrenches. Conventional tapered pipe threads are not suitable for flush couplings, and conventional square threads will often not sustain significant internal water pressure.

Open standpipe piezometers are generally considered to be more reliable than other types, and the reliability of unproven piezometers is usually evaluated on the basis of how well the results agree with those of adjacent open standpipe piezometers. Advantages and limitations are summarized in Section

9.12. A major limitation is their slow response to changes in piezometric head, because a significant volume of water must flow out of or into the soil or rock mass to register a change in head. This slow response is referred to as *hydrodynamic time lag* and is discussed further in Section 9.10. A second major limitation is caused by the existence of the standpipe: when embankment fill is placed around the standpipe, nearby compaction tends to be inferior, interruption to normal filling operations is costly, and the standpipe is subject to damage by construction equipment.

Open standpipe piezometers of the type described in Section 9.8 can be installed by pushing or driving into place. The term *push-in* piezometer is used in this book for piezometers that are connected to a pipe or drill rod and pushed or driven into place. A special type of open standpipe piezometer, the *heavy liquid piezometer,* is also described in Section 9.8.

Figure 9.4. Electrical dipmeter (courtesy of Geotechnical Instruments (U.K.) Ltd., Leamington Spa, England).

9.3.2. Methods of Reading Open Standpipe Piezometers

Various methods are available for reading open standpipe piezometers, most of which involve sounding the elevation of the water surface with a probe.

Electrical Dipmeter

The most commonly used probe is an *electrical dipmeter* (Figure 9.4), consisting of a two-conductor cable with a cylindrical stainless steel weight at its lower end. The weight is divided electrically into two parts, with a plastic bushing between, and one conductor is connected to each part. The upper end of the cable is connected to a battery and either an indicator light, buzzer, or ammeter. When the probe is lowered within the standpipe and encounters the water surface, the electrical circuit is completed through the water and the surface indicator is actuated. For small-diameter standpipes where the water level is no deeper than about 15 ft (5 m), a coaxial cable with bared ends can be used.

Capillary Reader

An alternative device is manufactured by Piezometer Research and Development and shown in Figure 9.5. The term *capillary reader* originates from an earlier device that relied on the capillary action of liquid in a small-diameter tube. The air valve is held open to allow the two surfaces of colored liquid to stabilize at the same level, and the nylon tubing is pushed down the standpipe until the lower end is submerged below the water surface. Submergence is indicated by a lowering of liquid level in the graduated sight tube. The nylon tubing is then slowly withdrawn and the level of colored liquid rises until the surface of the water in the standpipe is reached, and at this point the graduation on the tubing is read. A 0.125 in. (3 mm) outside diameter tubing is normally used, such that it can be inserted in standpipes with inside diameter as small as 0.2 in. (5 mm): an electrical dipmeter cannot normally be used in such small-diameter standpipes. When used in a large-diameter standpipe a small weight is usually added to the lower end.

Audio Reader

Sandroni (1980) describes an audio reader, shown in Figure 9.6, that can be assembled from readily available components. The arrows on the figure show the path of a noise created, for example, by a transistor radio. The graduated measuring tube is

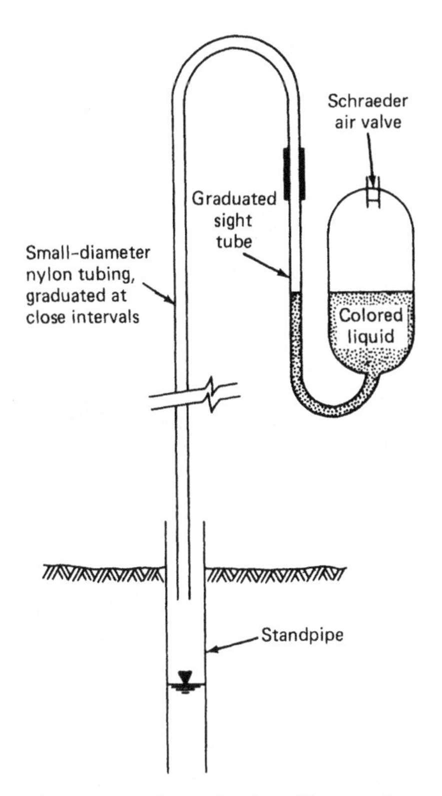

Figure 9.5. Schematic of capillary reader.

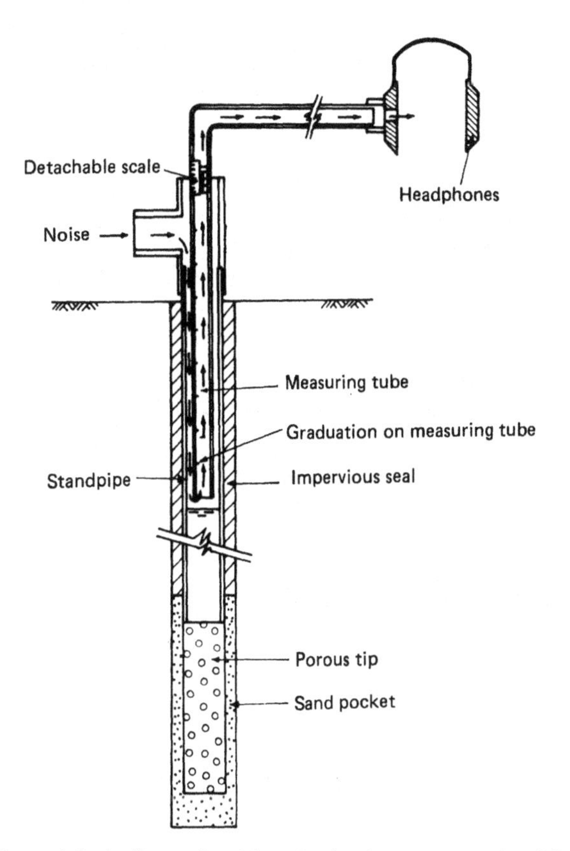

Figure 9.6. Audio reader (after Sandroni, 1980). Reprinted by permission of Institution of Civil Engineers, London.

inserted within the standpipe and, when the lower end of the tube touches water, the noise transmitted to the headphones ceases. The detachable scale is used to subdivide graduations on the measuring tube.

Survey Tape and Weight

If the standpipe diameter is large enough, a survey tape with a weight on its lower end can often be used by listening for the weight to contact the water surface. A small bell, with the striker removed, is sometimes used to increase the noise on contact.

Pressure Transducer

A pneumatic, vibrating wire, or electrical resistance strain gage pressure transducer can be inserted into the standpipe below the lowest possible piezometric level, thereby allowing readings to be made at a remote location. The transducer can be left hanging in place and recovered for periodic recalibration.

Alternatively, an open standpipe piezometer can be converted to a diaphragm piezometer by inserting a pressure transducer within the standpipe below the lowest possible piezometric level and sealing just above the transducer. A packer inflated with air can be used for the seal and allows recovery of the transducer for recalibration (e.g., Tao et al., 1980). This adaptation shortens the response time and reduces the possibility of clogging by repeated inflow and outflow of water through the porous tip.

Purge Bubble System

The *purge bubble* principle can be used to read an open standpipe piezometer, as shown in Figure 9.7, and this system also allows readings to be made at a remote location. A length of plastic tubing is inserted within the standpipe to a point below the lowest piezometric level, and the elevation of the lower end is recorded. A small controlled flow of air depresses the level of water within the plastic tubing until it falls to the lower end of the tubing, at which time air bubbles rise to the surface of water in the standpipe. The measured air pressure is then equal to the water pressure at the lower end of the plastic tubing, and the piezometric elevation is cal-

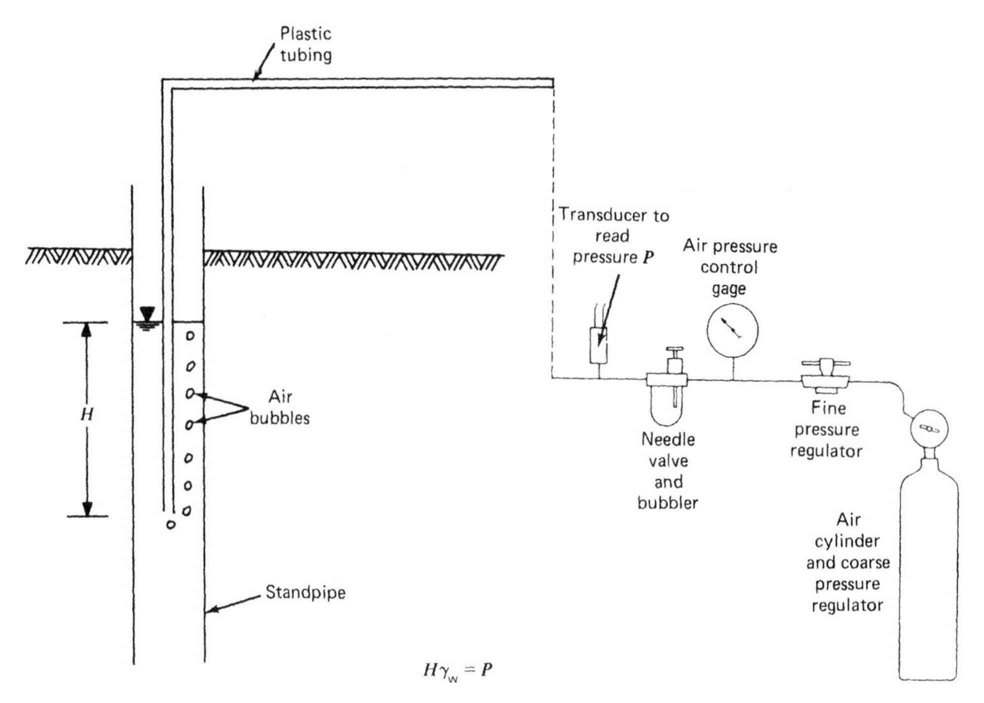

Figure 9.7. **Purge bubble principle (after Penman, 1982).**

culated as shown in the figure. Most readout units for pneumatic transducers can readily be adapted for this purpose.

The system shown in Figure 9.8 can be used for continuous recording by maintaining a constant small flow of air and recording pressure (e.g., Brand et al., 1983; Pope et al., 1982).

When the purge bubble system is used, the standpipe need not be vertical, provided that the elevation of the lower end of the plastic tubing is known. However, the system should not be used if the inside diameter of the standpipe is less than 0.3 in. (8 mm) because, as discussed in Section 9.3.3, the air bubbles may not rise to the top of the standpipe, resulting in a false reading.

Halcrow Bucket

Brand et al. (1983) describe the *Halcrow bucket* for recording peak water levels in standpipes. A chain of plastic cylinders, each with an intake hole, is fixed along a weighted nylon string at selected depth intervals above the normal base water level in the standpipe. When the string of buckets is withdrawn, the highest transient water level is indicated by the upper limit of water-filled buckets. This system is cheap and simple, and it can be used to provide important design information when expensive automatic monitoring systems cannot be justified.

Float and Recorder

Open standpipe piezometers are sometimes read with an automatic recorder by inserting a float and counterweight arrangement in the standpipe. However, this arrangement requires a minimum standpipe diameter of about 3 in. (76 mm) and therefore has a large response time.

Bourdon Tube Pressure Gage

If the piezometric level rises above the top of the standpipe, a Bourdon tube pressure gage can be attached.

9.3.3. Open Standpipe Piezometers in Unsaturated or Gaseous Soils

When an open standpipe piezometer is installed in unsaturated or gaseous soils, gas may enter the standpipe. If the inside diameter of the standpipe is less than about 0.3 in. (8 mm), gas bubbles may not rise to the top of the standpipe, and the water level in the standpipe will be elevated to give a false read-

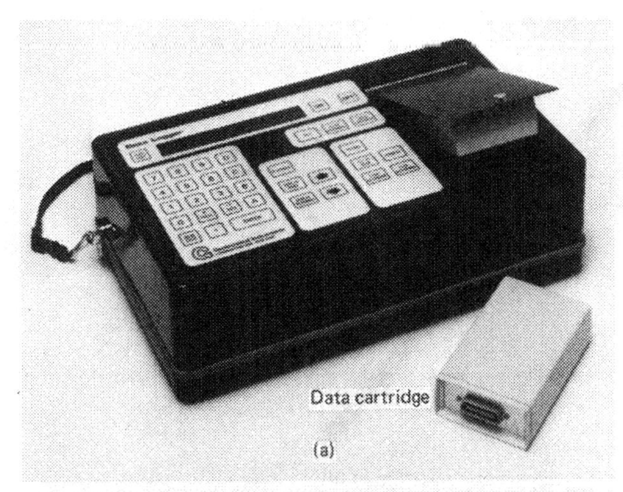

(a)

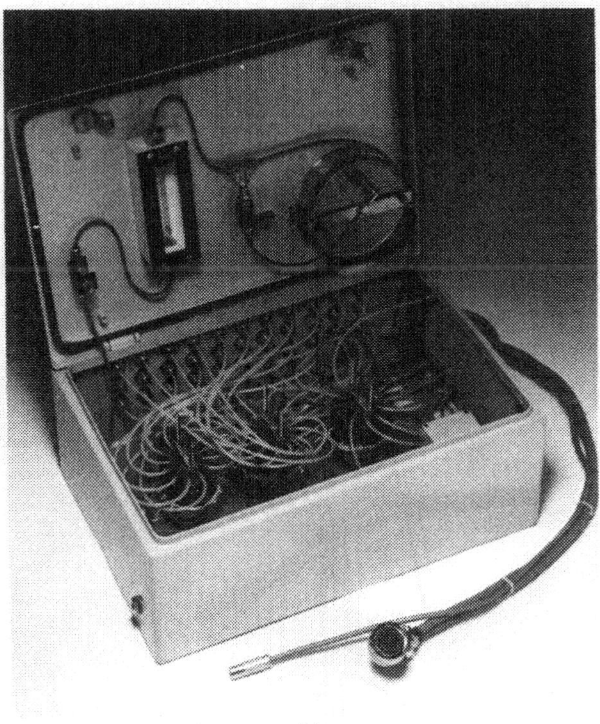
(b)

Figure 9.8. Continuous recording purge bubble system: (a) data logger and (b) scanner system used in conjunction with logger (courtesy of Geotechnical Instruments (U.K.) Ltd., Leamington Spa, England).

ing. In such soils the inside diameter of the standpipe must therefore be greater than 0.3 in. (8 mm).

9.3.4. Open Standpipe Piezometers in Consolidating Soils

When an open standpipe piezometer is installed in consolidating soils, the standpipe may buckle or shear, creating a leak or obstructing passage of the readout probe. Examples are shown in Figure 17.10. When this potential exists, either a telescoping pipe should be installed around the standpipe or an alternative type of piezometer should be used. When it is necessary to prevent leakage of water along the annulus between the pipes, the annulus should be filled with a slurry.

9.3.5. Freezing of Water in Open Standpipe Piezometers

If the piezometric level rises above the frost line, the water in the standpipe will freeze and the piezometer will become inoperable. However, the upper portion of water in the standpipe can be replaced with an antifreeze mixture with specific gravity less than unity. A mixture containing 4 parts methanol, 2 parts glycerol, and 3 parts water by volume has been used (Soderman, 1961). The mixture has a specific gravity of approximately 0.99 at 26°C (79°F) and a freezing point of −51°C (−60°F). However, methanol can damage some plastics, and it is safer to use a thin oil such as hydraulic oil, and to make a correction to the piezometric level because of the lower specific gravity. In this case an electrical dipmeter will not function, and one of the alternative reading methods must be used. In freezing situations, an alternative approach is use of the *heavy liquid piezometer*, described in Section 9.8.

9.3.6. Checking Integrity of Seal

A falling or constant head permeability test can be made through the piezometer in an attempt to check the integrity of the seal. The test should be made only after allowing dissipation of any pore water pressure generated during installation, as indicated by a stable water level in the standpipe. However, Vaughan (1969) points out that these tests are only of limited value because a seal has to be extremely bad before it will be detected by this method. Despite this limitation, the test appears to be worthwhile as an acceptance test.

Equations for calculating permeability from falling and constant head tests are given by Hanna (1985) and Hvorslev (1951).

9.4. TWIN-TUBE HYDRAULIC PIEZOMETERS

The twin-tube hydraulic piezometer (Figure 9.9), sometimes referred to as a *closed hydraulic piezometer,* was developed for installation in the

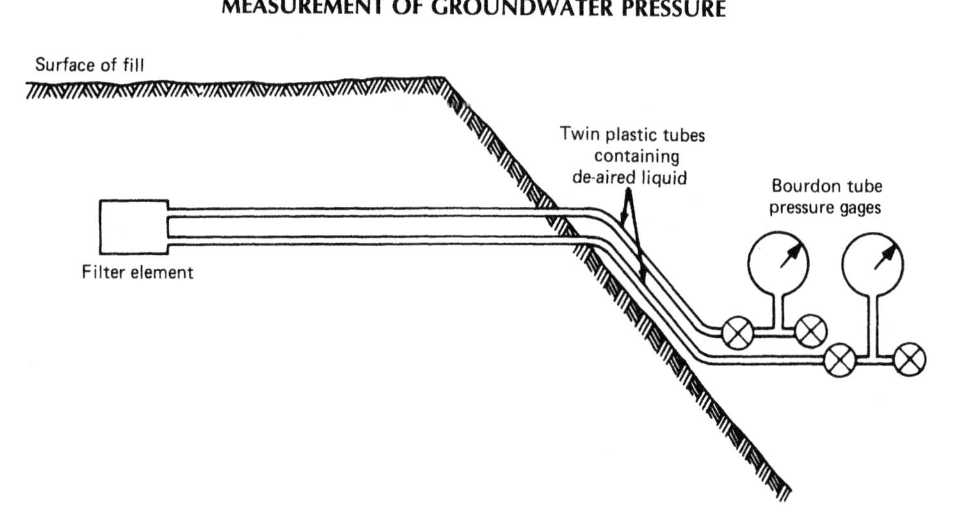

Figure 9.9. Schematic of twin-tube hydraulic piezometer installed in fill.

foundations and fill during construction of embankment dams. It consists of a porous filter element connected to two plastic tubes, with a Bourdon tube pressure gage on the end of each tube. U-tube manometers or electrical pressure transducers can be used instead of Bourdon tube pressure gages. The piezometric elevation is determined by adding the average pressure gage reading to the elevation of the pressure gages. If both plastic tubes are completely filled with liquid, both pressure gages will indicate the same pressure. However, if gas has entered the system through the filter, tubing, or fittings, the gas must be removed by flushing: this is the reason for requiring two tubes.

Figure 9.10 shows a method of converting an open standpipe piezometer to a twin-tube hydraulic piezometer. The standpipe is cut to a level such that the special head is protected by installing it slightly below the ground surface. The instrument is initially read by using the purge bubble principle. When the piezometric level rises to a level above the top of the standpipe, a valve at the upper end of the central tube can be closed permanently, and the instrument is read as a twin-tube hydraulic piezometer.

The application for twin-tube hydraulic piezometers is almost exclusively limited to long-term monitoring of pore water pressures in embankment dams, and therefore longevity is a primary need. Twin-tube hydraulic piezometer systems have been developed in the United States by the U.S. Bureau of Reclamation (Bartholomew et al., 1987; USBR, 1974) and in England by Imperial College, London (Bishop et al., 1960; Penman, 1960). The USBR piezometer tip is shown in Figure 9.11 and the Imperial College (Bishop) tip in Figures 9.12 and 9.13. Each system has been used widely in embankment dams throughout the world with very variable success. The system developed in England appears to have a better success record than the system developed in the United States, and successful long-term

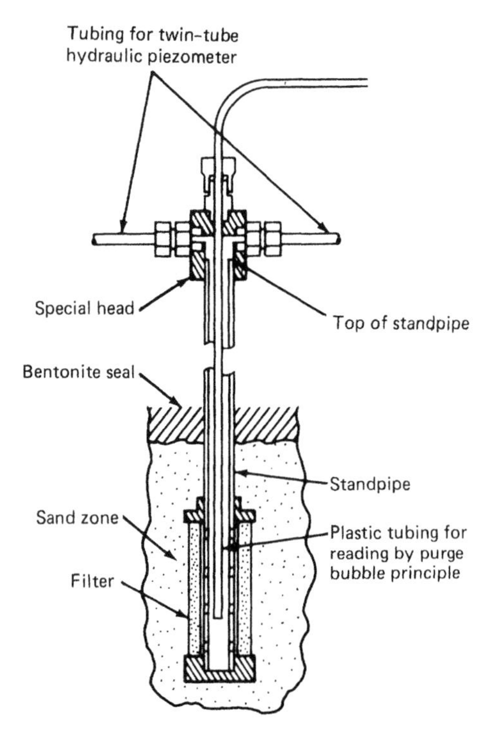

Figure 9.10. Conversion of open standpipe piezometer to twin-tube hydraulic piezometer (after Penman, 1982).

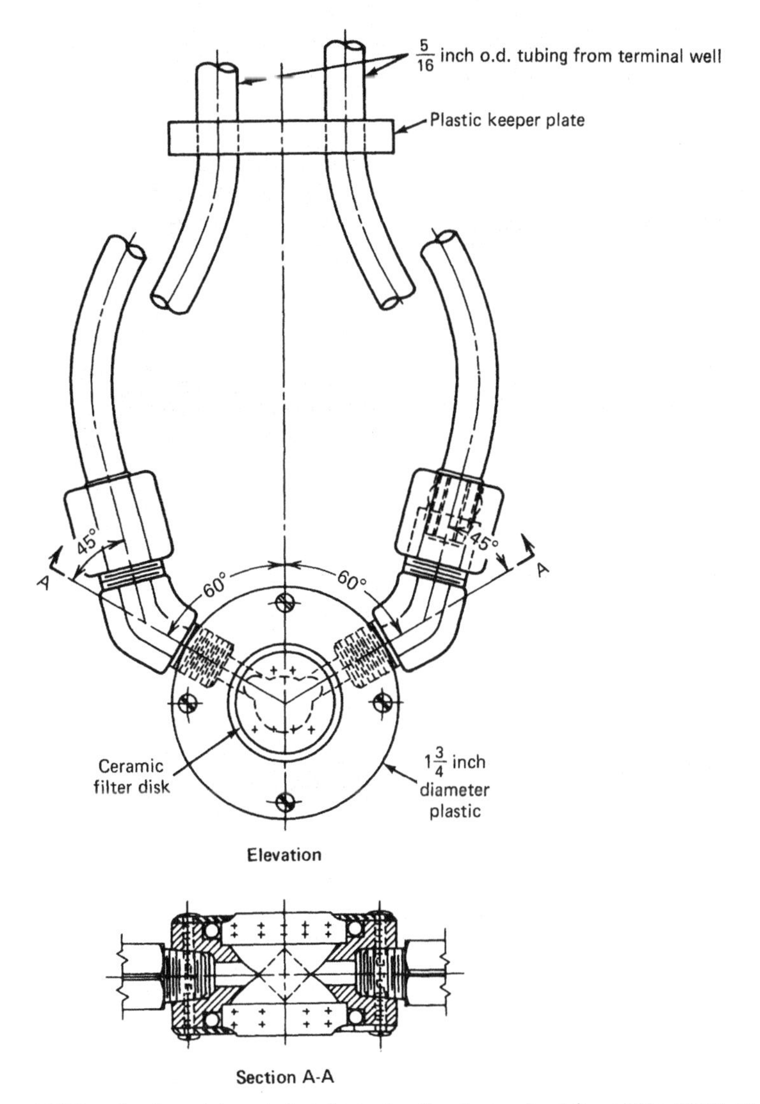

Figure 9.11. USBR embankment-type twin-tube hydraulic piezometer (after USBR, 1974). Courtesy of Bureau of Reclamation.

use of twin-tube hydraulic piezometers requires close adherence to many proven details. Some engineers believe that the twin-tube hydraulic piezometer should be superseded by diaphragm piezometers, but the author has a different view, and the arguments for and against both points of view are given in Chapter 21. Because the many proven details are not available in compact form in the literature, and because the author advocates the use of twin-tube hydraulic piezometers for long-term monitoring of pore water pressure in embankment dams, these details are given in Appendix E. Details relate to selection of components, method of installation, initial filling with liquid, and maintenance.

When compared with diaphragm piezometers, the major advantages of twin-tube hydraulic piezometers are the absence of inaccessible moving parts or electrical components and the capability of flushing the piezometer cavity. They are therefore very attractive for long-term monitoring. The system developed in England is reliable and has a long successful performance record. When installed in fill, the integrity of twin-tube hydraulic piezometers can be checked after installation by performing falling or constant head permeability tests. Automatic data acquisition systems are available, with a scanner allowing an electrical pressure transducer to be coupled hydraulically to each piezometer in turn.

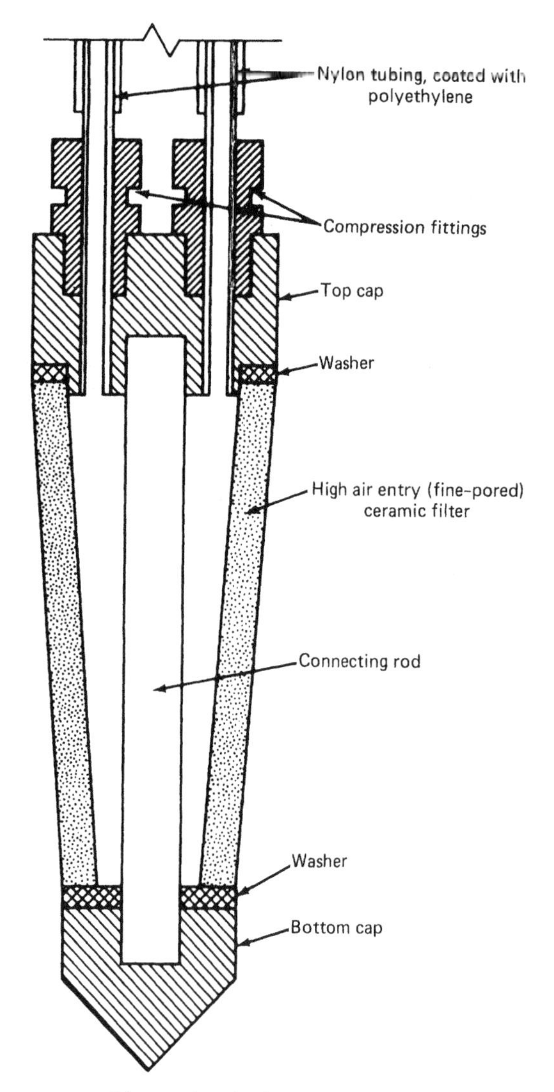

Figure 9.12. Bishop twin-tube hydraulic piezometer (after Penman, 1960).

Disadvantages include the need for a terminal enclosure to contain the readout and flushing arrangements; in addition, the enclosure must be protected from freezing either by heating or by constructing it below the frost line. Also, routing of tubing must be planned so that it does not rise significantly above the minimum piezometric elevation, otherwise the liquid will be required to sustain a subatmospheric pressure, and the liquid is likely to become discontinuous. However, if all components are selected to minimize gas entry, as described in Appendix E, and if high-quality de-aired water is used, the tubing can be installed up to 20 ft (6 m) above the minimum piezometric elevation.

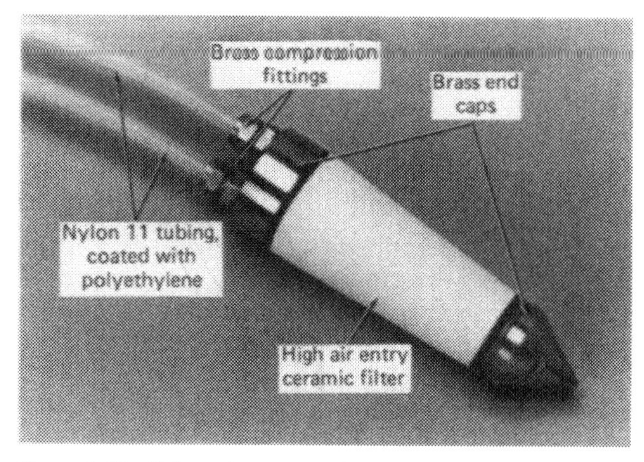

Figure 9.13. Bishop piezometer tip with attached tubing (courtesy of Soil Instruments Ltd., Uckfield, England).

Periodic flushing with de-aired water may be needed to remove any gas bubbles that have accumulated, and some engineers contend that this maintenance burden is unacceptable. However, the author believes that the maintenance burden has been exaggerated by improper selection of components, and flushing is seldom necessary in a properly designed and installed system when positive pore water pressures are being monitored.

9.5. PNEUMATIC PIEZOMETERS

The various types of *normally closed* and *normally open* pneumatic transducers and associated data acquisition systems are described and compared in Chapter 8. When these transducers are used for monitoring pore water pressure, a filter is added to separate the flexible diaphragm from the material in which the piezometer is installed (Warlam and Thomas, 1965). Figure 9.14 shows a schematic arrangement of a pneumatic piezometer, and a photograph of the version manufactured by Thor International, Inc. is shown in Figure 9.15. Similar piezometers, in which oil is used instead of gas, are available; however, they do not appear to offer any advantages over pneumatic piezometers.

Pneumatic piezometers can be packaged within push-in housings but are subject to the limitations discussed in Section 9.16. A special version of pneumatic piezometer, designed for pushing into place below the bottom of a borehole, is described in Section 9.8.

Some pneumatic piezometers can be used for

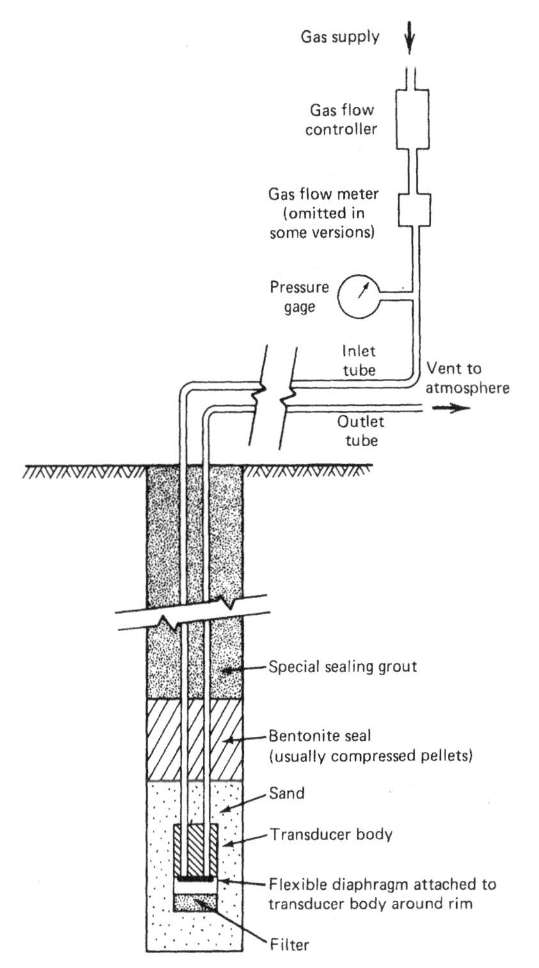

Figure 9.14. Schematic of pneumatic piezometer installed in a borehole. (*Normally closed* transducer, with two tubes, read as gas is flowing.)

monitoring negative (subatmospheric) pore water pressures by applying a vacuum to the outlet tube.

The *normally open* transducer is available from Terra Technology Corporation with one or two additional plastic tubes connected to the cavity between the filter and the flexible diaphragm. The intent of adding one tube is apparently to reduce the error caused by diaphragm displacement (Chapter 8) and also to allow unclogging a blocked filter by applying water pressure to this tube. If the piezometric level is below the end of the additional tube, and the end of the tube is capped, an air space is likely to exist above the water in this tube such that the error caused by diaphragm displacement is reduced, but the rapid response feature is impaired. If the piezometric level is above the end, and the end is capped, the error caused by diaphragm displacement remains. If the end is not capped, water will rise in the tube as the check valve closes during reading, possibly creating a false reading because of the increased head. Therefore, addition of this third tube appears to be a disadvantage in all cases. The intent of adding a fourth tube and connecting it to the cavity is apparently to permit flushing any accumulated gas from the cavity, thus minimizing response time and ensuring that pore water pressure rather than pore gas pressure is measured in unsaturated soils. However, the limitations described above for the single additional tube are applicable, and all the limitations of twin-tube hydraulic piezometers have been added. The author believes that a better solution to problems caused by large diaphragm displacement is use of an appropriately designed *normally closed* pneumatic piezometer.

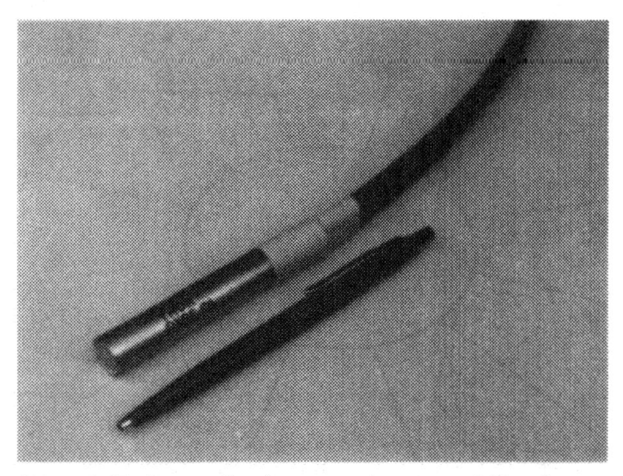

Figure 9.15. Pneumatic piezometer (courtesy of Thor International, Inc., Seattle, WA).

When selecting a pneumatic piezometer, a user must pay attention to the details discussed in Section 8.3. The author does not favor use of *normally open* pneumatic piezometers.

9.6. VIBRATING WIRE PIEZOMETERS

The vibrating wire piezometer has a metallic diaphragm separating the pore water from the measuring system. As shown in Figure 9.16, a tensioned wire is attached to the midpoint of the diaphragm such that deflection of the diaphragm causes changes in wire tension, and measurements are made by following the procedure described in Chapter 8. Figure 9.17 shows the version manufactured by Telemac.

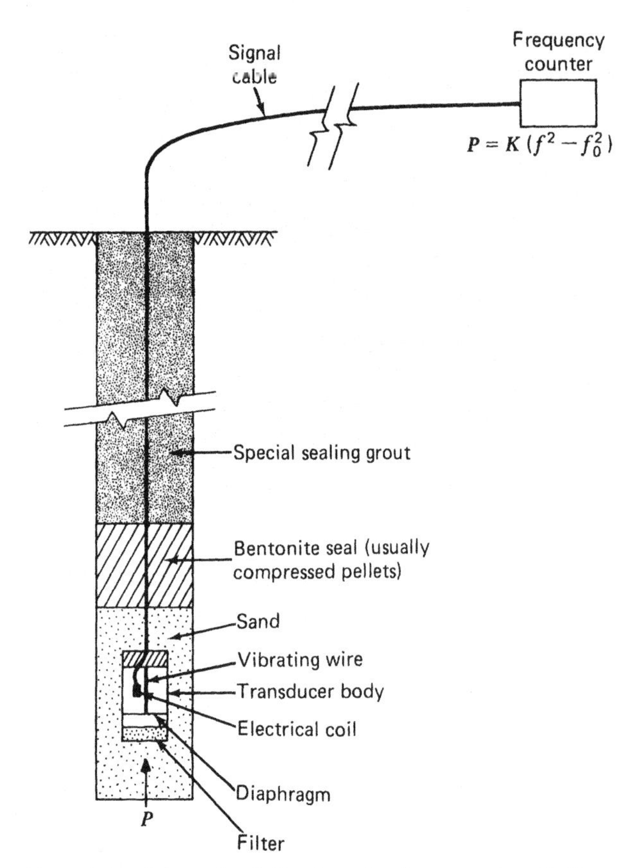

Figure 9.16. Schematic of vibrating wire piezometer installed in a borehole.

Chapter 8 describes advantages and limitations of vibrating wire transducers, identifies the major historical disadvantages as the potential for errors caused by zero drift and by corrosion of the vibrating wire, and indicates methods for minimizing errors. Most vibrating wire strain gage piezometers have a dried and hermetically sealed cavity around the vibrating wire, thereby minimizing the corrosion problem.

In recognition of the concern for zero drift, several versions of vibrating wire piezometer are provided with an in-place check feature whereby the zero reading and in some cases the calibration can be checked at any time during the life of the piezometer. These versions are described in Section 9.8.

There has been recent evidence in Norway that if standard vibrating wire piezometers are installed in compacted fill, they may show a reading change caused by total stresses acting on the piezometer body. The problem can be overcome at the manufacturing stage by constructing the piezometer

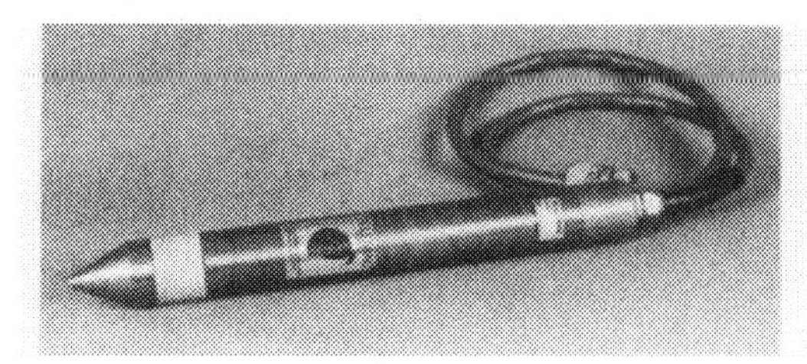

Figure 9.17. Vibrating wire piezometer (courtesy of Telemac, Asnières, France).

within a thick-walled cylinder (DiBiagio and Myrvoll, 1985) or within an outer protective shell, and this appears to be a worthwhile precaution for all vibrating wire piezometers that are to be installed within embankments. More details of this rubust instrument are given in Section 9.8.

Vibrating wire piezometers can be packaged within push-in housings, as shown on Figure 9.18. However, they are subject to the limitations discussed in Section 9.16.

9.7. ELECTRICAL RESISTANCE PIEZOMETERS

Operating principles and commercial versions of unbonded and bonded electrical resistance strain gage piezometers are shown in Figures 9.19–9.22.

The unbonded version uses a transducer invented by Roy Carlson in 1928 and patented by him in 1936.

Bonded versions generally include semiconductor resistance strain gages. Although highly stable bonded resistance strain gage transducers are available, their cost is high, at present they cannot be used in piezometers that are competitively priced, and commercially available piezometers have transducers for which long-term stability is uncertain. Bozozuk (1984) reports that numerous cases of long-term drift have been observed in the field. However, ongoing improvements in semiconductor

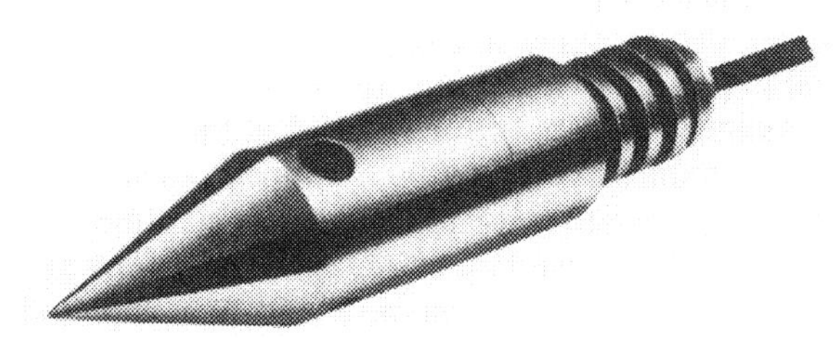

Figure 9.18. Push-in vibrating wire piezometer (courtesy of Geokon, Inc., Lebanon, NH).

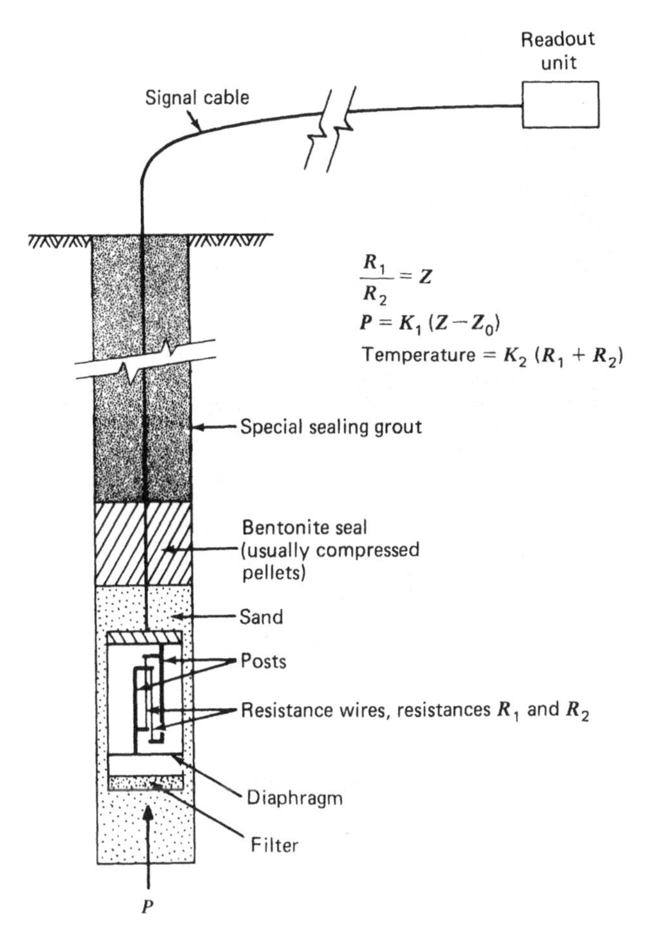

Figure 9.19. Schematic of Carlson unbonded electrical resistance strain gage piezometer installed in a borehole.

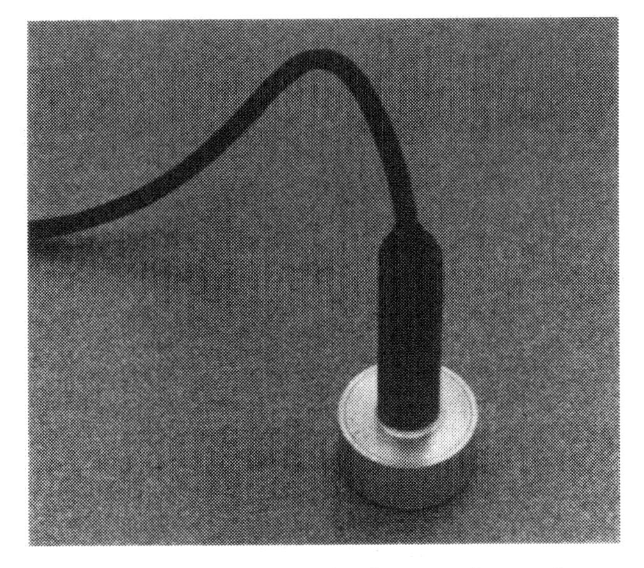

Figure 9.20. Unbonded electrical resistance strain gage piezometer (*pore pressure cell*) (courtesy of Carlson Instruments, Campbell, CA).

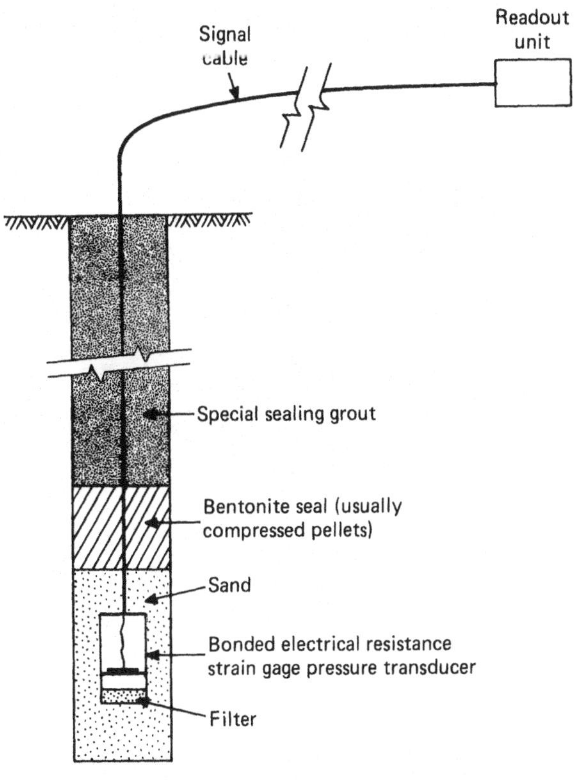

Figure 9.21. Schematic of bonded electrical resistance strain gage piezometer installed in a borehole.

strain gage technology may lead to development of more reliable and competitive piezometers of this type in the near future.

Bonded electrical resistance strain gage piezometers can be packaged within push-in housings (e.g., Torstensson, 1975; Wissa et al., 1975) and are used primarily for in situ determination of soil properties, where small size is critical and periodic recalibration is possible. They can, however, be left in place for monitoring pore water pressure, in which case they are subject to the limitations discussed in Section 9.16. They can also be used as *profiling piezometers* for measuring pore water pressure at many points, by pushing in increments and waiting for the reading to stabilize after each push.

A version of the bonded electrical resistance strain gage piezometer referred to as the *BAT piezometer* is arranged so that the transducer can be detached from the porous tip. The piezometer is described in Section 9.8.

When electrical resistance strain gage piezometers are installed in compacted fill they should, as described above for vibrating wire piezometers, be

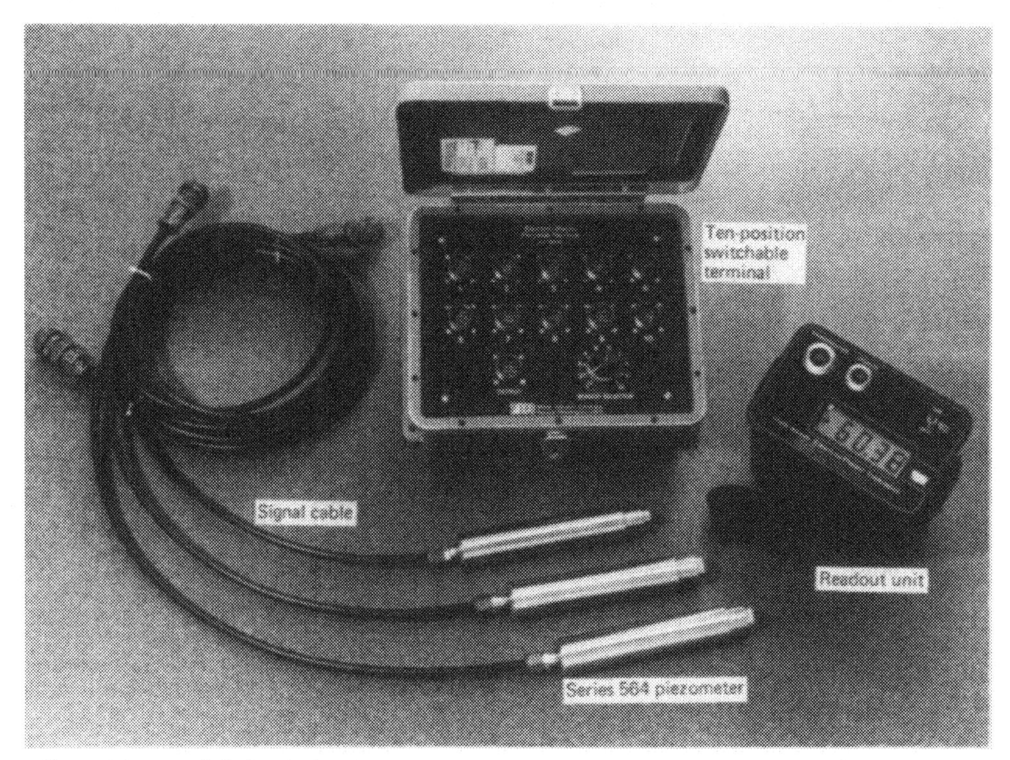

Figure 9.22. Bonded electrical resistance strain gage piezometers (courtesy of Slope Indicator Company, Seattle, WA).

housed within an adequately strong body so that reading changes are not caused by total stresses acting on the outside of the body.

9.8. MISCELLANEOUS SINGLE-POINT PIEZOMETERS

Numerous single-point piezometers are available, based in principle on one of the five types described above. The following sections describe examples that have a practical application.

9.8.1. Push-in Open Standpipe Piezometer: Wellpoint Type

A standard wellpoint (Section 9.2) is connected to galvanized pipe with threaded couplings. Nominal pipe diameter is typically 1.0 or 1.25 in. (25 or 32 mm). The author believes that use of this standard system will not result in an adequate seal above the piezometer and that it must be amended by adding a length of larger-diameter pipe above the wellpoint, as described in Section 9.16. Section 9.16 also describes other limitations of push-in open standpipe piezometers.

9.8.2. Push-in Open Standpipe Piezometer: Geonor Type

The Geonor *Model M-206 field piezometer* (Figure 9.23) was developed for use in soft sensitive clays in Norway and is pushed into place by attachment to EX-size drill rods.

EW-size drill rods will also fit on the piezometer, but the connection between an EX male and EW female thread will not be watertight, and excess pore water pressure is likely to bleed through this connection and dissipate along the inside of the drill rod. If EW rods are used, a special effort must be made to seal the connection. Sintered bronze filters are standard, and epoxy filters are used in corrosive environments. The outside diameters of EX and EW drill rods (Appendix F) are slightly larger than the outside diameter of the piezometer (1.30 in., 33.0 mm). The rods remain in place, and the piezometer and rods are recovered when measurements are no longer required, but economy can be achieved by using 1 in. (25 mm) steel pipe with threaded couplings above a sufficient length of EX rod for adequate sealing.

The piezometer has been used successfully as a *profiling piezometer* for measuring pore water pres-

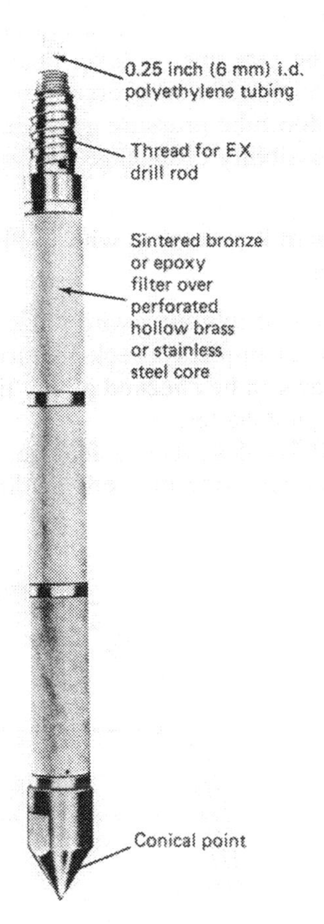

Figure 9.23. Geonor Model M-206 field piezometer (courtesy of Geonor A/S, Oslo, Norway).

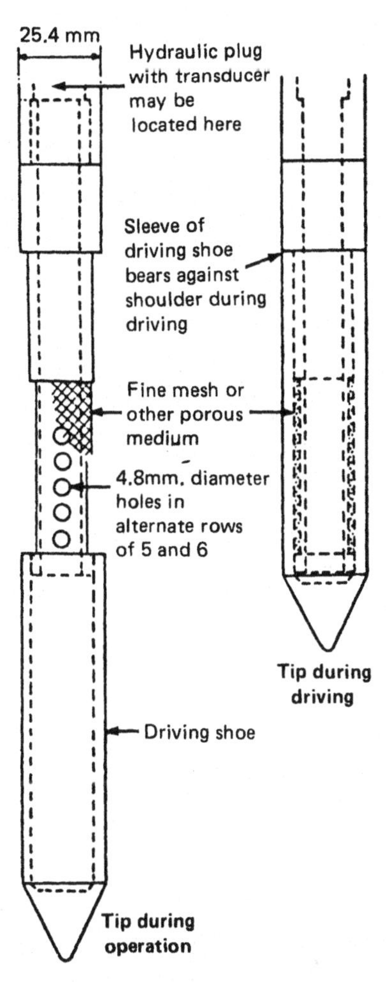

Figure 9.24. Cambridge drive-in piezometer (after Parry, 1971). Reprinted by permission of the Institution of Civil Engineers, London.

sure at many points in a very soft marine clay by pushing in 1.5 ft (0.5 m) increments and waiting for the water level in the standpipe to stabilize after each push, normally requiring an overnight wait (Handfelt et al., 1987). In addition to providing data at many points, this procedure can be used to check permanently installed piezometers. The version of piezometer with sintered bronze filters should be used for this purpose, because the version with epoxy filters has a smaller-diameter core that tends to break while pushing or withdrawing the piezometer.

9.8.3. Push-in Open Standpipe Piezometer: Cambridge Type

The *Cambridge drive-in piezometer* (Parry, 1971) is shown in Figures 9.24 and 9.25 and was developed for pushing or driving through soil containing stones and hard seams without damaging, smearing, or clogging the porous element. All parts of the tip except the gauze mesh are mild steel. The piezometer is attached with a flush joint to a length of steel pipe and driven into the ground with the sliding driving shoe covering the porous section as shown on the right side of Figure 9.24. When in place, a mandrel is inserted within the standpipe and the driving shoe tapped down approximately 900 mm (35 in.) relative to the standpipe, thereby exposing the porous element to the soil.

The piezometer is available from Geotechnical Instruments (U.K.) Ltd. and Soil Instruments Ltd.

9.8.4. Heavy Liquid Piezometer

The *heavy liquid piezometer* (Figure 9.26) is a special type of open standpipe piezometer, developed by Piezometer Research and Development. The

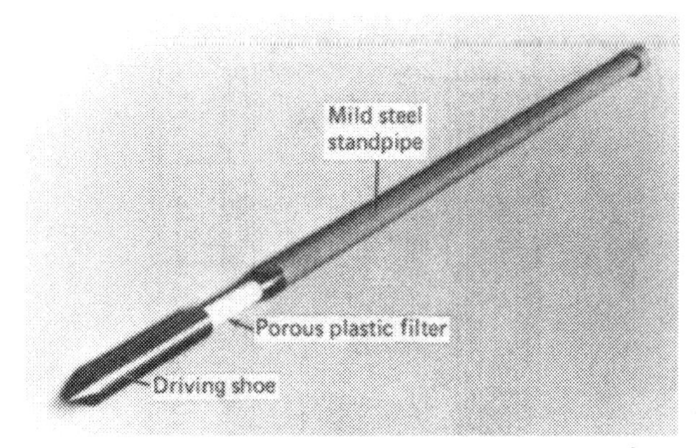

Figure 9.25. Cambrdge drive-in piezometer (courtesy of Soil Instruments Ltd., Uckfield, England).

standpipe and part of the piezometer tip are filled with acetylene tetrabromide, a liquid with a specific gravity of 2.96, and in the tip there is a stable interface between acetylene tetrabromide and water. The surface of liquid in the standpipe is therefore lower than the piezometric level, and the known specific gravity is used to convert measured liquid level to piezometric level.

When compared with a conventional open standpipe piezometer, the heavy liquid piezometer reduces response time and overcomes freezing problems. When the piezometric level rises above the top of the standpipe it also overcomes the need to attach a Bourdon tube pressure gage and therefore reduces the possibility of damage by vandals.

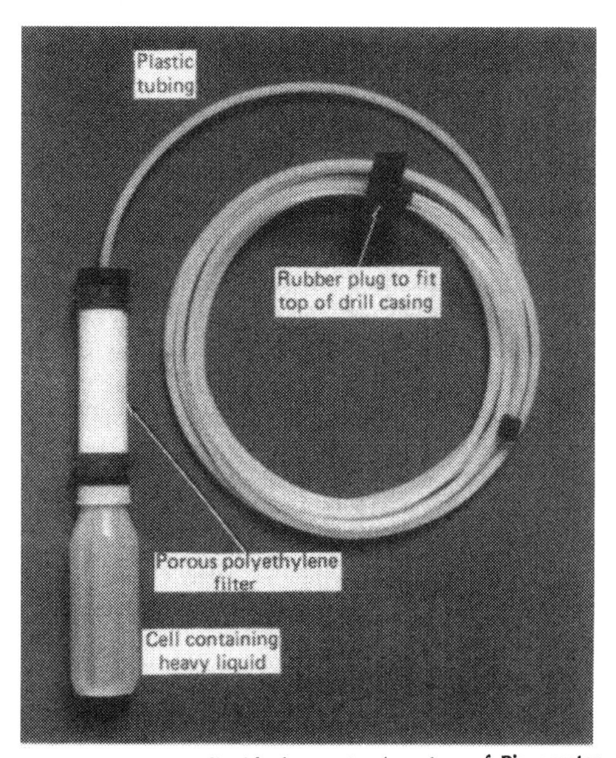

Figure 9.26. Heavy liquid piezometer (courtesy of Piezometer Research & Development, Bridgeport, CT).

9.8.5. Diaphragm Piezometers with In-Place Check Features

Several versions of vibrating wire piezometers are provided with an in-place check feature whereby the zero reading can be checked at any time during the life of the piezometer.

DiBiagio (1974) describes a feature, shown in Figure 9.27, for checking the zero reading and the

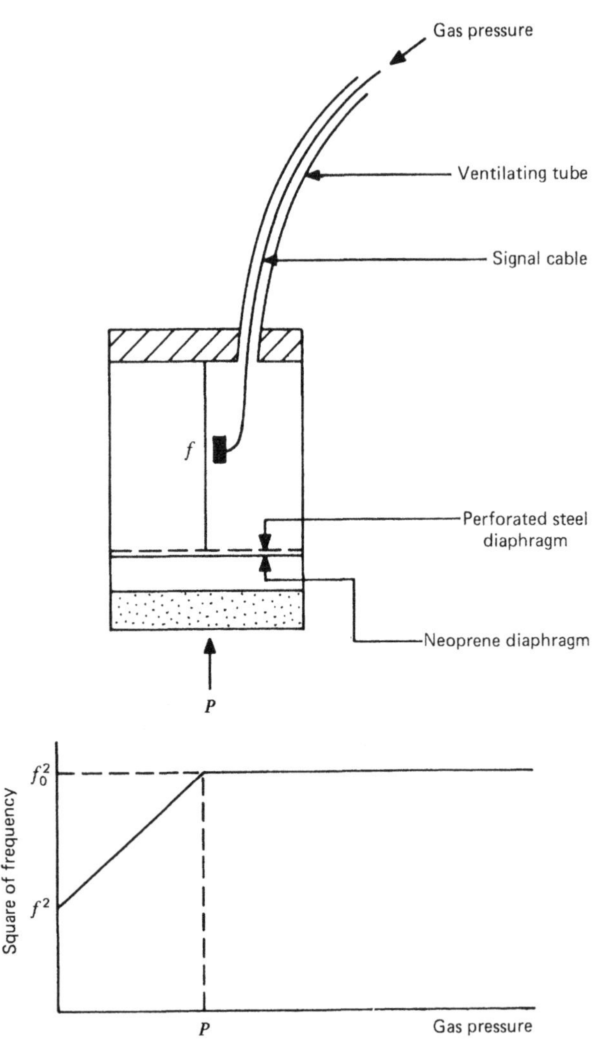

Figure 9.27. Schematic of Norwegian Geotechnical Institute vibrating wire piezometer with feature for checking calibration in place.

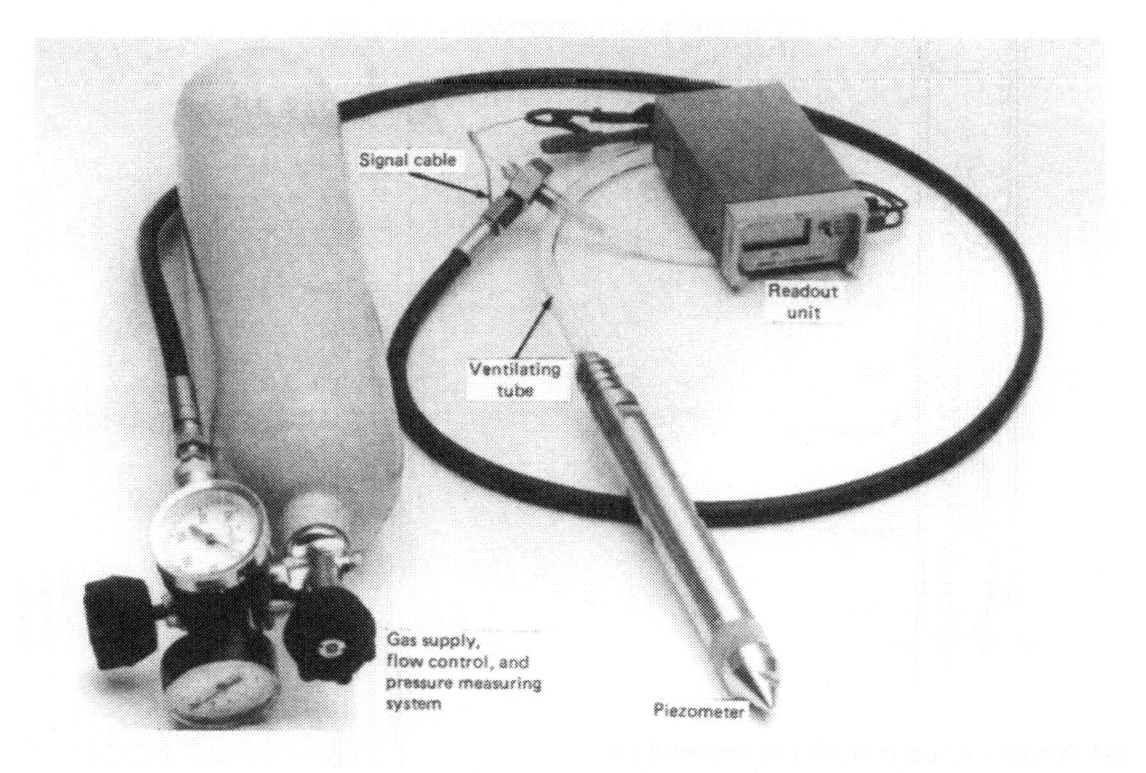

Figure 9.28. Vibrating wire piezometer with feature for checking calibration in place (courtesy of Geonor A/S, Oslo, Norway).

complete calibration in place. The instrument is available from Geonor (Figure 9.28). The body of the gage is ventilated through a tube, and the back of the neoprene diaphragm is normally at atmospheric pressure. Under normal working conditions the two diaphragms are in contact, and the instrument behaves as a conventional vented vibrating wire piezometer. To check the zero reading and calibration, gas pressure is applied to the back of the diaphragm through the ventilating tube. Readings of gage wire frequency are taken at various pressure increments until eventually, when the gas pressure is slightly greater than the pore water pressure, the neoprene diaphragm moves away from the metal diaphragm, leaving it with equal gas pressure on both sides, and the frequency of vibration of the gage wire reaches an asymptotic value. Any further increase in the applied gas pressure does not cause further change in the vibrating frequency of the gage wire, and this limiting frequency value is the zero frequency for the instrument. The calibration of the instrument is indicated by the slope of the pressure versus frequency plot.

Various manufacturers of vibrating wire piezometers now offer this feature as an option, and the feature could also be used for checking electrical resistance piezometers. Although an ingenious arrangement, it is incompatible with a hermetically sealed cavity within the piezometer tip and increases the potential for corrosion because any damage to the ventilating tube may provide a direct path for moisture to enter the cavity. Deterioration of the neoprene diaphragm may also result in malfunction. The instrument is therefore best suited to obtaining verified data of high accuracy over a short term, rather than as an in-place check over a long term.

The Swedish manufacturer Geotech AB has developed a method for checking the zero reading in place without the above limitation. As shown in Figure 9.29, bellows and a stem are added between the diaphragm and the filter. The zero reading can be checked (but not the slope of the calibration) by applying a voltage to the electromagnet to move the stem downward, thereby separating the stem from the diaphragm and returning the vibrating wire to its zero condition. The downward movement is approximately 0.05 mm (0.002 in.).

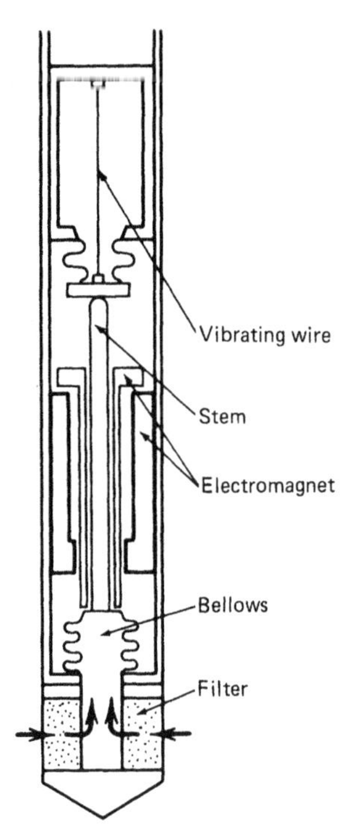

Figure 9.29. Schematic of Geotech AB vibrating wire piezometer Model PZ 4000, with feature for checking zero reading in place.

9.8.6. BAT piezometer

The *BAT piezometer* (Torstensson, 1984a) allows a bonded resistance strain gage transducer to be detached at any time from the porous tip, permitting servicing and recalibration of the transducer. The instrument is available from BAT Envitech, and a single transducer can be used for many piezometers. The porous tip has a short nozzle at its top, sealed with a rubber disk. The tip is saturated, attached to 1 in. (25 mm) galvanized pipe, and installed by the push-in method. Whenever a measurement of pore water pressure is required, a transducer unit is lowered down the standpipe to mate with the porous tip. A hypodermic needle at the base of the transducer unit pierces the rubber disk to create a hydraulic connection, and the disk is self-sealing when the transducer unit and needle are withdrawn. The arrangement is shown in Figure 9.30. Torstensson (1984a) reports that the rubber disk can be penetrated by the hypodermic needle several hundred times without loss of its self-sealing function. Separate units can be lowered down the standpipe for groundwater sampling and for in situ permeability testing.

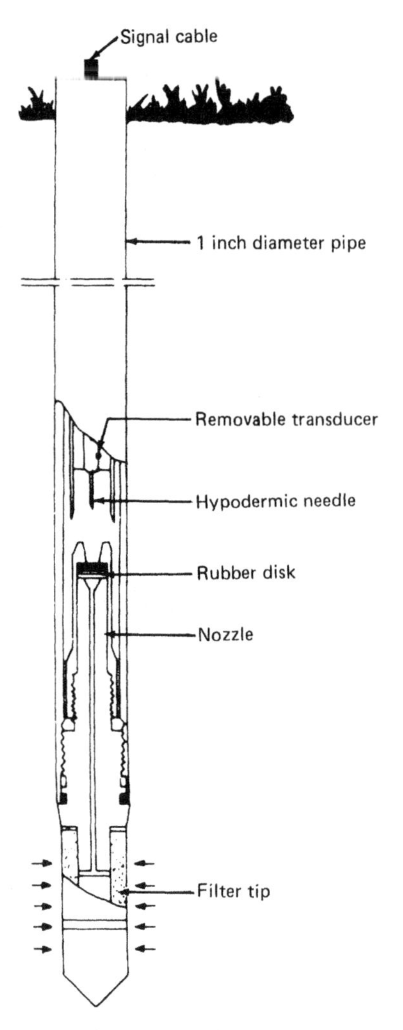

Figure 9.30. BAT piezometer (after Torstensson, 1984a). Reprinted by permission of *Ground Water Monitoring Review*, © 1984. All rights reserved.

9.8.7. Piezometer Designed to Be Installed by Pushing in Place Below the Bottom of a Borehole

A special version of pneumatic piezometer, manufactured by Slope Indicator Company, is designed for pushing into place below the bottom of a borehole (Handfelt et al., 1987). Figure 9.31 shows a version suitable for use in very soft clay (standard penetration test blowcount of less than about 2 blows per foot). Figure 9.32 shows a photograph of this version and also a more robust version suitable for pushing into stiffer clay (this has been installed in clay with a standard penetration test blowcount of 15 blows per foot, but is likely to be suitable for installation in stiffer clays). A similar packaging

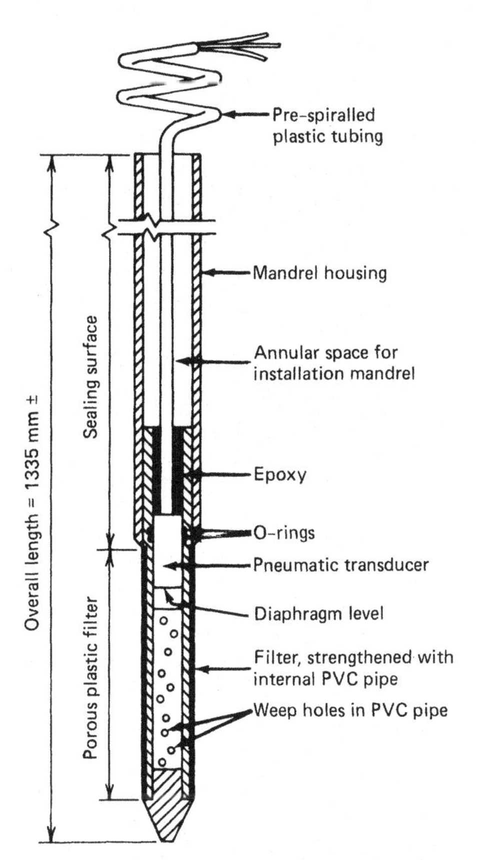

Figure 9.31. Pneumatic piezometer for installation by pushing in place below bottom of a borehole in very soft clay (courtesy of RMP Encon Ltd. and Dames & Moore).

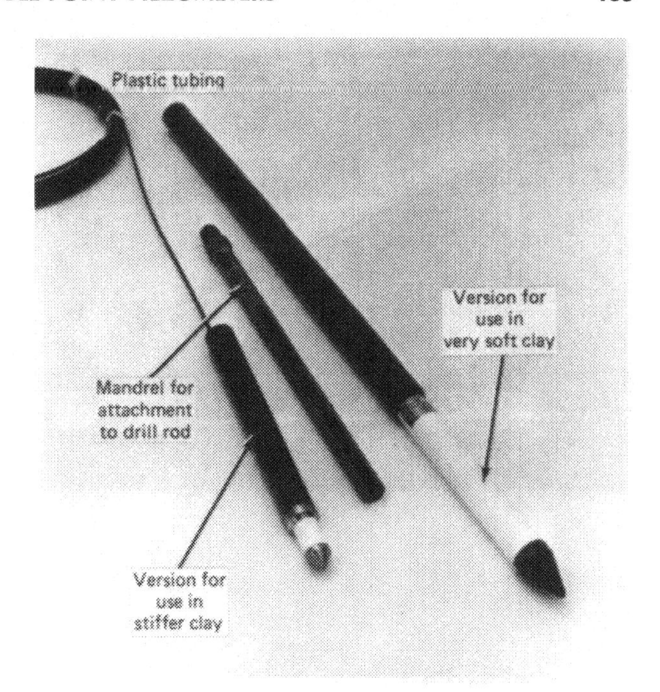

Figure 9.32. Pneumatic piezometers for installation by pushing in place below bottom of borehole (courtesy of Slope Indicator Company, Seattle, WA).

could be arranged for other types of diaphragm piezometer. The method of installation is described in Section 9.17.1.

9.8.8. Vibrating Wire Piezometers for Embankment Dams

DiBiagio and Myrvoll (1985) describe a vibrating wire piezometer specifically designed for installation in embankment dams. The piezometer is manufactured by Geonor and is shown in Figures 9.33 and 9.34. It has a thick-walled stainless steel housing to prevent the vibrating wire transducer from responding to total stresses acting on the housing, a very robust and water-blocked signal cable, heavy-duty seals, and long high air entry filters. The piezometer, and also a heavy-duty vibrating wire piezometer manufactured by Geokon, are good examples of instruments designed for long-term survivability.

9.8.9. Piezometers with Duplicate Transducers

Piezometers are available with one pneumatic and one vibrating wire transducer housed in the same body, so that one transducer acts as a backup to the other. An example is shown in Figure 9.35. Alternatively, two separate piezometers can be installed, but the combination arrangement simplifies installation of the tubing/signal cable. Vibrating wire piezometers and also pneumatic piezometers are

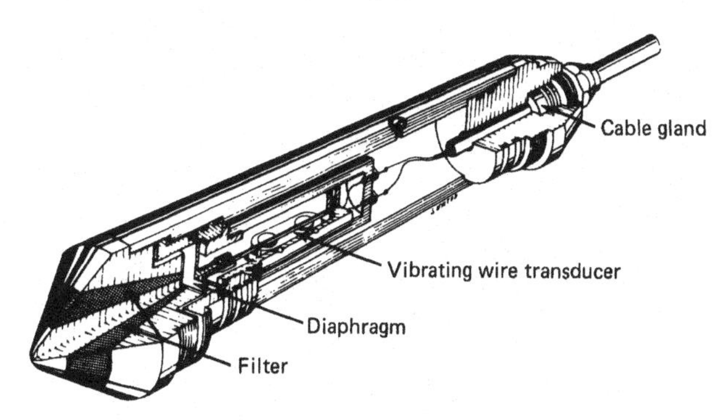

Figure 9.33. Vibrating wire piezometer for use in embankment dams (after DiBiagio and Myrvoll, 1985).

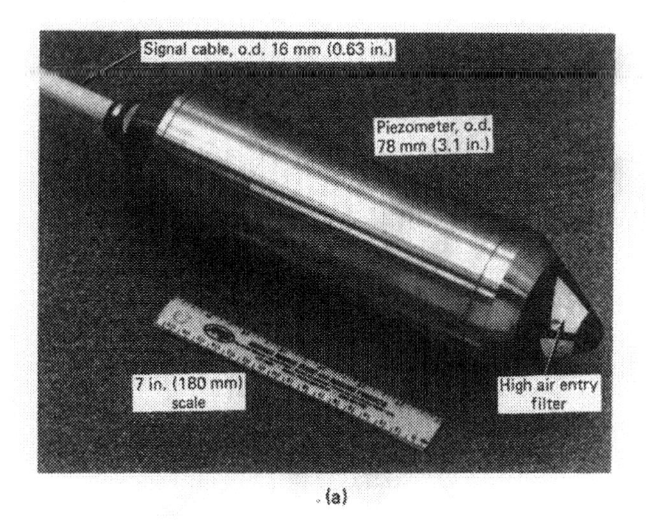

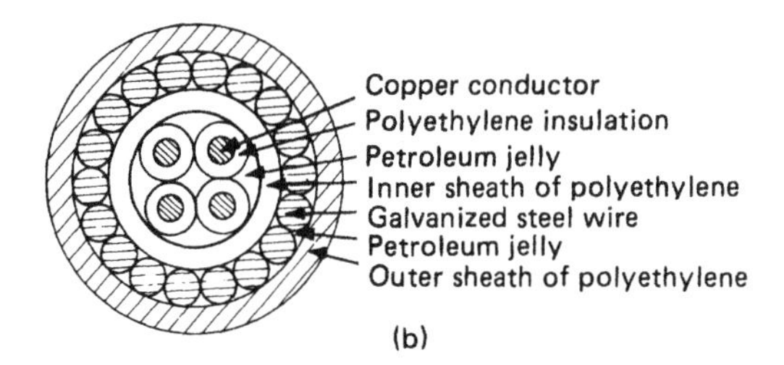

Figure 9.34. Vibrating wire piezometer for use in embankment dams: (*a*) Model S-411 piezometer and (*b*) cross section of P-430 signal cable (courtesy of Geonor A/S, Oslo, Norway).

available with two hydraulic tubes connected to the cavity between the filter and diaphragm, thereby adding a twin-tube hydraulic piezometer as backup, and this arrangement also allows for flushing any gas out of the cavity via the hydraulic tubes.

Whenever considering a piezometer with duplicate transducers, the combination will generally be subject to the combined limitations of both but will not necessarily be subject to the combined advantages of both. For example, addition of twin hydraulic tubes to a pneumatic piezometer requires that the tubes must not be installed significantly above the minimum piezometric elevation, a limitation that does not apply to pneumatic piezometers.

9.9. MULTIPOINT PIEZOMETERS

Multipoint piezometers can be created by four methods, described in turn.

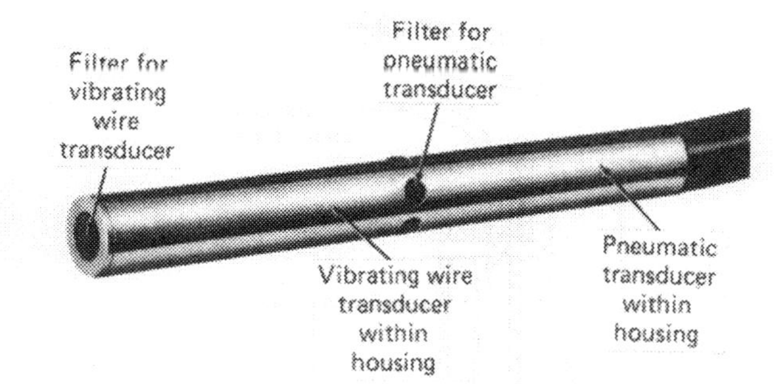

Figure 9.35. Piezometer with one pneumatic and one vibrating wire transducer (courtesy of Geokon, Inc., Lebanon, NH).

9.9.1. Multipoint Piezometers with Packers

Several piezometers can be assembled with a packer above and below each piezometer to create a seal.

Inflatable packers (e.g., Figure 9.36) are most frequently used and are filled with air, water, or epoxy. Commercial sources of inflatable packers, with multiple central holes for passage of leads from lower piezometers and packers, are given in Appendix D.

An innovative system has recently been developed by Solinst Canada Ltd., based on a design by the University of Waterloo (Cherry and Johnson, 1982). The arrangement, shown in Figure 9.37, uses Dowell chemical sealant (Section 9.17.8), which swells when in contact with water that is poured down the central PVC pipe. The figure shows standpipe piezometers, but alternatively pneumatic or vibrating wire piezometers can be used. The instrument can also be adapted for groundwater sampling. The system is supplied in modular form, with threaded connections in the PVC pipe between modules. An O-ring seal is included in each connection, as shown in Figure 9.37, to ensure that connections are watertight.

9.9.2. Multipoint Piezometers Surrounded with Grout

Several piezometers can be packaged within a length of pipe to create a multipoint piezometer, which is then inserted within a borehole in soil and entirely surrounded with grout.

Vaughan (1969) describes a multipoint open standpipe piezometer that is installed in this way. Vaughan reports that the response times of the piezometers in grout are adequate for most engineering purposes for all permeabilities of the soil.

Figure 9.36. Multipoint pneumatic piezometers with inflatable packers (courtesy of Thor International, Inc., Seattle, WA).

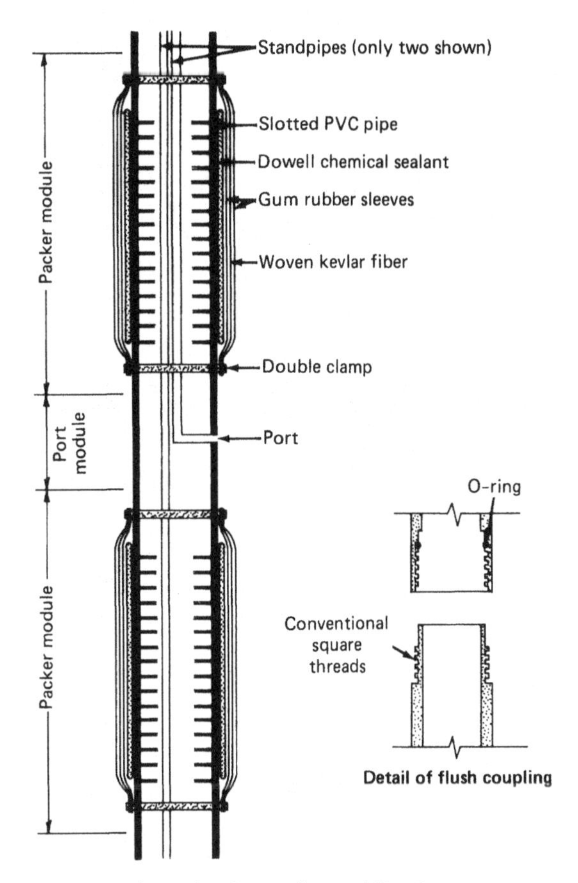

Figure 9.37. Schematic of Waterloo Multilevel System (courtesy of Solinst Canada Ltd., Burlington, Ontario, Canada).

When the permeability of the grout is greater than that of the soil the response time of the piezometer in grout is only slightly greater than that of a piezometer with a sand filter. When the soil has a permeability more than two orders of magnitude greater than the permeability of the grout, the response time is controlled by the grout annulus only. However, bearing in mind permeability and conformance criteria, the author believes that the applicability of this method is limited to use in a uniform clay of known properties, such as the core of an embankment dam. It is also limited by the difficulties of manufacturing and installing a grout so that its **in-place** compressibility and permeability are reasonably uniform.

9.9.3. Multipoint Push-in Piezometers

A multipoint open standpipe piezometer can be assembled from flush-coupled pipe with several filter zones and plastic tubes and pushed into place. Massarsch et al. (1975) report on good performance of a multipoint vibrating wire piezometer, with transducers 1 m (3 ft) apart and installed by pushing into clay, and indicate that leakage did not occur along the flush-coupled pipe. Carpentier and Verdonck (1986) describe a multipoint piezometer with electrical resistance strain gage transducers, used for measuring the influence of tides and waves on pore water pressures below the seabed.

Push-in piezometers are subject to the limitations discussed in Section 9.16.

9.9.4. Multipoint Piezometers with Movable Probes

A pipe is installed and sealed within a borehole and a movable probe periodically inserted within the pipe. The development of multipoint piezometers with movable probes has created a significant

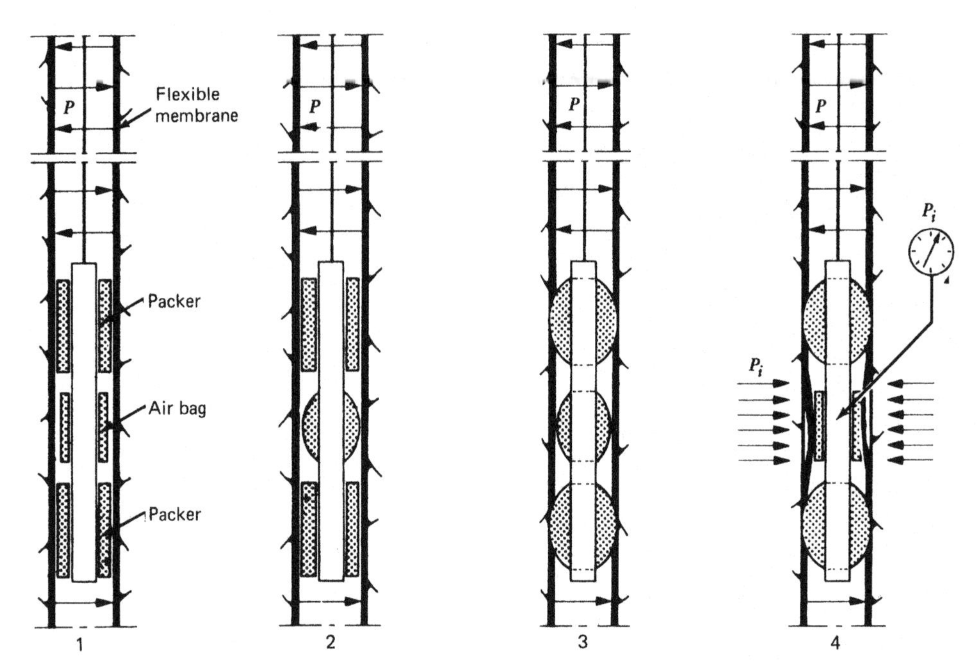

Figure 9.38. Operating procedure for Piezofor multipoint piezometer (courtesy of Telemac, Asnières, France).

breakthrough in the measurement of groundwater pressure. They allow a detailed pressure–depth profile to be defined, with an almost unlimited number of measurement points. Although they have a high initial cost, their use may result in substantial cost savings by ensuring that a project is designed with a reliable knowledge of the pressure–depth profile.

Three configurations are commercially available.

Piezofor System

The *Piezofor* system was developed in France for measurement of joint water pressure in rock (Hoek and Londe, 1974; Londe, 1982). A slotted pipe is grouted into a borehole with a brittle slurry that cracks as it shrinks. A soft waterproof tubular membrane is inserted into the pipe and pressurized to a higher value than the expected groundwater pressure.

Joint water pressure measurements are taken by following the steps shown in Figure 9.38. The Piezofor probe is lowered to the desired point, and water is forced out of the test section between the packers by inflating the air bag. The packers are inflated to isolate the test section and the air bag is deflated. The membrane is now forced inward by the outside pressure from the groundwater, causing the inside pressure to rise asymptotically until the two balance. The inside pressure is monitored with an electrical pressure transducer. The procedure can be repeated along the entire length of the borehole to provide the full piezometric profile, without affecting the natural seepage pattern. The rate at which equilibrium occurs between inside and outside pressures allows an estimate to be made of the permeability of the ground at that level. A pressure lock can be fitted at the top of the pipe if required, so that the probe can be inserted within the membrane without depressurizing the borehole.

The system has been used successfully in boreholes 330 ft (100 m) deep, and a system for use in 5000 ft (1500 m) boreholes is currently being tested.

Westbay Multiple Piezometer

More recently, a movable probe device has been developed by Westbay Instruments Ltd. in Canada (Patton, 1979; Rehtlane and Patton, 1982) and is referred to as the *MP (multiple piezometer) System*. Installations can be made in soil and rock.

Figure 9.39 shows the general arrangement and the probe. The system consists of pipe, couplings and packers permanently installed in a borehole, a portable pressure measurement probe, and installa-

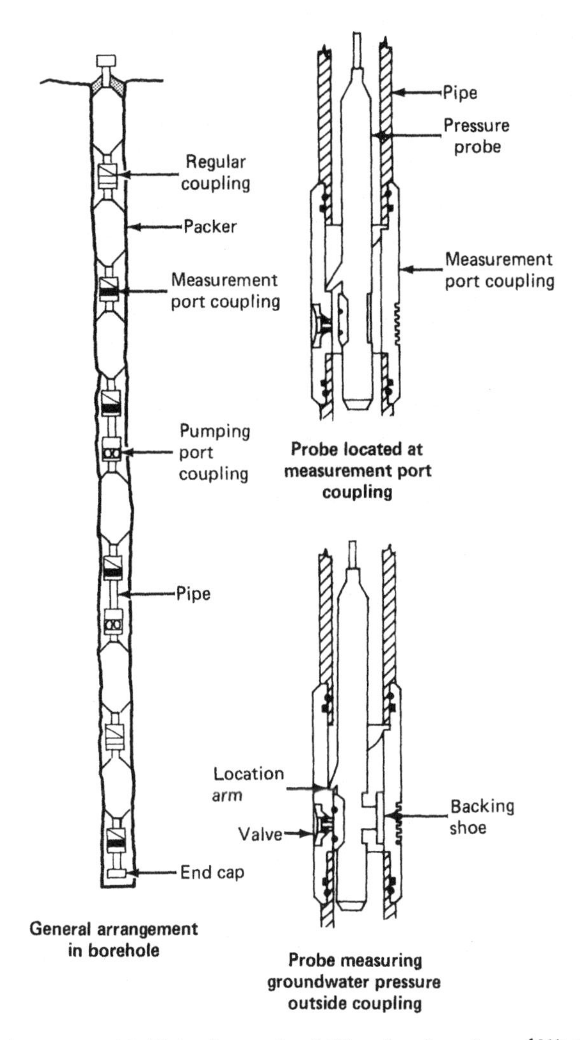

Figure 9.39. Multiple piezometer (MP) system (courtesy of Westbay Instruments Ltd., North Vancouver, BC, Canada).

tion tools. *Measurement port couplings* are installed in the pipe wherever groundwater pressure measurements are desired. Each of these couplings has a hole through its wall, with a filter on the outside and spring-loaded check valve on the inside. The remainder of the pipe is connected with sealed couplings, and a packer is installed around the pipe above and below each measurement port coupling. The assembly is lowered into the borehole, and the packers inflated one at a time with water, using a probe temporarily inserted within the pipe. To take readings, a pressure measurement probe is lowered within the pipe to locate a measurement port, jacked against the opposite wall of the pipe to open a check valve, and a measurement made. The procedure is repeated at each measurement port.

Pipe and couplings are manufactured in PVC or stainless steel. Pneumatic pressure probes are available for depths up to 250 ft (75 m) and standard electrical probes for depths up to 1000 ft (300 m). Recently a system has been used in boreholes 3000 ft (900 m) deep with 10–15 measurement ports in each and is designed for operation to a depth of 5000 ft (1500 m). Separate pneumatic and electrical probes are also available for taking pressurized or unpressurized samples of groundwater through the measurement ports. Additional valved couplings, referred to as *pumping port couplings,* can be included in the pipe for permeability testing or decontamination pumping. Grooved inclinometer casing can be used instead of smooth-walled pipe to create a combined piezometer and inclinometer system (Westbay Instruments Ltd., CPI system).

Piezodex System

The *Piezodex* system has recently been developed by the Federal Institute of Technology, Zürich, and Solexperts AG (Kovari and Köppel, 1987). Pressure transmitters are mounted in the wall of the access pipe and isolated with pneumatic packers. Each pressure transmitter consists of a piston with very low volumetric displacement. When the movable probe makes contact with a piston, the piston moves outward slightly (0.1–0.2 mm, 0.004–0.008 in.), and groundwater pressure is transmitted to a force transducer mounted in the probe. The depth capability is currently limited to 50 m (165 ft), but a system with twice this depth capability is under development.

9.10. HYDRODYNAMIC TIME LAG*

When a piezometer is installed and groundwater pressure changes, the time required for water to flow into or out of the piezometer to effect equalization is called the *hydrodynamic time lag*. It is dependent primarily on the type and dimensions of the piezometer and the permeability of the ground. Open standpipe piezometers have a much greater hydrodynamic time lag than diaphragm piezometers

*Sections 9.10–9.14 have been written with the assistance of Arthur D. M. Penman, Consulting Geotechnical Engineer, Harpenden, England.

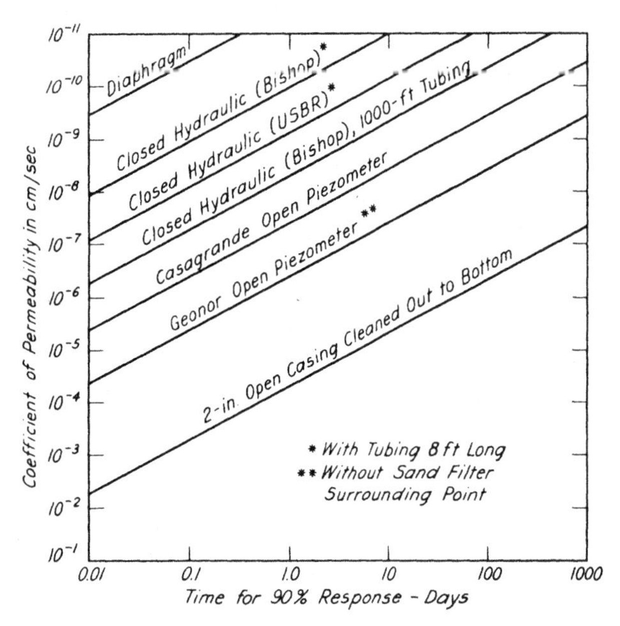

Figure 9.40. Approximate response times for various types of piezometer (after Terzaghi and Peck, 1967).

because a much greater movement of pore or joint water is involved. The term *slow response time* is used to describe a long hydrodynamic time lag.

Methods of estimating time lag are presented by Hvorslev (1951) and Terzaghi and Peck (1967). Brooker and Lindberg (1965) evaluate time lag effects for twin-tube hydraulic piezometers. Brand and Premchitt (1982), Penman (1960), Premchitt and Brand (1981), and Vaughan (1974a) report on the effects for various types of piezometers.

The order of magnitude of the time required for 90% response of several types of piezometer installed in homogeneous soils can be obtained from Figure 9.40, in which the *Geonor open piezometer* is the push-in instrument shown in Figure 9.23. The 90% response is considered adequate for many practical purposes, and of course the time for 100% response is infinite. The response time of open standpipe piezometers can be estimated from equations given by Penman (1960). For example,

$$t = 3.3 \times 10^{-6} \frac{d^2 \ln[L/D + \sqrt{1 + (L/D)^2}]}{kL}$$

where t = time required for 90% response in days,
d = inside diameter of standpipe in centimeters (cm),
L = length of intake filter (or sand zone around the filter) in centimeters (cm),
D = diameter of intake filter (or sand zone) in centimeters (cm),
k = permeability of soil in centimeters per second (cm/sec).

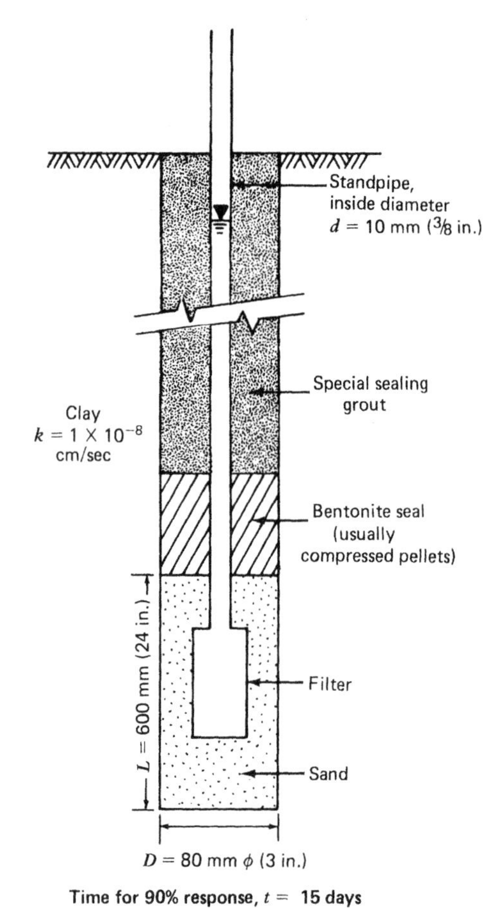

Figure 9.41. Example of time lag for open standpipe piezometer.

Figure 9.41 gives an example for an open standpipe piezometer surrounded by a sand zone. It should be noted that time lag can be minimized by using a minimum-diameter standpipe and a maximum-sized sand zone. However, as discussed in Section 9.3, a minimum-diameter standpipe can nullify the self-de-airing feature of the piezometer and will often create problems with insertion of the readout probe. For borehole installations, the diameter of the sand zone is limited by the economics of drilling large-diameter boreholes, and the length is often limited by the need to make measurements within a short length of the borehole.

The significance of hydrodynamic time lag depends primarily on the purpose of the measurements and on the anticipated fluctuations of the groundwater pressure. For example, if measurements are made to determine joint water pressure in a rock slope in which pressure fluctuations are not likely to be significant, an open standpipe piezometer may be suitable. If an embankment is being constructed on soft ground, and piezometers are used to monitor gain in strength, the rate of embankment placement and anticipated rates of pore water pressure dissipation will enter into judgments concerning time lag. If pore water pressure measurements are made with a push-in piezometer by leaving it in place for a short time and then pushing it deeper for additional measurements, a long time lag will not be acceptable. If the groundwater pressure is subject to daily fluctuations, for example, near the ocean or in an embankment or slope that forms part of a pumped storage hydroelectric project, a time lag of more than a few hours would obscure real variations in pressure and the measurements would have no value. Time lag criteria should be evaluated on a case-by-case basis.

9.11. TYPES OF FILTER

All piezometers include an intake filter. The filter separates the pore fluid from the structure of the soil in which the piezometer is installed and must be strong enough to avoid damage during installation and to resist the total stresses without undue deformation. Filters can be classified in two general categories: *high air entry* and *low air entry*.

Figure 9.42 shows a fine-pored filter placed between two parts of a container, one part containing gas, the other part containing water. The gas pressure is higher than the water pressure, but the fluids are in equilibrium because the pressure differential is balanced by surface tension forces at the gas/water interface. The smaller the radius of curvature of the menisci at the interface, the larger can be the pressure difference between water and gas. The minimum radius of curvature of the menisci is dictated by the pore diameter in the filter; thus, the finer the filter, the greater can be the pressure differential. The *air entry value* or *bubbling pressure* of the filter is defined as the pressure differential at which blow-through of gas occurs. Thus, a filter with a high air entry value (or high bubbling pressure) is a fine filter that will allow a high pressure differential before blow-through occurs. Because such a filter also has a low permeability, the terms can be confusing. Although the fluids in Figure 9.42 are in equilibrium, the longevity of the fluid separation is uncertain because gas may enter the water by diffusion.

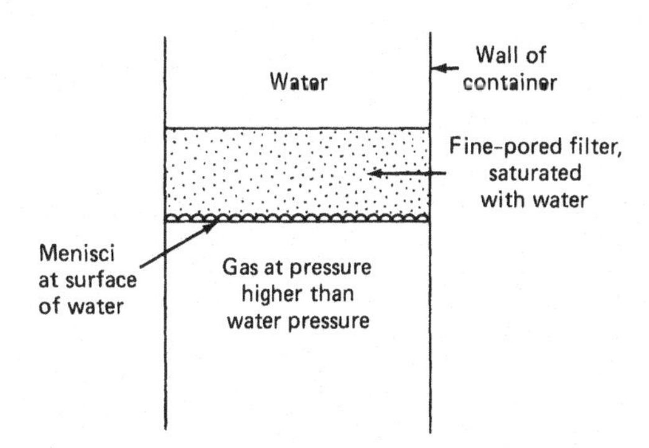

Figure 9.42. Separation of gas from water by a fine-pored filter.

Low air entry filters are coarse filters that readily allow passage of both gas and water. Typical low air entry filters have a pore diameter of 0.001–0.003 in. (0.02–0.08 mm, 20–80 microns). Typical air entry values range from 0.4 to 4.0 lb/in.^{2} (3–30 kPa). A filter with a pore diameter of 0.002 in. (50 microns) has a permeability to water of about 3×10^{-2} cm/sec.

High air entry filters are fine filters that can be used when piezometers are installed in unsaturated soil with the intent of measuring pore water pressure as opposed to pore gas pressure, in an attempt to keep gas out of the measuring system. Filters used for this purpose typically have a pore diameter of 4×10^{-5} in. (0.001 mm, 1 micron), an air entry value of at least 15 lb/in.^{2} (100 kPa), and a permeability to water of about 3×10^{-6} cm/sec.

9.12. RECOMMENDED INSTRUMENTS FOR MEASURING GROUNDWATER PRESSURE IN SATURATED SOIL AND ROCK

9.12.1. General

General guidelines on the selection of instruments are given in Chapter 4, and a recommendation is made to maximize reliability by using the simplest instrument that will achieve the purpose. Advantages and limitations of instruments for measuring groundwater level are summarized in Table 9.1.

Table 9.1. Instruments for Measuring Groundwater Pressure

Instrument Type	Advantages	Limitations[a]
Observation well (Figure 9.1)	Can be installed by drillers without participation of geotechnical personnel	Provides undesirable vertical connection between strata and is therefore often misleading; should rarely be used
Open standpipe piezometer (Figure 9.2)	Reliable Long successful performance record Self-de-airing if inside diameter of standpipe is adequate Integrity of seal can be checked after installation Can be converted to diaphragm piezometer Can be used for sampling groundwater Can be used to measure permeability	Long time lag Subject to damage by construction equipment and by vertical compression of soil around standpipe Extension of standpipe through embankment fill interrupts construction and causes inferior compaction Porous filter can plug owing to repeated water inflow and outflow Push-in versions subject to several potential errors: see Section 9.16
Twin-tube hydraulic piezometer (Figure 9.9)	Inaccessible components have no moving parts Reliable Long successful performance record When installed in fill, integrity can be checked after installation Piezometer cavity can be flushed Can be used to measure permeability	Application generally limited to long-term monitoring of pore water pressure in embankment dams Elaborate terminal arrangements needed Tubing must not be significantly above minimum piezometric elevation Periodic flushing may be required Attention to many details is necessary: see Appendix E
Pneumatic piezometer (Figure 9.14)	Short time lag Calibrated part of system accessible Minimum interference to construction: level of tubes and readout independent of level of tip No freezing problems	Attention must be paid to many details when making selection: see Section 8.3 Push-in version subject to several potential errors: see Section 9.16
Vibrating wire piezometer (Figure 9.16)	Easy to read Short time lag Minimum interference to construction: level of lead wires and readout independent of level of tip Lead wire effects minimal Can be used to read negative pore water pressures No freezing problems	Special manufacturing techniques required to minimize zero drift Need for lightning protection should be evaluated Push-in version subject to several potential errors: see Section 9.16
Unbonded electrical resistance piezometer (Figure 9.19)	Easy to read Short time lag Minimum interference to construction: level of lead wires and readout independent of level of tip Can be used to read negative pore water pressures No freezing problems Provides temperature measurement Some types suitable for dynamic measurements	Low electrical output Lead wire effects Errors caused by moisture and electrical connections are possible Need for lightning protection should be evaluated

Table 9.1. ***(Continued)***

Instrument Type	Advantages	Limitations[a]
Bonded electrical resistance piezometer (Figure 9.21)	Easy to read Short time lag Minimum interference to construction: level of lead wires and readout independent of level of tip Suitable for dynamic measurements Can be used to read negative pore water pressures No freezing problems	Low electrical output Lead wire effects Errors caused by moisture, temperature, and electrical connections are possible. Long-term stability uncertain Need for lightning protection should be evaluated Push-in version subject to several potential errors: see Section 9.16
Multipoint piezometer, with packers (e.g., Figures 9.36 and 9.37)	Provides detailed pressure–depth measurements Can be installed in horizontal or upward boreholes Other advantages depend on type of piezometer: see above in table	Limited number of measurement points Other limitations depend on type of piezometer: see above in table
Multipoint piezometer, surrounded with grout	Provides detailed pressure–depth measurements Simple installation procedure Other advantages depend on type of piezometer: see above in table	Limited number of measurement points Applicable only in uniform clay of known properties Difficult to ensure in-place grout of known properties Other limitations depend on type of piezometer: see above in table
Multipoint push-in piezometer	Provides detailed pressure–depth measurements Simple installation procedure Other advantages depend on type of piezometer: see above in table	Limited number of measurement points Subject to several potential errors: see Section 9.16 Other limitations depend on type of piezometer: see above in table
Multipoint piezometer, with movable probe (e.g., Figures 9.38 and 9.39)	Provides detailed pressure–depth measurements Unlimited number of measurement points Allows determination of permeability Calibrated part of system accessible Great depth capability Westbay Instruments system can be used for sampling groundwater and can be combined with inclinometer casing	Complex installation procedure Periodic manual readings only

[a] Diaphragm piezometer readings indicate the head above the piezometer, and the elevation of the piezometer must be measured or estimated if piezometric elevation is required. All diaphragm piezometers, except those provided with a vent to the atmosphere, are sensitive to barometric pressure changes.

Reliability and durability are often of greater importance than sensitivity and high accuracy. The fact that the actual head may be in error by 1 ft (300 mm) as a result of time lag may not matter in some cases provided the piezometer is functioning properly. If a malfunction occurs, it is of little importance that the apparent head can be recorded to 0.01 in. (0.25 mm). It is generally believed that if the instrument is installed correctly, that it is functioning, and that no time lag remains, the accuracy of all piezometers can be within 6 in. (150 mm) of water head. High-accuracy requirements of course necessitate selection of high-accuracy components, such as *test quality* pressure gages in pneumatic readout units.

9.12.2. Use of Observation Wells

As indicated in Section 9.1, applications for observation wells are very limited. They create a vertical connection between strata, and their **only** application is in continuously permeable ground in which groundwater pressure increases uniformly with depth. This condition can rarely be assumed. In the view of the author, observation wells are often misleading and should be used only when the groundwater regime is well known. If the groundwater regime is well known, measurements may be unnecessary: therefore, a practitioner must have a strong argument in favor of an observation well before selecting this option.

9.12.3. Use of Open Standpipe Piezometers

For measurements of pore or joint water pressure, an open standpipe piezometer is the first choice and should be used provided that the time lag and other limitations listed in Table 9.1 are acceptable. When these limitations are unacceptable, a choice must be made among the remaining piezometer types.

9.12.4. Short-Term Applications

Short-term applications are defined as applications that require reliable data for a period of a few years, for example, during the typical construction period. When open standpipe piezometers are unsuitable, the choice is generally between pneumatic and vibrating wire piezometers. The choice will depend on the factors in Table 9.1, on the user's own confidence in one or the other type, and on a comparison of cost of the total monitoring program.

When the economics of alternative piezometers are being evaluated, the **total** cost should be determined, considering costs of instrument procurement, calibration, installation, maintenance, monitoring, and data processing. The cost of the instrument itself is rarely the controlling factor and should never dominate the choice. Until recently, vibrating wire piezometers were preferable to pneumatic piezometers when automatic acquisition of data was required, but now both types can be read with automatic data acquisition systems, and signals can be transmitted using the various communication systems described in Chapter 8.

Twin-tube hydraulic piezometers are rarely chosen for short-term applications: their primary role is for long-term monitoring of pore water pressure in embankment dams. Resistance piezometers are the only type capable of monitoring pore water pressures that are varying with high dynamic frequency, such as during earthquakes or while piles are driven nearby. However, the author does not favor their use except under circumstances where frequent checks on zero reading and calibration can be made. They provide a convenient method for monitoring and recording pressures during short-term tests, such as pumping tests to determine permeability in situ, and in such cases frequent calibration presents no difficulties.

9.12.5. Long-Term Applications

For long-term applications, selection criteria are similar, but twin-tube hydraulic piezometers become an attractive option. General recommendations for long-term applications are given in Chapter 15. Recommendations specific to long-term monitoring of pore water pressure in embankment dams are given in Chapter 21.

9.12.6. Detailed Monitoring of Pressure–Depth Profile

When detailed monitoring of the pressure–depth profile is required, multipoint piezometers will normally be the instruments of choice. Comparative information is given in Table 9.1.

9.12.7. Filter Requirements

When selected for installation in saturated soil or rock, all piezometers should have low air entry filters. Open standpipe and twin-tube hydraulic piezometers that are sealed into boreholes below the water table require low air entry filters so that the seal can be checked after installation by performing permeability tests. The permeability of the filter should be at least ten times greater than the permeability of the ground. There is no benefit in using a high air entry filter on a diaphragm piezometer installed in saturated soil or rock and, if such a filter is used and not fully saturated, readings will be incorrect.

9.13. RECOMMENDED INSTRUMENTS FOR MEASURING PORE WATER PRESSURE IN UNSATURATED SOIL

If the pores in a soil contain both water and gas, the pressure in the gas will be greater than the pressure in the water (Chapter 2). In fine-grained soils the pressure difference can be substantial, and special techniques may be required to ensure measurement of pore water pressure rather than pore gas pressure. Applications for pore water pressure measurements in unsaturated soil are generally limited to two situations: first, measurements in compacted fills for embankment dams, and second, measurements in organic soil deposits, in which gas is generated as organic material decomposes. These applications are discussed in turn.

9.13.1 Measurements in Compacted Fills for Embankment Dams

When measurements of pore water pressure are required in compacted fills for embankment dams, piezometer selection criteria are similar to those given in Section 9.12. However, open standpipe piezometers are generally unsuitable for installation during construction, because they interfere with fill placement.

Twin-tube hydraulic piezometers and diaphragm piezometers should have high air entry filters to ensure that pore water pressure rather than pore gas pressure is measured.

Twin-tube hydraulic piezometers should have filters sized to minimize response time, and the size used in the Bishop piezometer (Figures 9.12 and 9.13 and Appendix E) is recommended. Penman (1960) has compared the measured response time of the filter used in the Bishop piezometer with that of a smaller filter in the form of a 2 in. (50 mm) diameter disk and reports on the significant advantage of the larger filter. Clearly, it is undesirable for a filter unit to present too small an area to the soil, but a diaphragm piezometer, because it requires a much smaller volume of water to operate its transducer, can tolerate a smaller contact area than a twin-tube hydraulic piezometer.

When required for installation in compacted fills for embankment dams, diaphragm piezometers should either have hollow cylindrical filters (e.g., the Telemac instrument shown in Figure 9.17) or filters emerging at the tapered nose of the piezometer (e.g., the Geonor instrument shown in Figure 9.34), so that the surface of the filter can be installed in intimate contact with the fill. Sherard (1980) reports on measurements with vibrating wire piezometers installed in fine clayey soil at Porto Colombia Dam in Brazil, where all piezometers read zero during a time when the true pressures were clearly subatmospheric. The piezometers, manufactured by Maihak, had disk-shaped filters on their ends. It is believed that, with this arrangement for the filter, it is difficult to avoid entrapping air between the filter and soil; therefore, subatmospheric pressures will not be monitored.

Despite use of saturated high air entry filters, the longevity of filter saturation is uncertain because gas may enter the filters by diffusion. The compacted fill in an embankment dam may remain unsaturated for a prolonged period after the reservoir is filled, and in fact the fill may never become permanently saturated by reservoir water. Increase of water pressure causes air to go into solution, and the air is then removed only when there is enough flow through the fill to bring in a supply of less saturated water. The pressure and time required to obtain saturation depend on the soil type, degree of compaction, and degree of initial saturation. Pore gas pressure may therefore remain significantly higher than pore water pressure for a substantial length of time, perhaps permanently. The vibrating wire piezometer shown in Figures 9.33 and 9.34 has a high air entry filter that is much longer than in other versions, so that filter saturation is maintained for a maximum time. However, even with this long filter, permanent saturation is not ensured.

Twin-tube hydraulic piezometers allow for flushing of the filter and cavity with de-aired liquid, thereby ensuring that pore **water** pressure continues to be measured, and are preferable to diaphragm piezometers when long-term measurements are required. A more detailed justification for this recommendation is given in Chapter 21.

As discussed in Section 9.15, no sand pocket should be placed around piezometers when they are installed in compacted fills for embankment dams.

9.13.2. Measurements in Organic Soil Deposits

When pore water pressure measurements are required in organic soil deposits, difficulties caused by the presence of gas are usually overcome by installing open standpipe piezometers fitted with low air entry filters and with standpipes of adequate diameter to ensure self-de-airing. The use of twin-tube hydraulic piezometers may be limited by a piezometric elevation that will be too far below the elevation of the tubing.

If the response time or other limitation of an open standpipe or twin-tube hydraulic piezometer is unacceptable, a diaphragm piezometer must be used. A short-term method entails use of a diaphragm piezometer with high air entry filter. The

Table 9.2. Filter and Saturation Requirements for Piezometers

Type of Piezometer	Method of Installation	Saturated Soil or Rock: Type of Filter[a]	Saturated Soil or Rock: Need for Saturation[b]	Unsaturated Soil: Type of Filter[a]	Unsaturated Soil: Need for Saturation[b]
Open standpipe	Push-in	Low	Yes	Low[d]	Yes
	Sealed within borehole[c]	Low	Yes[e]	Low[d]	Yes[e]
	Placed in fill	—	—	Low[d]	Yes[e]
Twin-tube hydraulic	Push-in	Low	Yes	High	Yes
	Sealed within borehole[c]	Low	Yes	High	Yes
	Placed in fill	—	—	High	Yes
Diaphragm	Push-in	Low	Yes[f]	High	Yes[f]
	Sealed within borehole[c]	Low	Yes[f]	High	Yes[f]
	Placed in fill	—	—	High	Yes[f]

[a] "Low" indicates low air entry. "High" indicates high air entry.
[b] "Yes" indicates that complete saturation is required, including any cavity between filter and diaphragm.
[c] Surrounded by zone of saturated sand.
[d] Minimum inside diameter of standpipe is 0.3 in. (8 mm).
[e] Saturation not required if permeability tests will not be made through the piezometer.
[f] See discussion in text on saturation of filters attached to diaphragm piezometers.

filter and the cavity between filter and diaphragm must be completely saturated before installation but, as described previously, the longevity of the fluid separation is uncertain. Piezometers with long high air entry filters have been made for this purpose, to prolong the time for gas to infiltrate. For longer-term applications, two hydraulic tubes can be connected to the cavity between the filter and the diaphragm (Section 9.8.9), to allow flushing any gas out of the cavity, using the procedure described in Appendix E.

9.14. SATURATION OF FILTERS

9.14.1. Reasons for Saturating Filters

Saturation of filters serves four major purposes. First, if the pores in the filter are filled with water, the chance of clogging is reduced. Second, if gas is removed from the filter the response time is minimized. Third, saturation allows permeability tests to be made through open standpipe and twin-tube hydraulic piezometers. Fourth, filter saturation is necessary if the filter is to separate water from gas in unsaturated soils.

9.14.2. Recommendations for Type of Filter and Need for Saturation

Recommendations for filter type and the need for saturation are summarized in Table 9.2.

Table 9.2 indicates that filters attached to diaphragm piezometers should always be saturated, but there is one exception to this rule. Saturation is not necessary for a diaphragm piezometer that is sealed within a borehole in saturated soil or rock if there is no interest either in minimizing the time after installation during which readings are perturbed by the installation process or in minimizing subsequent response time. Saturation of a diaphragm piezometer should include the cavity between the filter and the diaphragm and of course requires selection of a piezometer that **can** be saturated. Some diaphragm piezometers are available with permanently attached high air entry filters and are supplied by the manufacturers in a dry condition. Such piezometers will cause the measured pressure to be higher than the pore water pressure because of trapped air, and they should not be used.

9.14.3. Methods of Saturating Low Air Entry Filters

Low air entry filters can be saturated by forcing water through the pores.

A low air entry filter on an open standpipe piezometer can be saturated by immersing the filter in water and applying a suction to the standpipe, for example, with a bilge pump: when a piezometer is installed in a borehole, this can usually be done within the borehole. A low air entry filter on a twin-tube hydraulic piezometer can be saturated in the same way or the filter can be saturated before attachment to the tubing by plugging one tube and connecting a pressurized supply of water to the other tube. If a low air entry filter on a diaphragm piezometer can be removed from the piezometer, it can be saturated in a similar way. If the filter cannot be removed and saturation is required for minimizing response time, a vacuum should be applied to the dry filter (having determined from the manufacturer that the instrument will not be damaged by the vacuum) and the filter then allowed to flood with water as the vacuum is released. The cycle should be repeated two or three times until no gas bubbles appear from the filter as the vacuum is applied.

Filters should preferably be saturated at the site rather than before shipment. When filters are saturated before shipment and shipped in a sealed saturated container, there is a risk that they may be damaged if freezing occurs in transit. After saturation, filters should be stored under water. When transferring a saturated filter from one container to another or from a container to a water-filled borehole, a water-filled plastic bag or rubber sheath can be placed around the filter and later removed by tearing.

9.14.4. Methods of Saturating High Air Entry Filters

High air entry filters require a more intensive saturation procedure than low air entry filters. Some manufacturers recommend total immersion in water for 24 hours, but this is ineffective because air will not escape.

A high air entry filter for a twin-tube hydraulic piezometer can be saturated, as shown in Figure 9.43, by attaching short lengths of tubing and plac-

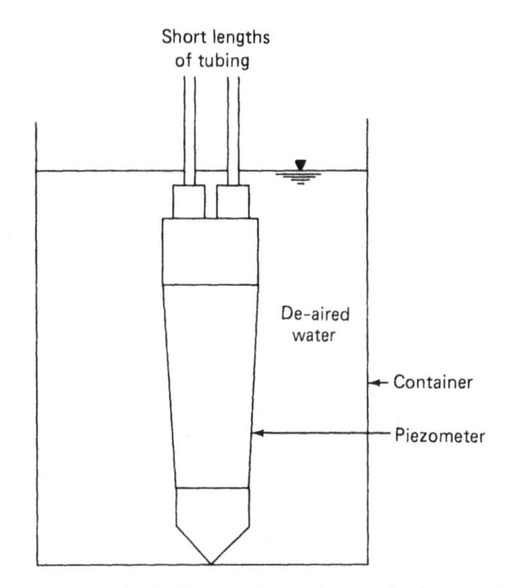

Figure 9.43. Method of saturating twin-tube hydraulic piezometer.

ing the dry filter in a container of de-aired water for 24 hours. The water will pass through the filter and drive out the air. Even though some air will enter the de-aired water through the water surface, the amount appears to be small during the 24 hour period provided that the container is not moved or the water stirred. Suction can be applied to the tubes to speed up the process, but this is usually unnecessary.

A high air entry filter for a diaphragm piezometer can be saturated by detaching it from the piezometer, placing the dry filter in a container, and applying a vacuum. The filter should then be allowed to flood with water, preferably warm de-aired water, allowing it to rise in the container so that it wets the lower side of the filter and rises slowly through it, thereby driving out the air. If the vacuum is high enough, the water will boil. The vacuum should then be released gradually and the filter kept submerged in the water for at least 1 hour. Attempts to saturate a high air entry filter by complete immersion in water and applying suction will often not be successful. Whatever the applied suction, air within the pores of the filter may be retained by the surrounding water and simply expand and contract as the pressure is lowered and raised.

There is no general rule about whether high air entry filters should be saturated at the manufacturer's facility or at the site. Because saturation requires significantly more care and effort than for low air entry filters, there is an argument favoring saturation at the manufacturer's facility. Bordes (1986) reports that, because of numerous problems in the field, Telemac saturates its filters in its workshop and, in addition, that freezing of submerged presaturated filters in a sealed container at −40°F (−40°C) has resulted in no damage. However, if appropriate facilities are available at the site, field saturation may be the more conservative approach.

9.15. INSTALLATION OF PIEZOMETERS IN FILL

As indicated in Section 9.6, there has been recent evidence that diaphragm piezometers that are installed in fill can show a reading change caused by total stresses acting on the piezometer body. Diaphragm piezometers that are to be installed in fill should therefore have thick-walled housings or an outer protective shell.

9.15.1. Compacted Clay Fill

A trench should be excavated to contain the piezometer tubes or cables, as described in Chapter 17, and piezometers are normally pushed into the wall of the trench. If the clay contains any material larger than 0.2 in. (5 mm), a pocket should be hand-excavated for each piezometer, alongside the trench, and backfilled with screened clay. The pocket is typically 18 × 18 × 30 in. (460 × 460 × 760 mm) deep. The backfill material should have a water content and density similar to the adjacent compacted clay and should be compacted with hand-operated equipment in 4 in. (100 mm) layers.

Because piezometers with high air entry filters will be used, it is essential that a good contact be obtained between the filter and the embankment material. Even a small thickness of loose, poorly compacted soil adjacent to the filter can prevent the piezometer from functioning correctly. Piezometers with conical filters, such as Bishop twin-tube hydraulic tips, are normally installed by pushing a mandrel approximately 12 in. (300 mm) into the clay soil and pressing the piezometer into the hole. The mandrel is exactly the same size as the piezometer and thus the piezometer makes close contact with the soil. As discussed in Section 9.13, diaphragm piezometers should have either hollow cylindrical filters as shown in Figure 9.17 or filters emerging at the tapered nose of the piezometer as shown in Figure 9.34. They should be installed by forming a cylindrical hole in the side of the trench, with diameter slightly smaller than the piezometer body, and forcing the piezometer into the hole. Smearing the filter with a thin coat of saturated soil paste, mixed to a consistency near the liquid limit, is probably desirable to aid in ensuring continuity between the water in the pores of the filter and the pore water in the embankment material (Sherard, 1981). The length of the hole above the piezometer is backfilled with tamped layers of embankment material, after removal of any larger stones.

No sand pocket should be allowed around the piezometer: sand would create a reservoir of air.

Guidelines for installation of piezometer tubing and cables are given in Chapter 17.

9.15.2. Granular Fill

Piezometers in granular fill are normally installed by excavating pockets alongside the trench, as described above for compacted clay, and placing the piezometer within backfill in the pocket. In rockfill or coarse granular material, it is necessary to place a zone of coarse clean sand around the piezometer to prevent damage. In rockfill, it may also be necessary to place a graded filter, from coarse gravel to sand, to prevent passage of material through the rockfill.

9.16. INSTALLATION OF PIEZOMETERS BY THE PUSH-IN METHOD

Three types of push-in open standpipe piezometer have been described in Section 9.8: wellpoint type, Geonor type, and Cambridge type. Pneumatic, vibrating wire, and bonded electrical resistance strain gage piezometers are also available as push-in versions. Multipoint push-in piezometers have been described in Section 9.9.3.

Installation of push-in piezometers is much less expensive than installation and sealing within a borehole, and therefore they tend to be used without recognition of their limitations. In addition to the limitations given in Sections 9.3–9.9 and summarized in Table 9.1, push-in piezometers are limited by the potential for an inadequate seal, for smearing or clogging, and for false readings caused by gas generation. These limitations are discussed in turn and installation guidelines are given.

9.16.1. Sealing

Sealing of a push-in piezometer requires an adequately intimate contact between the soil and pipe immediately above the piezometer.

The open standpipe configuration with wellpoint and steel pipe has a standard coupling immediately above the wellpoint, with a pipe of smaller outside diameter above the coupling. During installation, this coupling scores a hole of larger diameter than the outside of the steel pipe, and adequate sealing assumes that the soil will squeeze in to make intimate contact with the pipe. However, despite recommendations by AASHTO (1976) in favor of this procedure, the author believes that this may be an unwarranted assumption and that the configuration should not be used unless a length of pipe, with minimum diameter equal to the coupling diameter, is attached to the top of the wellpoint. This length of pipe will then be in contact with the soil for sealing purposes.

The drill rods attached to the Geonor field piezometer (Figure 9.23) have an outside diameter slightly larger than the outside diameter of the piezometer and have no protruding couplings; thus, they are in intimate contact with the soil. A minimum of 10 ft (3 m) of EX-size drill rod is generally used when the piezometer is installed in soft sensitive clays. Parry (1971) obtained good data by installing the Cambridge drive-in piezometer (Figures 9.24 and 9.25) in porous deposits underlying clay, using a 1 m (3 ft) sealing length of pipe. Torstensson (1984b) reports that, when using the BAT piezometer, a sealing length of 20 in. (500 mm) has been shown to be adequate for installations in clays, silts, and fine sands. The author has used a push-in open standpipe piezometer for falling head permeability measurements in the core of an embankment dam, with a 6 ft (2 m) sealing length of pipe. The core material was decomposed granite, a clay containing residual grains of angular coarse sand, and leakage occurred along the soil/pipe contact, presumably along channels created by the sand grains.

It is therefore evident that push-in piezometers are not appropriate for all soil conditions. No positive guidelines can be given on the length of the sealing pipe because the length depends on soil conditions, but it appears unwise to economize on the length: the longer the better. When a pipe surrounds the piezometer leads or standpipe, for use during pushing, connections in this pipe should also be sealed so that hydraulic short circuits are prevented. For example, when installing the Geonor field piezometer, all drill rod threads should be sealed. When an open standpipe piezometer is installed by the push-in method, the adequacy of the seal should be checked by allowing dissipation of any pore water pressure generated during installation and then performing a falling or constant head permeability test as described in Section 9.3.6.

9.16.2. Smearing and Clogging

Various approaches are taken to prevent excessive smearing or clogging of push-in piezometers during installation.

The Geonor field piezometer has a maximum pore size of 0.03 mm (0.001 in.) and is installed by first filling the standpipe totally with water and plugging the top, so that soil cannot enter the pores as the piezometer is pushed into place. Alternatively, water is pumped down the standpipe (a conventional garden spray tank, fitted with a pressure gage and flowmeter, is suitable) during all except the last 10 ft (3 m) of the push and a valve closed at the top of the standpipe for the remainder of the push. This alternative procedure requires that the pressure supply system is disconnected when each new length of pipe or rod is added during installation but ensures complete filling with water and minimum possibility of clogging. Pore water pressures are generated at the tip as the piezometer is pushed into clay and these must be allowed to dissipate before the standpipe plug is removed, to discourage clogging. The waiting time of course depends on the soil type: in the Norwegian soft sensitive marine clays for which the piezometer was developed, 24 hours is a typical waiting time.

The Cambridge drive-in piezometer is protected from smearing and clogging by use of the sliding driving shoe and by selection of an appropriate mesh size.

When a diaphragm piezometer is packaged within a push-in housing, the filter and cavity between the filter and diaphragm should be thoroughly saturated with water prior to installation, to prevent smearing and clogging.

9.16.3. Gas Generation

A further limitation of push-in piezometers results from the potential for gas generation if dissimilar metals are in contact.

DiBiagio (1977) reports that when piezometers were recovered after each use and reused at other sites, they sometimes provided good data and sometimes highly erratic data. The piezometers had vibrating wire transducers housed within stainless steel bodies and were attached to mild steel EX-size drill rod and installed by the push-in method. The drill rods remained in place, so that the mild and stainless steels were in contact. At sites where good measurements were obtained, the soil was quick clay, which has a low concentration of salt in the pores. At sites where erratic measurements were obtained, the clay had a high salt content in the pore water, such that the two dissimilar metals and the adjacent soil formed a galvanic cell (battery circuit), which can influence the measured pore water pressure in two ways. First, the corrosion process may include liberation of hydrogen gas at the stainless steel cathode (the piezometer). Second, the electric current generated by the corrosion process may have an electro-osmotic effect on the pore water. Both these effects tend to increase the pore water pressure around the stainless steel piezometer and also could account for fluctuations in measured pressures. A simple nylon bushing, inserted between the piezometer and the extension rods, was all that was needed to separate them electrically and eliminate the problem: this has been confirmed by numerous subsequent piezometer installations.

There is no reason to believe that this potential error is limited to vibrating wire piezometers. Presumably, the same problem could occur with any type of piezometer that is installed by the push-in method, if the push pipe is left in place.

9.16.4. Some Detailed Guidelines for Installation of Piezometers by the Push-in Method

Several details must be addressed when using push-in piezometers.

As a piezometer is pushed through soil, temporary changes in pore water pressure will generally occur around the piezometer. If a diaphragm piezometer is used, and these pore water pressures exceed the measuring range of the instrument, the piezometer may permanently be damaged. Piezometer readings should therefore be made continuously as the instrument is pushed and the rate of pushing controlled as appropriate. The feature shown in Figure 9.27, for checking the calibration of a diaphragm piezometer in-place, allows the application of a controlled backpressure during pushing: this type of piezometer can therefore often be installed more rapidly than a piezometer with a sealed cavity behind the diaphragm.

When installed in clays that are later subject to consolidation, the push pipe or rod will be subjected to downdrag forces, the piezometer tip will be pushed downward with respect to clay around it, and in extreme cases this may cause the piezometer tip to move from one stratum to another. It may be necessary to monitor the changing elevation by surveying methods on the top of the pipe and to interpret data accordingly. If the Cambridge drive-in piezometer is installed in consolidating clays, the downdrag forces will tend to close up the sliding arrangement at the tip, and the gap between the shoulder and driving shoe should be checked periodically by probing inside the standpipe and the driving shoe tapped further down if necessary.

When a piezometer is installed in very soft clay by the push-in method, the drill rod or pipe used for pushing may descend under its own weight after the piezometer is in place, and a steel plate should be clamped around the pipe to rest on the ground surface. This of course aggravates the problem just described, because the piezometer tip will then move downward by the full amount of vertical compression between tip and steel plate.

9.17. INSTALLATION OF PIEZOMETERS IN BOREHOLES IN SOIL

When it is not possible or satisfactory to install a piezometer below the ground surface by pushing or driving from the surface, a borehole method is required. The author is aware of seven methods that are in use for installing single piezometers in boreholes in soil, each of which is included in Table 9.3 and outlined in Sections 9.17.1–9.17.7.

The first method entails pushing from the bottom of a borehole, while the remaining six entail placement of the piezometer within a sand pocket and sealing above with bentonite pellets and/or grout. Recommendations and various practical details are given after the methods have been outlined.

9.17.1. Method 1. Push-in Below Bottom of Borehole

Installations in very soft to stiff clay can be made by using a variation of the push-in method, employing the instruments described in Section 9.8.7. A borehole is advanced to a short distance above the piezometer elevation and the piezometer pushed into place below the bottom of the borehole.

A borehole is drilled to the planned elevation of the top of the mandrel housing (Figure 9.31), the piezometer saturated, the tubing prespiraled as described in Chapter 17, and a mandrel inserted into the housing. The mandrel consists of a length of

Table 9.3. Methods for Installing Single Piezometers in Boreholes in Soil

Method	Method of Borehole Support	Casing With-drawn	Ability to Take Soil Sample	Sand Around Piezometer	Sealing Method	Advantages	Limitations
1. Push-in below bottom of borehole	Casing or bio-degradable drilling mud	Yes	Yes	No	Grout and contact between soil and mandrel housing	Rapid installation method	Possible only in relatively soft soils Requires special piezometer Precautions necessary to prevent excessive smearing or clogging
2. Bentonite pellet seal, casing with-drawn	Casing	Yes	Yes	Yes	Pellets and grout	High-quality seal	Requires significant time for installation Great care needed to clean inside of casing Spiraled tubing or cable cannot be used
3. Bentonite pellet seal, casing left in place (AASHTO method)	Casing	No	Yes	Yes	Pellets and casing/soil contact	None	Casing not recovered Relies on seal between soil and outside of casing Great care needed to clean inside of casing Risk of damage to piezometer if casing is dragged down Problems often caused by sticky bentonite pellets
4. Bentonite pellet seal, casing left in place (Piezometer R&D method)	Casing	No	No	Yes	Pellets and casing/soil contact	Use of bottom plug keeps casing wall clean and minimizes bridging problems Rapid installation method	Casing not recovered Relies on seal between soil and outside of casing
5. Grout seal, casing with-drawn	Casing	Yes	Yes	Yes	Pellets and grout	Rapid installation method	Requires skill to design grout mix Great care needed to clean inside of casing Layer of bentonite pellets recommended over sand to avoid contamination by grout Grouting pressure must be minimized
6. Bentonite pellet or grout seal, using hollow-stem augers	Augers	N/A	Yes	Yes	Pellets and grout	Convenient method if augers are more readily available than casing	Grout seal has same limitations as Method 5 Great care needed to clean inside of augers Requires large volume of backfill material Augers must not be with-drawn by rotation
7. Bentonite pellet or grout seal, using bio-degradable drilling mud	Biodegradable drilling mud	N/A	Yes	Yes	Pellets and grout	Convenient method if local practice favors use of biodegradable drilling mud for site investigation borings	Requires skill and experience to avoid borehole collapse and ensure enzyme breakdown Grout seal has same limitations as Method 5 Difficult to ensure that bentonite seal is in place if poured into mud-filled borehole Increases response time of diaphragm piezometers

appropriately sized drill rod, with a slot milled out of one side to allow passage of the tubing. A nylon line is attached to the top of the housing with a hose clamp and used to hold the piezometer on to the mandrel as more drill rods are added and the piezometer is lowered to the bottom of the borehole. On contacting the bottom of the borehole, the piezometer is pushed into place using the mandrel, the borehole filled with a cement/bentonite grout pumped down the drill rods as they are withdrawn, and the borehole topped up with grout. The borehole may be either cased temporarily, withdrawing casing without rotation, or supported by biodegradable drilling mud (Section 9.17.7).

9.17.2. Method 2. Bentonite Pellet Seal, Casing Withdrawn

The steps for this method are as follows:

1. Drive the casing, without a drive shoe, to 1 ft (300 mm) below the elevation of the bottom of the piezometer, taking a soil sample at the piezometer elevation, and wash until the water runs clear. The washing is extremely important because, if clay adheres to the inner wall of the casing, backfill materials will stick to the casing and form a plug. The *reverse circulation method,* described in Section 9.17.8, should be used.
2. Pull the casing 6 in. (150 mm) and pour in enough sand to fill the borehole below the bottom of the casing. Check the borehole depth.
3. Repeat step 2. Lower the piezometer to the top of the sand.
4. Pull the casing 6 in. (150 mm) and pour sand to fill the borehole below the bottom of the casing. Repeat until the sand and casing are 1 ft (300 mm) above the top of the piezometer, checking the depth after each pour.
5. Pull the casing 6 in. (150 mm) and pour compressed dry bentonite pellets to fill the borehole below the bottom of the casing. Repeat until a 4 ft (1.2 m) layer of pellets is in place, checking the depth after each pour.
6. Fill the casing with cement/bentonite grout, withdraw all casing, without rotation, and top up the borehole with grout.

9.17.3. Method 3. AASHTO Method: Bentonite Pellet Seal, Casing Left in Place

The method is described by AASHTO (1984) and is based on the procedure described by Casagrande (1949, 1958). For the reasons given later in this section and in Section 9.17.8, this method is not recommended for use but is included here for completeness. The steps for this method are as follows:

1. Drive 2 in. (50 mm) diameter casing, without a drive shoe and without a coupling in the lowest 10 ft (3 m), to 1 ft (300 mm) below the elevation of the bottom of the piezometer, taking a soil sample at the piezometer elevation. Do not wash in advance of the casing as the last 10 ft (3 m) are driven. After driving is complete, wash until water runs clear, using the *reverse circulation method.*
2. Pull the casing 1 ft (300 mm) while pouring in enough sand to fill the borehole below the bottom of the casing. Check the borehole depth. Lower the piezometer to the top of the sand.
3. If less than 3 ft (900 mm) of settlement is anticipated, pull the casing 1 ft (300 mm) while pouring in enough sand to fill to a level 30 in. (760 mm) above the bottom of the casing. Check the borehole depth. If more than 3 ft (900 mm) of settlement is anticipated, the casing is not pulled during this step.
4. Pour in a 1 in. (25 mm) layer of 0.5 in. (13 mm) rounded pebbles and tamp.
5. Pour in a 3 in. (76 mm) layer of bentonite pellets, followed by a 1 in. (25 mm) layer of pebbles, and tamp.
6. Repeat step 5 four times, to form a five-layer seal.
7. Pour in a 2 ft (600 mm) layer of sand, cover with pebbles, and tamp.
8. Repeat steps 5 and 6 to form a second five-layer seal.
9. Fill at least 10 ft (3 m) of the casing with sand.

The method suffers from several disadvantages. The cost of each installation includes the cost of casing left in place. The method relies on the adequacy of the seal between the soil and a 10 ft (3 m) length of casing. Although this is no doubt justified for installations in soft clay as described by Casagrande (1949), there may be doubts about adequacy in other soils. If the bottom of the casing is above the piezometer tip and the natural ground above the tip is subjected to vertical compression, the casing will be dragged down with respect to the tip, possibly resulting in damage to the installation. Blocked standpipes have been reported on several projects where significant vertical compression has occurred, indicating kinked standpipes. The provisions of step 3 are supposed to overcome this problem, but the author believes that the casing should not be pulled during step 3 even when as little as 6 in. (150 mm) of settlement is anticipated.

The method was developed before the advent of compressed dry bentonite pellets, and bentonite was prepared by mixing powdered bentonite with water to the consistency of putty, and rolling into small balls. Tamping was necessary to ensure a good seal, and pebbles were needed over each layer of bentonite to prevent bentonite from sticking to the tamping hammer, but use of pebbles risks damage to the piezometer leads while tamping. As discussed later, compressed dry bentonite pellets do not need tamping and do not need a layer of pebbles.

9.17.4. Method 4. Piezometer R&D Method: Bentonite Pellet Seal, Casing Left in Place

A variation on Method 3 is described by Piezometer R&D (1968, 1983). The major variations are as follows:

- Use of a temporary bottom plug in the casing, thus ensuring that the inside of the casing is clean. Sand can therefore be poured inside the casing before it is pulled, without risk of forming a "sand-grab." All casing ends are reamed to discourage bridging.
- Use of plastic centering fins around the piezometer to allow the piezometer to be installed half in and half out of the lower end of the casing, so that when casing is dragged downward as vertical compression occurs the piezometer and/or standpipe will not be damaged.
- Use of 2.5 in. (63 mm) instead of 2 in. (50 mm) casing.
- Use of compressed dry bentonite pellets, without pebbles.

The primary steps for this method are as follows:

1. Ensure that all casing ends are reamed and that the bottom section is 10 ft (3 m) long without threads on the bottom end. Insert a special steel and rubber plug in the bottom end. Drive casing to 2 ft (600 mm) below the bottom of the piezometer, adding water as each new section is connected.
2. Lower a tamping hammer to rest on the plug, and tap lightly. A solid "feel" indicates that no soil has bypassed the plug. Verify the casing depth. Pull the casing 3 in. (76 mm), and ensure that the plug leaves the casing by observing no upward movement of the tamping hammer cable. (If the plug follows the casing, it is likely that artesian conditions or excess pore water pressures exist, and the procedure described in Section 9.17.8 should be followed.) Remove the tamping hammer.
3. Pour in sufficient sand to fill 2 ft (600 mm) of the borehole.
4. With the casing water-filled, lower the piezometer, fitted with plastic centering fins, to rest on the sand.
5. Pour in sufficient sand to fill around the piezometer and 2 ft (600 mm) of borehole above the piezometer.
6. Pull the casing 2 ft (600 mm) plus half the length of the piezometer. The sand and piezometer will remain at their original elevation.
7. Pour in a 3 in. (76 mm) layer of bentonite pellets and tamp.
8. Repeat step 7 four times, to form a five-layer seal.
9. Pour in a 2 ft (600 mm) layer of sand.
10. Repeat steps 7 and 8 to form a second five-layer seal.
11. Fill at least 2 ft (600 mm) of the casing with sand. The remainder of the casing is not filled.
12. Install a rubber plug in the top of the casing, with the piezometer lead passing through a central hole and clamped above the plug.

Use of this method overcomes most of the problems associated with bridging of backfill, because the inside of the casing is flush and clean. However, no soil sample can be taken, and use of the bottom plug creates a closed-ended driving method and concern for unwarranted soil disturbance. However, Piezometer R&D (1983) report on many tests to evaluate the effect of the plugs, by installing pairs of piezometers 5 ft (1.5 m) apart, one by the bottom plug method and one by the more conventional method of driving casing and washing out thoroughly. Falling head permeability tests and piezometer readings as fill was placed over the piezometers indicated no difference in behavior.

9.17.5. Method 5. Grout Seal, Casing Withdrawn

The method is described by USBR (1974) and is essentially the same as Method 2, except that grout replaces the bentonite pellets.

The Corps of Engineers (1971), Fetzer (1982), and USBR (1974) suggest various grout mixes: details are given in Chapter 17. Lambe (1959) describes use of a chemical grout and, although the grout used by Lambe is no longer acceptable for environmental reasons, other chemical grouts are a viable option provided that the advice of grouting specialists is obtained and that the chemicals have no adverse effects on piezometer materials. Lambe recommends a thin layer of bentonite pellets above the sand to prevent the grout from invading the sand around the piezometer tip, and in general a thin grout should be avoided for this reason. A general and significant concern with any grout seal is the possibility of grout mixing with the sand around the piezometer as it flows under pressure from the end of the tremie pipe, and sand has on occasion been washed upward and out of the borehole. Grout pressure should therefore be minimized, and a grout pipe with a bottom plug and several side ports is preferable to an open-ended grout pipe.

Vaughan (1969) indicates that in relatively permeable material it is not difficult to select a grout with a lower permeability than the surrounding material, but in an impermeable material this may not be possible. For example, the permeability of a pumpable cement/bentonite grout may be on the order of 5×10^{-8} cm/sec, whereas the in situ permeability of an overconsolidated clay may be as low as 2×10^{-9} cm/sec. However, Vaughan describes a method for calculating the reading error when the grout has higher permeability than the ground and indicates that the error is minor if the two differ by one order of magnitude.

Whenever one is designing a grout mix for sealing piezometers in boreholes, the recommendations given in Chapter 17 should be followed.

9.17.6. Method 6. Bentonite Pellet or Grout Seal, Using Hollow-Stem Augers

The method is essentially the same as either Method 2 or 5, depending on the selection of sealing material, except that hollow-stem augers are used instead of drill casing.

9.17.7. Method 7. Bentonite Pellet or Grout Seal, Using Biodegradable Drilling Mud

Biodegradable mud* is composed of an organic polymer (no bentonite) that self-destructs through enzyme breakdown with time, leaving only the water used to mix the mud in the borehole. This type of drilling mud provides the necessary hole stabilization and cutting removal during drilling and then, upon reverting to water, leaves the borehole free of the organic polymer that was originally added to the water, providing a suitable environment for groundwater pressure measurement.

The primary steps for this method are as follows:

1. Mix the biodegradable mud with water until it is free from lumps.
2. Advance a borehole to 1 ft (300 mm) below the elevation of the piezometer.
3. Flush the borehole with "clean" mud, to remove any clay that has mixed with the mud while advancing the borehole. This clean mud should be as thin as possible, so that placement of backfill material is not obstructed, but not so thin that the borehole wall collapses. Clearly, this step requires judgment and experience.
4. Proceed with installing the sand, piezometer, and either a bentonite pellet or grout seal, generally as described above for Method 2 or 5.
5. If standpipe piezometers are used, "develop" the completed installation by clearing any filter cake of silt or clay from the borehole wall alongside the piezometer. Outward flow from the piezometer is usually not successful. The preferred method involves using compressed air to blow water out of the standpipe, allowing formation water to flow inward and break down the filter cake. If diaphragm piezometers are used, the filter cake cannot

*For example, Revert. See Appendix D.

be removed and hydrodynamic time lag is likely to be increased.

Where potable water supplies can be affected, a mud formulation that meets environmental protection standards must be used.

Biodegradable mud that is supplied in liquid form is easier to mix but is expensive and needs an additive to trigger breakdown. Quality control tends to be difficult, and this material is not recommended.

9.17.8. Recommendations for Installing Single Piezometers in Boreholes in Soil.

The author believes that, with the exception of Method 3, all of the above seven methods are viable. Each method may be the method of choice, the selection depending on the specifics of each application. Users are therefore advised to study the descriptions of the seven methods, Table 9.3, the many details included in Section 9.17.8, to answer the questions listed in Table 9.4, and to plan a method most suited to their particular case. The planning should include preparation of written step-by-step procedures, including a detailed listing of required materials and tools, following the guidelines given in Chapter 17.

Reservations about Method 3 stem from the possibility of damage to the piezometer when casing is dragged down, use of small-diameter casing and hand-rolled sticky pellets that aggravate the bridging problem, use of pebbles that may damage the piezometer leads, and reliance on the seal between the casing and soil. The author believes that Method 3 should not be used.

It appears that, if the soil is sufficiently clayey so that a seal can be assured between the soil and the outside of the casing and if the longer response time is acceptable, the push-in method using the Geonor field piezometer or Cambridge drive-in piezometer will perform as well as Method 4 and will be more economical. If adequacy of the seal is in doubt, none of these methods should be used.

Various practical details for installing piezometers in boreholes in soil are given below.

Role of Drillers and Specialist Personnel

Drilling crews should not be relied on to install piezometers in boreholes unless specialist personnel are intimately involved: the task should be a team effort.

Table 9.4. Some Questions to be Answered when Selecting a Method for Installing Single Piezometers in Boreholes in Soil

1. What is the soil type?
2. Can a piezometer be pushed from the bottom of the borehole, and is sufficient lead time available for procuring special piezometers?
3. Are there artesian conditions?
4. Are there excess pore water pressures?
5. What are the local drilling customs?
6. What drilling methods are available?
7. What skill/care/experience is available among personnel who will be responsible for installations?
8. Is adequate skill available for designing an appropriate grout mix?
9. How much vertical compression will occur in the soil above the piezometer?
10. Are soil samples required?
11. How can the borehole be supported?
12. Can the soil be relied on to seal against the outside of the casing?
13. How much time is available for installation?
14. How much time is available between completion of installation and the need to establish zero readings?
15. What are the requirements for response time?
16. What borehole diameter will be used?
17. How will the casing or augers be prepared?
18. How will the casing or augers be cleaned?
19. How will the casing or augers be backfilled?
20. How will the casing or augers be pulled?
21. If required, what will be the design of the sounding hammer?
22. Do bentonite pellets swell adequately, with or without a retarder?
23. What is the required waiting time for swell?
24. Is there a need to retard the swell and, if yes, what method will be used?
25. Do pellets behave satisfactorily when tested in flowing water?
26. Is one of the alternative bentonite seals preferable to compressed dry bentonite pellets?
27. Is there a need to form a seal below the piezometer?
28. If artesian conditions or excess pore water pressures exist, how will they affect the installation?
29. Which method of installation is most appropriate?

Borehole Diameter

Borehole diameter will be controlled primarily by the size of the piezometer or leads, the need for an annular space adequate for backfilling purposes, sampling needs, and the availability of drilling equipment. Boreholes drilled with hollow-stem augers may be as large as 10.2 in. (260 mm) in diameter, but casing as small as 2 in. (50 mm) has been used, for example, by Casagrande (1949), although this minimizes the space available for backfilling and is not recommended. A common borehole diameter in the United States is 3 in. (76 mm), and NW casing is often used.

Guidelines for Preparation and Cleaning of Casing and Augers

When casing is used with Method 2, 3, 4, or 5, all casing ends should have a flush inside diameter so that backfill material will not bridge. If standard pipe is used as casing, the inside wall will normally turn inward at the ends to form a sharp protrusion, and ends must be reamed.

When Method 2, 3, 5 or 6 is used, the casing or augers must be cleaned very thoroughly before sand is poured. If this is not done, sand and bentonite pellets are likely to stick to the casing or augers and prevent a good installation. The most effective method is to reverse the jetting pump and to use the jet pipe as the intake, with its lower end a few inches from the bottom of the borehole. The borehole is kept filled by pouring **clean** water in until all cloudy water is pumped out. This is referred to as the *reverse circulation method,* and in this way the velocity of the outgoing water is maximized.

Guidelines for Backfilling and Pulling Casing and Augers

General guidelines are given in Chapter 17.

When backfill is placed around piezometers as casing or augers are pulled, they should be pulled in short increments, to avoid collapse of the borehole prior to backfilling. They should be pulled without rotation, to avoid spiraling piezometer leads. The installation methods outlined above refer to increments of as little as 6 in. (150 mm), and this is a good starting point for each installation. However, experience with each installation will indicate whether larger increments are acceptable or whether smaller increments are necessary.

For all methods except Method 4, pulling of casing should precede backfilling with solid material, because if backfill is allowed to settle within the casing it may "grab" the piezometer leads, causing the piezometer to be lifted when casing is pulled. The same applies to the use of hollow-stem augers. An exception to this rule, at a site where borehole caving was of great concern, is described by Fetzer (1982). A standpipe piezometer with 3 in. (76 mm) diameter PVC flush jointed standpipe was installed through 6 in. (150 mm) casing. Sand was poured to a level as high as 3 ft (1 m) above the bottom of the casing before the casing was pulled, but the piezometer could be held in place by weighting the top. This would not be possible with a less robust standpipe or with another type of piezometer.

Volumes of solid backfill should be controlled by using a small container with a volume equal to a known depth of borehole, remembering to calculate required volumes using the **outside** diameter of casing or augers. Quantities of Ottawa sand and compressed bentonite pellets, required to fill various sizes of borehole, are given in Figure 17.9.

Sand should be saturated with water before pouring and should be poured slowly to avoid bridging in the borehole. Any washed and screened sand between U.S. standard sieve size #20 and #40 (approximately 0.034 and 0.017 in.; 0.9 and 0.4 mm) is suitable and is usually available from suppliers of sand blasting materials. Screening and washing concrete sand is usually impracticable. Ottawa sand is often used, but it should be noted that this term does not define the gradation, and gradation limits must be included in specifications. No sand finer than #40 sieve size should be included because this will take too long to settle and will mix with the sealing material. These gradation requirements are made in consideration of filter criteria and must be maintained for sand around open standpipe piezometers. However, the upper size limit can be relaxed for diaphragm piezometers, because no significant flow of water occurs. The lower size limit should be maintained to ensure that sand falls to the bottom of the borehole. Sand falls through water in the borehole at a rate of about 3 seconds per foot (10 seconds per meter).

Bentonite pellets should be poured very slowly, and the best way to avoid bridging is to drop them individually at a rate of about 3 per second and to allow enough time for their descent before measuring borehole depth. They fall at a rate of about 1 second per foot (3 seconds per meter). If given sufficient space to do so, compressed dry bentonite pellets will swell to between 10 and 15 times their original volume, and they should **not** be tamped. If they are tamped, the dry pellets may be forced into the soil beyond the walls of the borehole, where they are not required. Moreover, as discussed later in this section, if tamping causes the pellets to assume a denser packing, expansion of the pellets against the soil may create artificially high piezometer readings for a substantial time after installation. The rule is: *make sure that the pellets are where they should be, and leave them alone*. The adequacy of a seal created by this method can be demonstrated by placing one layer of pellets at the bottom of a glass filled with water, and observing how they swell and form a complete seal around the glass. No pebbles are needed over a layer of compressed bentonite pellets. It is absolutely essential that pellets are allowed to expand before overlying material is placed, otherwise overlying material may descend through the pellets and prevent a seal.

When pouring sand or bentonite pellets in the borehole, the piezometer leads should be kept taut to discourage bridging, and depth measurements should be made repeatedly to verify correctness of the backfill level.

When a grout seal is used (Method 5, 6, or 7), care must be taken to avoid contaminating the sand with grout. It appears worthwhile to place a 6 in. (150 mm) layer of bentonite pellets above the sand and to allow them to expand before grouting, for a time determined while testing pellets in water as described later in this section. If a stiff grout has been selected, with a slump of about 4 in. (100 mm) or less, the bentonite pellets are not needed. In either case, a grout pipe with a bottom plug and several side ports should be used, and care should be taken to pump grout slowly and to raise the grout pipe slowly.

Sounding Hammer

A *tamping hammer* is often used to tamp the bentonite seal and to center the piezometer leads or standpipe. In fact the tamping hammer serves two more purposes: to measure depths and to assist in dislodging any bridges of backfill material that may form in the borehole. As indicated above, compressed bentonite pellets should **not** be tamped. However, the device is required for the other three purposes: centering, depth measurement, and dislodging bridges. By persisting with the term *tamping hammer,* users are encouraged to use the device for tamping, and the author proposes the new term *sounding hammer*.

The need for centering the leads or standpipe within a bentonite pellet seal has been demonstrated during laboratory tests by Guarino (1985). When a single lead ran continuously along the inside wall of a pipe, filling the pipe with bentonite pellets and flooding with water did not result in a good seal.

A sounding hammer is required whenever placing sand and whenever using a bentonite pellet seal and is advisable for centering purposes when using a grout seal around all piezometer leads except exterior coupled standpipes.

When used around a single piezometer lead or a multiple-lead with a single jacket, the sounding hammer should be a steel cylinder with outside diameter 0.3–0.5 in. (8–13 mm) less than the inside diameter of the casing or augers, inside diameter approximately 0.2 in. (5 mm) larger than the outside diameter of the piezometer lead, a length of not less than 2 ft (600 mm), and a weight of approximately 15 lb (7 kg). The diameter limits are necessary so that backfill is forced to the bottom of the borehole, the length limit so that the hammer does not become angled and stuck in the borehole, and the weight limit for handling purposes. For small-diameter boreholes, the hammer can be made from solid steel, but usually it is necessary to weld one pipe inside another, with donut-shaped top and bottom plates, to keep the weight adequately low. If the required outside diameter cannot be achieved by using standard steel pipe, the next smaller pipe size can be used and the top and bottom plates made to protrude beyond the outside diameter by the required amount. All corners on the hammer should be thoroughly rounded. A bridle should be attached to the top of the hammer and a cable attached to the bridle with a secure smooth connection that will not damage the piezometer lead. A 0.125 in. (3 mm) stainless steel airplane cable, covered with a plastic sheath, is ideal for the bridle and cable. The cable should be graduated, using a hot stamping machine, with zero at the bottom of the hammer. Figure 9.44 shows a typical sounding hammer.

The arrangement shown in Figure 9.44 requires that the sounding hammer is inserted over the entire length of piezometer lead: a laborious task for long leads. This can be overcome by milling a slot along one side of the cylinder so that the lead can be inserted sideways. Removable top and bottom plates, each with a slot, are attached to the cylinder with their slots rotated 180 degrees with respect to the slot in the cylinder, so that they can also be inserted sideways but can retain the lead within the cylinder.

When a piezometer has two separate leads, for example, a pneumatic or twin-tube hydraulic piezometer for which a common tubing jacket has not

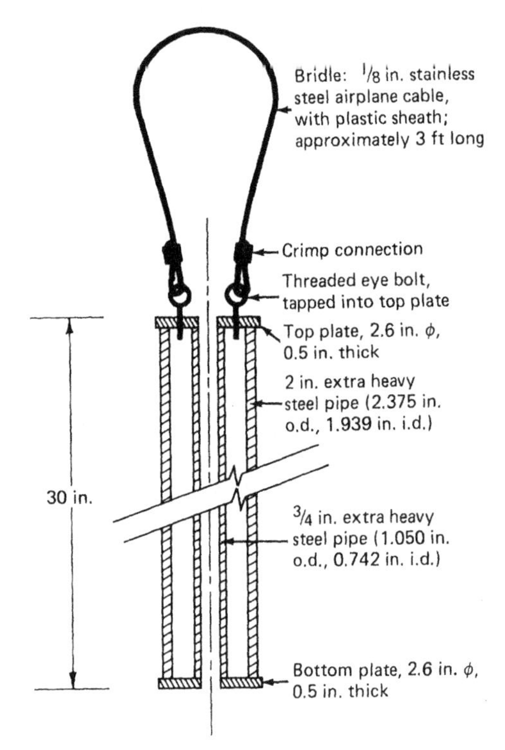

Figure 9.44. Typical sounding hammer for use inside NW flush-joint casing (casing i.d. 3.0 in.). Note: (1) plates and pipe welded together; (2) all corners well rounded; (3) weight approximately 15 lb (7 kg); (4) graduated cable attached to bridle with loop and crimp connection.

been used, the tubes must be separated by the sounding hammer. If the arrangement for a single lead is used, the two leads will be forced into close contact and a seal may not form along the contact. Separation is achieved by use of a bottom plate with two holes or by welding two separate small-diameter pipes within a larger pipe.

There is a temptation to use a sounding rod rather than a sounding hammer, because a rod is much easier to use. However, this practice is likely to push the standpipe or leads to one side of the borehole and may result in an inferior seal (Guarino, 1985). Bozozuk (1960) describes a sounding hammer consisting of an enlarged foot attached to pipe, and both foot and pipe are inserted over the piezometer lead. While a suitable method for shallow holes, the need for repeated coupling and uncoupling of pipes makes this a more laborious procedure.

Bentonite Pellets

Compressed dry bentonite pellets have been shown to form an adequate seal (Filho, 1976) provided that they can be inserted in the borehole at the required location. Typical pellets are shown in Figure 9.45.

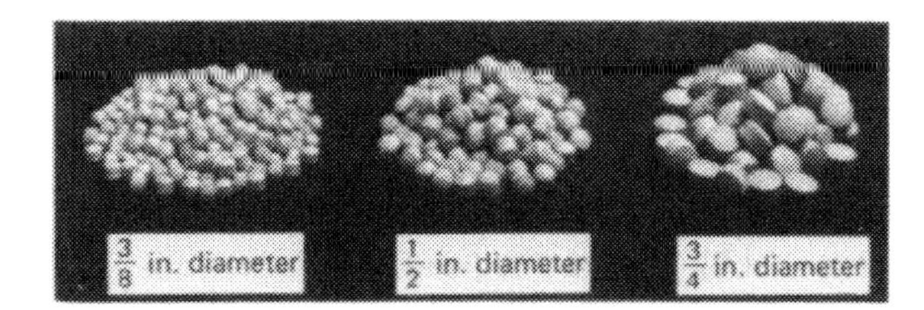

Figure 9.45. Bentonite pellets (courtesy of Piezometer Research & Development, Bridgeport, CT).

The swelling properties of bentonite pellets depend both on the properties of the constituents of the pellets and on the chemistry of the water in which they are immersed. Users should always verify that the pellets swell to at least five times their original volume when placed in a container of water for 24 hours. The water should be from the source that will be used in the field. During this test a judgment should be made on the length of waiting time required before overlying material can be placed without risk of the material descending past the bentonite. A typical waiting time is 30 minutes.

Most people who have installed piezometers in boreholes with bentonite pellets have experienced the problem of pellets becoming stuck part way down the borehole, and the problem is aggravated if pellets are poured too fast. Fetzer (1982) reports that, when installing an open standpipe piezometer with a 3 in. (76 mm) flush coupled standpipe within a 6 in. (150 mm) casing, pellets became lodged in the annular space. Some manufacturers add retarding agents to their pellets.* Users have tried a variety of methods for retarding the onset of swelling and stickiness. Brief soaking in diesel fuel, hydraulic oil, alcohol, or varnish has sometimes been successful but on other occasions has caused pellets to break apart or to become sticky very rapidly. Other methods include coating with hair spray, or soaking in a shellac solution and allowing to dry. Biodegradable drilling mud is also used: the powder is mixed with water, using no more than 0.1 lb per U.S. gallon (12 g/liter), which creates a very viscous liquid. The mixture is allowed to hydrate for at least 2 hours before briefly soaking the pellets and placing them in the borehole. Other users have tried freezing pellets before use. When there is a need to re-

*For example, Piezometer Research & Development "Non-stick" pellets.

tard the onset of swelling and stickiness, no unique procedure can be recommended for all pellets and all water, and users must therefore make their own tests.

When making such tests, pellets should be from the same shipment as those to be used in the field, because pellets do not always have uniform properties. The water should be from the field source. Although swelling can be demonstrated by placing pellets in a container of water, this is an inappropriate test for investigating the onset of swelling and stickiness when pellets are falling through water, because the action of flowing water has a marked effect on behavior. Piezometer Research & Development base their reported pellet behavior on tests in flowing water, and the author recommends this approach. A short length of transparent pipe, with diameter similar to the diameter of the borehole, is placed vertically and plugged at its lower end.* A water supply is connected to an opening in the plug, the pipe filled with water, pellets inserted at the top, and the water flow controlled so that pellets remain at about the midheight of the pipe. This arrangement is now a good model of field conditions, pellet behavior can be observed, and the effect of retarders can be evaluated. The time required to achieve a particular observed condition can be recorded and converted to an equivalent borehole depth by using the measured flow of water spilling from the top of the pipe.

If a retarder is employed, users should verify that pellets swell to at least five times their original volume when placed in a container of water for 24 hours.

Commercial sources of bentonite pellets are given in Appendix D.

Alternative Bentonite Seals

The above methods have referred to use of compressed dry bentonite pellets, installed by pouring into the borehole, but several other methods have been used, one of which may be the method of choice when planning an installation.

Casagrande (1949) describes use of bentonite balls of 0.5 in. (13 mm) diameter, formed by mixing powdered bentonite to a putty-like consistency, rolled in talcum powder to minimize sticking, and stored in glass jars to prevent drying. This procedure has in general been superseded by the advent

*The test apparatus is available from Piezometer Research & Development. See Appendix D.

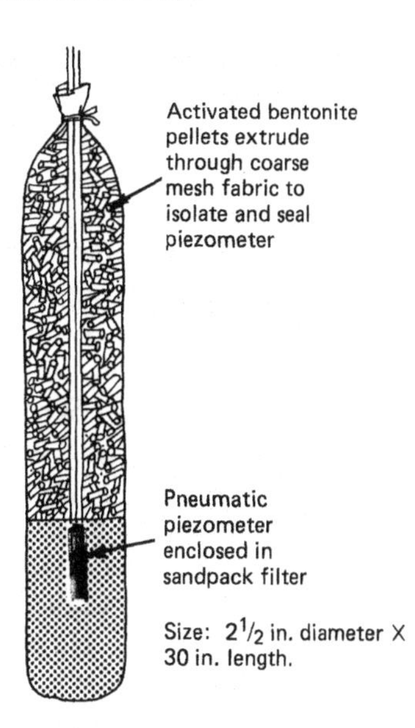

Figure 9.46. Piezometer with integral sand pocket and bentonite pellet seal (courtesy of Thor International, Inc., Seattle, WA).

of compressed dry bentonite pellets but may be used if the manufactured pellets are not available. However, as described in Section 9.17.3, tamping is needed, as also is a layer of pebbles to prevent sticking to the sounding hammer. Despite great care during installation, pellets tend to stick to the casing, and a heavy residue of bentonite tends to remain in the casing, obstructing the passage of pellets and reducing the quality of the seal. This method should be avoided wherever possible.

When piezometers are installed offshore or in other environments where groundwater is saline, conventional compressed bentonite pellets will not swell adequately unless special precautions are taken. Attempts to solve the problem with pellets made by compressing salt water bentonite appear to have been unsuccessful, and better success has been achieved by creating a local freshwater environment in which the conventional pellets can swell and remain swollen. This has been done successfully by using Method 4 and filling the casing with fresh water. Alternative approaches are to use a push-in method, Method 1, Method 5, 6, or 7 with a suitable grout, or a chemical sealant (Driscoll, 1986; Senger and Perpich, 1983).

Pellets can be prepackaged within a cylindrical bag as shown in Figure 9.46. The sand-filled part of the bag will normally be of canvas and the upper part of a coarser mesh. The arrangement must be sized to match the diameter of the borehole and the required lengths of sand and seal zone and substantially decreases the time and effort required for seal placement. As with all bentonite seals, sufficient waiting time must be allowed for swelling before placing overlying materials.

Pellets can be tremied to the bottom of the borehole by adding them to flowing water, using the method described in Chapter 17 for backfilling boreholes with sand or pea gravel, but with a larger-diameter pipe. A 1.5 in. (38 mm) Schedule 80 PVC pipe with flush threaded couplings is inserted within the borehole as the piezometer is lowered, and a 45 degree Y-branch is fitted to the pipe near its upper end. A water supply is connected to the branch and water circulated down the pipe until it spills out at the top of the borehole. Pellets are poured slowly into the top of the pipe and washed to the bottom of the borehole, and the pipe is gradually raised as backfilling progresses. The same tremie pipe arrangement can be used for placing sand.

Bentonite gravel (Appendix D) is often used in eastern Canada in preference to bentonite pellets. The gravel ranges in size from 0.3 to 0.8 in. (8–20 mm) and is significantly less expensive than compressed pellets.

Peters and Long (1981) describe the development of bentonite rings, installed with the aid of a modified tamping hammer around open standpipe piezometers in 6 in. (150 mm) diameter boreholes. Each bentonite ring was made by compacting bentonite at 75% water content in a 3.5 in. (90 mm) diameter by 6 in. (150 mm) long cylindrical mold and forcing a 0.8 in. (20 mm) rod through the center. Each ring was lowered around the standpipe with the special tamping hammer shown in Figure 9.47. To compact a ring, the hammer was lowered to the bentonite ring and the sliding steel weight raised and dropped. A series of four to six compacted rings constituted a seal. This method appears to be an attractive alternative to bentonite pellets but necessitates an effort by the user to prepare the rings and special hammer.

"Volclay sausages" are available from American Colloid Company (Appendix D), consisting of granular bentonite packaged within soluble plastic tubes 2 in. (50 mm) in diameter by 22 in. (560 mm) long. They are designed for curing loss of circulation in wireline drilling operations by using a downhole grouting technique. The author is not aware of their use for sealing piezometers, but smaller-diameter tubes appear to have a potential for this purpose.

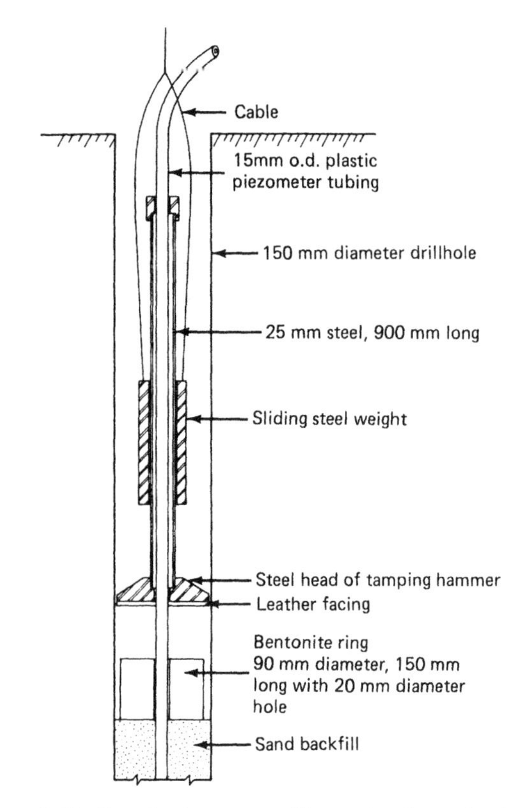

Figure 9.47. Tamping hammer and bentonite ring seal (after Peters and Long, 1981).

Chemical Sealant

A chemical sealant has been developed by the Dowell Division Laboratory of Dow Chemical Company, Tulsa, OK, for sealing groundwater monitoring wells where the groundwater is highly mineralized (Senger and Perpich, 1983). The sealant contains a polymer that swells in water and can be preformed by pouring into a mold of appropriate size and allowing it to set overnight to a consistency of soft rubber. Swelling time can be adjusted by changing the formulation of the sealant.

Dowell chemical sealant is used in the Waterloo Multilevel System (Figure 9.37) and is being considered as a borehole seal during studies relating to disposal of high-level nuclear waste.

This chemical sealant appears to have applications for sealing conventional piezometers in boreholes, although the author is not aware of practical experience to date. Because of the very substantial effort involved in ensuring that compressed dry bentonite pellets are placed in the borehole at the required location, the alternative of an easy-to-place chemical sealant is extremely attractive. Users are encouraged to conduct comparative tests between seals made with compressed dry bentonite pellets and with chemical sealant and to report their findings to the profession.

Change of Pore Water Pressure Caused by Swelling of Bentonite

The high-expansive property of bentonite pellets clearly has the potential for causing changes of pore water pressure. Depending on the soil conditions, confining conditions, and the bulk density of the pellets in place, there is a potential for an increase in pore water pressure as the pellets expand against the soil, but on the other hand there is a potential for a decrease in pore water pressure as the pellets hydrate and draw water from the pores.

Casagrande (1949) indicates that the purpose of the sand below, within, and above the bentonite seals in Method 3 is "to minimize the effect of swelling pressures of the overlying bentonite." The sand also increases frictional forces between backfill and casing, thereby discouraging a "blow-out" of the seal.

Binnie & Partners (1979) warn that "compressed bentonite pellets, which swell to many times their original size . . . can artificially reduce the pore pressures for long periods and should, therefore, only be used with caution." Guarino (1985) has made two sets of tests to examine this hypothesis. In the first set, a sand zone containing a piezometer was placed in the bottom of a closed-ended pipe, followed by a layer of pellets and water. Pressure was applied to the water and the piezometer monitored. In every case the piezometer reading decreased as hydration of the bentonite progressed. In the second set of tests, a piezometer was installed in a 12 in. (300 mm) diameter clay-filled pipe, in an attempt to model field conditions. A sand zone was provided around the piezometer and pellets above. The magnitude of pressure changes and the elapsed time before equilibrium pressure was established depended on the method of installation of the pellets. Use of a tamping hammer to pack the pellets in place caused excessively high readings of pore water pressure, in some cases for more than 1 month after installation. When pellets were poured into the "borehole" but not subsequently tamped, the piezometer readings decreased as hydration progressed, but equilibrium pressure was always established within 1 day. It is apparent from these tests that the rule is: *make sure that the pellets are where they should be, and leave them alone.*

In general, whenever bentonite pellets are used, seals should be installed well ahead of the need for measurements of pore water pressure so that any changes caused by the swelling of bentonite have had time to dissipate.

Sealing Below Piezometer

When a piezometer is installed at the bottom of a borehole, no seal is needed below the piezometer. However, if the borehole has been advanced below the elevation where measurements are required, a seal must be placed up to the bottom of the sand zone. Sealing options and procedures are the same as for the seal above the piezometer, but care must be taken to prevent sealing material from contacting the wall of the borehole where the sand zone will be placed, usually by protecting the zone with casing.

Installation Under Artesian Conditions

When a piezometer is to be installed under artesian conditions or where excess pore water pressures exist, arrangements must be made to place the seal while water is not flowing up the borehole and to overcome the problem of soil rising up inside the casing or augers. The most straightforward method is to extend the casing or augers to a level above the piezometric level and to counterbalance the flow with a static head of water, but this can only be done for a limited excess head. The author is aware of four methods for coping with the problem for larger excess heads.

First, in appropriate soil conditions a push-in piezometer can be installed.

Second, Method 4 can be used. Under artesian conditions the casing plug tends to follow the casing when the casing is first pulled. An extension is added to the casing, filled with water to minimize any flow of water around the casing plug, the lower sand, piezometer, and upper sand placed within the casing, the casing pulled to its final position, and the extension casing removed. The remainder of the procedure is as described in Section 9.17.4.

Third, Kinner and Dugan (1982, 1985) describe installation of piezometers offshore in sands and gravels in which the piezometric elevation ranged from 25 to 35 ft (8–11 m) above mean sea level. The arrangement is shown in Figure 9.48. The boreholes were advanced with heavy-weight conventional drilling mud. The wellpoint, riser pipe, and packer

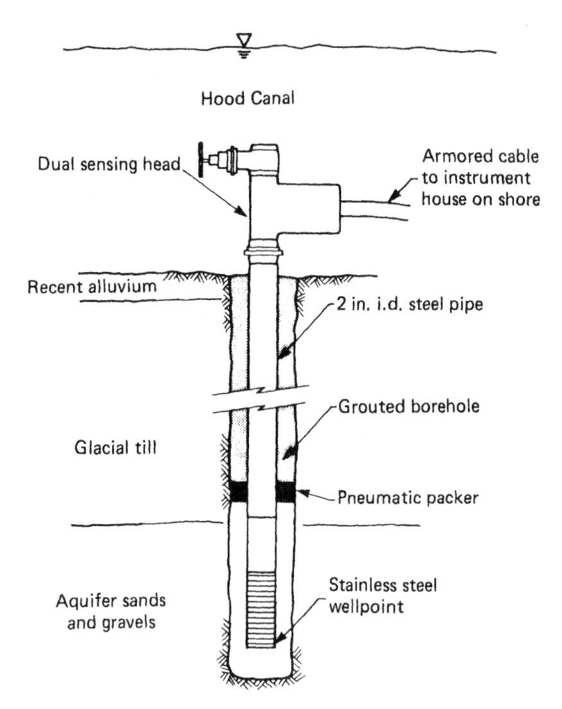

Figure 9.48. Piezometer installed in soil under artesian conditions (after Kinner and Dugan, 1982).

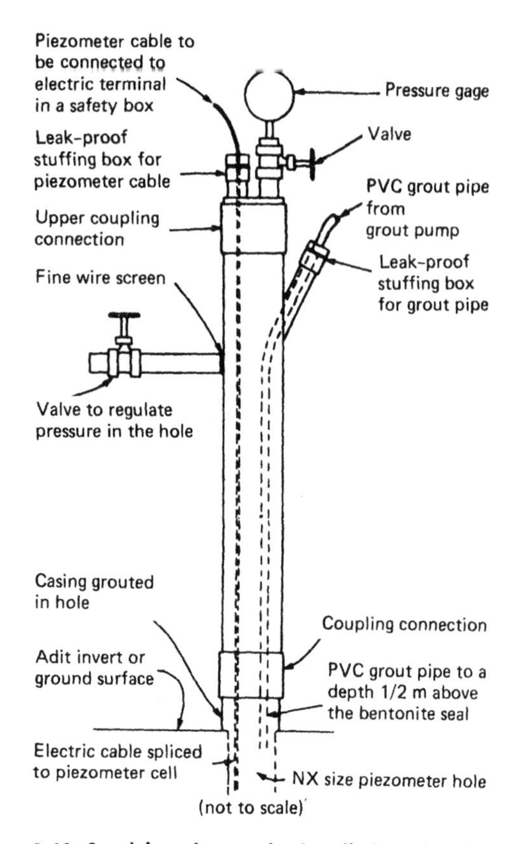

Figure 9.49. Special equipment for installation of a piezometer under artesian conditions (after Logani, 1983).

were inserted into the mud-filled borehole, the packer inflated, and the borehole grouted. After grout setup, the wellpoint and riser pipe were purged with water until artesian flow was generated, thereby cleansing the system of drilling mud. While the sensing head was being attached, the valve at the top of the assembly was opened to allow water flow and closed after making the connection. The dual sensing head contained both a pneumatic transducer and a single hydraulic tube and was packaged in this way to allow replacement of the head by a diver if damage or malfunction occurred. For a more conventional on-shore installation, a diaphragm piezometer could be set within the standpipe with an inflatable packer.

Fourth, Logani (1983) describes a method for installing piezometers under artesian conditions in rock, which could perhaps be adapted for use in soil provided that the borehole could be supported adequately during installation. The equipment shown in Figure 9.49, but without the grout pipe and piezometer cable, was attached to the borehole collar and the artesian pressure recorded. The pressure regulating valve was opened to allow free water flow and to release artesian pressure, and coarse sand was poured into the borehole by removing the upper coupling. The upper coupling was replaced and the pressure regulating valve closed, allowing the sand to settle under no-flow conditions. After again opening the valve and removing the upper coupling, the piezometer, placed within a saturated sand-filled pervious bag, was lowered into the borehole, followed by more sand, using the same procedure to allow the sand to settle. A 2 m (7 ft) layer of bentonite pellets was placed in increments in the same way and the grout pipe lowered within the borehole. The borehole was filled with a cement/bentonite grout under pressure to counter the artesian head. At this site the artesian pressure was as much as 40 ft (12 m) above the ground surface. Logani (1985) and Reyes (1985) both describe methods for installing piezometers under higher artesian pressures, by using a two-valve pressure loading chamber for introducing backfill material into the borehole under very low or no-flow conditions.

Some "Don'ts"

Never advance the borehole using conventional drilling mud (unless using the method shown in Figure 9.48 for an installation under artesian conditions), as this will seal the soil and inhibit piezometer response.

When casing is withdrawn after placing a bentonite seal, never fill the remainder of the borehole with sand or drilling cuttings. Always use a cement/bentonite grout. The practice of filling the remainder of the borehole with sand presumably has its origin in the procedure described by Casagrande (1949), in which the sand is bounded by casing. If the casing is withdrawn and sand is placed, conformance has been violated by creation of a pervious column that allows drainage of pore water pressure and may feed surface runoff into the zone around the piezometer.

When installing diaphragm piezometers, do not allow any damage to the outer sheath of the tubing or cable. Pneumatic piezometer tubing often consists of individual tubes with a common sheath, and in many cases water can travel between the tubes and the sheath. Electrical piezometers may have cable that allows "wicking" inside the sheath. If the sheath is damaged just above the piezometer, a pathway may be created for drainage of pore water pressure and/or for water ingress to the transducer.

Whenever planning to use a bentonite pellet seal, do not allow any subsurface couplings in plastic tubes, connections in electrical cables, or exterior couplings in steel or plastic pipes. Apart from creating weaknesses, these protrusions interrupt the passage of the pellets and sounding hammer.

9.17.9. Installation of More than One Piezometer in a Borehole in Soil

Adaptations of Methods for Installing Single Piezometers

It is generally agreed that if more than one piezometer is installed in a borehole in soil by adapting one of the seven methods described previously, seals may be inferior. Two piezometers can be installed successfully if great care is taken and if each piezometer has only a single lead, but more than two are not recommended. For this purpose multiple tubes encased within a common jacket are considered as single leads.

When two piezometers are installed in a borehole, clearly Methods 1, 3, and 4 are unsuitable. The lower piezometer is installed by using one of the remaining four methods, and a seal must then be formed up to a level just below the upper piezometer. Bentonite pellets can be used if the distance is not great, but more normally the intervening space is grouted, and it is generally necessary to allow the grout to set overnight. The upper piezometer is then installed in the same way, with special attention to the seal around and between the two leads. When a bentonite pellet seal is used, the sounding hammer must ensure separation between the two leads, as described in Section 9.17.8. When a grout seal is used, spacers should be placed between the leads to ensure that grout seals the intervening space: a short length of small-diameter PVC pipe can be taped every few feet to the lead for the upper piezometer. Because these spacers interfere with placement of a bentonite pellet layer between sand and grout, a low slump grout should be used to avoid contamination of the sand.

When more than one open standpipe piezometer is installed in a single borehole, acceptance testing should include a verification of seal integrity by adding water to each standpipe in turn and verifying that the others do not respond.

Installation of Multipoint Piezometers

Section 9.9 describes several methods for installing more than two piezometers in a borehole in soil, using multipoint piezometers. Methods include use of inflatable or chemical packers, piezometers surrounded with grout, push-in piezometers, and movable probes. Advantages and limitations of the methods are given in Table 9.1.

9.18. INSTALLATION OF PIEZOMETERS IN BOREHOLES IN ROCK

As indicated in Chapter 2, measurements of joint water pressure in rock must be made within zones of open jointing. The core should be examined carefully, and additional in-hole studies such as water pressure tests using packers, sonic velocity profiles, borehole camera photographs, and impression packers may be required. Piezometers must be installed within a section of the borehole of sufficient length to intersect several discontinuities, and the borehole should be flushed thoroughly with water to clean out the fractures prior to placing backfill.

9.18.1. Installation of Single Piezometer in a Borehole in Rock

If support is required for the borehole, a single piezometer can be installed by using Method 2, 5, or

7, as described previously for boreholes in soil, but with longer sand zones. Simplified versions of Methods 2 and 5 are suitable for installing a single piezometer when support is not required.

9.18.2. Installation of More than One Piezometer in a Borehole in Rock

Adaptations of Methods for Installing Single Piezometer

As described in Section 9.17.9 for boreholes in soil, two piezometers can be installed successfully in a borehole in rock by using Method 2, 5, or 7, with longer sand zones, if each piezometer has only a single lead and if great care is taken.

More than two piezometers have been installed by using alternate layers of granular backfill and grout. Although not acceptable in soil because seals beween three or more leads are likely to be inferior, in rock the interval between piezometers will tend to be long, allowing use of long grout seals and overcoming concerns for leakage along the seals. Two methods are possible.

First, if the borehole requires support, biodegradable drilling mud and an adaptation of Method 7 can be used, with a grout seal. An example of this approach is described by Deardorff et al. (1980), who installed three pneumatic piezometers per hole in 6.5 in. (165 mm) diameter boreholes up to 900 ft (275 m) deep. A single 1.5 in. (38 mm) diameter steel pipe was used both as a grout pipe and for circulation of drilling mud. Gravel was poured into the borehole to create the pervious zone around each piezometer, and the grout was stiff enough to provide a firm base to support successive gravel zones and grout without waiting for individual seals to harden. The writers indicate that grout penetration into the gravel was negligible provided that the gravel was given adequate time to settle through the drilling mud, and provided that care was taken to pump grout slowly and to raise the pipe slowly after completion of a grout stage. Deardorff et al. (1980) also describe installation of pneumatic piezometers in boreholes drilled downward at an angle of 40 degrees to the horizontal and upward at an angle of 45 degrees. Packers were used as plugs while grouting the upward holes.

Second, if the borehole does not require support, an adaptation of Method 5 can be used. An example of this approach is described by Lang (1983), who installed four open standpipe piezometers per hole in 1200 ft (365 m) deep boreholes drilled with HQ wireline equipment. A 0.5 in. (13 mm) diameter PVC pipe was used to deliver alternate layers of sand and grout down the borehole.

Both of the above procedures are painstaking and laborious, and use of a multipoint piezometer is likely to be more economical.

Installation of Multipoint Piezometers

Section 9.9 describes several methods for installing more than two piezometers in a borehole in rock, using multipoint piezometers. Methods include use of inflatable or chemical packers, piezometers surrounded with grout, and movable probes. Advantages and limitations of the methods are given in Table 9.1. In almost every case where more than two piezometers are required in a single borehole in rock, a multipoint piezometer system with inflatable or chemical packers, using either several single piezometers or a movable probe, is the system of choice.

9.18.3. Installation Under Artesian Conditions

When a single piezometer is to be installed in a borehole in rock under artesian conditions, it may be possible to set an inflatable packer in an aquaclude above the piezometer, thereby stopping the flow and allowing the borehole above the packer to be grouted. Alternatively, either the third or fourth method described in Section 9.17.8 (Kinner and Dugan, 1982, 1985; Logani, 1983, 1985; Reyes, 1985) can be used. Multipoint piezometers with inflatable packers can also be used in this application.

CHAPTER 10

MEASUREMENT OF TOTAL STRESS IN SOIL

10.1. INSTRUMENT CATEGORIES AND APPLICATIONS

Total stress measurements in soil fall into two basic categories: measurements within a soil mass and measurements at the face of a structural element. Instruments are referred to as *earth pressure cells, soil stress cells,* and *soil pressure cells,* and in this book the terms *embedment earth pressure cells* and *contact earth pressure cells* will be used for the two basic categories.

Embedment earth pressure cells are installed within fill, for example, to determine the distribution, magnitude, and direction of total stress within an embankment dam or within fill overlying a culvert. Applications for contact earth pressure cells include measurement of total stress against retaining walls, culverts, piles, and slurry walls and beneath shallow foundations.

The primary reasons for use of earth pressure cells are to confirm design assumptions and to provide information for the improvement of future designs; they are less commonly used for construction control or other reasons. When concerned with stresses acting on a structure during construction or after construction is complete, it is usually preferable to isolate a portion of the structure and to determine stresses by use of load cells and strain gages within the structure. For example, this approach has been used successfully in the determination of earth pressures in braced excavations from measurements of support loads.

Most earth pressure cells are designed to measure static or slowly varying stresses only. When cells are required for seismic or large-scale dynamic loading studies, they must be designed to have a sufficiently rapid response time.

10.2. EMBEDMENT EARTH PRESSURE CELLS

Attempts to measure total stress within a soil mass are plagued by errors resulting from poor conformance, because both the presence of the cell and the installation method generally create significant changes in the free-field stress. It is difficult and expensive to match the elastic modulus of the earth pressure cell to that of an individual soil. It is also very hard to place the cell under field conditions so that the material around the cell has the same modulus and density as the surrounding soil or rockfill and with both faces of the cell in intimate contact with the material. It is also very difficult and costly to perform a truly representative calibration in the laboratory to determine the cell response or calibration factor. Therefore, it is usually impossible to measure total stress with great accuracy.

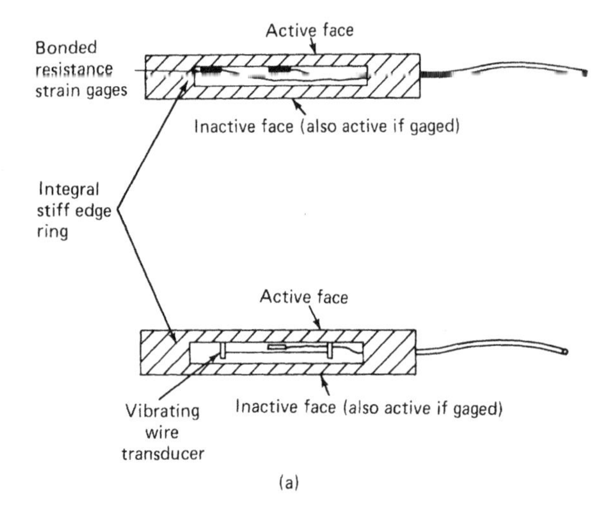

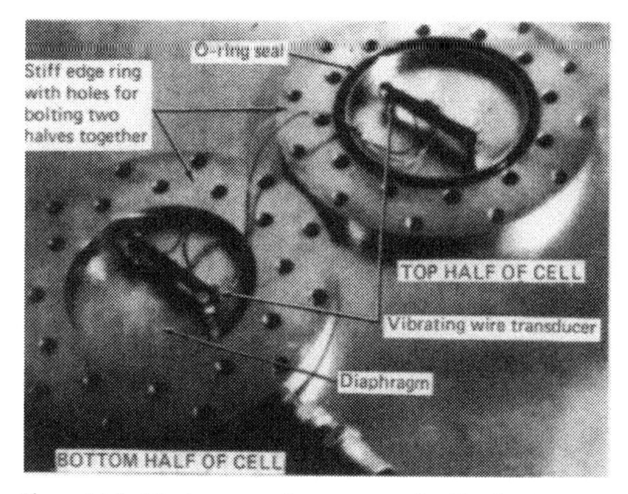

Figure 10.2. Diaphragm earth pressure cell with vibrating wire transducers (courtesy of Soil Instruments Ltd., Uckfield, England).

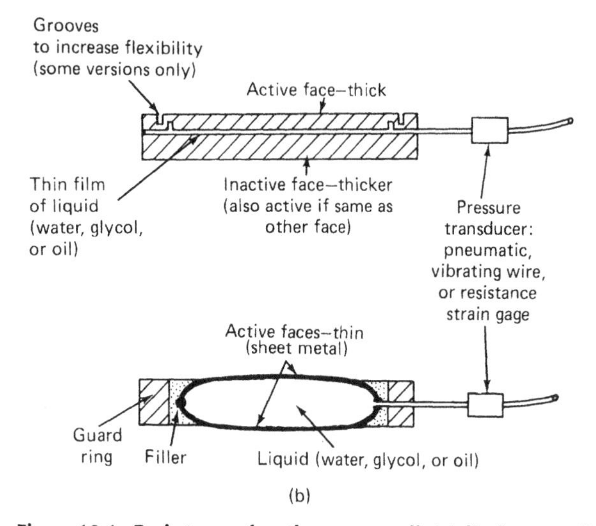

Figure 10.1. Basic types of earth pressure cell: (a) diaphragm cells and (b) hydraulic cells.

10.2.1. Types of Embedment Earth Pressure Cells

There are two basic types of embedment earth pressure cells: *diaphragm cells* and *hydraulic cells*. Examples of the two types are shown in Figure 10.1.

In the diaphragm type, a stiff circular membrane, fully supported by an integral stiff edge ring, is deflected by the external soil pressure. The deflection is sensed by an electrical resistance strain gage transducer bonded directly on the interior face of the cell or by a vibrating wire transducer, as shown in Figure 10.2. The vibrating wire transducer is usually mounted on two posts located at the points of contraflexure of the diaphragm, causing maximum rotation of the posts, magnifying the strain, and creating adequate sensitivity. A diaphragm cell may have one or two independent active faces. In the latter case, the two independent measurements provide an important check on the quality of the installation, in particular whether both faces are in similar contact with the surrounding soil. Thomas and Ward (1969) describe the design, construction, and performance of a diaphragm cell with two active faces and vibrating wire transducers. Induction coil transducers of the soil strain gage type (Chapter 8) have also been used in diaphragm cells.

The hydraulic type of cell consists of two circular or rectangular steel plates, welded together around their periphery, with liquid filling the intervening cavity and a length of high-pressure steel tubing connecting the cavity to a nearby pressure transducer. Total stress acting on the outside of the cell is balanced by an equal pressure induced in the internal liquid. It is essential that the cell is filled with de-aired liquid and that no gas bubbles are trapped within the cavity during filling.

Two versions of hydraulic cell are shown in Figure 10.1. In the first version, also shown in Figure 10.3, the active face is relatively thick (typically 0.1–0.25 in., 2.5–6 mm). Grooves are sometimes machined around the edge to increase edge flexibility, so that the active face tends to work as a piston. In addition, the layer of liquid is thin (0.02–0.08 in., 0.5–2 mm) so that the stiffness of the cell is high and more closely matches that of the surrounding soil, and the installed cell experiences minimal effects caused by thermal expansion and contraction of the liquid. In the second version of the hydraulic cell,

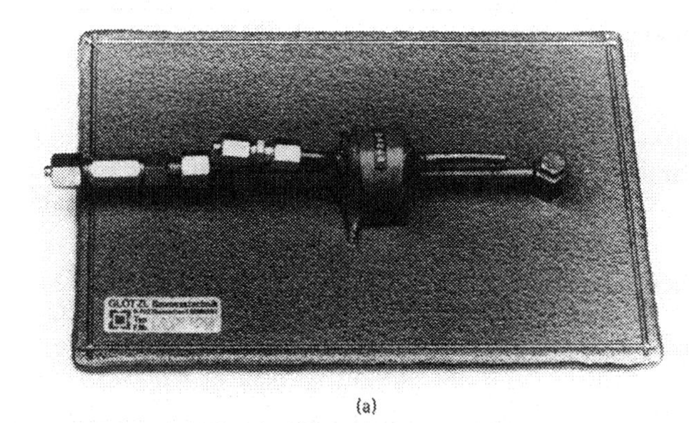

(a)

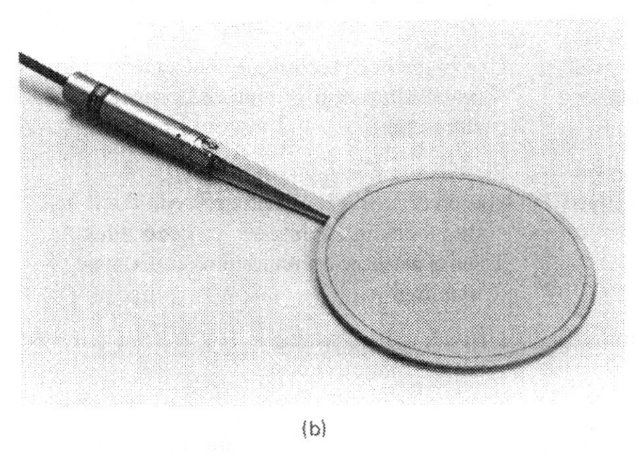

(b)

Figure 10.3. Hydraulic earth pressure cells with grooved thick active faces. (a) Cell with pneumatic transducer, (courtesy of Glötzl GmbH, Karlsruhe, West Germany and Geo Group, Inc., Wheaton, MD). (b) Cell with vibrating wire transducer (courtesy of Soil Instruments Ltd., Uckfield, England).

both faces are fabricated from relatively thin sheet metal, usually with a rolled edge. The layer of liquid is much thicker (0.1–0.4 in., 2.5–10 mm), so that the cell is less stiff. It is therefore subject to conformance errors in stiffer soils and is more susceptible to temperature effects. A guard ring may be provided to protect the thin sheet metal cell from radial edge loads. The active faces are usually domed outward, which makes installation against a prepared surface more difficult than a flat-faced cell.

Pressure transducer options for both versions of hydraulic cell are the same as for diaphragm piezometers: pneumatic, vibrating wire, and bonded or unbonded resistance strain gage type, with the advantages and limitations given in Chapter 8 and 9.

Comprehensive evaluations of various types of commercially available earth pressure cells, along with criteria for cell design and manufacture, are given by Brown (1977), Corbett et al. (1971), Hvorslev (1976), O'Rourke (1978), Reese et al. (1968), Selig (1964), State of California (1968, 1971), and Weiler and Kulhawy (1978, 1982). A report by the International Society for Rock Mechanics (ISRM, 1981c) includes a comprehensive description of the hydraulic type of cell with a pneumatic transducer, in addition to guidelines on installation, reading, calculation, and reporting procedures.

10.2.2. Limitations Imposed by Soil Environment

Measurement of total stress at a point within a soil mass requires the following:

1. An earth pressure cell that will not appreciably alter the state of stress within the soil mass because of its presence (conformance).
2. A large enough sensing area to average out local nonuniformities.
3. Minimum cell sensitivity to nonuniform bedding.
4. A method of installation that will not seriously change the state of stress.

The last requirement generally limits these measurements to fills and other artificial soil conditions. However, some success has been achieved in measuring horizontal stress in soft soils by pushing specially designed earth pressure cells downward into natural ground (e.g., Massarsch, 1975). Other instruments used for this purpose include the stepped blade (Handy et al., 1982), pressuremeters (Baguelin et al., 1972, 1977; Wroth and Hughes, 1973), and the flat plate dilatometer (Marchetti, 1980; Schmertmann, 1982). Earth pressure measurements have also been made in natural ground by hydraulic fracturing through hydraulic piezometers (e.g., Penman, 1975). Wroth (1975) discusses and compares the various methods. Attempts to measure stress in soil by advancing a large-diameter borehole, inserting earth pressure cells, and backfilling around the cells are generally subject to gross conformance errors.

10.2.3. Factors Affecting Measurements

Hvorslev (1976), Selig (1964), and Weiler and Kulhawy (1978, 1982) report on studies of factors affecting measurements with embedment earth pressure cells. Table 10.1 is based on a table presented by Weiler and Kulhawy (1982), with substantial re-

Table 10.1. Major Factors Affecting Measurements with Embedment Earth Pressure Cells

Factor	Description of Error	Correction Method[a]
Aspect ratio (ratio of cell thickness to diameter)	Cell thickness alters stress field around cell	Use relatively thin cells ($T/D < 1/10$)
Soil/cell stiffness ratio (ratio of soil stiffness to cell stiffness)	May cause cell to under- or overregister Error will change if soil stiffness changes	Design cell for high stiffness and use correction factor
Size of cell	Very small cells subject to scale effects and placement errors Very large cells difficult to install and subject to nonuniform bedding	Use intermediate size of cell: typically 9–12 in. (230–300 mm) diameter[b]
Stress–strain behavior of soil	Measurements influenced by confining conditions	Calibrate cell under near-usage conditions[b]
Placement effects	Physical placement and backfilling causes alteration of material properties and stress field around cell	Use placement technique that causes minimum alteration of material properties and stress field[b]
Eccentric, nonuniform, and point loads	Soil grain size too large for cell size used Nonuniform bedding causes nonuniform loading	Increase active diameter of cell[b] Use hydraulic cells with grooved thick active faces in preference to other types[b] Take great care to maximize uniformity of bedding[b]
Proximity of structures and other embedded instruments	Interaction of stress fields near instruments and structure causes errors	Use adequate spacing
Orientation of cell	Changing orientation while placing fill over cell causes reading change	Use placement methods that minimize orientation changes Attach tiltmeters to cell
Concentrations of normal stress at edges of cell	Causes cell to under- or overregister, depending on stiffness of cell relative to soil	For diaphragm cell, use inactive stiff edge ring to reduce sensitive area ($d/D \approx 0.6$)[b] For hydraulic cell, use grooved thick active face and thin layer of liquid
Deflection of active face	Excessive deflection of active face changes stress distribution around cell by arching	Design cell for low deflection: for diaphragm cell, diaphragm diameter/diaphragm deflection at center > 2000–5000; for hydraulic cell, use thin layer of liquid[b]
Placement stresses	Overstressing during soil compaction may permanently damage cell	Check cell and transducer design for yield strength (hydraulic cells with pneumatic transducers have high overload capacity)[b]
Corrosion and moisture	May cause failure of cell by attacking cell materials	Use appropriate materials and high-quality waterproofing[b]
Temperature	Temperature change causes change of cell reading	Design cell for minimum sensitivity to temperature; if significant temperature change is likely, measure temperature and apply correction factor determined during calibration[b]
Dynamic stress measurements	Response time, natural frequency, and inertia of cell cause errors	Use appropriate type of cell and transducer, together with dynamic calibration[b]

[a] D = cell diameter; T = cell thickness; d = diaphragm diameter.
[b] Applies also to contact earth pressure cells. See Section 10.3.2.

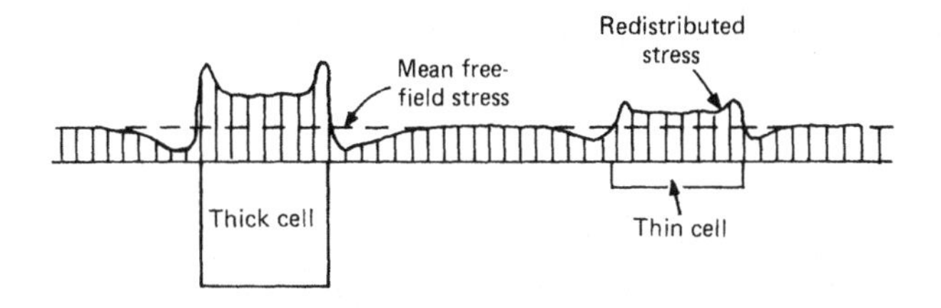

Cell stiffer than soil (overregistration)

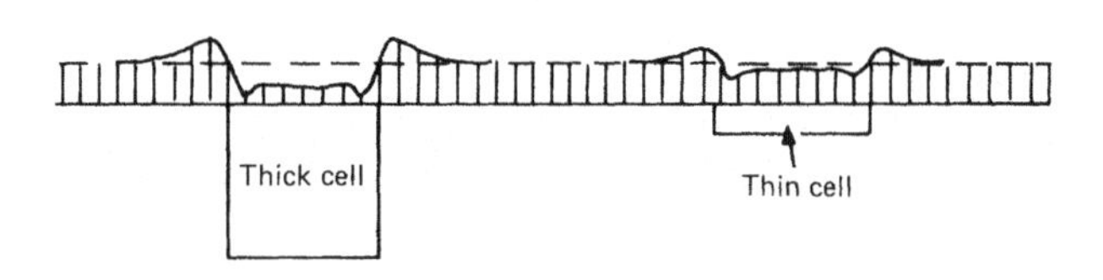

Cell less stiff than soil (underregistration)

Figure 10.4. Effect of embedment earth pressure cell aspect ratio and soil/cell stiffness ratio (after Selig, 1964).

visions by the author, and summarizes the major factors and correction methods. Various factors that affect measurements are discussed in the following subsections.

Aspect Ratio and Soil/Cell Stiffness Ratio

The effects of aspect ratio (ratio of cell thickness to diameter) and soil/cell stiffness ratio (ratio of soil stiffness to cell stiffness) are illustrated in Figure 10.4. The error resulting from these causes can be minimized by designing for high stiffness and an aspect ratio of less than 1 : 10. It can be seen from Figure 10.4 that if the transducer is mounted above the cell, the error will tend to be larger than if it is mounted at the end of a steel tube emerging through the rim of the cell as shown in Figure 10.3b. Mounting the transducer above the cell provides greater protection for the transducer, but makes compaction of soil around the cell more difficult.

Size of Cell

The typical diameter of cells available for field use is 9–12 in. (230–300 mm). Small cells of 2–3 in. (50–75 mm) diameter are available, but they are not recommended for general field use because scale effects and placement problems are likely to cause greater errors than for larger cells. Very large hydraulic cells (e.g., Alberro and Borbón, 1985; Sparrow, 1967), up to 3 or 4 ft (1 or 1.2 m) in diameter, have been used in limited numbers but, although they may provide measurements of the average stress over a large area and thus be more accurate than cells of conventional size, they are very expensive and awkward to handle and install, and it is very difficult to create a uniform bedding over such a large area.

Figure 10.5 shows stresses measured at Chicoasén Dam in Mexico by installing a group of embedment hydraulic earth pressure cells of different sizes and shapes. Sizes and shapes are identified on

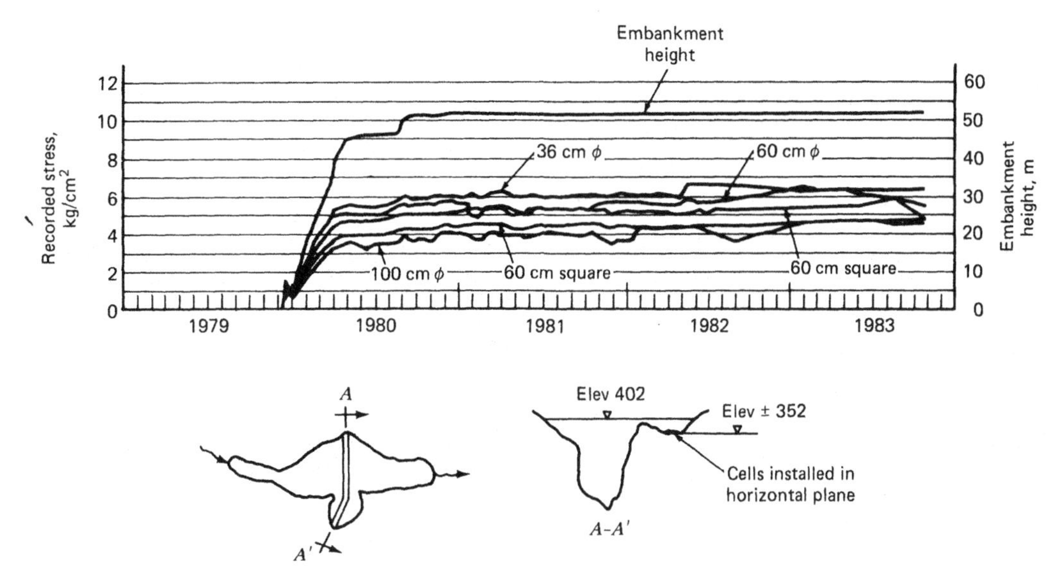

Figure 10.5. Measurements with embedment earth pressure cells installed in Chicoasén Dam (after Alberro and Borbón, 1985).

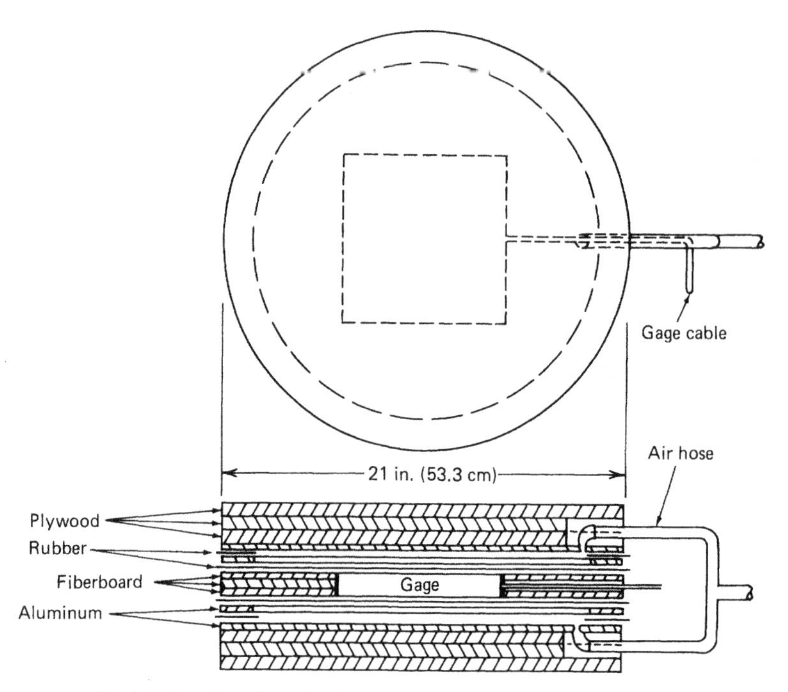

Figure 10.6. Assembled fluid calibration chamber for earth pressure cells (after Selig, 1980). Reprinted with permission from ASTM *Geotechnical Testing Journal*. Copyright ASTM, 1916 Race Street, Philadelphia, PA 19103.

the plots. The cells have thin active faces and guard rings, and the layer of liquid is 0.375 in. (9.5 mm) thick. The data show a scatter of about ±25% with respect to the mean value and generally indicate that the measured stress decreased with increasing size of cell.

Laboratory Calibrations

Each cell should be calibrated under fluid pressure to be sure that it is functioning correctly and not leaking, and most manufacturers of commercial earth pressure cells provide a calibration chart based on all-round fluid pressure loading, using air or water. Selig (1980) describes a simple and inexpensive calibration chamber for use with a laboratory load reaction frame or compression testing machine for performing all-round fluid pressure calibrations. The chamber is constructed of easily fabricated layers of metal, plywood, and rubber and assembled as shown in Figure 10.6.

Unless installations are to be made in soft clay, fluid pressure calibrations are insufficient. If measurement accuracy must be maximized, each cell should be calibrated in a large calibration chamber, using the soil in which it will be embedded. Cell design and soil placement details have a very substantial influence on measurements, and during laboratory calibrations it is most important that installation procedures represent the intended field methods as closely as possible. Calibration within large chambers is an expensive and difficult undertaking, and chambers that have been used are described by Alberro and Borbón (1985), Bozozuk (1970, 1972b), Hadala (1967), and Selig (1980). The chamber should be at least three times, and preferably five times, the diameter of the cell.

The chamber shown in Figure 10.7 was used by Hadala at the U.S. Army Corps of Engineers Waterways Experiment Station at Vicksburg, Mississippi, to calibrate 2 in. (50 mm) diameter cells in both sand and clay. The chamber would also be suitable for calibrating the 9–12 in. (230–300 mm) diameter cells that are more typically used in the field. Hadala reports on the effect of the placement method used. The cells satisfied the Table 10.1 recommendations for stiffness and diaphragm deflection, and the aspect ratio was 1:9, yet Hadala concludes the following:

- The placement method used has a definite influence on the mean over-registration ratio and the scatter of the data about the mean. In some cases, over-registration ratio changed as much as 40 per-

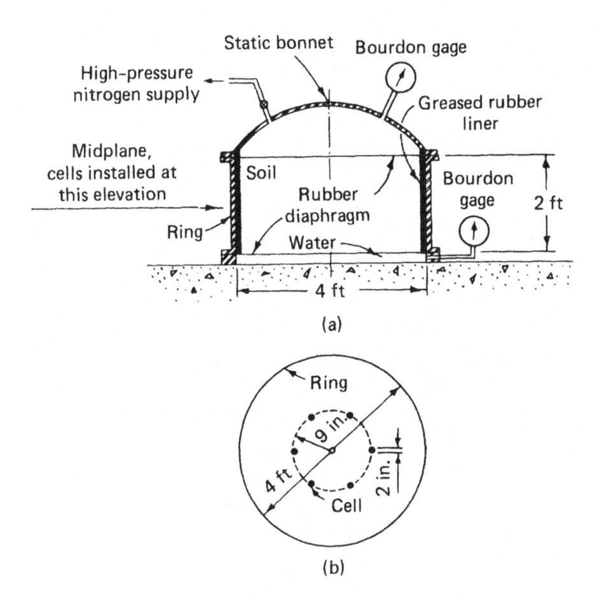

Figure 10.7. (a) Test chamber for calibration of embedment earth pressure cells and (b) array of cells at midplane (after Hadala, 1967).

cent due to placement method changes. In the case of scatter, it was noted that the simpler procedures resulted in less data scatter.

- Both over- and under-registration of this very stiff gage are possible, depending on the soil in which it is embedded and the way it is placed.
- The tests in clay generally exhibited lower over-registration ratios than those in sand.
- Of the sand test placement methods, the set-on-surface method [cells simply set on the surface, followed by normal construction procedures to complete the fill] was determined to be the best, while of the clay placement methods examined, the cut/no-cover method [cells placed in a shallow excavation of the same diameter and depth as the cell] was selected as the best.
- The data scatter noted in this study suggests that the use of a single soil stress gage to measure the magnitude of stress in a soil mass with any reasonable degree of confidence is a fruitless effort. When attempting to measure earth pressure magnitudes in virgin-loaded, compacted specimens similar to those used in this study, the average of at least three soil stress gage measurements should be used if 20 percent accuracy is required nine times out of ten.

It is likely that the small size of Hadala's cells aggravated placement effects, and it is believed that the effects are not so severe for more conventional field-sized cells.

Selig (1980) describes a chamber similar to Hadala's and provides practical guidelines on its use for calibrating field-sized cells. Bozozuk's chamber consisted of a heavily reinforced plywood box constructed in two sections, one fitting on top of the other. The cell was installed in the same plane as the junction between the two sections. As each loading increment was applied, the two parts of the box were jacked vertically apart to create a thin, continuous crack separating the two parts of the box. This crack forced all of the applied vertical load to be transmitted past the cell in grain-to-grain contact, since none of the load was lost in wall friction.

The Comision Federal de Electricidad at their Experimental Laboratories in Mexico City have constructed a large laboratory facility to test the response of embedment earth pressure cells to applied loads (Alberro and Borbón, 1985). The diameter of the test chamber is 3 m (10 ft) and the height 3.18 m (10.4 ft). Preliminary calibrations have been made to date in sand, and additional tests are planned to examine the effects of placement procedure, geometry and dimensions of the cells, and cell location and orientation within the chamber.

Field Placement Effects

The above comments on placement effects refer to laboratory calibrations, and field placement effects add an additional source of error that may well be of even greater magnitude. As discussed in Section 10.2.4, the accepted field installation procedure involves compacting fill with heavy equipment, installing the cells in an excavated trench, and backfilling around and over them by hand tamping or light machine. The probability is high that cells are therefore surrounded by a zone of soil with greater compressibility than the remainder of the fill, that imposed stresses are therefore redistributed by arching, and that substantial underregistration occurs.

Binnie et al. (1967) report that when diaphragm cells with two active faces and vibrating wire transducers were installed in washed gravel fill at Mangla Dam, no cell recorded more than half the calculated added vertical stress. Opposite faces generally recorded different pressures, even for initial readings. The initial difference persisted as fill was placed, and in some cases increased so that one face showed a pressure less than half that of the other face. The writers comment that the initial differ-

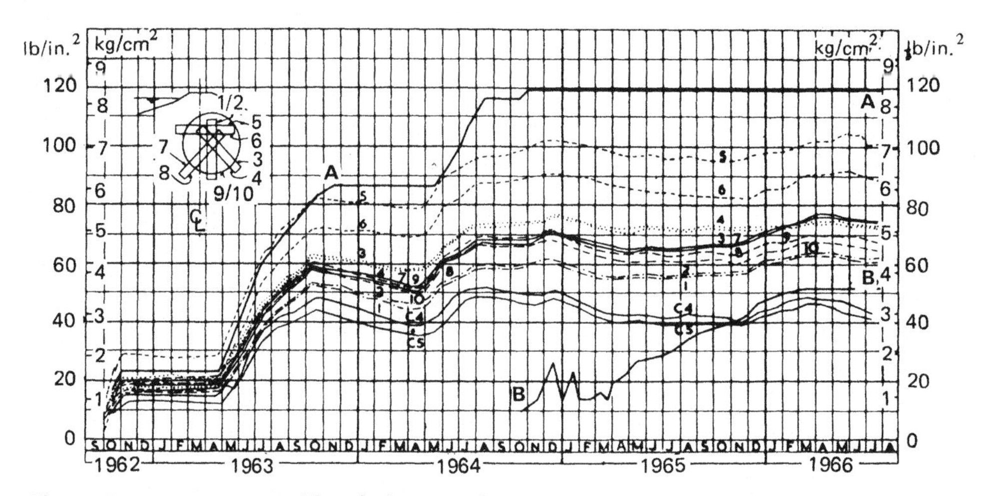

Figure 10.8. Measurements with embedment earth pressure cells in the clay core at Balderhead dam: A—overburden pressure; B—reservoir pressure (after Kennard et al., 1967).

ences were probably caused by nonplanar trimming of fill surfaces or uneven backfilling, whereas a part of the later differences may be due to stress variations in the fill caused by the cells themselves.

Kennard et al. (1967) and Thomas and Ward (1969) report on use of diaphragm cells with two active faces and vibrating wire transducers in the clay core of Balderhead Dam. Cells placed vertically showed good agreement between the two faces, as might be expected. During the first two filling seasons the average reading from a cell placed horizontally coincided approximately with the calculated overburden pressure, but thereafter the readings from the upper face increased at a greater rate than those from the lower face. On completion of the dam the average reading was about 80% of the overburden pressure. Data are shown in Figure 10.8.

Wilson (1984) comments that when earth pressure cells are installed in a horizontal plane in compacted fills for embankment dams, the cells typically register only 50–70% of the calculated added vertical stress as embankment construction continues. Wilson also comments that, because it is often difficult to shape the bottom of the excavation to the exact shape of the cell, bedding may be uneven and the application of vertical stress may deform the cell and cause additional error.

It appears to the author that two general observations can be made from a review of reported field placement effects. The first relates to the method of installation, the second to selection of type of cell.

First, although the conventional installation procedure usually prevents damage to the cells, its limitations are clear, and further research on the behavior of embedment earth pressure cells is needed to establish an improved procedure for installation within compacted fills. There is a need to develop a controlled method of field compaction around the cells that prestresses the soil to match the prestress in the remainder of the fill that is compacted by heavy equipment, without damaging the cells. It is hoped that improved installation techniques will result from the tests now in progress in Mexico (Alberro and Borbón, 1985).

The second general observation relates to the effect of nonuniform bedding and leads to a recommendation for selection of type of cell. Diaphragm cells are designed and calibrated for a uniformly distributed load on the active faces, and point loads, stress nonuniformities, or arching will cause significant errors. Hydraulic cells are also subject to errors from these causes but to a lesser extent than diaphragm cells. The best choice appears to be a flat hydraulic cell with thick active face, preferably with grooves to increase flexibility, and a thin layer of liquid.

Thomas and Ward (1969) state that their diaphragm cell, with two active faces and vibrating wire transducers, was designed for use in clay, therefore the effect of non-uniform bedding was not considered important. Uff (1970) demonstrated the high sensitivity of the same cell to uniformity of bedding by installing it as a contact cell, as shown in Figure 10.9, and loading the outer face with a 50 lb (23 kg) point load. Uff found that measurements at the outer face were highly dependent on the point of application of the load, whereas measurements at

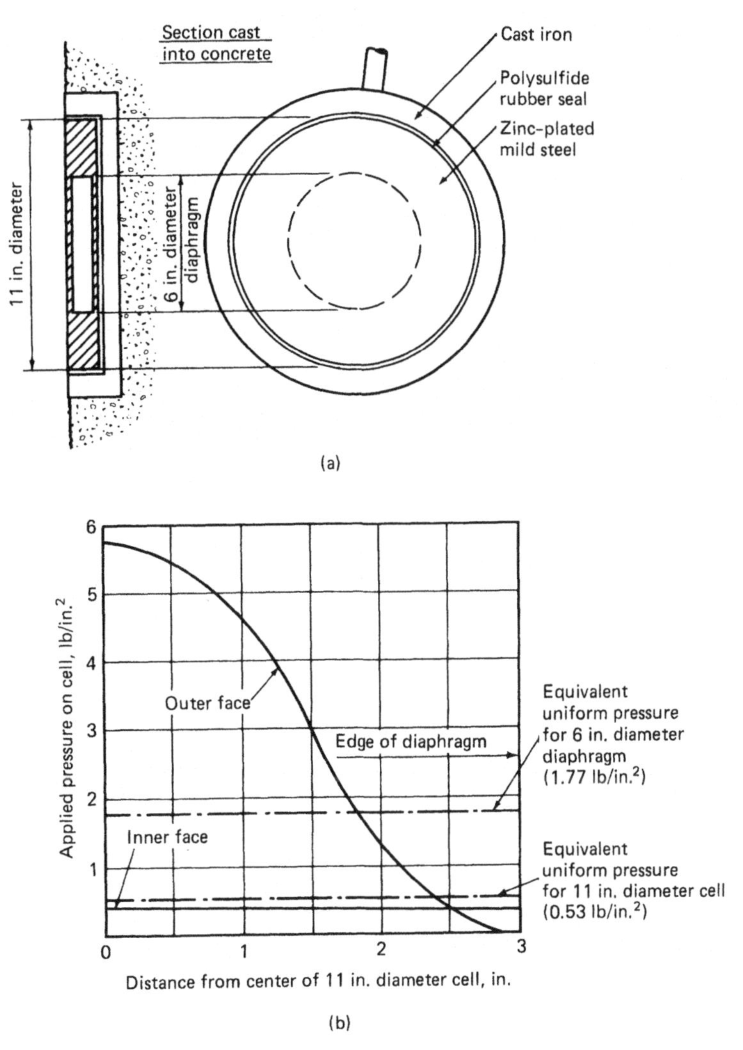

Figure 10.9. Response of diaphragm cell to nonuniform loading: (a) installation arrangement and (b) response of cell to moving a 50 lb point load across the diaphragm (after Uff, 1970).

the inner face were relatively uniform. The "equivalent uniform pressures" are calculated by applying the 50 lb load over circular areas having 6 and 11 in. diameters. Figure 10.9 shows that the response of the inner face is independent of the position of the point load on the outer face, and that the measured pressure is close to that calculated assuming the point load is uniformly distributed over a circular area of 11 in. diameter. Binnie et al. (1967) report on substantial underregistration and differences between the two faces when the same diaphragm cells were installed as embedment cells in gravel. Thomas and Ward (1969) report on smaller underregistration and smaller differences between the two faces when the cells were installed as embedment cells in clay. Vaughan and Kennard (1972) installed the same cells at the contact between concrete and compacted clay, using Uff's arrangement, and in four out of five cases the pressure measured by the outer face was higher than the pressure measured by the inner face. In most cases the pore water pressure was about 80% of the measured total stress, and therefore the two faces indicated very significant differences in effective stress. The writers comment that the inner face may well give the more accurate measurement.

This review of reported placement effects on diaphragm cells with vibrating wire transducers makes it clear that they are very sensitive to nonuniformity of bedding.

In contrast, Figure 10.10 shows the results of tests by McRae and Sellers (1986), similar to those

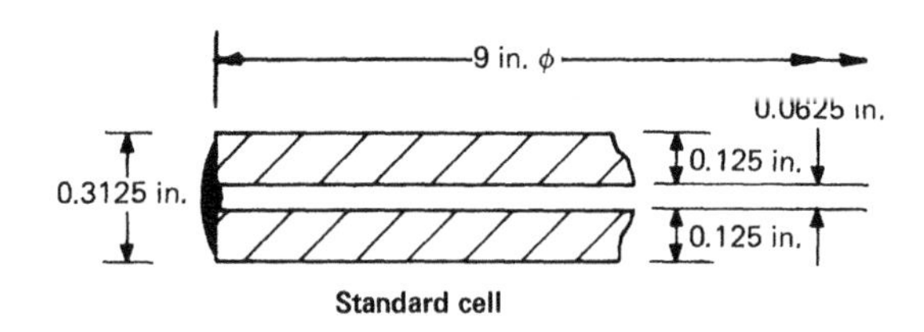

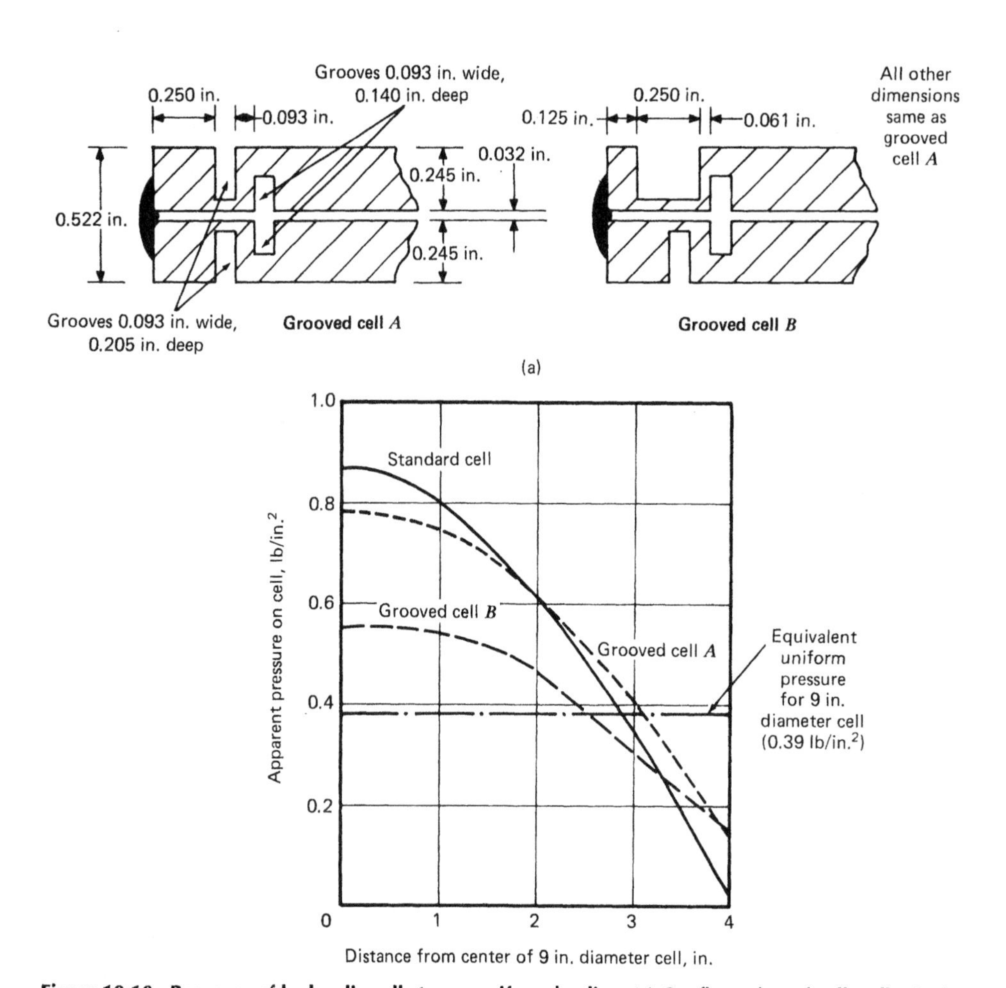

Figure 10.10. Response of hydraulic cells to nonuniform loading. (a) Configuration of cells: all 9 in. in diameter, filled with antifreeze. Vibrating wire transducer is connected at the rim. (b) Response of cells to moving a 25 lb point load across the upper faces of cells (after McRae and Sellers, 1986).

by Uff, made on hydraulic cells with thick active faces and vibrating wire pressure transducers. Cells *A* and *B* were grooved, following the procedure first adopted by Glötzl. The cells were placed on a 0.1 in. (2.5 mm) thick rubber pad, resting on a firm flat horizontal surface and loaded on the upper face with a 25 lb (11 kg) point load. When compared with Uff's test results on diaphragm cells, all the hydraulic cells showed less deviation from the "equivalent uniform pressure" when the point load was applied at the center of the cell. McRae and Sellers made similar tests by omitting the rubber pad, by changing the loading from a point load to a 25 lb (11 kg) load applied over an area of 0.5 in.^{2} (320 mm^2), and by using point loads of 13 and 67 lb (6 and 30 kg), all with similar results to those shown in Figure 10.10. The wide groove in the face of cell *B* evidently reduces bending when the cell is subjected to nonuniform loading and clearly is a worthwhile feature. A reasonable minimum thickness of the active face, when grooves are included, appears to be 0.25 in. (6.3 mm).

It is appreciated that all the point loading tests described above are extreme tests and that in practice great efforts are made to minimize nonuniformity of bedding by appropriate field installation procedures. However, since uniformity of bedding cannot be assured, it is apparent that hydraulic cells with grooved thick active faces are preferable to diaphragm cells. More tests are needed to define the optimum grooving pattern. In practice this observation limits the application of diaphragm cells to installations in clay. Even for installations in clay it cannot be said with certainty that the two independent measurements with a diaphragm cell should be preferred to the single measurement with a hydraulic cell. Comparative tests are needed to establish this preference. The cost of a hydraulic cell is typically less than half the cost of a diaphragm cell. Until further comparative data are available, the author favors selection of hydraulic cells with grooved thick active faces for all applications.

Green (1986) suggests that a "best of both worlds" type of hydraulic cell could be created with two back-to-back hydraulic chambers, each with its own pressure transducer. Each outer face would be grooved, and the stiff disk between the two hydraulic chambers would have flat faces. The cell would therefore provide two independent measurements and would have minimum sensitivity to nonuniform bedding.

Design of Cell Rim: Diaphragm Cells

For a diaphragm cell it is important to minimize the effect on the diaphragm of stress concentrations normal to the cell near the rim. An inactive integral stiff edge ring is used at the rim. Peattie and Sparrow (1954) recommend that the ratio of the sensitive area to the total facial area should be less than about 0.45. Typical diaphragm cells with vibrating wire transducers, such as the cell shown in Figure 10.9a, have a ratio of about 0.3.

The diaphragm must deflect, so that a measurable strain is achieved, and thus the effective modulus of the center of a diaphragm cell is much lower than that of the solid steel edge ring. Stress therefore tends to be concentrated on the edge ring.

Design of Cell Rim: Hydraulic Cells

The two versions of hydraulic cell shown in Figure 10.1 will be considered separately.

The cell with a grooved thick active face has a uniform modulus across most of its face, and if the liquid film is thin the modulus can be close to that of the soil. Under these conditions there is no need for an edge ring, and in fact an edge ring would distort the stress normal to the face of the cell.

The cell with thin active faces is generally provided with an edge ring (also called a *guard ring, stiffening ring,* or *confining ring*) for three reasons. First, it adds strength to an otherwise weak and flexible cell. Second, it tends to reduce the sensitivity to temperature changes. Third, it reduces the effect of inward radial stresses on the edge of the cell and the possibility of buckling the thin sheet metal faces.

Orientation of Cell

An additional factor affecting measurements arises if the orientation of a cell changes as overlying fill is placed. DiBiagio (1977) reports on changing orientation of cells installed in a vertical plane in the moraine core of a dam, as indicated by biaxial tiltmeters mounted on the cells. As overlying fill was placed and compacted, the inclination of a cell to vertical typically varied up to 18 degrees by the time the fill surface was 11 feet (3.4 m) above the cell.

Consequence of Factors Affecting Measurements: An Overview

Weiler and Kulhawy (1982) conclude:

> A stress cell which performs well mechanically as installed does not guarantee that the measured stress is representative of the correct free-field stress. In-soil calibration of the stress cells under the conditions expected in the field combined with an understanding of the cell–soil system behavior is essential to achieve good results. The time and expense [and difficulty] of in-soil calibration make the procedure unattractive to most engineers, so the number and accuracy of stress measurements are not high. Further research is needed before the understanding of stress cell behavior is complete, but it is believed that successful stress measurements may be made even now, in soils placed by man, if sufficient time and care are taken in making the stress measurement and in interpreting the results.

The author agrees that successful stress measurements can be made in clayey soils but has doubts about success in sands and rockfill. In contrast to

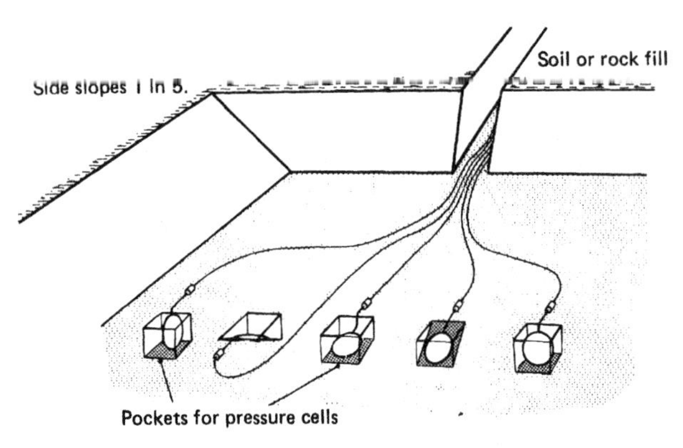

Figure 10.11. Typical layout of embedment earth pressure cells (courtesy of Soil Instruments Ltd., Uckfield, England).

the view of Weiler and Kulhawy, Wilson (1984) believes that the error caused by arching of stresses around embedment earth pressure cells can be so great that the very expensive in-soil calibrations (which cannot match site conditions) are often of marginal value.

In summary, the many factors that affect measurements can result in substantial errors, so that measurements with embedment earth pressure cells can rarely be made with high accuracy.

10.2.4. Installation of Embedment Earth Pressure Cells

A report by the International Society for Rock Mechanics (ISRM, 1981c) includes detailed recommendations for installation of embedment earth pressure cells. The report recommends the following procedure:

> An excavation to accommodate a cluster of five cells should be as shown in Figure 10.11, with dimensions not less than 13 × 13 × 6 ft deep (4 × 4 × 2 m), with side slopes not steeper than 1 vertical in 5 horizontal. The base of the excavation should be compacted and level. The cells should be individually installed in small pockets at the base of the excavation, each being approximately twice the size of the cell. Pockets should be separated from each other by at least 3 ft (1 m). Cell locations should be marked out, the pockets hand dug very carefully, and trimmed. Protruding stones should be removed and the holes filled with compacted stone-free soil. Each cell should be positioned in its pocket and checked for correct functioning and for alignment and level. The pocket should then be backfilled, whenever possible using the excavated soil, stone-free and at unchanged water content, to a density similar to that of the surrounding soil. The main excavation should then be backfilled with embankment material at unchanged water content, having removed rocks larger than the size of the cell. Three lifts of 4–8 in. (100–200 mm) each should first be placed and compacted by hand-operated equipment before completing the backfill with light mechanical equipment. No heavy vibratory rollers should be used until at least 6 ft (2 m) of fill has been so placed. When installing cells in rockfill, the pocket for each cell should be larger than in soil and should be backfilled with thoroughly compacted material of progressively smaller size, until the soil in contact with the cell is of a grain size less than 0.2 in. (5 mm).

Although the above procedure appears to prevent damage to the cells and to be accepted by many users, the author believes that several improvements are warranted in an attempt to increase accuracy. First, a 3 ft (1 m) deep excavation is preferable to an excavation twice that depth, and vibratory rollers can generally be used after 3 ft (1 m) of fill has been placed over the cells if they have been designed for high short-term overload capacity. Second, the procedure of installing cells in small pockets at the base of an excavation clearly runs the risk of decreasing measurement accuracy. Although no clear recommendations can be made, perhaps a procedure similar to that described in Chapter 17 (Clements, 1982) for installing horizontal tubes and cables in fill by mounding rather than excavating may be workable. Cells would be required to withstand temporary stress caused during compaction, and a robust hydraulic cell appears to be the best choice. The transducer should be housed within a thick-walled steel cylinder and a thick-walled annealed steel tube used to connect the transducer to the cell. The tests currently planned in Mexico (Alberro and Borbón, 1985) may result in improved placement procedures. Third, Wilson (1984) suggests that if a layer of plaster of Paris or weak cement is used as bedding for each cell and the cell worked into the bedding, errors caused by nonuniform bedding may be reduced, but the improvement, if any, resulting from this modification has not yet been verified. Fourth, where knowledge of cell orientation is critical, consideration should be given to attaching two electrolytic levels to the cell, mounted 90° apart.

The author believes that, for installations in rockfill, the free-field stress may be so perturbed by installation that measurements made with cells as small as 9 in. (230 mm) in diameter may not be worthwhile. Larger cells up to 3 or 4 ft (1 or 1.2 m) in diameter appear to be preferable, despite their substantial installation difficulties and high cost.

A variation on the procedure described by ISRM (1981c) involves placing the earth pressure cell within a hole in the middle of a plate approximately 2 ft (600 mm) square. Diaphragm cells with a vibrating wire transducer mounted on one active face, and with a diameter to thickness ratio of 4.5, are customarily installed in this way in Norway (DiBiagio and Myrvoll, 1985). Each cell is mounted in the middle of either a steel or a segmented concrete plate, flush front and back, as shown in Figure 10.12. The segmented concrete plate creates flexibility and best possible conformance with the prepared surface of the fill, both immediately after installation and when compaction equipment passes over the plate. The diameter to thickness ratio is increased to 15, and it is claimed that because the edge effects shown in Figure 10.4 now occur near the edges of the plate rather than the cell, the stress against the cell will be more uniform and representative of free-field conditions. However, it appears to the author that use of an appropriately designed hydraulic cell, of uniform stiffness across the face of the cell, would make the plate unnecessary.

As indicated in Section 10.2.2, some success has been achieved in measuring horizontal stress in soft soils by using specially designed "push-in" earth pressure cells, stepped blades, pressuremeters, and flat plate dilatometers. Installation procedures for these instruments are described in the references cited in Section 10.2.2.

10.3. CONTACT EARTH PRESSURE CELLS

Measurements of total stress against a structure are not plagued by so many of the errors associated with measurements within a soil mass, and it is possible to measure total stress at the face of a structural element with greater accuracy than within a soil mass. However, cell stiffness and the influence of temperature are often critical.

(a)

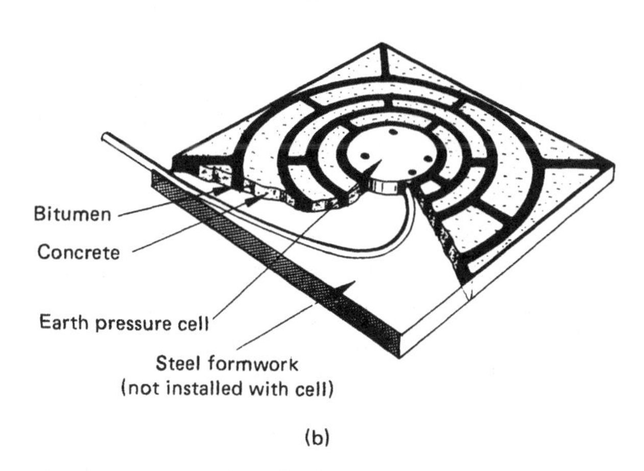

(b)

Figure 10.12. Mounting a diaphragm earth pressure cell (a) in the middle of a large steel plate and (b) in segmented concrete plate to increase diameter to thickness ratio (courtesy of Geonor A/S, Oslo, Norway).

10.3.1. Types of Contact Earth Pressure Cells

Standard Types of Cell

Some embedment earth pressure cells can be used directly as contact cells. All the cells shown in Figure 10.1 are suitable, with two exceptions. First, a hydraulic cell with a thin active face is unsuitable, because it is not readily possible to install this type of cell completely flush with the surface of a structure. Its stiffness is too low, and is also sensitive to the temperature changes that often occur at the locations of contact cells. Second, a diaphragm cell with a single active face is not likely to be suitable

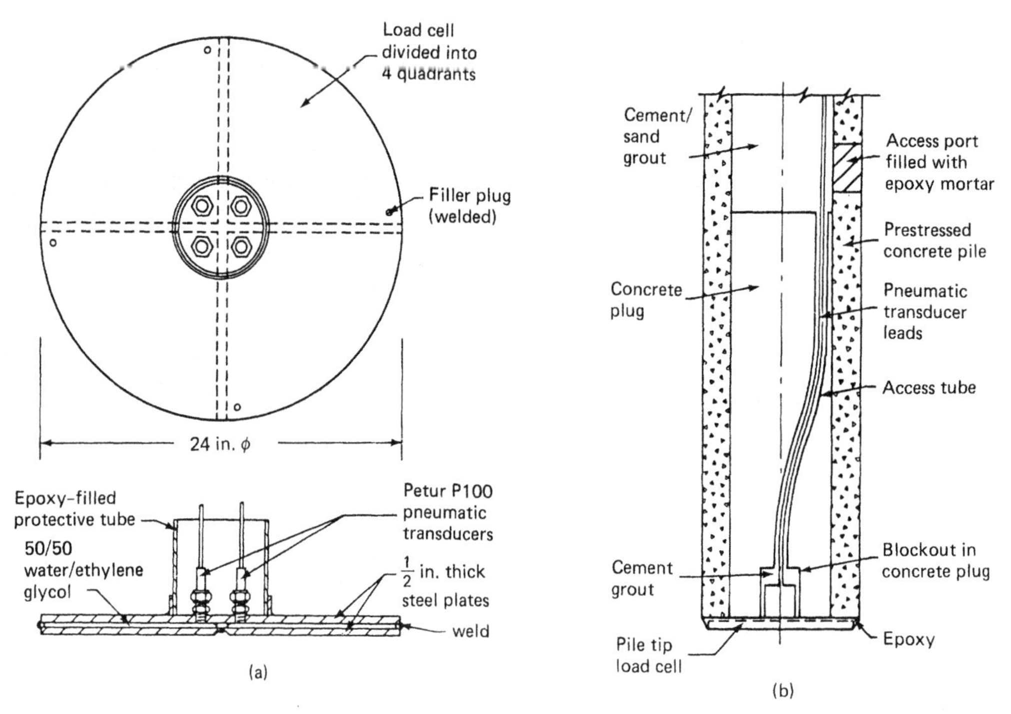

Figure 10.13. Contact earth pressure cell for measurement of pile tip load: (a) construction details of cell and (b) cell mounted on pile (after Green et al., 1983).

because, as indicated by Uff's (1970) test results, measurements on the outer face are subject to significant errors caused by uneven loading. A diaphragm cell with two active faces can be used, with the mounting arrangement shown in Figure 10.9, placing more reliance on measurements at the inner face.

When installing a hydraulic cell to measure contact stress between soil and concrete, there is a possibility of uncoupling between the cell and concrete when temperature rises and the liquid in the cell expands during concrete cure (Section 13.6). It is therefore preferable to construct the cell with one steel plate approximately 0.5 in. (13 mm) thick and to place that plate against the concrete, thereby ensuring that any expansion occurs outward. The layer of liquid should be as thin as possible.

Cells for Measurement of Load at Tips of Driven Piles or Drilled Shafts

Various types of cells have been developed for measurements of load at the tips of driven piles and drilled shafts. The cells can be considered as load cells but are classified in this book as contact earth pressure cells.

A hydraulic cell with pneumatic transducers is described by Green et al. (1983) for measurement of tip load on driven prestressed concrete piles. The cell is divided into four independent quadrants as shown in Figure 10.13, each with a pneumatic pressure transducer, and covers almost the entire area of the pile tip. Design criteria included adequate robustness to withstand driving forces and a thin enough active face to transmit tip pressure to the liquid in the cell. High-quality welding and leak testing procedures were used. The piles were 24 in. (610 mm) octagonal, with a 15 in. (380 mm) diameter hollow core. A 5 ft (1.5 m) long solid concrete plug was cast into the tip of each instrumented pile and the pneumatic transducer tubing carried up through the core. The writers discuss the difficulties of calibrating such a cell and justify their adopted check calibration procedure, whereby the pile was first lifted into a vertical position and cell readings taken in air, then lowered on to four wood blocks, one centered on each quadrant. The measured load was within 10% of the dead weight of the pile. Three instrumented piles were driven as part of a bridge foundation, through loose to medium dense sand into dense to very dense sand. Tubing to one of the 12 quadrants was damaged during driving, but the remaining quadrants have provided consistent and apparently reliable data during a 5 year monitoring period.

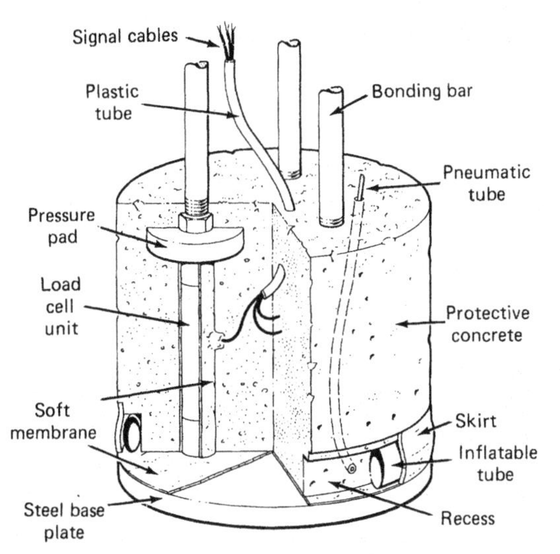

Figure 10.14. Contact earth pressure cell for measurement of load at the tip of a drilled shaft (after Price and Wardle, 1983). By courtesy of Building Research Establishment, Crown copyright.

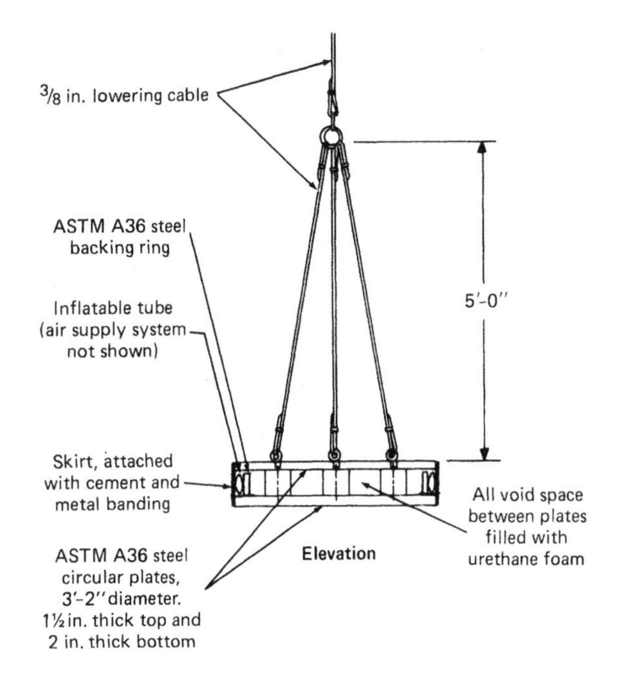

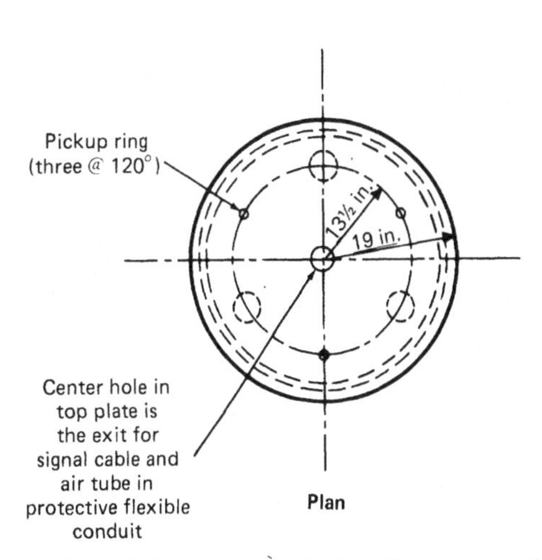

Figure 10.15. Schematic of contact earth pressure cell for measurement of load at the tip of a drilled shaft: vibrating wire load cells mounted between two thick circular plates.

Figure 10.14 shows a cell designed at the Building Research Establishment in England (Price and Wardle, 1983) for measurement of tip loads in drilled shafts. The load cell units are steel tubes with internal strain gages, usually the vibrating wire type. The steel base plate is isolated from the concrete shaft by a soft membrane sheet, so that the tip load is transferred entirely into the load cells. The load is prevented from bridging the cell by the inflated tube. Load from the load cells is then transferred to the concrete shaft through pressure pads and bonding bars, located above the load cells. The cell is lowered into place on a small bed of concrete, the inflatable tube pressurized to form a seal, and the shaft concrete poured. After 24 hours the air pressure is released.

Barker and Reese (1969) describe a "bottomhole cell," also designed for measuring tip loads in drilled shafts, consisting of electrical resistance strain gaged load cells mounted between two thick circular steel plates. The cell is similar to one designed by Whitaker (1964) at the Building Research Station, which has been superseded by the Price–Wardle cell.

Nowack and Gartung (1983) used a hydraulic cell with a pneumatic transducer for measuring tip loads in a drilled shaft. The cell was bedded on a concrete pad and the annular space around the cell filled with a soft material to ensure that the entire tip load was transmitted through the cell. Horvath (1985) used two flatjacks for the same purpose, positioned in series with intervening bearing plates. The second flatjack served as a backup in the event of malfunction.

The author has recently been involved in planning an instrumented load test of a drilled shaft in sand and gravel, with cobbles and boulders, to be conducted by the State of Ohio Department of Transportation, that requires measurement of tip load. Steel casing and drilling mud will be used to drill the shaft, and the bottom cannot be inspected. The selected cell is shown in Figure 10.15, and will be manufactured by Geokon. The circular steel plates will be held together by three steel bolts, one through the hollow-core of each load cell, with heads flush with the bottom face of the lower steel plate. The bolts will be tightened only hand tight, and steel domed covers (not shown in the figure) will be attached to the top plate over each bolt to ensure that no load will be carried by the bolts. This provision allows use of a calibration based on the sum of the three separate load cell calibrations. The skirt and inflatable tube ensure that no load can bridge the cell during the load test. The cell will be placed on a 6 in. (150 mm) layer of either stiff grout or leveled and compacted sand, the inflatable tube pressurized to center the cell, the tube depressurized, the reinforcing cage inserted over the lowering cable and conduit, the shaft concrete poured and casing raised. Early during pouring concrete and raising casing the inflatable tube will again be pressurized to form a seal, and the air pressure will be released after 24 hours.

Cells for Measurement of Stress on Sides of Driven Piles

Various types of cells are commercially available for attachment to sheet and H-piles, but measurement difficulties are great in both cases. If a cell is attached on the face of a pile, the stress field is likely to be disturbed significantly, with the possible exception of piles driven through soft clay. If the cell is attached within a cutout in an attempt to place the sensitive face in the same plane as the face of the pile, the cutout will usually weaken the section excessively. Even if an accurate measurement could be made, stresses on the surfaces of the sheet and H-piles are likely to be irregular, and a few point measurements are therefore of limited value.

Measurements on the sides of pipe piles are also difficult, because of the need to mount the sensitive face flush with the outside of the pile. Possibly the best approach is to manufacture special hydraulic cells with curvature to match the outside of the pipe and to use a thick grooved outer face. If the grooves are omitted, the curved face is likely to prevent flexing. Alternatively, rectangular cells could be used, with the long side parallel to the axis of the pipe, but the face would not be flush with the outside of the pile. For either option, the cells should be installed in cutouts, and measurement accuracy is likely to increase with increasing pile diameter.

Measurements on the sides of driven concrete piles can be made satisfactorily (e.g., Clemente, 1979), provided that a flat face (piles square or octagonal) is available.

Hydraulic cells with pneumatic transducers have been shown to survive pile driving and are generally preferred.

10.3.2. Factors Affecting Measurements

Overall requirements for contact earth pressure cells are the same as given in Section 10.2.2 for embedment cells: good conformance, adequate sensing area, minimum sensitivity to nonuniform bedding, and a method of installation that will not seriously change the state of stress. These requirements can usually be accomplished by installing a flat-faced cell of adequate size with its sensitive face absolutely flush with the surface of the structure and by attention to correction methods for minimizing various sources of error. These methods are identified by a superscript *b* in Table 10.1. Various additional factors that affect measurements are discussed in the following subsections.

Number of Cells

Although contact stresses may be reasonably uniform for the structure as a whole, stresses measured over areas the size of most contact earth pressure cells may be very irregular, owing to local variations in soil conditions. Measurements with contact earth pressure cells therefore often show considerable scatter. The more cells the better, but it is usually difficult to determine whether scatter results from real variations in stress or from measurement errors.

As indicated in Section 10.1, it is sometimes preferable to isolate a portion of the structure and to determine stresses by use of load cells and strain gages within the structure.

Laboratory Calibrations

Each cell should be calibrated under fluid pressure to be sure that it is functioning correctly and not leaking, and the calibration chamber shown in Figure 10.6 can be adapted for performing fluid calibrations of contact earth pressure cells. Selig (1980) presents details of the procedure, whereby the bottom half of the chamber is replaced with a concrete block containing the cell cast in place. The cell can be removed and the concrete block reused by casting an oversized cavity in the block and seating each cell in the cavity with a thin layer of high-modulus gypsum plaster.

Selig also describes calibrations in contact with soil, by placing the cell and concrete block within a 3 ft (910 mm) diameter chamber used for soil calibration of embedment cells. A similar procedure is described by Felio and Bauer (1986), but it is believed that their 2 ft (610 mm) diameter chamber is too small for calibrating their 12 in. (300 mm) diameter cells.

Temperature Effects on Cell

Embedment earth pressure cells are not usually subject to temperature changes, but contact cells are often near the atmosphere or near concrete pours that create changes in temperature.

As discussed in Section 10.3.1, hydraulic cells used to measure contact stress between soil and concrete should have an inner thick inactive face, and the layer of liquid within the cell should be as thin as possible. The active face should also be thick and grooved. Examples of the significant influence of temperature on hydraulic cells are given by Coyle and Bartoskewitz (1976) and Felio and Bauer (1986), but the cells in both examples did not satisfy the above guidelines. Calibrations to examine temperature sensitivity under unconfined conditions, for example, in a heated water bath, are sometimes quoted but are of limited value. When confined, the cell is likely to be much more susceptible to temperature effects that cause expansion or contraction of the layer of liquid, and this condition is very difficult to simulate correctly in a laboratory test. In both the laboratory and in the field there may also be real stress changes occurring at the interface, owing to temperature effects on the soil and the structure. The best solution to this dilemma is to design the cell for minimum sensitivity to temperature.

A properly designed diaphragm cell is not significantly affected by temperature changes and, if temperature changes are great, diaphragm cells may be more appropriate than hydraulic cells. However, the cells should have two active faces, the mounting arrangements shown in Figure 10.9 should be used, care should be taken to minimize nonuniform loading, and more reliance should be placed on measurements at the inner face.

Stiffness of Cell

As for embedment cells, errors will normally be caused by excessive diaphragm displacement. When installed at a contact between concrete and all materials other than soft clay, it is recommended that the same modulus criterion should be adopted as for concrete stress cells (Chapter 13): the modulus of elasticity of the cell should be at least one-half that of the concrete.

Irregularity of Surface of Structure

When contact earth pressure cells are installed at an irregular surface, significant measurement errors can be expected. As an example, DiBiagio (1977) describes measurements on a sheet pile wall. Because of the corrugations in the wall, the cells must be placed either on the protruding or indented corrugation, as shown in Figure 10.16. Cells were placed at both locations at the same depth in soft clay. Measurements were almost equal immediately after pile driving but later followed the pattern shown in Figure 10.16, presumably because of sheeting movements or consolidation and arching. A "best estimate" of the average total stress per unit length of the wall can be made by averaging P_1 and P_2, but clearly this average may be significantly in error.

10.3.3. Installation of Contact Earth Pressure Cells

Installation at Interface Between Soil and Concrete

As indicated in Section 10.3.1, when a hydraulic cell is to be installed at an interface between soil and concrete, a thick steel plate should be used for the face of the cell adjacent to concrete.

ISRM (1981c) indicates that, when placing cells to measure contact stress between soil and concrete, the cells may be installed by any one of three methods: (1) attached to the formwork and placed in the structure before concreting, (2) fastened to the structure after concreting and prior to backfilling, or (3) embedded in the backfill a short distance away from the structure. In the view of the author, the overriding need is to ensure that the sensitive face of the cell is **absolutely** flush with the interface. The first method is therefore suitable, and also the second if the cell is grouted into a blockout. The second method is not suitable if the cell protrudes outside the profile of the structure, because

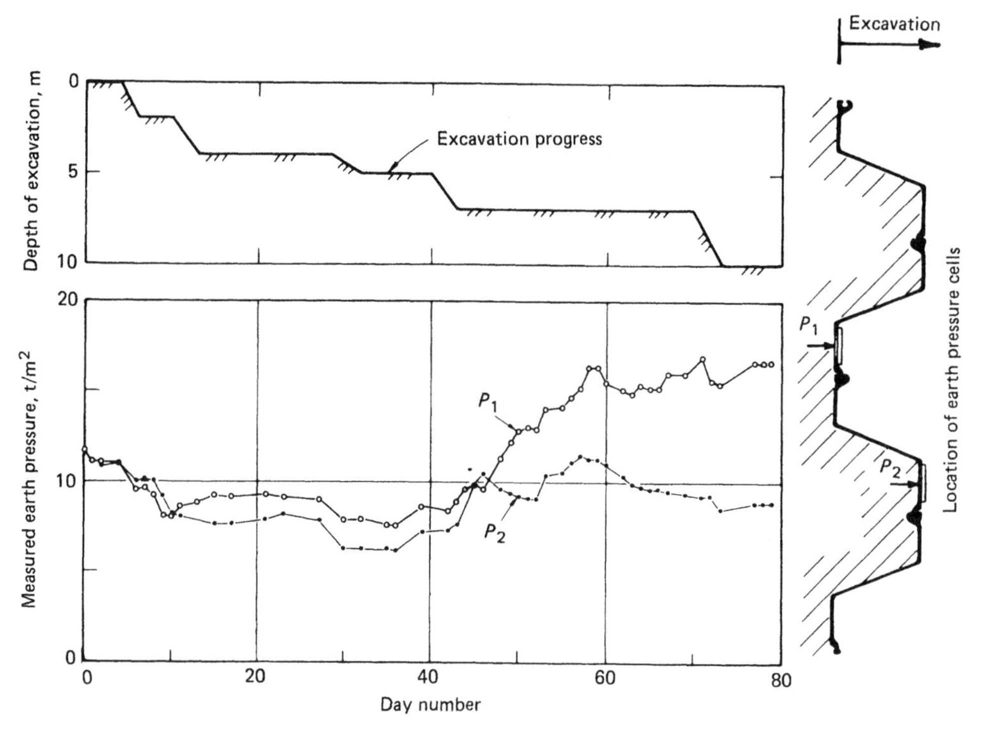

Figure 10.16. Variation in measured earth pressure on the corrugated surface of a sheet pile wall (after DiBiagio, 1977).

conformance errors may be significant. The third method creates the potential for all the errors described above for embedment cells. Also, leads may be sheared if they are installed through the concrete, and leads are exposed to damage if they are installed in the soil. It is therefore not recommended.

The first method, whereby cells are attached to the formwork and placed in the structure, can generally be used for measurements against retaining walls, culverts, precast concrete piles, and precast concrete tunnel linings. The cell can sometimes be held in place during concreting by light bracing against the form at the opposite face, or alternatively, the cell can be attached by bolts with heads on the outside of the form. The back face of the cell should be cleaned and degreased to ensure good contact between cell and concrete. A ring of soft sealing material, about 0.1 in. (2.5 mm) thick, around the edge of the cell is useful to prevent concrete fouling the outside face of the cell and to limit the influence of stresses that may act on the edge of the cell as the concrete cures. The same soft sealing material should be used during calibration. When using diaphragm cells, the mounting arrangement shown in Figure 10.9 is suitable. Concrete vibrators should not be allowed to touch the cell. In cases where cells cannot be installed in this way, a blockout can sometimes be made in the concrete and the cell subsequently embedded within the blockout.

Where concrete is cast directly against soil, for slurry walls, drilled shafts, base slabs, spread footings, cast-in-place tunnel linings, and other structures, installation is less straightforward, and a method must be devised whereby the cell is held against a suitably flat surface of soil and protected during concreting. DiBiagio and Roti (1972) describe installation of cells in a slurry trench by using an expendable hydraulic jack to hold the cell in position against the soft clay. Each cell was first mounted in the center of a flat steel plate such that its sensitive face was flush with the surface of the plate. The cell and plate were attached to one end of the hydraulic jack with a flexible coupling and a reaction plate of similar size attached to the other end. The cells and hydraulic jacks were attached to the reinforcing cage before lowering into the trench and, to avoid exerting forces on the cage during installation of the cells, the body of each jack was placed inside a short length of oversized pipe welded to the cage. Actuation of jacks forced the cells in contact with the sides of the trench; pressure was maintained during concreting and released after the concrete had set. A similar installation procedure is described by Uff (1970), and this approach can be used in other applications where concrete is cast directly against a vertical soil face. Bierschwale et al. (1981) describe use of steel pins to attach contact earth pressure cells to the soil in the wall of a dry hole for a drilled shaft.

Where there is concern for damage or disturbance to cells during concreting, each cell may be cast in a briquette of concrete identical to the parent concrete. The sensitive face of the cell should be **absolutely** flush with the face of the briquette that will be in contact with the soil. It is **essential** that the briquette be cast not more than 24–48 hours ahead of the main concrete pour, otherwise, accuracy will be severely reduced because of poor conformance. When cells are installed against a horizontal soil surface, for example, at the base of a foundation slab or spread footing, it is best to use a briquette over the entire cell, and to cast it in place so that good contact with the soil is ensured. A typical briquette size for a 9 in. (230 mm) diameter cell with transducer attached to a tube emerging from the rim of the cell is 2 ft × 1 ft × 3 in. (600 × 300 × 76 mm). When cells are installed in situations where the briquette cannot be cast in place, it should be 6 in. (150 mm) thick to ensure adequate strength for moving.

Wherever possible, grains larger than about 0.2 in. (5 mm) should not be allowed in contact with the cell, and if necessary a thin layer of grout or fine-grained material should be placed against the face of the cell.

Installation at Interface Between Soil and Steel

Wherever possible, the cell should be mounted such that its face is absolutely flush with the surface of the steel structure. If a hole can be cut in the structure, the cell can be attached to studs welded on the back face, and any irregularities in the front face can be smoothed with a suitable filler material such as epoxy resin.

Installation at Tips of Driven Piles or Drilled Shafts and on Sides of Driven Piles

Installation for these special applications is discussed in Section 10.3.1.

Installation at Interface Between Soil and Rock

Three approaches are possible, depending on the irregularity of the rock surface at and near the place of installation and the strength of the rock.

First, if the surface at and near the place of installation is generally planar and the rock is strong, a location should be selected for the cell where the surface is flat within about ± 0.4 in. (± 10 mm), and any loose material should be removed. A cement mortar or epoxy resin pad should be troweled on to the rock surface and the cell pushed into the pad, squeezing out mortar or resin until a layer no thicker than 0.2–0.4 in. (5–10 mm) remains beneath the cell. Entrapment of air bubbles must be avoided, and the cell may need to be secured in position by tying to pins in the rock. The area around the cell should be blended into the surrounding rock surface by troweling cement or epoxy after the pad has hardened.

Second, if the surface at and near the place of installation is generally planar and the rock is weak, a hollow can be excavated in the rock and the cell installed with its sensitive face outward, flush with the surface of the rock, as described for installation at an interface between soil and concrete.

Third, if the surface of the rock is irregular, such that either of the above methods might result in measurement anomalies, an embedment earth pressure cell should be used and installed entirely within the soil as described in Section 10.2.4.

Installation at Interface Between Rock and Concrete

The three approaches decribed above for an interface between soil and rock are generally applicable. However, additional precautions are necessary to minimize the problem of cell/concrete uncoupling when temperature rises and falls during concrete cure (Section 13.6), and a stiff concrete stress cell rather than a contact earth pressure cell should be used.

CHAPTER 11

MEASUREMENT OF STRESS CHANGE IN ROCK*

11.1. APPLICATIONS

The primary application for measurement of stress change in rock is to monitor the stability of pillars and walls formed during underground mining.

Similar measurements are occasionally made during excavation of underground openings for civil engineering purposes, but most civil engineers prefer to monitor opening stability by measuring deformation. The mining engineer is accustomed to working with much lower factors of safety than the civil engineer and is therefore generally more concerned with critical stress conditions. However, the current and frequent use of design tools such as finite and boundary element modeling techniques has greatly increased the importance of monitoring stress change in rock. The primary output from these design aids are estimates of rock deformation and stress, and the effectiveness of an adopted design can only be confirmed by monitoring one or both of these variables in situ.

Stress change monitoring is used to diagnose critical stability in situations where the rock approaches a failure condition caused either by an increase or decrease in stress. A decrease in stress can lead to failure by reducing confinement, thereby causing rock blocks to slide or unravel. It is important to realize that the actual stress within a rock mass may vary significantly from point to point, depending on geologic features and the local stress concentrating effect of an opening. A single point measurement of stress may therefore be misleading, and if reliable data are required it is usually necessary to make measurements at numerous points throughout the rock mass of interest.

Extensive measurements of stress change are made during thermomechanical testing in underground excavations for studies relating to the disposal of high-level nuclear waste. This application requires devices capable of withstanding the high temperatures created to simulate anticipated repository conditions and is not discussed in this book.

11.2. INSTRUMENT CATEGORIES

The following three methods can be used for monitoring stress change in rock.

*Coauthored by Robert J. Walton, Senior Experimental Scientist, Commonwealth Scientific and Industrial Research Organization (CSIRO), Victoria, Australia. J. Barrie Sellers, President, Geokon, Inc., Lebanon, NH, and Frank S. Shuri, Senior Engineer, Golder Associates, Inc., Seattle, WA, have also assisted in preparation of this chapter.

Repeated Measurements of in Situ Stress

Repeated measurement of rock stress can be made in situ by using one of the absolute stress measurement techniques such as borehole overcoring. However, this approach is not often used because the cost is high, absolute accuracy is low, and there is a possibility that important data may be missed between measurements. Even when the most accurate borehole overcoring measurements are made, the variation between tests is generally at least ±25% for magnitude and ±15 degrees for orientation of the principal stresses. These variations are caused both by measurement inaccuracy and by real variations in stress from one point in the rock mass to another.

Geophysical Techniques

Several geophysical techniques use the relationship between deformability and stress—hence the propagation of seismic waves—in an attempt to determine in situ stress. For example, the time can be measured for an energy pulse to travel through the rock, or the rise time of a similar pulse can be measured. Geophysical techniques for monitoring stress change in rock are still at an early stage of understanding and development and are not discussed in this book.

Measurements in a Borehole

An instrument can be installed in a borehole to measure displacement, strain, or pressure caused by changes in the near-field stress. This approach is the most viable method for monitoring stress change and is the subject of the remainder of this chapter.

Instruments can be divided into two general categories: *soft* and *hard (rigid) inclusions*.

Soft inclusion gages have small stiffness relative to the host rock, and stress determination requires knowledge of rock properties and constitutive behavior. In practice, a gage having an intact modulus of elasticity less than that of the surrounding rock by a factor of 3 or more is considered to be a soft inclusion gage. These gages therefore offer minimum resistance to rock deformation, so that gage readings depend on rock stress and elastic properties but are independent of the modulus of the gage.

Rigid inclusion gages are designed to be stiff relative to the host rock, and stress determination requires knowledge of rock properties and constitutive behavior only within broad bounds. For practical purposes, a rigid inclusion gage is a gage having a modulus of elasticity exceeding that of the surrounding intact rock by a factor of at least 3. Rock deformations are resisted by the gage so that strains in the gage have minimum sensitivity to rock modulus. Rigid inclusion gages are usually referred to as *stressmeters*.

It is evident that an instrument that is a soft inclusion when installed in one type of host rock could be deemed a rigid inclusion when installed in rock of lower modulus.

Comprehensive descriptions of measurement methods are given by IECO (1979), LBL (1982), Lemcoe et al. (1980), Lingle et al. (1981), RKE/PB (1984), and Schrauf and Pratt (1979). Methods in practical use are described and compared in Sections 11.3 and 11.4, and recommendations are given in Section 11.5.

11.3. SOFT INCLUSION GAGES

As discussed in Section 11.2, soft inclusion gages have small stiffness relative to the host rock, and stress determination requires knowledge of rock properties and constitutive behavior. This definition clearly places borehole deformation gages and biaxial and triaxial strain cells in the category of soft inclusions, but the categorization may be less appropriate for borehole pressure cells. Borehole pressure cells have been used to measure stress change in a wide variety of rock types; in some cases stress determination requires knowledge of rock properties, while in other cases it does not. Some borehole pressure cells can be filled with mercury specifically to increase their stiffness so that they can be considered as rigid inclusions when used in low-modulus rocks. Despite this qualification, borehole pressure cells are classified in this book as soft inclusion gages.

11.3.1. Flat and Cylindrical Borehole Pressure Cells

A borehole pressure cell consists of a flat or cylindrical metal chamber, filled with a liquid and fitted with pressure regulation and monitoring capabilities. The term *borehole pressure cell* (BPC) is normally used for a flat cell and the term *cylindrical pressure cell* (CPC) for a cylindrical cell.

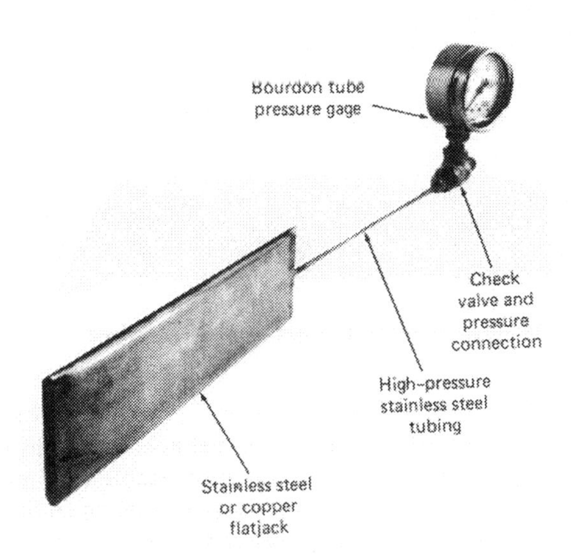

Figure 11.1. Flat borehole pressure cell (courtesy of Geokon, Inc., Lebanon, NH).

Flat cells are essentially miniature flatjacks and are typically about 2 in. wide, 0.25 in. thick, and 8 in. long (50 × 6 × 200 mm). A BPC is shown in Figure 11.1. A remote reading pressure transducer can replace the Bourdon tube pressure gage.

The U.S. Bureau of Mines BPC is described by Panek and Stock (1964) and Smith (1972). Embedment earth pressure cells with pneumatic transducers have been used as BPCs (e.g., Sauer and Sharma, 1977).

A BPC can be installed in any one of four ways. First, the cell can be inserted into a borehole already filled with grout. Second, grout can be pumped into the borehole after the cell is in place. Third, the cell may be preencapsulated inside an appropriately sized cylinder of mortar or epoxy, or sandwiched between two curved aluminum platens, inserted into a borehole, and pumped up to create adequate contact with the wall of the borehole. Fourth, a preencapsulated cell can be grouted within a borehole. A method that includes grouting is highly recommended, because it results in more intimate contact between the cell and borehole wall. After the grout has set, the cell is pumped up until the hydraulic pressure is somewhat greater than the estimated rock stress, a valve in the pressure line is closed, and the system is allowed to stabilize.

In elastic (the terms *elastic* and *viscoelastic* are defined in Chapter 2) rocks the relationship between change in BPC pressure gage reading and change in rock stress can be determined by using the pump-up procedure described by Sellers (1970), but rock deformability values must be known. The BPC responds primarily to stresses acting perpendicular to the plane of the flatjack but also has a small sensitivity to stress changes in the plane of the flatjack. Two BPCs oriented at right angles to each other in the same borehole will, provided the orientations of the principal stresses in the plane perpendicular to the axis of the borehole are known, allow determination of changes in these stresses. It is generally assumed that in viscoelastic rocks the BPC will achieve an equilibrium pressure that is close to the actual rock stress.

CPCs are typically 1.5 in. (38 mm) in diameter and 8 in. (200 mm) long. The U.S. Bureau of Mines CPC (Hall and Hoskins, 1972; Panek et al., 1964; Smith, 1972) consists of a cylindrical copper jacketed inflatable probe connected to a screw pump for pressure application. The Colorado School of Mines dilatometer, which is 3 in. (76 mm) in diameter and has a flexible plastic jacket (Hustrulid and Hustrulid, 1973), can also be used as a CPC.

For a CPC the diameter and smoothness of the borehole are critical, and every effort should be made to keep the tolerance within approximately 0.02 in. (0.5 mm). The borehole should preferably be drilled with a rotary diamond bit, and if percussion drilling is used a reaming shell should be inserted behind the bit. The CPC is inserted into the borehole and pressure is applied with the screw pump to seat the cell against the wall of the borehole. As for the BPC, the cell is then pumped up until the hydraulic pressure is somewhat greater than the estimated rock stress, a valve in the pressure line is closed, and the system is allowed to stabilize. The procedure for calculation of stress change is described by Sellers (1970). It should be noted that CPC data provide a measurement of the **average** change of the two principal stresses in the plane perpendicular to the axis of the borehole, and this may not be appropriate for strongly anisotropic stress conditions.

A typical range of commercially available BPCs and CPCs is 0–10,000 lb/in.^{2} (0–70 MPa), with a sensitivity of 40 lb/in.^{2} (300 kPa).

If biaxial stress measurements are required, three BPCs can be installed in the same borehole. Lu (1981) describes determination of stress changes in viscoelastic rock by using one CPC and two BPCs in the same borehole.

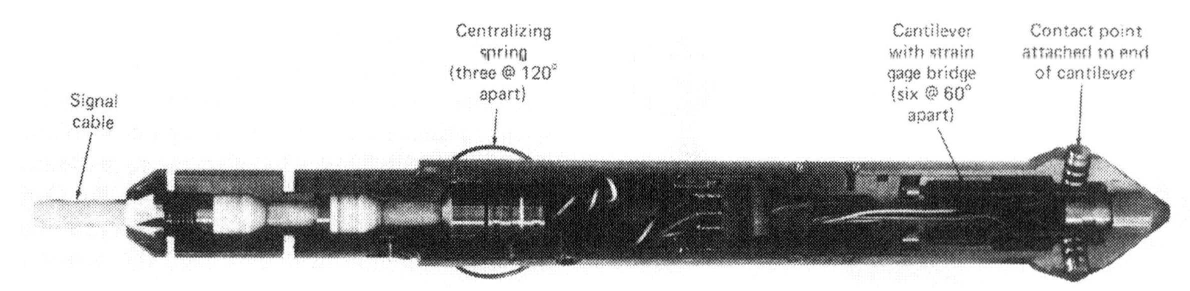

Figure 11.2. U.S. Bureau of Mines borehole deformation gage (courtesy of Rogers Arms and Machine Company, Inc., Grand Junction, CO).

11.3.2. Borehole Deformation Gages

A borehole deformation gage (BDG) is designed to measure diametral changes in a small-diameter borehole, using remotely read electrical transducers.

U.S. Bureau of Mines Gage

The most well-known BDG is the U.S. Bureau of Mines (USBM) instrument, shown in Figure 11.2, allowing measurement of changes in three diameters 120 degrees apart. The gage has three pairs of contact points, the outer ends contacting the borehole wall and the inner ends bearing against cantilevers, each of which is clamped to the gage body. Electrical resistance strain gages are bonded to both faces of each cantilever and the strain gages on opposite cantilevers are wired into a full Wheatstone bridge configuration. A conventional strain indicator is used as a readout unit. Changes in borehole diameter cause bending of appropriate cantilevers and changes in strain gage resistances. Borehole deformations are related, via an assumed or measured deformability and Poisson's ratio, to stress changes in the plane perpendicular to the axis of the borehole. Most versions of the USBM BDG are designed for measurements within 1.5 in. (38 mm) diameter boreholes, but larger and smaller sizes are available.

The USBM BDG was developed for determination of absolute in situ stress by using the overcoring procedure (Hooker and Bickel, 1974; Hooker et al., 1974; Obert, 1966), a test that involves reading the gage for less than 30 minutes. Most commercially available USBM BDGs have transducers that are insufficiently stable for monitoring stress change over a longer period, and the design makes it very difficult to prevent ingress of moisture into the transducer area. The version shown in Figure 11.2 is believed to be the most stable version available and has been used for stress change monitoring during in situ tests related to underground high-level nuclear waste repositories. However, even this version has been proved reliable for stress change monitoring only for short-term applications at ambient temperature.

The USBM gage is designed as a reusable device and is held in place in the borehole by frictional forces. For long-term monitoring purposes, the instrument would have to be held in place by grouting to prevent movement caused by blasting or other vibrations. However, the cost of the instrument usually prevents grouting in place.

The USBM BDG is not recommended for monitoring stress change in rock.

CSIRO Yoke Gage

The yoke BDG has been designed by the CSIRO (Commonwealth Scientific and Industrial Research Organization) in Australia specifically for stress change monitoring. It has been used to monitor stress changes in the roof of a coal mine (Walton and Fuller, 1980) and in pillars of metalliferous mines (Walton and Worotnicki, 1986). As shown in Figure 11.3, three strain gaged yokes are used to measure changes in three diameters 120 degrees apart. As for the USBM gage, output is related, via an assumed or measured deformability and Poisson's ratio, to stress changes in the plane perpendicular to the axis of the borehole.

The instrument is held in place in a borehole by epoxy cement, but the cement merely fills the annular space and plays no part in transmitting rock deformations to the yokes. The tips of the yokes are rounded, so that they slide freely along the wall of the borehole during installation. A front-to-rear

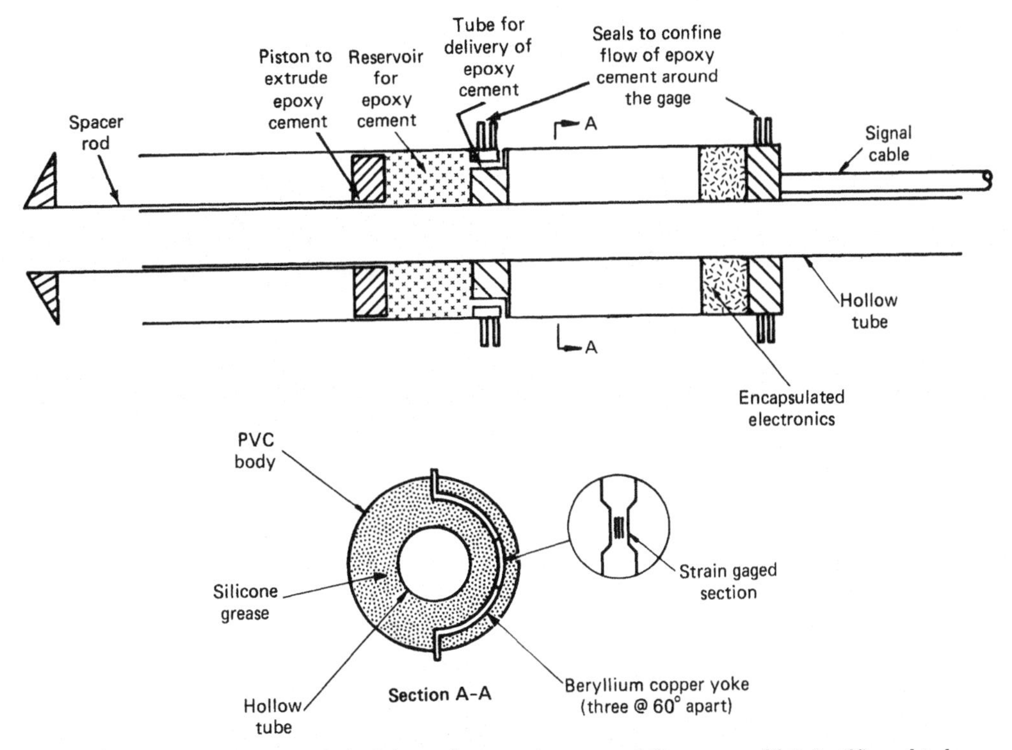

Figure 11.3. CSIRO yoke borehole deformation gage (courtesy of Commonwealth Scientific and Industrial Research Organization, Australia).

opening allows passage of signal cables from instruments installed deeper in the borehole.

To date, instruments have been installed in 56.5 mm (2.22 in.) diameter boreholes. The instrument has a large range in both tension and compression. For example, the tensile range is 100 MPa (1.4×10^4 lb/in.^{2}) in rock of modulus 70 GPa (10×10^6 lb/in.^{2}). The compressive range is greater than the tensile range.

11.3.3. Biaxial and Triaxial Strain Cells

Biaxial and triaxial strain cells are designed to measure strains on the wall or end of a borehole. The cells include electrical resistance strain gage transducers and are bonded to the rock. Their primary application, as for the USBM borehole deformation gage, is the determination of absolute in situ stress by using the overcoring procedure, and five versions are in practical use for this very short-term application.

The South African CSIR (Council for Scientific and Industrial Research) doorstopper biaxial strain cell (Gregory et al., 1983; Leeman, 1971; Stickney et al., 1984) is bonded to the end of a borehole. Four versions of triaxial cell are bonded to the wall of a borehole: CSIR, South Africa (Herget, 1973; Leeman, 1971); Luleå, Sweden (Gregory et al., 1983; Hiltscher et al., 1979; Stillborg and Leijon, 1982); LNEC, Portugal (Rocha and Silverio, 1969; Rocha et al., 1974); and CSIRO, Australia (Jagger and Enever, 1978; Walton and Worotnicki, 1978; Worotnicki and Walton, 1976).

The CSIRO cell has been used for stress change monitoring on more occasions than the other four cells. Likely reasons are its high degree of waterproofing of the strain gages, use of a hard-wired cable, which eliminates the need for downhole connectors, and ready commercial availability.

The CSIR doorstopper and triaxial strain cells and the Luleå cell all require bonding of strain gages directly to the rock surface. For analytical reasons it is preferable for the strain gages to be on the rock surface, but moisture can cause severe degradation of the resistance signal, and in practice it is better to ensure complete waterproofing by full encapsulation of the strain gages, accepting that a correction factor is required to account for the annulus of en-

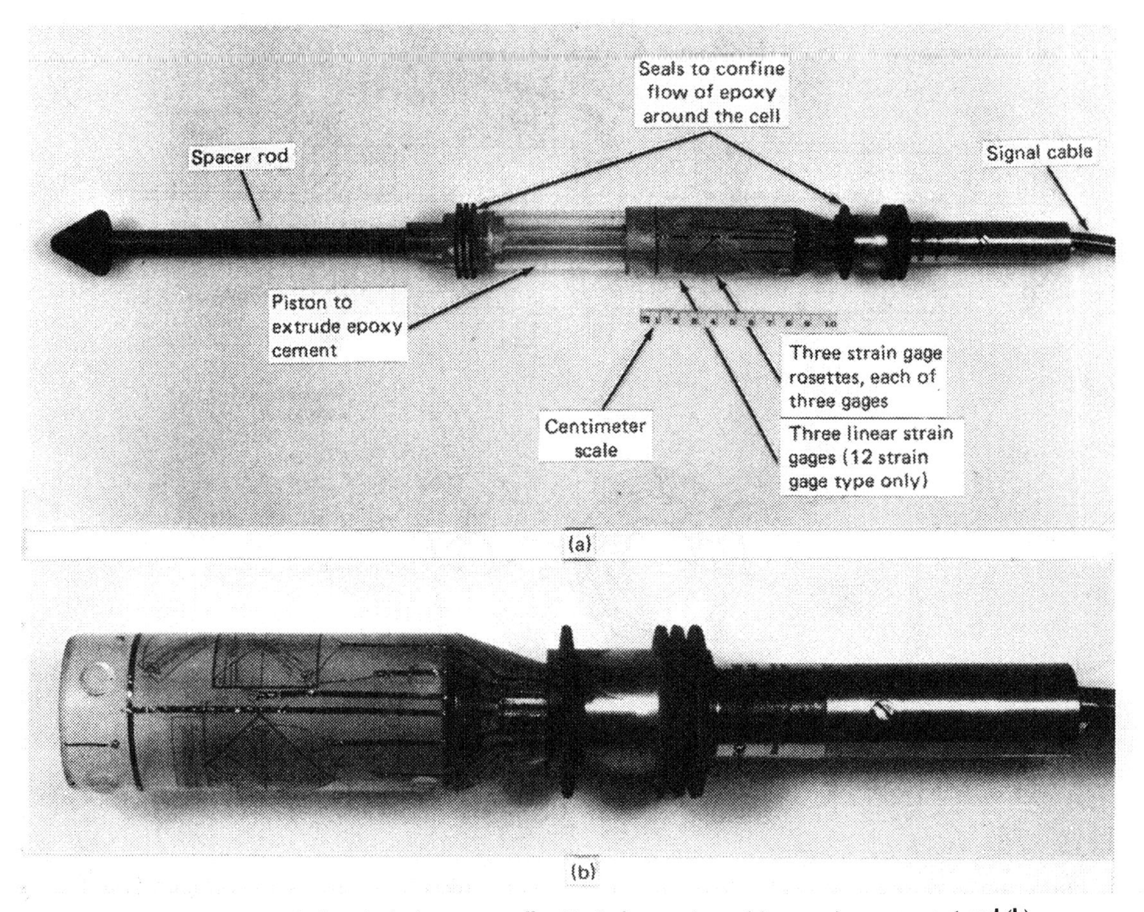

Figure 11.4. CSIRO hollow inclusion stress cell—12 strain gage type: (a) general arrangement and (b) detail of epoxy body of cell, containing strain gages (courtesy of Commonwealth Scientific and Industrial Research Organization, Australia).

capsulating material. The LNEC and CSIRO cells have strain gages that are fully encapsulated in epoxy resin.

The CSIRO cell, referred to as a *hollow inclusion (H.I.) cell,* is shown Figure 11.4. It consists of a thin epoxy tube, with three strain gage rosettes, each of three gages, embedded within the epoxy. The strain gage layout provides for some duplication of gages, so that in the event of malfunction of two and sometimes three of the nine gages it is still possible to determine the full stress tensor. When all nine gages are functional, the extra readings are a significant benefit, in that they allow data evaluation by least-squares fitting methods. Gages are wired in three-wire quarter bridge circuits. A version with 12 strain gages is available for use in anisotropic rock or to give greater redundancy of measured strains.

The cell is installed in a 1.5 in. (38 mm) diameter borehole by filling the hollow core with specially formulated epoxy cement, attaching the piston and spacer rod, and pushing with a setting tool to the end of the borehole. The cement is extruded between the cell and rock and allowed to set before initial strain readings are taken.

Methods of converting measured strain to rock stress are described by Duncan Fama and Pender (1980) and by Worotnicki and Walton (1976). An example of successful measurements is given by Kohlbeck and Scheidegger (1986).

The cell can be used to monitor changes in compressive stress up to the point where nonlinear conditions occur in the rock around the borehole. The ultimate sensitivity of the instrument is dependent on the resolution and accuracy of the readout equipment used: between 1 and 5 microstrain is common.

Despite the advantages of the CSIRO cell when compared with the other four versions, the user should be aware of four sources of error.

First, the cell may exhibit an output that implies a radial biaxial compression, which is in fact caused by a slight volumetric expansion of the epoxy as moisture is absorbed. The effect is largest soon after the cell is installed and decreases significantly after 30–100 days (Walton and Worotnicki, 1986). Windsor and Worotnicki (1986) state that cells used for long-term monitoring should preferably be installed at least 1 month before reliable data are required. The writers also present more than 1 year of data for cells that were installed 6–12 months before the monitoring period commenced, indicating apparent stability of the cells.

Second, the cell is sensitive to temperature changes, particularly in the circumferential direction, but the magnitude of the temperature response is known (Walton and Worotnicki, 1986). The cells are supplied with a thermistor to monitor temperature; therefore, a correction can be applied, but this is often unnecessary because many installed locations are deep underground where the rock undergoes little change in temperature.

Third, the epoxy body of the cell could continue to polymerize after the cell has been installed, which would cause a small volume change of the epoxy to occur. This effect is significantly reduced by curing the instrument at a temperature in excess of the field temperature. All commercially available instruments are cured by the manufacturer at 50°C (122°F) for 6 days, and most experience with long-term use of the cell has been at temperatures below 40°C (104°F). Long-term performance above this temperature is unknown, although the cell has been used for overcoring at temperatures up to 60°C (140°F).

Fourth, the body of the cell and the cement may creep. Walton and Worotnicki (1986) have demonstrated that, over a 1 year period and at a temperature of 18–25°C (64–77°F), the creep rate is approximately 0.3 microstrain/day radial compression and almost zero in the axial direction. They have also carried out a long-term laboratory test of a conventional and prototype thin-wall version of the cell, showing that the thin-wall version had superior long-term stability. However, no field experience was reported.

11.3.4. Miscellaneous Soft Inclusion Gages

Blackwood (1977), Rocha and Silverio (1969), and Worotnicki and Walton (1976) all describe variations of a solid soft inclusion cell for use when measuring in situ stress by the overcoring procedure. When used as stress monitoring instruments, solid soft inclusion cells appear to offer no advantages over hollow soft inclusion cells and in fact have a disadvantage when monitoring reduction in compressive stress. In this application the solid cells require a more effective cell-to-rock bond than the hollow cells, because higher tensile stresses must be maintained across the interface.

11.3.5. Comparison Among Soft Inclusion Gages

Comparative information is summarized in Table 11.1.

11.4. RIGID INCLUSION GAGES

Section 11.2 indicates that rigid inclusion gages, usually referred to as *stressmeters,* are designed to be stiff relative to the host rock and that stress determination does not require accurate knowledge of rock properties and constitutive behavior. Clearly, this is an advantage when comparing rigid gages with soft inclusion gages. The definition of a rigid inclusion gage, also given in Section 11.2, sets a practical upper limit of intact rock modulus for which rigid inclusions remain "rigid" as one-third the modulus of the gage. Steel, which has a modulus of approximately 30×10^6 lb/in.^{2} (200 GPa), is the most practical material from which to manufacture rigid inclusion gages. If a steel inclusion totally fills the cross section of the borehole, the upper limit of intact rock modulus for which a rigid gage can be considered "rigid" is therefore 10×10^6 lb/in.^{2} (70 GPa). When used in rock with a higher modulus, the gage must be calibrated in the host rock if maximum accuracy is required.

11.4.1. Vibrating Wire Stressmeters

The vibrating wire transducer, together with methods for minimizing errors caused by zero drift and by corrosion of the vibrating wire, is described in Chapter 8.

Uniaxial Stressmeter

The most commonly used and commercially available rigid inclusion gage is the uniaxial version of the vibrating wire stressmeter, which was developed for measuring stress change in underground

Table 11.1. Soft Inclusion Gages

Method	Advantages	Limitations[a]
Flat (e.g., Figure 11.1) and cylindrical borehole pressure cells	Economical	Small measurement scale[b] Rock deformability needed for pressure/stress conversion in elastic rocks
USBM borehole deformation gage (e.g., Figure 11.2)	Recoverable Suitable for monitoring reductions in compressive stress	Small measurement scale[b] Low electrical output; lead wire effects; errors resulting from moisture, temperature, and electrical connections are possible Not proved for other than short-term monitoring Movement of gage causes false readings: expensive to grout in place Rock deformability needed for strain/stress conversion Not recommended for use
CSIRO yoke borehole deformation gage (Figure 11.3)	Suitable for monitoring reductions in compressive stress	Small measurement scale[b] Low electrical output; lead wire effects; errors resulting from moisture, temperature, and electrical connections are possible Rock deformability needed for strain/stress conversion Limited field experience
CSIRO hollow inclusion triaxial strain cell (Figure 11.4)[c]	Triaxial data Strain gages encapsulated in epoxy resin Suitable for monitoring changes in tensile stress	Small measurement scale[b] Low electrical output; lead wire effects; errors resulting from moisture, temperature, and electrical connections are possible Rock deformability needed for strain/stress conversion Cementing difficulties in wet holes Potential errors caused by creep and moisture absorption of the epoxy Should be installed at least 1 month before requiring data

[a] Requires knowledge of rock properties and constitutive behavior.
[b] Limitation of small measurement scale can be minimized by installing several instruments in the rock mass of interest.
[c] When compared with other four versions of biaxial and triaxial strain cell, CSIRO cell is preferred: see Section 11.3.3. Other versions are therefore not included in this table.

coal mines (Hawkes and Bailey, 1973; Sellers, 1977). As shown in Figures 11.5 and 11.6, the stressmeter consists of a thick-walled steel cylinder with a vibrating wire transducer mounted across a diameter at approximately midlength. The cylinder is wedged in the borehole and acts as a proving ring. Figure 11.5 shows the vibrating wire in line with the wedge, a configuration adopted during the original development of the stressmeter. This configuration causes a range limitation, because the tension in the wire decreases as stress across the proving ring increases, and the wire can eventually become too

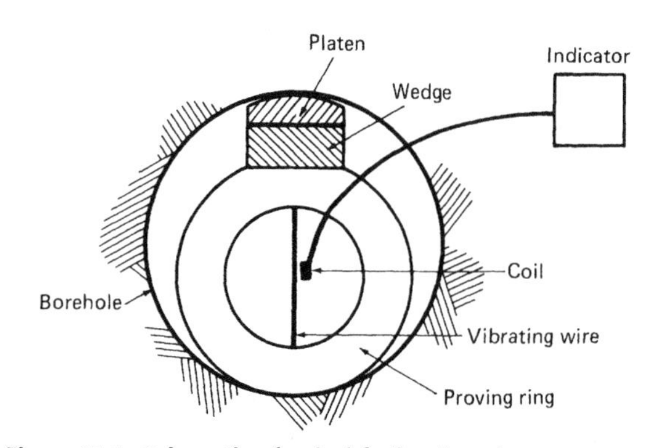

Figure 11.5. Schematic of uniaxial vibrating wire stressmeter.

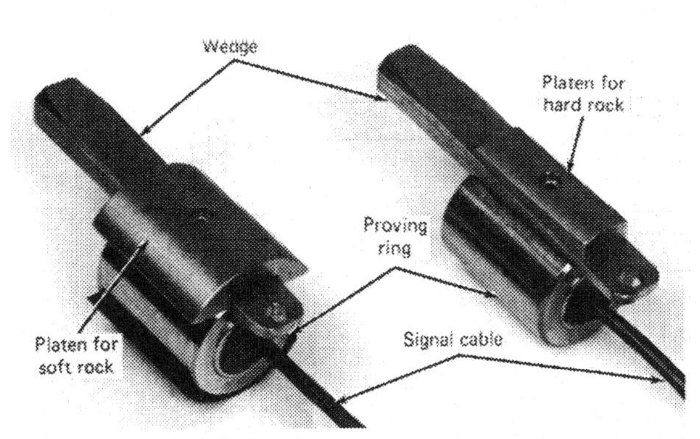

Figure 11.6. Uniaxial vibrating wire stressmeter (courtesy of Irad Gage, a Division of Klein Associates, Inc., Salem, NH).

slack. The device is now available in an alternative configuration, with the vibrating wire perpendicular to the alignment shown, such that the tension in the wire increases as stress increases. The range of this version is 15,000 lb/in.^{2} (100 MPa) compression and, by setting the wedge to create a high preload, it can be used to monitor a stress reduction of about 5000 lb/in.^{2} (35 MPa). Sensitivity is 0.5–5 lb/in.^{2} (3–30 kPa).

The stressmeter is available for installation in boreholes of nominal diameter 1.5, 2.4, and 3.0 in. (38, 60, and 76 mm), using special setting tools, and is preloaded across a diameter by pulling the wedge between the body and platen. A high preload is necessary so that both positive and negative stress changes can be monitored. Various platen sizes are available, and in soft rocks wide platens are used so that stress at the borehole wall is minimized. If this stress is excessive, local crushing of the rock can be caused at the line of contact between platen and rock. For rock moduli less than about 3×10^6 lb/in.^{2} (20 GPa), the soft rock platens should be used. The borehole should preferably be drilled with a rotary diamond bit, because if the contact between the stressmeter and the wall of the borehole is not uniform, the in-place loading will not match the loading during stressmeter calibration. If percussion drilling is used, a reaming shell should be inserted behind the bit. Installation details are given by Hawkes and Bailey (1973), Sellers (1977), and manufacturers of the instrument.

Early work with the stressmeter suggested that readings are insensitive to rock deformability when deformability of the intact rock is less than about 2×10^6 lb/in.^{2} (15 GPa), but recent tests indicate that the device is somewhat sensitive to rock deformability throughout its range. Dutta et al. (1981) and others have determined that the amount of surface contact between the gage platen and the borehole wall has a major influence on gage readings, and it is therefore important to control borehole size, shape, and surface condition at gage locations. When these factors are controlled carefully, the sensitivity to rock deformability approximately halves when intact rock deformability changes from 1 to 10×10^6 lb/in.^{2} (7–70 GPa). Rock deformability should therefore be measured or estimated and the calibration determined from data supplied by the instrument manufacturer.

When using the stressmeters in rocks with intact modulus higher than about 2×10^6 lb/in.^{2} (15 GPa), two cautions are applicable. First, if the modulus of the gage is taken as about 7.5×10^6 lb/in.^{2} (50 GPa) (Dutta et al., 1981), the definition given in Section 11.2 for a rigid inclusion indicates that the gage is no longer a true *stressmeter* when used in a host rock with an intact modulus greater than about 2.5×10^6 lb/in.^{2} (17 GPa). The higher the modulus of the host rock, the greater is the need for calibrations specific to the rock type in which the gage will be used. Second, Dutta et al. (1981) have established that, in stiffer rock, it is not always possible to reproduce the amount of surface contact between the gage and borehole for each installation. Because the amount of surface contact has a major influence on gage readings, it is desirable to install several gages in one location if maximum accuracy is required.

When a uniaxial vibrating wire stressmeter is installed in rock subject to biaxial or triaxial stress changes, it may not give a correct indication of stress changes in the direction of measurement. When a complete evaluation is required of stress change in the plane normal to the borehole axis, three uniaxial stressmeters can be set at known orientations to each other (Pariseau, 1985). Alternatively, a biaxial stressmeter can be used.

Biaxial Stressmeters

A biaxial stressmeter, shown in Figure 11.7, has been developed by the Soil and Rock Instrumentation Division of Goldberg-Zoino and Associates, Inc., Newton, MA, in conjunction with Geokon, Inc., Lebanon, NH, for thermomechanical testing in an underground excavation for studies relating to disposal of high-level nuclear waste in salt rock (Shuri and Green, 1987).

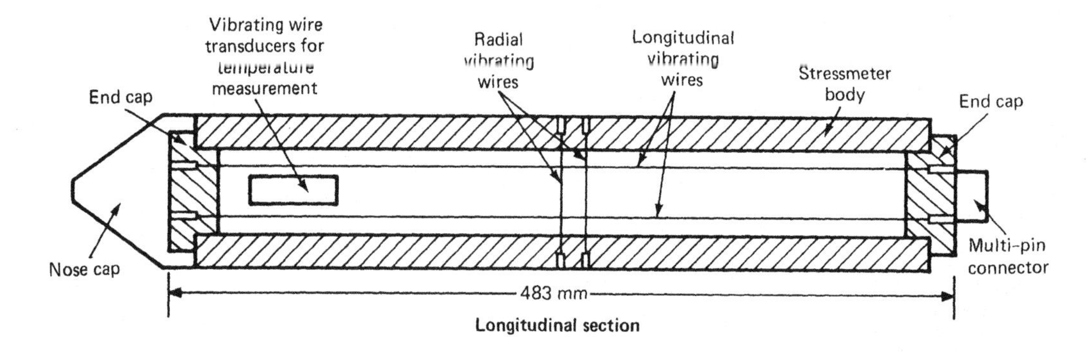

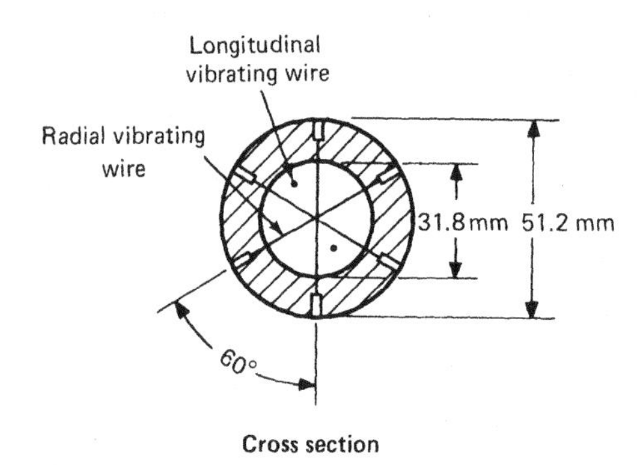

Figure 11.7. Biaxial vibrating wire stressmeter (courtesy of Soil & Rock Instrumentation Division, Goldberg-Zoino & Associates, Inc., Newton, MA, and Geokon, Inc., Lebanon, NH).

Radial deformation is measured using three vibrating wire transducers mounted at 60 degree intervals, and two such sets are included to provide redundancy. Two longitudinal transducers are also included to allow for correction of extensional effects, and two additional vibrating wire transducers are incorporated to provide data for thermal corrections. The stressmeter is grouted into the borehole using a cement/water mix. Extensive large-scale laboratory testing has been performed under triaxial stress conditions and at elevated temperatures.

There is currently no field experience with this stressmeter, but a similar instrument has been used successfully in the field to monitor stress in ice. The scale and extent of the laboratory testing program are both large, and the development appears promising. The results of the testing program indicate that, when installed in salt rock, the stressmeter has only minor dependence on characteristics of the rock and of the grout between stressmeter and rock. Test results were also satisfactory under compressive loading in granite with an intact modulus up to 10×10^6 lb/in.^{2} (70 GPa).

11.4.2. Photoelastic Stressmeters

A photoelastic stressmeter (Hall and Hoskins, 1972; Roberts and Hawkes, 1979) consists of a cylindrical glass plug with an axial hole, together with a light source and polarizing filter. The meter is bonded around its periphery into a borehole as shown in Figure 11.8. When the glass plug is subjected to stress, light and dark areas are visible when the plug is illuminated with polarized light and viewed through a hand analyzer. The light and dark areas are referred to as *photoelastic interference fringes,* and the change in the number of fringes is proportional to shear strain in the glass. Methods for installation and reading photoelastic stressmeters are described by Roberts and Hawkes (1979).

Use of photoelastic stressmeters is limited by the need for experienced reading personnel, and they are not viable instruments for general usage.

11.4.3. Tapered Plugs

Several solid metal instruments with a tapered outer surface have evolved in an attempt to create a posi-

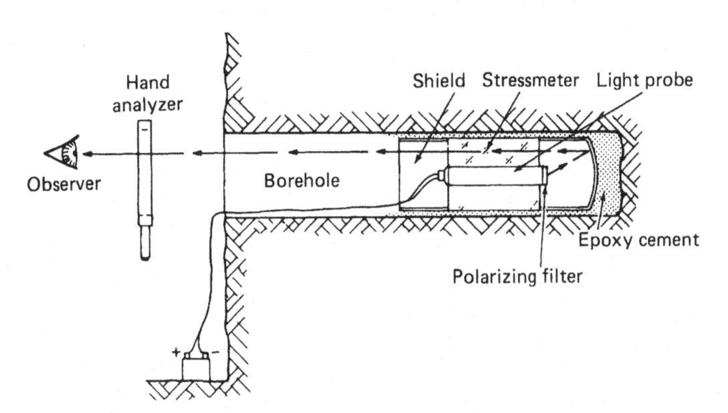

Figure 11.8. Photoelastic stressmeter (after Roberts and Hawkes, 1979).

tive contact between the instrument and the rock. They are forced into a matching tapered socket drilled at the bottom of a borehole.

Most designs are variations on those of Potts and Tomlin (1960) and of Wilson (1961). Wilson's gage consists of a tapered brass plug, split longitudinally to form a plane to which resistance strain gages are bonded, then rejoined. The instrument is designed for measuring stress change in coal and is forced into a matching tapered socket drilled at the bottom of a borehole. It can be used for determination of stress change across one diameter of the borehole only; that is, it is a uniaxial device. Enever et al. (1977) describe a similar uniaxial instrument, manufactured from steel, for determining stress changes in rock of higher modulus. Variations of this instrument, with strain gage rosettes mounted on planes transverse to the axis of the instrument to create biaxial gages, are described by Truong (1977) and Walton and Matthews (1978).

The main disadvantage of tapered plugs is their requirement for nonstandard boreholes and the consequent increased drilling cost. Walton and Matthews (1978) report that it is sometimes difficult to achieve adequate and reproducible transfer of stress from the rock to the gage, owing to uneven contact with the rock. It is also difficult to install these gages into a borehole in a way that produces moderate radial compression within the gage. They can therefore be used to monitor increase in stress but not reduction in stress, unless first radially compressed by a stress increase.

11.4.4. Miscellaneous Rigid Inclusion Gages

A rigid inclusion stressmeter with mechanical components similar to the vibrating wire stressmeter but with a bonded electrical resistance strain gage transducer is described by Cook and Ames (1979). Peng et al. (1982) describe a biaxial stressmeter consisting of a thick steel pipe with a transverse central portion on which a resistance strain gage rosette is mounted. It is bonded into a borehole with epoxy resin.

Spathis and Truong (1987) have analyzed the effect of placing a layer of in-fill material between a cylindrical plug and the wall of the borehole. A biaxial inclusion design that is bonded to the rock using a thick layer of high-modulus cement grout has been proposed, but no laboratory or field trials have been reported. Percussion drilled boreholes could be used with this design.

11.4.5. Comparison Among Rigid Inclusion Gages

Comparative information is summarized in Table 11.2.

Several of the gages fill the borehole completely. This is considered an advantage in all rock types, because when compressive stress is high, the support provided by the gage will help to minimize stress concentrations around the borehole. In contrast, with the "open" borehole condition that exists with most soft inclusion gages, failure of the rock around the borehole may occur because of concentration of stress. In viscoelastic rocks, if the gage does not fill the borehole completely, the rock will creep around the gage and results are likely to be incorrect.

11.5. RECOMMENDED PROCEDURES FOR MEASUREMENT OF STRESS CHANGE IN ROCK

Methods for the determination of stress change in rock masses have been a major area of research for

Table 11.2. Rigid Inclusion Gages

Method	Advantages[a]	Limitations[b]
Uniaxial vibrating wire stressmeter (Figure 11.5)	Lead wire effects minimal Installation is relatively simple Can be used to monitor reduction in compressive stress of up to about 5000 lb/in.^{2} (35 MPa)	Calibration dependent on contact geometry and initial preload Calibration somewhat dependent on rock deformability in high-modulus rock Gage does not fill borehole completely Single measurement axis Risk of wedge slippage if blasting nearby
Biaxial vibrating wire stressmeter (e.g., Figure 11.7)	Biaxial Lead wire effects minimal Gage fills borehole completely	Cannot be used to monitor large reductions in compressive stress[c] No field experience in rock
Photoelastic stressmeter (Figure 11.8)	Biaxial Gage fills borehole completely Poor cementing is immediately apparent Shows principal stress directions very well	Cannot be used to monitor large reductions in compressive stress[c] Lack of wide commercial availability Readings subjective Not amenable to remote readout
Tapered plugs	Gage fills borehole completely	Lack of wide commercial availability Custom drilling equipment required Low electrical output; lead wire effects; errors resulting from moisture, temperature, and electrical connections are possible Cannot be used to monitor large reductions in compressive stress[c] Not recommended for use

[a] Requires knowledge of rock properties and constitutive behavior only within broad bounds.
[b] Small measurement scale, but limitation can be minimized by installing several instruments in the rock mass of interest.
[c] Unless first subjected to a significant increase in compressive stress.

many years. Because it is impossible to measure stress or stress change directly, indirect techniques are required, and theoretical and practical difficulties are common. Monitoring stress changes involves the accurate measurement of small quantities over a long time period, often in a harsh environment, and at this stage of development there is no universal technique that can be applied to every rock type, rock mass condition, and measurement situation.

As indicated in Section 11.1, it is important to realize that the actual stress within a rock mass may vary significantly from point to point, depending on geologic features and the local stress concentrating effect of an opening. A single point measurement of stress may therefore be misleading, and if reliable data are required it is usually necessary to make measurements at numerous points throughout the rock mass of interest. Whenever possible, attempts should be made to determine the locations of discontinuities and to install the instruments in zones of intact rock.

When deciding on a technique to monitor stress change, consideration must be given to the type and condition of the rock mass, its stiffness, and whether uniaxial, biaxial, or triaxial measurements are required. The problem of attaching a measuring instrument to the rock and ensuring that the instrument experiences the correct stress is a general problem, common to nearly all instruments. Firm recommendations are impossible and the following recommendations are intended as guidelines only, to be used while bearing in mind the advantages and limitations given earlier. The guidelines are summarized in Table 11.3.

When considering accuracy of measurements, the authors concur with the following views expressed by IECO (1979):

> Evaluation of the accuracy of in situ stress-change measurement devices is difficult. Many types of gages . . . have produced data for periods of a year or more after emplacement. In most cases, the consistency of the stress change readings has led inves-

Table 11.3. Guide to Recommended Procedures for Measurement of Stress Change in Rock[a]

Number of Measurement Axes	Elastic Rock	Viscoelastic Rock
Uniaxial	Uniaxial vibrating wire stressmeter	Borehole pressure cell
Biaxial	Three uniaxial vibrating wire stressmeters Biaxial vibrating wire stressmeter CSIRO H.I. triaxial strain cell CSIRO yoke gage	Three borehole pressure cells Biaxial vibrating wire stressmeter
Triaxial	CSIRO H.I. triaxial strain cell	No instrument available

[a] Comparative information among options is given in Sections 11.3 and 11.4, and summarized in Tables 11.1 and 11.2.

> tigators to assume the results are reliable to at least ±25 percent. . . . But no detailed effort has been reported to document either gage accuracy or the rock and gage characteristics that affect the gage accuracy.

11.5.1. Uniaxial Measurements in Elastic Rock

The uniaxial vibrating wire stressmeter is well suited to applications in this category. However, the guidelines given in Section 11.4.1 should be followed carefully.

11.5.2. Biaxial Measurements in Elastic Rock

The choice for biaxial measurements is among four gages: a group of three uniaxial vibrating wire stressmeters, the biaxial vibrating wire stressmeter, the CSIRO H.I. cell, and the CSIRO yoke gage.

Provided that the Section 11.4.1 guidelines are followed, three uniaxial vibrating wire stressmeters, installed at different but known orientations in a single borehole, can be used in an attempt to make biaxial measurements. The dependence on deformability characteristics is within reasonable limits, and the vibrating wire transducers provide a stable signal that cannot be degraded readily. However, recognizing that contact geometry and initial preload cause uncertainties in calibration, substantial errors may arise from the combination of data from three stressmeters. Also, the capability for measuring large reductions in compressive stress depends greatly on the amount of preload created during installation.

The biaxial vibrating wire stressmeter, although having no field experience in rock, is an option for use in elastic rock with an intact modulus up to about 10×10^6 lb/in.^{2} (70 GPa), provided that the stress change is known to be compressive. As for the uniaxial stressmeter, the vibrating wire transducers provide a stable signal.

The CSIRO H.I. cell has been used successfully for measuring stress change, but there is a potential for errors caused by creep and moisture absorption. The errors can be minimized by installing the cell at least 1 month before requiring data. Measurement accuracy depends on the accuracy of deformability estimates or measurements, and electrical resistance transducers are subject to lead wire effects, and to the possibility of errors caused by moisture, temperature, and electrical connections.

The CSIRO yoke gage, although having limited field experience, has also been used successfully for monitoring stress change. As for the H.I. cell, measurement accuracy depends on the accuracy of deformability estimates or measurements, and the electrical resistance transducers can be sources of error.

It is important that readers should not make a selection among these four options, based merely on a reading of this summary section. The factors discussed earlier and summarized in Tables 11.1 and 11.2 should also be studied. Any one of the four options, depending on the specifics of each case and on comparative costs, may be the instrument of choice.

11.5.3. Triaxial Measurements in Elastic Rock

When monitoring stress changes in elastic rock, use of nontriaxial instruments involves assumptions for the magnitude of stress changes that occur in directions or planes in which there is no measurement. A triaxial instrument is required to measure any rotation of the principal stresses and to monitor stress changes in a situation where a borehole cannot be drilled perpendicular to the stress direction or plane of interest. Although the CSIRO hollow inclusion triaxial strain cell was not originally designed for monitoring stress change, with due care and consideration it can be used in elastic rocks when triaxial data are required. Because of the limitations reviewed in the previous section, it is best suited to monitoring three-dimensional stress changes that occur over short time periods (approximately 1–2 months), rather than a slow buildup or decrease in stress levels.

11.5.4. Measurements in Viscoelastic Rock

Borehole pressure cells are suitable for measurement of stress change in viscoelastic rock. The biaxial vibrating wire stressmeter, although having no field experience in rock, is an attractive alternative for biaxial measurements, and of course it can also be used if only uniaxial data are required.

CHAPTER 12

MEASUREMENT OF DEFORMATION

12.1. INSTRUMENT CATEGORIES

Instruments for measuring deformation can be grouped in the categories listed in Table 12.1. Definitions of each category, together with an indication of typical applications, are given in later sections of this chapter. It can be seen that there is a vast array of instruments for monitoring deformation, but Peck (1972) warns:

> An instrument too often overlooked in our technical world is a human eye connected to the brain of an intelligent human being. It can detect most of what we need to know about subsurface construction. Only when the eye cannot directly obtain the necessary data is there a need to supplement it by more specialized instruments. Few are the instances in which measurements by themselves furnish a sufficiently complete picture to warrant useful conclusions.

12.2. SURVEYING METHODS*

Surveying methods are used to monitor the magnitude and rate of horizontal and vertical deformations of structures, the ground surface, and accessible parts of subsurface instruments in a wide variety of construction situations. Frequently, these methods are entirely adequate for performance monitoring, and geotechnical instruments are required only if greater accuracy is required or if measuring points are inaccessible to surveying methods, as is the case for subsurface measurements. In general, whenever geotechnical instruments are used to monitor deformation, surveying methods are also used to relate measurements to a reference datum.

Surveying methods are described briefly in the following subsections, and comparative information is given in Table 12.2. Reference datums and measuring points for monitoring surface deformation are also described in this section.

Surveyors who work on construction sites often have little experience with the accuracies required for deformation monitoring, and a well-trained survey crew is essential when maximum accuracy is required. Measurement accuracy is controlled by the choice and quality of surveying technique and by characteristics of reference datums and measuring points. Survey instrument technology is well established, and most reputable manufacturers include a statement of accuracy in their instrument specifications, which can be relied on if the instrument is calibrated and operated in accordance with instructions.

Discussions of surveying methods by Cording et al. (1975), Gould and Dunnicliff (1971), and Senne

*Written with the assistance of Thomas S. McGrath, Land Surveyor, Upper Montclair, NJ, and Joseph H. Senne, Professor and Chairman, Department of Civil Engineering, University of Missouri–Rolla, MO.

Table 12.1. Categories of Instruments for Measuring Deformation

Category	Type of Measured Deformation						Section
	↔	↕	⤢	⟲	∸	⨪	
SURVEYING METHODS	●	●	●		●		12.2
Optical and other methods							
Benchmarks							
Horizontal control stations							
Surface measuring points							
SURFACE EXTENSOMETERS	●	●	●		●		12.3
Crack gages							
Convergence gages							
TILTMETERS				●	●	●	12.4
PROBE EXTENSOMETERS	●	●	●			●	12.5
Mechanical heave gage							
Mechanical probe gages							
Electrical probe gages							
Combined probe extensometers and inclinometer casings							
FIXED EMBANKMENT EXTENSOMETERS	●	●	●			●	12.6
Settlement platform							
Buried plate							
Mechanical gage with tensioned wires							
Gages with electrical linear displacement transducers							
Soil strain gage							
FIXED BOREHOLE EXTENSOMETERS	●	●	●			●	12.7
Single-point and multipoint extensometers							
Subsurface settlement points							
Rod settlement gage							
INCLINOMETERS	●	●	●	●		●	12.8
TRANSVERSE DEFORMATION GAGES	●	●	●			●	12.9
Shear plane indicators							
Plumb lines							
Inverted pendulums							
In-place inclinometers							
Deflectometers							
Borehole directional survey instruments							
LIQUID LEVEL GAGES		●				●	12.10
Single-point and multipoint gages							
Full-profile gages							
MISCELLANEOUS DEFORMATION GAGES							12.11
Telltales	●	●	●		●	●	
Convergence gages for slurry trenches	●					●	
Time domain reflectometry	●	●	●		●	●	
Fiber-optic sensors	●	●	●		●	●	
Acoustic emission monitoring	●	●	●			●	

Key: ↔ horizontal deformation; ↕ vertical deformation; ⤢ axial deformation (↔ or ↕ or in between); ⟲ rotational deformation; ∸ surface deformation; ⨪ subsurface deformation.

Table 12.2. Surveying Methods

Method	Advantages	Limitations	Approximate Accuracy
Elevations by optical leveling	Fast, particularly with self-leveling instruments Uses widely available technology	First-order leveling requires high-grade equipment and careful adherence to standard procedures	Third order: ± 0.05 ft $\times \sqrt{mi}$ (± 12 mm $\times \sqrt{km}$)[a] Second order: ± 0.025–0.033 ft $\times \sqrt{mi}$ (± 6–8 mm $\times \sqrt{km}$) First order: ± 0.012–0.020 ft $\times \sqrt{mi}$ (± 3–5 mm $\times \sqrt{km}$)
Distance measurements by taping	Direct measurements	Requires clear, relatively flat surface between measuring points and reference datum Tape should be checked against a standard frequently	Third order: $\pm$ 1/3000–1/6000 of distance Second order: $\pm$ 1/20,000–1/50,000 of distance First order: $\pm$ 1/300,000 of distance
Offsets from a baseline using theodolite and scale	Direct measurements	Requires baseline unaffected by movement	± 0.001–0.005 ft (± 0.3–2 mm)
Traverse lines	Usable where direct measurements are not possible	Accuracy decreases as number of legs in the traverse line increases If possible, traverse should be closed	$\pm$ 1/30,000–1/150,000 of distance
Triangulation	Usable where direct measurements are not possible	Requires accurate measurement of angles and baseline length Very slow when compared with trilateration by EDM	$\pm$ 1/30,000–1/1,000,000 of distance
Laser beam leveling and offsets	Faster than conventional optical methods Readings can be made by one person	Seriously affected by air turbulence, humidity, and temperature differential Requires curvature and refraction corrections beyond about 650 ft (200 m) Limited to about 0.25 mile (0.4 km)	± 0.01–0.03 ft (± 3–10 mm)
Electronic distance measurement (EDM)	Long range Fast and convenient Very accurate	Accuracy is influenced by atmospheric conditions	For distance: ± 0.001–0.03 ft (± 0.3–10 mm) ± 2–5 ppm For lateral position change by trilateration: ± 0.005–0.03 ft (± 2–10 mm) ± 2–5 ppm
Trigonometric leveling	Long range Fast and convenient Can be done simultaneously with traversing	Accuracy is influenced by atmospheric conditions Requires a very accurate measurement of zenith angle	Third order: ± 0.05 ft $\times \sqrt{mi}$ (± 12 mm $\times \sqrt{km}$)[a] Second order: ± 0.025–0.033 ft $\times \sqrt{mi}$ (± 6–8 mm $\times \sqrt{km}$)
Photogrammetric methods	Can record movement of hundreds of potential points at one time for determination of overall deformation pattern	Weather conditions can limit use Interpretation requires specialist skill For good accuracy the baseline should be not less than one-fifth of the sight distance	$\pm$ 1/5000–1/100,000 of distance
Global positioning system	Operates with little attention from personnel Can be set to trigger a warning device Very accurate	Very expensive Availability very limited at present Requires special ephemeris and computer software	± 0.0005–3 ft (± 0.2 mm to 1 m); average accuracy is ± 0.03–0.1 ft (± 10–30 mm) with 1½ hours of observation per point

[a] mi = distance in miles; km = distance in kilometers.

Figure 12.1. Second-order automatic level, Kern Model GK2-A (courtesy of Kern Instruments, Inc., Brewster, NY).

(1980) have provided substantial material for preparation of this section. General texts describing surveying methods include Bouchard and Moffitt (1987) and Davis et al. (1981): these should be consulted by the reader who needs more detail.

12.2.1. Elevations by Optical Leveling

Construction site settlement surveys are usually carried out using engineer's levels (e.g., Figure 12.1) at second- or third-order accuracy.

Second-order leveling requires limiting sight distances, balancing foresight and backsight, carefully plumbing the rod, and reading on well-defined marks and stable turning points. The circuit should be closed on a benchmark and the apparent closing error distributed. A *two-peg test* should be made prior to measurements and the engineer's level adjusted if necessary.

First-order leveling requires an optical micrometer attachment and a pair of invar survey rods. Requirements for first-order leveling include very-high-quality equipment, careful adherence to standard procedures, a minimum of atmospheric disturbance, and minimum temperature variation.

12.2.2. Distance Measurements by Taping

Taping over distances greater than about 200 ft (60 m) has largely been superseded by electronic distance measurements (EDM). However, if no EDM is available, distance measurements can be made by taping. A thorough treatment of sources of error and methods of improving accuracy of tape measurements is given by Milner (1969).

12.2.3. Offsets from a Baseline Using Theodolite and Scale

Measurements are made from a baseline by simple right angle offset, using a scale or steel tape.

12.2.4. Traverse Lines

A traverse to determine change in lateral position is a survey made by measuring successive distances and angles. If the traverse returns to the starting point it is called a *closed traverse,* in which the sum of interior angles of the polygon can be calculated and adjusted. The sides can also be adjusted for error of closure, which gives a good indication of the overall precision of the traverse.

The accuracy of angular measurements depends on the theodolite, sighting target, and atmospheric turbulence. Theodolites reading to 1 arc-second generally do not have the resolving power to align on a target to that accuracy, but six to eight positions will yield a standard error within 1 arc-second.

12.2.5. Triangulation

Where direct taping is impracticable, triangulation can be used to determine change in lateral position. A fixed baseline is measured accurately by EDM or precise taping techniques, and two angles are determined between the baseline and the measuring points. It is vital that the reference baseline is established on stable ground outside the zone of movement. Once the baseline has been established, it is necessary only to determine angles to the measuring points and to calculate their positions and propagated errors. Use of equi-angular figures increases accuracy. Greatest accuracy, for example, when monitoring movements of the downstream face of an arch dam, requires first-order triangulation. Figure 12.2 shows a theodolite suitable for first-order triangulation.

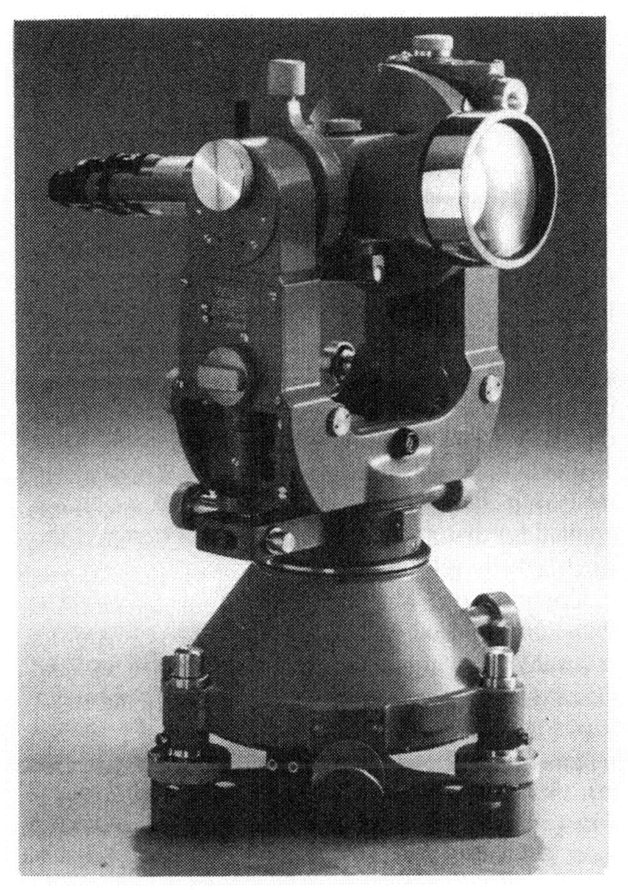

Figure 12.2. Precision theodolite, Wild Model T3 (courtesy of Wild Heerbrugg Instruments, Inc., Farmingdale, NY).

Most triangulation work is done at night or on cloudy days, since atmospheric turbulence and temperature variations are the limiting factors in making accurate observations.

12.2.6. Laser Beam Leveling and Offsets

The word *laser* is an acronym for *l*ight *a*mplification by *s*timulated *e*mission of *r*adiation. Lasers (e.g., Figure 12.3) can be used for alignment measurements and leveling, but the beam is deflected by air turbulence, humidity, and temperature differential. When these factors are small, measurement accuracy over distances of up to about 1000 ft (300 m) is about ±0.01 ft (±3 mm). Attempts have been made to increase accuracy by using split photocell detectors, but under typical field conditions accuracy is usually no better than ±0.01 ft (±3 mm) even over distances as small as 100 ft (30 m).

Figure 12.3. Electronic level, Spectra-Physics Model EL-1 (courtesy of Spectra-Physics, Dayton, OH).

12.2.7. Electronic Distance Measurement

Electronic distance measurement (EDM) equipment is used for measurement of distances, either for direct determination of distance change or for determination of lateral position change by trilateration. It is also used for trigonometric leveling, as discussed in the next section. Over the last 20 years the availability of increasingly reliable and accurate EDM equipment has radically changed conventional surveying practices. EDM devices require fewer personnel than conventional optical instruments, are faster to use, and are more accurate.

EDM equipment makes use of the velocity of electromagnetic radiation to measure the distance between the instrument and a reflector prism that is placed at the measuring point. Some equipment uses microwave radiation while others use infrared or visible light. Since air density has an effect on the velocity of light, air temperature, pressure, and humidity must be monitored. In the modern EDM, these factors are monitored and processed internally with a microcomputer.

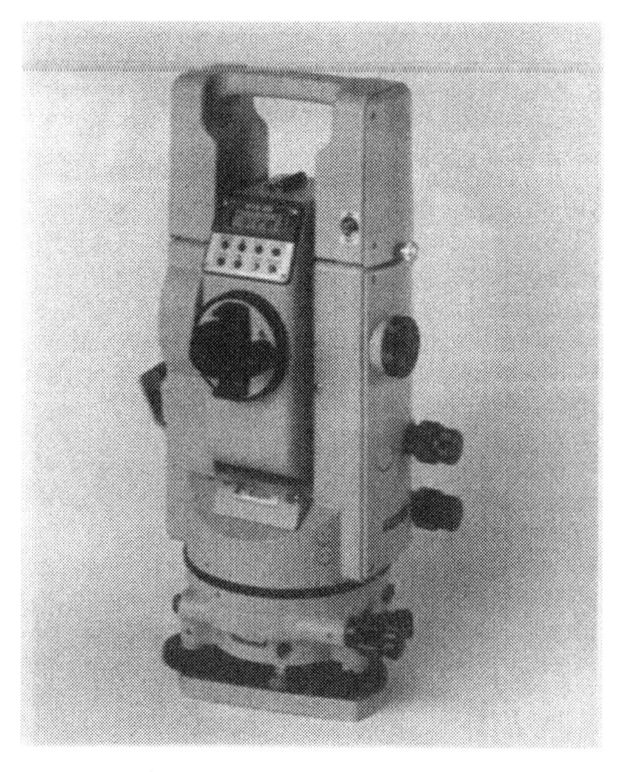

Figure 12.4. Electronic distance measurement equipment, Topcon Model GTS-2B (courtesy of Topcon Instrument Corporation of America, Paramus, NJ).

Depending on the model, an EDM can have a range of a few feet to several miles. Most instruments have two components of error: a random error plus a small percentage of the sight length. As an example of a highly accurate EDM, the Mekometer, developed in England in the late 1960s, has a range of 50–10,000 ft (15–3000 m) and an accuracy of ±0.001 ft (±0.3 mm) ±2 parts per million (ppm). The Mekometer has been used for monitoring movements of embankment dams (Penman and Charles, 1973; Penman and Mitchell, 1970). A number of instruments, for example, the instrument shown in Figure 12.4, are widely available and in use at a more modest price and are capable of measuring distances to within ±0.015 ft (±5 mm). A list of available instruments is given by Hanna (1985).

K. Robertson (1977, 1979) describes the operating principle of EDM and discusses the usefulness of precision EDM in detecting small movements of large dams. He reports that surveys can detect movements within 0.01 ft (3 mm) when measurements are made during daylight hours, requiring about 5 minutes per measurement. In contrast, theodolite observations are usually made at night to minimize refraction effects and require the averaging of eight to twelve positions for each angle, involving at least ½ hour at each station.

12.2.8. Trigonometric Leveling

Trigonometric leveling uses EDM equipment to measure the slope distance from the instrument to a prism placed at the measuring point. The vertical angle between this sloping line and horizontal (the *zenith angle*) is measured with either a semi-precise (6 arc-seconds) or precise (2 or 1 arc-second) theodolite. The elevation difference between the instrument and measuring point is calculated from the measured distance and angle, and corrections are applied for distortion caused by the curvature of the earth and by refraction.

Trigonometric leveling is much more economical than conventional optical leveling when third-order accuracy is adequate and can be used when measuring points are physically inaccessible. Second-order trigonometric leveling requires use of special targets, sight distances not exceeding 1000 ft (300 m), more expensive equipment, and longer procedures. However, it is generally more economical than second-order optical leveling, especially in hilly terrain.

12.2.9. Photogrammetric Methods

Precise photography for measuring structural movements employs phototheodolites to take successive photographs from a fixed station along a fixed baseline. Movements are identified in a stereocomparator by steroscopic advance or recession of pairs of photographic plates in relation to stable background elements. The procedure defines components of movement taking place in the plane of the photograph. The method is similar to triangulation and involves the intersection of lines of sight, as shown in Figure 12.5. The film planes are usually parallel to each other and as nearly perpendicular as possible to a line joining the midpoints of the baseline and observation points. To calculate the position of a point, the focal length and orientation of the camera must be known, as well as the elevation of the ends and the length of the baseline. Once the control stations are established for the phototheodolite, the field work is minimal. The photogrammetric method has the advantage that hundreds of potential movements are recorded on a

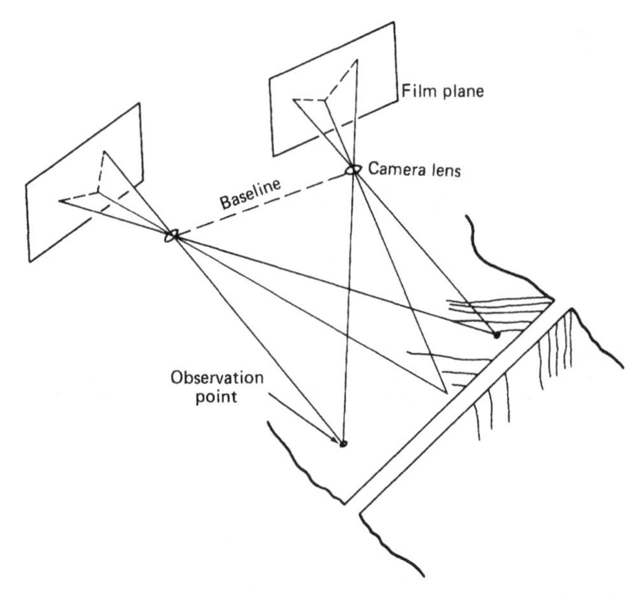

Figure 12.5. Photogrammetric arrangement using stereo pairs (after Senne, 1980).

single stereo photographic pair, allowing an appraisal of the overall displacement pattern in a minimum time. Figure 12.6 shows a stereoplotter system.

Measurement accuracy depends on many factors. Phototheodolites designed for this purpose should be used. The baseline should be as long as topography permits, not less than one-fifth the sight distance and nearly perpendicular to the line of sight. Stereocomparator measurements should be made to micron accuracy. In general, the standard error of measurement may vary from $\pm 1/5000$ to $1/100{,}000$ of the sight distance, and accuracies as good as ± 0.02 ft (± 6 mm) have been reported for sight distances less than 200 ft (60 m). For longer sights up to 1600 ft (500 m), with baselines near 330 ft (100 m), accuracies of ± 0.16 ft (± 50 mm) have been obtained.

Moore (1973) describes photogrammetric measurements to determine deformations of a rockfill dam. A Wild P-30 phototheodolite was used and the data processed using a Wild A-7 Autograph. The standard deviations of the mean coordinates were in eastings ± 20 mm (± 0.07 ft), in northings ± 40 mm (± 0.13 ft), and in elevation ± 10 mm (± 0.03 ft).

Bozozuk et al. (1978) report on the use of photogrammetry for measuring pile head movements as surrounding piles were driven in a large pile group. Accuracy of measured horizontal deformations is reported as ± 0.07 ft (± 20 mm), of vertical deformations ± 0.02 ft (± 6 mm).

McVey et al. (1974) describe experiments with

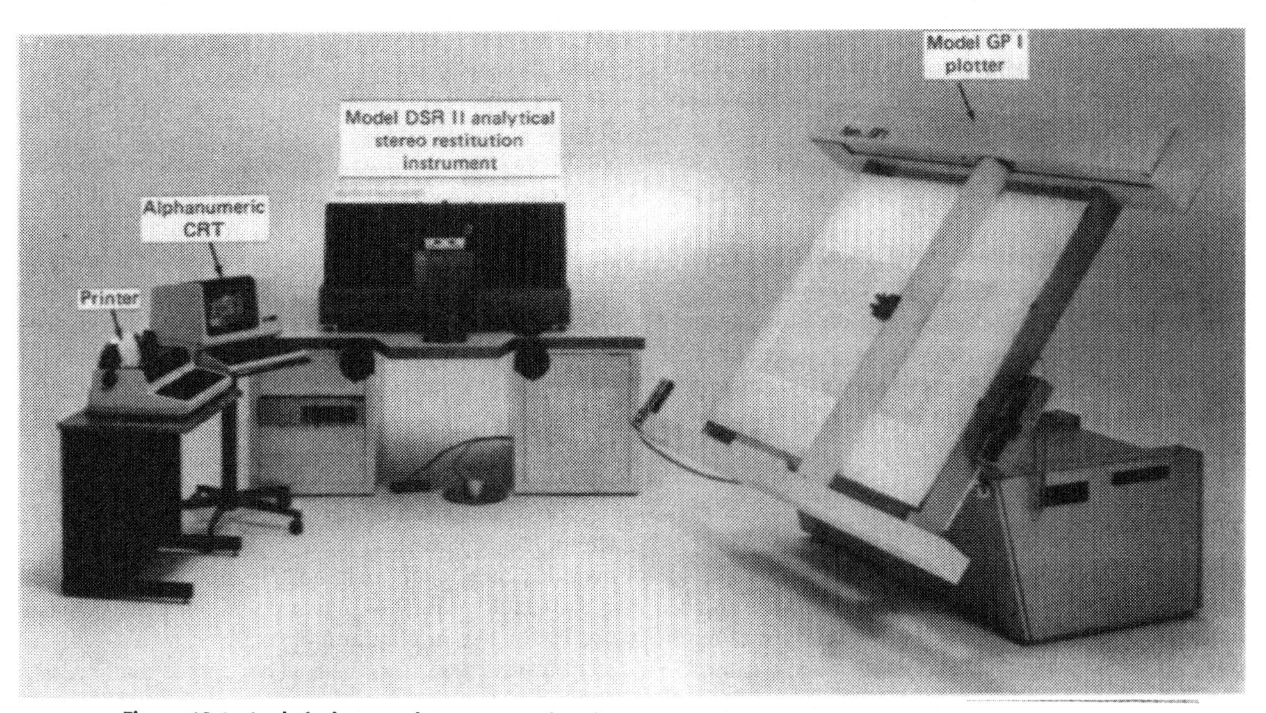

Figure 12.6. Analytical stereoplotter system for photogrammetric measurements (courtesy of Kern Instruments, Inc., Brewster, NY).

photogrammetric methods for monitoring deformation of underground openings, report on the diffi culty of maintaining a suitable baseline, and suggest that an accuracy of ±0.002 ft (±0.6 mm) might be possible.

12.2.10. Global Positioning System

The *NAVSTAR Global Positioning System* (GPS) was originally conceived as a tool for military sea, land, and air navigation. The method is described by Hoar (1982) and Laurila (1983).

The system consists of three parts: satellites, a ground control network, and user equipment. Radio signals are used in an interferometric mode. Two or more GPS receivers simultaneously receive signals from the same set of satellites, and the resulting observations are subsequently processed to obtain the interstation difference in position. If one of the receivers is placed at a known position, the three-dimensional position of the second receiver may be determined, and the number of stations determined simultaneously is limited only by the number of receivers available. Calculations require use of a book of satellite positions as a function of time and earth position (an *ephemeris*) and special computer software. Accuracy of measurements is highly dependent on the time allowed for observations, and thus on the available funds, and submillimeter accuracy is possible in the extreme case. For civil engineering purposes, such as monitoring the deformation of dams, a more typical accuracy is ±0.03–0.1 ft (±10–30 mm), requiring about 1½ hours of observation per point, at a cost of U.S. $500–2500 per point.

At present there are seven GPS satellites in orbit, and availability for civil engineering purposes is somewhat limited. Original plans called for 18 satellites to be operational by 1989, so that availability would be unlimited, but these plans are currently delayed by the interruption of the U.S. space program.

12.2.11. Reference Datums

A stable reference datum is required for all survey measurements of absolute deformation. A reference datum for measurements of vertical deformation is referred to as a *benchmark*. A reference datum for measurements of horizontal deformation is generally referred to as a *horizontal control station* or a *reference monument*.

Benchmarks

Benchmarks established on substantial stable permanent structures ordinarily do not contribute error to settlement observations. However, a verification should first be made that the structure is not moving vertically owing to conditions such as groundwater lowering, seasonal thermal effects, or loading on piled foundations by negative skin friction.

The author suspects that many "benchmarks" used on construction projects are not as stable as the user thinks. If no suitable permanent structure is available that is **known** to be stable and remote from all possible vertical movement, a deep benchmark should be installed to a depth below the seat of vertical movement. Benchmarks placed at shallow depths in soil probably move to some extent and the movement may be sufficient to interfere with the desired accuracy of a survey. Apart from effects of frost heave, seasonal moisture changes, and nearby trees, construction activities may cause a near-surface benchmark to settle by subsoil densification from blasting or pile driving, by consolidation from nearby loading or drawdown, or as a result of extension strains directed toward an excavation.

A deep benchmark consists of a pipe or rod, anchored at depth, surrounded by and disconnected from a sleeve pipe. The sleeve pipe protects the inner pipe or rod from drag caused by soil movement. Details are similar to single-point fixed borehole extensometers, as described later in this chapter. The anchor may be mechanical, hydraulic, or grouted. Figure 12.7 shows an arrangement for a benchmark installation in rock, with a grouted anchor and the inner rod centered in the casing with nylon spacers. The space between the rod and casing may be filled with heavy oil or bentonite slurry to minimize friction.

Unless protected, exposed benchmarks are affected by thermal expansions and contractions from temperature changes. In high-precision surveys where these movements cannot be tolerated, the top 10–15 ft (3–4.5 m) of the deep benchmark rod should be replaced with a temperature compensating alloy steel such as invar (Bozozuk, 1984).

Bozozuk et al. (1962) give details of deep benchmarks for use in clay and permafrost areas, with a steel foot driven to refusal. The Borros anchor, as described later in this chapter, may be used for a benchmark if anchored in unyielding strata. Kjellman et al. (1955) describe a benchmark that is

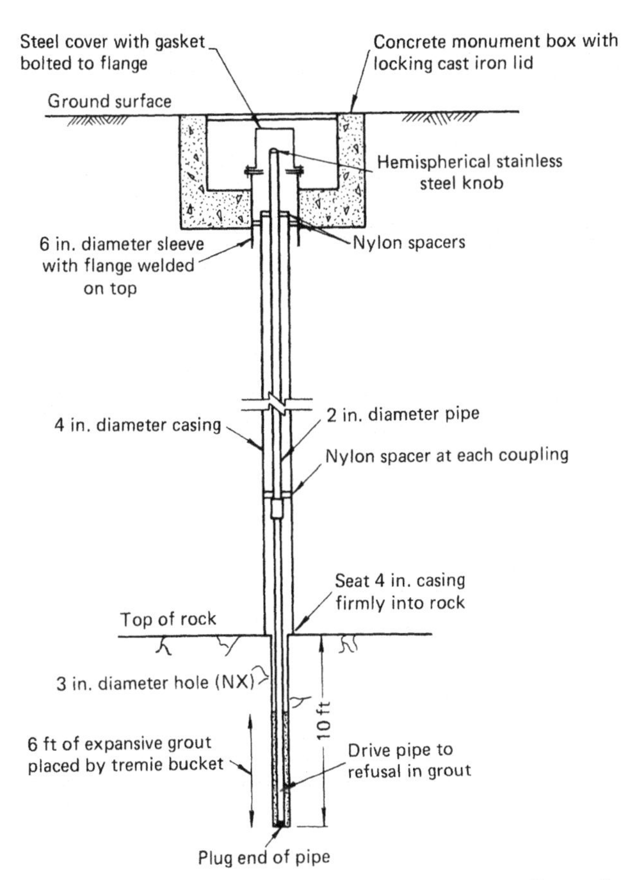

Figure 12.7. Benchmark installation in rock (after Cording et al., 1975).

isolated by using a coating of asphalt or silicon grease wrapped with aluminum foil, instead of casing, to break the soil/rod bond. However, casing appears to provide a more positive method of bond breaking.

It is good practice to install three benchmarks and to survey between them on a regular basis, thereby identifying any vertical movement of a particular benchmark. If existing deep "benchmarks" are being considered, their details should be evaluated carefully, since they may not be designed to isolate the inner pipe or rod from soil movements.

Horizontal Control Stations

Reference datums for horizontal movements require a pedestal with a force centering device in the top for attaching a target and theodolite or electronic distance measuring device. The pedestal should be designed to prevent tilting and should be located below the seasonal movement zone. Figure 12.8 shows a horizontal control station with a force

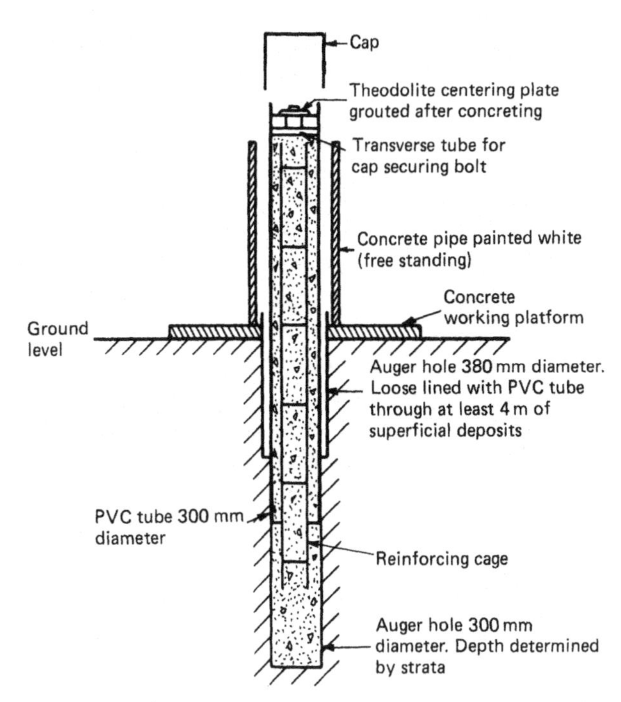

Figure 12.8. Horizontal control station for a theodolite (after Burland and Moore, 1974).

centering plate for the theodolite. The white concrete pipe around the upper part of the pillar shields it from the sun, since differential heating of the pillar can cause considerable apparent movement.

It is generally wise to use a horizontal control station that is at about the average elevation of the project. This eliminates the need for datum reduction computations.

There have been cases where horizontal control stations installed on river banks have moved inward; thus, the distance between such stations should be checked occasionally.

Inverted pendulums (Section 12.9.2) can also be used as horizontal control stations.

12.2.12. Measuring Points for Monitoring Surface Deformation

The term *measuring point* is used in this book for a point that may move. The term *observation point* can also be used. The word *benchmark* is sometimes used for a point that may move, but this use is incorrect: as discussed in the previous section, a benchmark is an immovable reference datum.

Measuring points may be located on structures or on the ground surface. The primary requirement is for stable and robust points that will survive throughout their required life. Temporary points that are later replaced by permanent points should be avoided, as they tend to be easily damaged and errors may occur when converting to the permanent points. Measuring points must have a clearly identified mark or surface to which each measurement is made. For example, for optical leveling, a machined hemispherical corrosion-free seat should be provided for the rod. Measurements of horizontal deformation may require filed cross-lines on the head of the measuring point. For procedures using a theodolite, force centering trivets may be used to produce repeatable centering of the instrument.

Various surface measuring points are described in this section. Measuring points for monitoring subsurface deformation are described in later sections of this chapter.

Measuring Points on Structures

Measuring points on structures may be simple 0.375 in. (10 mm) diameter by 2 in. (50 mm) long nickel plated steel carriage bolts cemented into shallow holes, using an epoxy-bonded patching compound. They can be used for both horizontal and vertical deformation measurements. Alternatively, bolts can be attached by using one of the proprietary powder actuated or expansion anchoring systems.

Figure 12.9 shows two types of measuring point for use while optical leveling on vertical surfaces. Permanent scales may also be attached to structures for elevation or offset measurements.

Measuring Points on Surface of Ground

When measurements of ground surface movement are required, it is generally necessary to seat the measuring point below the zone of frost heave and seasonal moisture changes. Cycles of shrinkage and swell in clay under grass-covered fields have been observed to cause vertical movements to depths of 15 ft (5 m) or more in severe climates and in expansive soils. Bozozuk et al. (1962) indicate that in the plastic Leda clay near large elm trees, vertical movements have been measured varying from 3 in. (80 mm) at the ground surface to 0.5 in. (13 mm) at a depth of 14 ft (4 m).

If the purpose is to monitor surface settlement caused by excavation, and the surface has structural strength, the measuring point must be seated below the surface. For example, when monitoring the settlement trough caused by tunneling below a

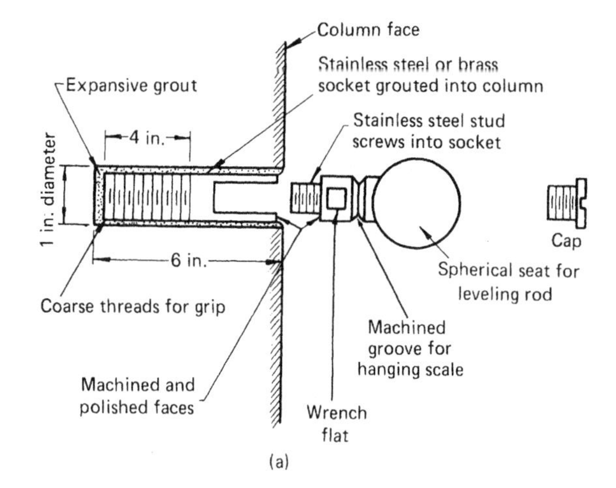

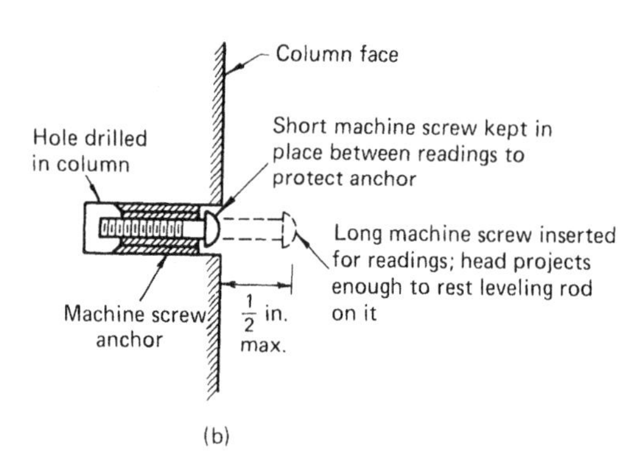

Figure 12.9. Measuring points on structures, suitable for use with optical leveling: (a) for precise work and (b) for less precise work (after Cording et al., 1975).

concrete pavement, use of paint marks on the pavement can be misleading.

A typical measuring point for monitoring settlement of the ground surface is shown in Figure 12.10. Alternatively, a shallow subsurface settlement point (Section 12.7.6) can be used. Figure 12.11 shows a typical measuring point for monitoring settlement on the surfaces of embankment dams. Wilson (1967) describes a measuring point (Figure 12.12) with a vernier gage and movable target for use with a theodolite when measuring horizontal deformation of the surfaces of embankment dams. Other ground surface measuring points are described by Burland and Moore (1974).

Control stakes, also referred to as *alignment stakes, displacement stakes, toe stakes, reference stakes,* and *slope monitoring stakes,* are used in highway and other work for monitoring offsets from a baseline. An example is shown in Figure 12.13. A more rugged steel construction can be used in an

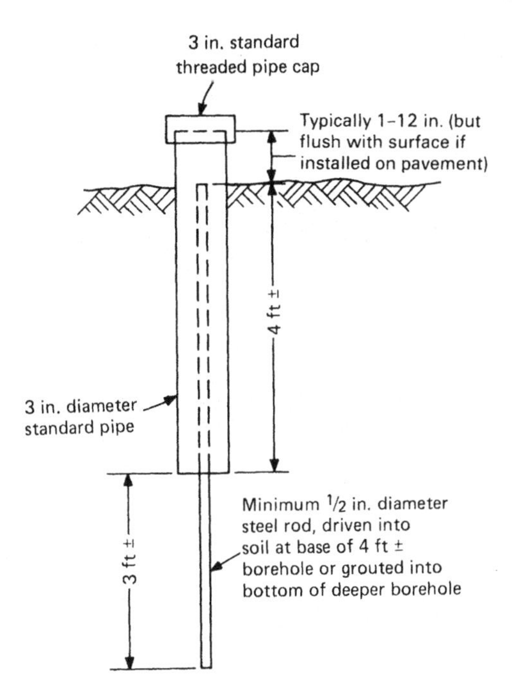

Figure 12.10. Measuring point on ground surface, suitable for use with optical leveling.

attempt to resist vandalism, and stakes can be set in concrete below frost penetration depth to minimize frost-induced movements. Wherever feasible, stakes should be installed in a straight line, so that significant movements can be detected easily and approximate measurements of offsets made by reading the scale with binoculars or a survey instru-

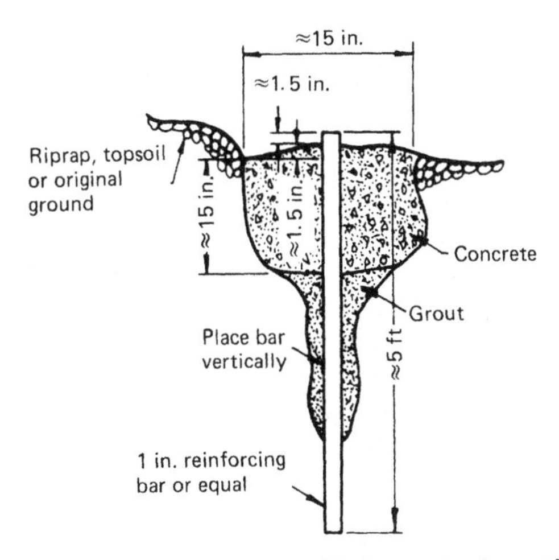

Figure 12.11. Measuring point suitable for monitoring settlement on surfaces of embankment dams (after USBR, 1974; courtesy of Bureau of Reclamation).

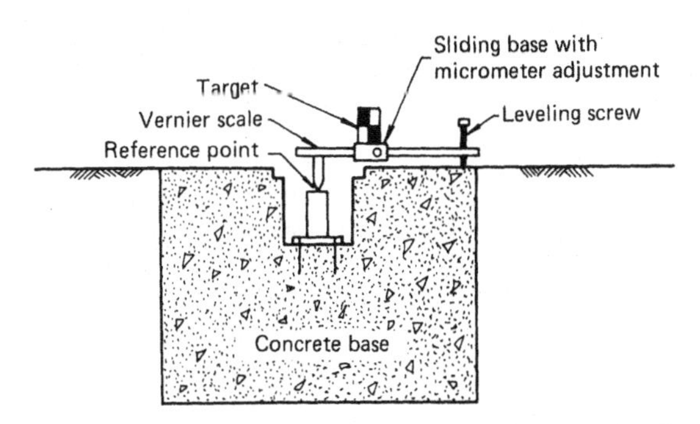

Figure 12.12. Measuring point on surface of embankment dam, with vernier gage and target, for monitoring horizontal deformation (after Wilson, 1967).

ment. A scale can also be marked on or attached to the vertical lumber for monitoring settlement. Problems in using control stakes include vandalism, disturbance owing to construction activities, and lack of precision. In addition, when movement is discerned it is often "too late." Therefore, control stakes should be used with caution.

12.3. SURFACE EXTENSOMETERS

Surface extensometers are defined in this book as devices for monitoring the changing distance between two points on the surface of the ground or a structure. They may also be used for monitoring changing distances between two points on the surface of an excavation. Surface extensometers can be divided into two categories: *crack gages* and *convergence gages*.

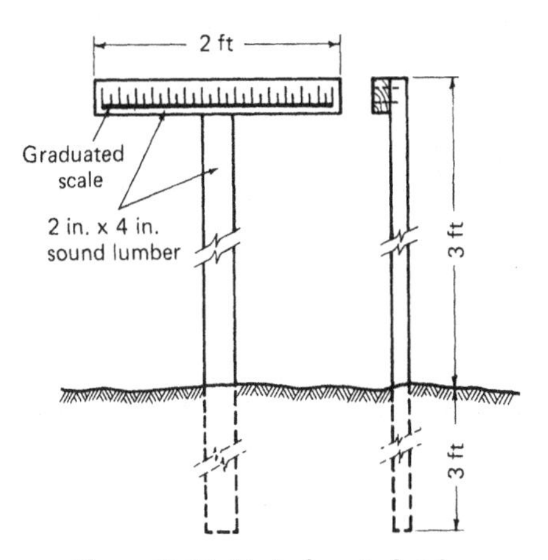

Figure 12.13. Typical control stake.

Table 12.3. Mechanical Crack Gages

Method	Advantages	Limitations	Approximate Precision
Wooden wedges	Inexpensive Can be used by all personnel	Qualitative only Interpretation unreliable	Crude
Glass plates or plaster patches	Inexpensive	Usually qualitative only	±0.05 in. (±1 mm)
Pins and tape	Inexpensive		±0.1 in. (±3 mm)
Pins and steel rule	Inexpensive	Span limited	±0.02 in. (±0.5 mm)
Pins and calipers	Inexpensive	Span limited	±0.01 in. (±0.3 mm)
Pins and tensioned wire, using weight	Inexpensive Can be adapted to trigger alarm		±0.1 in. (±3 mm)
Pins and mechanical extensometer			±0.005 in. (±0.1 mm)
Convergence gages	See Section 12.3.3	See Section 12.3.3	±0.001–0.1 in. (±0.03–3 mm)
Grid crack monitor	Inexpensive	Span limited	±0.05 in. (±1 mm)
Mechanical strain gage		Span limited	±0.0001–0.002 in. (±0.003–0.05 mm)
Dial indicator	Inexpensive	Span limited	±0.0001–0.001 in. (±0.003–0.03 mm)

Crack gages (sometimes called *jointmeters* or *strainmeters*) are typically used for monitoring tension cracks behind slopes and for monitoring cracks in concrete or other structures, pavements or tunnel linings, or joints or faults in rock. Observations of fracturing on a rock surface can provide useful information concerning behavior at depth, and surficial monitoring is usually far less expensive than borehole instrumentation or other techniques for subsurface monitoring.

Convergence gages are typically used for monitoring convergence within tunnels, braced excavations, and mines.

12.3.1. Mechanical Crack Gages

There are numerous mechanical devices available for monitoring the width of discontinuities (e.g., cracks, faults, and joints). The most common methods are described in the following subsections, and comparative information is given in Table 12.3. All methods require access to the gage location for monitoring.

Wooden Wedges

The use of wooden wedges driven into open fractures, and observing when wedges loosen, has been a traditional method used by mining personnel to indicate deformation, but more definitive techniques are usually required.

Glass Plates

The method consists essentially of cementing glass plates across discontinuities and observing breakages. Glass plates, usually about 3 × 1 × 0.1 in. (76 × 25 × 2.5 mm) thick, are cemented to the cleaned and roughened surface using epoxy resin adhesive. The observations are supplemented as necessary by measurement and recording of crack separation and direction of relative movement. Details of the procedure are described by ISRM (1984).

Plaster Patches

Gypsum plaster, which is brittle, is applied across discontinuities with a flat trowel, and observations are made as described above for glass plates.

Pins and Tape

A pin is set on each side of a discontinuity and separation monitored using a steel tape. The type and dimensions of the pins and the fixing system to be used should be appropriate to the condition of the ground or structure to be monitored, to ensure that the pins are rigidly attached to the surface and will remain attached throughout the monitoring program.

When the surface is strong rock or concrete, unaffected by local cracking, pins are typically 1 in. (25 mm) long and 0.25 in. (6 mm) in diameter with a tapered point at one end and a welded base at the other. Epoxy resin is used to fix the base to the surface, or holes can be drilled in the base and a concrete rivet gun used.

For soils or soft rocks, pins 20 in. (500 mm) long and 0.5 in. (13 mm) in diameter are typically used where they are to be driven into the formation. Alternatively, pins may be installed by grouting into a drilled hole. The exposed ends should be pointed or alternatively have filed cross-lines on flat heads.

Details of the installation methods are given by ISRM (1984), including equipment description, installation of pins, and reading and calculation procedures.

Pins and Steel Rule or Calipers

Pins are set as described previously and separation monitored using a steel rule or calipers. Measurements are more precise than when using a tape, but the span is limited.

Pins and Tensioned Wire, Using Weight

Figure 12.14 illustrates the use of a pin and tensioned wire gage for monitoring tension cracks at the top of a rock slope. A wire is stretched across the discontinuities between an anchor on one side and a pulley mounted on a measurement station on the other side. A weight on the wire below the pulley maintains tension. A scale is attached to the measurement station behind the wire, and a measurement block is fixed to the wire. Observation of the position of the measurement block with respect to the scale provides movement data. A trip block can be added to the wire, arranged to contact a trip switch on the scale when a predetermined movement occurs. This can be connected to an alarm if required.

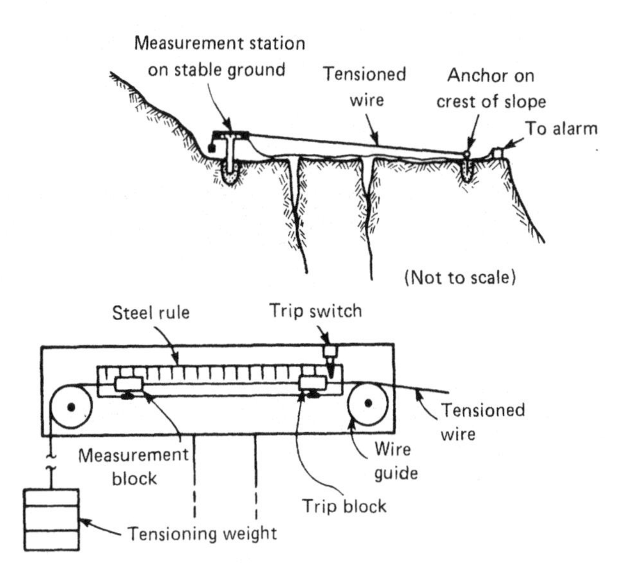

Figure 12.14. Mechanical crack gage, using pins and tensioned wire (after Hoek and Bray, 1981).

Pins and Mechanical Extensometer

Where greater precision is required and the span is too large for a steel rule or calipers, a *slack-wire* mechanical extensometer indicator (Section 12.7.4) can be used to measure the distance between pins, as shown in Figure 12.15. A 100 ft (30 m) steel tape with punched holes 1.5 in. (38 mm) apart can be used for measurement of any span. Alternatively, a length of 0.05 in. (1.3 mm) diameter stainless steel wire can be used for measuring a designated span, with a stainless steel button attached at each end of the wire by stainless steel set screws.

Convergence Gages

Mechanical tape, wire, rod, or tube convergence gages (Section 12.3.3) can be used as mechanical crack gages by attaching anchors to either side of the discontinuity.

Grid Crack Monitor

The *grid crack monitor,* also called a *calibrated crack monitor* and *calibrated telltale,* consists of two overlapping transparent plastic plates, one mounted on each side of the discontinuity. As shown in Figure 12.16, crossed cursor lines on the upper plate overlay a graduated grid on the lower plate. Movement is determined by observing the position of the cross on the upper plate with respect to the grid.

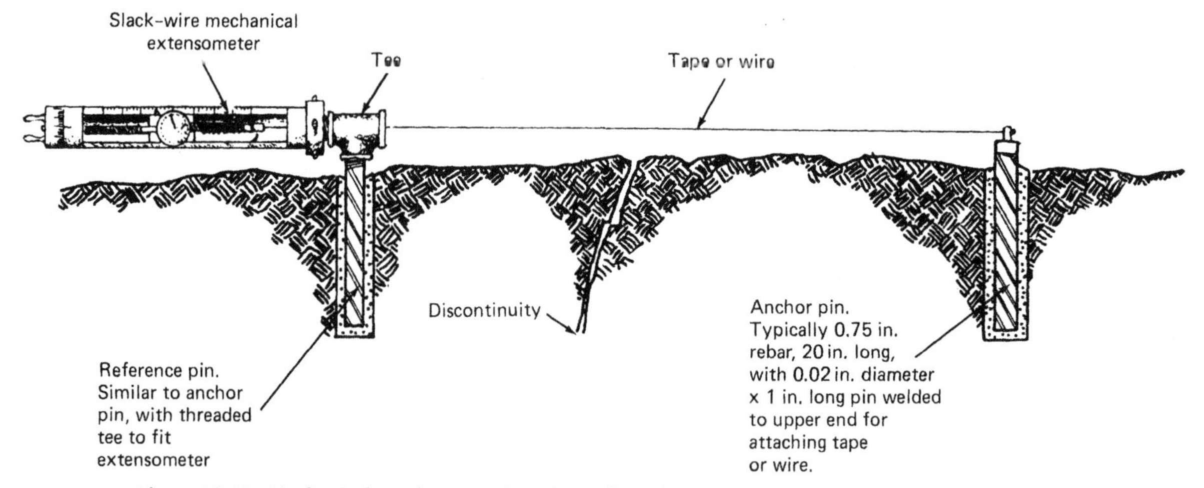

Figure 12.15. Mechanical crack gage, using pins and mechanical extensometer (after Yu, 1983; courtesy of *Canadian Mining Journal* and Kidd Creek Mines Ltd.).

Mechanical Strain Gage

Surface-mounted mechanical strain gages (Chapter 13) can be used for measurements across discontinuities.

Dial Indicator

A dial indicator can be attached temporarily or permanently to a bracket on one side of the discontinuity and arranged to bear against a machined reference surface on the other side. Three-axis versions are also available for measurements in orthogonal directions. Portable micrometers can be used instead of dial indicators.

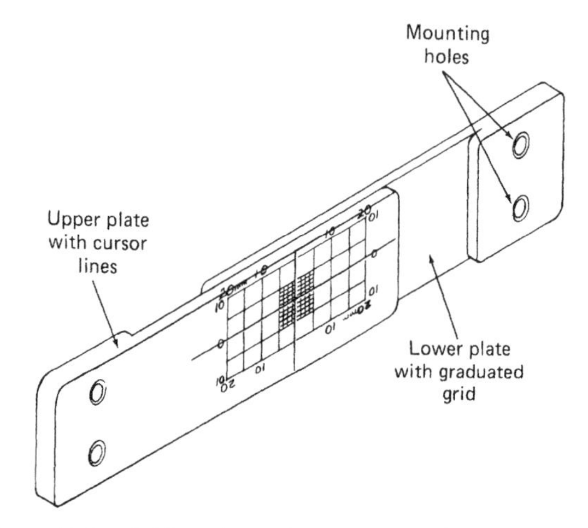

Figure 12.16. Grid crack monitor (courtesy of Avongard Products, U.S.A. Ltd., Waukegan, IL).

12.3.2. Electrical Crack Gages

When access to the gage location is not available for monitoring, or when continuous monitoring is needed, a remote reading electrical gage is required.

Three general arrangements are possible. First, an electrical linear displacement transducer can be attached to a bracket on one side of the discontinuity and arranged to bear against a machined reference surface on the other side. Second, anchor points can be located on either side of a discontinuity and the transducer attached to the anchor points via ball joints, as shown in Figure 12.17. Details of this arrangement, including instrument description, anchor point installation procedure, and reading and calculation procedures, are given by ISRM (1984). Third, an electrical linear displacement transducer can be incorporated in the mechanical system shown in Figure 12.14.

Available transducers (Chapter 8) include linear potentiometers, linear variable differential transformers (LVDTs), direct current differential transformers (DCDTs), variable reluctance transducers (VRTs), vibrating wire transducers, bonded and unbonded resistance strain gage transducers, and transducers with frequency-displacement induction coils. A magnetostrictive convergence gage, described in Section 12.3.3, can also be used as an electrical crack gage.

A single gage can be installed to monitor deformation perpendicular to the discontinuity, or a three-axis version can be used for measurements in orthogonal directions.

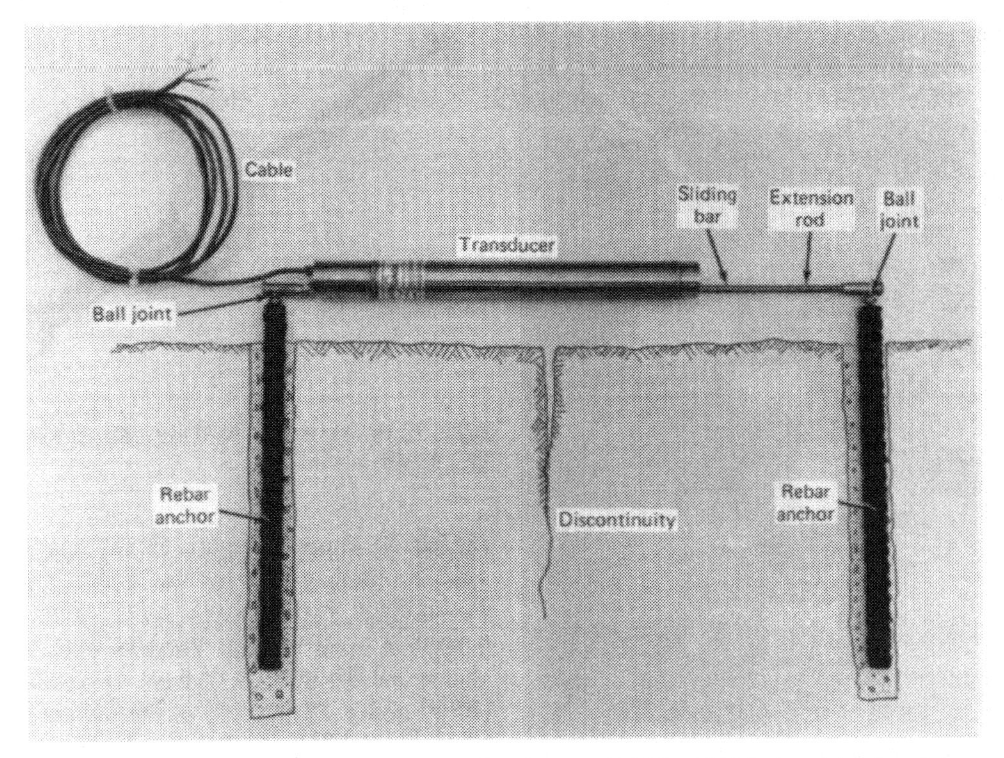

Figure 12.17. Electrical crack gage (courtesy of Irad Gage, a Division of Klein Associates, Inc., Salem, NH).

Electrical crack gages are more expensive than mechanical gages, and their range is limited. However, range can usually be extended by resetting. Precision is between ±0.0001 and 0.005 in. (±0.003 and 0.13 mm), depending on the transducer. Also depending on the transducer, readings can be affected by lead wire changes and by temperature and other environmental conditions.

12.3.3. Convergence Gages

A *convergence gage* usually consists of a tape, wire, rod, or tube in series with a deformation indicator. The gage is usually portable and is attached at the time of reading to permanent anchors mounted at each end of the measuring span. The most common gage types are described below.

Protective covers should be provided for anchors. The stability of all gages should be checked on a regular schedule, either by using a calibration frame supplied by the manufacturer or by reading a span that is known to be stable. The latter is preferable, because it provides a check on the complete gage rather than merely the part containing the deformation indicator.

Tape Convergence Gages

A typical *tape convergence gage,* often called a *tape extensometer,* is shown in Figure 12.18. The tape has punched holes at 2 in. (50 mm) intervals. Grouted rebar and expansion shell anchors are shown, but anchors can be welded or bolted. The tension in the tape is controlled by a compression spring, and to standardize tension the collar is rotated until the scribed lines are in alignment. After attachment of the extensometer to the anchors and standardizing the tension, readings of distance are made by adding the dial indicator reading to the tape reading. Precision is typically ±0.005 in. (±0.13 mm) in a 30 ft (10 m) span, decreasing with increasing span. Maximum span is approximately 200 ft (60 m).

Wire Convergence Gages

Conventional portable *wire convergence gages,* or *wire extensometers,* are similar to portable tape convergence gages. Either a separate wire can be used for each span or the wire can be equipped with a series of collars to accommodate variation in span. Precision with conventional steel wires is

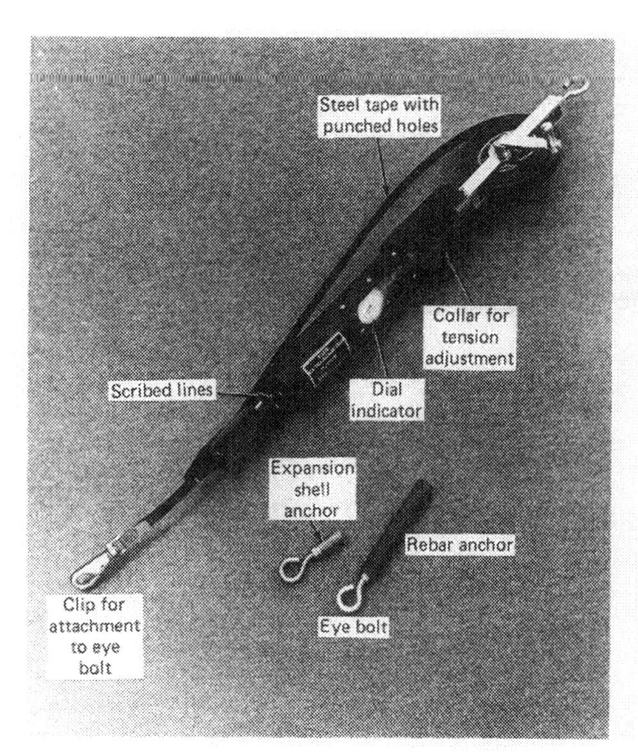

Figure 12.18. Tape extensometer (courtesy of Slope Indicator Company, Seattle, WA).

similar to precision of tape convergence gages and can be improved by using an invar wire: a 10°C temperature increase causes an expansion in a 10 ft (3 m) invar wire of approximately 0.002 in. (0.05 mm).

A precise invar wire convergence gage, the

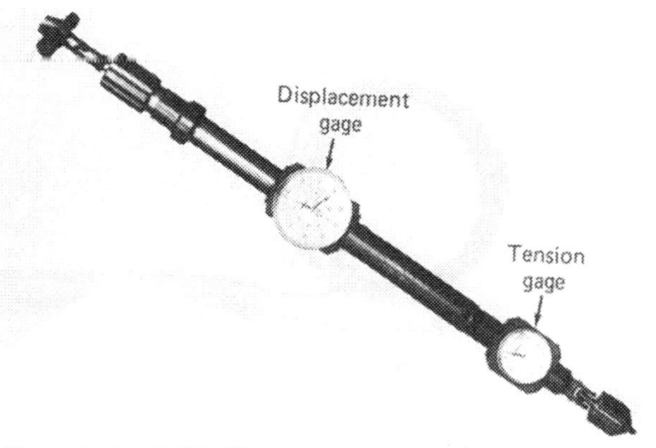

Figure 12.19. ISETH Distometer (courtesy of Kern Instruments, Inc., Brewster, NY).

ISETH Distometer (Figure 12.19), has been developed in Switzerland by the Federal Institute of Technology, Zürich (Kovari et al., 1974). Individual lengths of invar wire are used for each span. Precision is ±0.001 in. (±0.03 mm) for spans up to 60 ft (20 m) and ±1/1,000,000 of the distance for longer spans up to 150 ft (50 m). An alternative invar wire convergence gage, the *Distomatic* (Figure 12.20), is widely used in France (Londe, 1977). The wire is tensioned by an electric motor, actuating an electronic limit switch at the required standard tension. The motor is geared to a counter and the reading displayed digitally. Precision is reported to be ±0.001 in. (±0.03 mm) over a 20 ft (6 m) span and ±0.002 in. (±0.05 mm) over an 80 ft (25 m) span.

All the wire convergence gages described above

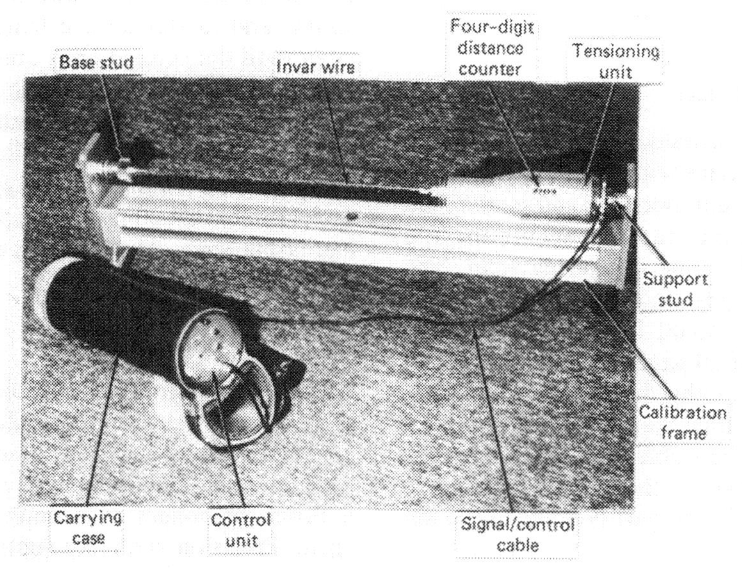

Figure 12.20. *Distomatic* convergence gage (courtesy of Telemac, Asnières, France).

Figure 12.21. Invar tube convergence gage (courtesy of Soiltest, Inc., Evanston, IL).

are portable. The arrangement shown in Figure 12.14, using a tensioned wire and a pulley, can be adapted for use as a convergence gage, using either a mechanical or electrical transducer.

Rod and Tube Convergence Gages

Rod and *rigid tube convergence gages* generally consist of telescoping rods or rigid tubes, a dial indicator or micrometer, and contact seats that mate with anchors. Some gages have invar rods or tubes, others have aluminum or galvanized or stainless steel for which a temperature correction can be applied to maximize precision. Range of span, depending on the model, is 6 in. to 25 ft (150 mm to 8 m). The telescoping arrangement is often spring loaded. Figure 12.21 shows a gage with telescoping invar tubes, placed in its calibration frame. Stacey and Wrench (1985) describe a gage with telescoping stainless steel tubes and a vernier scale.

These gages are alternatives to tape or wire gages for vertical spans in tunnels and mines where access to the upper anchor is inconvenient, and for vertical spans precision is typically ±0.005 in. (±0.13 mm).

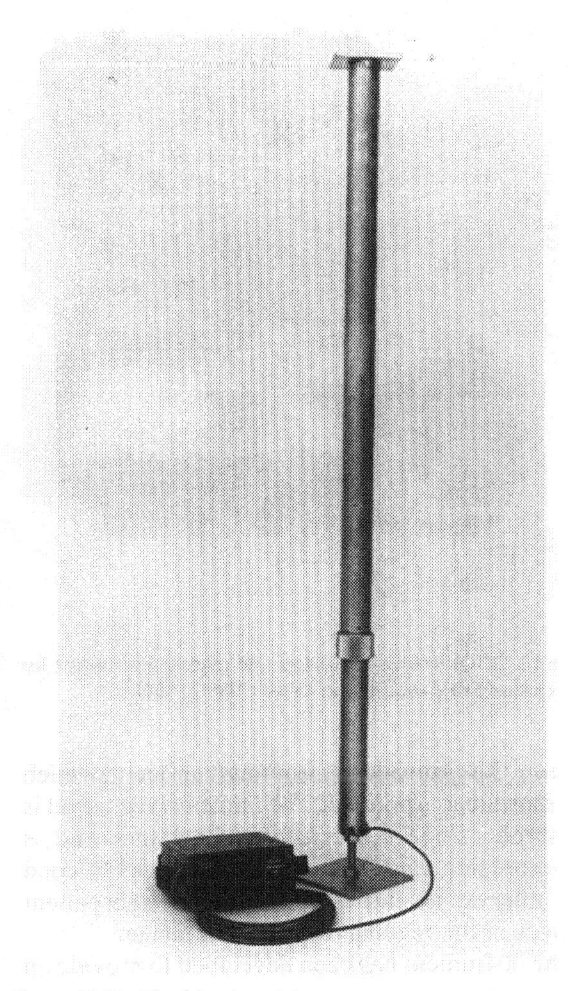

Figure 12.22. Fixed-in-place tube convergence gage (courtesy of Roctest Ltd., Montreal, Canada).

However, precision of horizontal or inclined spans is reduced by sag and for spans greater than 10 ft (3 m) may be as low as ±0.1 in. (±3 mm).

Rod and tube convergence gages are also available with electrical displacement transducers for remote reading, both as fixed-in-place and portable instruments. Figure 12.22 shows a fixed-in-place instrument.

Ultrasonic Convergence Gage

USBM (1984) reports on the recent development of an *ultrasonic convergence gage*. As shown in Figure 12.23, the gage consists of a transducer and a readout device. The small ultrasonic transducer, similar to an audio speaker diaphragm, is plucked with an ultrasonic frequency every 6 seconds. This frequency wave is sent through the air and reflected

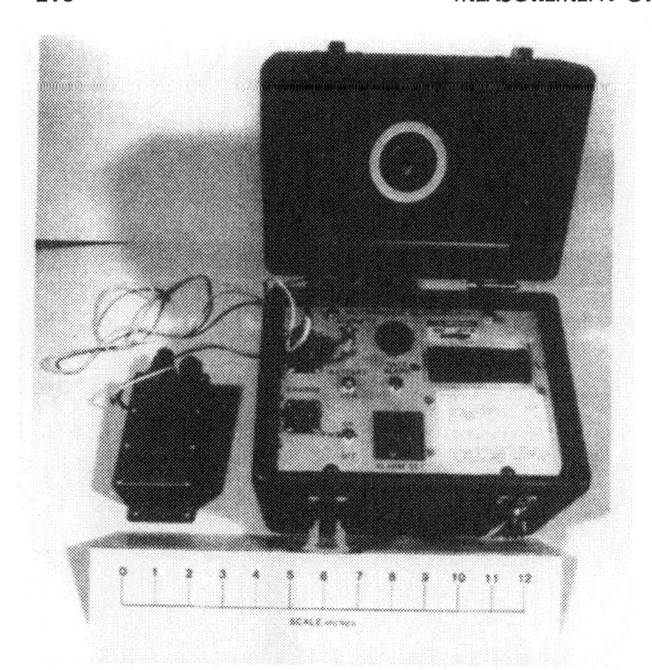

Figure 12.23. Ultrasonic transducer and readout instrument for mine closure rate measurement (after USBM, 1984).

back to the transducer from any surface at which the transducer is pointed. The time of wave travel is measured using a precise oscillator/timer and is converted into a measure of distance. The 6 second time interval is used to convert any subsequent changes in the reading to a rate of change.

The instrument has been developed to provide an unobstructive means of measuring changes in distance from roof to floor or from rib to rib of a mine opening and to indicate the rate of closure so that operators of underground mines can detect hazardous ground conditions. Based on use to date, USBM comments that a convergence of 0.5 in. (13 mm) can readily be measured, but the instrument is not capable of measuring a convergence of less than 0.05 in. (1.3 mm). Range is 1–35 ft (0.3–11 m), and alarm settings can be included. The device is therefore most applicable for measuring convergence during longwall mining, pillar robbing, and other mining operations resulting in large ground movements.

12.4. TILTMETERS

Tiltmeters, also referred to as *clinometers,* are used to monitor the change in inclination (rotation) of points on or in the ground or a structure.

A tiltmeter consists of a gravity-sensing transducer within an appropriate housing, and housings are available for installation either on or below the surface of the ground or structure. Surface versions may be either fixed-in-place or arranged as portable devices by mating with reference points permanently attached to the surface. Subsurface versions are usually fixed-in-place within boreholes.

Applications of tiltmeters include monitoring tilt of retaining walls and concrete dams and monitoring landslides in which the failure mode can be expected to contain a rotational component. Very precise tiltmeters can sometimes be used during a short time period to provide a rapid indication of deformation trends. They are also used for monitoring ground subsidence over mined areas, and tiltmeters are available for detecting earthtides and other geodetic or seismic events. Tiltmeters have been used for monitoring safety of buildings alongside excavations in an attempt to provide forewarning of distress, but unless a rotational component of deformation is expected, settlement measurements are likely to be more meaningful.

The precision of a tiltmeter is normally expressed in radians, arc-seconds, or gons. Conversion factors are given in Appendix H.

12.4.1. Mechanical Tiltmeters

Figure 12.24 shows a mechanical tiltmeter consisting of a beam and bubble level, similar to a builder's level, with a leveling adjustment at one end of the beam. The beam is located on two anchored reference balls, the leveling adjustment turned until the bubble level indicates that the beam is in a horizontal plane, and a dial indicator or micrometer inserted through the leveling adjustment to bear on the reference ball. Relative movements in the vertical plane are recorded as changes in dial indicator readings. Errors caused by zero shift are accounted for by reversing the tiltmeter, repeating the readings, and averaging. Precision is approximately ± 0.0005 in. (± 0.013 mm) for beams up to 8 in. (200 mm) long, decreasing to ± 0.005 in. (± 0.13 mm) for beams 3 ft (900 mm) long. These figures correspond to ± 13 and 29 arc-seconds. Similar tiltmeters are available for monitoring tilt of near-vertical surfaces.

12.4.2. Tiltmeters with Accelerometer Transducer

A portable tiltmeter for measuring tilt in both a horizontal and vertical plane is shown in Figure 12.25.

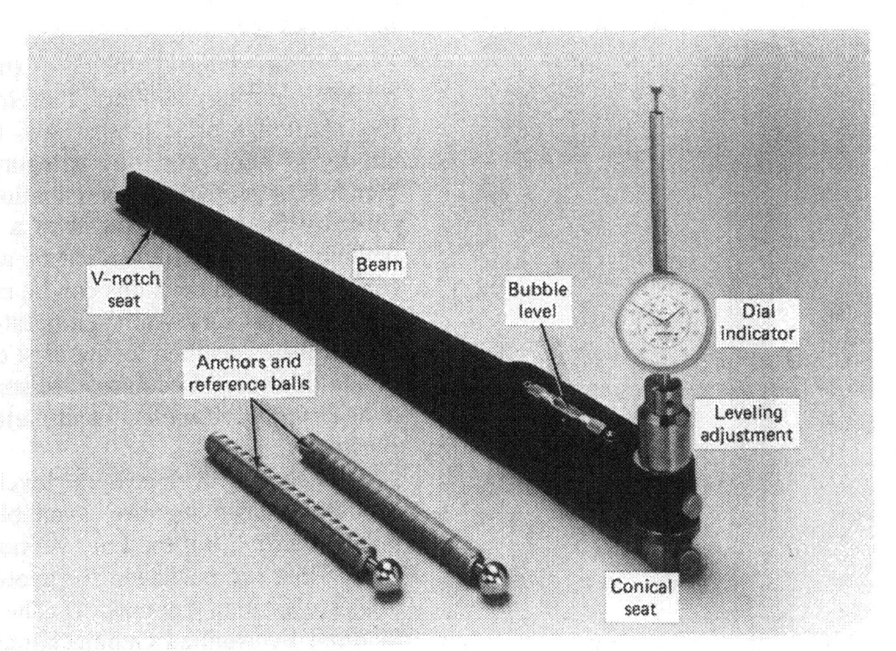

Figure 12.24. Mechanical tiltmeter (*portable clinometer*) (courtesy of Soil Instruments Ltd., Uckfield, England).

A measurement is made by placing the tiltmeter in an exactly reproducible position on a reference plate, taking a reading, turning the tiltmeter 180 degrees, and again taking a reading. This method allows use of the *check-sum* procedure described for inclinometers in Section 12.8. The reference plate is either metallic or ceramic and must be securely bonded or bolted to the monitored surface. Details of the equipment and procedure for use of this instrument are given by ISRM (1981a). Typical range is ±30 degrees from the horizontal or vertical, and precision is typically ±50 arc-seconds. Temperature sensitivity is typically 2–3 arc-seconds/°F.

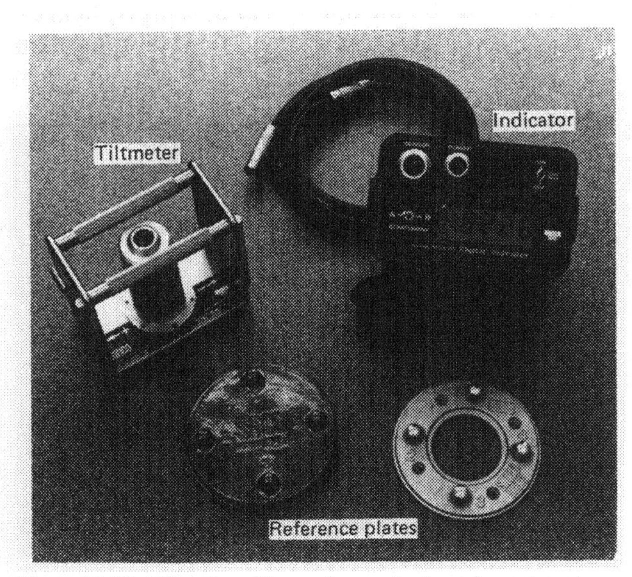

Figure 12.25. Tiltmeter with accelerometer transducer (courtesy of Slope Indicator Company, Seattle, WA).

A more precise portable tiltmeter with accelerometer transducer, referred to as a *clinometer,* is available from Solexperts AG. The instrument is used for measuring tilt in both horizontal and vertical planes, and measurements are made using the *check-sum* procedure. The standard range is approximately ±3 degrees from the horizontal or vertical, and tests by Thut (1987) indicate a precision of less than ±1 arc-second. It is believed that this high precision can be attributed to the highly repeatable mating arrangement between the portable instrument and the reference plate.

12.4.3. Tiltmeters with Vibrating Wire Transducer

Pendulum-actuated vibrating wire transducers are available for attachment to the face of a structure or for embedment below the surface of the ground or structure.

Two configurations are available. First, a pendulum is rigidly attached to the top of the tiltmeter such that tilt causes bending strains in the pen-

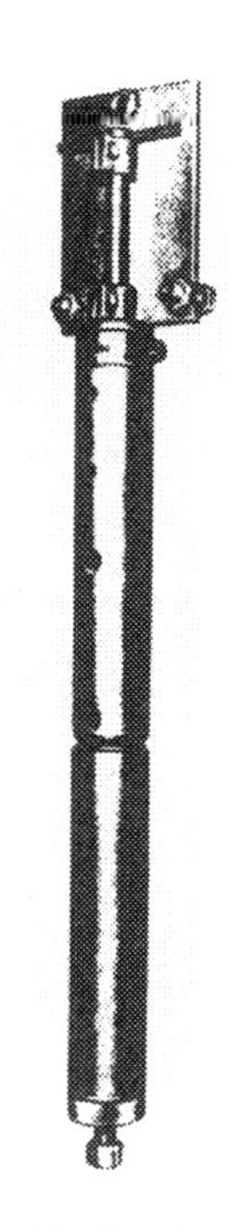

Figure 12.26. Tiltmeter with vibrating wire transducers (courtesy of Telemac, Asnières, France).

dulum, and the strains are monitored by two vibrating wire transducers, one on each side of the pendulum in the plane of tilting. Figure 12.26 shows an example. Alternatively, four transducers can be spaced around the pendulum for monitoring tilt in two vertical planes 90 degrees apart. In the second configuration the vibrating wire transducers span between the pendulum and the instrument housing. Whenever transducers are mounted on opposite sides of the pendulum, in either configuration, tilting causes equal and opposite strains in the two transducers, and temperature effects are eliminated. However, the instruments are usually bulky and heavy. Ranges of available tiltmeters typically vary from ± 0.1 to 1 degree, with precision approximately 0.5% of range, that is, ± 2–20 arc-seconds

12.4.4. Tiltmeters with Electrolytic Level Transducer*

Electrolytic level transducers (Chapter 8) can be grouped into two categories. In the first category the glass vial is made simply by sealing the ends of a standard glass tube, softening the glass, and bending to the required radius. Any irregularities that exist in the inside diameter of the tube are compounded during bending; therefore, accuracy is low. Although price is also low, temperature sensitivity is high, and this category is not recommended for geotechnical applications. In the second category the vial is made with a vacuum forming technique that greatly increases accuracy because exact internal dimensions can be controlled and reproduced in every vial. Temperature sensitivity is significantly less than for the first category. Tiltmeters in this second category are sometimes referred to as *earthtide tiltmeters* or tiltmeters with *geodetic sensitivity*.

Tiltmeters with electrolytic level transducers and geodetic sensitivity are available from several manufacturers, but the only versions known to the author that are packaged for geotechnical applications are listed in Appendix D. The tiltmeters manufactured by Applied Geomechanics, Inc. (for surface and downhole use), Sperry Corporation (for surface use), and G + G Technics (for surface and undersea use) all incorporate a transducer manufactured by Spectron Glass and Electronics, Inc., Hauppauge, NY. The *Sperry Tilt Sensing System,* shown in Figure 12.27, is manufactured with ranges from ± 20 arc-minutes to ± 45 degrees. A console and recorder are available for monitoring multiple tiltmeters, and threshold levels can be set individually to provide hazard warning. Use of the Sperry Tilt Sensing System is described by Cape (1984) and ENR (1984).

Tests made on the Sperry Tilt Sensing System by New York State Department of Transportation

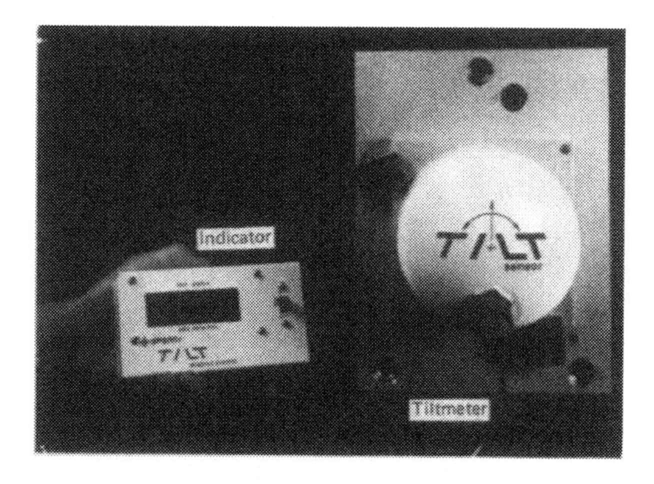

Figure 12.27. Sperry Tilt Sensing System (courtesy of Sperry Corporation, Phoenix, AZ).

*Written with the assistance of Robert S. Marshall, President, Spectron Glass and Electronics, Inc., Hauppauge, NY.

(Gemme, 1984) indicate that the version with a range of ±20 arc minutes has a repeatability in the laboratory of ±0.3 arc-seconds. Holhauzen (1985) states that in stable underground vault environments, the tiltmeter manufactured by Applied Geomechanics, Inc. consistently achieves a stability of ±0.3 arc-seconds or better during tests of 3 or 4 days. Firm figures for longer-term repeatability in a field environment do not appear to be available.

Temperature compensation of tiltmeters with electrolytic level transducers and geodetic sensitivity can be provided by including a thermistor with the same thermal coefficient as the electrolytic level transducer, but compensation is applicable only to uniform temperature conditions. When the temperature of the transducer and mounting arrangements is nonuniform, temperature sensitivity is greatly increased, and mounting arrangements should therefore be designed to average any nonuniformity of temperature or to allow temperatures to equalize quickly.

12.5. PROBE EXTENSOMETERS

Probe extensometers are defined in this book as devices for monitoring the changing distance between two or more points along a common axis, by passing a probe through a pipe. Measuring points along the pipe are identified mechanically or electrically by the probe, and the distance between points is determined by measurements of probe position. For determination of absolute deformation data, either one measuring point must be at a location not subject to deformation or its position with respect to a reference datum must be determined by surveying methods. The pipe may be vertical, providing measurements of settlement or heave, may be horizontal, providing lateral deformation measurements, or may be inclined.

Typical applications of probe extensometers are monitoring vertical compression within embankments or embankment foundations, settlement alongside excavations, heave at the base of open cut excavations, and lateral deformation of embankments. In general, they are alternatives to fixed borehole extensometers (Section 12.7), allowing for more measuring points and minimizing the cost of permanently installed materials, but generally measurements are less precise than fixed borehole extensometer measurements.

Various mechanical and electrical probe extensometers are described below, and comparative information is given in Table 12.4. Table 12.4 includes a column for accuracy, which refers to accuracy of deformation measurements, assuming that one measuring point is at a location not subject to deformation. If a surveying method is used to determine absolute deformation data, accuracy of deformation measurements may depend on the surveying method.

12.5.1. Mechanical Heave Gage

The gage is used for monitoring heave at the bottom of braced or other open cut excavations. As shown in Figure 12.28, a conical steel point is installed, facing upward, in a slurry-filled borehole just below the eventual bottom of the excavation. At any time during excavation, a probing rod of known length is lowered down the borehole to mate with the conical point. The elevation of the top of the rod is determined by surveying methods, giving the elevation of the conical point. The borehole can be located readily during construction by coloring the slurry with ethrysene dye. The gage is described by Swiger (1972). Bozozuk (1970) gives details of a similar gage, using a four-bladed steel vane at the bottom of the borehole.

12.5.2. Crossarm Gage

The crossarm gage was developed by the U.S. Bureau of Reclamation (Bartholomew et al., 1987; USBR, 1974) for installation during construction of embankment dams. As shown in Figure 12.29, it consists of a series of telescoping pipe sections with alternate sections anchored to the embankment by horizontal steel channel crossarms at 5–10 ft (1.5–3m) intervals. The crossarms ensure that the pipes move together an amount equal to compression of the intervening fill. Depths to the measuring point at the lower end of each interior pipe are sounded to the nearest 0.01 ft (3 mm) by a probe with spring-loaded sensing pawls, lowered on a steel tape. The probe is lowered just beyond each interior pipe in turn and raised until the pawls latch against the lower end. On reaching the bottom of the pipes, the pawls retract and lock within the body of the probe.

Installation details are described by USBR (1974) and require that most of the soil immediately surrounding the pipes be excavated and replaced by hand-compacted backfill. As each successive crossarm is placed, the elevation of a reference point on

Table 12.4. Probe Extensometers

Method	Advantages	Limitations	Approximate Accuracy[a]
Mechanical heave gage (Figure 12.28)	Inexpensive	Requires survey crew Risk of borehole caving during excavation	±0.2–1.0 in. (±5–25 mm)
Crossarm gage (Figure 12.29)	Crossarms ensure that axial deformations of pipe conform with deformations of soil Can accommodate large compression	Pawls in probe may jam Cannot be installed in boreholes Steel pipes can corrode Probe may become wedged if large lateral deformations occur Laborious to read Compaction difficult[b]	±0.05–0.2 in. (±1–5 mm)
Mechanical probe within inclinometer casing (e.g., Figure 12.30)	Simple When installed in fill, collars provide positive driving force to slide pipe Both vertical and horizontal deformations may be monitored[c]	Pawls in probe may jam If used in boreholes, axial deformations of casing may not conform with deformations of soil Laborious to read Compaction difficult[b]	±0.05–0.2 in. (±1–5 mm)
Sliding micrometer (Figure 12.31)	Very accurate Installation can be vertical or horizontal[c,d]	Very expensive Laborious to read	±0.0001 in. in 39 in. (±0.002 mm in 1 m)
Gage with current-displacement induction coil (e.g., Figures 12.32, 12.34)	Versions available for all site conditions Various installations possible[c,d,e]	Readings somewhat subjective Accuracy reduced by stray electric currents Compaction difficult[b]	±0.02–0.2 in. (±0.5–5 mm) for vertical installations ±0.02–1.0 in. (±0.5–25 mm) for horizontal installations
Gage with frequency-displacement induction coil (e.g., Figure 12.37)	Very accurate Various installations possible[c,d,e]	Compaction difficult[b]	±0.001 in. (±0.03 mm)
Magnet/reed switch gage (Figure 12.38)	Versions available for all site conditions Can be used as a heave gage at base of open cut excavations without interference to excavation Various installations possible[c,d,e]	Compaction difficult[b]	±0.02–0.2 in. (±0.5–5 mm) for vertical installations ±0.1–1.0 in. (±3–25 mm) for horizontal installations
Bellow-hose gage		Readings subjective No positive anchorage at measuring points Compaction difficult[b]	±0.1–0.3 in. (±3–8 mm)
Magnetostrictive gage (e.g., Figure 12.41)	Suitable for boreholes at all inclinations, including upward[d]	Reading depth limited to 25 ft (7.6 m) Requires careful handling	±0.002–0.01 in. (±0.05–0.3 mm)

[a] Accuracy refers to accuracy of deformation measurements, assuming that one measuring point is at a location not subject to deformation. If a surveying method is used to determine absolute deformation data, accuracy of deformation measurements may depend on the surveying method.
[b] When installed vertically through fill, large compaction equipment cannot be used near the pipe; thus, compaction tends to be inferior. Interruption to normal filling is costly, and damage by construction equipment is possible. Liquid level gages (Section 12.10) sometimes provide a preferable measurement method.
[c] Vertical installations can be combined with an inclinometer casing (Section 12.8) for monitoring both vertical and horizontal deformations.
[d] The instrument can be installed either horizontally or vertically.
[e] Horizontal installations can be combined with a full-profile liquid level gage (Section 12.10) for monitoring both horizontal and vertical deformations.

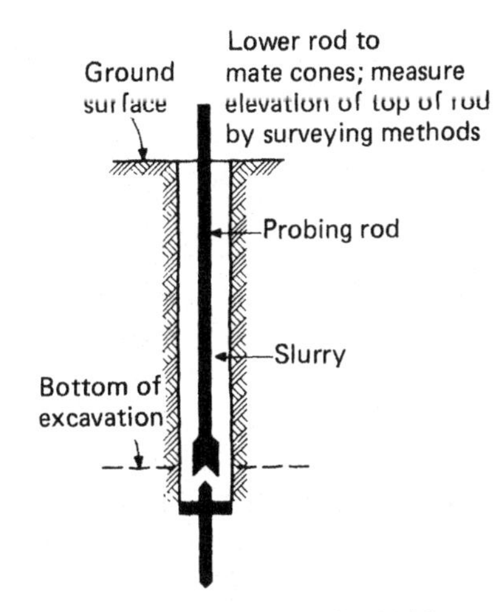

Figure 12.28. Schematic of mechanical heave gage.

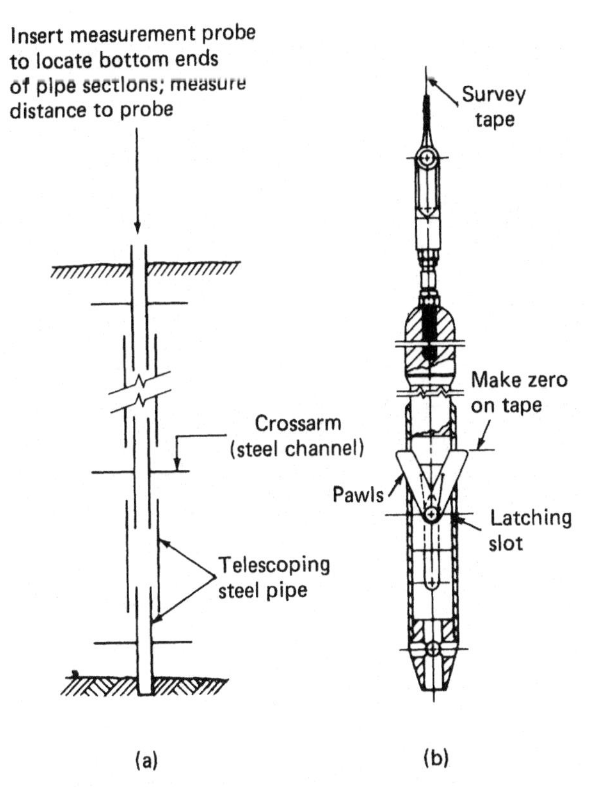

Figure 12.29. Crossarm gage: (a) Schematic of pipe arrangement and (b) measurement probe (after USBR, 1974; courtesy of Bureau of Reclamation).

the uppermost pipe section is determined to the nearest 0.01 ft (3 mm) by optical leveling.

For determination of compression between crossarms, the optical leveling error is eliminated and a precision of ±0.05 in. (±1.3 mm) can be obtained by applying a constant tension to the tape, using a pulley and counterweight.

The original version of the gage used 1.5 and 2 in. (38 and 51 mm) pipes, requiring a slender probe somewhat prone to malfunction and to becoming wedged in the pipes. Recent improvements include the use of 3–4 in. (76–102 mm) pipes, thus permitting the passage of a larger and more rugged probe than the original. Corrosion of the steel pipes has caused problems in some older installations, usually near the phreatic surface and especially in warmer climates and where soil conditions produce corrosive seepage water. Use of plastic pipe with a stainless steel ring at the measuring point on the lower end of each interior pipe could obviate the difficulty. Poor alignment of pipes or large lateral deformations have caused malfunctioning (Peters and Long, 1981), but a shorter probe can be tried if the standard probe will not pass down the pipes.

12.5.3. Mechanical Probe Within Inclinometer Casing

Measurements with a mechanical probe can be made in inclinometer casing connected with telescoping couplings (Section 12.8.3).

Vertical deformations are measured either with a probe similar to the crossarm gage probe or with a special hook supplied by some manufacturers of inclinometer casing (Figure 12.30). In embankments, the casing is forced to follow the pattern of soil compression by attaching settlement collars to its outside. If these are omitted, the telescoping action relies solely on friction between soil and the surface of the casing, and precision will be decreased if soil slips with respect to the section of casing it surrounds (Rosati and Esquivel, 1981). Settlement collars are not possible for borehole installations; thus, the telescoping of casing will not necessarily conform with soil compression, and data may not be correct.

12.5.4. Sliding Micrometer

A recent development by the Federal Institute of Technology, Zürich (Kovari and Amstad, 1982; Kovari et al., 1979), allows very accurate measurements of axial deformation to be made within a pipe. Measurements in soil and rock do not often merit such accuracy, and the primary application is for determination of axial and bending strains in

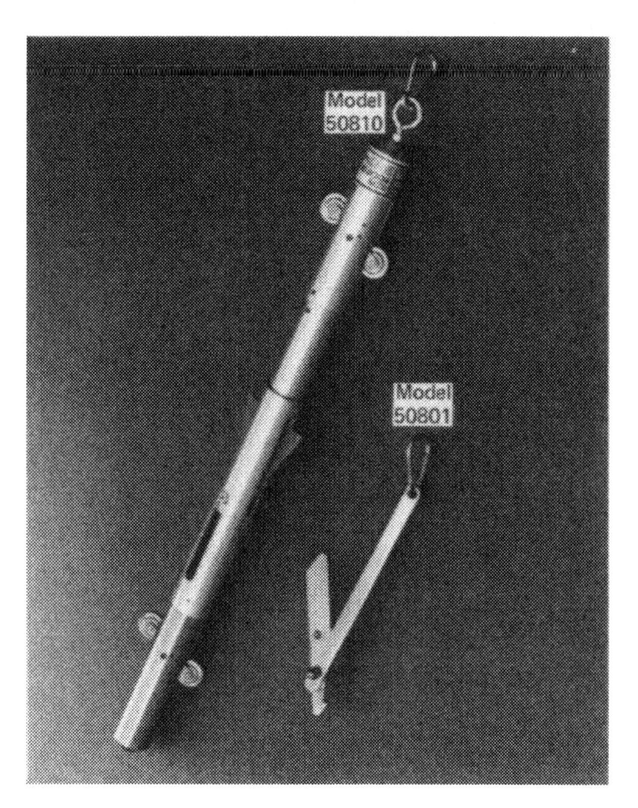

Figure 12.30. Mechanical probes for use within inclinometer casing (courtesy of Slope Indicator Company, Seattle, WA).

slurry walls, concrete piles, drilled shafts, and other concrete structures.

Figure 12.31 shows the basic arrangement. The pipe casing is either embedded directly within the concrete or grouted within a borehole. Cone-shaped measuring points are attached to the pipe casing at 1 m (39 in.) intervals, and the cones are fluted to allow passage of the probe. The probe (the *micrometer*) consists of a waterproof spring-loaded protective sleeve with fluted spherical ends and an invar tube and linear displacement transducer for measuring the distance between spherical ends. To take a set of measurements, the probe is attached to an installation rod, inserted within the pipe casing, the spherical ends mated with the first two cone-shaped measuring points, and a reading made. The probe is then rotated 45 degrees to allow the fluted spheres to pass through the fluted cones and relocated on the next pair of measuring points. The procedure is repeated to the end of the pipe casing, and a further set of readings is made during withdrawal, thus providing a check.

Precision of measurement between adjacent measuring points 1 m (39 in.) apart is reported to be ±0.0001 in. (±0.002 mm, ±2 microns), corresponding to ±2 microstrain. Temperature change problems have reportedly been eliminated through

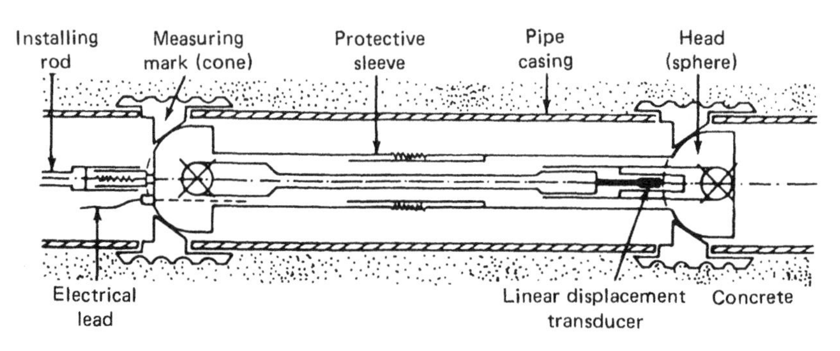

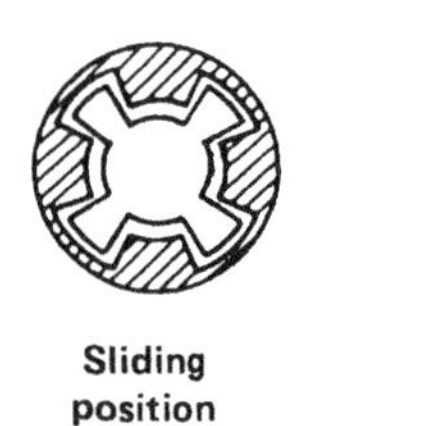

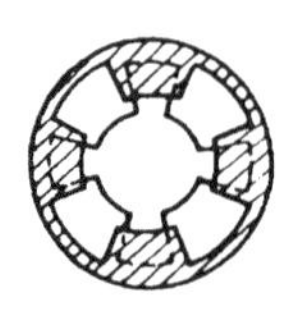

Figure 12.31. Sliding micrometer (after Kovari and Amstad, 1982). Reprinted by permission of the Institution of Civil Engineers, London.

self-temperature compensation, and the system is usable at any pipe orientation. A portable invar steel calibration frame is available to check the proper functioning of the instrument and its long-term stability.

The sliding micrometer is also available with a gravity-sensing transducer mounted within the probe, so that the system can also be used as an inclinometer (Section 12.8) for determination of the three orthogonal displacement vector components along a near-vertical borehole (Köppel et al., 1983). The instrument is supplied by Solexperts AG under the trade name *Trivec*.

12.5.5. Gage with Current-Displacement Induction Coil

Induction coil transducers with current output, described in Chapter 8, are used by several manufacturers of probe extensometers. The embedded part of the instrument consists of a telescoping pipe surrounded by steel rings or plates at the required measuring points. The reading device consists of a primary coil housed within a probe and an attached signal cable connected to a current indicator.

Depth measurements are made either with a survey tape attached to the probe or by using graduations on the signal cable: the former is very preferable because the cable can experience significant change in length during its life (Robinson, 1985). Use of a composite survey tape and signal cable is preferable (similar to the composite tape and cable shown in Figure 9.4 for use with an electrical dipmeter when reading open standpipe piezometers), allowing accurate measurements to be made without the need for dual lines running along the pipe. Readings are made by traversing the probe along the pipe and noting the tape graduation when output current is a maximum. Various arrangements for traversing the probe along horizontal pipes are described in Chapter 17. Changes in reading with time provide deformation or strain data, either in the horizontal or vertical direction.

For determination of absolute deformation data, either one steel ring or plate must be at a location not subject to deformation or its position with respect to a reference datum must be determined by surveying methods. The first option maximizes precision and is preferable. For borehole installations, the lowest measuring point should be placed at sufficient depth to serve as the reference datum, and precision can be maximized by installing three such measuring points and averaging readings.

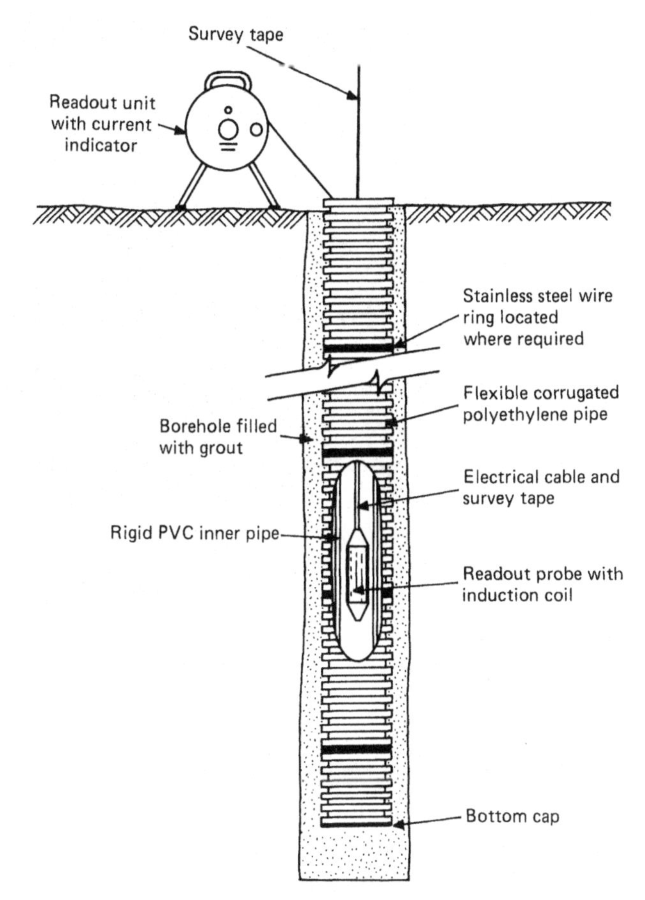

Figure 12.32. Schematic of Slope Indicator Company *Sondex* probe extensometer, installed in a borehole.

Two arrangements are possible for the embedded part of the instrument and are described in the following subsections.

Instrument with Corrugated Pipe and Steel Wire Rings

A corrugated polyethylene pipe, typically of 3 or 4 in. (76 or 102 mm) inside diameter, is surrounded by rings of stainless steel wire. A schematic of a borehole installation is shown in Figure 12.32 and a photograph in Figure 12.33. This arrangement can be installed either in a borehole or in fill but, as discussed later, the alternative arrangement with rigid plastic pipe, telescoping couplings, and steel plates is preferable for installations in fill.

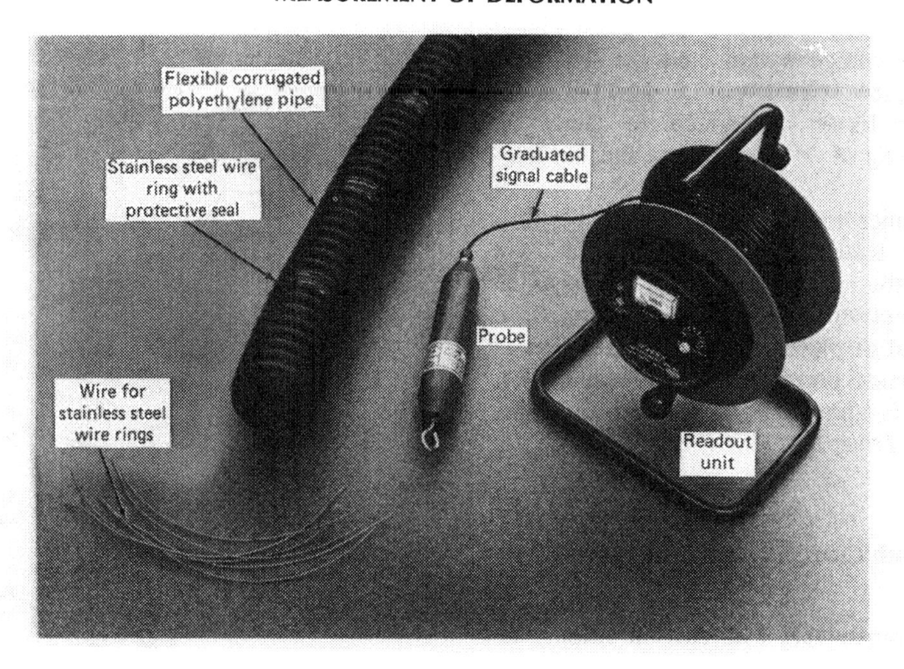

Figure 12.33. *Sondex* probe extensometer (courtesy of Slope Indicator Company, Seattle, WA).

Instrument with Rigid Plastic Pipe, Telescoping Couplings, and Steel Plates

Rigid PVC pipe, typically of 2.5 in. (63 mm) nominal diameter with telescoping couplings, is surrounded by steel plates when installed horizontally or vertically in fill or by expanding anchors when installed in a borehole. The steel plates are typically 12 in. (300 mm) square and 0.2 in. (5 mm) thick, each with a central hole. The version for horizontal installation in fill is described by Penman and Charles (1973), who refer to it as a *horizontal plate gage* (Figure 12.34).

For horizontal or vertical installations in fill, this arrangement is preferable to corrugated polyethylene pipe, because PVC pipe is more resistant to crushing and allows passage of the probe more easily. Steel plates are preferable to steel wire rings, as conformance with soil deformation is ensured.

A version of the gage, the *Idel Sonde*, uses a radio transmitter housed within the probe to transmit a signal when the probe passes a steel ring or plate. The primary use of the device has been in embankment dams, and it has generally not performed as well as the simpler and less expensive conventional version.

Precision

The gage is electrically very sensitive and can locate a steel ring or plate to within ±0.02 in. (±0.5 mm).

Figure 12.34. Horizontal plate gage (courtesy of Building Research Establishment, Watford, England; Crown copyright, 1987).

Table 12.5. Methods for Borehole Installation of Probe Extensometer with Current-Displacement Induction Coil

Method	Applications and Typical Minimum Borehole Diameter[a]	Method for Attempting Conformance	Outer Pipe	Inner Pipe	Anchors	Installation Sequence
1	Widely used in United States Uniform soils Predicted vertical compression up to about 2% 4 in. (100 mm)	Grout	Corrugated polyethylene, continuous or coupled	Flush-coupled or belled-end PVC	Steel wire rings mounted on corrugated pipe	Drill, insert double pipe system length-by-length, grout; *or* drill, insert corrugated pipe, insert inner pipe, grout
2	Uniform soils Predicted vertical compression up to about 5% 4 in. (100 mm)	Grout	Corrugated polyethylene, with telescoping couplings	Flush-coupled or belled-end PVC	Steel wire rings mounted on corrugated pipe	Drill, insert double pipe system length-by-length, grout
3	Widely used in England All soils Small and large predicted vertical compression 8 in. (200 mm)	Positive anchorage	None	PVC with telescoping couplings	Pneumatically actuated arrowheads	Drill, grout, insert coupled pipe with surrounding anchors, actuate anchors
4	Very large predicted vertical compression (up to 30%) 6 in. (150 mm)	Positive anchorage	PVC with telescoping couplings	PVC with standard socket couplings or belled ends	Spring-actuated anchors	Drill, insert outer coupled pipe with surrounding anchors, insert inner coupled pipe, grout, actuate anchors, withdraw inner pipe

[a] If drill casing is used, borehole diameter is inside diameter of casing.

When a vertical installation is read by using a survey tape and noting the tape reading at which output current is a maximum, precision with respect to the deep measuring point will generally be ±0.1–0.2 in. (±3–5 mm). If greater precision is required, a long-range dial height gage or micrometer reading head can be placed on a reference surface at the top of the installation, the sliding arm inserted within one of a series of holes punched in the survey tape, and precision with respect to the deep measuring point can be ±0.02 in. (±0.5 mm). If the deep measuring point is not a reference datum and optical leveling is required, precision will also depend on the accuracy of the optical leveling.

Measurement precision in horizontal installations, with respect to a steel plate at one end, generally ranges from ±0.1 to 1.0 in. (±3–25 mm). If greater precision is required, a rod containing a primary induction coil near the position of each steel plate can be installed permanently within the telescoping pipe (Penman and Charles, 1982). Each plate can then be located horizontally by switching to the appropriate coil along the rod and inching the rod along with a rack and pinion mechanism until the coil coincides exactly with the plate. With this system, it is only necessary to move the rod over a short distance, and a precision of ±0.02 in. (±0.5 mm) is possible.

Installation in Boreholes

Installations in boreholes should follow the general guidelines given in Chapter 17. Four methods are in use, the preferred method depending on the predicted vertical compression, the stratigraphy, other site-specific conditions and needs, instrument availability, and experience of installation personnel. The four methods are described in the following subsections and summarized in Table 12.5.

Installation in Boreholes, First Method: Corrugated Plastic Pipe and Steel Wire Rings. Conventional Method

Corrugated pipe and steel wire rings are arranged as shown in Figure 12.32: this arrangement is widely used in the United States for installing the *Sondex* instrument, manufactured by Slope Indicator Company.

The rigid PVC inner pipe is included to prevent the possibility of the corrugated pipe becoming crushed, and it also centers the probe for maximum reading precision. This installation method relies on conformance between the soil and corrugated pipe via the grout and, as discussed in Chapter 17, very careful attention must be paid to selection of grout mix. The grout should ideally have a modulus and undrained shear strength as similar as possible to that of the subsoil. If the grout is either excessively stiff or soft, or if measurements are required in strata having differing compressibilities, conformance is questionable. The method has been used satisfactorily in uniform soils where vertical compression has not exceeded about 2%, but conformance is open to question for greater compressions. Pea gravel has been used as a backfill instead of grout, but a granular backfill should only be used if it satisfies the conformance criteria discussed in Chapter 17: it should never be used for installations in clay.

The system can be installed by working with 10 ft (3 m) lengths of corrugated and rigid pipe, coupling them and installing the two together, length-by-length, filling the annular space between the two pipes with a bentonite slurry to minimize friction, and finally grouting the space outside the corrugated pipe. If this method is used, the couplings in the corrugated pipe **must** be sealed against intrusion of grout, normally by generous use of mastic filler and sealing tape.

As an alternative for installations less than 100 ft (30 m) deep, a continuous length of corrugated pipe can be installed first, followed by installation of the inner pipe. A typical installation procedure for this alternative, which supplements the guidelines given in Chapter 17, is as follows:

1. Drill a vertical hole using HW drill casing to the depth required, allowing length for the weight on the bottom cap (required to overcome buoyancy while grouting), and flush with clean water.
2. Stretch out the corrugated pipe and securely connect the weighted bottom cap with its attached male check valve.
3. Lower the corrugated pipe into the cased borehole, using a safety line, always keeping the pipe filled with clean water. Tie the top of the corrugated pipe to the top of the drill casing.
4. After the weight comes to rest on the bottom of the borehole, install the rigid inner pipe within the corrugated pipe until it rests on the bottom cap. The rigid pipe may be flush coupled Sch. 40 PVC or belled end Class 200 PVC, installed with female down if settlement is anticipated.
5. Lower 1 in. (25 mm) Sch. 40 steel pipe inside the rigid inner pipe, with female check valve attached, filling with water prior to mating with the bottom cap check valve. Pump a small amount of water to ensure proper mating.
6. Mix grout and pump enough grout to fill approximately 30 ft (10 m) of annular space between the corrugated pipe and drill casing. Pull 30 ft (10 m) of drill casing.
7. Continue step (6) until all drill casing is removed and an undiluted grout return is observed at the top of the borehole.
8. Remove the steel grout pipe and flush the inside of the rigid inner pipe with clean water.
9. Install a protective top cap and secure the corrugated pipe while the grout sets.

The above procedure uses the double shutoff check valve arrangement for grouting, as described for inclinometers in Section 12.8. Alternatively, the annular space can be grouted via a tremie pipe lowered to the bottom of the borehole. The single shutoff arrangement cannot be used for installation of corrugated pipe by this method; because the inner pipe is installed after installation of a continuous length of corrugated pipe, there is no seal between the bottoms of the two pipes, and grout might enter the annular space between the two pipes at the bottom of the borehole.

Installation in Boreholes, Second Method: Corrugated Plastic Pipe with Telescoping Couplings and Steel Wire Rings

The second installation method is similar to the first, installing the two pipes length-by-length, except that telescoping couplings are provided in the corrugated pipe. The couplings are typically made from PVC, with O-rings, and they encourage conformance. This method appears suitable for vertical compressions up to about 5%, but for larger compressions a positive soil anchorage is recommended.

Installation in Boreholes, Third Method: Rigid Plastic Pipe with Telescoping Couplings and Pneumatic Anchors

The third installation method, developed by the Building Research Establishment in England, uses rigid PVC pipe with telescoping couplings, and expanding anchors at the measuring points.

The expanding anchors have two arrowheads attached to pistons in two small cylinders and are placed around the PVC pipe as it is lowered into the borehole. They are later driven into the sides of the borehole by applying pneumatic pressure to the cylinders via small-bore nylon tubing. Use of expanding anchors requires an 8 in. (200 mm) diameter borehole. The annular space between the soil and pipe should be filled with a grout designed to keep the borehole open but soft enough so that it does not impede telescoping of the pipe.

Installation in Boreholes, Fourth Method: Rigid Plastic Pipe with Telescoping Couplings and Spring Anchors

The fourth installation method has recently been developed by Slope Indicator Company for use in very soft clays with their *Sondex* probe extensometer. It is based on anchor systems developed earlier by the Building Research Establishment in England (Marsland, 1974a), which use explosive cutters to release spring anchors. The system uses rigid plastic pipe with telescoping couplings and spring anchors. An anchor consists of a length of PVC pipe, sized to fit over the telescoping pipe, cut, heated, and allowed to cool as shown in Figure 12.35. A stainless steel wire ring is mounted around the anchor to form the measuring point. Shortly before installation, anchors are held in a retracted position by a nylon line with slip knots, passing through a cutting block as shown in Figure 12.36. The system is installed in a cased borehole as follows:

1. Place anchors over the telescoping pipe and prevent them from slipping along the pipe by tying with a second nylon line. This second line passes through the cutting block and anchor and partly through the wall of the telescoping pipe.
2. Install the telescoping pipe and anchors through the drill casing as described above for the alternative first installation method, with grout pipe and double shutoff arrangement. Insert a trip rod with bottom cutter head through the cutting blocks as installation proceeds. Great care is needed to avoid premature anchor tripping as the system is lowered, and the inside profile of the drill casing must be smooth and the anchor ends beveled.
3. After lowering to the bottom of the borehole, verify correct wire ring locations by using the probe, and insert a rigid PVC stiffening pipe within the telescoping pipe to bear on the bottom cap. The purpose of this stiffening pipe is to ensure that all telescoping joints remain in their open position during grouting. Fasten the stiffening pipe and telescoping pipe together at the top, and grout.
4. Withdraw all drill casing entirely, top up with grout, and withdraw grout pipe.

Figure 12.35. Spring anchor for use with Slope Indicator Company *Sondex* probe extensometer.

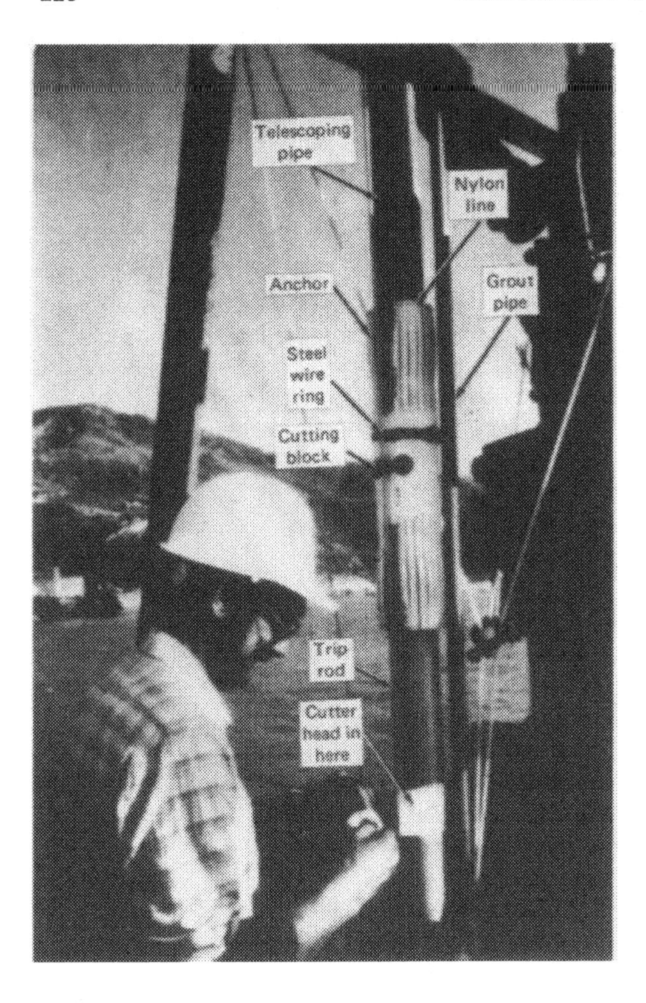

Figure 12.36. Installation of spring anchors for use with Slope Indicator Company *Sondex* probe extensometer.

5. Pull the trip rod to cut the pair of nylon lines at each anchor such that anchors positively grip the soil and are free to slide with respect to the telescoping pipe.
6. Allow the grout to set, and remove the stiffening pipe.

It must be emphasized that this fourth installation method requires great skill and care and should not be undertaken by inexperienced personnel. The system has recently been installed in very soft marine clays that form the foundation of a test embankment for a new airport in Hong Kong. Consolidation of the clays was accelerated by using vertical drains, and predicted vertical compression was greater than 30% (Handfelt et al., 1987). To date, maximum vertical compression has reached this value, and settlement data appear excellent. Plastic inclinometer casing with special long-range telescoping couplings was used instead of conventional telescoping pipe, so that both settlement and horizontal deformation could be measured (Section 12.5.10). Since the work in Hong Kong, Syncrude Canada Ltd. has simplified the installation procedure by eliminating the external cutting block and trip rod. The first and second nylon lines were replaced by two identical lines, one at each end of the anchor. Each line passed around the end of the anchor and through small-diameter holes drilled at opposite ends of the diameter through the anchor and telescoping pipe. Each line therefore held the anchor in its retracted position and also prevented it from slipping along the telescoping pipe. After installation in the borehole, the nylon lines were cut by lowering a dummy probe with a cutting blade fastened to its lower end, and the lines were drawn out of the drilled holes by the anchor spring action. The holes were not large enough to allow grout to pass through. This modification, although a simplification, prevents use of the internal stiffening pipe, and the author is reluctant to recommend it for all applications unless it can be shown that the stiffening pipe is unnecessary.

Installation in Fill

Installations in fill should follow the guidelines given in Chapter 17. Installations in boreholes that are later extended upward through fill should be converted to the telescoping pipe and steel plate system at the ground surface.

12.5.6. Gage with Frequency-Displacement Induction Coil

Induction coil transducers with frequency output, described in Chapter 8, are used by Telemac in their *Extensofor* probe extensometer (Bellier and Debreuille, 1977; Bordes and Debreuille, 1983), shown in Figure 12.37. The instrument consists of a 6 ft (2 m) long probe with a primary coil at each end. Steel rings or plates are installed around telescoping PVC pipe, generally as described previously for the gage with a current-displacement induction coil, at about 6 ft (2 m) spacing. The center-to-center distance between adjacent pairs of steel rings or plates is derived from the frequency output, using a conversion table, with a measurement precision of ±0.001 in. (±0.03 mm). The probe carries two inflatable packers to hold it steady while readings are

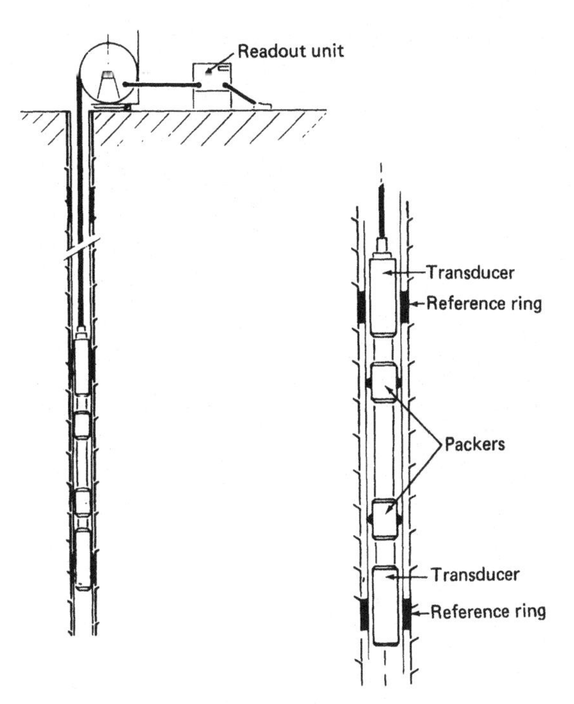

Figure 12.37. ***Extensofor*** **probe extensometer (courtesy of Telemac, Asnières, France).**

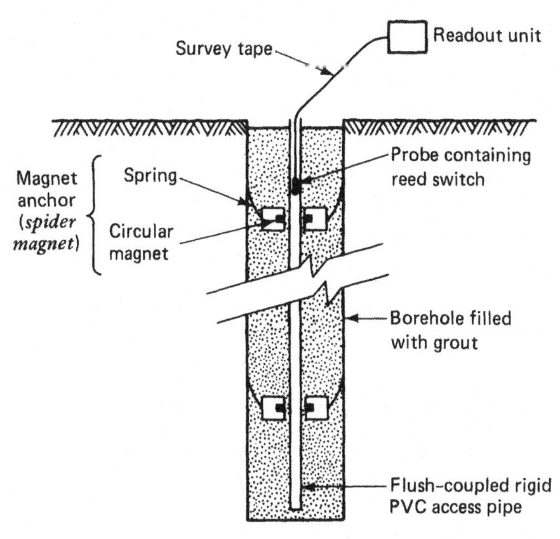

Figure 12.38. Schematic of probe extensometer with magnet/reed switch transducer, installed in a borehole.

taken, and the measurement range is about ±1.5 in. (±38 mm). The primary application is for monitoring axial deformations along boreholes in rock.

12.5.7. Magnet/Reed Switch Gage

The *magnet/reed switch gage* was developed by the Building Research Establishment in England (Burland et al., 1972; Marsland, 1974a) and incorporates the magnet/reed switch transducer described in Chapter 8. When used as a probe extensometer, the device consists of a series of circular magnetic anchors surrounding a rigid or telescoping plastic access pipe and is referred to as a *magnetic probe extensometer* or *magnetic extensometer*. A schematic of a borehole installation is shown in Figure 12.38 and various components in Figure 12.39.

An improved probe has been developed in Brazil (Figueiredo and Negro, 1981) that uses a different method of signal transmission when the reed switch is closed by a magnet. When closure occurs, a coupled oscillator in the probe is activated, generating square waves that are conducted through a steel survey tape up to the surface. The oscillator signal is received and amplified at the surface, enabling the operator to detect the signal through headphones. This improved probe avoids the use of a

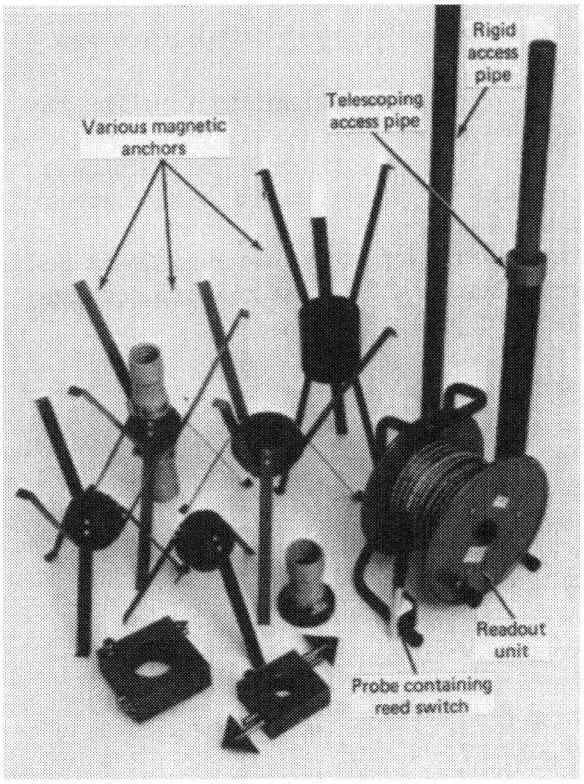

Figure 12.39. Magnet/reed switch probe extensometer (*magnetic extensometer*) (courtesy of Geotechnical Instruments (U.K.) Ltd., Leamington Spa, England).

Table 12.6. Methods for Borehole Installation of Probe Extensometer with Magnet/Reed Switch Transducer

Method	Applications and Typical Minimum Borehole Diameter[a]	Method for Attempting Conformance	Outer Pipe	Inner Pipe	Anchors	Installation Sequence
1	Predicted vertical compression up to about 1% 3.5 in. (90 mm)	Positive anchorage (but may be inhibited by grout)	None	PVC with flush couplings	*Spider magnet* with three leaf springs	Drill, grout, insert coupled pipe, insert anchors
2	All soils Small and large predicted vertical compressions 6 in. (150 mm)	Positive anchorage	None	PVC with telescoping couplings	Pneumatically actuated arrowheads	Drill, grout, insert coupled pipe with surrounding anchors, actuate anchors
3	All soils Small and large predicted vertical compressions 3.5 in. (90 mm)	Positive anchorage	Corrugated, coupled to anchors	PVC with flush couplings	*Spider magnet* with six leaf springs	Drill, grout, insert double pipe and anchor system length-by-length, actuate anchors
4	Rock 3.5 in. (90 mm)	Contact between leaf springs and rock, *or* grout	None	PVC with flush couplings	Either attached to pipe; *or spider magnet* with three leaf springs	Drill, insert coupled pipe, insert *spider magnet; or* drill, grout, insert coupled pipe with attached magnets

[a] If drill casing is used, borehole diameter is inside diameter of casing.

separate signal cable and greatly simplifies the reading task.

As an alternative to the single circular magnet, several small bar magnets can be set in a circle to form a measuring point.

Arrangements for reading are generally as described in Section 12.5.5 for the gage with a current-displacement induction coil.

Precision

Precision of reed switch closure is between ±0.001 and 0.01 in. (±0.03 and 0.3 mm), depending on guidance arrangements of the probe. System precision for vertical installations is similar to that of the gage with a current-displacement induction coil. Long probes are available with two reed switches, one near each end, such that readings can be made at two adjacent magnets in vertical installations by operating a height gage or micrometer and switching from one reed switch to the other. This is the usual procedure for installations in rock, where high precision is often required, and the entire system can be traversed by reading magnets in pairs. For maximum precision in horizontal installations, the twin reed switch probe can also be used. Alternatively, a rod containing a reed switch near the position of each magnet can be installed permanently within the telescoping pipe, making micrometer measurements of the movement of the rod that is required to align switches and magnets.

Installation in Boreholes

Installations in boreholes should follow the general guidelines given in Chapter 17. Four methods are in use, the preferred method depending on the predicted vertical compression, the stratigraphy, other site-specific conditions and needs, instrument availability, and experience of installation personnel. The four methods are described in the following subsections and summarized in Table 12.6

Installation in Boreholes, First Method: Rigid Plastic Pipe with Conventional Spider Magnets

Rigid flush-coupled PVC access pipe and *spider magnets* are installed as shown in Figure 12.38. Each anchor has three upward leaf springs, mounted at 120 degrees around the anchor. The borehole is usually cased and filled with a suitable grout, the access pipe is greased and inserted, the casing is raised to a level just above the lowest spider magnet, and a magnet is pushed downward over the access pipe until the leaf springs snap out of the casing bottom and bite into the soil. The procedure is repeated until all anchors are installed and casing withdrawn. While this is a simple installation method, there appears to be a possibility that anchors may become grouted to the access pipe, thereby inhibiting conformance. The method should therefore be used only for monitoring small vertical compressions, up to about 1%.

Installation in Boreholes, Second Method: Rigid Plastic Pipe with Telescoping Couplings and Pneumatic Anchors

The second method uses expanding anchors with arrowheads, as described in Section 12.5.5 for the induction coil gage. Because the reed switch probe can be of smaller diameter than the induction coil probe, arrowhead anchors can be installed in a smaller diameter borehole, and 6 in. (150 mm) is typical. This method overcomes the conformance concerns raised for the first method but entails more expensive anchors and a larger diameter borehole.

Installation in Boreholes, Third Method: Rigid and Corrugated Plastic Pipes and Spring Anchors

The third method consists of *spider magnets* around a rigid PVC access pipe, with lengths of plastic corrugated pipe linking adjacent pairs of spider magnets. The spider magnets are set with a pneumatic or explosive cutter, and the corrugated pipe ensures that spider magnets are free to slide along the rigid pipe, thereby ensuring conformance. The system is assembled and inserted in the borehole length-by-length and can be installed in a 3.5 in. (90 mm) borehole.

Three additional leaf springs are mounted on each *spider magnet,* at 120 degrees, pointing downward, and the six springs are held in a retracted position by loops of nylon twine. A pneumatic cutter is placed on each anchor such that when anchors are in position they can be actuated from the ground surface by applying pneumatic pressure via 0.19 in. (5 mm) air lines. The lowest anchor is actuated first, the air line pulled in an attempt to recover the lowest cutter, and the other anchors actuated in turn. Typically about 75% of cutters are recovered.

Installation in Boreholes, Fourth Method: Boreholes in Rock with Rigid Plastic Pipe

The magnet/reed switch gage can also be used for monitoring axial deformation in boreholes in rock. Magnets are either attached directly to the rigid PVC access pipe and the system grouted in place, or the rigid PVC access pipe/*spider magnet* arrangement is used.

Installation in Fill

For installations in fill, magnets are attached to rectangular or circular PVC or aluminum plates and installed around rigid PVC pipe with telescoping joints.

Installation for Monitoring Heave at Bottom of Open Cut Excavations

The magnet/reed switch gage can be arranged to provide measurements of heave at the bottom of braced or other open cut excavations, in a way that does not interfere with excavation work. As shown in Figure 12.40, the installation is made prior to the start of excavation, using one of the first three borehole installation methods. After taking initial readings, the access pipe is sealed 5–10 ft (1.5–3 m)

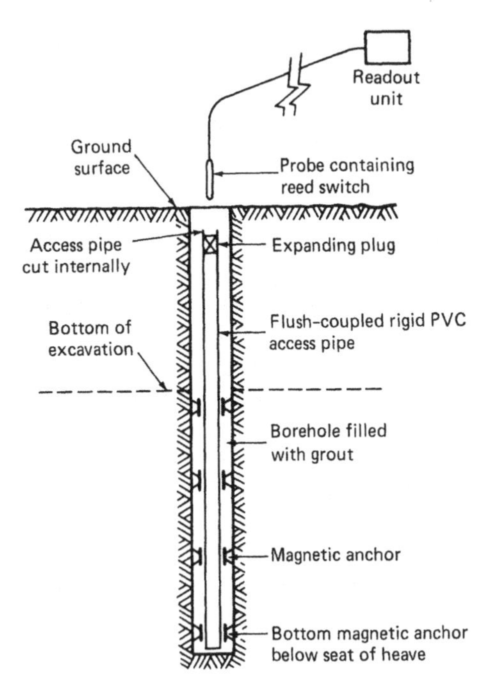

Figure 12.40. Schematic of probe extensometer with magnet/reed switch transducer, arranged for monitoring heave at the bottom of open cut excavations.

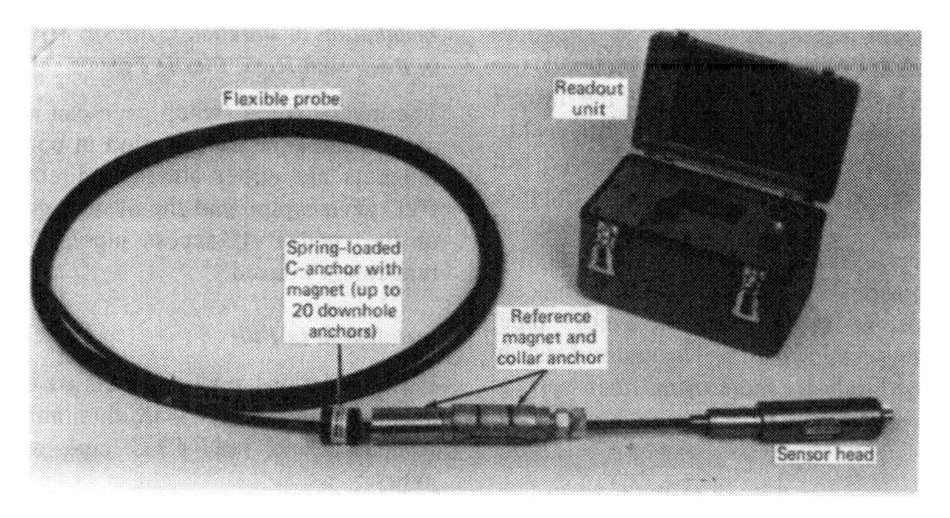

Figure 12.41. Flexible sonic probe (courtesy of Irad Gage, a Division of Klein Associates, Inc., Salem, NH).

below the ground surface, using an expanding plug set with an insertion tool, and the pipe is cut with an internal cutter just above the plug. A good survey fix is made on the plan position and, just before general excavation reaches the top of the pipe, the pipe is carefully located, a reading made, and the pipe again sealed and cut. The procedure is repeated until excavation is complete. Clearly, vigilance on the part of supervisory personnel is required, but the method is far less prone to damage and malfunction than use of fixed borehole extensometers with electrical transducers.

12.5.8. Bellow-Hose Gage

The *bellow-hose gage* (Bozozuk and Fellenius, 1979) consists of flexible polyethylene tube with short lengths of copper pipe mounted at predetermined locations to serve as measuring points. A probe with three protruding steel spring contact arms is lowered within the tube on the end of a survey tape. The steel springs are electrically insulated from each other, but two are connected to a cable leading from the probe to a voltmeter. When the probe contacts a copper pipe, the electrical circuit is closed, activating the voltmeter, and a tape reading is made alongside the top of the tube. The system is installed in a borehole by grouting, using the first method described above for the gage with a current-displacement induction coil (Section 12.5.5). When compared with the induction coil gage, readings are more subjective, no positive anchor arrangement is available to ensure conformance, and there appear to be no advantages.

12.5.9. Magnetostrictive Gage

The *magnetostrictive gage (sonic probe)* described in Section 12.7.4 is available in a removable flexible probe version and is shown in Figure 12.41. A magnet is attached directly to each anchor and a central access tube (not shown in the figure) is installed throughout the borehole. The flexible probe is inserted to the deepest magnet, and distances between magnets are displayed at the readout unit. The system has its primary application in underground rock excavations, where more elaborate permanently installed components would be subject to damage, and can be used in upward holes. The current version has an anchor depth limitation of 25 ft (7.6 m).

12.5.10. Combined Probe Extensometers and Inclinometer Casings

As indicated in Table 12.4, several of the probe extensometers can be used in conjunction with inclinometer casing in a vertical borehole, thereby obtaining both horizontal and vertical deformation data from one installation.

If a mechanical probe is used within inclinometer casing, no additional permanently installed features are required. However, use of this combination

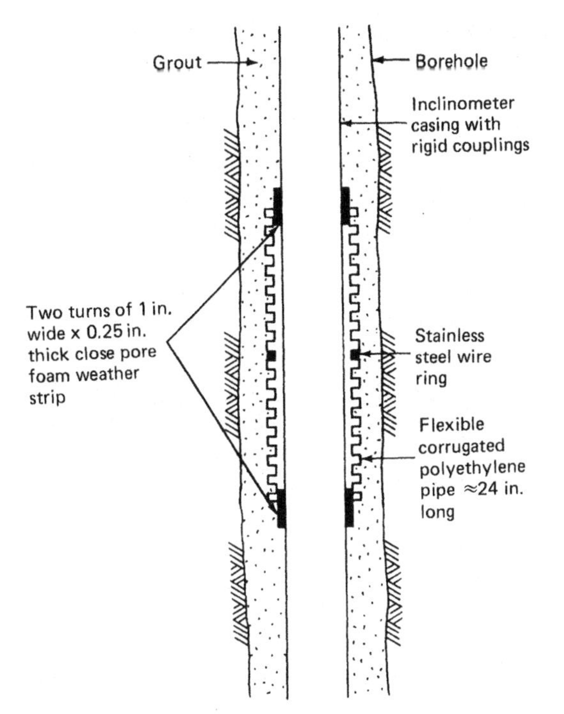

Figure 12.42. Schematic of induction coil ring combined with inclinometer casing.

should be limited to installations in fill, when settlement collars can be added. If the induction coil gage or the magnet/reed switch gage is installed vertically in fill or in a borehole, inclinometer casing with appropriate couplings can replace the inner pipe. If inclinometer casing is surrounded by corrugated plastic pipe, the intervening space must be well lubricated, and careful relative sizing of casing and pipe is required so that the casing is not free to move laterally with respect to the pipe. It is also important to fix the inclinometer casing so that it is not free to rotate.

Available installation methods are similar to those summarized in Tables 12.5 and 12.6, with the exception of the fourth method in Table 12.6. If the gage with current-displacement induction coil is used, the arrangement shown in Figure 12.42 provides an alternative simpler installation method for applications where less than about 2% vertical compression is anticipated. The thickness of weather strip is selected so that the polyethylene pipe can be pushed over the strip. The strip is intended to hold the corrugated pipe in place during installation, to prevent grout entering the annular space between the casing and pipe and to allow the corrugated pipe to break free as needed when settlement occurs.

If the user is concerned that the combined arrangement may reduce the reliability of either vertical or horizontal deformation measurements, the author recommends separate installations.

12.5.11. Combined Probe Extensometers and Open Standpipe Piezometers

The induction coil and magnet/reed switch instruments can also be combined with the standpipe of an open standpipe piezometer, so that both groundwater pressure and settlement can be monitored.

12.5.12. Recommendations for Choice of Probe Extensometer

It is not possible to make definitive recommendations for the choice of a probe extensometer. The author does not favor use of the bellow-hose gage, but all the others may on occasion be the instrument of choice. The selection depends on the application, the predicted axial compression, the ground conditions, other site-specific conditions and needs, the skill and experience of installation personnel, availability of hardware, the general factors for selection of instruments that are given in Section 4.9, and the more specific factors given in this section.

When reliable deformation data are essential, the author recommends use of a positive anchorage, rather than reliance on grout for ensuring conformance.

12.6. FIXED EMBANKMENT EXTENSOMETERS

Fixed embankment extensometers are defined in this book as devices placed in embankment fill as filling proceeds for monitoring the changing distance between two or more points along a common axis without use of a movable probe. They are used for monitoring settlement, horizontal deformation, or strain.

12.6.1. Settlement Platform

Settlement platforms are typically used for monitoring settlement below embankments on soft ground.

A settlement platform consists of a square plate of steel, wood, or concrete placed on the original ground surface, to which a riser pipe is attached (Figure 12.43). Optical leveling measurements to

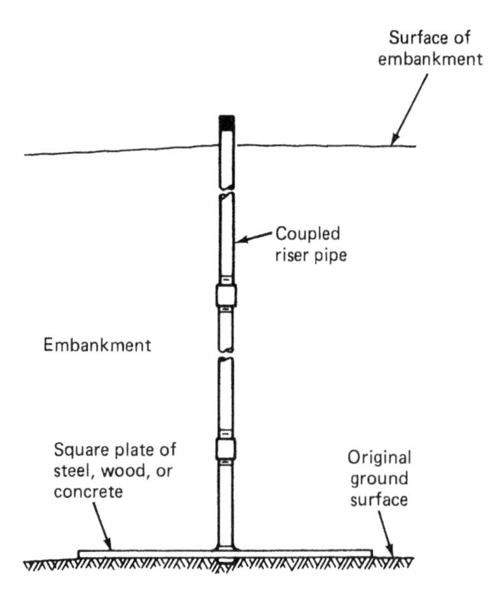

Figure 12.43. Typical settlement platform.

the top of the riser provide a record of plate elevations. The plate is typically 3 or 4 ft (1 or 1.2 m) square, and the riser pipe is typically 2 in. (50 mm) standard black iron pipe with threaded couplings. Either the pipe and plate are welded together or a floor flange is bolted to a wooden plate or embedded in a concrete plate. A sleeve pipe is sometimes placed around the riser pipe, with a gap between the bottom of the sleeve pipe and the plate to prevent downdrag forces on the riser pipe from being transmitted to the plate. Reasonable practice appears to require a sleeve pipe only if the embankment is over about 25 ft (8 m) high or if the plate is seated on highly compressible material such that downdrag forces might punch the plate below the original ground surface.

Care must be taken to maintain pipe verticality, and a record of added pipe length must be made as fill is placed. At the time of plate installation, an initial reference mark should be scribed on the riser pipe and its elevation determined and recorded. The pipe should be scribed to an accuracy of 0.01 ft (3 mm) at maximum intervals of 5 ft (1.5 m), measured from the initial mark. Immediately after the pipe is scribed, the graduations should be numbered to reflect the distance from the initial scribe. When pipe lengths are added, the extensions should be scribed in a similar way by tape measurement from scribes on the lower pipe. An adjustable hose clamp can be placed around the pipe at the highest scribe mark to provide a place to rest the survey rod when determining the elevation of the scribe mark. The clamp is loosened and moved upward as fill is placed.

The primary advantage of settlement platforms is their simplicity. Limitations include their tendency to be damaged by construction equipment and the difficulty in compacting around the riser pipe: cases have occurred in which a highway pavement has settled directly over a settlement platform, indicating inadequate compaction. These problems can sometimes be avoided by using liquid level gages (Section 12.10) instead of settlement platforms. Other limitations are the potential for measurement errors caused by additions of pipe lengths and caused by non-vertical sections of pipe, and the requirement for a survey crew when taking readings. However, with proper care, they can provide reliable data of adequate precision, generally in the range ± 0.1–1.0 in. (± 3–25 mm).

Where an unyielding layer exists at shallow depth, economy can sometimes be achieved by installing a subsurface settlement point (Section 12.7.6) through a hole in the plate as shown in Figure 12.44. The arrangement allows measurements

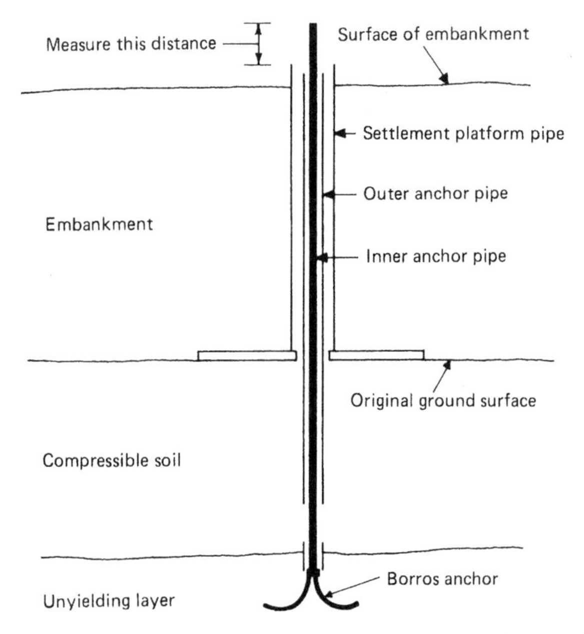

Figure 12.44. Schematic of combined settlement platform and Borros anchor.

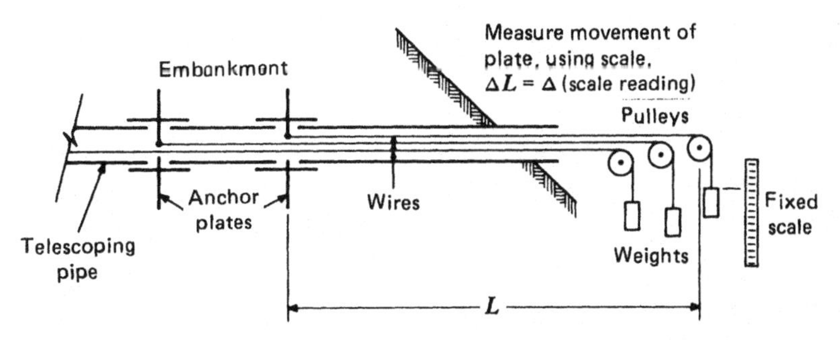

Figure 12.45. Schematic of typical mechanical gage with tensioned wires.

of absolute settlement to be made—without using a survey crew—by using a scale or mechanical gage to measure the distance shown.

12.6.2. Buried Plate

A *buried plate* is identical to the steel or concrete base plate of a settlement platform, and its use overcomes problems associated with the coupled riser pipe. To take an elevation reading on the plate, a vertical borehole is drilled, jetted, or augered from an accurately surveyed surface position, the plate located, and a depth measurement made. An accurate record must of course be made of the initial location of the plate in plan and elevation, and the plate must be large and level. The primary application for buried plates is for monitoring settlement below embankments on soft ground in cases where accuracy requirements are not great and where a boring rig is readily available.

Rózsa and Vidacs (1983) illustrate the application of buried plates, installed below an embankment on soft ground, in a row across an entire cross section.

12.6.3. Mechanical Gage with Tensioned Wires

Tensioned wire gages are used horizontally within an embankment or along the ground line at the base of an embankment to measure absolute horizontal deformation or horizontal strain.

As shown in Figure 12.45, the gage consists of vertical anchor plates attached to couplings in telescoping pipe laid horizontally. Steel wires or cables attached to each anchor plate are brought through the pipe to a measuring point where movement of a mark on the wire is observed relative to a fixed reference, while a standard tension is applied to the wire. Hosking and Hilton (1963) used at least ten wires in each pipe and 20 lb (9 kg) tension. Two different but standard weights can be used, as described in Section 12.7 for fixed borehole extensometers, to examine downhole friction characteristics. Additional guidelines are given in Section 12.7.

The pipe is normally installed in a shallow trench, as described in Chapter 17, and a terminal enclosure is required to house the pulleys, weights, and scale. Precision is generally ±0.2–0.8 in. (±5–20 mm). When absolute deformation data are required, the system must be located with respect to a horizontal control station, using surveying methods, and precision may be reduced.

12.6.4. Gage with Electrical Linear Displacement Transducers

These instruments consist of an electrical linear displacement transducer (LVDT, DCDT, linear potentiometer, vibrating wire transducer, variable reluctance transducer, or induction coil transducer with frequency output) mounted in line with a rod that connects two anchors, and cabled to an accessible location (Figure 12.46). An oil-filled telescoping

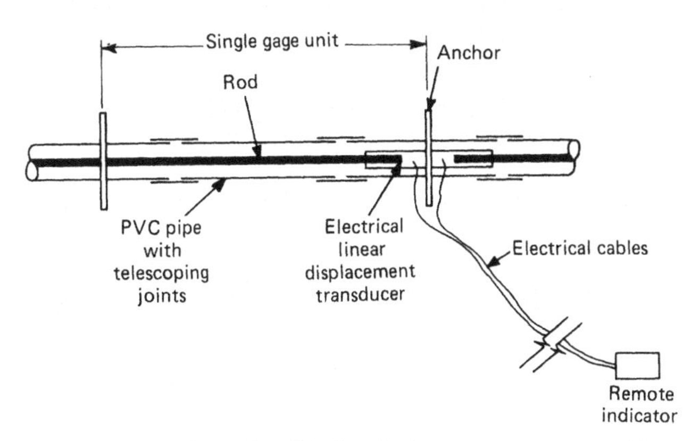

Figure 12.46. Schematic of fixed embankment extensometer with electrical linear displacement transducers.

Figure 12.47. Installation of fixed embankment extensometer, with linear potentiometer transducers, in embankment dam.

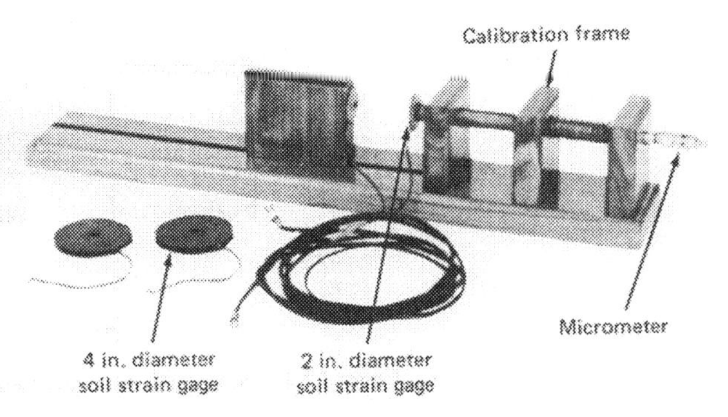

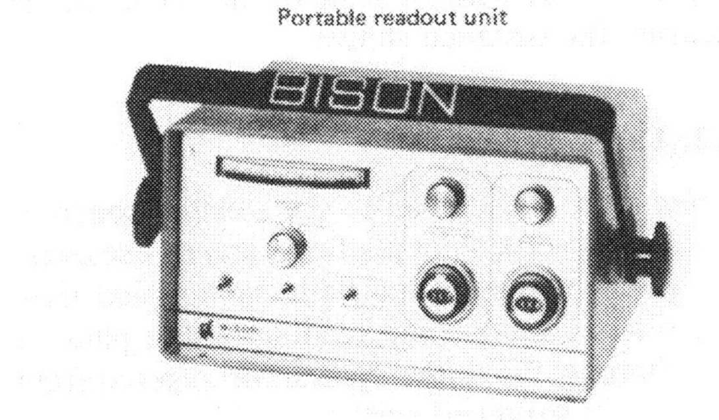

Figure 12.48. Soil strain gage (courtesy of Bison Instruments, Inc., Minneapolis, MN).

PVC pipe usually surrounds the rod and transducer, with O-rings in the couplings, to provide physical protection and waterproofing.

Their use is primarily for measuring horizontal strain in embankment dams, and the instruments are sometimes referred to as *horizontal strainmeters*. If the distance between anchors is too small, local variations may produce nonrepresentative data, whereas too long a distance will integrate true variations into an "average" value (Wilson, 1967). An appropriate anchor spacing is usually 10–20 ft (3–6 m). Gages can be installed singly or several can be coupled in series as shown in Figure 12.47. They can be grouped in different alignments to provide complete strain data in a horizontal plane.

Typical precision is ±0.01 in. (±0.3 mm), corresponding to a strain of ±0.005% with an anchor spacing of 16 ft (5 m).

12.6.5. Soil Strain Gage

The soil strain gage (Selig, 1975a) is described in Chapter 8 and is used for measurement of strain in any direction in earth fills. Components are shown in Figure 12.48.

Gages have been used extensively to measure vertical and horizontal strain beneath moving vehicles, as input to research studies on highway base and subbase materials, and on railroad ballast. Selig (1975b) describes strain determination around large buried culverts. They have also been used as convergence gages for monitoring the width of the trench during a full-scale test of a slurry trench excavation: this application is described in Section 12.11.2.

As shown in Figure 12.49, the two coils may be related to each other in an orthogonal, coaxial, or coplanar configuration. Strain is determined from the change in spacing between the coils after installation. The system is designed to operate with the two coils separated at a distance of less than five coil diameters, and coils are commercially available with diameters from 1 to 4 in. (25–100 mm). How-

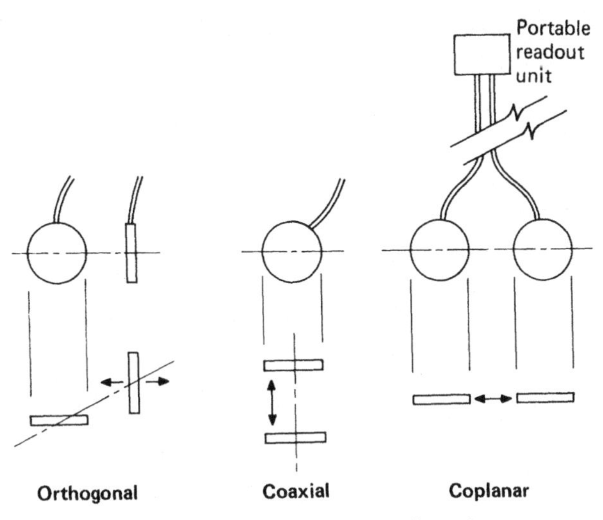

Figure 12.49. Soil strain gage configurations.

ever, larger coils can readily be manufactured to allow measurement over larger distances. Lord and Koerner (1977) describe a modified soil strain gage that uses a commercially available metal detector.

The primary advantage is the lack of any mechanical linkage between the coils; therefore, placement is simplified and conformance with strain of the soil is excellent. These devices are applicable for measuring both static and dynamic strains and are not affected by soil composition, moisture content, or temperature changes in the soil between the coils. Long-term precision is approximately $\pm 0.05\%$ strain, but dynamic values as small as $\pm 0.001\%$ strain can be detected. The primary disadvantage is sensitivity to the presence of metal objects, and movement of metal objects can cause a change in the electromagnetic field. This can create significant measurement errors unless adequate electrical shielding is installed. Stationary metal objects, however, generally do not cause a problem. The gages are also sensitive to changes in relative orientation of the coils.

12.7. FIXED BOREHOLE EXTENSOMETERS*

Fixed borehole extensometers are defined in this book as devices installed in boreholes in soil or rock

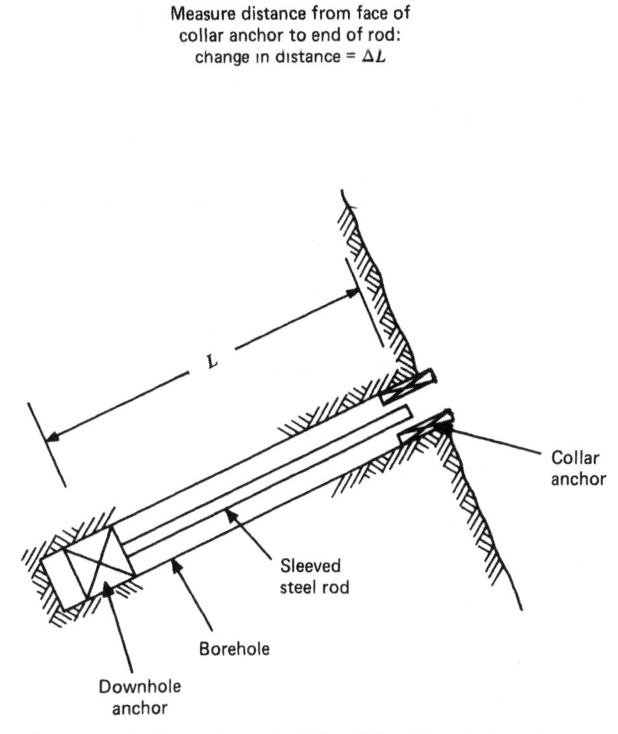

Figure 12.50. Operating principle of fixed borehole extensometer.

for monitoring the changing distance between two or more points along the axis of a borehole, without use of a movable probe. When the location of one measurement point is determined with respect to a fixed reference datum, the devices also provide absolute deformation data.

Typical applications are monitoring deformations around underground excavation in rock and behind the faces of excavated slopes. Fixed borehole extensometers are also used for monitoring consolidation settlements in soil, bottom heave in open cut excavations, and strain in concrete structures.

The operating principle is shown in Figure 12.50. The distance from the face of the collar anchor to the end of the rod is measured using either a mechanical or an electrical transducer. The device shown is a single-point borehole extensometer (SPBX), but several downhole anchors can be located in a single borehole, each with an attached rod from the downhole anchor to the collar anchor, to create a multipoint borehole extensometer (MPBX). MPBXs are used to monitor the deformation or strain pattern along the axis of an appropriately oriented borehole, for example, so that potential failure zones can be located and dangerous deep

*Written with the assistance of J. Barrie Sellers, President, Geokon, Inc., Lebanon, NH, and Howard B. Dutro, Vice President, Slope Indicator Company, Seattle, WA.

seated movements separated from surface spalling. Several SPBXs of different lengths, installed near each other, can serve the same purpose as an MPBX. Tensioned wires can be used instead of rods.

Many types of fixed borehole extensometer are available, the primary variables being choice of rods or wires, anchor type, SPBX or MPBX, transducer type, and extensometer head. These variables are discussed in the following sections. Alternative instrument types and installation, reading, data calculation, and processing procedures are described in ISRM (1981b). O'Rourke and Ranson (1979) provide comprehensive data on instrument performance.

12.7.1. Choice of Rods or Wires

In general, rod extensometers are of a simpler design than wire types and are more easily installed, especially if only one anchor is installed per borehole. They are generally preferred for extensometers up to about 300 ft (90 m) long, but the advantages of rods over wires are reduced as the length of the extensometer increases.

Wires

Wires are typically single-strand stainless steel, 0.02–0.05 in. (0.5–1.3 mm) in diameter. When wires are used, wires **must** be straightened by use of a wire straightener before extensometer assembly and, if this is not done, precision will be greatly reduced.

When the tension in the wire may vary, for example, when coil springs or cantilevers are used, wire stretch must be taken into account. Wire tension may vary as movement occurs along the extensometer, and wire stretch factors are usually supplied by the extensometer manufacturer. These factors can sometimes be verified after the extensometer is installed, by moving the extensometer head a known amount to and fro in line with the borehole axis and measuring the corresponding reading change. Some extensometer designs attempt to overcome wire stretch corrections by using constant-tension devices. Negator springs designed to exert a constant force are appropriate when very long range is required, but they should not be used when requiring high accuracy over a small range, because accuracy can be limited by hysteresis. Despite the cumbersome aspects of pulleys and suspended weights, similar to the system shown in Figure 12.45, they provide good tension control and are a good choice for permanently tensioned wire extensometers in applications where the head arrangements are acceptable.

Rods

Rods are typically 0.2–0.5 in. (5–13 mm) in diameter and are available in mild steel, stainless steel, aluminum alloy, fiberglass, and invar. A carbon fiber/vinyl ester composite material has been developed (Beloff, 1986) for use in high-temperature applications, such as studies relating to disposal of high-level nuclear waste, and can be manufactured to create a near-zero longitudinal thermal coefficient.

Rods are shipped either in coils or in straight lengths typically 10 ft (3 m) long. The straight lengths are generally connected with threaded flush couplings. Coiled steel and aluminum alloy rods may need to be straightened before installation, using a tool supplied by the manufacturer, but this is not necessary for coiled fiberglass rods. Rods should be encased in individual plastic sleeves to minimize frictional effects, and sleeves can be filled with oil if the borehole is inclined downward.

Rods are sometimes allowed to remain slack within their sleeves but sometimes are tensioned by coil springs, and there is no unanimity of opinion on the advantages of tensioning the rods. Proponents of tensioning argue that tensioning minimizes slack in the system. For some types of extensometer head, tensioning also allows the head to be moved outward a known distance to examine frictional effects. Opponents of tensioning argue that, because no borehole is perfectly straight, tensioning will increase normal stresses between a rod and its sleeve, thus increasing frictional effects and reducing precision. Opponents also argue that use of untensioned rods allows frictional effects to be examined at any time, by incorporating a bayonet disconnect fitting (Figure 12.53a) between the rod and downhole anchor. The rod can then be turned at the collar anchor to disconnect it from the downhole anchor and moved within its sleeve. The author has had good success with this arrangement and generally favors untensioned rods, but there is a need for comparative evaluations.

When using the bayonet disconnect fitting, thread sealing compound must be used on all threaded rod connections, and the rods must protrude beyond the face of the collar anchor so that they can be gripped and turned. The arrangement also simplifies installation, because rods can generally be inserted within sleeves as a separate step, after anchors and sleeves have been installed. When using the bayonet fitting in upward installations, a stop should be provided near the upper end of each rod so that rods cannot fall out of sleeves and cause injury to reading personnel, and rods must be installed concurrently with sleeves.

12.7.2. Choice of Downhole Anchor

ISRM (1981b) recommends a criterion for anchor adequacy, based on an applied load of 220 lb (100 kg) or five times the rod or wire tension, whichever is greater. ISRM recommends that, when subjected to this load, the anchor should not move in either direction by an amount greater than the sensitivity of the instrument. Hawkes (1978) describes many of the available downhole anchors; these and others in common use are discussed in the following subsections.

Applications for each anchor are suggested, but the generalizations will not apply to all cases. Several manufacturers of geotechnical instruments have wide experience in performance of their anchors under a variety of conditions, and their advice should be sought before selecting anchors. Factors that affect selection include soil or rock type and quality, borehole depth, borehole diameter, inclination and wall roughness, number of downhole anchors, use of rods or wires, and type of extensometer head.

Expanding Wedge Anchors

Two *expanding wedge anchors* are shown in Figure 12.51.

The *rockbolt expansion shell anchor* consists of a one-piece cylindrical split shell and a wedge nut. Rotation of the extensometer rod causes the wedge nut to move axially relative to the shell, expanding it into close contact with the walls of the borehole. To prevent the shell from rotating, it must initially be in rubbing contact with the borehole wall. Similar anchors are available with a split shell and wedge nut at each end. A steel block can be welded to each part of the split shell to allow gripping in a large-diameter borehole. The large setting force of the rockbolt expansion shell anchor enables it to be used at locations affected by blasting.

The *flat wedge spring anchor* consists of a flat wedge-shaped body with two expansion jaws, connected by a bent wire and spring loaded against the body as shown. The extensometer wire is attached to the body. When the anchor is pushed down the borehole on the end of a setting rod, the jaws rub against the sides of the borehole under the action of the spring. The anchor body is then pulled back by the extensometer wire to lock it into place. The anchor is designed for multipoint installations and the wires from deeper anchors are threaded through the bail prior to setting.

The rockbolt expansion shell anchor is generally the preferred anchor for single- and double-point rod extensometers installed in rock. The flat wedge spring anchor is useful in upward boreholes in rock, where multiple anchors are required. Both anchors can be installed rapidly, have wide expansion capabilities, and thus are suitable for boreholes with rough uneven walls.

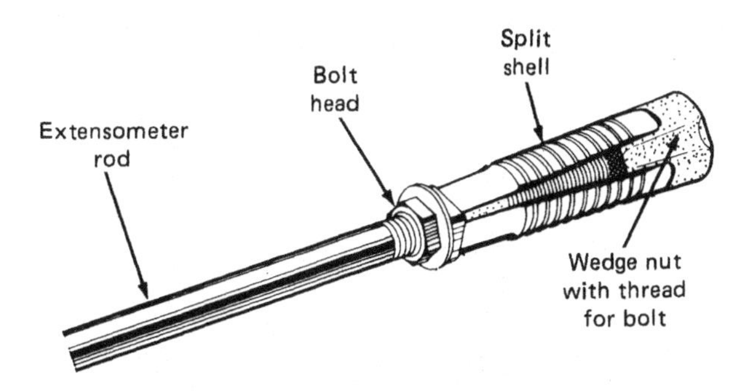

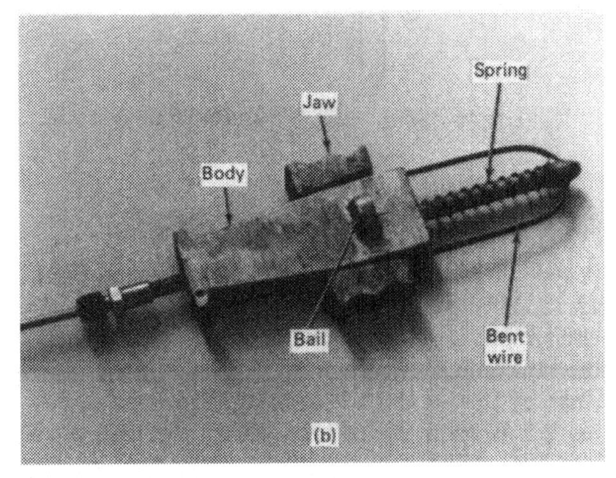

Figure 12.51. Expanding wedge anchors: (a) rockbolt expansion shell anchor (courtesy of Irad Gage, a Division of Klein Associates, Inc., Salem, NH) and (b) flat wedge spring anchor (courtesy of Slope Indicator Company, Seattle, WA).

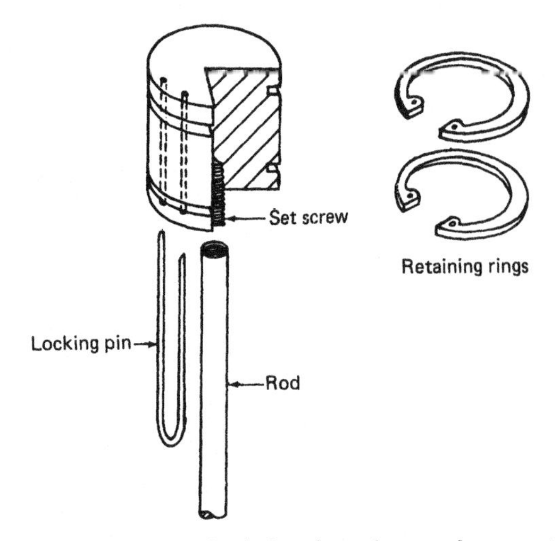

Figure 12.52. Spring-loaded anchor (after Hawkes, 1978).

Spring-Loaded Anchors

A typical *spring-loaded anchor* is shown in Figure 12.52. The anchor is also referred to as a *C-anchor* and *snap-ring anchor.* It is installed by pushing into the borehole to the required depth and tripping the rings by pulling out a U-shaped locking pin, allowing the rings to spring out against the wall of the borehole. These anchors are useful in hard or competent rock where smooth uniform boreholes can be drilled. Under these conditions, they offer the ultimate in installation speed and simplicity, where several anchors are installed in the same borehole. Dutta (1982) reports on pull tests within 3 in. (76 mm) diameter steel pipe, indicating a resistance to pullout in excess of 200 lb (90 kg). Other tests by Dutta indicate that the anchors are not likely to be disturbed by nearby blasting.

Groutable Anchors

Groutable anchors (Figure 12.53) are simple to install and are usually the preferred anchors for boreholes in rock that are inclined downward. Nominally horizontal boreholes can often be inclined slightly downward to allow their use. They are suitable for use at locations affected by blasting but not for use in soil, as the grout may inhibit conformance. The measurement rods or wires are sleeved within plastic pipes or tubes to isolate them from the grout.

Extensometers incorporating groutable anchors are preassembled at the surface and pushed into the borehole, with tubes for grout outlets and air vents as discussed in Chapter 17. Boreholes directed upward require a stage grouting procedure, and an alternative anchor is usually preferable.

Hydraulic Anchors

Two types of *hydraulic anchor,* which can be used with rods or wires, are shown in Figure 12.54.

The *expanding tube type* is suitable for use in soil or rock. A soft metal tube is flattened, its ends sealed, and wrapped around the body of the anchor. The tube is pressurized by hydraulic oil to expand it beyond the yield point of the metal and in contact with the walls of the borehole. When hydraulic pressure is released, the tube maintains its deformed shape.

The *prong type* is used primarily in soft ground, where the anchor must be forced outward into the soil to spread the anchorage forces and prevent slippage. Hydraulic oil pumped to the anchor forces the prongs outward into the walls of the borehole. Single- and double-acting versions are available.

12.7.3. Choice of Single-Point or Multipoint Extensometer

In general, the more anchors installed in one borehole, the greater are installation difficulties and duration, with possible consequent delay to construction schedules. MPBXs typically have between four and eight anchors and generally require more skilled installation personnel than SPBXs.

Several SPBXs installed side by side in boreholes of different lengths can simulate an MPBX and may result in overall economy and reduced construction delay. However, the applicability of this option is affected by many site-specific conditions, including borehole length, type and number of anchors, and type of extensometer head. When a cluster of SPBXs is used and core recovery is required for planning the installation and interpreting data, only the deepest SPBX hole need be cored, and the shallower holes can be percussion drilled if that is more economical. If the collar anchors can be set in the same block of intact rock, anchor movement data are calculated as for an MPBX. However, if collar anchors may move with respect to each other and if downhole relative deformation data are required, their relative movement must be measured. If the SPBX heads are installed in a vertical line, the mechanical tiltmeter described in Section 12.4.1 can sometimes be used for this purpose.

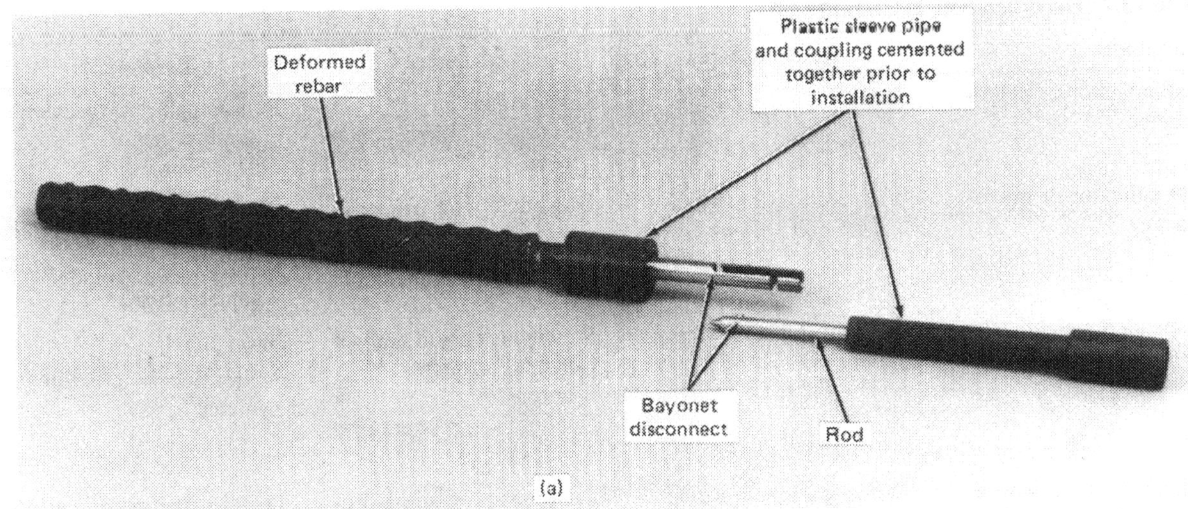

Figure 12.53a. Groutable anchor for rod extensometer (courtesy of Geokon, Inc., Lebanon, NH).

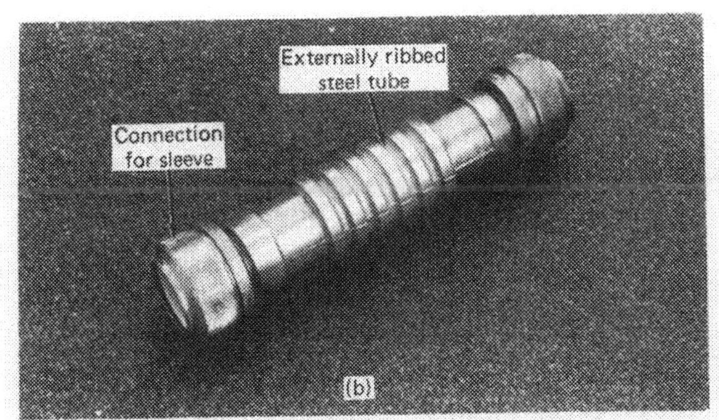

Figure 12.53b. Groutable anchor for multipoint wire extensometer (courtesy of Slope Indicator Company, Seattle, WA).

12.7.4. Choice of Transducer and Extensometer Head

Mechanical and electrical transducers are available for reading fixed borehole extensometers. Mechanical transducers are less expensive and are generally more reliable and more resistant to damage. When access to the collar anchor is available, mechanical transducers should be the first choice unless automatic or remote monitoring is required. In other cases, electrical transducers must be used, but arrangements should be made within the measuring head for backup mechanical reading without disturbing the electrical transducers, so that a periodic check can be made on the electrical system by creating temporary access.

Mechanical and electrical transducers are described and compared in Chapter 8. The following subsections provide additional information relevant

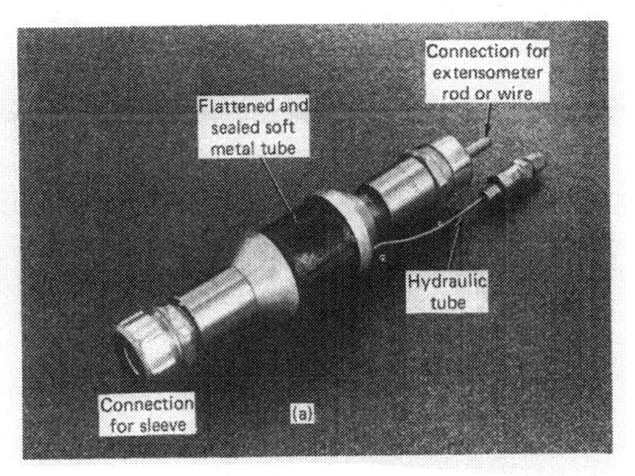

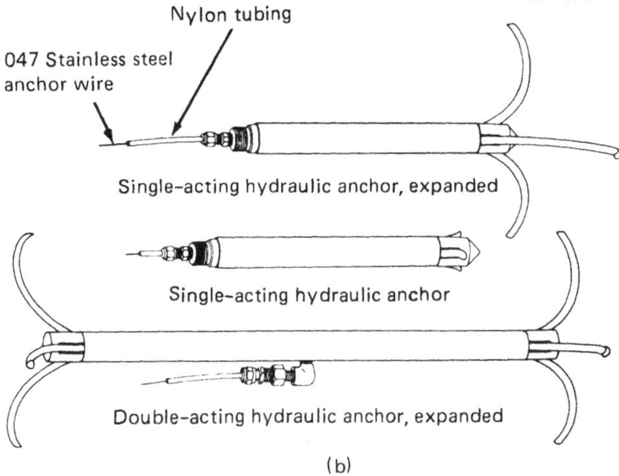

Figure 12.54. Hydraulic anchors: (a) expanding tube type and (b) prong type (courtesy of Slope Indicator Company, Seattle, WA).

Table 12.7. Fixed Borehole Extensometers[a]

Transducer	Rods or Wires	Advantages	Limitations
Dial indicator or micrometer	Rods	Large range Disconnect fittings can be used[b]	Requires access to extensometer head
Dial indicator or micrometer	Variable-tension wires with coil springs	Downhole wire friction can be examined[c]	Requires more skill to install than rod system Requires access to extensometer head
Suspended weights (similar to Figure 12.45)	Constant-tension wires	Excellent tension control Reliable for deep boreholes Dual tension reading provides check	Bulky head Requires access to extensometer head
Indicator for slack-wire extensometer (e.g., Figure 12.55)	Wires tensioned during reading	Dual tension reading provides check	Tedious reading procedure Requires access to extensometer head
Electrical resistance strain gage (e.g., Figure 12.56)	Variable-tension wires	Downhole wire friction can be examined[c] Can be read with dial indicator or micrometer to check electrical reading	Limited range Low electrical output Lead wire effects Errors owing to moisture, temperature, and electrical connections are possible Need for lightning protection should be evaluated
LVDT, DCDT, or linear potentiometer (e.g., Figures 12.57, 12.58, 12.59)	Rods, or variable-tension wires with coil springs	Versions available with transducers mounted in series between anchors for improved physical protection Versions available for use as heave gages in open cut excavations Disconnect fittings can be used[b] or downhole wire friction can be examined[c]	LVDTs have unwanted lead wire effects Linear potentiometers require perfect water seal at rod entry Need for lightning protection should be evaluated
Rotary potentiometer or rotary transformer	Constant-tension wires with constant-force springs	Long range Can be read with dial indicator or micrometer to check electrical reading Can be read initially with dial indicator or micrometer and later converted to electrical transducer	Precision likely to be reduced by hysteresis in springs Need for lightning protection should be evaluated
Vibrating wire	Rods, or variable-tension wires with coil springs	Lead wire effects minimal Disconnect fittings can be used[b] or downhole wire friction can be examined[c]	Special manufacturing techniques required to minimize zero drift Need for lightning protection should be evaluated

Table 12.7. ***(Continued)***

Transducer	Rods or Wires	Advantages	Limitations
Induction coil (e.g., Figures 12.60, 12.61)	Rod	No need for strong mechanical connection between transducers and rock Lead wire effects minimal Radio telemetered SPBX available	Need for lightning protection should be evaluated
Magnetostrictive (sonic probe) (e.g., Figure 12.62)	Rods	MPBX requires only one transducer MPBX requires only a single six-conductor cable Disconnect fittings can be used[b] Can be read with dial indicator or micrometer to check electrical readings Can be read initially with dial indicator or micrometer and later converted to electrical transducer	Signal requires amplification if lead wires are longer than about 600 ft (200 m) Need for lightning protection should be evaluated

[a]For additional comparative data between use of rods or wires, downhole anchor types, and SPBX or MPBX configurations, see Sections 12.7.1.–12.7.3. Mechanical transducers should be used in preference to electrical transducers wherever access is available. Assuming that the deepest anchor or the head can serve as an immovable reference point, precision of all extensometers is generally ±0.001–0.005 in. (±0.03–0.13 mm).

[b]If untensioned rods are used, or if rods are temporarily detensioned, downhole friction can be examined by using bayonet disconnect fittings between rods and downhole anchors. See Section 12.7.1.

[c]This can sometimes be done by moving the head outward a known distance. See Section 12.7.1.

to their use as transducers for fixed borehole extensometers. In general, each transducer can be used with any of the anchor types, and with a single-point or multipoint system. Table 12.7 provides comparative data on various combinations of transducers, rods, and wires. Selection of a particular arrangement depends on site-specific conditions and needs, instrument availability, and experience of installation personnel.

Dial Indicator or Micrometer

Depth micrometers are more rugged than dial indicators and thus are preferred; micrometers with digital counters capable of reading movements to 0.001 in. (0.03 mm) are preferable. A minimum of two should be available in case one is damaged. Dial indicators and micrometers should be checked regularly against a standard to verify that there has been no change in the zero reading, and most geotechnical instrumentation manufacturers will supply a pipe with one end closed for this purpose.

Suspended Weights

The *suspended weight* arrangement shown in Figure 12.45 can be used for reading fixed borehole extensometers. Excellent tension control is provided, and the arrangement is well suited for multipoint wire extensometers where access is available and where the bulky head arrangement can be tolerated. Whittaker and Woodrow (1977) describe the design and performance of an extensometer with suspended weights.

Hedley (1969) describes a useful technique to examine downhole friction characteristics and to verify precision, whereby readings are always taken at two different but standard wire tensions. The difference in reading at these two tensions should be the same every time a set of readings is taken. A small

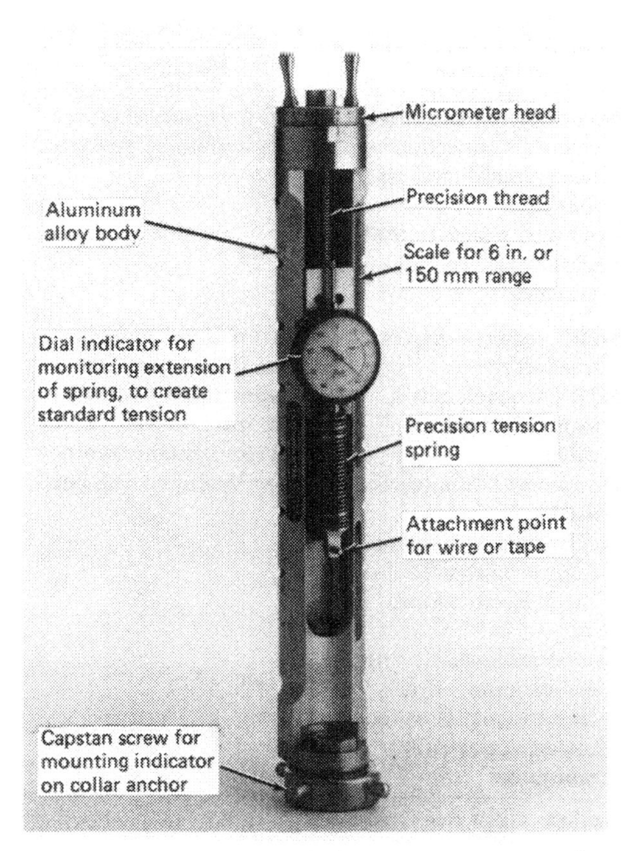

Figure 12.55. Indicator for slack-wire extensometer: Mark 2 model (courtesy of Terrascience Systems Ltd., Vancouver, Canada).

change could mean that friction in the borehole has altered, while a major change indicates that the wire is obstructed or the borehole has sheared and clamped the wires. The readings at two tensions provide a means of determining the location of any obstruction.

Indicator for Slack-Wire Extensometer

The *slack-wire* extensometer indicator (Potts, 1957) allows application of standard tension to a wire and includes a micrometer for reading deformation. As shown in Figure 12.55, the capstan screw of the Mark 2 version is threaded on to the collar anchor, the extensometer wire is connected temporarily to the attachment point, tension is applied to the wire by rotating the micrometer head until the dial indicator reads a standard value, and the micrometer is read. When used with a multipoint system, the micrometer head is retracted, the first wire removed from the attachment point and replaced with a second wire, and the procedure repeated. The indicator can be used with wires up to 600 ft (180 m) long and has a reading range of 6 in. (150 mm.)

The lighter Mark 1 version is limited to extensometer wire lengths up to about 100 ft (30 m), and tension is applied directly by turning the micrometer. The thread within the micrometer must therefore sustain the full tension; there is a risk of stripping the thread, and the author does not favor this version.

Readings are usually taken at two different but standard wire tensions as described previously for the suspended weight system, and the difference in the two micrometer readings should always be the same.

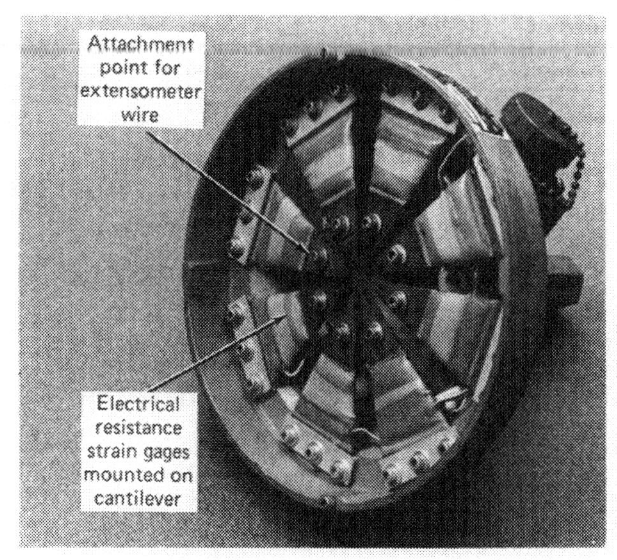

Figure 12.56. Extensometer head with electrical resistance strain gage transducers (courtesy of Slope Indicator Company, Seattle, WA).

Electrical Resistance Strain Gage Transducer

Figure 12.56 shows an eight-point extensometer head with electrical resistance strain gage transducers, using cantilevers to exert wire tension. Strain gages are connected in half bridge networks, but full bridge networks are available as an option. Relative deformation between the collar anchor and a downhole anchor causes bending of the corresponding cantilever and change in strain gage reading. Backup mechanical readings can be made by inserting a dial indicator through holes in the cover plate, to bear against each cantilever near its tip. Wire tension varies as movement occurs, and wire

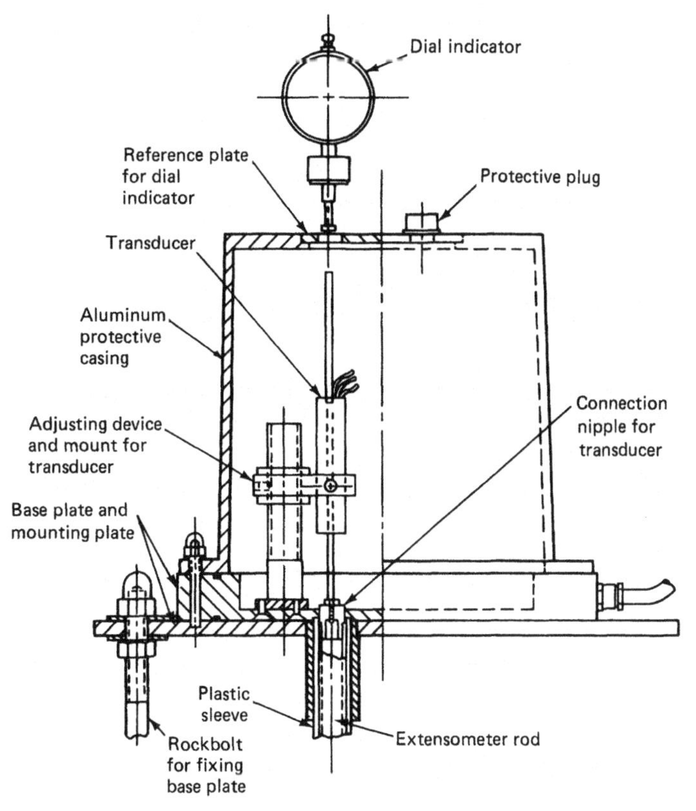

Figure 12.57. Head for fixed borehole extensometer (courtesy of Interfels GmbH, Bentheim, West Germany).

stretch factors must be accounted for by calibration or computation. This type of head is not currently used extensively.

LVDT, DCDT, and Linear Potentiometer

LVDTs, DCDTs, and linear potentiometers are all used as transducers for single-point and multipoint borehole extensometers. Comparisons among the three types are given in Chapter 8.

A single-point version is shown in Figure 12.57, and a similar arrangement is in use for multipoint versions. The arrangement allows for remote readings and also for backup mechanical readings.

Normally, all transducers are housed in an extensometer head as shown in Figure 12.57, but an alternative multipoint configuration is available with the transducers in series. In the series configuration, a transducer is mounted between each anchor, using appropriate lengths of extensometer rod. An example is shown in Figure 12.58 and is installed by inserting each anchor/transducer/rod module in turn and setting the mechanical anchor with an installation tool. Each module is spring loaded so that the downhole end of each rod can contact the adjacent anchor at any point on its surface. The system is retrievable. The series configuration gives a more direct measurement of relative deformation between adjacent anchors and increases reliability where the extensometer head is subject to mechanical damage. However, the transducers are less readily accessible for any necessary checking and maintenance, and this is generally a disadvantage. In very deep holes the series configuration may be the system of choice, because it overcomes the need for very long rods.

Rotary transducers can be used when a long range is required. When downhole wires are used, each wire passes over a pulley in the head and is attached to a constant-force spring. A single-turn or multiturn rotary potentiometer or rotary transformer is mounted on the axle of the pulley, so that relative movement between the collar anchor and

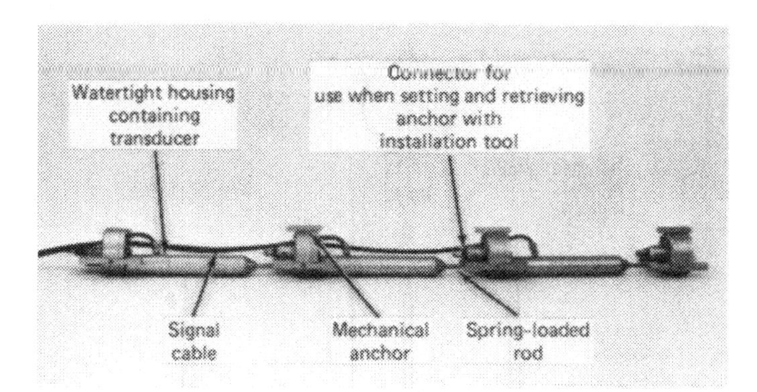

Figure 12.58. Fixed borehole extensometer, with LVDTs, DCDTs, or linear potentiometers mounted in series: Roctest Model BOF-EX (courtesy of Roctest Ltd., Montreal, Canada).

the downhole anchor causes an output change. When downhole rods are used, a flexible wire is attached to the end of each rod and passes over a pulley. Shuri and Green (1987) describe an extensometer head of this type.

Versions of fixed borehole extensometers with LVDT, DCDT, and linear potentiometer transducers are available for monitoring bottom heave at the base of open cut excavations. The arrangement is shown in Figure 12.59, requiring that a borehole is drilled to below the anticipated seat of heave, and setting a deep anchor. A sleeved rod spans between the anchor and an electrical linear displacement transducer set below the eventual bottom of the excavation, such that any change in distance between the transducer and deep anchor causes an identical movement within the transducer itself. An electrical cable passes up the borehole to the ground surface, and arrangements are made for damage protection and cable retrieval as described in Section 12.5.7 for the probe extensometer with a magnet/reed switch transducer. The upper connector on the coiled electrical cable is attached to the expanding plug, so that a gage reading can be made whenever the plug is retrieved. Several transducers can be set in the same borehole, each with a rod attached to the deep anchor, to provide a pattern of heave measurement with depth. If the transducers are functioning on completion of excavation, a multiconductor cable can be spliced to the separate coiled cables and routed to a suitable remote location for subsequent monitoring of recompression settlements.

Vibrating Wire Transducer

Conventional vibrating wire transducers have insufficient range for direct use in fixed borehole extensometers, but if a coil spring is added in series with the extensometer wire and vibrating wire, adequate range is created.

Induction Coil Transducer

A transducer with a frequency-displacement induction coil, described in Chapter 8, is used by Telemac in their *Distofor* fixed borehole extensometer (Bellier and Debreuille, 1977; Bordes and Debreuille, 1983). As shown in Figure 12.60, steel rings are mounted within telescoping PVC pipe couplings, and the system is grouted within a borehole. A single central rod is inserted through all rings, with a primary coil mounted on the rod alongside each ring. Changes in frequency output are con-

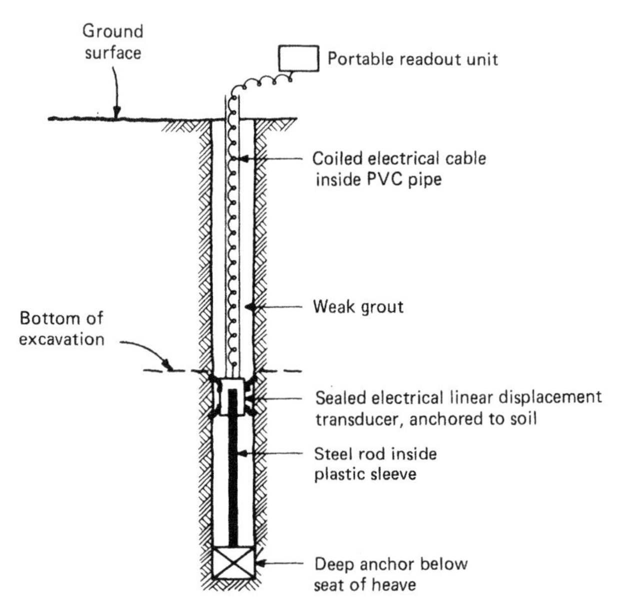

Figure 12.59. Schematic of fixed borehole extensometer arranged for monitoring heave at the bottom of open cut excavations.

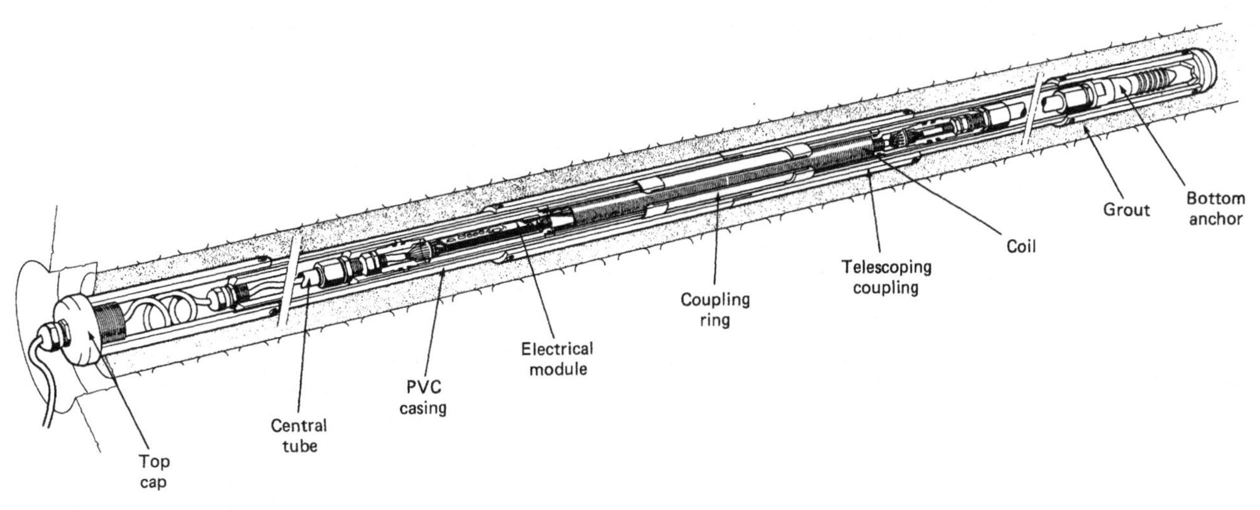

Figure 12.60. ***Distofor*** **fixed borehole extensometer (courtesy of Telemac, Asnières, France).**

verted to deformation, using conversion tables, with a measurement precision of about ±0.001 in. (±0.03 mm) and a range of ±2.5 in. (±60 mm). Londe (1982) indicates that the fundamental advantage of this instrument over conventional fixed borehole extensometers using rods or wires is the absence of any strong mechanical connection between the transducers and the rock.

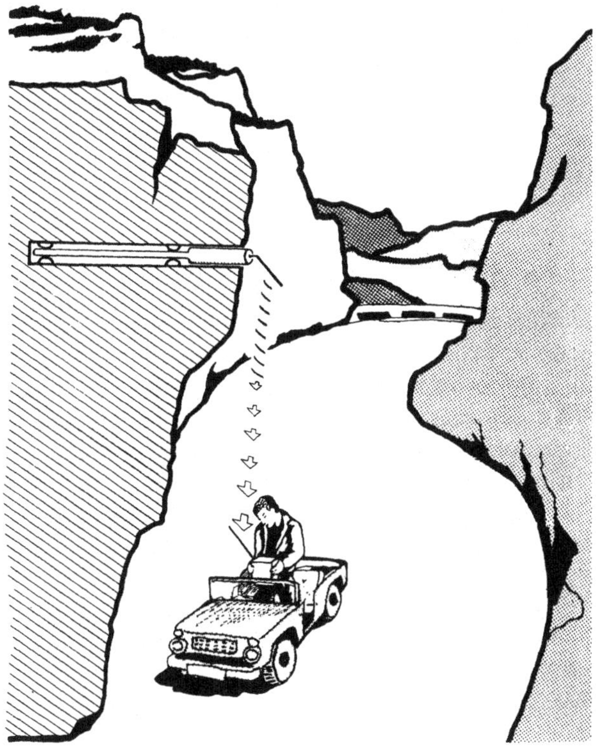

Figure 12.61. ***Radiofor*** **single-point fixed borehole extensometer (courtesy of Telemac, Asnières, France).**

The single-point version shown in Figure 12.61, the *Radiofor,* can be read remotely with a portable battery-operated indicator. The frequency output is transmitted as a radio signal, thus overcoming the need for lead wires. Operating range for the signal in tunnels is reported as 60 ft (20 m), elsewhere 1000 ft (300 m). The device was originally developed for monitoring tunnel convergence during driving but is also suitable for monitoring deformations of rock slopes. The transmitter and sensor are recoverable.

Magnetostrictive Transducer

The *magnetostrictive transducer (sonic probe),* described in Chapter 8, can be either left in place at the extensometer head or used as a portable "wand." The head arrangement is shown in Figure 12.62. A bar magnet is mounted on each rod, near the collar anchor, and the distance is measured between each magnet and a reference magnet mounted in the head.

12.7.5. Installation of Fixed Borehole Extensometers

When planning installation procedures, one should follow the Chapter 17 guidelines. Particular attention should be given to whether groundwater must be prevented from passing along the borehole. If an open borehole is unacceptable, it should be filled with grout, and downhole components must be se-

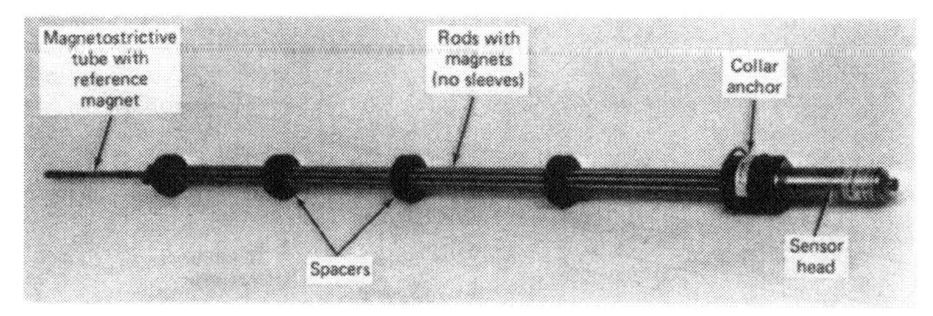

Figure 12.62. **Head arrangements for fixed borehole extensometer with magnetostrictive transducer (*sonic probe*) (courtesy of Irad Gage, a Division of Klein Associates, Inc., Salem, NH).**

lected accordingly. Additional points are given in the following subsections.

Borehole Requirements

Most fixed borehole extensometers have minimum and maximum allowable borehole diameters, and selection of the instrument should be influenced by available methods for drilling the borehole. Sometimes the allowable diameter variation is very small, and *go* and *no-go* gages may be required to check whether the borehole falls within the allowable tolerance.

If possible, boreholes for multipoint extensometers should be arranged such that either the collar anchor or deepest downhole anchor can be used as an immovable reference point. This assessment should be based on the geology, the geometry of the excavation or structure, and on other site-specific features. If fixity is in doubt, provision should be made to monitor absolute deformation of the collar anchor on a regular basis, using surveying methods or a convergence gage. Anchor locations should also be based on geology and geometry, and clearly cored holes are helpful. If not determined from rock cores, discontinuities can sometimes be located by borehole television or borescope surveys.

Where excavation will occur close beneath the deepest anchor, for example, a tunnel, a "telltale" consisting of a colored plastic tube may be attached to the anchor to protrude into the path of the future excavation. Subsequent exposure of the telltale indicates the borehole location.

Installation of Downhole Components

When multipoint borehole extensometers are assembled, the various rods or wires must not become entangled. Correct relative positions must be maintained at all times. Color coding is often helpful, with corresponding colored marks around the mouth of the borehole.

When using groutable anchors, the guidelines given in Section 17.5.6 should be followed. A thick cement grout is normally used, and no sand should be included in the mix. When used in rock where predicted deformations are compressive, grout strength should be weakened by use of additives. Buoyancy forces must be overcome, and grout pressure must be controlled to avoid collapse of the sleeves protecting extensometer rods or wires. If flexible plastic tubing is used for the sleeves, risk of collapse can be avoided by filling the tubes with oil and pressurizing the oil with a hand pump until grout sets. More details of these procedures are given in Chapter 17.

12.7.6. Subsurface Settlement Points

Subsurface settlement points are categorized in this book as fixed borehole extensometers but are used for monitoring absolute deformation rather than relative deformation between a collar anchor and a downhole anchor. Typical applications are for monitoring settlement below embankments and surcharges, above soft ground tunnels or adjacent to braced excavations, and for monitoring uplift during grouting operations.

The device consists essentially of a riser pipe anchored at the bottom of a vertical borehole and an outer casing to isolate the riser pipe from downdrag forces caused by settlement of soil above the anchor. Settlement of the anchor is determined by measuring the elevation of the top of the riser pipe, using surveying methods. Three arrangements are

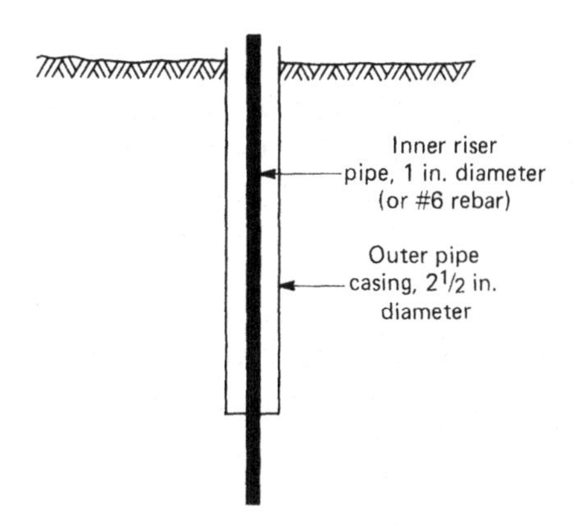

Figure 12.63. Schematic of subsurface settlement point with driven anchor.

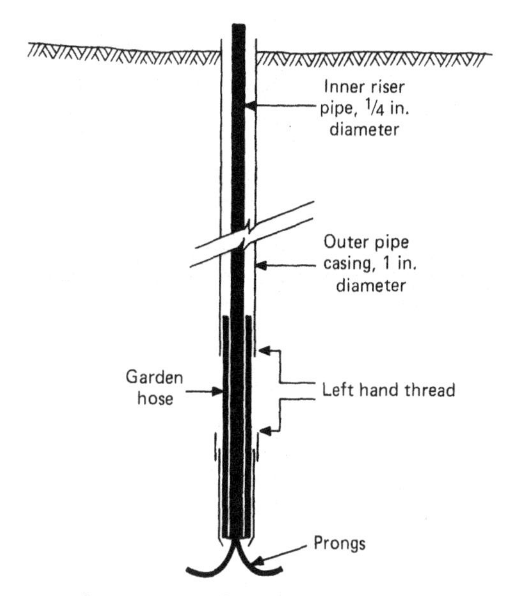

Figure 12.64. Schematic of Borros anchor.

in common use and are described in the following subsections.

Driven or Grouted Anchor

A typical arrangement, sometimes referred to as a *deep settlement point,* is shown in Figure 12.63. Outer pipe casing is driven to the required depth and cleaned out. If the casing is set in a predrilled hole, the annular space between the casing and borehole should be backfilled with sand, pea gravel, or grout, and any grout should be allowed to set before the riser pipe is installed. The riser pipe is then inserted and driven 1–3 ft (300 mm to 1 m) below the bottom of the casing. A rounded reference surface is often attached to the top of the riser pipe and the arrangement protected by a surface cover. Details are given by Cording et al. (1975).

When a more secure anchor is required, a measured quantity of grout can be tremied to the bottom of the borehole before inserting the riser pipe, the riser pipe driven through the grout, and the outer casing bumped back so that its bottom remains about 1 ft (300 mm) above the top of the grout.

Borros Anchor

The *Borros anchor,* or *Geonor settlement probe,* is shown in Figure 12.64. The anchor consists of three steel prongs housed within a short length of 1 in. (25 mm) steel pipe, with points emerging from slots in a conical drive point. The upper end of the 1 in. (25 mm) pipe has a left-hand thread, and 0.25 in. (6 mm) steel pipe is welded to the tops of the prongs.

A borehole is advanced to a few feet above the planned anchor depth and the anchor inserted by attaching extension lengths of riser and outer pipe. All threads are wrench-tight, except the left-hand thread, which is greased and hand-tight. When the point reaches the bottom of the borehole, it is driven 1–3 ft (300 mm to 1 m) by driving on the top of the outer pipe. The prongs are then ejected by driving on the riser, the left-hand thread opened by turning the outer pipe clockwise, and the outer pipe bumped back a distance larger than the anticipated vertical compression of the soil above the anchor. Any drill casing used to advance the borehole is then withdrawn.

The Borros anchor provides a more positive anchorage than the driven anchor. However, although a frequently used and simple device, a problem can arise owing to binding of the riser where it exits from the bottom of the outer pipe, such that downdrag forces cause settlement of the prongs. The problem can be minimized by installing an O-ring bushing or a length of greased garden hose in the annular space at the bottom of the pipes. Settlement of the prongs in soft clays can also be caused by the weight of the riser, particularly during the

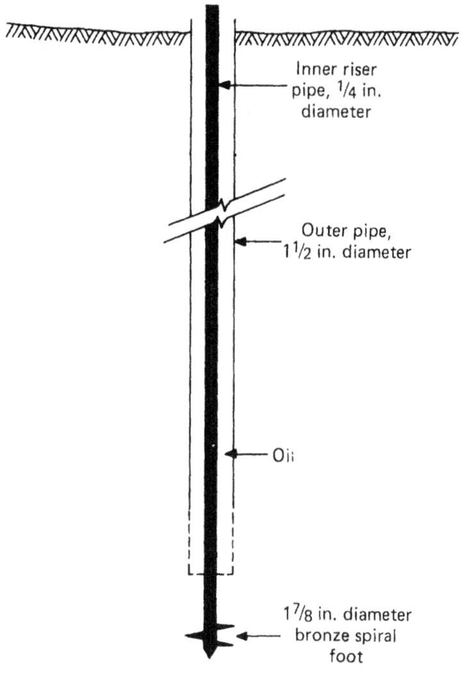

Figure 12.65. Schematic of spiral-foot subsurface settlement point.

period soon after installation, when soil strength is reduced by high pore water pressures that are created as prongs are ejected. The top of the riser should be supported to avoid settlement of the prongs during this period. Longer-term settlement of the prongs with respect to surrounding soft clay can be avoided by counterweighting the riser pipe at the top with a rope, pulley, and weight.

Spiral-foot Anchor

The *spiral-foot anchor,* shown in Figure 12.65, overcomes some of the problems described above for the Borros anchor.

The anchor consists of one or more turns of a bronze helical auger (Bozozuk, 1968) and is connected to a 0.25 in. (6 mm) steel riser pipe. A borehole is advanced to a few inches above the planned anchor depth and the spiral-foot, riser pipe, and outer casing inserted. The spiral-foot is screwed down to the required elevation, and the outer casing is raised and the drill casing withdrawn as for the Borros anchor. Oil is pumped down the riser pipe, out through holes provided just above the foot, and up into the outer casing, providing protection from corrosion and against damage from freezing. Increased anchorage can be obtained by using additional auger turns or a larger diameter auger, but the arrangement appears to suffer from the same potential problem of pipe binding, discussed previously for the Borros anchor. Again, use of garden hose or an O-ring bushing should overcome this problem, provided that the holes for oil outlet are drilled sufficiently high up the riser pipe.

Assuming that the riser pipe is properly anchored and sleeved, measurement accuracy of all three arrangements depends on accuracy of surveying methods, and ±0.1 in. (±3 mm) is typical.

When the riser pipe is carried upward through fill, compaction around the pipe cannot be made with large compaction equipment; thus, compaction tends to be inferior. Interruption to normal filling is costly, and damage by construction equipment is possible. Other limitations are the potential for cumulative errors owing to the addition of pipe lengths and the requirement for a survey crew when taking readings. Liquid level gages (Section 12.10) sometimes provide an alternative measurement method.

12.7.7. Rod Settlement Gage

A single-point fixed borehole extensometer can be used for precise monitoring of ground surface settlement or heave, by using a deep benchmark arrangement similar to the one shown in Figure 12.7 and attaching a dial indicator to the top sleeve. The dial indicator stem bears on the top of the riser pipe to provide data for direct determination of absolute settlement without the need for a survey crew. The system is referred to in this book as a *rod settlement gage* but is also called a *precision settlement gage.* Accuracy is generally ±0.002–0.005 in. (±0.05–0.13 mm). The gage can also be used as a benchmark for surveying methods.

12.8. INCLINOMETERS*

Inclinometers fall within the category of *transverse deformation gages* (Section 12.9) but, because they are used so much more widely than alternative transverse deformation gages, they merit a separate section in this book.

*Written with the assistance of P. Erik Mikkelsen, Vice President, Slope Indicator Company, Seattle, WA.

Inclinometers are defined as devices for monitoring deformation normal to the axis of a pipe by means of a probe passing along the pipe. The probe contains a gravity-sensing transducer designed to measure inclination with respect to the vertical. The pipe may be installed either in a borehole or in fill, and in most applications is installed in a near-vertical alignment, so that the inclinometer provides data for defining subsurface horizontal deformation. Inclinometers are also referred to as *slope inclinometers, probe inclinometers,* and *slope indicators.*

Typical applications include the following:

1. Determining the zone of landslide movement.
2. Monitoring the extent and rate of horizontal movement of embankment dams, embankments on soft ground, and alongside open cut excavations or tunnels.
3. Monitoring the deflection of bulkheads, piles, or retaining walls.

Some probes can be operated within a horizontal pipe, for monitoring settlement of embankments, oil tanks, and other structures on soft ground, and in this application are an alternative to full-profile liquid level gages (Section 12.10). Measurements within an inclined pipe are also possible, for example, when monitoring deformation of the upstream face of concrete face rockfill dams. In addition to their use for deformation monitoring, inclinometers can be used for absolute determination of position, for example, in borehole directional surveys, and for determining the alignment of piles and slurry trenches. Inclinometers have also been used to estimate bending moments: this application is discussed in Section 12.8.11.

Most inclinometer systems have four major components:

1. A permanently installed guide casing, made of plastic, aluminum alloy, fiberglass, or steel. When horizontal deformation measurements are required, the casing is installed in a near-vertical alignment. The guide casing usually has tracking grooves for controlling orientation of the probe.
2. A portable probe containing a gravity-sensing transducer.
3. A portable readout unit for power supply and indication of probe inclination.

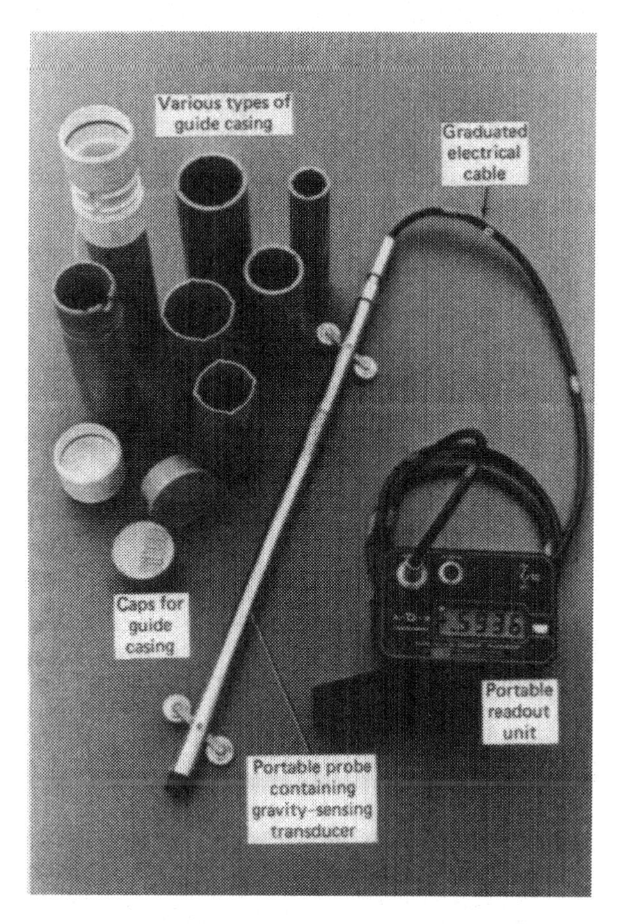

Figure 12.66. Inclinometer system: Slope Indicator Company Digitilt® system (courtesy of Slope Indicator Company, Seattle, WA).

4. A graduated electrical cable linking the probe to the readout unit.

Examples of these components are shown in Figure 12.66.

Figure 12.67 shows the normal principle of inclinometer operation for near-vertical guide casings. After installation of the casing, the probe is lowered to the bottom and an inclination reading is made. Additional readings are made as the probe is raised incrementally to the top of the casing, providing data for determination of initial casing alignment. The differences between these initial readings and a subsequent set define any change in alignment. Provided that one end of the casing is fixed from translation or that translation is measured by separate means, these differences allow calculation of absolute horizontal deformation at any point along the casing.

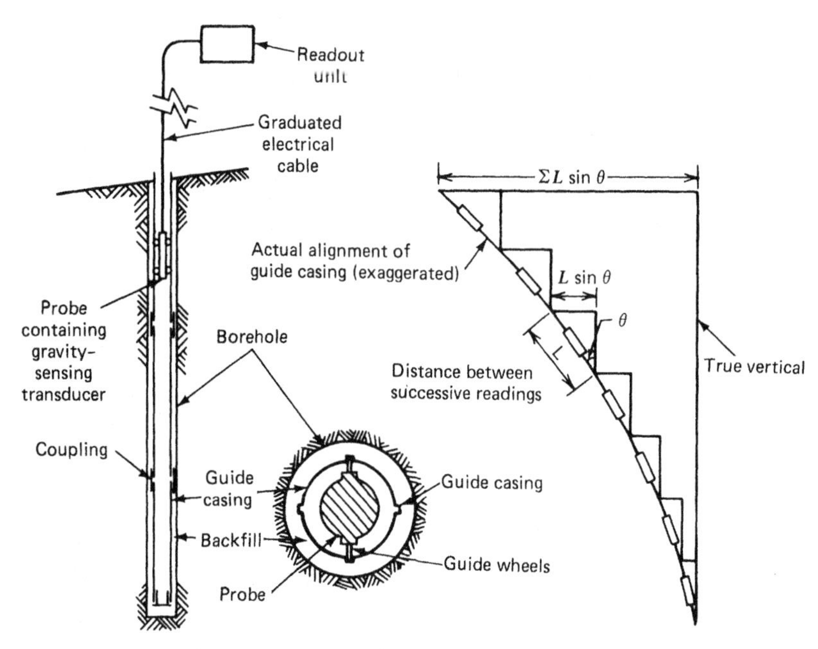

Figure 12.67. Principle of inclinometer operation.

Inclinometers are described in many publications, including Gould and Dunnicliff (1971), Green and Mikkelsen (1986), ISRM (1981a), and Wilson and Mikkelsen (1978).

12.8.1. Types of Inclinometer

Inclinometer types in common use are described in the following subsections, and comparative information is given in Table 12.8. Most inclinometers provide casing inclination data in two mutually perpendicular near-vertical planes. Thus, horizontal components of movement, both transverse and parallel to any chosen direction, can be computed from the measurements.

Inclinometer with Force Balance Accelerometer Transducer

A force balance accelerometer is mounted in the probe such that voltage output is proportional to inclination. The biaxial version includes two transducers, one mounted below the other, with sensing planes 90 degrees apart. The force required to balance the mass, and thus the output voltage, is directly proportional to sin θ (Figure 12.67); therefore, the digital readout is used directly in calculations. The inclinometer with force balance accelerometer transducer has become the most widely used type.

Manual and automatic readout units are available. Manual units are the simplest and least expensive and consist of a power supply, controls, and one or two digital displays. An example is shown in Figure 12.66. The readings are recorded on field data sheets, for later calculation by hand or electronic calculator or for input to a computer program.

Automatic readout units contain a power supply, controls, one or two digital displays, and either magnetic cassette tape or solid-state recording capability. Weeks and Starzewski (1986) describe an automatic readout unit developed in England by Geotechnical Instruments (U.K.) Ltd. and shown in Figure 12.68. Mikkelsen and Wilson (1983) describe a *Recorder–Processor–Printer (RPP),* developed in the United States by Slope Indicator Company and shown in Figure 12.69, that has field checking, editing, data reduction, and plotting capability. Thomas (1985) presents time and cost comparisons for manual and automatic readout units manufactured by Slope Indicator Company and concludes that when the volume of inclinometer reading work exceeds 600 ft (180 m) of casing per month, the *RPP* is

Table 12.8. Inclinometers

Type of Inclinometer	Advantages	Limitations	Typical Range	Approximate Precision[a]
Force balance accelerometer transducer (e.g., Figure 12.66)	Long successful experience record Most widely used type Version available with automatic readout, recording, data reduction, and plotting provisions Versions available for use in 1.5 in. (38 mm) inside diameter grooved casing Versions available for use in horizontal casing for monitoring settlement		±30°, optional to ±90°	±0.05–0.5 in. in 100 ft (±1–13 mm in 30 m)
Slope Indicator Series 200B (Figure 12.70)	Long successful experience record	Standard version is uniaxial No provision for automatic readout No longer manufactured	±12°, optional to ±25°	±0.3–1.0 in. in 100 ft (±8–25 mm in 30 m)
Bonded resistance strain gage transducer	Version available for use in smooth 1.5 in. (38 mm) inside diameter pipe	Errors owing to moisture, temperature, and electrical connections are possible Abandoned by most manufacturers	±20°	±0.02–1.0 in. in 100 ft (±0.5–25 mm in 30 m)
Vibrating wire transducer	Long successful experience record	Special manufacturing techniques required to minimize zero drift Bulky transducer results in large probe Abandoned by most manufacturers	±20°	±0.1–0.5 in. in 100 ft (±3–13 mm in 30 m)
Electrolytic level transducer		Size of transducer limits its use to near-horizontal holes Short experience record	±40°	±2 in. in 100 ft (±50 mm in 30 m)
Shear probe (*poor man's inclinometer*) (e.g., Figure 12.78)	Simple Inexpensive	Poor precision Does not measure inclination Cannot determine curvature below point of smallest curvature	±30°	Very crude

[a] Defined as repeatability with which the instrument can determine the horizontal position of one end of a near-vertical casing with respect to the other (±1 in. in 100 ft corresponds to about ±25 mm in 30 m, $\pm 8 \times 10^{-4}$ radian, or ±170 arc-seconds). Repeatability in near-horizontal casings is similar. Repeatability is decreased in inclined casing: see Section 12.8.2.

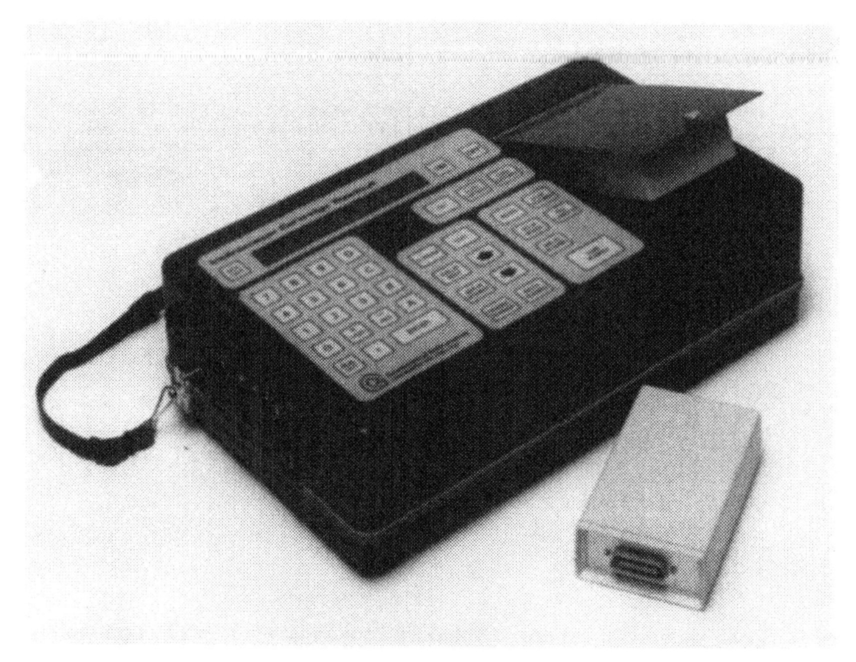

Figure 12.68. Solid-state data logger for inclinometer (courtesy of Geotechnical Instruments (U.K.) Ltd., Leamington Spa, England).

more economical than alternative readout units available from the same manufacturer.

Slope Indicator Series 200B

The Slope Indicator Series 200B uniaxial inclinometer (Figure 12.70), manufactured until recently by Slope Indicator Company, has a potentiometric transducer for measuring inclination. Major components of the transducer are a free-swinging pendulum and an arc-shaped resistance coil, mounted so that the center of the arc corresponds with the pivot of the pendulum. The tip of the pendulum acts as a wiper on the resistance coil, subdividing the coil into two resistances that form one-half of a balance bridge circuit. Resistance output depends on the position of the tip on the resistance coil and thus on the inclination of the probe. The other half of the bridge, including switches, batteries, and potentiometer indicator, is enclosed in the readout unit. Cornforth (1974) and Green (1974) give performance details.

Inclinometer with Bonded Resistance Strain Gage Transducer

Bonded resistance strain gages are mounted around a pendulum that is not free to pivot at its upper end. Tilt causes bending strains in the pendulum and a change in strain gage output. A Wheatstone bridge circuit is used for monitoring.

Green (1974) compares the performance of a bonded resistance strain gage inclinometer with the Slope Indicator Series 200B. Kallstenius and Ber-

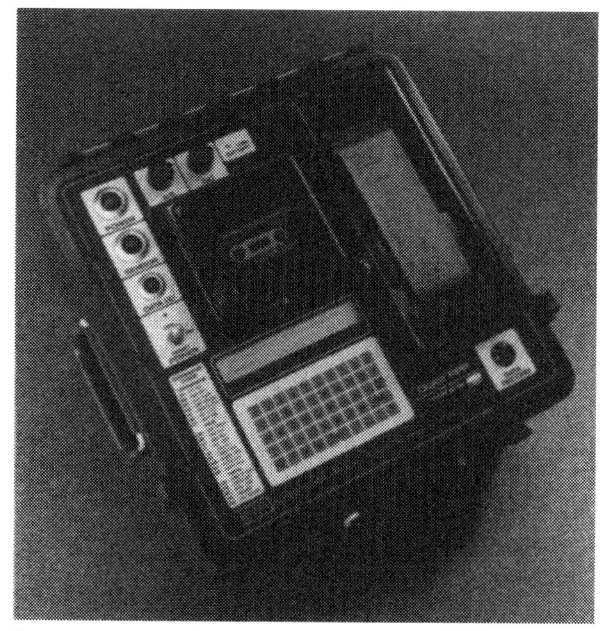

Figure 12.69. Recorder–Processor–Printer (RPP) for Digitilt® inclinometer (courtesy of Slope Indicator Company, Seattle, WA).

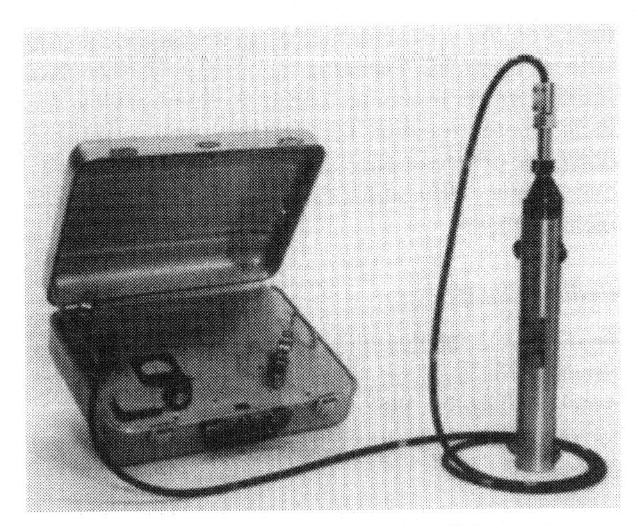

Figure 12.70. Slope Indicator Series 200B inclinometer (courtesy of Slope Indicator Company, Seattle, WA).

gau (1961) describe a bonded resistance strain gage inclinometer, developed by the Swedish Geotechnical Institute, that operates within a standard 1.5 in. (38 mm) inside diameter PVC pipe. The inclinometer is attached to orientation rods, and precision is reported as ±0.4 in. in 100 ft (±10 mm in 30 m).

Inclinometer with Vibrating Wire Transducer

Vibrating wire transducers are mounted on a stiff pendulum in a similar configuration to the inclinometer with a bonded resistance strain gage transducer. Two transducers are required for uniaxial tilt monitoring, four for biaxial monitoring. The readout unit houses power supply, controls, and a frequency counter.

Inclinometer with Electrolytic Level Transducer

An inclinometer with electrolytic level transducer is described by Gravina and Carson (1983) for operation within 1.4 in. (36 mm) inside diameter drill rods in near-horizontal holes. The instrument has been designed for borehole directional surveying in coal mines as drilling of gas drainage holes progresses. A self-orienting mechanism, including a DC motor, slip rings, and mercury switch, is provided to orient the electrolytic level transducer into a vertical plane prior to each reading. A piston is mounted on the front of the probe to allow the probe to be driven along the drill rods under water pressure, while the instrument cable passes through a stuffing box mounted on the outer end of the drill rods.

Cooke and Price (1974) report on the use of inclinometers with electrolytic level transducers, but with their version it was not possible to locate the instrument in exactly the same positions for each set of readings. An in-place version (Section 12.9.3), with a train of transducers, was found to be more satisfactory.

Shear Probe

Although it does not measure inclination, the simple *shear probe* described in Section 12.9.1 is also referred to as a *poor man's inclinometer*.

12.8.2. Factors Affecting Precision of Inclinometer Data

Factors affecting precision of inclinometer data are discussed in the following subsections. However, the overriding factor within control of the user is the skill and care of personnel responsible for all phases of the measurement program.

Precision of Gravity-Sensing Transducer

Manufacturers normally specify the precision of the gravity-sensing transducer, and a systematic check should be made on a regular basis to ensure that the transducer is functioning satisfactorily (Section 12.8.6).

Errors can be grouped into three categories: *scaling, zero offset,* and *azimuth rotation*. The scaling of a transducer is the relationship between input and output and defines the slope of the calibration plot. The zero offset (also called *bias*) of an inclinometer is the reading when the probe is in a true-vertical alignment. Azimuth rotation error results from the difference in orientation between the axis of the transducer and the wheel assembly on the probe, and this cannot be set during manufacture to closer than about ±0.5 degree. All three sources of error can be created by rough handling, wear of the wheel carriage, electronic aging, and temperature change.

Design and Condition of Wheel Assembly

Wilson and Mikkelsen (1977) indicate that the probe should have a well-designed wheel carriage with little or no side-play in the wheels, axle, and swing-arm assemblies. Preferably, wheels should have sealed double ball bearings, and the wheel profile should be compatible with the geometry of the grooves in the casing. During extensive use, these parts wear out more than other parts and should be easy to replace when worn excessively.

Casing Alignment

Table 12.8 includes approximate values for precision with which various inclinometers can be used to determine the horizontal position of one end of a near-vertical casing with respect to the other end. When an inclinometer probe is to be used for vertical deformation measurements in near-horizontal casing, the gravity-sensing transducer is mounted with its axis perpendicular to the long axis of the probe. Precision in near-horizontal casings is similar to near-vertical casings and may on occasion be better because greater control of groove orientation is usually possible when casing is installed horizontally in a trench. Precision degrades as the alignment of near-vertical casings deviates from vertical, and as the alignment of near-horizontal casings deviates from horizontal.

When an inclinometer probe is to be used in inclined casing, two approaches are possible. The first approach uses a conventional "vertical" probe in casings inclined within 45 degrees of vertical and a conventional "horizontal" probe in casings inclined within 45 degrees of horizontal. With this approach the transducer will not be working in the most sensitive part of its range, but this is a small price to pay for the ability to use the *check-sum* procedure, described in Section 12.8.9. The second approach uses the conventional probes in casings inclined within approximately 25 degrees of vertical and horizontal and a special probe for the remaining inclinations. In this special probe the transducer is mounted so that its axis is approximately vertical when the probe is within the inclined casing. Although the transducer is now working in the most sensitive part of its range, the *check-sum* procedure cannot be used, and the second approach is not recommended.

When measurements are made in inclined casings, changes in azimuth can cause substantial errors. For example, when the casing is 5 degrees off vertical, an error of 2 in. per 100 ft (50 mm per 30 m) is caused if the azimuth changes by 1 degree (Wilson and Mikkelsen, 1978). A detailed discussion of errors in inclined casings is given by Mikkelsen and Wilson (1983).

Penman and Hussain (1984) describe an ingenious method for making deformation measurements on the upstream face of an embankment dam with an upstream asphaltic membrane. Rather than running an inclinometer within inclined casing, the inclinometer remains near-vertical and is tracked down the upstream face on a trolley device, thereby overcoming difficulties associated with inclined measurements.

Casing Diameter

Precision can be maximized by using large-diameter casing. For a given wheel thickness and groove width, rotational "play" decreases as casing diameter increases, and thus azimuth rotation error also decreases.

Borehole Backfilling Procedure

Poor-quality backfilling around the casing may cause scatter in readings shortly after installation, but usually the backfill will stabilize with time. If maximum precision is required, the boring procedure should be planned to minimize disturbance to surrounding ground, and backfill should fill the annular space completely. Grout backfill is generally more effective than a compacted granular backfill. Installation procedures are described in Section 12.8.4.

Spiraling of Casing

When casing is installed in boreholes, the orientation of the casing grooves at depth is not necessarily the same as at the surface. Extrusion of aluminum casing may cause a spiral as large as 1 degree per 10 ft (3 m) length, and similar deviations have been noted in extruded and machine-grooved plastic casing. Green (1974) measured a spiral of 18 degrees in an 80 ft (24 m) length of coupled extruded plastic casing, and others have reported similar measurements. Exposure to hot sunlight prior to installation will often cause spiraling of initially true plastic casing, and lengths should always be stored in the shade and supported adequately to avoid bending.

The spiral of each length can be measured before installation. For example, lengths of casing can be set on V-blocks on a bench and a plumb line of nylon filament hung alongside each end. Spiral can be estimated to about 1 degree by setting one end parallel to the plumb line, making sure the casing is not twisted by placement in the blocks, and observing by eye at the other end. Alternative methods for measuring the spiral of individual lengths are available from casing manufacturers: for example, Slope Indicator Company has a method that uses a standard biaxial inclinometer within casing held in a near-horizontal alignment.

If possible, lengths of casing should be selected so that successive spirals cancel out each other.

Spiraling can also be created by using poor installation techniques. When lengths of casing are assembled, alternate couplings should be twisted left and right before fixing, because manufacturing tolerances usually allow for some rotational movement. If grooves turn out of the planned orientation during installation, they should never be forced back merely by rotating the casing top. If the bottom of the casing is free to rotate, the top should be raised and lowered repeatedly as the orientation is gradually corrected. Drill casing and hollow-stem augers should be withdrawn without rotation.

When casing is installed in fill, groove orientation is easier to control, and spiraling usually does not degrade precision in near-vertical and near-horizontal installations. However, as discussed above, errors in inclined casings can be large.

A survey of spiral after casing installation is recommended when difficulties have occurred during installation and for casings deeper than about 200 ft (60 m), particularly if it is necessary to know the exact direction of ground movements at depth. A survey can be made in one of two ways. First, most manufacturers supply a spiral survey instrument (e.g., the *spiral checking sensor* shown in Figure 12.71), typically consisting of a 5 ft (1.5 m) long rod between upper and lower guide wheels, with a rotary transducer mounted at one end. Second, a system can be assembled from a dummy inclinometer probe or appropriately sized rectangular plate, connected via a universal joint to orientation rods, fitted with high-precision tongue-and-groove connections (Figure 12.71). The assembly is lowered within the casing, and orientation of the rods with respect to the top of the casing is recorded as more rods are added and the probe or plate passes down the grooves within the entire casing. Orientation can be measured with a circular protractor mounted on the top of the casing.

For a particular installation, groove spiraling does not change with time, and thus a single set of spiral survey data can be used with biaxial inclinometer data to determine true direction of deformation, or deformation in any predetermined plane. Adjustment of data to correct for groove spiral entails graphic or computerized data reduction.

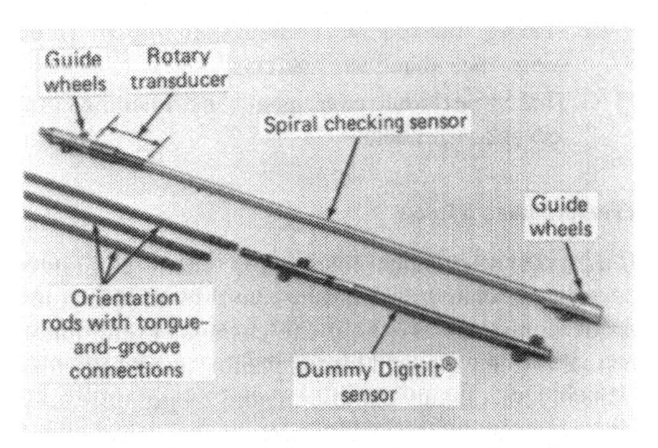

Figure 12.71. Instruments for measuring spiral in inclinometer casing after installation (courtesy of Slope Indicator Company, Seattle, WA).

Repeatability of Reading Position

Any lack of care in repeating depth measurements to the inclinometer probe with each set of readings will markedly reduce precision. The cable should not be subject to stretch or slip between the components of the cable and should be marked permanently and accurately. This source of error increases with increasing length of casing and is aggravated if the depth of casing grooves is irregular or if installed casing is not straight, and its significance increases with increase in precision of the gravity-sensing transducer. Methods of reading in casing equipped with telescoping couplings are discussed in Section 12.8.8.

Depth Interval Between Reading Positions

Maximum precision is achieved by using a reading interval equal to the spacing between wheels on the inclinometer probe. If the interval is greater than the wheelbase of the probe and deformations of the casing are not composed of smooth curves or straight lines between measuring points, significant errors may arise. Reading with a 2 ft (610 mm) probe at depth intervals as great as 5 ft (1.5 m) could miss shear zones of vital interest entirely. However, in cases where localized movement (e.g., at a shear zone) is **known** to be absent, for example, as is often the case when monitoring the bending of piles during a lateral load test, reading effort can be minimized by the following method:

1. Taking readings initially at the wheelbase interval.

2. Taking subsequent readings at two or three times the wheelbase interval.
3. Taking periodic readings at the wheelbase interval as a check.

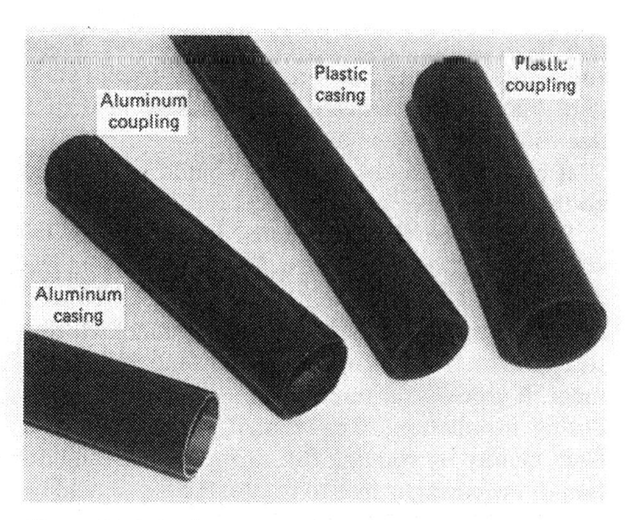

Figure 12.72. Inclinometer casing (courtesy of Geotechnical Instruments (U.K.) Ltd., Leamington Spa, England).

Temperature Effects

If the gravity-sensing transducer is sensitive to temperature, a change in reading may be noted as the probe enters water within the casing. Potentiometer, vibrating wire, and force balance accelerometer transducers do not exhibit major temperature effects, but bonded resistance strain gage transducers may show a greater variation in reading with change in temperature. However, in all cases a waiting period of at least 10 minutes should be allowed for temperature stabilization after the probe has been lowered to the first reading position, before readings are taken. Most readout units also show some reading variation with temperature, and for maximum precision the readout unit should not be used in extreme temperatures. Errors caused by temperature variation of the readout unit increase with increasing departure of the probe from vertical.

Handling of Probe

Shock to the probe can cause a zero shift of the transducer. Errors can be minimized by careful handling and by operating the probe only when connected to the readout unit with power switched on. The bottom of the probe should be provided with a rubber cushion, and during operation the probe should never be allowed to contact the bottom of the casing suddenly.

12.8.3. Types of Inclinometer Casing

Plastic, aluminum alloy, and fiberglass casings are available, with either rigid or telescoping couplings. Examples are shown in Figures 12.72 and 12.73. Steel casing is also available but is used less frequently.

Plastic Casing

ABS (acrylonitrile/butadiene/styrene) is the most commonly used plastic casing and is suitable for most applications. Alternative plastics, such as PVC (polyvinylchloride), tend to be more brittle, especially at low temperatures. Most plastic casing is manufactured by broaching the grooves in extruded pipe, but in some cases the pipe is grooved during extrusion. Outside diameters range from 1.9 to 3.5 in. (48–89 mm).

Rigid couplings are available with self-aligning grooves or keys and are preferable to couplings requiring use of a special aligning tool. A self-aligning coupling developed by Westbay Instruments Ltd. has O-rings for sealing, and nylon shear wires to lock the coupling and casing together. These couplings are very convenient since they minimize installation time and necessary skill and allow for easy disassembly if difficulties are encountered during installation or when casing is used for borehole directional surveys, but they have a larger outside

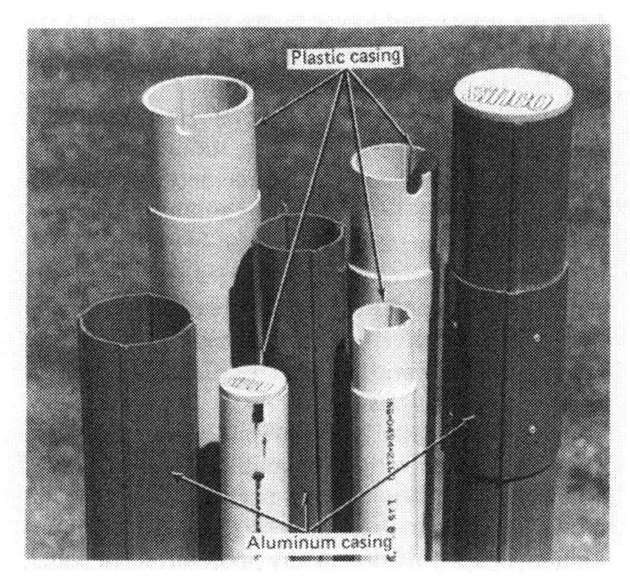

Figure 12.73. Inclinometer casing (courtesy of Slope Indicator Company, Seattle, WA).

diameter than conventional couplings. The Westbay coupling is also available with a spring-loaded port for use as a multipoint piezometer (Section 9.9) and can be installed on inclinometer casing to allow measurements of both transverse deformation and groundwater pressure.

Telescoping plastic couplings also have self-aligning grooves or keys, and standard versions typically have a telescoping range of up to 6 in. (150 mm), allowing for 9% compression when used with 5 ft (1.5 m) casing lengths. Long-range telescoping ABS couplings have recently been developed by Slope Indicator Company for use in very soft clays subjected to large vertical compression and allow for up to 30% compression (Handfelt et al., 1987).

Aluminum Casing

Aluminum alloy casing is grooved during extrusion, the groove also appearing on the outside profile as a protrusion. Outside diameters, measured across opposite protrusions, range from 2.4 to 3.4 in. (61–86 mm), and slightly larger-diameter extrusions of the same shape are used for rigid and telescoping couplings. The maximum range of telescoping couplings is typically 6 in. (150 mm).

Aluminum casing is subject to corrosion, either by groundwater or by free lime in cement grout used during installation, and several cases of total corrosion within a period of a few months have been reported. In most applications casing should therefore be treated both on the inside and outside with a suitable coating (e.g., baked-on epoxy paint), but corrosion potential remains at any cut ends or rivet holes.

Steel Casing

Seamless welded square steel tubing, available from suppliers of steel pipe, can sometimes be used as inclinometer casing, and most inclinometer probes allow the wheels to ride in the tubing corners. Possible applications include measurements on driven steel piles, applications where cost must be minimized and where lower accuracy is acceptable, and special applications when other instruments (e.g., strain gages in piles) are attached to the outside of the casing.

A 2 in. (50 mm) outside, 1.75 in. (44 mm) inside size tubing is typically used, but a larger size should be used when significant transverse deformations are expected. Alternatively, the inclinometer probe can be arranged such that wheels ride on the flat surfaces, requiring a larger size of tubing (e.g., DiBiagio and Myrvoll, 1985). Extruded steel tubing usually has excessive twist and is not recommended, but seamless welded tubing is not twisted excessively. Couplings are made from suitably sized seamless welded square steel tubing and often have a very loose fit.

Selection of Inclinometer Casing

The type of casing and coupling, together with the installation procedure, should be selected to ensure that the casing conforms with soil or rock movements. Selection among the inclinometer casing and coupling options depends on the answers to the following questions:

1. Will the casing be subject to axial compression or extension? If axial compression or extension in excess of about 1% is expected, telescoping couplings should be used to prevent damage to the casing and to ensure conformance. If a probe extensometer will be used within the inclinometer casing to monitor axial movements, the guidelines given for combined probe extensometers and inclinometer casings in Section 12.5.10 should be followed.
2. What is the predicted transverse deformation? Users sometimes express concern that inclinometer casing may be too stiff to follow transverse ground movements in soft soils. This concern implies that the soil might flow around the casing, but available evidence suggests that casing movement conforms with soil movement. However, when installed in soft soil, deformations at a distinct shear zone are likely to cause local nonconformance, such that shear measurements are more gradual than actual deformation. Deformation at a distinct shear zone will cause bending of the casing, and excessive localized bending will prevent passage of the probe. If substantial deformation is predicted at distinct shear zones, a large-diameter casing and small-diameter probe should be used. In extreme cases the casing can be installed in a large-diameter borehole and surrounded by a soft grout, as described in Section 12.8.4. Manufacturers will provide bending limits for their various probe casing combinations.

3. Is maximum precision required? As indicated in Section 12.8.2, precision can be maximized by using large-diameter casing.
4. What will be the alignment of the installed casing? When the gravity-sensing transducer is mounted with its axis at an angle to the long axis of the inclinometer probe, for example, for measurement of settlement in near-horizontal casing, the diameter of the probe will generally be larger than standard probes, and larger-diameter casing will be required.
5. What will be the climatic conditions during installation? When plastic casing is exposed to hot sunlight, groove spiraling may increase and lengths may become warped. Use of plastic casing in hot temperatures entails shaded storage. Solvent cement for plastic casing cannot be used in very cold conditions: options are therefore plastic casing with Westbay Instruments Ltd. self-aligning couplings or aluminum, fiberglass or steel casing.
6. Are there limitations on acceptable casing diameter? For example, drilling costs usually increase with borehole diameter. Where casing is installed in concrete, structural considerations may limit diameter. Where casing is installed within pipes attached to H-piles, allowable pipe diameter may be the controlling factor. If the diameter is limited, couplings that do not protrude beyond the outside diameter of the casing will usually be preferred.
7. Will installation personnel be skilled and careful? If not, Westbay Instruments Ltd. self-aligning couplings may be preferred.
8. Is longevity required? Major factors are groundwater alkalinity or free lime in cement grout used during installation and the existence of stray ground currents, which may be present in urban environments. If pH is greater than about 10 or if stray ground currents are suspected, plastic casing should be used. When aluminum casing is used in alkaline conditions, both the inside and outside should be treated with a suitable coating to prevent corrosion.
9. What is the depth of the installation? Users sometimes favor aluminum casing in very deep installations, fearing that plastic casing may be damaged by external grout pressure or axial stresses during installation. These concerns can be overcome by use of appropriate installation procedures, sometimes including stage grouting, and should not be a reason for rejecting plastic casing.
10. What backfill will be used? Plastic couplings are easier to seal against intrusion of backfill than aluminum, fiberglass or steel couplings. If grout backfill is used, a good seal is imperative. If a granular backfill is used in a borehole installation, the outside profile at couplings should be flush, otherwise backfill may bridge above the couplings.
11. What are the costs of alternative types of casing?

12.8.4. Installation of Inclinometer Casing

When planning installation procedures, one should follow the guidelines given in Chapter 17. Additional guidelines for installation of inclinometer casing are given in the following subsections.

Installation procedures vary widely, according to the type of casing and couplings and to site-specific conditions. Procedures are described by AASHTO (1978), ISRM (1981a), Wilson and Mikkelsen (1977,1978), and in the instruction manuals of various manufacturers of inclinometers.

When installing inclinometer casing for use in combination with a probe extensometer, one should follow the guidelines given in Section 12.5.10.

Coupling Requirements

Care must always be taken to seal inclinometer couplings and bottom caps against intrusion of backfill. Couplings with O-rings do not require additional sealing, but others usually require sealing mastic and tape. Where pop-rivets are intended to shear to allow telescoping movement, sealing mastic and tape should be used over the rivet heads. Even with rigid riveted couplings, the stem of the rivet occasionally pulls out during installation, and rivet heads should be filed smooth and sealed. Solvent cement is used on most rigid plastic couplings for sealing and tensile strength, and usually one or more pop-rivets are installed on each side of the coupling to provide strength while the cement sets. When maximum tensile strength is required across cemented couplings in ABS or PVC casing, a primer should be used before the solvent cement is applied. The primer etches the surfaces and allows proper adhesion. Rivets should not be installed in casing grooves or any other location that would interfere with tracking of the probe.

When assembling couplings, one must be careful to avoid creating spiraled casing. Alternate couplings should be twisted left and right before fixing.

Installation in Fill

Inclinometer casing is installed in fill by using the methods described in Chapter 17. When a mechanical probe is used within inclinometer casing to monitor compression of the fill, settlement collars must be attached to the casing (Section 12.5.3).

Borehole Requirements

Inclinometer casing can occasionally be installed in an unsupported borehole. However, where there is risk of borehole collapse during installation, the borehole should be supported by drilling mud, drill casing, or hollow-stem augers. Drill casing or hollow-stem augers should be used if any doubt exists about the ability of drilling mud to support the borehole. If maximum precision is required, the boring procedure should be planned to minimize disturbance of the surrounding ground, and boreholes for vertical installations should be as near as possible to true vertical.

In general, the bottom of the inclinometer casing should be fixed from translation so that absolute deformation data can be calculated by assuming base fixity, and thus the borehole should be advanced to stable ground. This assessment should be based on the geology, the geometry of the excavation or structure, and other site-specific factors. A depth of 10–20 ft (3–6 m) below the expected active deformation zone is suggested for most installations. Additional borehole length should be allowed for any added weight required at the casing bottom to overcome buoyancy during installation. As described in Section 12.8.6, it is often convenient to continue one borehole to a greater depth than required for deformation data and to use the bottom length of inclinometer casing for checking the instrument.

Samples are normally taken to define stratigraphy, as input to the decision on borehole depth and to assist with interpretation of data.

Inclined boreholes are not recommended for inclinometer casings, because groove orientation is difficult to control and will typically have a spiral on the order of 3–10 degrees per 100 ft (30 m) (Mikkelsen and Wilson, 1983).

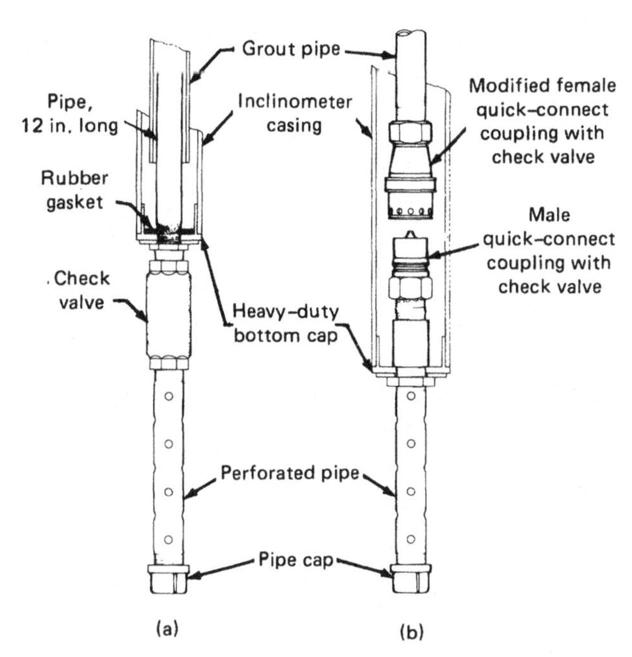

Figure 12.74. Arrangements for grouting through a pipe within inclinometer casing: (a) single shutoff arrangement with gasket seal and (b) double shutoff arrangement with quick-connect (courtesy of Slope Indicator Company, Seattle, WA).

Installation in Boreholes

The annular space between the borehole wall and inclinometer casing should be backfilled with grout, sand, or pea gravel. Grout backfill is more likely than granular backfill to fill the annular space completely but cannot be used if grout would be lost into the surrounding ground.

When a grout backfill is used, grout can be tremied via a pipe inserted outside the inclinometer casing, but the arrangements shown in Figure 12.74 allow installation in a smaller-diameter borehole.

When the arrangement in Figure 12.74a is used, steel pipe or drill rod is lowered over the 12 in. (300 mm) long pipe to seal against the rubber gasket and is used for grouting. After the annular space between the borehole wall and inclinometer casing is filled with grout, a measured volume of water is pumped to displace grout from most of the grout pipe, the grout pipe is raised a short distance, and the check valve closes. The remaining grout in the grout pipe is removed by flushing thoroughly with water, under low flow to avoid reopening the check valve. Two methods are available for overcoming buoyancy until grout sets and for ensuring that the casing remains as straight as possible in the borehole. First, an adequate bottom weight can be attached to the bottom cap. Second, the weight of the grout pipe can be used, temporarily weighting the top of the inclinometer casing while the grout pipe is raised to flush with water. The second method is preferable for deep holes, for which the magnitude of the bottom weight would become excessive. The first method must be used if the inclinometer casing has telescoping couplings.

When the arrangement in Figure 12.74b is used, both check valves are opened when the two fittings mate under the weight of the grout pipe and close after grouting when the grout pipe is raised, thereby preventing spillage of grout into the inclinometer casing. It is important to fill the grout pipe with water prior to mating the two fittings, otherwise particles from the bottom of the borehole are likely to be washed into the check valves and prevent later closure. A conservative installation will include a check valve below the bottom cap, of the type shown in Figure 12.74a, and a spare female quick-connect should always be on hand for use if grout is allowed to set in the fitting attached to the grout pipe. Although the arrangement in Figure 12.74b prevents spillage of grout when the grout pipe is raised, the weight of the grout pipe cannot readily be used to overcome buoyancy; thus, a bottom weight must be used.

When an outside tremie pipe is used for grouting, buoyancy can be overcome by use of a bottom weight, the weight of a temporary internal pipe, or a first stage of grout can be allowed to set around the bottom 10 ft (3m) of the casing so that no buoyancy forces remain.

For all methods of overcoming buoyancy, the attachment between the bottom cap and casing must be adequately strong. The manufacturer should be consulted for information on the capacity of the standard attachment and for methods of reinforcement.

Drill casing and hollow-stem augers must be withdrawn without rotation.

An example of a procedure for installing inclinometer casing in a borehole, including a required materials and equipment list, is given in Appendix G. This example procedure should not be used as a "cookbook" procedure: each installation has its own criteria, dependent on site-specific conditions and needs, available materials and equipment, and the skill of installation personnel.

Installation when Large Movements Are Expected on Thin Shear Zones

As indicated in Section 12.8.3, deformation at a distinct shear zone will cause bending of the casing, and excessive localized bending will prevent passage of the probe. If substantial deformation is predicted at distinct shear zones, a large-diameter casing and small-diameter probe should be used. It is also helpful to use a large borehole and soft backfill, so that the casing causes the backfill to shear, thereby prolonging the life of the installation. In this case the localized large shear movements will be redistributed over a larger casing length, but this is preferable to shearing the casing or blocking probe access.

Careful monitoring, by using the *shear probe* technique described in Section 12.9.1, can provide a forewarning of blocking probe access and can indicate the timeliness of converting the system to a *slope extensometer* (Section 12.9.1). A single anchor can be installed below the shear zone, usually at the bottom of the casing, and subsequent deformation monitored. If the single wire is left slack between slope extensometer readings, inclinometer readings can usually be obtained until blockage occurs.

Installation on Piles

When installing casing on solder piles or driven steel piles, two methods are possible. First, square steel tubing can be welded directly to the piles and used as inclinometer casing. Second, a steel pipe can be welded to the pile and inclinometer casing installed within the pipe, using a grout backfill and one of the arrangements shown in Figure 12.74. This second method allows drilling through the steel pipe after pile installation, to create base fixity by setting the inclinometer casing below the tip of the pile, and is usually preferable.

12.8.5. General Guidelines on Use of Inclinometers

General guidelines on instrument calibration and maintenance and on data collection, processing, presentation, interpretation, and reporting are given in Chapters 16 and 18. Additional guidelines for inclinometer use are given in manufacturers' instruction manuals, by Green and Mikkelsen (1986), ISRM (1981a), and Wilson and Mikkelsen (1978), and are summarized in the following sections.

12.8.6. Calibration

An inclinometer can be returned to the manufacturer for regular calibration checks, but it is usually inconvenient to do this on a frequent schedule. There are three available methods for making field checks of an inclinometer for use in near-vertical casings and additional methods for near-horizontal and inclined casings. These are described in turn.

All the field checks described in the following subsections can only be used to examine variations in field readings with time, and these values cannot be evaluated by referencing to similar data obtained during factory calibration. For comparison with the original factory calibration, the probe must be returned to the manufacturer. An inclinometer should be subjected to one of the field checks regularly and repaired by the manufacturer when appropriate. If this is done, there should be no need to send it to the manufacturer on a regular schedule for routine checking.

Near-Vertical Test Casing

A short length of test casing can be installed in suitably stable ground at a location where the temperature remains as constant as possible. The unheated basement of a building founded on bedrock is an ideal location. If drilling is inconvenient, the casing can be embedded in a concrete block formed by a 55 gallon (208 liter) oil drum founded on bedrock. The casing should be a 4 ft (1.2 m) length of standard inclinometer casing, installed so that one pair of grooves is in a vertical plane and the other pair 10–15 degrees inclined from the vertical.

Readings are taken, after waiting for temperature stabilization, with the wheels oriented in the four grooves at 90 degrees apart, with the biaxial probe hanging on the cable at a standard position within the test casing. These eight readings (i.e., transducers *A* and *B* for each wheel orientation) provide a check on *azimuth rotation, scaling,* and *zero offset (bias)* errors and provide data for systematic error corrections if necessary. More details on the procedure are available from some manufacturers of inclinometers.

Test in Bottom of Near-Vertical Field Casing

One inclinometer casing can be installed in the field to a greater depth than required for deformation data and the bottom length used as a stable reference for checking purposes. The bottom 20 or 30 ft (6 or 9 m) should be installed in rock or firm soil that, without doubt, will not move throughout the instrument use period. Prior to taking field readings on any day, check measurements should be made within this zone, following the same procedure described previously for a test casing. However, because the casing will not have significant inclination, this method does not provide as good a check on scaling as the previous method.

Figure 12.75. Inclinometer test stand, master type (courtesy of Geotechnical Instruments (U.K.) Ltd., Leamington Spa, England).

Test Stand

A test stand is provided by some manufacturers for checking the health of the instrument, as described previously for a test casing. A highly accurate *master test stand* (Ohdedar and Dawes, 1983; Figure 12.75) can be used or, alternatively, a less accurate *field test stand* (Figure 12.76). When this method is

Figure 12.76. Inclinometer test stand, field type (courtesy of Geotechnical Instruments (U.K.) Ltd., Leamington Spa, England).

employed, it is essential that the test stand remains absolutely stable, both as the set of readings is taken and between sets of readings.

Checks on Inclinometer for Use in Near-Horizontal Casings

When an inclinometer probe is to be used for vertical deformation measurements in near-horizontal casing, it can be checked in a similar way to a conventional inclinometer. The procedure requires two test casings, the first near-horizontal with the grooves oriented 10 degrees off vertical and the second inclined 10 degrees off horizontal but with the grooves oriented vertically.

Checks on Inclinometer for Use in Inclined Casings

When an inclinometer probe is to be used for measurements in inclined casing, the method of checking depends on the configuration of the transducers, as described in Section 12.8.2. If a conventional "vertical" or "horizontal" probe is used, the instrument can be checked as described above. If a special probe is used, with its transducer mounted so that its axis is approximately vertical when the probe is within the inclined casing, the instrument can be checked by taking readings in a test casing installed at the average inclination of casings in the field. However, the probe cannot be rotated 180 degrees or turned end-for-end, so that the check is less reliable than for a conventional probe.

12.8.7. Maintenance

The inclinometer probe should be checked frequently and wheel fixtures and bearings tightened and replaced as necessary. After each casing has been read, guide wheels should be cleaned and oiled.

For long-term installations it is particularly important to minimize wear of the casing grooves. Primary needs are clean and oiled guide wheels and clean casing. If any solid material enters the casing, it will settle to the bottom, possibly stick to the guide wheels, and abrade the grooves. It is usually worthwhile to flush the casing occasionally with clean water, and a large bottle brush helps to remove any solid material. The need applies to casings at all alignments, and horizontal casings are particularly prone to deposition of material in the lowest groove.

12.8.8. Data Collection

Data collection generally requires two trained technicians.

Although they may fit other manufacturers' casing, instruments supplied by different manufacturers should not be used interchangeably. In the interest of accuracy, even the interchangeable use of probes supplied by the same manufacturer should be avoided (Wilson and Mikkelsen, 1978). Use of the same technicians, instrument, and cable for all measurements on a particular project is highly desirable, and mandatory if high precision is required.

Field Check on Inclinometer

Prior to collecting data on any day, a field check should be made to ensure that the inclinometer system is functioning correctly, preferably by using one of the methods described in Section 12.8.6.

If experienced users consider this recommendation to be unnecessarily conservative they should, as a minimum, use the following procedure:

1. Lower the probe to a point within the casing, preferably about 20 ft (6 m) below the water level.
2. Wait for temperature stability (usually at least 10 minutes).
3. Take repeated readings.
4. Remove the probe, rotate it 180 degrees, lower it to the same point.
5. Wait for temperature stability and take repeated readings.
6. Examine the stability of *check-sums* (Section 12.8.9).

By using this procedure, any tendency for drift in the check-sums can be detected before data are collected. Good stability of check-sums during each data set is essential for good accuracy.

Measurement Method

The probe should be inserted to the bottom of a vertical or inclined casing or the "far" end of a horizontal casing. A measurement traverse is made by holding the probe stationary at each depth interval throughout the casing and recording depth and inclination data. Maximum precision is achieved by use of a reading interval equal to the inclinometer wheelbase.

When reading in near-vertical casings, readings in one vertical plane are normally taken with the probe at one orientation and then repeated with the probe turned through 180 degrees. Measuring locations must be identical to those in the first traverse. If a uniaxial probe is used and deformation data in both vertical planes are required, two further traverses are made, with the probe guide wheels in the second pair of grooves, 90 degrees from the first. By reducing data based on the differences between readings, 180 degrees apart, systematic instrument errors and errors caused by casing irregularities are minimized. An excellent check on the reliability of each measurement is provided by calculating the algebraic sum of readings 180 degrees apart, and this should be done **in the field** while collecting the data. These are referred to as *check-sums* and are discussed further in the following section on data processing.

When reading in near-horizontal casings, the same check-sum procedure is used by turning the inclinometer probe end-for-end, and a cable connection is provided at each end for this purpose.

When reading in inclined casings, the check-sum procedure is possible if the axis of the transducer is mounted parallel or perpendicular to the long axis of the inclinometer probe, but this is not possible with the inclined mounting arrangement described in Section 12.8.2.

Initial Readings

Measurements of the initial profile, to which all subsequent data will be related, should follow the guidelines for initial readings given in Chapter 18. When taking initial readings, a fixed orientation reference for the probe should be established and recorded and consistently used for all subsequent reading sets. For a biaxial inclinometer, generally accepted practice is to orient the *A* transducer so that it will register deformation in the principal plane of interest as a positive change.

At the time of initial readings, survey measurements should, if practicable, be made on the top of the casing to establish its lateral position to within the accuracy required of inclinometer measurements. If base fixity is in doubt, provision should be made to monitor absolute transverse deformation of the top of the casing on a regular basis, using surveying methods. Even with confidence in base fixity, it is good practice to check calculated data occasionally by survey measurements on the top of the casing.

Readings Within Telescoping Casing

If the inclinometer casing incorporates telescoping couplings, the depth below ground surface of each element in the stratigraphic profile and of the corresponding casing section will not remain constant. In this situation, two methods of data collection and processing are possible.

First, inclinometer readings are made at constant locations with respect to the stratigraphic profile. Telescoping coupling locations are recognized by feeling a "bump" as inclinometer guide wheels pass into or out of a coupling, and inclination readings are taken at constant distances from the nearest coupling. Length changes across couplings are determined by using a mechanical probe extensometer within the inclinometer casing, and these data are used during data processing.

Second, readings are taken at uniform depth intervals, settlement data are obtained either by using a probe extensometer or by alternative method, and a computer program is used to interpolate the inclination of the casing at the same elevations as the initial set of data.

The first method is the more accurate, because in the second method the guide wheels may rest on a corner at the end of a section of casing, causing poor repeatability. However, the second method requires less overall effort and is suitable if the reduced accuracy is acceptable.

Automatic Recording

Although it increases data collection speed, an automatic readout unit may be subject to a significant limitation.

With manual data recording, data collection personnel can scan the data in the field for variations in the check-sums and make corrections or repeat readings immediately. Automatic readout units with field checking and editing capability also allow these checks to be made. However, some automatic units do not allow the check-sums to be examined in the field, and data must be scanned for errors after being printed out in the office and before computer processing. Additional field work may be needed if errors are found, and use of an automatic readout unit may not necessarily increase overall efficiency.

12.8.9. Data Processing

The first step in data processing should be a review of the check-sums, and **this should be done while**

collecting the data. The check-sum is usually equal to twice the zero offset (bias) of the transducer.

Check-sums are used to examine data for errors, and ideally the check-sums should remain constant for all depth intervals in a given data set. In reality, the check-sums vary according to casing conditions, instrument performance, and operator technique. If opposite walls of the casing are not parallel, if wheels are influenced by the uneven inside profile of telescoping couplings, or if depth control is not precise, the check-sum may vary randomly about a mean value. Small variations do not usually indicate a problem.

Some inclinometer manufacturers will indicate the magnitude of normal variations in check-sums. For example, for its Digitilt® inclinometer, Slope Indicator Company (1987) states that the check-sum should remain within ± 10 or 20 units of the mean of all check-sums for that data set. Typical standard deviations for the *A* axis are 1–10 and are usually double for the *B* axis. In general, a standard deviation greater than 10–20 units indicates problems with casing irregularities, the instrument, or operator technique.

After reviewing check-sums, data processing consists of four steps: 180 degree differences, "change" at each reading depth by subtracting each 180 degree difference from the initial value, "cumulative change" by cumulating change data from the bottom of the traverse upward, and conversion of cumulative change to deformation units via a calibration constant.

Manual reduction of inclinometer data is tedious and time consuming and, as a minimum, a small programmable calculator should be used. Because the calculations are based on a cumulative process, it only requires one error to produce completely misleading results. Consequently, manual reduction of inclinometer data is practicable only where a small number of casings are involved and when they are read at infrequent intervals. Powerful IBM-PC compatible software is available for reduction and plotting of inclinometer data and for error detection and correction, and the author strongly recommends its use. For example, the PC-SLIN data reduction program, available from Slope Indicator Company, has error detection routines that make a statistical evaluation of variations in check-sums from the mean value and provide an expedient method for detecting mistakes and evaluating errors in the data. Green and Mikkelsen (1986) present a good summary of data processing procedures and sources of error.

If an automatic readout unit has been used, data can be entered into a computer, via a telephone modem if required, after the scan for errors. Use of a *Recorder–Processor–Printer* allows data reduction in the field.

When preparing to plot data, special attention should be paid to the orientation sign convention. Some manufacturers provide guidelines in their instruction manual.

The two plots shown in Figure 12.77 are in common use. The "change" plot is useful to dramatize the location of deformation zones. The "cumulative change" plot gives a more graphic representation of the actual deformation pattern and is more readily understood by personnel unfamiliar with the data reduction procedures. A third plot, of deformation at a particular depth versus time, is particularly useful in studying deformation trends and making predictions. For example, such plots might be prepared for the Figure 12.77 data at depths of 16 and 46 feet.

If a groove spiral survey has been made, spiral data can be used with biaxial inclinometer data to determine true direction of deformation or deformation in any predetermined plane.

12.8.10. Data Interpretation

The normal purpose of inclinometer measurements is to define the location of any deforming zone and to allow an evaluation of that zone as time progresses, rather than to survey an exact profile of the casing. Often the deforming zone is only a few feet thick, and the sum of the changes over a few adjacent reading depths will often be representative of the magnitude and rate of the entire movement. Thus, the most useful plots are generally plots of deformation at a few selected depths versus time, and the Figure 12.77 plots are merely steps in visualizing what is occurring and in developing the deformation–time plots. The cumulative change plot may in fact be misleading because, although the instrument may be operating within its range of precision, over a period of time it may suggest tilting back and forth. If a small kink begins to develop somewhere in the plot, primary concern should lie with the developing kink and not with the overall tilt.

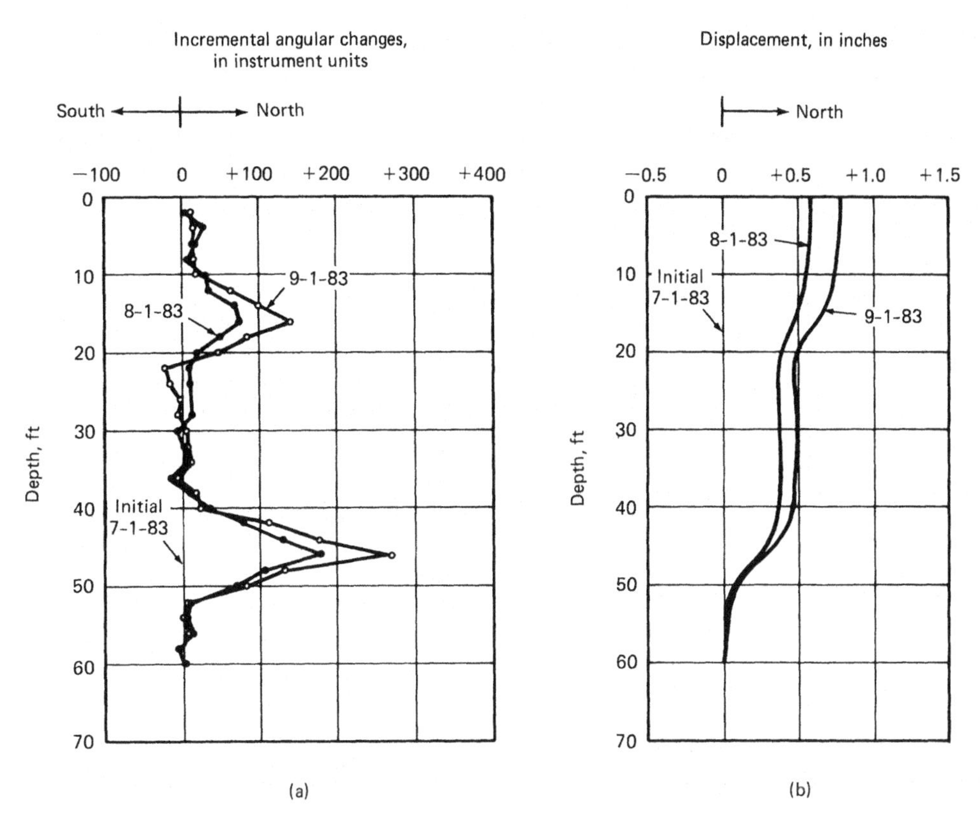

Figure 12.77. Typical plots of inclinometer data: (a) "change" plot and (b) "cumulative change" plot (courtesy of Slope Indicator Company, Seattle, WA).

12.8.11. Use of Inclinometer Data to Estimate Bending Moments*

Moments and corresponding stresses in structures can sometimes be back-calculated from inclinometer measurements. The moment diagram can be obtained from

$$M_x = \frac{d\theta_x}{dx} \cdot EI_x,$$

where: θ_x = angle measured by inclinometer at section x,
M_x = moment at section x,
E = elastic modulus,
I_x = moment of inertia of section x.

If inclinometer readings are taken at 2 ft (610 mm) depth intervals, the above equation can be rewritten as

$$M_x = \left(\frac{\theta_x - \theta_{x-1}}{24}\right) EI_x = \frac{\Delta\theta_x}{24} \cdot EI_x.$$

As indicated in Table 12.8, the best possible precision for inclinometer data, when using a force balance accelerometer transducer, is ±0.05 in. in 100 ft (±1 mm in 30 m, $\pm 4 \times 10^{-8}$ radian, ±9 arc-seconds). This high precision is possible only when maximizing the quality of **all** factors discussed in Section 12.8.2 and by taking at least three sets of readings for each data set, so that repeatability can be examined.

Inclinometer data can give good results when used with flexible steel structures such as sheet or steel piles (e.g., Boissier et al., 1978), where deflections are large and the section modulus is known. However, they are of limited use for stiff composite structures such as drilled shafts, reinforced slurry walls, and reinforced concrete retaining walls, both because angular changes are likely to be very small and because the moment of inertia is not known accurately (Wolosick and Feldman, 1987). Because

*Written with the assistance of Alex I. Feldman, Senior Engineer, Shannon & Wilson, Inc., Seattle, WA.

the behavior of concrete is inelastic, these zones also vary with time. For such structures, inclinometer data should be used for estimating bending moments only if other measurements can be used for checking, such as measurement of internal stresses or external load. Examples of measurements in a sheet pile wall, slurry wall, and cylinder pile wall are given by Gould and Dunnicliff (1971). Saxena (1974) and Soares (1983) describe measurements in slurry walls.

12.9. TRANSVERSE DEFORMATION GAGES

Transverse deformation gages are defined in this book as devices installed within a borehole or pipe for monitoring deformation normal to the axis of the borehole or pipe. Inclinometers fall within this category but are described separately in Section 12.8.

Typical applications are determination of depth and extent of sliding zones in natural and excavated slopes and earth fills, and measurement of the pattern of horizontal deformation within embankment dams and around braced excavations. Transverse deformation gages include shear plane indicators, plumb lines, inverted pendulums, in-place inclinometers, and deflectometers.

Borehole directional survey instruments are also discussed in this section.

12.9.1. Shear Plane Indicators

Shear plane indicators range from crude and inexpensive rupture stakes to more precise and expensive slope extensometers.

Rupture Stakes

In soft clays simple wooden stakes can be pushed or driven into the ground to a depth beyond the anticipated shear plane. Shearing will break the stakes, and the depth to the shear plane can be determined by pulling out the upper part of each stake. Stakes can be 2 × 1 in. (50 × 25 mm) softwood, or hardwood beading, without knots. Eide and Holmberg (1972) made saw cuts halfway through each stake at 2 in. (50 mm) intervals to ensure breakage. Stakes are usually installed by first pushing or driving a steel pipe with a loose end cap, inserting the stake within the pipe, raising the pipe while pushing on the stake to dislodge the end cap, and withdrawing the pipe. This is an economical procedure if installations can be made by hand, but if a drill rig is required, the shear probe described in the following subsection may be the preferred approach. Because there is a significant risk of breakage when removing stakes, a large number should be used so that false data can be discarded.

Shear Probe

The *shear probe,* also referred to as a *poor man's inclinometer, slip indicator,* and *poor boy,* consists of plastic tubing or thin-wall polyvinylchloride (PVC) pipe, installed in a nominally vertical borehole. The depth to the top of the shear zone is determined by lowering a rigid rod within the tubing or pipe and measuring the depth at which the rod stops at a bend. The depth to the bottom of the shear zone can be measured by leaving a rod with an attached graduated nylon line at the bottom of the tubing or pipe and pulling on the line until the rod stops. Curvature can be determined by inserting a series of rigid rods of different lengths and noting the depth at which each rod will not pass further down the tubing or pipe. Curvature is given by:

$$R = \frac{L^2}{8\,(D_1 - D_2)},$$

where: R = radius of curvature of tubing or pipe,
D_1 = inside diameter of tubing or pipe,
D_2 = outside diameter of rod,
L = length of rod.

Components of the system typically used in England are shown in Figure 12.78. The PVC pipe is used as temporary sleeving around the plastic tubing, while the borehole is backfilled with sand, and is withdrawn as backfilling proceeds. In the United States, thin-wall PVC pipe is normally used instead of tubing, and therefore the temporary sleeving is unnecessary. Typically, the pipe is 2 in. (50 mm) SDR 21 PVC with belled ends, and a typical set of four reading rods consists of 1 in. (25 mm) pipe, 6, 15, 30, and 40 in. (150, 380, 760, and 1020 mm) long, each arranged for separate attachment to a 100 ft (30 m) long graduated 0.125 in. (3 mm) diameter steel cable. The shortest rod will stop at a bend of 6 in. (150 mm) radius, the longest at a bend of 20 ft (6 m) radius.

In soft clays the pipe can usually be installed by attaching a strong bottom plug and pushing inside

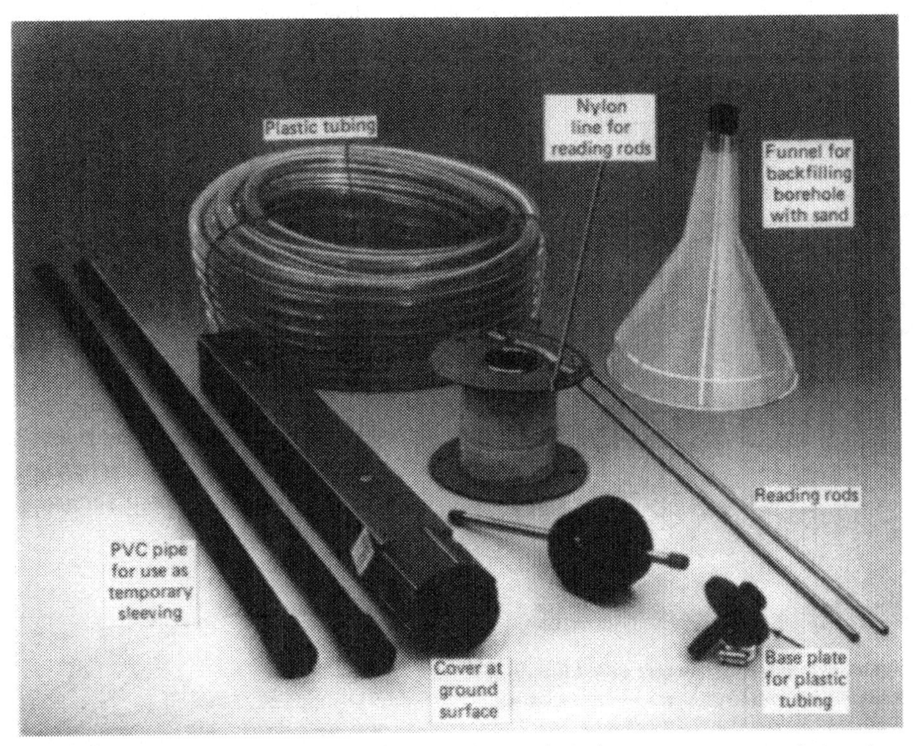

Figure 12.78. Slip indicator components (courtesy of Geotechnical Instruments (U.K.) Ltd., Leamington Spa, England).

with drill rods. In stiffer soils it may be necessary to drive casing, clean out, insert the belled-end pipe, and withdraw the casing. Clearly, the pipe should be installed as straight as possible.

McGuffey (1971) describes use of the shear probe in observation well pipes, allowing monitoring of both groundwater level and horizontal deformation. A similar arrangement would be possible in open standpipe piezometers.

Shear Strip

The *shear strip* consists of a parallel electrical circuit made up of resistors that are mounted on a brittle backing strip and waterproofed. As shown in Figure 12.79, the locations of up to two breaks in the strip are determined by measuring resistances at the top and bottom of the strip. Resistors can be spaced at any interval, but 3 ft (1 m) is typical, and the maximum number of resistors per strip is about 100.

The device is generally installed in soil by drilling a 3 in. (76 mm) diameter borehole, inserting 2 in. (50 mm) PVC or polyethylene pipe, inserting the shear strip with a polyethylene grout tube, grouting with cement grout, and withdrawing the grout tube. Installation in rock is similar, but the protective pipe is usually unnecessary. The shear strip can be connected to an automatic recording system and also arranged to sound an alarm if the strip breaks.

Slope Extensometer

The *slope extensometer* (Kirschke, 1977; Müller et al., 1977) is a multipoint fixed borehole extensometer with tensioned wires, arranged for monitoring

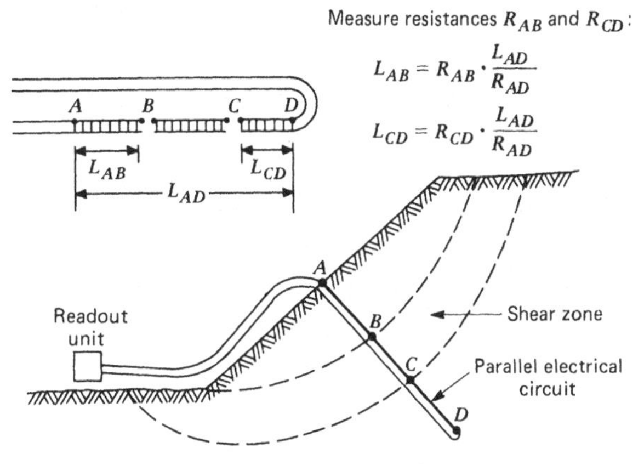

Figure 12.79. Schematic of shear strip.

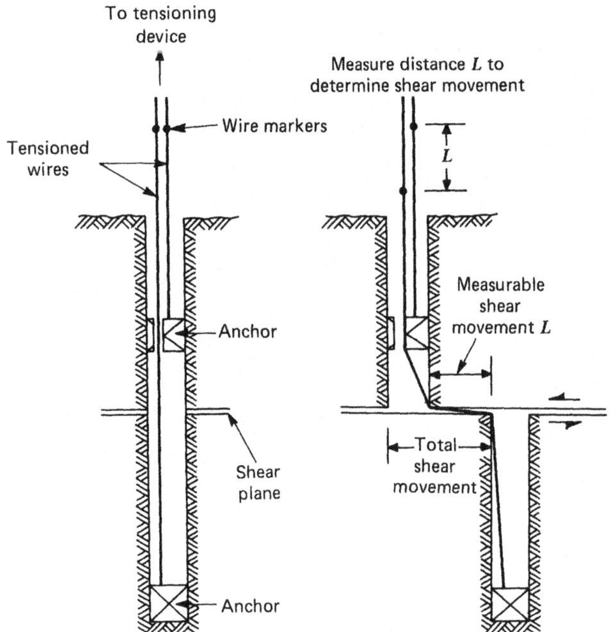

Figure 12.80. Schematic of slope extensometer (after Kirschke, 1977). Note: Additional anchors and tensioned wires not shown.

deformations normal to the axis of the borehole (Figure 12.80). Up to about 10 anchors and wires can be installed in a borehole.

Initial shear deformation will not cause an equivalent reading change, owing to lateral movement of the wires within the borehole, but after the borehole has been separated completely, the reading change will equal the shear deformation. Wires are tensioned either by coil springs or by pulleys and suspended weights, and deformations between wire markers are read with a ruler. Measurement precision can be increased by using alternative mechanical or electrical transducers.

When compared with more conventional inclinometer measurements, advantages of the slope extensometer include a simple and rapid reading procedure, the option to provide an alarm by inclusion of limit switches, and the ability to monitor much larger shear deformations. However, Kirschke (1977) comments that the device is suitable only for monitoring distinct shear planes or thin shear zones.

12.9.2. Plumb Line and Inverted Pendulum

Plumb lines, or *hanging pendulums,* can be used for monitoring horizontal displacements of concrete dams, dam abutments, shafts, and tall buildings. A typical arrangement is shown in Figure 12.81.

Inverted pendulums are used for the same purposes as plumb lines and are applicable where access is not available to the bottom of the system. They can also be used for accurate measurements of absolute ground surface deformation and as horizontal control stations for surveying methods (Marsland, 1974b). A typical arrangement is shown in Figure 12.82. The float, which is free to move in a water tank, tensions the wire and keeps it vertical.

Clearly, both systems require a near-vertical duct when installed as construction progresses or a near-vertical borehole when installed after completion of construction or in original ground. The instruments can be read to an accuracy of ± 0.02 in.

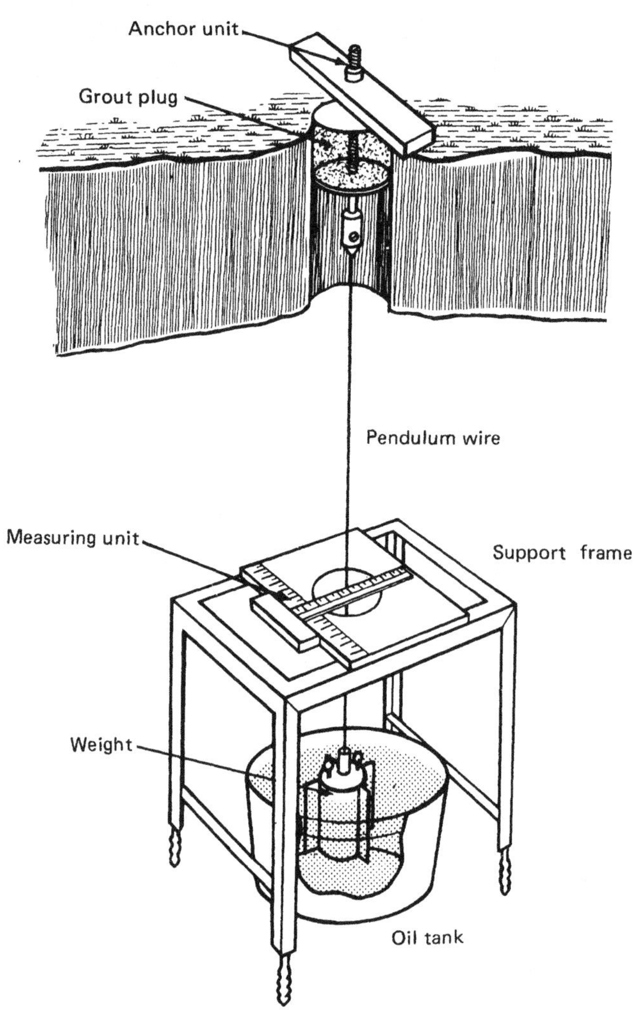

Figure 12.81. Plumb line (courtesy of Soil Instruments Ltd., Uckfield, England).

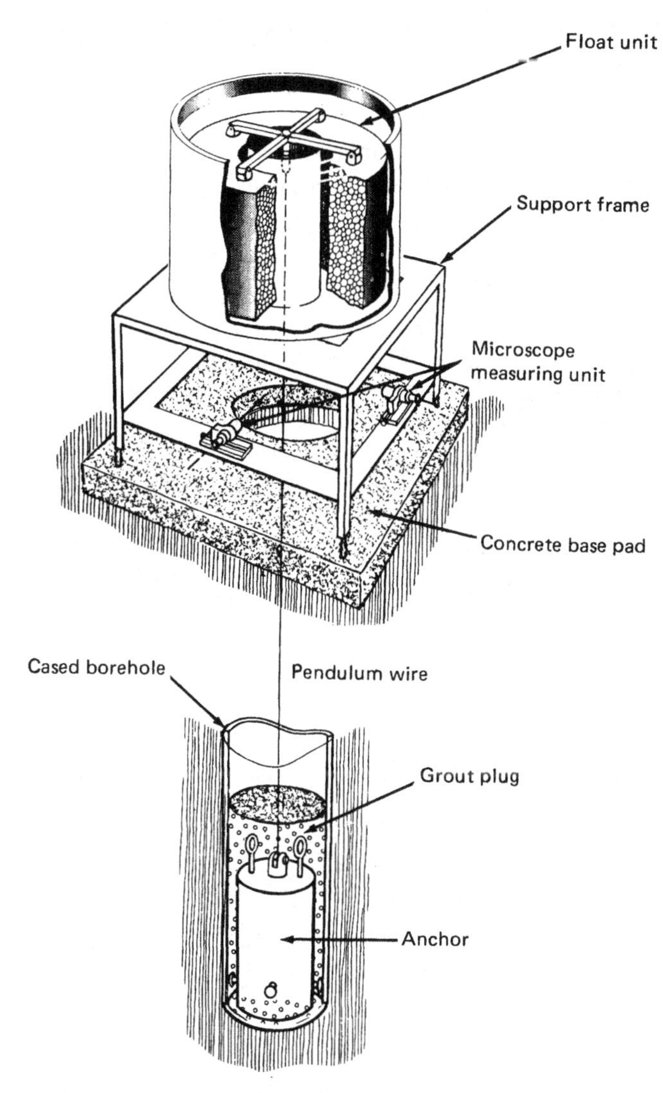

Figure 12.82. Inverted pendulum (courtesy of Soil Instruments Ltd., Uckfield, England).

(±0.5 mm) by using a steel measuring scale or to ±0.001 in. (±0.03 mm) by using traveling vernier microscopes mounted to sight the wire in orthogonal directions. Carpenter (1984a) describes a method for remote monitoring of plumb lines that have been installed in four dams in the Colorado River Storage Project. The system was designed around an optical vision system used for control of industrial robots. Rays from light-emitting diodes are used to sense the shadow of the pendulum wire, and signals are projected on to a linear photodiode array that is scanned by a microprocessor. The commercial supplier of the system is given in Appendix D.

Advantages of plumb lines and inverted pendulums include their simplicity and longevity. Their main disadvantage is the difficulty in creating a straight vertical duct or borehole, and borehole installations may require specialized and expensive drilling. Debreuil and Hamelin (1974) describe a technique for drilling 3 in. (76 mm) diameter holes using standard equipment, directional drilling techniques, and an inverted pendulum for monitoring verticality as drilling proceeds.

12.9.3. In-Place Inclinometers and Multiple Deflectometers*

In-place inclinometers and *multiple deflectometers* are typically used for monitoring subsurface deformations around excavations or within slopes, when rapid or automatic monitoring is required.

In-Place Inclinometers

An *in-place inclinometer* is generally designed to operate in a near-vertical borehole and provides essentially the same data as a conventional inclinometer (Section 12.8). The device, shown schematically in Figure 12.83, consists of a series of gravity-sensing transducers joined by articulated rods. Uniaxial or biaxial transducers can be used. The transducers are positioned at intervals along the borehole axis and can be concentrated in zones of expected movement. Movement data are calculated using the same methods as for conventional inclinometers.

Figure 12.84 shows an in-place inclinometer with force balance accelerometers as gravity-sensing transducers. Londe (1982) describes a system with a pendulum blade and induction transducers, and Cooke and Price (1974) report on development of a system with electrolytic level transducers.

The device generally uses standard inclinometer casing as guide pipe and can be removed for repairs. However, data continuity will be interrupted when the device is removed and replaced. When compared with conventional inclinometers, advantages include more rapid reading, an option for continuous automatic reading, and an option for connection to a console for transmission of data to remote locations or for triggering an alarm if deformation exceeds a predetermined amount. Disadvantages include greater complexity and expense of the

*Written with the assistance of Howard B. Dutro and P. Erik Mikkelsen, Vice Presidents, Slope Indicator Company, Seattle, WA.

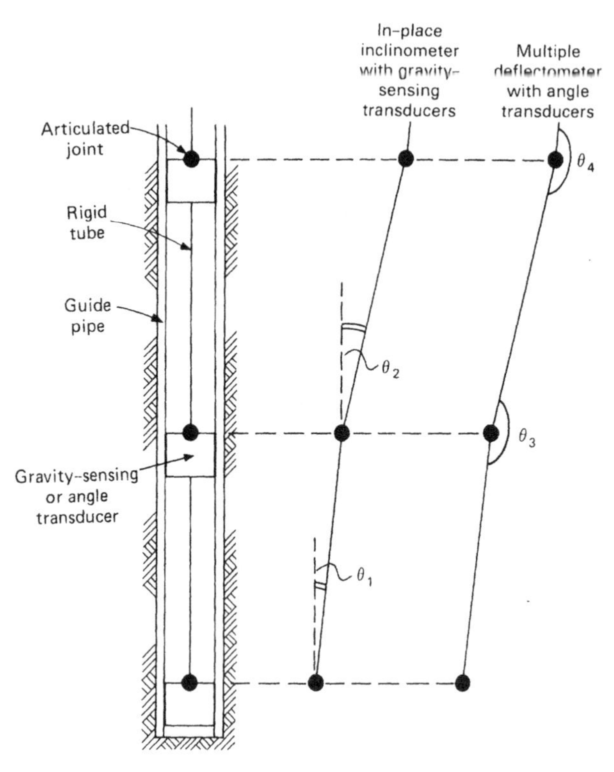

Figure 12.83. Schematic of in-place inclinometer and multiple deflectometer.

hardware. When conventional inclinometers are read, any long-term drift of the gravity-sensing transducers is removed from calculations by taking a second set of readings with the inclinometer rotated 180 degrees (the *check-sum* procedure), but this is not possible with the in-place version. Although the transducers are generally stable, there is always the possibility of a "rogue transducer," and this possibility should be recognized if one is planning to use an in-place inclinometer for long-term applications where high precision is required.

In-place inclinometers can be used effectively in combination with a conventional inclinometer. An in-place version can first be installed to define the location of any transverse deformation, with minimal labor costs for reading. If deformation occurs, the in-place system can be removed and the moving zone monitored with a conventional inclinometer. Alternatively, a conventional inclinometer can be used first to indicate any deformation and an in-place version later installed across a critical deforming zone to minimize subsequent effort and perhaps to provide an alarm trigger.

Typical field precision is ±0.02–0.04 in. over a 10 ft gage length (approximately ±0.5–1.0 mm in 3 m, ±34–69 arc-seconds), and it is believed that greater precision can be obtained by careful use of a conventional inclinometer with a force balance accelerometer transducer.

The in-place inclinometer, together with installation, calculation, data processing, and reporting procedures, is described by ISRM (1981a).

Multiple Deflectometers

Multiple deflectometers, also referred to as *chain deflectometers,* operate on a similar principle to in-place inclinometers, but rotation is measured by angle transducers instead of tilt transducers (Figures 12.83 and 12.85).

Two versions are commercially available: articulated rods with full bridge bonded resistance strain gage transducers attached to cantilevers and a tensioned wire passing over knife edges with induction transducers.

Multiple deflectometers are usually installed within inclinometer casing. The system can usually

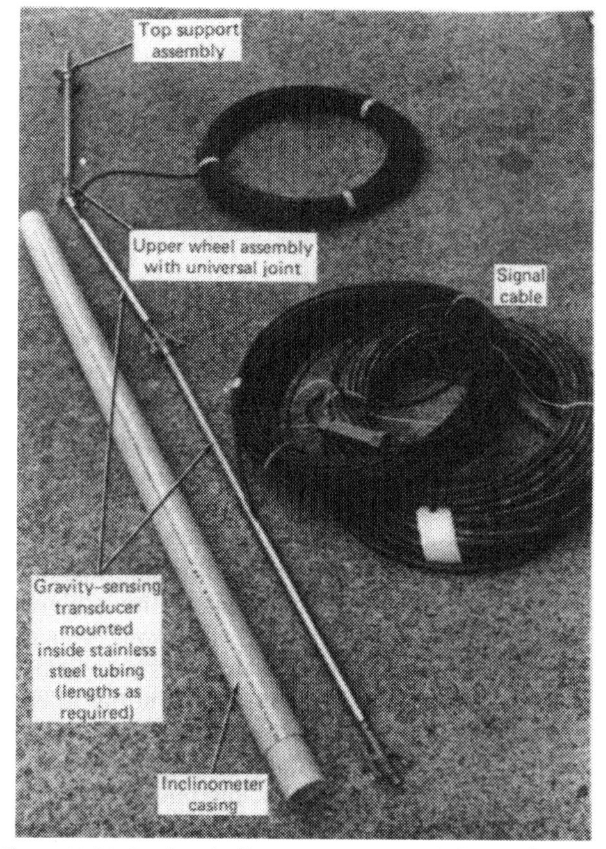

Figure 12.84. In-place inclinometer (courtesy of Slope Indicator Company, Seattle, WA).

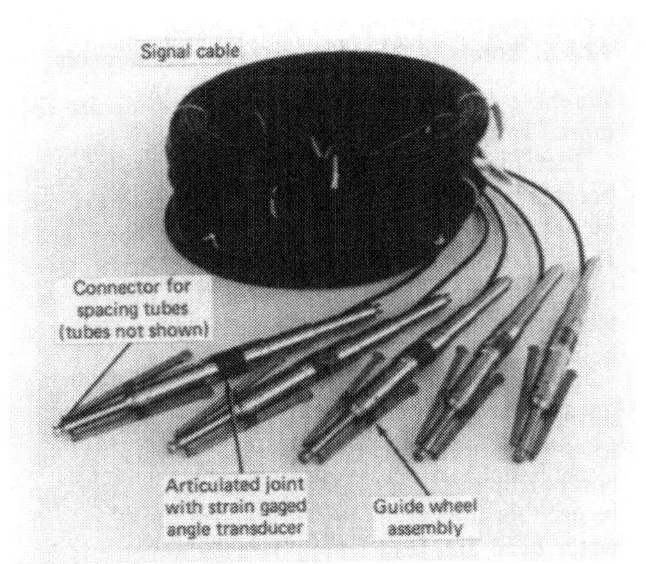

Figure 12.85. Downhole components of multiple deflectometer (courtesy of Slope Indicator Company, Seattle, WA).

be removed from the borehole at any time for maintenance and calibration, but data continuity will be interrupted when the device is removed and replaced.

Although advantages and limitations are generally the same as for in-place inclinometers, there are three important differences between the two systems. The first favors multiple deflectometers, while the second and third favor in-place inclinometers. First, multiple deflectometers are not limited by borehole inclination, because their transducers are not referenced to gravity. They can therefore be used to sense horizontal deformation in a horizontal borehole. Second, because deflectometer data are calculated by determining the position of one arm of the instrument relative to another, and **not** with respect to gravity, the device has no means of sensing rotation of the system as a whole. Third, deflectometer errors accumulate exponentially, whereas inclinometer errors accumulate arithmetically.

Multiple deflectometers have not been used widely and performance experience is sparse. Müller and Müller (1970) describe a tensioned wire version with induction transducers and quote a precision of ±0.0015 in. over a 15 ft gage length (approximately ±0.04 mm in 4.6 m). This appears to be a case of unusually high precision, perhaps when not operating through mechanical zero (Dutro, 1985). The author believes that, under field conditions, users should assume a typical precision similar to that suggested previously for in-place inclinometers.

12.9.4. Portable Borehole Deflectometers

The deflectometer is also available in a portable version, allowing similar use as a conventional inclinometer but without the limitation of borehole inclination. It can therefore be used to provide horizontal deformation data in a horizontal hole as well as for borehole directional surveys.

A typical instrument consists of two beams of equal length connected by an articulated joint, with an angle transducer arranged to sense angular rotation between the two beams. Some systems operate within grooved inclinometer casing, and others require use of insertion rods for orientation control. Data reduction procedures are similar to procedures used in conventional surveying by open traverse lines and are described by Dutro (1977).

The most accurate instrument known to the author was developed by the Federal Institute of Technology, Zürich, (Kovari et al., 1979), and is available from Solexperts AG. It uses the same sphere/cone-shaped measuring points as the sliding micrometer (Figure 12.31), spaced 1.5 m (4.9 ft) apart, has linear displacement transducers, and is attached to insertion rods. For a borehole 30 m (100 ft) long, the repeatability of transverse deformation measurement is reported as ±2–3 mm (±0.08–0.12 in.) (Thut, 1987). Figure 12.86 shows the basic arrangement, including transducers for measuring

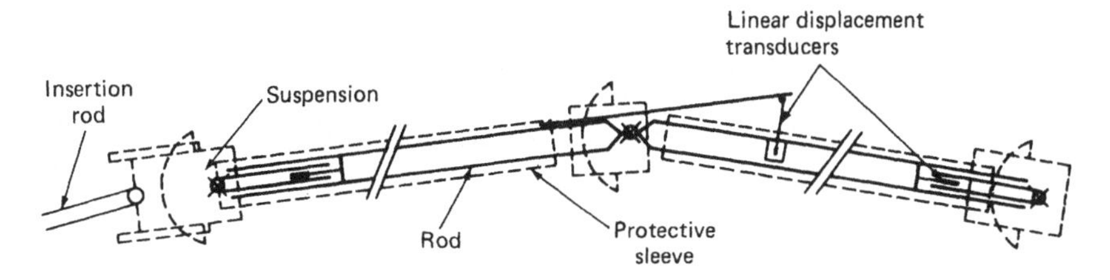

Figure 12.86. ***Extenso-Deflectometer*** **(after Kovari et al., 1979). Reprinted from ASCE 4th RETC, Atlanta, June 1979, "New Developments in the Instrumentation of Underground Openings."**

axial deformation, so that the instrument functions as a combined portable borehole deflectometer and probe extensometer. The portable borehole deflectometer is commercially available, but the combined instrument shown in Figure 12.86 has not been developed.

Instruments manufactured in the United States by Slope Indicator Company have strain gaged cantilevers to sense angular rotation between the two beams and operate either within grooved inclinometer casing or smooth-walled pipes. O'Rourke and Kumbhojkar (1984) provide guidelines on the measurement procedure and sources of error when using these instruments. The major sources of error are mechanical interference in the articulation and positioning uncertainties in successive legs of a traverse, and both can be minimized by carefully following established reading procedures and by repeated traverses. Dutro (1984) indicates that, when following these procedures within a 100 ft (30 m) long smooth-walled clean pipe, the instrument enables the true position of one end of the pipe with respect to the other end to be determined to within ±6 in. (±150 mm). Robinson et al. (1985) report on laboratory and field test measurements, indicating that the vertical profile for a 15.2 m (50 ft) near-horizontal grooved casing could be measured to an accuracy of ±12 mm (±0.5 in.), while the horizontal profile could be measured to an accuracy of ±4 mm (±0.15 in.). The particular instrument used tended to change its zero value between readings, and this resulted in measurement errors of ±50 mm (±2 in.) over the length of the casing. The use of a developed data correction constant was found to produce accurate measurements to within ±8 mm (±0.3 in.). It is believed that the above accuracies were limited by mechanical shortcomings in the design of the instrument then available from Slope Indicator Company. Design of an improved version is in progress.

The accuracies reported above should not be converted to a proportion of the length of pipe or casing and used to estimate accuracy for pipes and casings of different lengths, because the error accumulates with the number of measurements according to the power function given by Dutro (1977).

12.9.5. Fiber-Optic Sensor

The *fiber-optic microbending sensor* has a potential use as a transverse deformation gage and is described in Section 12.11.4.

12.9.6. Borehole Directional Survey Instruments

Borehole directional surveys are occasionally required before installing instruments.

Portable borehole deflectometers can be used in boreholes at any inclination, and inclinometers can be used to survey the vertical profile of boreholes. However, both methods are cumbersome because they usually require either orientation rods or the temporary insertion of inclinometer casing and may be unnecessarily precise.

The vertical profile of a borehole that is inclined downward can be surveyed with a full-profile liquid level gage (Section 12.10). Alternatively, if the borehole holds water, a diaphragm piezometer can be used for the survey, by making measurements of water head and lead length as a piezometer is inserted into the borehole. If the borehole will not hold water, a closed-ended PVC pipe can be inserted temporarily and filled with water.

A comprehensive description of borehole directional surveying methods, ranging from the simple acid etch technique to more complex photographic and gyroscopic methods, is given by Cumming (1956). Two methods described by Cumming are used occasionally for directional surveys of boreholes prior to installing geotechnical instruments and are outlined in the following subsections. Commercial sources are included in Table D.7 of Appendix D.

Photographic Method

Inclination is measured by photographically recording the position of the tip of a free-swinging pendulum, and orientation is indicated by a magnetic compass. *Single-shot* and *multiple-shot* versions are available, a timing mechanism controlling the photographic exposures (Figure 12.87). Inclination angles can be read to the nearest 0.25 degree and compass bearings to the nearest 0.5 degree. Survey services are provided by several companies specializing in oil field operations.

Pajari Method

The device used in the Pajari method contains a pendulum, magnetic compass, and timing mechanism (Figure 12.88). A single reading is taken for each insertion, by setting the timing mechanism, inserting the instrument to the required measurement point, waiting for the timing mechanism to lock the pendulum and compass, and retrieving and reading the instrument.

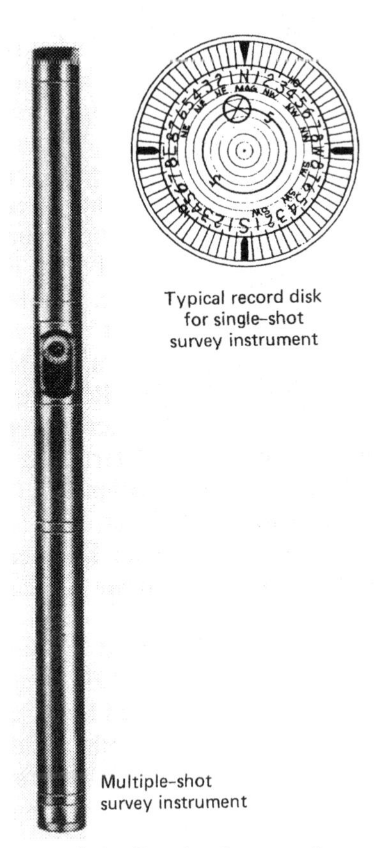

Figure 12.87. Borehole directional survey instrument, photographic type (courtesy of Eastman Christensen Company, Houston, TX).

Figure 12.88. Pajari borehole directional survey instrument (courtesy of Pajari Instruments Ltd., Orillia, Ontario, Canada).

12.10. LIQUID LEVEL GAGES

Liquid level gages are defined in this book as instruments that incorporate a liquid-filled tube or pipe for determination of relative vertical deformation. Relative elevation is determined either from the equivalence of liquid level in a manometer or from the pressure transmitted by the liquid.

The primary application for liquid level gages is monitoring settlements within embankments or embankment foundations. In general, they are alternatives to vertical probe extensometers, settlement platforms, and subsurface settlement points, allowing installation to be made without frequent interruption to normal fill placement and compaction and minimizing the potential for instrument damage. Most liquid level gages also allow measurements to be taken at a central reading location. Certain types of liquid level gage can also be used where more precise measurements are required, such as when monitoring the settlement of structures.

The gages only provide a means of measuring **relative** elevations between two or more points. If **absolute** settlement or heave is required, as is usually the case, data must be referenced to a benchmark. If one end of the gage cannot be mounted directly on a benchmark, a surveying method will normally be used, and accuracy may be dependent on accuracy of the surveying method.

In general, liquid level gages are sensitive to liquid density changes caused by temperature variation, to surface tension effects, and to any discontinuity of liquid in the liquid-filled tube. These three sources of error are discussed in detail in Section 8.2.3, which also provides guidelines on selection of tubing material and diameter, tubing fittings, liquid, and routing of liquid-filled tubes. The greatest potential source of error is discontinuity of liquid caused by the presence of gas, and great care must always be taken to ensure absence of gas. Precision claimed for these instruments is sometimes unrealistic, and users are encouraged to study this section and Section 8.2.3 before selecting liquid level gages.

A general caution about the use of mercury in these systems is appropriate. In the United States mercury is considered to be a hazardous material, and environmental restrictions prevent its use in many applications. In addition, if mercury remains in plastic tubing in the long term, there is evidence that it can leach gas from the tubing and create discontinuities in the mercury. Also, oxidation can cause contamination and blockage of the tubing. When mercury is used, it should be triple distilled.

Single-point, multipoint, and full-profile liquid level gages are available. Single-point gages can be grouped into three categories, depending on the relative elevations of the two ends of the system: both ends at the same elevation, readout unit higher than the cell, and readout unit lower than the cell. Gages are described in turn, and comparative information is given in Table 12.9.

All tubing diameters referred to in this discussion of liquid level gages are **inside** diameters, because the **inside** diameter impacts on the behavior of fluid within the tubing. The reader is cautioned about a possible confusion: industry standards use the **outside** diameter when referring to tubing sizes.

12.10.1. Single-Point Gages with Both Ends at Same Elevation

Hose Levels

A simple liquid level manometer was used in ancient times for establishing relative levels during construction of canals in Mesopotamia, between the Tigris and Euphrates rivers in what is now Iraq. Modern versions are usually called *hose levels* and are used primarily for precise long-term measurement of differential settlement within buildings, where operating conditions can be well controlled.

The hose level, also called the *Terzaghi water level meter,* consists of two burettes connected by a length of transparent water-filled tube, normally of 0.38 in. (10 mm) **inside** diameter. The burettes are hung on a pair of wall-mounted observation pins, and micrometer spindles are advanced to touch the water surfaces simultaneously. Relative elevations of pins are determined from micrometer readings, and absolute elevations are determined by mounting one of the pins over a benchmark. The version manufactured earlier by Soiltest, Inc. is shown in Figure 12.89, but this product has been discontinued, and the author is not aware of a current commercial source.

The arrangement is described by the Corps of Engineers (1980) and Terzaghi (1938), and revisions to Terzaghi's design are detailed by Casagrande et al. (1967). Other versions—together with methods of minimizing errors resulting from discontinuity of the liquid, temperature differences along the hose, and differences in atmospheric pressure between

Table 12.9. Liquid Level Gages

Gage	Advantages	Limitations[a]	Approximate Precision[b]
Hose level (e.g., Figure 12.89)	Very precise	Great care needed to minimize errors Both ends must be at same elevation and barometric pressure Accuracy reduced significantly if forced air ventilation systems cause different air pressures at ends	±0.001–0.5 in. (±0.03–13 mm)
Multistation hose level (Figure 12.90)	Simple	Both ends must be at same elevation and barometric pressure	±0.1–0.5 in. (±3–13 mm)
Single-point overflow gage with both ends at same elevation (Figure 12.91)	Cell can be attached to a subsurface settlement point to allow readout above measuring point	Unless cell is attached to a subsurface settlement point, both ends must be at same elevation Both ends must be at same barometric pressure Only single point is monitored	±0.02–0.8 in. (±0.5–20 mm)

Table 12.9. ***(Continued)***

Gage	Advantages	Limitations[a]	Approximate Precision[b]
Single-point gage with pressure transducer in cell, with readout unit higher than cell (Figure 12.93)	Cell can be installed in a borehole Backpressured version available so that liquid-filled tube can readily be checked for continuity of liquid	Design and operation of transducer requires close attention to many details to ensure precision; see Section 12.10.2 Version with vibrating wire transducer may be subject to damage by overranging the transducer if initial filling with liquid is completed before shipment; see Section 12.10.6. If transducer cavity is vented, potential for corrosion and reading errors. If transducer cavity is not vented, gage sensitive to changes in barometric pressure Regular de-airing of liquid required (but backpressured version greatly reduces frequency) Only single point is monitored Requires accurate knowledge of liquid density Readout unit must be above cell Use of mercury creates possible environmental hazard[c]	With pneumatic transducer and aqueous solution: ±0.5–1.5 in. (±13–38 mm) With vibrating wire transducer and mercury: ±0.1–1.0 in. (±3–25 mm)
Single-point gage with pressure transducer in readout unit, with readout unit higher or lower than cell (Figure 12.96)	Liquid-filled tube can readily be checked for continuity of liquid by increasing backpressure No buried transducer Readout can be above or below cell Cell can be installed in a borehole	Only single point is monitored Requires accurate knowledge of liquid density	With electrical pressure transducer and aqueous solution: ±0.25–1.0 in. (±6–25 mm)
Single-point overflow gage with readout unit higher or lower than cell (Figure 12.98)	Readout can be above or below cell Cell can be installed in a borehole	Regular de-airing of liquid required when readout is higher than cell Both ends must be at same barometric pressure Only single point is monitored Requires accurate knowledge of liquid density	±0.4–0.8 in. (±10–20 mm)
Multipoint gage with interconnected chambers (e.g., Figure 12.99)	Precise Automatic recording	All chambers must be at same elevation and barometric pressure	±0.004–0.1 in. (±0.1–3 mm)

Table 12.9. (*Continued*)

Gage	Advantages	Limitations[a]	Approximate Precision[b]
Full-profile overflow gage (similar to Figure 12.91)	No limit to number of measuring points No delicate or expensive parts buried Can be used with probe extensometer to measure both horizontal and vertical deformations Can be used for elevation surveys along a near-horizontal or inclined borehole, pipeline, or culvert	Both ends must be at same elevation and barometric pressure Buried pipe must slope upward from access point Longitudinal position of measuring point must be controlled very carefully	±0.1–0.8 in. (±3–20 mm)
Full-profile gage with pressure transducer in probe and attached liquid-filled tube (similar to Figure 12.93; also Figure 12.100)	No limit to number of measuring points No delicate or expensive parts buried Can be used with probe extensometer to measure both horizontal and vertical deformations Can be used for elevation surveys along a near-horizontal or inclined borehole, pipeline, or culvert	Cumbersome Design and operation of transducer requires close attention to many details to ensure precision; see Section 12.10.2 Versions with vibrating wire or electrical resistance transducer are subject to overranging while handling. If transducer cavity is vented, potential for corrosion and reading errors. If transducer cavity is not vented, gage sensitive to changes in barometric pressure Version with bladder has limited range, and great operator care is required Prone to temperature errors if external air temperature is very different from temperature in pipe Regular de-airing of liquid required Liquid-filled tube must be transparent Requires accurate knowledge of liquid density Readout unit must be above probe Use of mercury creates possible environmental hazard[c] Longitudinal position of measuring point must be controlled very carefully	With pneumatic transducer and aqueous solution: ±0.5–1.5 in. (±13–38 mm) With vibrating wire or electrical resistance transducer and mercury: ±0.1–1.0 in. (±3–25 mm) With bladder: ±0.3–1.5 in. (±8–38 mm)
Full profile gage with pressure transducer in readout unit and attached liquid-filled tube (similar to Figure 12.96)	Liquid-filled tube can be checked for continuity of liquid by increasing backpressure	Cumbersome Prone to temperature errors if external air temperature is very different from temperature in pipe	With electrical pressure transducer and aqueous solution: ±0.25–1.0 in. (±6–25 mm)

Table 12.9. (*Continued*)

Gage	Advantages	Limitations[a]	Approximate Precision[b]
	No limit to number of measuring points No delicate or expensive parts buried Can be used with probe extensometer to measure both horizontal and vertical deformations Can be used for elevation surveys along a near-horizontal or inclined borehole, pipeline, or culvert	Requires accurate knowledge of liquid density Longitudinal position of measuring point must be controlled very carefully	
Full-profile gage with pressure transducer in probe but without attached liquid-filled tube (Figure 12.103)	Can be used for elevation surveys along downward inclined boreholes No limit to number of measuring points No delicate or expensive parts buried	Great care needed to ensure pipe is completely filled with liquid Buried pipe must not leak Sensitive to barometric pressure if pressure transducer is not vented Requires accurate knowledge of liquid density Readout unit must be above probe Longitudinal position of measuring point must be controlled very carefully	±0.05–20 in. (±1.3–510 mm)
Double fluid full-profile gage (DFSD) (Figure 12.104)	Very long profiles can be monitored No open pipe required Automatic data acquisition system available No limit to number of measuring points No delicate or expensive parts buried	Complex control system Excessive pressure can burst tube Tube cannot be more than 20 ft (6 m) below readout unit Requires accurate knowledge of liquid densities Readout unit must be above tubing Use of mercury creates possible environmental hazard[c] Longitudinal position of measuring point must be controlled very carefully	±0.1–1.5 in. (±3–38 mm)

[a] All gages are sensitive to liquid density changes caused by temperature variation, to surface tension effects, and to any discontinuity of liquid in the liquid-filled tube. See this section and Section 8.2.3 for details and recommendations. They are also subject to freezing problems.

[b] Precision refers to *relative* elevations betwen two or more parts of the gage and is highly dependent on factors discussed in Section 8.2.3. If *absolute* settlement or heave is required, data must be referenced to a benchmark, and accuracy will usually be dependent on the referencing method.

[c] If left in tubing long term, mercury can leach gas from tubing and create discontinuities in liquid. Also, oxidation can cause contamination and blocking of tubing.

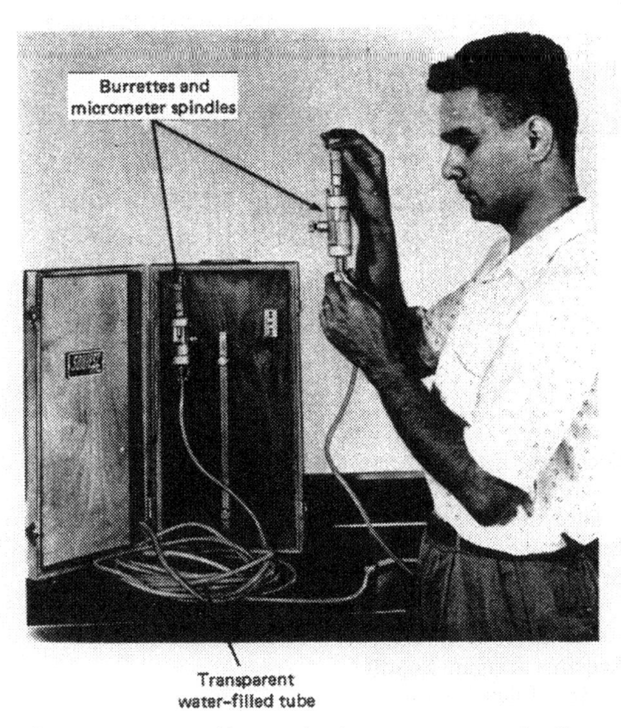

Figure 12.89. Terzaghi water level meter (courtesy of Soiltest, Inc., Evanston, IL).

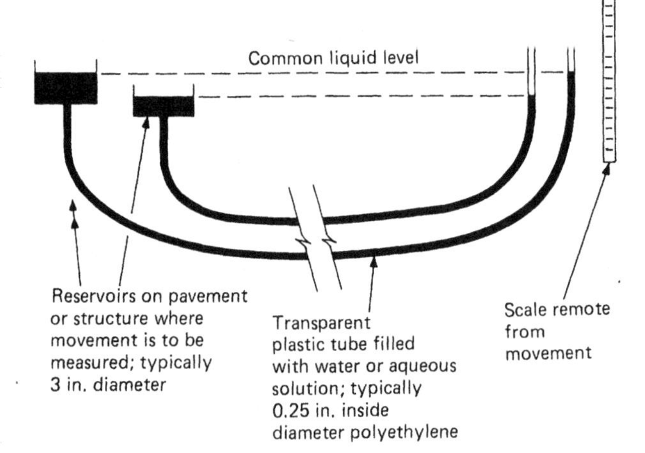

Figure 12.90. Schematic of multistation hose level.

the two water surfaces—are described by Gould and Dunnicliff (1971).

A multistation hose level, Figure 12.90 (Warner, 1978), is used for monitoring compaction grouting and slab jacking of highway pavements or buildings. The reservoirs are fixed at locations where monitoring is required, and the tubes are routed to a terminal panel and scale alongside the grout pump for direct observation by the pump operator. A single hose can also be used for contouring a structure prior to compaction grouting or slab jacking, by setting the reservoir at a fixed location and placing the other end at successive points on the structure.

Overflow Gages

The most frequently used instruments with both ends at the same elevation are called *overflow gages,* or alternatively *hydraulic leveling devices, water overflow pots,* or *overflow weirs.* They are commonly used for settlement measurements in embankment dams and as alternatives to settlement platforms during construction of embankments on soft ground, overcoming the need to extend a riser pipe through the embankment.

The instrument is described by Penman et al. (1975) and shown schematically in Figure 12.91. The gage is normally read by adding liquid to the liquid-filled tube at the readout station, causing overflow in the cell such that the visible level at the readout station stabilizes at the same elevation as the overflow point. The vent tube is essential to maintain equal pressure on both surfaces of liquid, and the drain tube is needed to allow overflowed liquid to drain out of the cell. A four-tube version, with duplicate liquid-filled tubes, provides a verification of reading correctness and is the preferred instrument. Penman et al. (1975) indicate that it is necessary to flush the liquid-filled tube with at least its own volume of de-aired liquid before reproducible readings can be obtained. Forsyth and McCauley (1973) and the New York Department of Transportation (1979) indicate that liquid is not normally added for highway applications, because liquid overflows as settlement occurs. However, this procedure optimistically assumes no discontinuity of liquid and no loss of liquid by evaporation, and

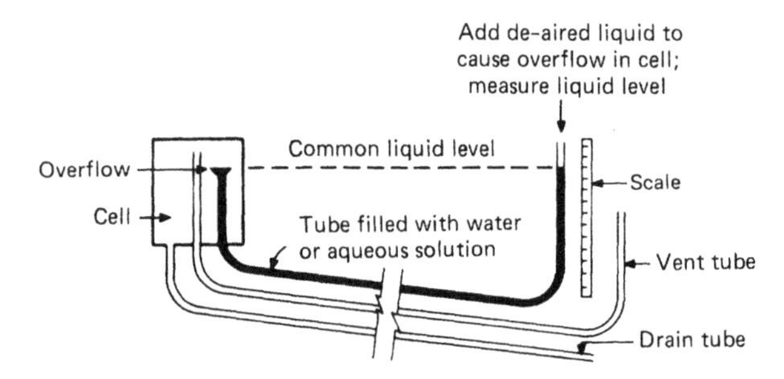

Figure 12.91. Schematic of overflow gage with both ends at same elevation.

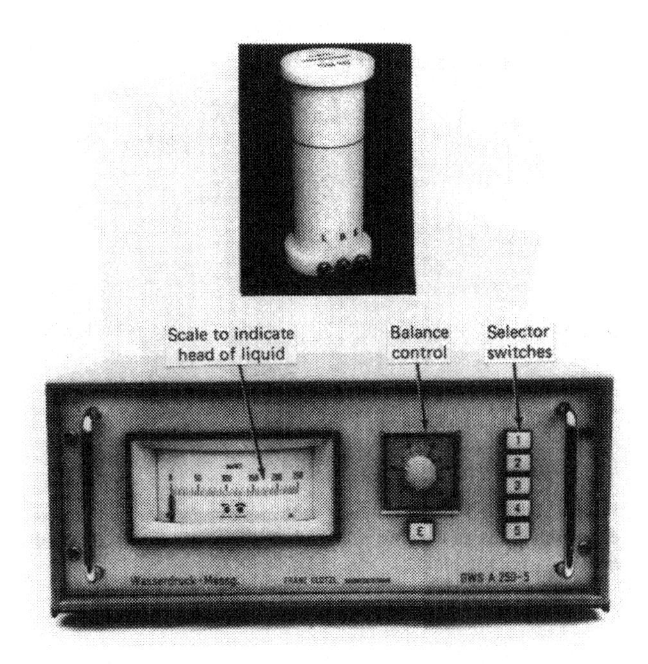

Figure 12.92. Overflow liquid level gage: (top) cell and (bottom) readout unit (courtesy of Glötzl GmbH, Karlsruhe, West Germany and Geo Group, Inc., Wheaton, MD).

good practice involves establishing a regular flushing schedule, reading before and after, and revising the schedule in accordance with changes caused by flushing.

Figure 12.92 shows the cell and readout unit manufactured by Glötzl. The liquid-filled tubes from a maximum of five cells are connected to the readout unit, and selector switches allow connection to each tube in turn. An electrical pressure transducer in the readout unit is used to indicate the head of liquid.

Forsyth and McCauley (1973) describe a gage without a drain tube, in which overflowing liquid drains directly into the embankment. Assumptions when using this gage are first that the embankment is sufficiently pervious to allow free draining, second that barometric pressure has free connection to the buried gage, and third that there is no chance of a free water surface rising above the level of the overflow at any time during the active life of the gage. This version of the overflow gage is not recommended.

As indicated in Section 8.2.3, the optimum **inside** diameter of the liquid-filled tube is 0.25 in. (6 mm).

The diameter of the vent tube should be large enough to create air pressure equilibrium along the tube but small enough so that any water that enters the tube can be blown out. Penman et al. (1975) indicate that when 0.6 in. (15 mm) **inside** diameter vent tubes were used to satisfy the first criterion, and they become flooded, it was extremely difficult to remove all the water. Water traps formed and destroyed the accuracy of the gages. An **inside** diameter of 0.25 in. (6 mm) appears to be a maximum for ensuring that water can be blown out, and even with this diameter a steady continuous flow of dry gas is needed to remove the film of water remaining in the tube. If the film is not removed after blowing out water, it will collect at a low point and again form a water trap. With the compromise of 0.25 in. (6 mm) **inside** diameter vent tubes, air pressure equilibrium can be achieved with an error of less than 0.005 in. (0.1 mm) head of water in 5 minutes over 600 ft (180 m). The inside diameter of the drain tube is usually the same as the vent tube.

The optimum **inside** diameter of all three (or four) tubes is therefore 0.25 in. (6 mm). However, available commercial versions have liquid-filled and vent tubes ranging from 0.2 to 0.4 in. (5–10 mm) **inside** diameter. The volume within the cell, between the overflow point and the drain tube connection, should be significantly larger than the volume of the liquid-filled tube, so that the overflow point does not become submerged during flushing. Backup of liquid in the drain tube can cause false readings and can be overcome by applying air pressure to the vent tube until liquid ceases to flow from the drain tube. It is advisable to install the drain tube on a continuous slope toward the readout unit.

The overflow gage can be used for monitoring vertical deformation below the elevation of the readout unit by attaching the cell to the top of a subsurface settlement point (Section 12.7.6). If this arrangement is used beneath an embankment, a protective cover must be provided around and over the cell, with sufficient internal height to allow settlement of the cover without contacting the cell (New York Department of Transporation, 1979). Special precautions must also be made to protect the tubes near the cell from damage as settlement proceeds.

12.10.2. Single-Point Gages with Readout Unit Higher than Cell

Gages that allow the readout unit to be higher than the cell include either a pressure transducer or a method of applying a suction or backpressure to the overflow gage described in the previous section.

$P = H\gamma$; thus, determine elevation of cell

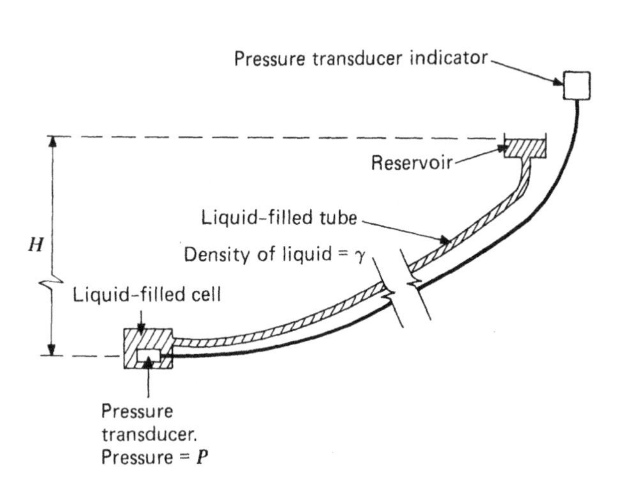

Figure 12.93. Schematic of liquid level gage with pressure transducer in cell, with readout unit higher than cell.

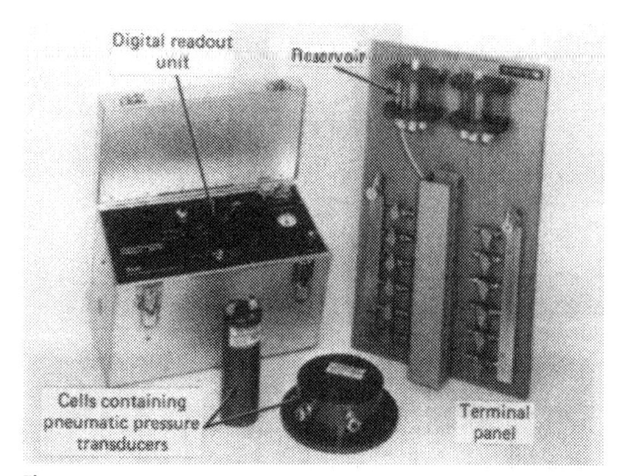

Figure 12.95. Liquid level gage with pneumatic pressure transducer in cell, with readout unit higher than cell (courtesy of Soil Instruments Ltd., Uckfield, England).

Gages with Pressure Transducer in Cell

The arrangement is shown schematically in Figure 12.93. The pressure transducer can be either a pneumatic (e.g., Figures 12.94 and 12.95) or vibrating wire type. The upper surface of the liquid column is at a known elevation at the readout location; therefore, relative elevation of the transducer and reservoir can be determined from the pressure measurement and liquid density.

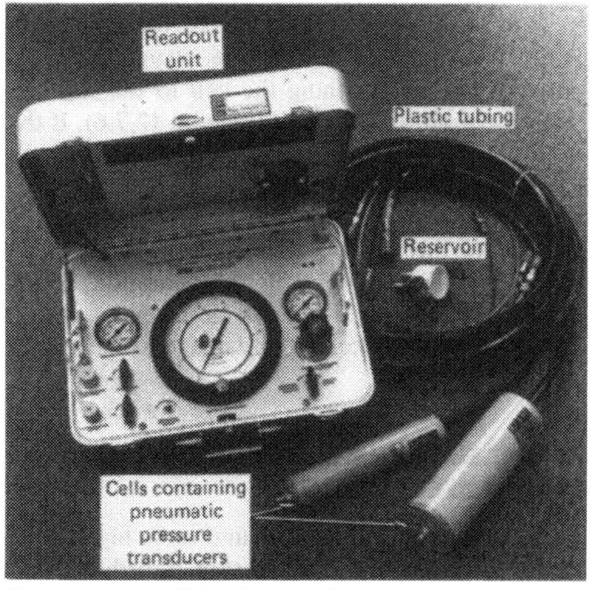

Figure 12.94. Liquid level gage with pneumatic pressure transducer in cell, with readout unit higher than cell (courtesy of Slope Indicator Company, Seattle, WA).

Figure 12.93 shows only one liquid-filled tube, but in practice it is preferable to have two tubes so that flushing is possible and independent measurements can be made on each tube as a check.

Precision of these gages is dependent on two major factors. First, the pressure transducer must read the liquid head correctly and second, the liquid must transmit static head correctly from the free surface at the reservoir to the diaphragm of the pressure transducer. The discussion of the first major factor will consider pneumatic and vibrating wire transducers separately.

When using pneumatic pressure transducers, one must pay close attention to the factors discussed in Section 8.3. When using the transducer for monitoring settlement, precision requirements are much greater and much more difficult to achieve than when the transducer is used for monitoring pore water pressure or total stress. If the transducer is read under a condition of no gas flow (Figure 8.6), **very** careful control of gas pressure is essential. If the transducer is read as gas is flowing (Figure 8.7 or 8.8), the rate of gas flow should be as low and constant as possible, and a flow meter should preferably be included in the system. The flow controller should be insensitive to variations in temperature, and the accuracy of the flow meter should be within 0.5 cm^3/min. The types of flow controllers and flow meters that are adequate when monitoring pore water pressure and total stress may not be adequately precise when monitoring settlement. Also, it is very important to use a pneumatic transducer with a very low volumetric displacement of the diaphragm (Section 8.3), because large volumetric displacement may cause a pressure surge in the liquid-filled tube and errors in the liquid part of the system.

When using vibrating wire pressure transducers, one must pay close attention to the factors discussed in Section 8.4.9. If the cavity behind the diaphragm is hermetically sealed, the gage is sensitive to changes in barometric pressure. On the other hand, if the cavity is vented to atmosphere to avoid this limitation, the transducer is subject to corrosion, and if any water enters the vent tube significant reading errors may be caused by the air/water interfaces (Penman, 1978). These are significant disadvantages. In addition, if initial filling of the liquid-filled tubes is completed before the gage is installed, care must be taken while handling the system to avoid overranging the vibrating wire transducer. Bearing in mind these various limitations, the author prefers pneumatic transducers when using this type of liquid level gage.

The second major factor that affects precision is the liquid part of the system. The system must transmit static head correctly from the free surface at the reservoir to the diaphragm of the pressure transducer, and close attention must be given to the factors discussed in Section 8.2.3. The selection of tubing diameter is critical. An upper limit is necessary to ensure that any gas in the tube can readily be displaced, and a lower limit is necessary to ensure that equilibrium is achieved in an acceptably small time. When aqueous solutions are used, the **inside** diameter of the liquid-filled tube should be between 0.17 and 0.25 in. (4.3–6 mm). If a larger diameter is used, gas cannot readily be displaced during initial filling and subsequent flushing. If a smaller diameter is used, the equilibrium time is likely to be too long, and any gas/liquid interfaces may create significant errors (Penman, 1978). An **inside** diameter of 0.25 in. (6 mm) is strongly recommended if a pressure surge is caused by movement of the liquid column at the time of reading, for example, by using a pneumatic transducer with significant volumetric displacement. When mercury is used, the **inside** diameter of the tubing normally ranges from 0.07 to 0.2 in. (2–5 mm).

The instrument shown in Figure 12.95 is available with a system for backpressuring both the reservoir and pneumatic transducer with air pressure. The magnitude of the air pressure is not used in calculations, because it is applied to both ends of the system. This version allows a check to be made for continuity of liquid, by taking readings as the air pressure is increased. In fact, if an initially de-aired system **does** become discontinuous, provided the amount of free gas in the liquid is not excessive, it can often be driven into solution under the backpressure.

The author believes that the precision of gages with pneumatic transducers and aqueous solutions is generally limited to ±0.5 in. (±13 mm). This precision is possible provided that the above guidelines for the diameter of the liquid-filled tube are followed and that temperature variations in the liquid-filled tube are not great (Section 8.2.3). Gages with vibrating wire transducers and mercury can be used to obtain higher precision but are subject to the limitations given in Table 12.9. Until a sufficient bank of data has been obtained to demonstrate precision, the author recommends that users conduct full-scale calibrations prior to field installation.

Gages with Pressure Transducer in Readout Unit

The arrangement is shown in Figures 12.96 and 12.97.

Figure 12.96 shows only two tubes, but in practice it is preferable to have two liquid-filled tubes and two tubes for gas, so that flushing is possible and independent measurements can be made on each liquid-filled tube as a check. The rubber bladder is made slightly larger than the rigid case of the cell and is therefore never in tension. Initially, the liquid-filled tubes are connected to a reservoir to ensure that the rubber bladder is expanded to fill the rigid case. Sufficient gas pressure is then applied to overcome the liquid head H, thereby compressing the liquid slightly, ensuring that all liquid is at a pressure greater than atmospheric pressure and that gas and liquid pressures across the bladder are equal. The magnitude of the gas pressure is not used in calculations, because it is applied to both ends of the system. The change in pressure transducer reading, divided by the specific gravity of the liquid, gives vertical deformation directly.

The gage shown in Figure 12.96 has two major advantages when compared with the gage shown in Figure 12.93. First, a check can readily be made for continuity of liquid, by taking readings as the gas pressure is increased. If two liquid-filled tubes are provided, separate readings provide an additional check. In fact, if an initially de-aired system **does** become discontinuous, provided the amount of free gas in the liquid is not excessive, it can often be driven into solution under the backpressure. Second, the transducer is accessible for checking and recalibration if necessary. Additionally, there is no

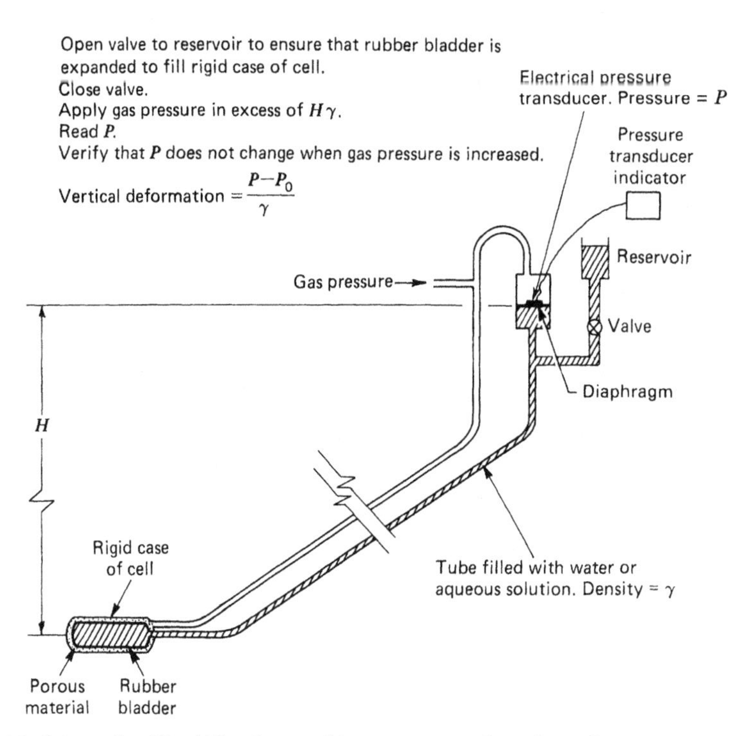

Figure 12.96. Schematic of liquid level gage with pressure transducer in readout unit, with readout unit higher than cell.

error caused by the pressure surge associated with using a pneumatic transducer in the system shown in Figure 12.93.

Precision of these gages is dependent on the same two major factors discussed for the gage shown in Figure 12.93: the pressure transducer must read the liquid head correctly and the liquid must transmit static head correctly. These conditions are easier to achieve when the system is backpressured and when the transducer is accessible, and therefore the gage shown in Figure 12.96 is preferred. A liquid-filled tubing inside diameter of between 0.17 and 0.25 in. (4.3–6 mm) is a good choice when aqueous solutions are used. A precision of ±0.25 in. (±6 mm) appears to be possible, provided that the above guidelines for the diameter of the liquid-filled tube are followed and that temperature variations in the liquid-filled tube are not great (Section 8.2.3).

Overflow Gages

The *overflow gage* shown in Figure 12.91 can be converted for use with the readout unit higher than the cell, by applying either a measured suction to the readout end of the liquid-filled tube or a measured backpressure to the vent tube.

The former version is shown in Figure 12.98 and described by Penman (1982) and Penman et al. (1975). The 1982 paper describes a revised method of reading, and this should be used in preference to the 1975 method. This gage is limited to an elevation difference between the cell and readout unit of about 15 ft (4.6 m): at a greater distance the liquid tends to become discontinuous. Precision is typically ±0.4 in. (±10 mm).

The latter version is described by Dunnicliff (1968) and has duplicate air tubes both to allow circulation of air and to give two independent means of measuring air pressure. This system allows measurements to be made at up to 100 ft (30 m) elevation difference between the cell and readout unit, and precision is ±0.8 in. (±20 mm).

12.10.3. Single-Point Gages with Readout Unit Lower than Cell

The backpressured gage shown in Figures 12.96 and 12.97 can be used with the readout unit lower than the cell. As an alternative, the *overflow gage* shown

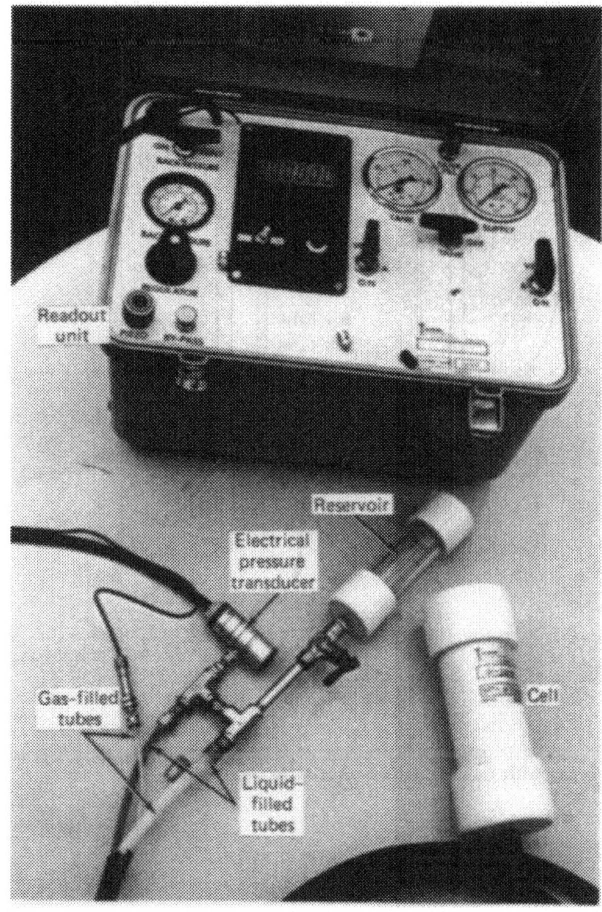

Figure 12.97. Liquid level gage with electrical pressure transducer in readout unit (courtesy of Thor International, Inc., Seattle, WA).

in Figure 12.98 can be used with the readout unit lower than the cell by applying a measured pressure to the readout end of the liquid-filled tube (Penman, 1982).

12.10.4. Multipoint Gages

Various *multipoint gages* have been developed, but most users prefer to install several single-point gages, usually because failure of a single gage in a multipoint system may result in loss of the entire system. However, a multipoint gage that consists of a series of interconnected liquid-filled chambers, placed at a similar elevation, is available and useful and is described in the following paragraphs.

Figure 12.99 shows a version manufactured by Geokon, Inc., recently developed for monitoring vertical deformation along a bench on the slope of an open pit mine. The DCDTs are connected to a console and can be arranged to sound an alarm in the event that a predetermined settlement is exceeded at any chamber. There are 11 chambers and interconnecting pipework, all covered by about 3 ft (1 m) of fill to minimize inaccuracy caused by temperature variations. The liquid is a 3 : 2 ethylene glycol/water mix, and precision over a 1000 ft (300 m) long system is about ±0.02 in. (±0.5 mm). The version is based on an original design by the Building Research Establishment in England (Ward et al., 1968), which achieved a precision of ±0.004 in. (±0.1 mm) during a loading test on chalk, by using a magnifying lever on each float. Precision of all versions is affected by temperature variations, which were greater at the open pit mine than the loading test on chalk.

Readers may question why this system used such a large-diameter liquid-filled pipe (nominal diameter 3 in., 76 mm), whereas for single-point systems 0.25 in. (6 mm) **inside** diameter tubes are recommended so that continuity of liquid is ensured. The problem of filling a near-horizontal large-diameter pipe with liquid was demonstrated during recent tests in Colorado. A 1.5 in. (38 mm) nominal diameter PVC pipe was laid on reasonably level ground around a building, with the two ends terminating in vertical risers alongside each other. Pouring water in one riser, supposedly to fill the pipe, resulted in unrepeatable relative water levels in the two risers, and levels were up to 1.5 in. (38 mm) different, indicating errors resulting from breaks in continuity of liquid. If a pipe or tube with an **inside** diameter larger than 0.25 in. (6 mm) is used, trapped air must be excluded in one of two ways. First, outlets for air bleeding must be provided on the top of the pipe at any point where a crest may form, and in any event no further apart than about 100 ft (30 m). Second, the pipe should be installed with a pronounced slope toward each end (i.e., in a vertical U or V shape). The first approach is more reliable and is used in the arrangement shown in Figure 12.99, the connectors to the floats acting as air bleeds. The author believes that the arrangment shown in Figure 12.99 would be satisfactory during the short-term if a 0.25 in. (6 mm) **inside** diameter liquid-filled tube had been used instead of the larger-diameter pipe, but that occasional flushing with de-aired liquid would be needed to maintain continuity of liquid in the long-term. In summary, if air bleeds or pronounced slopes can be provided (with confidence that their effectiveness will not be reduced by shape

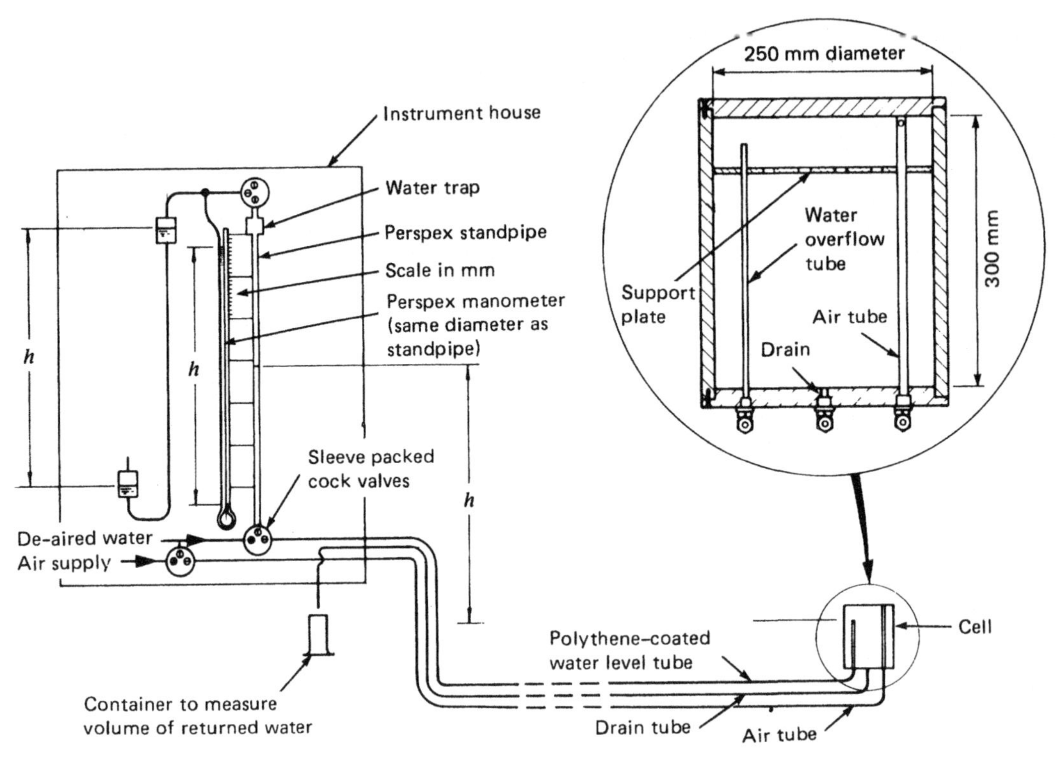

Figure 12.98. Schematic of liquid level gage, overflow type, with readout unit higher than cell (after Penman, 1982).

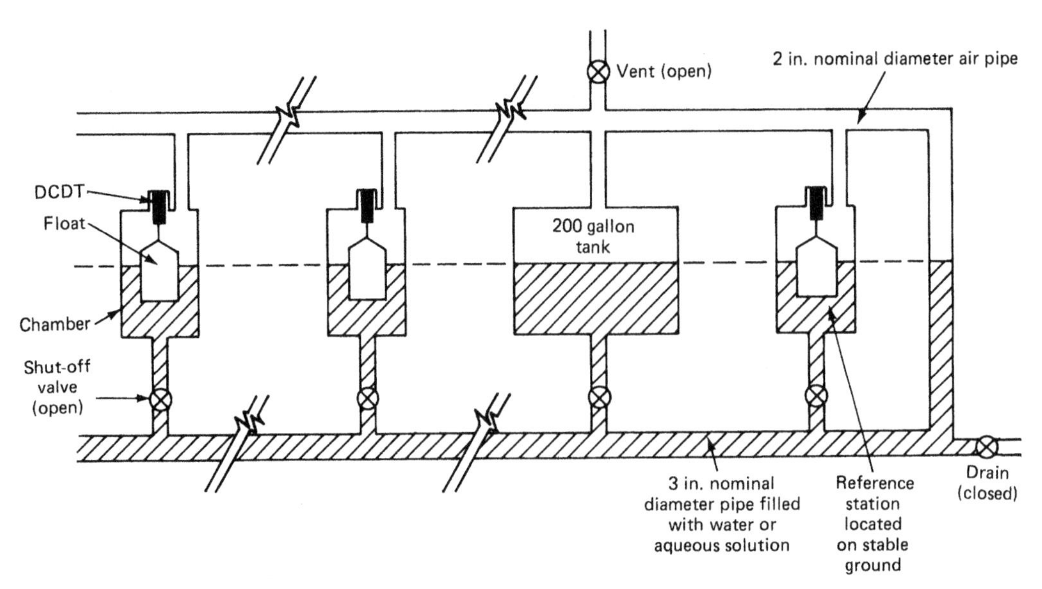

Figure 12.99. Schematic of multipoint gage (courtesy of Geokon, Inc., Lebanon, NH).

changes as vertical deformation occurs), a large-diameter pipe can be used. If not, a 0.25 in. (6 mm) **inside** diameter tube should be used, with occasional flushing.

Müller et al. (1977) describe a system similar to the arrangement shown in Figure 12.99, incorporating a capacitance transducer. The instrument is manufactured by Interfels. Geonor manufactures a system that uses a vibrating wire transducer to measure the buoyancy of a partially submerged float.

Two similar vibrating wire/float systems have been used satisfactorily in Russia for many years: first, the arrangement with separate air- and liquid-filled pipes as shown in Figure 12.99 and second, a single pipe of about 12 in. (305 mm) diameter, with connections to float chambers at intervals along the pipe. The connection for air enters the pipe at the top, the connection for liquid at the bottom, and the pipe remains partly filled with liquid.

12.10.5. Full-Profile Gages

Most *full-profile gages* consist of a near-horizontal plastic pipe and an instrument that can be pulled along the pipe. Readings are made at points within the pipe, and the entire vertical profile can be determined. Differences in vertical profile with time provide data for determination of vertical deformation.

These gages are particularly appropriate where vertical deformation is likely to be nonuniform, such that many single-point gages would otherwise be required. They provide the same data as an inclinometer used within horizontal inclinometer casing, and in fact an inclinometer may often be the instrument of choice if high accuracy is required. Most gages can also be used for surveying elevations along a near-horizontal or inclined borehole, or along the invert of a pipeline or culvert. Survival records are generally excellent, since no delicate parts are buried, and the instruments can be checked on a day-to-day basis and any malfunctions corrected in the laboratory. The expensive and calibrated part of the system is portable and can be used at several locations on one project or on several projects.

The distance of the instrument from one end of the pipe is established from graduations on a traction line and must be controlled carefully. For example, if part of the pipe is inclined at an angle of 10 degrees to the horizontal when using an aqueous solution in the system, a longitudinal positioning error of 1 in. (25 mm) will cause a measurement error of about 0.17 in. (4.3 mm). If maximum precision is required, the pipe should be as horizontal as possible. If the pipe is also arranged as a probe extensometer, both vertical and horizontal deformations can be monitored, and probe extensometer data can be used to control longitudinal positioning of the full-profile gage.

Various types of full-profile gage are described in the following subsections.

Overflow Gage

The cell of the overflow gage shown in Figure 12.91 is shaped as a probe (Penman, 1982; Penman and Charles, 1982). The pipe, typically 2.5 in. (63 mm) Sch. 40 PVC, is inclined slightly upward away from the readout and is free-draining.

Gages with Pressure Transducer in Probe and Attached Liquid-Filled Tube

Most of the single-point gages based on the arrangement shown in Figure 12.93 can be manufactured as full-profile gages using pneumatic, electrical resistance strain gage, or vibrating wire pressure transducers, with aqueous liquids or mercury. Typical pipe diameter is 1.5 in. (38 mm) nominal; requirements for tubing, transducer, and readout unit are as discussed previously for single-point gages, and precision is similar. When used as a full-profile gage, the liquid-filled tube should be transparent, so that regular inspections can be made for continuity of liquid. If any gas is observed in the liquid, the gage should be flushed with fresh de-aired liquid. Reading correctness should be verified on each day prior to any field readings, by using the gage to measure a known elevation difference.

Bergdahl and Broms (1967) describe a full-profile gage, shown schematically in Figure 12.100. With the readout unit higher than the probe, air pressure is slowly applied to the inside of the bladder until a small quantity of liquid returns to the readout unit, indicating an air/liquid pressure balance across the bladder. In the original version the air tube was contained coaxially within the liquid-filled tube, thus creating a large wetted surface for the liquid. This feature created a long time lag between application of air pressure and stabilization of the free liquid surface, and precision was poor when tubes were longer than about 100 ft (30 m). An improved version, manufactured by Water Nold Company, Inc. under the trade name *Aquaducer*™ (Figure 12.101), uses separate tubes and a 0.25 in. (6 mm)

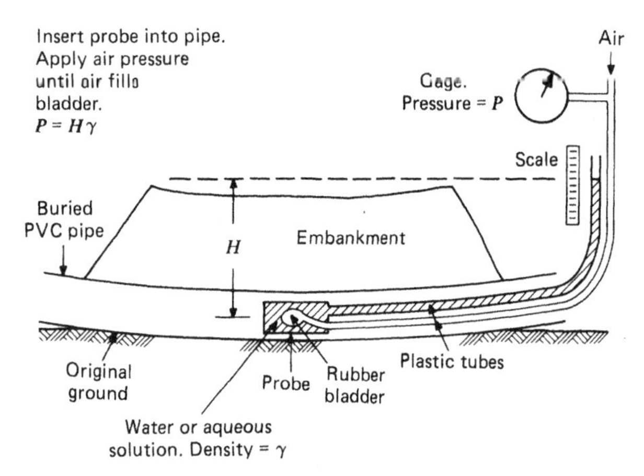

Figure 12.100. Schematic of full-profile gage, with air/liquid pressure balance across a bladder.

inside diameter liquid-filled tube. Standard tubing length is 500 ft (150 m), allowing profiles of up to 1000 ft (300 m) to be surveyed if access is available to both ends of the pipe. The elevation difference between the probe and readout is limited by the strength of the bladder to about 10 ft (3 m).

Gages with Pressure Transducer in Readout Unit and Attached Liquid-Filled Tube

The single-point gage shown in Figure 12.96, with the pressure transducer in the readout unit, can be manufactured as a full-profile gage, as illustrated in Figure 12.102. Because the liquid is backpressured, this version is preferable to the version with a pressure transducer in the probe.

Gages with Pressure Transducer in Probe and without Attached Liquid-Filled Tube

The types of full-profile gages described previously have attached liquid-filled tubes and therefore tend to be cumbersome. Also, the liquid in the liquid-filled tube may be subject to large temperature changes during insertion and withdrawal from the pipe, with consequent thermal errors. In an effort to overcome these limitations, several gages have been developed whereby a pressure transducer is pulled along a liquid-filled pipe or tube. The arrangement is shown in Figure 12.103.

A version described by Bozozuk (1969) used an unbonded resistance strain gage pressure transducer within a 1 in. (25 mm) nominal diameter pipe. The instrument is subject to significant errors (Tao, 1979), including error caused by temperature sen-

Figure 12.101. ***Aquaducer*™ full-profile liquid level gage (courtesy of Water Nold Company, Inc., Natick, MA).**

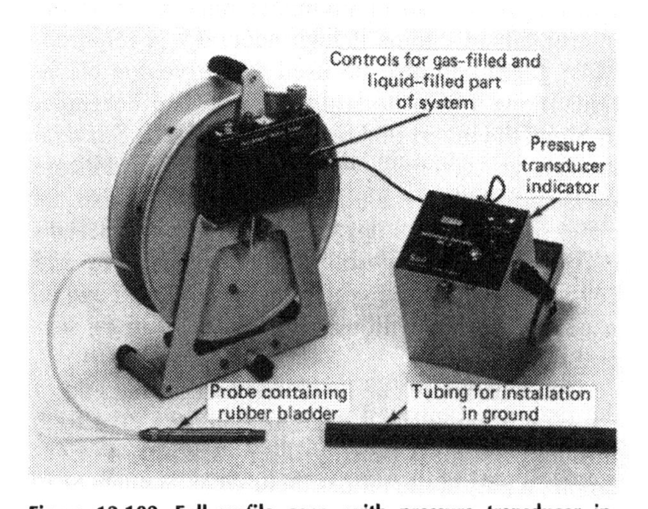

Figure 12.102. Full-profile gage, with pressure transducer in readout unit, and attached liquid-filled tube (*hydrostatic profile gauge*) (courtesy of Soil Instruments Ltd., Uckfield, England).

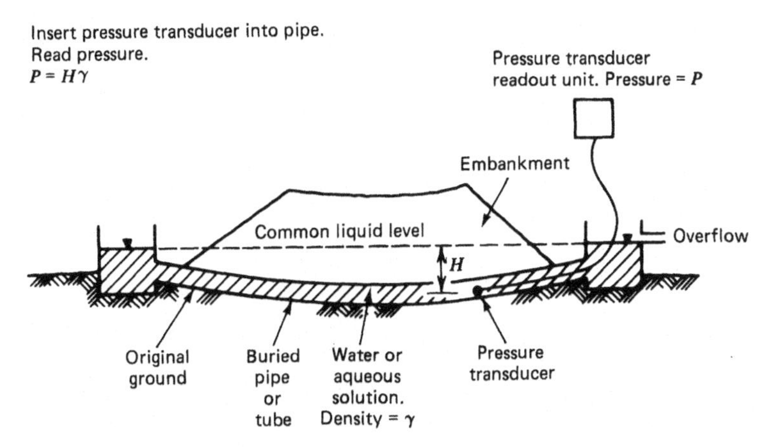

Figure 12.103. Schematic of full-profile gage, without attached liquid-filled tube.

sitivity of the pressure transducer, and it has not been widely used. However, with the advent of small stable bonded and unbonded resistance strain gage, vibrating wire, and pneumatic pressure transducers, this arrangement should be applicable in cases where access is available to both ends of the pipe or tube, where temperature variations are not great, where the barometric pressure at each end of the system is substantially the same (i.e., where air currents are not great), and where complete liquid-filling can be maintained. As discussed earlier, the last condition requires outlets for air bleeding, pronounced upward slopes toward the ends of the pipe, or a 0.25 in. (6 mm) **inside** diameter tube.

The author is not aware of commercially available versions with stable transducers, but Audibert (1985) reports on the successful development and use, by the Houston, TX, office of The Earth Technology Corporation, of a system using a miniature piezoresistive pressure transducer within 0.25 in. (6 mm) **inside** diameter thick wall tubing. The pressure transducer was Model No. 8507-2, manufactured by Endevco Corporation, San Juan Capistrano, CA, with an outside diameter of 0.092 in. (2.3 mm), a range of 0–2 lb/in.^{2} (0–14 kPa), a burst pressure of ±40 lb/in.^{2} (275 kPa), and a rated temperature sensitivity of 0.003 lb/in.^{2}/°F (0.037 kPa/°C). The system was used to make settlement measurements within a block of soil during a large-scale model test, conducted in a non-air-conditioned warehouse, and achieved a precision of ±0.05 in. (±1.3 mm). This high precision indicates that the transducer was in a very uniform temperature environment: the rated temperature sensitivity corresponds to an error of 0.83 in. per 10°F (38 mm per 10°C) and illustrates the difficulty of achieving high precision in a field environment. However, future improvements in pressure transducers may make this system more applicable for general field use.

It is interesting to note that the **inside** diameter (0.25 in., 6 mm) chosen by Audibert for the tubing is the same as the inside diameter recommended by the author in Section 8.2.3 and that therefore Audibert had no difficulties caused by discontinuity of liquid.

The author recommends that, whenever the type of gage shown in Figure 12.103 is used, instrument data should be compared with elevation data determined by surveying methods before backfilling over the pipe or tube and that repeatability tests should be made before accepting data. In most cases it may be more prudent to use the type of gage with an attached liquid-filled tube.

The gage without an attached liquid-filled tube is particularly suitable for surveying elevations along downward inclined boreholes. If the borehole is watertight, the survey merely involves lowering the pressure transducer through the water on a graduated cable and recording and plotting pressure and cable graduation at intervals to the bottom of the borehole. If the borehole is not watertight, a plastic pipe with a bottom cap can be inserted temporarily and filled with water.

Double Fluid Settlement Gages

A *double fluid settlement gage* has been developed by the Road Research Laboratory in England (Irwin, 1964). A buried reservoir containing mercury is pressurized by air during reading and the mercury driven along a 0.11 in. (2.8 mm) **inside** diameter nylon tube. The position of the mercury is detected by

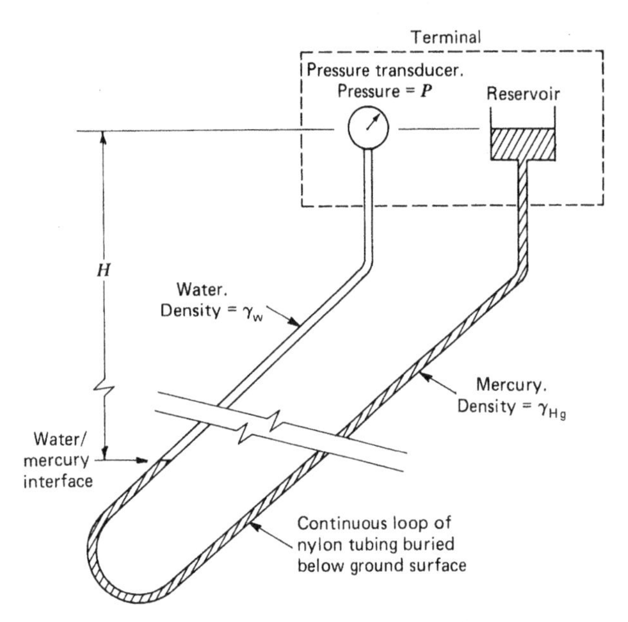

Figure 12.104. Schematic of double fluid settlement gage.

electrical contacts in the tube and settlement is determined by measuring the air pressure. Later, the design was modified to create a multipoint system with intermediate measuring points (Irwin, 1967). Estimated errors were less than ±0.1 in. (±2.5 mm) over distances up to 200 ft (60 m) between the reading station and the mercury reservoir.

Subsequently, a double fluid settlement gage was developed by the project engineers during construction of Tarbela Dam in Pakistan and subsequently named the *TAMS double fluid settlement device (DFSD)*. The gage is described by Clements (1978) and Clements and Durney (1982), and some measurement results are given by Szalay and Marino (1981). The operating principle is shown in Figure 12.104.

A continuous loop of 0.11 in. (2.8 mm) **inside** diameter polyethylene-sheathed nylon tubing is installed in a near-horizontal trench, and the ends of the loop are terminated at a common accessible point. Prior to measurement the tubing is filled with de-aired water. During the measuring phase a water/mercury interface is formed, and mercury is allowed to bleed into the tubing to advance the interface. By maintaining the free surface of the mercury at a constant level and monitoring the pressure on the top of the water column as shown in Figure 12.104, a continuous record of the elevation of the interface is obtained. The position of the interface is determined by measuring the volume of water forced out of the tubing, and a confirmation of the interface position can be obtained by creating small discrete risers at several points in the tubing during installation and using these as indicators. On completion of the measuring sequence, the mercury is removed from the installed tubing and replaced by water.

At Tarbela Dam the length of individual tubes exceeded 4000 ft (1220 m), and clearly no other gage could have provided a continuous settlement profile for this application. Since its original development, the system has been improved and used successfully in several embankment dams, notably for determining the settlement profile across zones of different compressibilities, for example, between the core and downstream filter. Clements and Durney (1982) and Clements (1984) describe three versions of the DFSD:

1. A manually operated system similar to the original development, designed to operate over tubing lengths up to approximately 4000 ft (1220 m) and to accommodate elevation differences up to 20 ft (6 m). The terminal cannot be higher than 20 ft (6 m) above the lowest part of the tube, otherwise pressures become excessive. Precision is dependent on three factors. First, the elevation differences in the system: the smaller the differences, the greater the precision. Second, the length of tubing: the longer the tubing, the less is the precision, owing to inertia and wall friction. Third, the expertise and patience of the operator: under average conditions a precision of ±0.4–1.5 in. (±10–38 mm) is typical.
2. A portable manual system, designed to operate over tubing lengths of up to 650 ft (200 m) and to accommodate elevation differences less than 4 ft (1.2 m). A precision of approximately ±0.1 in. (±3 mm) is achievable.
3. An automatic system (Figure 12.105), designed to accommodate tubing lengths up to 3300 ft (1000 m) and elevation differences in excess of 10 ft (3 m). A precision of ±0.4 in. (±10 mm) is achievable.

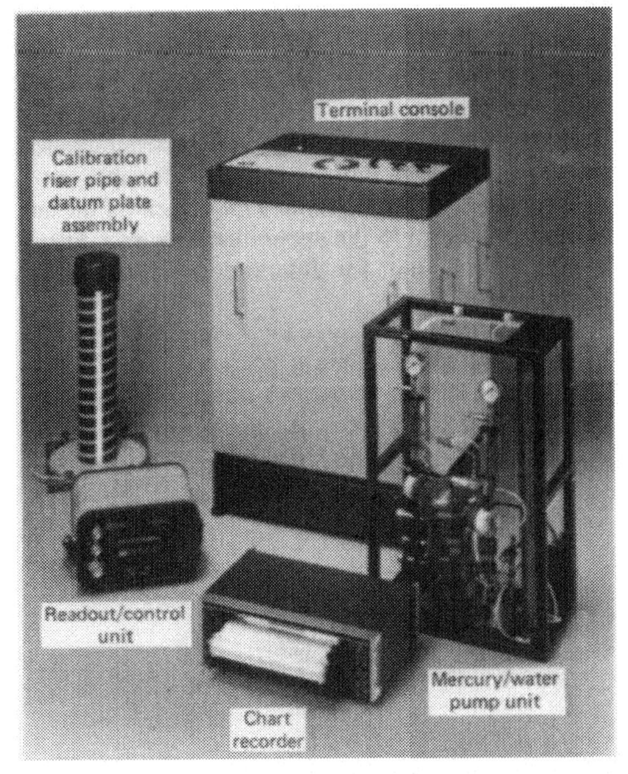

Figure 12.105. Automatic plotter system for double fluid settlement gage (courtesy of Soil Instruments Ltd., Uckfield, England).

Clements (1984) recommends that the slope of the tubing should normally not exceed 1 in 10, to prevent the possibility of interface breakup during the stop/start transitions when the instrument is read manually. However, interface breakup is normally encountered only when the operator allows an excessive rate of mercury feed. This concern does not exist with the continuously feeding automatic system, and it has been demonstrated that no breakup occurs even when the interface is forced to "loop the loop" in a vertically coiled tube.

When installing tubing for the DFSD, a duplicate length should be installed alongside the primary tube for use in the event of damage by application of excessive pressure.

12.10.6. Filling and Flushing Liquid-Filled Tubes

Two accessible liquid-filled tubes, with appropriate fittings and valves, are strongly recommended for all systems that are left in place, so that they can be flushed if discontinuity of liquid is suspected. Recommendations for tubing material, fittings, and liquid are given in Section 8.2.3.

In general, initial filling of liquid-filled tubes is best done by applying a vacuum at one end and allowing liquid to enter from the other end. This procedure reduces the amount of air in the tube, therefore reducing the chance of breaking continuity of the liquid, and minimizes the time required. During subsequent flushing, it is usually best to blow all liquid out of the tube under air pressure and introduce fresh liquid as described above.

In the United States many manufacturers complete the initial filling of liquid-filled tubes before shipment to the user. This practice runs the risk of the liquid becoming discontinuous during shipment and may require that special measures be taken to prevent damage to pressure transducers or other components subjected to the pressure of the liquid. In Europe it is more common practice to complete the initial filling in the field. This practice requires appropriately competent field personnel and may require filling and subsequent emptying at the factory so that factory calibrations can be made. Clearly, there are points in favor of both approaches, but in general the author recommends the European practice.

Nylon tubes are likely to require more flushing than polyethylene tubes during their initial use. Nylon tends to absorb water and leach gas until it becomes fully saturated with water, and therefore one or two additional flushes are often needed before the system can be fully commissioned.

12.10.7. Recommendations for Choice of Liquid Level Gage

It is not possible to make definitive recommendations for choice of a liquid level gage, and any one of the types described above may on occasion be the instrument of choice. The selection depends on the application, site-specific conditions and needs, the number of measuring points required, the relative elevations of measuring points and readout unit, availability of and familiarity with hardware, required precision, and the general factors given in Section 4.9 and the more specific factors given in this section and in Table 12.9.

12.10.8. Installation of Liquid Level Gages

Installation of liquid level gages in boreholes and in fill should follow the guidelines given in Chapter 17. When installed in boreholes, tubing can be prespiraled, as described in Section 17.5.1, to avoid damage caused by large vertical compression of the surrounding soil.

12.11. MISCELLANEOUS DEFORMATION GAGES

Various deformation gages do not fit readily into previously described categories: telltales, convergence gages for slurry trenches, time domain reflectometry, fiber-optic sensors, and acoustic emission monitoring.

12.11.1. Telltales

When a sleeved rod or wire is attached to an inaccessible point, routed to an accessible point, and used with a transducer for monitoring the changing distance between the two points, the device is often referred to as a *telltale*. When a telltale is installed in the ground, it is the same as a single-point fixed borehole extensometer, described in Section 12.7. Telltales can also be installed in or on structures for monitoring relative deformation, for example, on a tieback anchor for determining movement of the anchor with respect to the anchor head or on a driven pile or drilled shaft for determining tip settlement during a load test.

As described in Chapter 13, *multiple telltales* can be used for determination of strain and load in structural members. Figure 12.106 shows one of the telltales used for this purpose during a recent drilled shaft load test with which the author was involved.

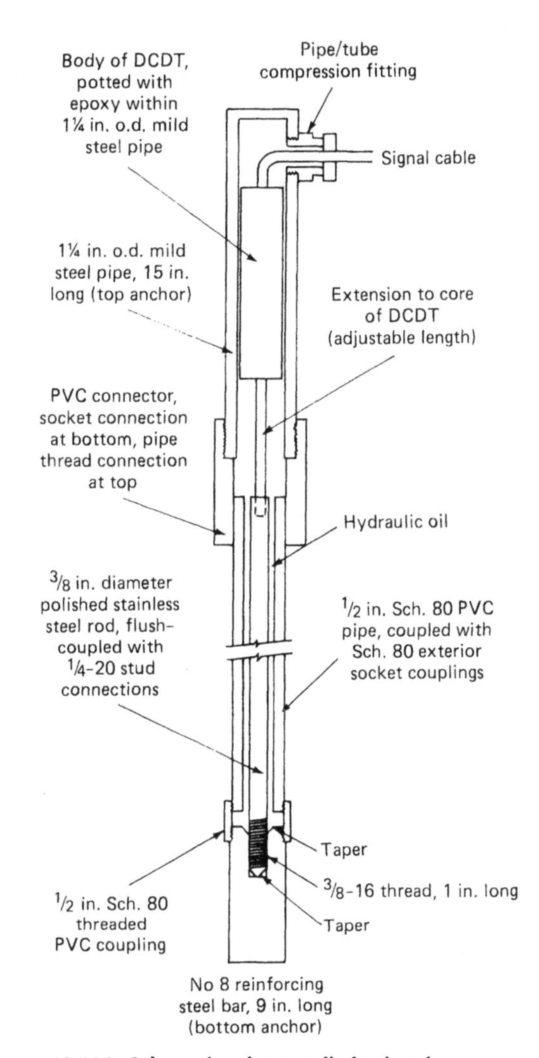

Figure 12.106. Schematic of one telltale that forms part of a remotely-read multiple telltale system for a drilled shaft load test.

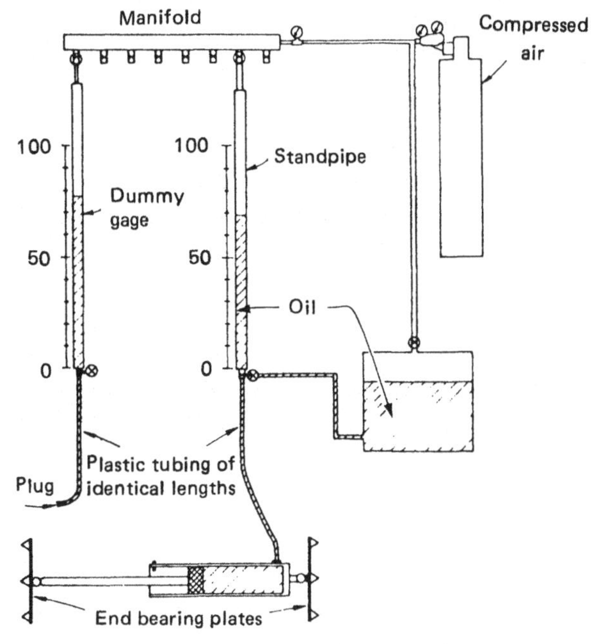

Figure 12.107. Schematic of hydraulic gage used to measure convergence of slurry trench (after DiBiagio and Myrvoll, 1972).

The top of the top anchor was installed about 12 in. (305 mm) below the shaft butt.

12.11.2. Convergence Gages for Slurry Trenches

Methods of monitoring stability in relation to time, while conducting full-scale tests of slurry trench excavations, are outlined in Chapter 19. The methods include monitoring closure of the trench. Two types of gage are available: the first gage is applicable if reinforcing steel is installed in the excavation, the second gage if reinforcing steel is not installed.

A *hydraulic gage* is described by DiBiagio and Myrvoll (1972) and shown in Figure 12.107. It consists of a piston within an oil-filled piston chamber, set horizontally across the trench. End bearing plates contact opposite walls of the trench, one attached to the end of the piston rod, the other to the opposite end of the assembly. A standpipe rises vertically from the piston chamber. A reduction in the width of the trench causes movement of the piston and an upward flow of oil into the standpipe; thus, the level of oil in the standpipe can be related to the width of the trench. The gage is attached to the reinforcing cage prior to installation in the trench.

An alternative instrument was used successfully in a recent full-scale test with which the author was involved and is shown in Figure 12.108. The instrument incorporated a *soil strain gage,* as described in Section 12.6.5. Coils 15 in. (380 mm) in diameter were embedded in opposite sides of the trench using a double-acting hydraulic jack, supported on orientation rods. Prior to installation five plastic tent pegs were attached to the back of each coil, the jack retracted, the coils supported on opposite ends of the jack, the jack and coils attached to the orientation rods and lowered into the trench, the jack actuated to drive the tent pegs into the soil until the coils were at the surface of the trench wall, and the jack retracted and withdrawn. Measurements were taken while the trench was filled with slurry, after concrete pouring, and after concrete set, with a pre-

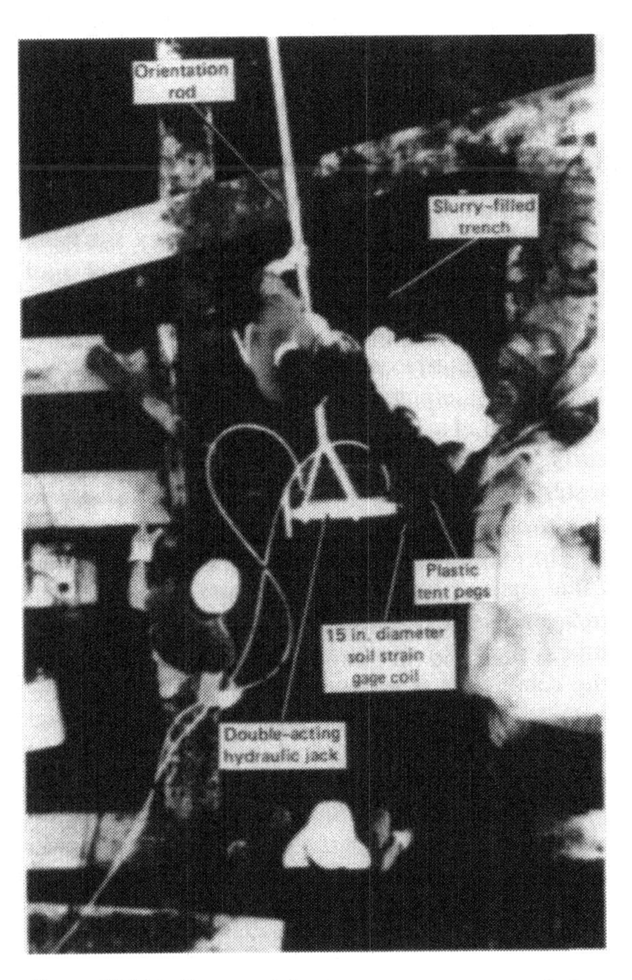

Figure 12.108. Gage used to measure convergence of slurry trench, based on *soil strain gage* transducer.

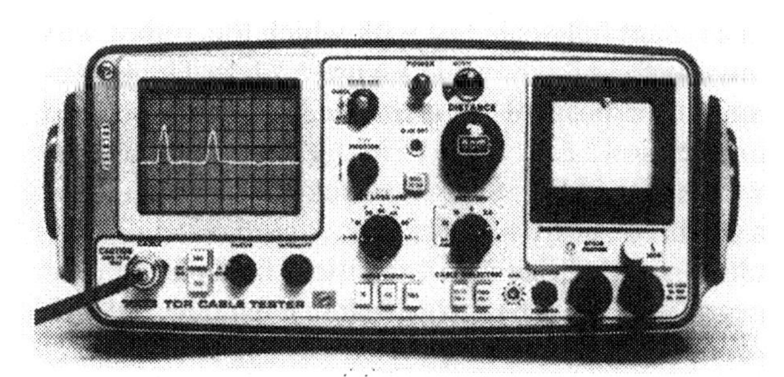

Figure 12.109. Time domain reflectometry (TDR) cable tester: Tektronix Model 1503 (courtesy of Tektronix, Inc., Beaverton, OR).

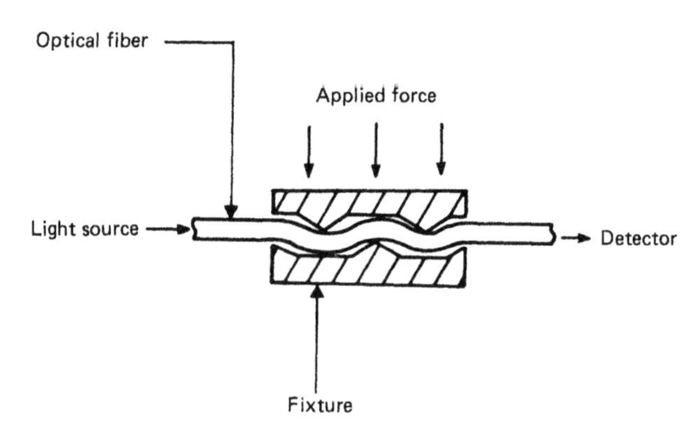

Figure 12.110. Fiber-optic microbending sensor (after Krohn, 1983). Reprinted by permission. Copyright © 1983, Instrument Society of America.

cision of approximately ±0.1 in. (±3 mm). Data were not influenced by the presence of slurry or concrete. This alternative would not be possible in a trench with reinforcing steel, because the steel would influence the output of the induction coil transducers.

12.11.3. Time Domain Reflectometry

Time domain reflectometry (TDR), originally developed to locate breaks in power line cables, has been used to monitor the successive collapse of roof strata after underground mining and the propagation of the resulting cave toward the surface (O'Connor and Dowding, 1984; Wade and Conroy, 1980). The equipment consists of a coaxial electrical cable grouted in a vertical borehole from the ground surface to the mine and a standard TDR cable tester* (e.g., Figure 12.109). The cable tester is used to transmit an electrical pulse along the cable and to monitor the return signal, and faults in the cable such as crimps, short circuits, or breaks are indicated as characteristic signals on a cathode ray tube screen and on a paper record. The distance to the cable fault is proportional to the elapsed time between transmission and arrival of a reflected signal. Accuracy is about 2% of the distance between the tester and the cable break. This can be improved by precrimping the cable at 10 ft (3 m) intervals: the crimps distort the signal and act as markers on the arrival waveform, to which any further distortions or breaks may be related.

*Commercial sources are given in Appendix D.

12.11.4. Fiber-Optic Sensors*

Fiber-optic sensors (Davis, 1985; Davis et al., 1982; Kersten and Kist, 1984; Krohn, 1983) depend on the ability of the fibers to carry light from a source to a photosensitive detector. Fiber-optic sensors can be used to sense the relative position between an object and the end of a fiber or the distance between two points along a fiber; they can also indicate bending. They are unaffected by temperature or humidity extremes and are immune to electrical noise. Small size and ability to transmit light along curved paths can provide access to normally inaccessible areas, and reliability is high because most sensors are passive.

The author is not aware that fiber-optic sensors have been used in geotechnical applications, but they appear to have good potential for monitoring deformation both along and transverse to the fiber. When optical fibers bend, small amounts of light are lost through the walls. By using the microbending sensor shown in Figure 12.110, changes in the intensity of received light can be related to the magnitude of transverse deformation. By using a pulsed system, axial deformation and bending can be monitored at **all** points along a fiber. Fiber-optic sensors therefore have the potential for providing the same information as a combination of an inclinometer together with a probe or fixed borehole extensometer, although precision has not yet been proved. A real-time *continuous fiber-optic strain monitoring sys-*

*Written with the assistance of Richard W. Griffiths, G2 Consultants, Pacific Palisades, CA.

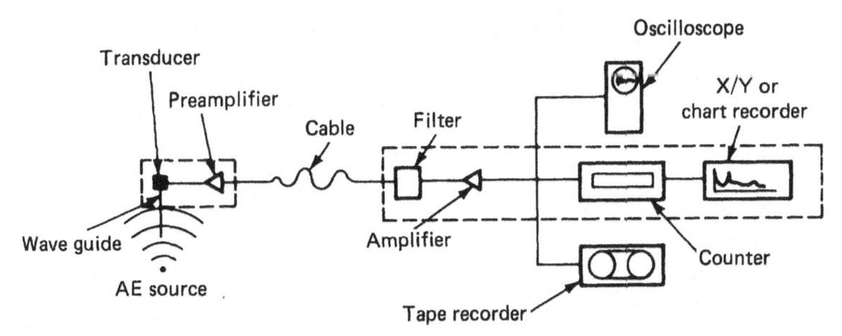

Figure 12.111. Schematic of basic single-channel acoustic emission monitoring system for recording total counts or count rate (after Koerner et al., 1981). Reprinted with permission from ASTM STP 750. Copyright ASTM, 1916 Race Street, Philadelphia, PA 19103.

tem is currently being developed (McKeehan and Griffiths, 1986; McKeehan et al., 1986; *Oil & Gas Journal,* 1985), and has been patented. Once the system is developed, geotechnical applications are likely to follow.

12.11.5. Acoustic Emission Monitoring*

Acoustic emissions are sounds generated within a soil or rock material that has been stressed and subsequently deforms. Sometimes these sounds are audible, for example, wood cracking, ice expanding, or soil and rock particles abrading against one another, but more often they are not, owing to their low amplitude or high frequency or both.

A piezoelectric transducer is generally used as a "pickup" to detect the acoustic emissions and produces an electrical signal proportional to the amplitude of sound or vibration being detected. The signal is then amplified, filtered, and counted or recorded in some quantifiable manner. Unwanted machine and environmental noise are electronically filtered from the signal or separately quantified and subtracted from the measurements. The counts or recordings of the emissions are then correlated with the basic material behavior to determine empirically the relative stability of the given material. Usually, if no acoustic emissions are present, the material is in equilibrium and therefore stable. If emissions are observed, the material is not in equilibrium and may be in a condition that eventually leads to failure. The technique was originated by the U.S. Bureau of Mines in the 1930s to detect mine pillar, wall, and roof instability. The method is sometimes referred to as *microseismic detection* and *subaudible rock noise monitoring,* but the term *acoustic emission,* or simply *AE,* is becoming the accepted term.

As shown in Figure 12.111, the components of an acoustic emission monitoring system consist of a waveguide to bring the signals from within the ground to a convenient monitoring point, a transducer (geophone, accelerometer, or hydrophone) to convert the mechanical wave to an electrical signal, a preamplifier to amplify the signal if long cable is being used, filters to eliminate undesirable portions of the signal, an amplifier to amplify the signal further, and a quantification system. The dashed lines in Figure 12.111 indicate components that are grouped in self-contained boxes.

Field monitoring efforts in AE have been directed to the following topics:

- Stability assessment of earth embankments
- Standup time of excavations in soil and rock
- Stability of bracing and anchors in ground support systems
- Providing an early warning of subsidence
- Determination of preconsolidation and prestress values in soil
- Detection of seepage, grout penetration, and hydrofracturing

AE is most effective when the amplitude of the signals is high and thus is more effective for rock and cohesionless soil than for cohesive soil. Guidelines for practical use of AE are given by Drnevich and Gray (1981), Hardy and Leighton (1975, 1978, 1981), and Koerner et al. (1976, 1977, 1978).

*Written with the assistance of Robert M. Koerner, Professor, Drexel University, Philadelphia, PA.

CHAPTER 13

MEASUREMENT OF LOAD AND STRAIN IN STRUCTURAL MEMBERS*

13.1. INSTRUMENT CATEGORIES AND APPLICATIONS

Instruments for measuring load and strain in structures fall into two groups: *load cells* and *strain gages*. In each case, the transducers are used to measure small extensions and compressions. Load cells, sometimes also called *dynamometers,* are interposed in the structure in such a way that structural forces pass through the cells, and strain gages are attached directly to the surface of the structure or are embedded within the structure to sense the extensions and compressions in the structure itself.

Typical load cell applications include load testing of piles, drilled shafts, tiebacks and rockbolts, and long-term performance monitoring of tiebacks and end-anchored rockbolts. Load cells are also used for monitoring temporary and permanent loads in prestressing and other cables and in supports for underground excavations, for example, under vertical steel or wood posts.

Strain gages are used where load cells cannot be interposed in the structure for reasons of geometry, capacity, or economy and where load and stress can be calculated with adequate accuracy from knowledge of the relationship between strain and stress. Strain gages are used, for example, to determine temporary or permanent stresses or loads in struts across braced excavations, in rockbolts, bridge members, tunnel linings, retaining walls, slurry walls, sheet piles, and nuclear containment vessels.

Additional applications of load and strain measurements in structural members are given in Part 5.

*Written with the assistance of J. Barrie Sellers, President, Geokon, Inc., Lebanon, NH.

13.2. LOAD CELLS

The types of load cell in general use are described in Sections 13.2.1–13.2.7 and comparative information is given in Table 13.1.

13.2.1. Mechanical Load Cells

Mechanical load cells usually contain either a torsion lever system or an elastic cup spring (Bellville washer) that is deformed during load application. Deformation is sensed by a dial indicator and calibrated to load. A hollow-core load cell with a torsion lever system is manufactured by Proceq SA and is shown in Figure 13.1. The compression of four points around a cylinder is amplified, cumulated, and linked to a single built-in dial indicator. The less accurate cup spring arrangement is available in a hollow-core load cell and is usually referred to as a *disk load cell.* The cell is designed primarily for use on rockbolts.

A mechanical load measuring arrangement,

Table 13.1. Load Cells

Type of Load Cell	Advantages	Limitations	Approximate Accuracy
Mechanical (e.g., Figure 13.1)	Robust and reliable	Requires access to cell	±2–10%[a]
Telltale (Figure 13.19)	Simple Inexpensive Calibrated in place	Requires access to telltale (but can be converted to remote reading device)	±2–10%
Hydraulic (Figure 13.2)	Low profile Remote readout is possible	Requires large-area rigid bearing plates	±2–10%[a]
Electrical resistance (Figure 13.4)	Remote readout Readout can be automated	Low electrical output Lead wire effects Errors owing to moisture and electrical connections are possible Need for lightning protection should be evaluated	±2–5%[a]
Vibrating wire (e.g., Figure 13.6)	Remote readout Lead wire effects minimal Readout can be automated Single-gage versions available for in-line use in tension	Special manufacturing techniques required to minimize zero drift Need for lightning protection should be evaluated	Single-gage versions for in-line use in tension, better than ±2% Multigage versions for general use, ±2–5%[a]
Photoelastic (Figure 13.8)	Robust and reliable	Requires access to cell Limited capacity Requires skill to read Most users prefer more direct numerical reading	±2–5%
Calibrated hydraulic jack (e.g., Figure 13.10)	Readily available	Low accuracy Error usually on unsafe side for load testing Should not be used alone for load measurement	±10–25%
Cable tension meter (e.g., Figures 13.12, 13.13, 13.14)	Removable versions available for use without need to unload cable; one meter can be used on many cables	Removable versions require calibration for each cable type and size	±2–5%

[a] These are accuracies of the load cells when used with adequately designed and installed mounting arrangements. However, because of misalignment of load, off-center loading, and end effects, system accuracy is often no better than ±5–10%, and may be significantly worse if mounting arrangements are inadequate. Guidelines on mounting arrangements are given in Section 13.2.8.

sometimes referred to as a *telltale load cell,* is categorized in this book as a mechanical strain gage and described in Section 13.3.1.

13.2.2. Hydraulic Load Cells

As shown in Figure 13.2, a hydraulic load cell consists of a flat liquid-filled chamber connected to a pressure transducer. The edges of the cell have either a bonded rubber/metal seal or a welded joint. A central hole can permit use with tiebacks or rockbolts. The pressure transducer can be a Bourdon tube pressure gage, requiring visual access to the cell for reading, or an electrical or pneumatic pressure transducer can be used to allow remote reading (Figure 13.3).

Figure 13.1. Mechanical load cell (courtesy of Proceq SA, Zürich, Switzerland).

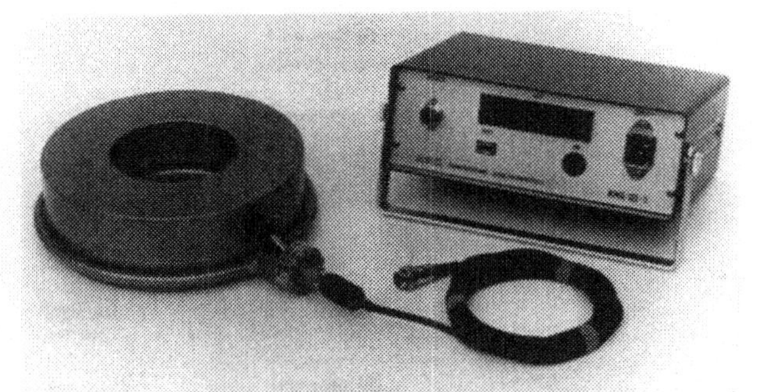

Figure 13.3. Hydraulic load cell with electrical pressure transducer (courtesy of Glötzl GmbH, Karlsruhe, West Germany, and Geo Group, Inc., Wheaton, MD).

Good design results in linear output and relative insensitivity to temperature fluctuations, particularly when installed on a system that is not stiff, such as a tieback. The cell requires very little headroom. Bourdon gage versions used for tieback load cells can easily be read with binoculars from a 100 ft (30 m) distance, so that remote pressure transducers are avoided. Since the cell is flexible, it must be mounted between rigid bearing plates; typical bearing plates used for tiebacks flex excessively, and thicker plates should be used where hydraulic load cells are installed. The bearing pads shown in Figure 13.2 account for any uneven contact between the cell and bearing plates. Non-embossed vinyl floor tile can be used for bearing pad material.

Experience with hydraulic load cells is variable, largely reflecting mounting techniques and detailed design at the edges of the cell. Green (1985) reports that two 600 ton (5.3 MN) cells, manufactured by Glötzl and installed on permanent tendons in a tunnel, have indicated a constant load to within ±1% over a 5 year period. Fifty 70 ton (620 kN) cells, manufactured by Thor International, Inc. to a controlled design, were used to monitor tieback loads on a 150 ft (46 m) deep temporary tieback wall in Seattle. The cells had an outside diameter of 10 in. (254 mm) and a hollow-core diameter of 4 in. (102 mm) and were mounted between bearing plates with an outside diameter of 10 in. (254 mm) and a thickness of 2.5 in. (63 mm). They showed ±5% variation in load over a 6 month period. Green suggests that the load variation was real, caused by thermal and construction-related effects.

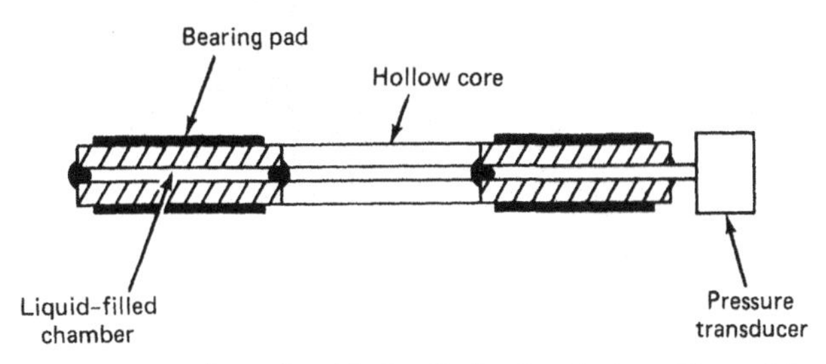

Figure 13.2. Schematic of hydraulic load cell. Note: The cell is circular in plan.

Specially fabricated hydraulic load cells can be used for measurement of load at the tips and sides of driven piles and at the tips of drilled shafts and are classified in this book as contact earth pressure cells. They are described in Chapter 10.

13.2.3. Electrical Resistance Load Cells

Most electrical resistance load cells used in geotechnical applications consist of a cylinder of steel or aluminum alloy, with electrical resistance strain gages bonded to the outer periphery of the cylinder at its midsection as shown in Figures 13.4 and 13.5. Several strain gages are used at regular intervals around the periphery. Half the gages are oriented to measure tangential strains and half to measure axial strains. They are connected to form a single full bridge network, thereby integrating individual strain gage outputs and reducing errors that result from load misalignment and off-center loading. In critical situations, the gages may be duplicated to provide two full bridge networks, one serving as a backup.

Alternative electrical resistance load cell configurations consist of gages bonded inside transverse holes drilled through the cylinder at its midsection. Others consist of three or more gaged cylinders mounted between two steel plates, allowing the height of the cell to be minimized.

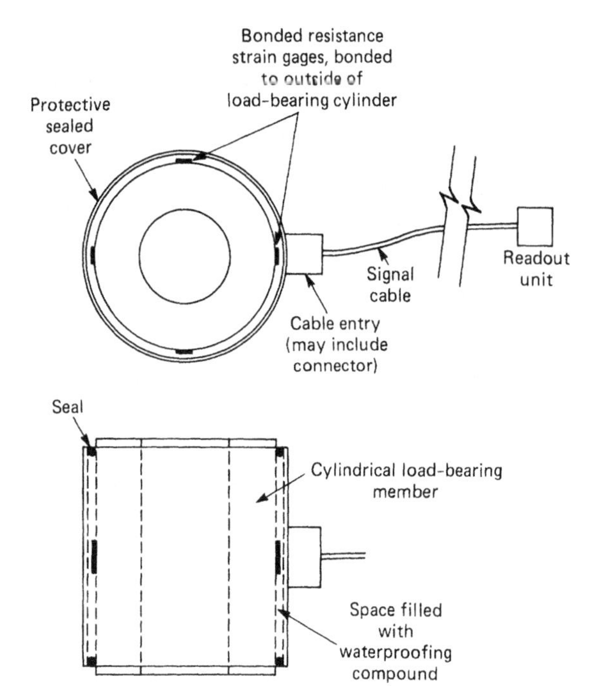

Figure 13.4. Schematic of electrical resistance load cell.

Figure 13.5. Electrical resistance load cells (courtesy of Geokon, Inc., Lebanon, NH).

			Load-Bearing Member		
Cell	Capacity (kips)	Material	Inside Diameter (in.)	Outside Diameter (in.)	Height (in.)
A	400	Steel	4.5	6.5	3.5
B	400	Aluminum	3.0	5.5	4.0
C	300	Aluminum	2.0	4.5	4.0

Various manufacturers of strain gages and process control equipment provide load cells that are not waterproofed for field use, and these should be avoided. Strain gages should be protected from mechanical and water damage by an outer protective steel cover, sealed at the ends with O-rings, and filled with waterproofing compound.

Specially fabricated electrical resistance load cells can be used for measurements of tip load on drilled shafts. These are classified in this book as contact earth pressure cells and are described in Chapter 10.

13.2.4. Vibrating Wire Load Cells

In most vibrating wire load cells, deformation of the load-bearing member is measured by using three or more vibrating wire transducers, and outputs from each transducer must be measured separately and averaged. Models are available with and without a central hole. The arrangement is similar to the electrical resistance load cell shown in Figure 13.4, but with vibrating wire transducers instead of bonded resistance strain gages. Figure 13.6 shows the version manufactured by Maihak. Methods for minimizing errors caused by zero drift and by corrosion of the vibrating wire are discussed in Chapter 8.

Three manufacturers have recently developed load cells for permanent attachment to rockbolts, each containing a single vibrating wire transducer. The Telemac instrument (Bellier and Debreuille, 1977; Figure 13.7) is screwed by the user on to the rockbolt near the free face. The cell is a hollow, high-grade steel cylinder, designed for use with end-anchored rockbolts. Bellier and Debreuille also describe a cell for use with fully grouted rockbolts, in which the normal solid bolt is replaced by a hollow

Figure 13.6. Vibrating wire load cell (courtesy of Maihak AG, Hamburg, West Germany).

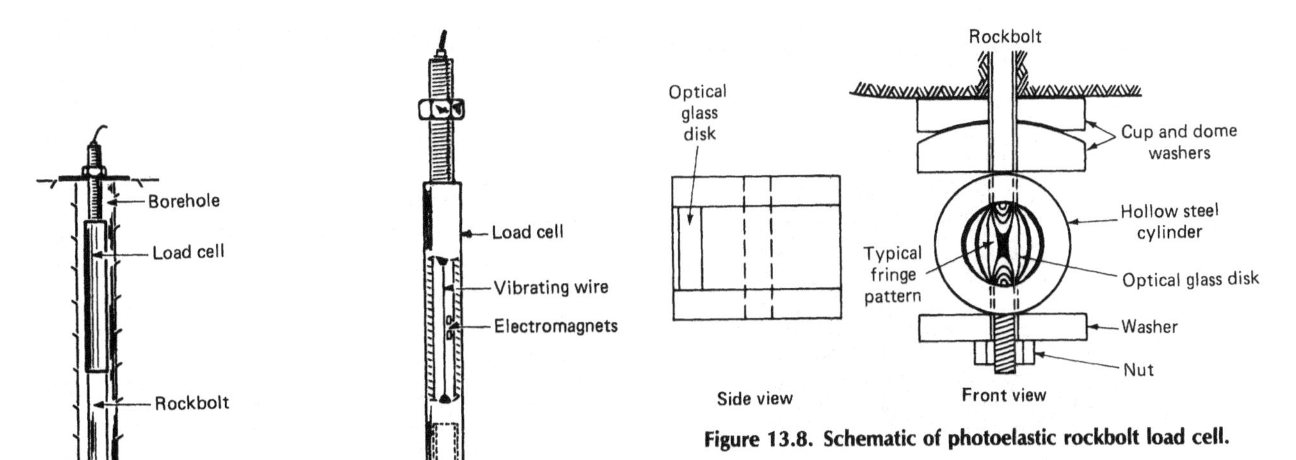

Figure 13.8. Schematic of photoelastic rockbolt load cell.

Figure 13.7. Schematic of vibrating wire rockbolt load cell (after Bellier and Debreuille, 1977).

bolt of similar cross-sectional area, and miniature vibrating wire transducers are fixed within the core at required measuring points, using set screws or spot welds. The instrument manufactured by Irad Gage and Geokon is a complete expansion shell type rockbolt with a vibrating wire transducer mounted in a central longitudinal hole in the bolt near the free face. The gage is available with a contact point on the head of the bolt so that readings can be made with a hand-held probe, but difficulties are sometimes experienced with the contact, and hard-wiring to either a nearby or a remote terminal is preferable. When using any of the three versions of this instrument, individual calibrations are required.

Specially fabricated vibrating wire load cells can be used for measurement of load at the tips of drilled shafts and are classified in this book as contact earth pressure cells. They are described in Chapter 10.

Fellenius and Haagen (1969) describe a vibrating wire load cell capable of withstanding pile-driving forces. A patented feature was adopted to ensure that dynamic load was not transmitted directly through the wire clamping points during the large number of impacts while pile driving.

13.2.5. Photoelastic Load Cells

Photoelastic load cells (Roberts and Hawkes, 1979) have been used, primarily in Europe, for monitoring loads in tiebacks, rockbolts, piles, mine pillars, scaffolding, cables, and chains. They consist of a disk of optical glass, normally enclosed within the end of a hollow steel cylinder. When the cylinder is loaded diametrically, a stress is applied to the glass disk, and strain in the disk produces light and dark areas when the disk is illuminated with polarized light and observed through a hand viewer. The light and dark areas are referred to as *photoelastic interference fringes*. The change in the number of fringes is directly proportional to the change in load on the steel cylinder. Figure 13.8 shows a schematic of a photoelastic load cell, designed for use on rockbolts, that is also suitable for measuring loads in stranded tendon tiebacks (Liu and Dugan, 1972). Figure 13.9 shows the version manufactured by Perard Torque Tension Ltd.

Although a versatile technique, use of photoelastic load cells has been hampered by awkwardness of reading and lack of wide commercial availability, most cells being manufactured on a custom basis. It is suspected that most users prefer a more direct numerical reading method, thus reducing the commercial market for photoelastic cells. However, they are relatively inexpensive, robust, and reliable.

13.2.6. Calibrated Hydraulic Jacks

Calibrated hydraulic jacks (Figure 13.10) are used for the application of load to rockbolts, tieback anchors, piles, drilled shafts, cross-lot struts, and other structural elements. They are also used for later determination of load in cross-lot struts and tieback anchors by *lift-off* testing, by determining the load required to free a loaded strut or anchor from its bearing surface.

Reliance on the measured pressure in the jack

Figure 13.9. Photoelastic load cell (courtesy of Perard Torque Tension Ltd., Worksop, England).

fluid as the sole method for load determination can often lead to significant inaccuracies. Misalignment of load, off-center loading, nonparallel bearing plates, and transverse relative movement of bearing plates all cause friction between the piston and cylinder, so that hydraulic pressure does not give a good indication of load. Temperature changes and pressure gage inaccuracy may cause additional errors. During laboratory calibration, the jack will usually be inserted in a high-quality testing machine, fitted with a spherical seating in the loading head, and rigid bearing plates. The jack will be placed with its axis in line with the loading axis. In the field, however, the jack bearing plates may flex and will rarely be parallel, loading will usually be misaligned and off-center, and consequently friction between the piston and cylinder will be significant. When used as the active member while load testing piles or tiebacks, errors will be on the unsafe side during loading, because the actual load will be less than the load implied from the laboratory calibration. When unloading is accomplished by releasing jack fluid pressure, the actual load will be higher than the load implied from the laboratory calibration.

Figure 13.10. Calibrated hydraulic jack: 150 ton capacity, 6 in. stroke, with center hole (courtesy of Richard Dudgeon, Inc., Stamford, CT).

Davisson (1966) reports on pile load testing using calibrated jacks equipped with swivel heads, where actual load was determined by using high-quality electrical resistance load cells. Jack fluid pressure was also recorded. Error in using the jack calibration was erratic and ranged up to 10%. Davisson concludes: "Even under relatively good conditions, using a swivel head, we must conclude that the probable error is approximately 5%, and varies from zero to 10% erratically. Without the swivel head, the error is probably higher. The foregoing behavior probably explains the [erratic] load–settlement curves often observed for pile tests." Davisson recommends use of both a swivel head and a load cell. Fellenius (1980, 1984) reports on the use of a load cell and a calibrated hydraulic jack during pile load testing and indicates that during loading the jack typically overregisters by 10–25%, and during unloading it typically underregisters by 5%. Figure 13.11 shows typical errors.

Whenever comparisons are made between the two methods of load measurement, it should be remembered that load cells can be in error by 5–10% because of load misalignment, off-center loading, and end effects; therefore, the disagreement between hydraulic jack and load cell does not necessarily result entirely from inaccuracies in hydraulic jack data. Load cell inaccuracy should be minimized by following the guidelines given in Section 13.2.8.

A further source of error in the use of calibrated hydraulic jacks results from incorrect calibration. If the fluid pressure in the jack must be used for load determination, the calibration procedure should model the method of field loading, in which the jack

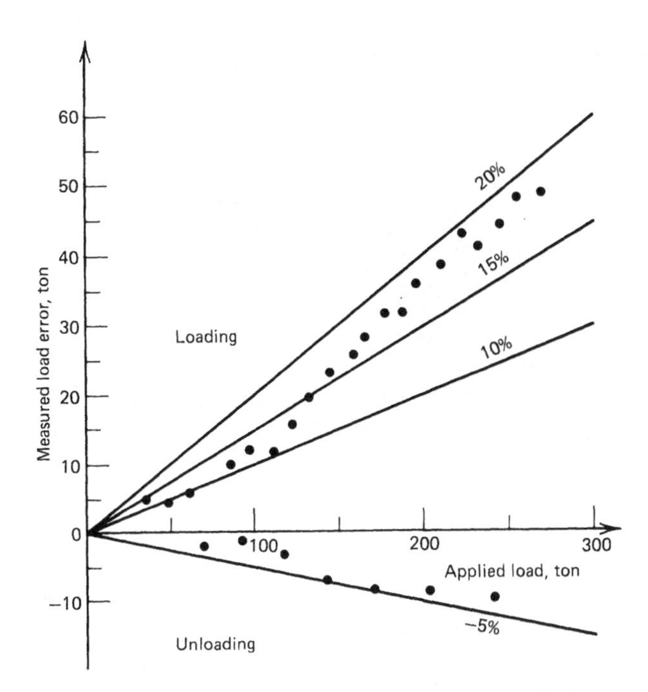

Figure 13.11. Typical error in load determined from fluid pressure in hydraulic jack (after Fellenius, 1984). Reprinted from *Geotechnical News*, Vol. 2, No. 4, December 1984.

is usually in the active mode as the load is both increasing and decreasing. For example, when calibrating a jack for tieback proof testing, the distance between the loading plates of the testing machine should initially be set equal to the expected jack height at alignment load, the jack activated to alignment load, and the testing machine read as the passive member. The distance between loading plates should then be increased to model the expected piston travel to the next load increment, the jack activated to that load, and a calibration reading taken. The procedure should be repeated in increments to the maximum load. Calibration during load decrease should be made in a similar way, with the jack in the active mode.

The following is a useful procedure for checking whether a hydraulic jack is functioning correctly:

1. Calibrate the pressure gage against a dead weight tester.
2. Measure the jack piston diameter (or circumference) and if possible check against the manufacturer's specification or general assembly drawing.
3. Determine the pressure/load calibration from steps 1 and 2.
4. Compare this calibration with the loads indicated by a load cell or testing machine, on loading and unloading cycles. The pressure/load calibration determined in step 3 should lie within the hysteresis loop from this step 4, usually nearer the readings made during the unloading cycle. If it does not, something is wrong.

In summary, load measurements based on jack fluid pressure may be significantly in error, may be on the unsafe side, and are usually unacceptable. Good practice when load testing piles, drilled shafts, and tiebacks includes use of a load cell in series with the hydraulic jack. When piles and drilled shafts are being load tested, a swivel head should also be used. When lift-off tests to determine load in a structural element are being performed, a load cell should be used in series with the jack.

13.2.7. Cable Tension Meters

Vibrating wire, electrical resistance, and photoelastic load cells can be manufactured in a configuration suitable for measuring tension in cables and chains. Figure 13.12 shows a version with an electrical resistance transducer. Two additional devices are manufactured specifically for measuring tension in cables. Both are portable devices for use on cables that are already under tension, and measurements are made without detensioning the cable.

First, the Fulmer tension meter (Hanna, 1985)

Figure 13.12. Cable tension meter with electrical resistance transducer (courtesy of Strainsert, West Conshohocken, PA).

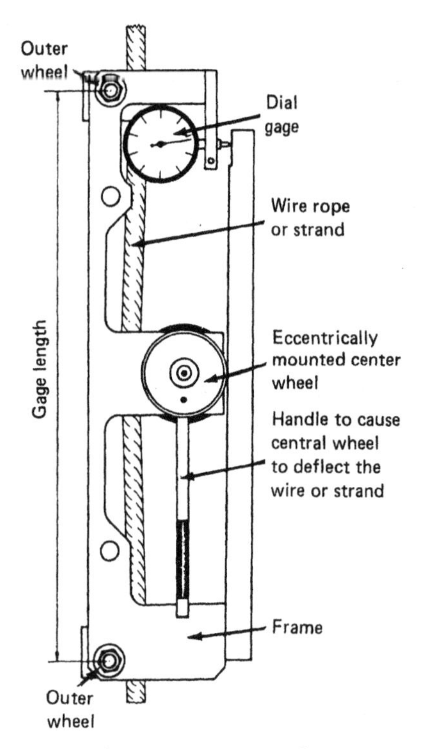

Figure 13.13. Fulmer tension meter (after Hanna, 1985).

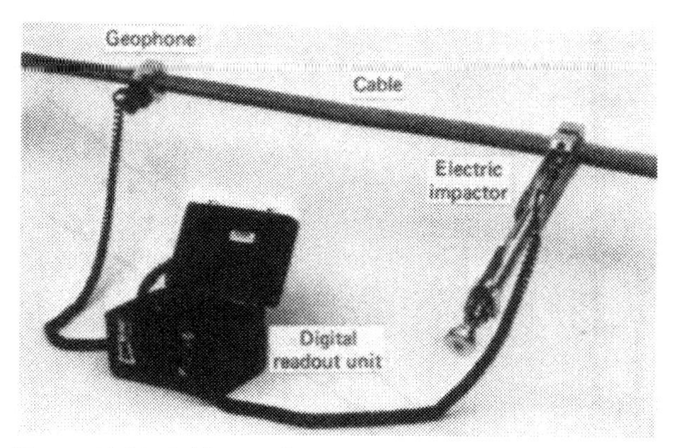

Figure 13.14. Cable Tensiometer (courtesy of Roctest Ltd., Montreal, Canada).

depends on the relationship between the tension in the cable and the normal force required to cause a transverse deflection. As shown in Figure 13.13, a deflecting force is applied via an eccentric wheel, causing the frame of the meter to bend in proportion to the applied force. Bending of the frame is measured with a dial indicator and calibrated to tension in the cable. Calibration charts are required for each cable type and size. Accuracies of ±5% on cable and ±3% on wire are claimed. Second, the Roctest cable tensiometer (Figure 13.14) depends on the relationship between the tension in the cable and the travel velocity of a wave generated by impact. An impactor and geophone are attached to the cable 5 ft (1.5 m) apart, and the time for wave travel is measured. Given the dimensions and mechanical properties of the cable, tension can then be determined from a calibration chart. Accuracy is claimed to be better than ±2% when the tension is at least 15% of the yield strength of the cable, but accuracy decreases rapidly when tension is lower.

13.2.8. Guidelines on Use of Load Cells

General guidelines on instrumentation use are given in Part 4. The following additional guidelines apply specifically to load cell use.

Cell Type, Configuration, and Capacity

Table 13.1 provides guidelines for selecting among the different types of load cell.

The major sources of inaccuracy in hollow-core load cells are off-center loading, misalignment of load, and end effects. Many general-purpose load cells for measuring compressive loads have a solid center, the loads being applied to the cell through a raised center section incorporating a spherical seat to allow for initial lack of fit in the loading setup. This arrangement is not possible with hollow-core load cells, and lack of fit and eccentric loading must be minimized by using swivel bearing washers and/or soft washers. Vibrating wire and electrical resistance load cells must be gaged at several intervals around the periphery so that strains can be averaged to compensate for eccentric loading. End effects can be minimized by using large height/diameter ratios, but if the height is large the cell becomes cumbersome and difficult to use. Most commercially available hollow-core cells have a height approximately equal to the outside diameter of the hollow-core load member and appear to be free from significant end effects. However, there are many load cells on the market that have much smaller height/diameter ratios, their compact design being claimed as an advantage.

The working capacity of the selected load cell should be equal to at least the yield load of the structural member, unless the application or previous experience indicates that substantially less load is expected. The size of the load-bearing member within the cell should be selected so that its elastic limit is at least twice the working capacity. Unwanted capacity of course reduces sensitivity and should be avoided.

Pretreatment of Cell

Residual stresses are usually created in the load-bearing portion of the cell during fabrication. During subsequent stressing, these residual stresses may readjust such that measured strains do not reflect applied stresses, and measurements will be in error. The load-bearing member should preferably be stress relieved by heat treatment before gaging.

Calibration

Load cell calibrations should model field conditions as far as possible, using similar bearing plates, washers, and range of possible off-center loading and load misalignment. Specific factors relating to calibration of hydraulic jacks are discussed in Section 13.2.6. Littlejohn (1981) recommends that, when basic characteristics of a load cell are being established, consideration should be given to the following rigorous tests:

(i) Routine calibration using centric loading and rigid flat platens at 20°C, say.
(ii) As in (i) but using (a) concave inclined platens, (b) convex inclined platens, and (c) 0.3 mm [0.012 in.] thick shims with irregular spacing to simulate uneven bedding [Figure 13.15].
(iii) Eccentric loading between rigid flat platens, with eccentric distance up to 10% of cell diameter.
(iv) Inclined platens up to 1 degree with centric loading.
(v) On completion of the appropriate series of tests, the cell should finally be subjected to a repeat routine calibration (i).

It is emphasized that these rigorous tests are not suggested for routine calibration but as a means of evaluating an unproven design. Littlejohn (1981) makes recommendations for routine calibration of load cells, including the following:

- For routine calibration the load cell should be delivered to the laboratory at least one day before the test to permit sufficient time for the cell to attain the correct ambient temperature (20°C). The cell should be subjected to centric loading between rigid flat platens using a testing machine with an absolute accuracy better than 0.5%.
- Bearing in mind that the load cell may not have been used for some time, it may be prudent to load cycle the cell two or three times over its full loading range until the zero and maximum readings are consistent. [This is particularly important for electrical resistance load cells.] The load increments and decrements should not exceed 10% of the rated capacity of the cell, and short pauses at these intervals need only be long enough to take careful readings.
- To measure the specific effects of temperature, a centric loading test using rigid flat platens should be carried out at temperatures above and below ambient (20°C), say 40°C and 0°C, respectively.
- [When used to monitor load while stressing a structural member], load cells should be calibrated every 200 stressings or after every 60 days use, whichever is the more frequent.

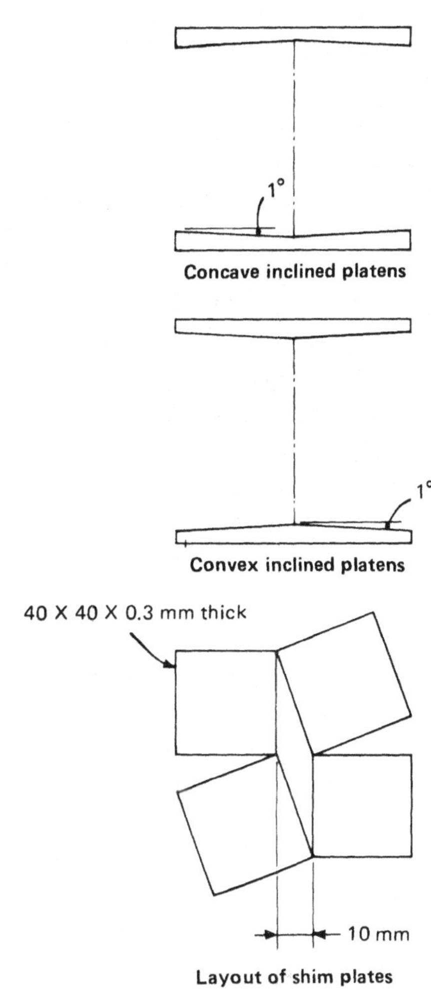

Figure 13.15. Typical types of platen to simulate uneven loading (after Littlejohn, 1981). Reprinted with permission from *Ground Engineering* (1981), Vol. 14, No. 3, p. 29.

Prior to accepting load cells for applications subjected to dynamic forces at some point in time, such as measurements of static load within a driven pile, full-scale tests should be made to ensure that zero readings will not be changed by the dynamic forces.

When load cells are used for long-term measurements, provision should be made, if possible, for periodic checks on the zero reading. In many applications a *lift-off* test can be made by measuring the load required to free the cell from its bearing surface, using a hydraulic jack with a second load cell in series. Special arrangements for periodic check-calibrations of load cells installed on permanent tiebacks are discussed in Chapter 19.

Installation

Accuracy of load cell data is highly dependent on the design of the mounting arrangements, and good load cells can fail to provide adequate data if insufficient attention is given to these arrangements. Bearing plates should be sufficiently smooth, flat, and stiff. Load should be axial and concentric, and features to reduce the effect of misalignment should be adequate. Cells and any swivel bearings or soft washers should be aligned with the load-bearing member to within ⅛ in. (3 mm) if possible. Centralizer bushings are often useful for centering hollow-core load cells around tiebacks and rockbolts. A quick-set mortar pad can be used to provide a firm and square bearing surface for a rockbolt load cell. Keil and Hellwig (1987) present practical experience compiled over many years of load cell installation and indicate the most effective mounting methods available.

Where load cells are used on steel tunnel sets or on struts across braced excavations, the standard rib or strut will usually require modification to accommodate the cell. The cell may break the continuity of the structural member, and arrangements should be made to ensure that the cell will not kick out or allow buckling or shear deformations of the member. The load cell should be installed concurrently with the structural member rather than inserted later, thereby simplifying erection and alignment and ensuring that the total load history will be recorded.

Where load cells are used for temporary measurements in permanent supports, arrangements must be made for jacking to remove the load cell and insert permanent spacers, thereby allowing reuse of the cell. Of course, cells should be calibrated before reuse.

13.3. SURFACE-MOUNTED STRAIN GAGES

Strain gages in general use for measurements on structural surfaces are described in Sections 13.3.1–13.3.4 and compared in Table 13.2.

13.3.1. Surface-Mounted Mechanical Strain Gages

Portable gages with dial indicators are the most common mechanical gages for measurement of surface strain in geotechnical applications. Scratch gages are occasionally used. Telltales are also used as mechanical strain gages.

Portable Gages with Dial Indicators

Portable gages with dial indicators, also called *mechanical extensometers,* are available in various configurations. The *Demec* (*de*mountable *mec*hanical) gage (Base, 1955; Morice and Base, 1953) was developed by the Cement and Concrete Association in England and is shown on Figure 13.16. The instrument consists of an invar bar with a conical locating point at each end—one fixed to the bar and the other pivoting on a knife edge. The pivoting movement is transmitted to a dial indicator via a lever arm. A 200 mm (approximately 8 in.) gage length is typically used for field measurements. A separate invar bar is supplied with the gage for use as a reference standard. Small stainless steel disks with central indentations are either cemented directly to the structure or mounted on anchor pins set in drilled holes. A *setting-out bar* is used to ensure correct initial disk spacing. Under ideal laboratory conditions a measurement accuracy of ±5 microstrain is possible, but under field conditions ±25–50 microstrain accuracy is more common. Details of the instrument, with procedures for disk installation, reading, and data calculation are given by ISRM (1984). Operating guidelines are also given by Cording et al. (1975). If maximum accuracy is required, extreme care must be taken during installation and reading. Cording et al. (1975) describe a temperature-compensated standard bar (Figure 13.17) in which the lengths of invar and steel are inversely proportional to their respective linear coefficients of thermal expansion. The separation of the reference points therefore does not vary with temperature, and the bar can be used to examine temperature and zero drift effects on the portable gage.

Table 13.2. Surface-Mounted Strain Gages

Gage Type	Advantages	Limitations	Typical Gage Length	Typical Range (microstrain)	Sensitivity (microstrain)	Approximate Accuracy (microstrain)
Portable dial indicator (e.g., Figure 13.16)	Simple Inexpensive Waterproofing not required Calibration can be checked at any time No delicate parts attached to structure	Requires access to structure Requires extreme care to read	2–80 in. (50–2000 mm)	Up to 50,000	3–50	± 5–200
Scratch (e.g., Figure 13.18)	Inexpensive Self-recording Waterproofing not required	Requires access to structure Requires skill to read Strains must be dynamic and large	3–48 in. (75–1200 mm)	Up to 6000	30–300	±25–200
Multiple telltales	Simple Inexpensive	Requires access to telltales (but can be converted to remote reading device)	Unlimited	Unlimited	Depends on application	±25–400
Vibrating wire (e.g., Figure 13.20)	Remote readout Lead wire effects minimal Readout can be automated Factory waterproofing Arc welded or bolted version is reusable	Limited range Cannot be used to measure high-frequency dynamic strains Special manufacturing techniques required to minimize zero drift Need for lightning protection should be evaluated	2–14 in. (50–350 mm)	3000	0.2–2	±5–50
Electrical resistance (weldable and bonded foil) (e.g., Figures 13.22, 13.23, 13.24)	Remote readout Readout can be automated Suitable for monitoring dynamic strains	Low electrical output Lead wire effects Errors owing to moisture, temperature, and electrical connections are possible Installation of bonded gages requires great skill and experience Need for lightning protection should be evaluated	0.01–6 in. (0.25–150 mm)	20,000	1–4	±1–100

A gage with accuracy similar to that of the Demec gage is described by Kovari et al. (1977). The *Whittemore gage*, used primarily for strain measurements during laboratory testing of concrete, is significantly less accurate than the Demec gage.

Kovari et al. (1977) also describe a *curvometer*, designed to monitor the change in curvature of a straight or curved beam at a given point, so that bending strains can be determined. The device consists of a mechanical reference bar, with a dial gage mounted perpendicular to the bar at its midpoint. Use of the curvometer to monitor bending strains in

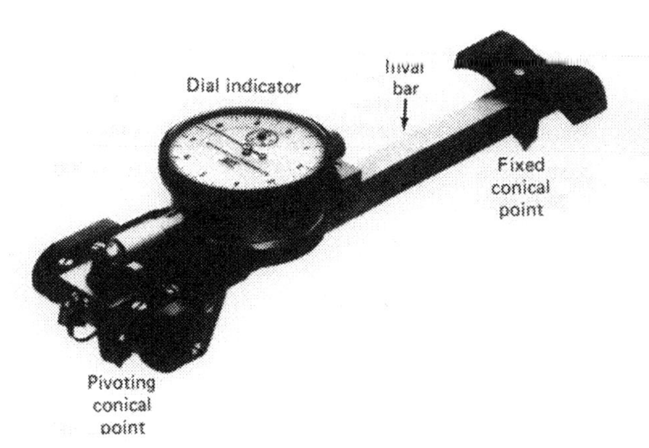

Figure 13.16. Demec strain gage (courtesy of W. H. Mayes & Son, Windsor Ltd., Windsor, England).

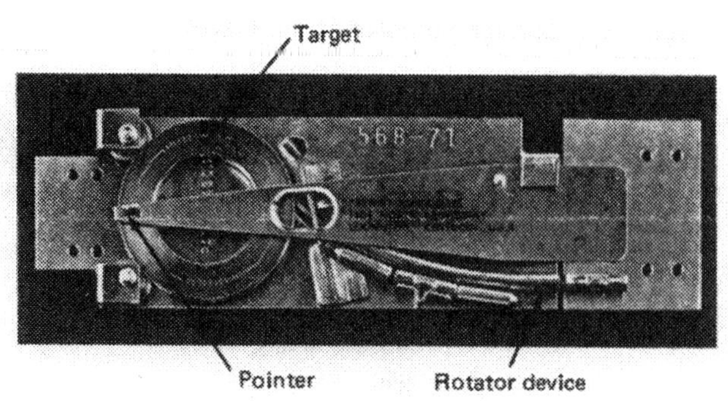

Figure 13.18. Prewitt scratch gage (courtesy of Eastlex Machine Corporation, Lexington, KY).

steel ribs and concrete segments for tunnel linings is described by Robinson et al. (1985).

Scratch Gages

Scratch gages provide a record of strain by scribing a line on a target. The target is attached to one end of the gage length, and a pointer attached to the other end bears on the target. A mechanical linkage causes the target to advance or rotate as strain occurs, so that the line scribed on the target by the pointer does not overlap a previously scribed line. The scribed lines are usually examined under a microscope, and irregularities are converted to strain.

The *Prewitt scratch gage,* shown in Figure 13.18, has a disk-shaped brass target. The two ends of the gage are clamped, bolted, or glued to the structure. The rotator device contacts the rim of the disk and causes it to rotate when compressive strain occurs, and a new disk can be inserted at any time. Scribe lines are interpreted by using a calibrated microscope or special optical comparator. The device is only suited to recording dynamic strains, and use is limited to situations in which strains are sufficiently large and dynamic so that the disk is turned automatically by the rotator device. The author attempted to use the gage for recording dynamic strains in struts across a braced excavation as a train moved along a railroad immediately adjacent to the excavation, but the disk did not rotate.

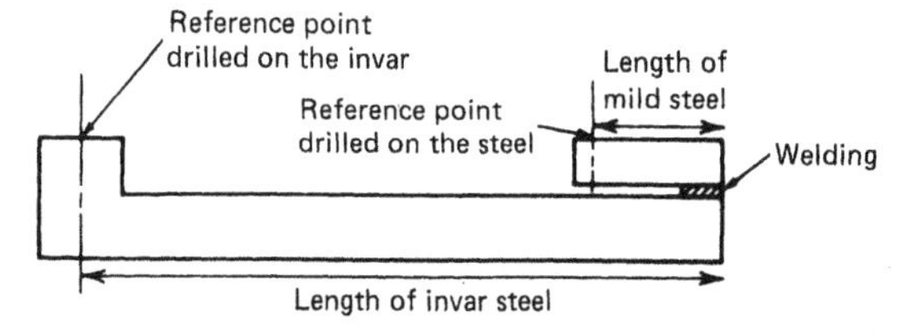

Figure 13.17. Temperature-compensated standard bar for portable strain gage with dial indicator (after Cording et al., 1975).

The scratch gage manufactured by Leigh Instruments Ltd. records strain data on metallic tape, stored in an interchangeable cassette attached to the gage. The primary application has been on aircraft structural components, and data are processed using a video display unit.

Telltales

A *telltale* is an unstressed wire, rod, or pipe, mounted alongside or within a structural member to indicate change in length of the member. Although a telltale is generally considered as an instrument for measuring deformation, change in length can be converted to strain, and the device is also classified in this book as a mechanical strain gage.

Figure 13.19 shows a telltale arranged to monitor load in a tieback anchor. This arrangement is also referred to as a *telltale load cell.* The telltale rod is inserted within a pipe sleeve, and one end is attached to the tieback near the lower end of the stressing length. The distance d between the upper end of the telltale rod and the anchor locking plate is measured, normally using a portable mechanical gage. The change in distance d is equal to ΔL, the change in stressing length L. Because the stressing length is acting as an elastic member under tension, strain is equal to $\Delta L/L$, and load change can be calculated since cross-sectional area and modulus of the tieback anchor are known. In practice, the relationship between ΔL and load change is determined in the field while proof testing the anchor, using a hollow-core load cell in series with the

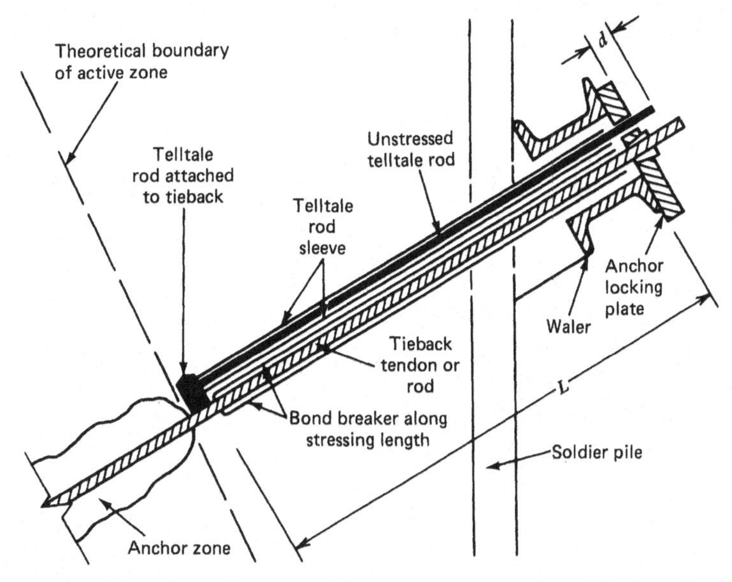

Figure 13.19. Schematic of telltale arranged for load measurement in a tieback anchor (*telltale load cell*).

stressing jack. During proof testing, the telltale rod can be shortened so that deformation measurements are made with respect to the waler. Alternatively, the rod can be extended outward through the jack and hollow-core load cell and deformation measurements made with respect to the pulling head, applying a correction for travel of the jack piston. The hollow-core load cell is also used to determine load at lock-off, and load changes are referenced to that datum. The fact that load is derived from deformation measurements sometimes raises concerns that slippage of the anchor or movement of the soldier pile will create errors in load measurement, but this is not so. The arrangement can be used to monitor change of load resulting from all causes and thus is an alternative to a hollow-core load cell at the anchor head. A similar arrangement, installed on the surface of a structure, can be used to monitor change of load.

A pair of telltales can be used to monitor strain. The arrangement is essentially the same as a multipoint fixed borehole extensometer (Section 12.7). For example, a second sleeved telltale can be attached to the tieback shown in Figure 13.19, within the anchor zone. The average strain between the two attachment points is given by:

$$\frac{\Delta L}{L} = \frac{(\text{change in } d_2) - (\text{change in } d_1)}{L},$$

where: d_1 = distance d for first telltale,
d_2 = distance d for second telltale,
L = distance between attachment points of the two telltales.

Therefore, average strain between the two attachment points can be monitored by measuring changes in the distances d_1 and d_2, and strains can be converted to approximate values of stress by using an estimate of the composite modulus of the tieback and grout within the anchor zone. By using multiple telltales, attached to the tieback at different points within the anchor zone, the pattern of load transfer in the tieback can be estimated. It is necessary, however, to assume the shape of the load transfer plot between each pair of attachment points: if a straight line is assumed, the calculated stress will be attributed to the location midway between the telltale attachment points. If the load at one attachment point is known, the load at the other can be calculated if the same straight line assumption is made, but if the line is not straight, there may be a large error in the calculated load. Fellenius (1980, 1987) provides guidelines on the interpretation of telltale data. When planning multiple telltale measurements in driven piles and drilled shafts, one should recognize that any telltale measurement may be in error by as much as ± 0.02 in. (± 0.5 mm). Errors in load transfer data can be minimized by installing telltales at more than two attachment points, so that data from each pair can be evaluated in light of data from adjacent pairs. Because load transfer calculations are extremely sensitive to any error in telltale data, it is advisable to install duplicate telltales at each attachment point. If they agree, they provide confidence in the data, and if they disagree the adjacent telltales are used to evaluate which is likely to be more correct.

Electrical linear displacement transducers can be mounted at the accessible ends of the telltales to permit remote monitoring. An example is shown in Figure 12.106. Multiple telltales can also be used for determining the pattern of load transfer in driven piles and drilled shafts and for monitoring strain in other structures.

13.3.2. Surface-Mounted Vibrating Wire Strain Gages

The vibrating wire transducer is described in Chapter 8, and a schematic of a surface-mounted gage is shown in Figure 8.22. Chapter 8 also includes a discussion of techniques for minimizing errors caused by zero drift and by corrosion of the vibrating wire, and guidelines on cable selection. The surface-mounted gage is available in the two basic configurations illustrated in Figure 13.20.

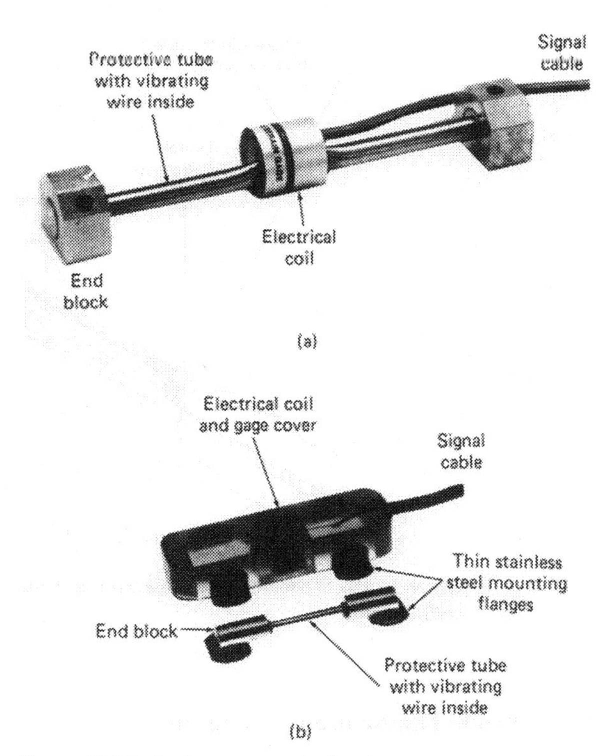

Figure 13.20. Surface-mounted vibrating wire strain gages: (a) gage installed by arc welding (gage can also be fitted with different end blocks for installation by bolting) and (b) gage installed by spot welding (courtesy of Geokon, Inc., Lebanon, NH).

The configuration designed for installation by arc welding (Figure 13.20a) or bolting is generally between 3 and 14 in. (75–350 mm) long. The end blocks are normally installed by arc welding to the structure, using a solid spacer bar to hold the blocks at their correct spacing. Alternatively, the end blocks can be bolted to the structure, requiring appropriate threaded studs or holes in the blocks. The remainder of the gage is then fixed into the end blocks and bolted in place. Light tapping of the end blocks with a hammer will improve the initial stability of the gage. Most versions have O-ring seals between the protective tube and the gage end blocks to provide waterproofing and to allow the tube to remain unstressed. All versions permit adjustment of the initial wire tension, and some contain internal springs such that the vibrating wire is set by the manufacturer in midrange. This arrangement promotes gage stability, allows for easy acceptance testing on receipt from the manufacturer, and eases the task of setting the initial tension. Gage range is typically 3000 microstrain, corresponding to a stress change of approximately 90,000 lb/in.^{2} (620 MPa) in steel, and the initial tension should therefore be set with care so that the wire will not become too slack or too tight under the anticipated tensile or compressive strains. Some versions allow the electrical coil assembly to be detached from the protective tube to allow for easier repair in the event of damage to the signal cable. After attachment of all gage components, a robust protective cover should be attached, and the signal cable should be protected, for example, by conduit.

The Norwegian Geotechnical Institute in Oslo has recently developed a method of installing the gage on the inside of a closed-ended pipe pile, using welded end blocks, thereby minimizing the chance of damage during pile driving. Four holes are drilled in the pile, on a straight line parallel to its axis. A setting tool is used to insert the gage within the pile and to locate end block stems through the first and last holes. The stems are held in place temporarily with a positioning jig and welded to the pile. The other two holes are used to set the initial wire tension, by using a screwdriver and allen wrench.

The arc welded gage is also available with two vibrating wires. The wires are at different distances

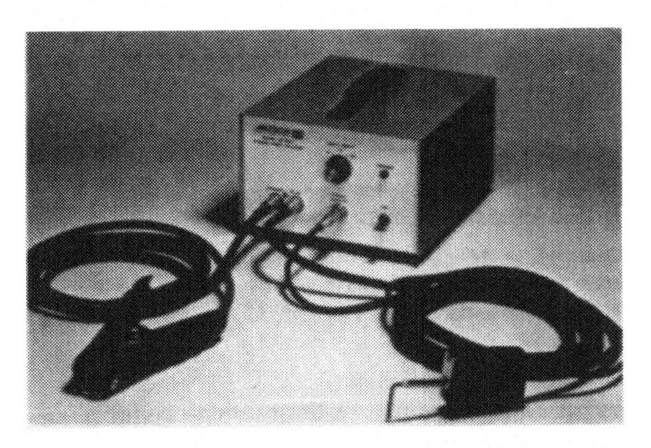

Figure 13.21. Capacitive discharge spot welder for installing weldable strain gages (courtesy of Eaton Corporation, Los Angeles, CA).

from the surface of the structure, so that bending strains can be monitored by attaching the gage to one surface only.

The spot welded gage is typically 2 in. (50 mm) long. The version shown in Figure 13.20b has O-ring seals between the protective tube and end blocks, so that the tube remains unstressed. An internal spring holds the wire at an initial tension, normally preset at midrange, but the initial tension can be set during installation, allowing for maximum range in compression or tension as required. Another version has the protective tube welded to a thin stainless steel mounting flange along the entire length of the gage, such that the tube is stressed as strain occurs, and in this version no field adjustment of wire tension is made. All versions are installed on steel by using a small capacitive discharge spot welder (Figure 13.21) to weld the mounting flange to the structure, with weld points at approximately 0.06 in. (1.5 mm) spacing. The surface is first prepared by removing surface paint, rust, and scale using a disk sander, with progressively finer grits, to expose the bare metal and remove pits. When welding, the weld current must be controlled carefully, following the manufacturer's recommendations. It is useful to practice welding and to demonstrate the suitability of the selected weld current by using test strips of the same material as the mounting flange. After welding, the weld points must be sealed with a waterproof coating to prevent later corrosion and gage slippage. Where access is available, the electrical coil and gage cover can be placed on the gage temporarily whenever a reading is required. However, they are normally spot welded in place and covered with a mastic compound as corrosion protection. A robust protective cover is usually installed over the gage, and the signal cable is normally installed in conduit.

When compared with the arc welded gage, advantages of the spot welded gage are small size, an installation procedure not requiring arc welding, and minimum errors that result from bending of the structure because of the close proximity between the vibrating wire and the structure surface.

Most manufacturers attempt to match the linear coefficient of thermal expansion of the vibrating wire to that of structural steel, so that the gage is insensitive to temperature change when installed on steel. However, remaining differences in coefficients may be significant enough to require correction, particularly where the structural member is exposed to large temperature variations. Temperature should therefore be measured at the time of each reading, and calculated stresses should be corrected on the basis of correction factors, developed as discussed in Section 13.3.9. Most manufacturers provide gages with integral thermistors for temperature measurement, or alternatively the resistance change of the plucking coil can be calibrated to temperature change and used for temperature measurement.

Where gages or signal cables are installed in regions of high thunderstorm activity, lightning protection (Chapter 8) should be considered. If gages will be subjected to vibration, the user should ensure that zero drift will not be caused. Guidelines are given in Chapter 8.

13.3.3. Surface-Mounted Electrical Resistance Strain Gages*

Five basic types of electrical resistance strain gage are described in Chapter 8: bonded wire, unbonded wire, bonded foil, semiconductor, and weldable. Chapter 8 also includes a description and comparison of Wheatstone bridge circuits for use with electrical resistance strain gages, a discussion of cable selection, data acquisition and communication systems, gage and cable integrity testing, and general guidelines for the use of electrical transducers and data acquisition systems.

Both bonded foil (Figure 13.22) and weldable

*Written with the assistance of Given A. Brewer, Consultant, Brewer Engineering Laboratories, Teledyne Engineering Services, Marion, MA.

Figure 13.22. Bonded foil electrical resistance strain gage (courtesy of Measurements Group, Inc., Raleigh, NC).

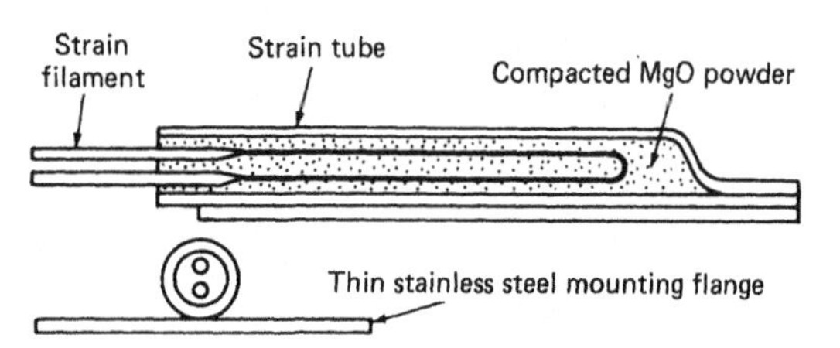

Figure 13.24. Weldable electrical resistance strain gage with strain filament encased in a small tube (courtesy of Eaton Corporation, Los Angeles, CA).

(Figures 13.23 and 13.24) gages are used for surface strain measurements in geotechnical field applications. Bonded wire, unbonded wire, and semiconductor gages are used for special applications only and will not be discussed further in this chapter. When one is selecting between weldable and bonded foil gages, the primary issues are environmental conditions, accuracy, cost, and size. Environmental conditions often override all other considerations, leading to the selection of weldable gages.

Environmental Conditions

When instruments must be installed in the field, weldable gages are usually preferable to bonded gages. Weldable gages can be installed successfully by relatively inexperienced personnel, but installation of bonded gages in the field requires personnel highly experienced in field procedures. Even with experienced personnel, installation of bonded gages is limited to warm-weather conditions. Success depends on many painstaking steps, including surface preparation, bonding, waterproofing, connection, and physical protection, and under field conditions success is difficult to attain. Installation steps are discussed later in this section. Successful installation is easier to achieve in a controlled indoor environment, but this also requires great experience. When bonded gages are installed by experts, they can be stable and reliable, but there are many case histories of unsuccessful installations by geotechnical personnel, and a *do-it-yourself* approach is not likely to lead to success.

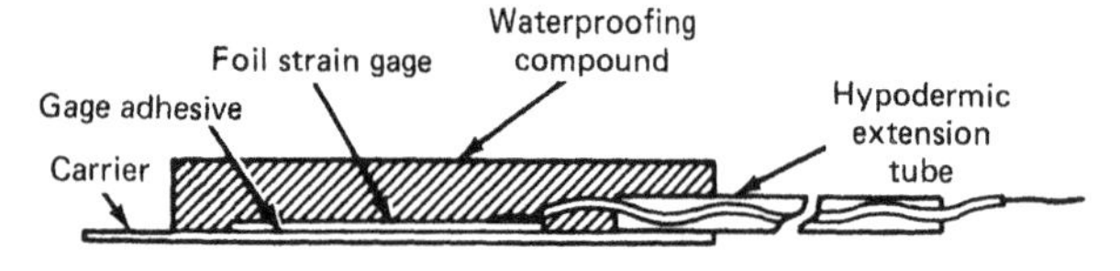

Figure 13.23. Weldable electrical resistance strain gage with bonded foil transducer (courtesy of HITEC Products, Inc., Ayer, MA).

Accuracy

When bonded foil strain gages for field use are installed in a controlled indoor environment by experts, they can have a short-term accuracy of ±1 microstrain. However, this accuracy will be downgraded severely if skill is inadequate or if installation and field conditions are other than ideal. Accuracy of weldable gages depends on the resistance element. When a bonded foil transducer is used (Figure 13.23), typical accuracy is ±5 microstrain; for the arrangement shown in Figure 13.24, typical accuracy is ±15 microstrain.

Cost

A typical weldable strain gage costs 10–15 times as much as a bonded gage. However, installation costs are generally significantly less, and cost comparison depends on the application.

Size

Typical weldable strain gages are about 1 in. (25 mm) long, but special gages as small as 0.2 in. (5 mm) long are available. Bonded gages range in length from 0.01 to 6 in. (0.25–150 mm) and may therefore be preferred when available space is very limited.

Selection and Installation of Bonded Gages

Key factors when using bonded gages in a field environment include gage selection, selection of bridge circuit, and installation procedure.

When one is selecting the type of gage, the following must be considered: gage material, backing material, resistance value, strain limits, fatigue life, temperature range, temperature compensation, static or dynamic application, and use environment. Gages with integral terminals are preferred, so that the user solders lead wires to a tab rather than to the gage wires themselves.

Guidance on selection of a bridge circuit is given in Chapter 8. Many factors influence this decision, but the primary consideration is that gages must detect the desired strains and properly cancel or eliminate the unwanted strains. In all cases, the lead wire shield should be isolated from the structural member to which the gages are attached.

Selection of the installation procedure for bonded gages depends primarily on usage and gage type. Installation typically involves the following five steps:

- *Surface Preparation.* The surface usually must be degreased, sanded, etched, neutralized, and cleaned with a solvent and a chemical cleaner.
- *Bonding.* Bonding entails extreme care in handling the gage, application of a layer of suitable bonding cement with careful control over glue thickness, careful positioning, clamping under a controlled pressure, and usually curing under a controlled temperature. When bonded in humid environments, the surface of the structure must be heated to prevent formation of a moisture film. The *easy-to-use* cyanoacrylate contact cements, which cure at room temperature and do not require application of controlled pressure throughout the bonding period, are usually not suitable. When they are used, they should be purchased only from strain gage manufacturers and applied by experienced personnel.
- *Connection.* Interconnect wires are soldered to create the required Wheatstone bridge circuit, connections cleaned with a solvent, the gage and connection area coated with a protective and vapor barrier, the connectors heat treated, and lead wires soldered in place. Rosin core solder should be used.
- *Waterproofing.* The type, number of coatings, vapor barrier and lead wire sealer depend on the field environment and duration of the monitoring program. For a one to two week outdoor monitoring program a single layer of waterproofing compound may be sufficient. For a one to two month outdoor monitoring program the layer should be covered by aluminum foil and in turn covered by a second layer of waterproofing compound. For longer term programs, more comprehensive waterproofing is required, and a multi-layered patented procedure has been proved (Brewer, 1972): waterproofing compound, allowed to cure; waterproofing compound extending beyond the first layer; stainless steel foil spot welded through unset waterproofing compound; second layer of waterproofing compound allowed to cure; third layer of waterproofing compound extending beyond the stainless steel foil.
- *Physical Protection.* Protection will generally be a steel cover plate or box or a fiberglas coating.

As the installation of gages proceeds, they should be tested functionally and checked for insulation, and if possible a leak detection test should be made after waterproofing is complete. Gage resistance testing, insulation testing, and leak detection testing are described in Chapter 8. All data should be recorded and maintained as initial test records. Strain gage manufacturers will normally provide standard installation procedures but will not necessarily have sufficient experience with field installations. If in doubt, the guidance of field specialists should be sought.

When using bonded strain gages for measuring surface strains on concrete or rock, moisture will cause gage deterioration unless special precautions are taken. All loose surface material should first be removed and the surface cleaned with a wire brush and a chemical cleaner. A room-temperature curing epoxy should be applied liberally, thus ensuring the absence of air bubbles, and allowed to cure. The epoxy should then be sanded, to create a smooth surface, and checked for absence of air bubbles; the gage should be bonded to the epoxy and waterproofed as described above.

Bonded strain gages can be attached to a spring steel strip to form a *clip-on* arrangement that is reusable. Fang and Koerner (1977) describe a clip-on gage for measurement of crack openings in structures while the structures were subjected to vibration.

Selection and Installation of Weldable Gages

Weldable gages are restricted to installation on steel, and the installation procedure is similar to the weldable vibrating wire gage procedure described in Section 13.3.2. Gages with integral leads and factory waterproofing are most suitable. Quarter bridge gages are available with integral three-wire leads. Half bridge gages are also available, with one active arm and one dummy.

13.3.4. Miscellaneous Surface-Mounted Strain Gages

Cragg (1984a) describes various strain gages for use on geotextiles in reinforced embankments. Simple gages were used by the U.S. Department of Agriculture, Forest Service (Steward et al., 1977; Vischer, 1975), consisting of different lengths of electrical wire attached to a standard gage length on the geotextile. Wire lengths were chosen such that breakage occurred at strain levels in the geotextile of 1, 3, 5, 10, and 20%, and an ohmmeter was used to monitor circuit continuity. Similar gages, using platinum wire and a copper tube, were used in Holland (Leeuwen and Volman, 1976), by clamping the wire in the tube so that a force of 2 lb (9 N) was needed to break the connection. Neither of these gages provided continuous strain readings. Cragg (1984a) and Rowe et al. (1984a) describe a hydraulic strain gage developed to provide continuous strain readings. The gage (Figure 13.25) is similar to a large-diameter, liquid-filled hypodermic syringe, connected to a small-diameter sight tube. A small movement of the large-diameter piston produces a large movement of the liquid level in the sight tube. The magnification factor is about 12. The gages were fitted into wood blocks that were both clamped and glued to the geotextile, to monitor strains over a gage length of 8 in. (200 mm). Dummy sight tubes were installed to correct for temperature differences between the gages and the remote monitoring station. The gages are applicable where strain is expected to be up to about 5% and have been used successfully for monitoring tensile strains up to 20%. Performance data are given by Rowe et al. (1984a, 1984b). The soil strain gage (Section 12.6.4) was also used, but Cragg (1984b) comments that

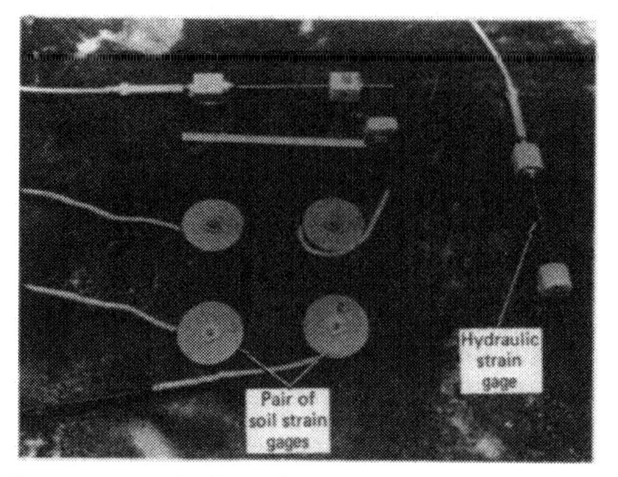

Figure 13.25. Hydraulic strain gages and soil strain gages attached to woven geotextile fabric (after Rowe et al., 1984a).

because the gage is sensitive to tilting out of the initial coplanar arrangement, it should not be used for monitoring strains in geotextiles of less than 1%.

A demountable electrical gage, called a *demountable extensometer,* is manufactured by Cambridge Insitu in England. The gage is similar to the Demec mechanical strain gage (Section 13.3.1) but with an electrical output. Precision is claimed to be ±5 microstrain but, as for the Demec gage, under field conditions ±25–50 microstrain is likely to be more common.

A special gage has been used successfully to monitor axial and bending strains in timber piles. The arrangement consists of electrical resistance strain gages bonded to the outside of a 0.25 in. (6 mm) aluminum alloy tube, which is then embedded in epoxy in longitudinal grooves routed along the piles.

Photoelastic strain gages (Roberts and Hawkes, 1979) depend on the principle that polarized light incident on certain types of plastics and glass will produce patterns of light and dark areas, referred to as *photoelastic interference fringes*. The number and position of these fringes changes in response to the strain in the material to which the gage is attached, and the magnitude and direction of strain can be determined from an analysis of the fringe patterns. Photoelastic materials are used extensively for model studies, and bonded photoelastic strain gages are available for surface mounting. However, these gages are subject to substantial errors because of creep or deterioration of the bonding cement and are rarely used for geotechnical applications.

Various electrical crack gages (Section 12.3.2) can be used for measurement of surface strain.

13.3.5. Selection of Strain Gage Length, Cable, and Connectors

The selection of strain gage length for measurements on steel depends on the purpose of measurements, gage type, available space, and access. In general, unless very localized measurements are required, gages should be as large as possible. For measurements on concrete the gage length should be long enough to average out strain differences in the constituents of the concrete, as aggregate will generally strain less than the cement matrix. If possible, the gage length should be at least five times the maximum size of aggregate.

Selection of cable and connectors should follow the guidelines given in Chapter 8.

13.3.6. Locations of Surface-Mounted Strain Gages

The location of strain gages depends on the purpose of the measurements, available access, and on protection requirements during and after installation. If stresses at particular points are of interest, the gages should be located at those points. However, measurements are often required at locations that are not influenced by uneven or eccentric loadings or by the proximity of irregularities such as holes in the structural member. For example, if end effects are to be avoided, gages should not be located near the ends of the structural member or near irregularities. When gaging struts across braced excavations, gages are normally located a minimum distance from the end of the strut equal to three times the depth of the strut and a similar distance from any interbracing connections and points of future welding or flame cutting. On tunnel ribs, gages should preferably be a minimum distance from connections, tie-rod holes, and blocking points equal to three times the depth of the rib.

Figure 13.26 shows typical strain gage locations on the cross section of various structural steel members. If the member is small and not subjected to bending (e.g., a threadbar tendon for tiebacks), a single gage may be sufficient, although two are usually more prudent because in most applications the member will be subjected to both axial and bending stresses. If axial stress alone is required, a minimum of two gages must be used, equidistant from and on opposite sides of the neutral axis. If maximum stress is required and if the member may be subjected to bending about any axis, more than two gages must be used. On a pipe pile or pipe strut, a minimum of three gages spaced 120 degrees apart is recommended for determination of maximum stress. Four gages spaced 90 degrees apart provide a worthwhile redundant measurement. On a driven H-pile, four gages are recommended, back-to-back on each side of the web and as near the flanges as possible. Gages at these locations can usually be protected adequately, using welded channels, angles, or corner strips. On large I-beams, four gages can be located as for an H-pile, but the accuracy of strain estimates in the flanges is often limited by the relative flimsiness of the web. It is usually better to install four gages on the inner faces of the flanges as far from the web as possible, providing a good compromise between a protected location and the zone of maximum stress. For sheet piles in tension, for example, for cofferdams, a pair of gages midway between the clutches is appropriate. For sheet pile walls in bending, pairs of gages on each of the outer portions of the Z-shape are recommended.

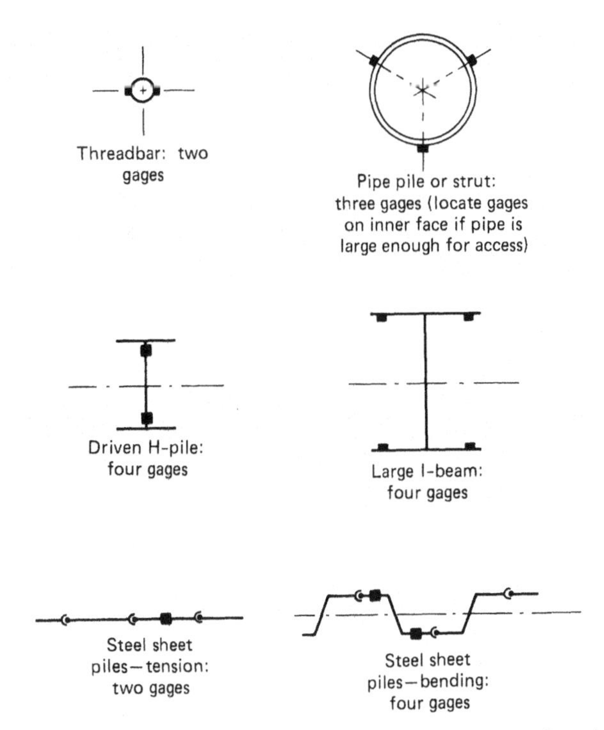

Figure 13.26. Typical locations of strain gages on structural steel members.

It should be remembered that the stress at any point in the cross section is the sum of the axial stress and the stress caused by bending, that stresses at the outer edges of the member can therefore be very much higher than stresses at the neutral axis, and that failure can consequently be initiated at the outer edges. Thus, two strain gages are rarely sufficient for providing a forewarning of failure.

Where a structural member is subjected to bending and only one surface is accessible—for instance, a steel tunnel lining or the outside of sheet piling—bending strains can be monitored by installing the type of vibrating wire gage with two vibrating wires at different heights above the surface of the member.

13.3.7. Installation of Surface-Mounted Strain Gages

General guidelines on the installation of instruments are given in Chapter 17. Methods of strain gage installation have been outlined in the above sections describing the various types of gage. The following additional guidelines apply to the installation of all types of surface-mounted strain gage.

Wherever possible, gages should be installed when the structural member is in an environment conducive to maintaining the quality of work, remote from pressures of construction progress, such as in a steel storage yard or warehouse. The work area should be protected from adverse weather conditions. Gages should be installed a minimum of 4 days before erection of the structural member to allow time for stabilization prior to taking initial readings.

Most strain gages and cables require protection from water and from mechanical damage caused by normal construction activities, vehicular traffic, and vandalism. Gages usually require protective cover plates, welded, bolted, or epoxied to the structure. Cover plates should not stiffen the structure and should not be tightened or welded so as to distort the structure and cause spurious strains at the gage location. Use of a stiff cover plate, attached to the structure at one end only, with mastic between the structure and cover plate, usually overcomes this problem. Cutting or welding can create significant stress changes in steel and may damage electrical cables; therefore, welding for attachment of cover plates should be minimized.

Gage cables are usually vulnerable to damage and must be protected by careful selection of location, by placement in conduit, or by embedment.

13.3.8. Initial Reading

General guidelines on initial readings are given in Chapter 18. Selection of initial readings depends on the purpose of the measurements. As an example, if actual stress in the structural member is required, initial readings should be made when the member is under no load, perhaps when supported evenly on a flat surface in a steel storage yard or warehouse. However, if gage readings are to be used for determination of load imposed on the structural member, initial readings should be made after erection of the member.

13.3.9. Relationship Between Strain and Stress*

Strain data are rarely of interest; the data are merely a step in the determination of stress. When making measurements on steel, provided that modulus is known and temperature is measured, conversion from strain to stress is straightforward. However, stress determination in concrete is by no means straightforward, and accurate results should not be expected.

Causes of Strain in Concrete

Strain in concrete can be caused by several factors other than stress change, and a strain gage responds to all causes of strain. Strain other than that caused by stress may be due to creep (strain under constant stress), shrinkage and swelling (moisture content change), temperature change, and the progress of autogenous volume change (dimensional change that is self-generated and not due to temperature, moisture change, or stress); other strain may also occur as concrete cures. Strain due to creep is generally more extensive in highly stressed prestressed concrete than in conventional reinforced concrete; therefore, for structural members such as prestressed concrete piles, creep strains are particularly significant.

*Written with the assistance of Stanley L. Paul, Associate Professor, Department of Civil Engineering, University of Illinois at Urbana–Champaign, IL.

The importance of creep and shrinkage is related to the duration of the measurement program. For a program lasting a few days, their influence may be small, but for a duration of several weeks or months, they will be much more important. Creep and shrinkage are influenced by mix design parameters, type of aggregate, fineness of the cement, and conditions during and after set, such as moisture content and temperature. The importance of creep and shrinkage is also related to stress magnitude. When stresses are greater than about 60% of the ultimate strength of the concrete, their importance increases substantially. A general discussion of the influences of factors that cause strain in concrete, and the conditions under which they are important, is given by Neville (1981).

Methods of Converting Strain to Stress

To avoid repetition in the later section describing embedment strain gages, the scope of this subsection includes strain measurements made with surface-mounted and also embedment strain gages.

The behavior of a concrete member is often predicted by measuring strains on the surface of the member or on reinforcement within the member. Stresses are then calculated by using the method of transformed areas, requiring knowledge of cross-sectional areas and moduli of steel and concrete, and these stresses are used to calculate axial loads and moments. This approach suffers from two major limitations. First, significant error may result from uncertainty in the modulus of the concrete. The modulus may vary from point to point because of variations in concrete properties that are caused by air voids, trapped moisture, and variation in compaction. This is particularly true of cast-in-place concrete. Second, the approach does not account for strains other than those caused by stress. A special effort must therefore be made to determine the relationship between strain and stress. The options are as follows:

1. Subject a specimen of the same concrete to the same influences as the prototype, in a known stress field, and measure resulting strain. This is sometimes attempted in the laboratory, but it is usually not possible to match laboratory and prototype conditions. A better approach is to install strain gages on a part of the prototype that is subjected to a known stress, thereby ensuring that the *specimen* behaves in a similar way to the prototype. For example, while load testing a concrete pile or drilled shaft, an *unconfined compression test specimen* can be created near the pile or shaft butt, either above the ground surface or within a sleeved section just below the ground surface. It is important to model moisture conditions correctly, because strains caused by moisture content change can be a significant part of total strain. The *specimen* at a pile or shaft butt should therefore be kept wet if the subsurface gages are below the water table.
2. Interpose a load-measuring device across at least one entire cross section and relate the known load to strain measured at nearby strain gage locations. The same relationship is applied to strains measured elsewhere. For example, a contact earth pressure cell can be installed at the tip of a pile or drilled shaft, or a load cell can sometimes be installed at a splice in a driven pile.

If neither of the above options is possible, three other approaches can be considered:

1. Use of creep curves developed for each particular project (e.g., England and Illston, 1965; Jones, 1961). These curves enable creep strains to be estimated from the age of the concrete at the time of loading and the duration and magnitude of loading. These strains are then subtracted from measured strains to obtain strains that result from stress.
2. Use of averaged empirical data that relate creep and shrinkage to environmental and mix parameters (e.g., ACI, 1982).
3. If the cross-sectional area is large enough, measure stress directly by using *concrete stress cells,* as described in Section 13.6.

It should be noted that the *no-stress* method, often used in large mass concrete structures, is not recommended. In this method a strain gage is installed in a cube of concrete and a gap is formed around the four sides and the top, so that the cube is in contact with surrounding concrete only at its base. Strain at the gage is assumed to result from all causes other than stress and is subtracted from strain readings of the active gages. However, this does not account for strain of the active gages caused by creep under stress. It also does not account correctly for shrinkage, because the size of the cube is so different from the structure that migration of moisture is entirely different.

Temperature Effects

Temperature effects on the gages themselves have been discussed in earlier sections describing the various types of strain gage. However, as discussed in Chapter 14, temperature change can cause a real stress change if the structural member is subjected to end restraint, and it may be necessary to apply a correction factor to account for temperature effects. Two cases will be considered: thermally matched strain gages and gages that indicate actual length change.

First, consider a strain gage that is thermally matched to the structural member: for example, a temperature-compensated electrical resistance strain gage (one for which the change in resistance caused by thermal strain of the member is compensated for by change in resistance caused by temperature change in the strain gage material) or a thermally matched vibrating wire strain gage (one for which the thermal coefficients of the vibrating wire and structural member are equal) attached to steel. In these cases the gage will respond only to strain changes that result from stress changes, and the stress changes can be calculated from the change in strain gage reading multiplied by modulus of elasticity. This rule is independent of end restraint conditions and applies for all causes of stress in an elastic member: change in applied loading, stress caused by temperature change in the member, and a combination of both. Figure 13.27 shows the example of thermally matched strain gages installed on a steel strut. For both conditions illustrated, and for any intermediate condition, stress changes can be calculated from the change in the strain gage reading, without the need for temperature corrections.

Second, consider a strain gage that is used to indicate actual length change, as would be the case with a mechanical strain gage referenced to the temperature-compensated standard bar illustrated in Figure 13.17. The thermal characteristics of the gage are now not matched with the member. Actual length change results both from temperature change and stress change, and the measurement must be

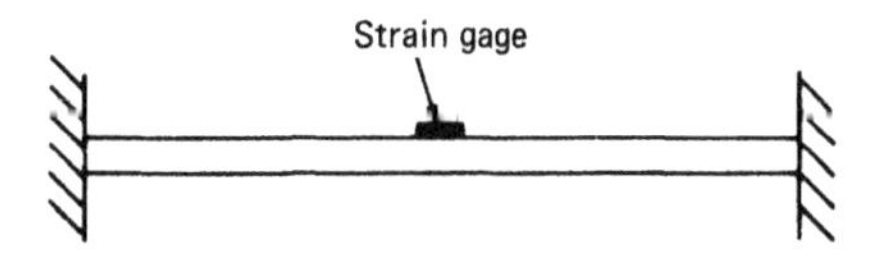

Complete end restraint. Temperature decrease causes decrease in compressive stress but no strain in the strut. Gage reading will change. Correct decrease in stress calculated from measured strain and known modulus of elasticity.

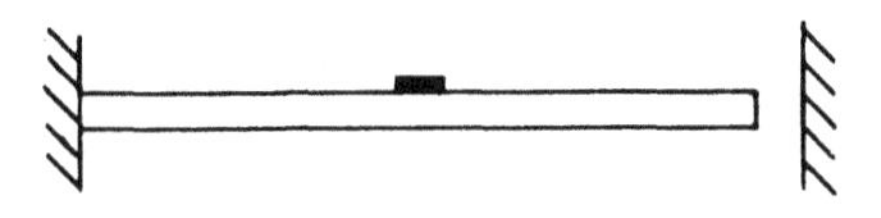

No end restraint. Temperature decrease causes strain in the strut but no change in stress. No change in strain gage reading.

Figure 13.27. Behavior of strain gages as temperature changes: gages thermally matched to structural steel member.

corrected for any strain caused by temperature change:

$$\text{total strain} = \frac{\Delta\sigma}{E} + \varepsilon\Delta T$$

where: $\Delta\sigma$ = stress change,
ΔT = temperature change,
E = modulus of elasticity of steel,
ε = linear coefficient of thermal expansion of steel.

Therefore, in this case it is necessary to measure temperature of the steel and apply a correction when calculating stress change. As an example of the importance of such corrections, the author has used a mechanical strain gage for stress determination in 24 in. (610 mm) diameter, 0.5 in. (13 mm) wall thickness pipe struts across an internally braced excavation. Design load was 520 kips (2.3 MN). An error of 2°C in temperature measurement (corresponding to an error of 23 microstrain) caused an error of 25 kips (110 kN) in the calculated strut load.

In cases where the user is uncertain whether the gage is thermally matched, whether the gage indicates actual length change, or whether an intermediate condition exists, a temperature correction factor can be developed by making strain measurements at various temperatures in the laboratory (Cording et al., 1975). The gage should be mounted on the same type of material as it will be mounted on in the field, and the test must be conducted in a way that avoids bending and other strains that are not related to temperature. If the gages are mounted on concrete, the moisture conditions must be maintained the same as in the field. Corrections to field readings can then be made by measuring temperature in the field and applying the correction factor. Alternatively, a temperature correction factor can sometimes be obtained by use of the *dummy gage* procedure, in which a gage is mounted on a specimen of the same material as the structural member and the specimen placed near the active strain gages in the field so that it will experience the same temperature but no stress. The dummy gage is read at the same time as the active strain gages, and its change in strain is used to correct the active gage readings. The dummy gage also serves as an indicator of any strain caused by factors other than stress and temperature. Several dummy gages should be used to obtain average corrections, since gages are unlikely to have identical characteristics.

An additional temperature error can occur if the strain gage and the structural member are at different temperatures. If possible, readings should be taken when temperature is uniform throughout the member. If the member is subjected to direct sun, readings should be taken on cloudy days whenever practicable. Shading or other insulation is often worthwhile and should extend to 1 ft (300 mm) on either side of the gage. White paint over gage covers and the nearby area can reduce the problem. When readings must be taken on sunny days, they are best taken in the early morning.

Zero Drift

A discussion of the zero drift potential of vibrating wire transducers is included in Chapter 8.

Zero drift of vibrating wire and electrical resistance strain gages can be examined by either of two methods. First, the dummy gage procedure, described above, can be used. When using this procedure, readings should ideally be taken with the dummy gage at the same temperature as the structure, so that temperature strains are not a factor. However, this is often not possible, and attempts must be made to correct for temperature strains by measuring temperature of the dummy gages and applying temperature correction factors developed in the laboratory. Second, a *check gage* can be placed in a constant-temperature environment.

If the gaged structural member is used for temporary ground support, *no-load* gage readings should be taken after removal of the structural member and compared with the initial no-load readings. This procedure of course does not provide data for correcting gage readings at the time they are taken, nor does it provide data on any variation of drift with time, but it can give an idea of any total drift that has occurred. The data may not be exact, because any permanent yielding of the member will affect the measurements.

In cases where access for mechanical readings is available and where accuracy of a mechanical strain gage is adequate, an estimate of zero drift at any time can be made by making duplicate measurements with a mechanical strain gage.

Zero drift can be assumed to be common to a gage type, model, and batch number and can be measured or estimated and applied to all gages. However, any abnormal behavior of individual gages owing to their own peculiarities will invalidate this approach.

Although zero drift can be examined by use of the above procedures, the overriding issue is to minimize the problem by gage selection and installation methods that promote low drift characteristics.

Summary of Guidelines for Converting Measured Strain to Stress

The following outline summarizes the guidelines given in this section.

All Measurements

- Determine whether gages are thermally matched to structure by using laboratory tests. Develop any necessary temperature correction factors, either in the laboratory or by using the dummy gage procedure. If temperature correction factors are required, measure temperature at each gage location.
- Wherever possible, take readings when temperature is uniform throughout the member.
- Use measurement systems with low zero drift characteristics. If zero drift is judged to be a possible problem, use *dummy gage* procedure, *check gage* at constant temperature, and/or check the readings with a mechanical gage.

Measurements on Steel

- Use modulus of steel and any necessary temperature correction factors.

Measurements in or on Concrete

Use one of the following six options.

- Subject specimen of same concrete to same influences as prototype, in known stress field, and measure strain. Preferably, the specimen should be in the field, but sometimes measurements are possible in the laboratory. Concrete must have the same moisture conditions as active gages in the field. Use the strain/stress relationship determined from the specimen and any necessary temperature correction factors.
- Interpose a load-measuring device across at least one cross section, relating load to measured strains.
- Use creep curves developed for each particular project.
- Use averaged empirical data.
- Use concrete stress cells.
- Use combinations of the above options.

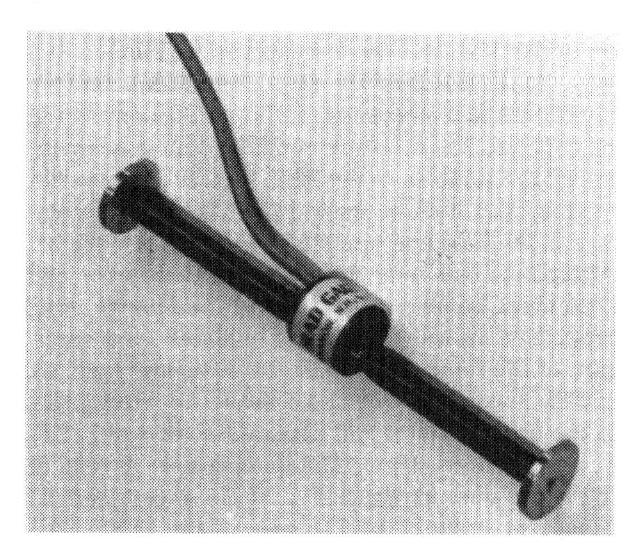

Figure 13.28. Vibrating wire strain gage for embedment in concrete (courtesy of Irad Gage, a Division of Klein Associates, Inc., Salem, NH).

13.4. EMBEDMENT STRAIN GAGES

The primary application for embedment strain gages is measurement of strain in concrete. Embedment strain gages in general use are described in Sections 13.4.1–13.4.4 and compared in Table 13.3.

When using embedment strain gages in concrete, most geotechnical applications involve simple cross sections subjected to compression and some bending, for example, driven piles and drilled shafts. We are not well able to monitor strains in complex cross sections involving rapid strain gradients, large tensile stresses, and/or cracked tensile sections.

13.4.1. Embedment Strain Gages: Mechanical Type

The mechanical type of embedment strain gage is identical in principle to the multiple telltales described in Section 13.3.1. They are sometimes used for monitoring strain when load transfer data are required while load testing driven piles and drilled shafts.

13.4.2. Embedment Strain Gages: Vibrating Wire Types

There are two types of embedment strain gage with a vibrating wire transducer. The first type is similar to the arc welded surface-mounted gage described in Section 13.3.2, except that large end flanges (Figure 13.28) replace the end blocks. Most gages are supplied with a preset tension, which can be specified by the user.

The second type of gage consists of a vibrating wire transducer mounted in the central portion of a length of steel bar, such as reinforcing steel. The use of a length of steel bar, embedded in concrete to measure strain, is sometimes referred to as a *sister bar,* presumably because the gaged bar exists alongside and behaves similarly to the reinforcing steel in the structural member. It is also referred to as a *rebar strainmeter.* Figure 13.29 shows a sister bar suitable for installation in a drilled shaft, to provide strain data during a load test. Guidelines for minimizing the *inclusion effect* of sister bars are given in Section 13.4.7.

13.4.3. Embedment Strain Gages: Electrical Resistance Types

Embedment strain gages with electrical resistance transducers can be grouped into five categories:

Bonded Foil or Weldable Resistance Gages Attached to Central Portion of Length of Steel Bar

These gages are also referred to as *sister bars*. Two or four foil or weldable gages are installed on the central portion of a length of steel bar, such as reinforcing steel, and connected in a half or full bridge

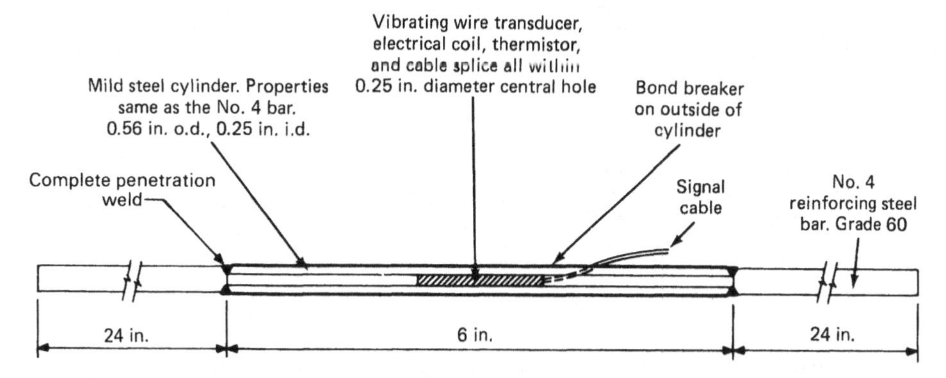

Figure 13.29. Schematic of *sister bar* with vibrating wire transducer.

network. Guidelines on minimizing the *inclusion effect* of sister bars are given in Section 13.4.7.

Unbonded Gages

The primary unbonded strain gage transducer for geotechnical applications is the Carlson *elastic wire strain meter,* shown in Figure 13.30. The instrument is stabilized by heat treating the resistance wire during manufacture, by filling with oil so that the resistance wire is totally immersed at all times, and by hermetically sealing to prevent the entrance of moisture. Accuracy is maximized and temperature effects are reduced by using two separate lengths of elastic wire, one of which increases in resistance with strain, while the other decreases, and by

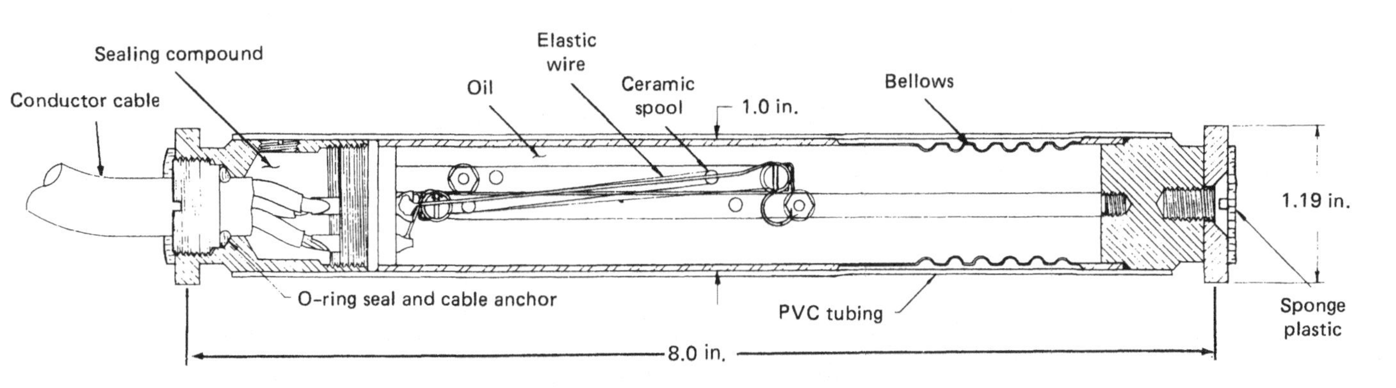

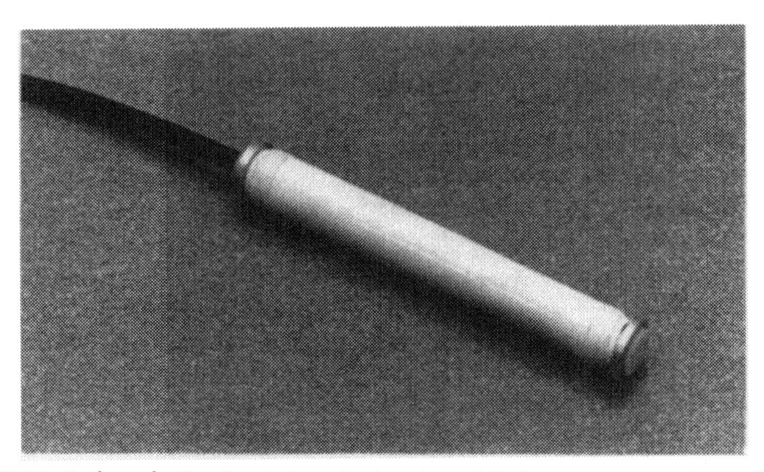

Figure 13.30. Carlson elastic wire strain meter (courtesy of Carlson Instruments, Campbell, CA).

Table 13.3. Embedment Strain Gages

Gage Type	Advantages	Limitations	Typical Gage Length	Typical Range (microstrain)	Sensitivity (microstrain)	Approximate Accuracy (microstrain)
Multiple telltales	Simple Inexpensive	Requires access to telltales (but can be converted to remote reading device)	Unlimited	Unlimited	Depends on application	±25–400
Vibrating wire; type similar to arc welded surface-mounted gage (e.g., Figure 13.28)	Lead wire effects minimal No conformance problem Remote readout Readout can be automated Factory waterproofing	Cannot be used to measure high-frequency dynamic strains Special manufacturing techniques required to minimize zero drift Need for lightning protection should be evaluated	5–10 in. (130–250 mm)	3000	0.2–2	±5–50
Vibrating wire; *sister bar* type (Figure 13.29)	Robust Easy to install Lead wire effects minimal Remote readout Readout can be automated Factory waterproofing	Cannot be used to measure high-frequency dynamic strains Special manufacturing techniques required to minimize zero drift Special design features required to minimize inclusion effects Sister bar must be small relative to size of structural member Need for lightning protection should be evaluated	Debonded length, plus 100× bar diameter	3000	0.2–2	±5–50
Bonded foil or weldable resistance gage; *sister bar* type (similar to Figure 13.29)	Robust Easy to install Suitable for monitoring dynamic strains Remote readout Readout can be automated Factory waterproofing	Special design features required to minimize inclusion effects Low electrical output Lead wire effects Errors owing to moisture, temperature, and electrical connections are possible Need for lightning protection should be evaluated	Debonded length, plus 100× bar diameter	20,000	1–4	±1–100

Table 13.3. (*Continued*)

Gage Type	Advantages	Limitations	Typical Gage Length	Typical Range (microstrain)	Sensitivity (microstrain)	Approximate Accuracy (microstrain)
Unbonded resistance (e.g., Figure 13.30)	Long good performance record No conformance problem Provides temperature measurement Remote readout Readout can be automated Factory waterproofing Suitable for monitoring dynamic strains but only up to about 25 Hz	Low electrical output Lead wire effects Errors owing to moisture, temperature, and electrical connections are possible Need for lightning protection should be evaluated	4–20 in. (100–500 mm)	3000	1.5–6	±20–75 (±2% full scale)
Mustran cell (Figure 13.31)	Robust Modulus matched Remote readout Readout can be automated	Large size Not available commercially Low electrical output Lead wire effects Errors owing to moisture, temperature, and electrical connections are possible Need for lightning protection should be evaluated	6–10 in. (150–250 mm)	1000	1–4	±10–50
Plastic encased gage (e.g., Figure 13.32)	Low cost Robust Easy to install No conformance problem Remote readout Readout can be automated Suitable for monitoring dynamic strains	Unstable in long term Low electrical output Lead wire effects Errors owing to moisture, temperature, and electrical connections are possible Need for lightning protection should be evaluated	0.5–10 in. (13–250 mm)	20,000	1–4	±10–50
Eaton Corporation gage (Figure 13.33)	No conformance problem Remote readout Readout can be automated Factory waterproofing Suitable for monitoring dynamic strains	Must be cast in a briquette before installation Low electrical output Lead wire effects Errors owing to moisture, temperature, and electrical connections are possible Need for lightning protection should be evaluated	2–6 in. (50–150 mm)	20,000	1–4	±10–50

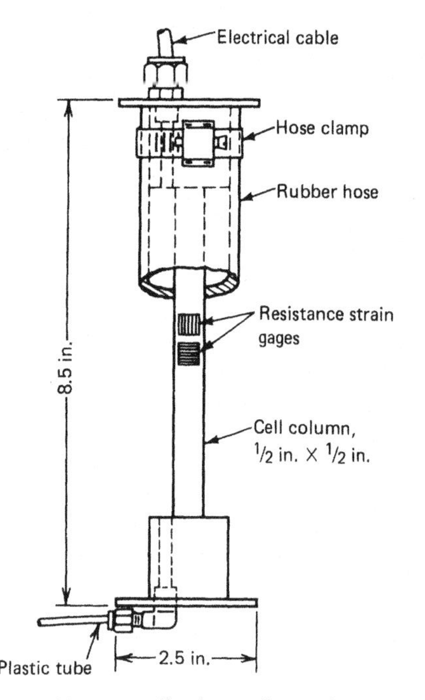

Figure 13.31. Mustran cell (after Barker and Reese, 1969).

Figure 13.32. Polyester mold strain gage for embedment in concrete (courtesy of Texas Measurements, Inc., College Station, TX).

measuring the ratio of the two resistances. It has a more than 40 year good performance record and is the first choice of many users for measuring strain within concrete. Variations of the Carlson elastic wire strain meter are the *joint meter,* for measuring movements across joints in concrete, and the *miniature strain meter,* for use in model studies. Details of the instruments, including installation and reading procedures, are given by Carlson (1975). Wiring diagrams are given by the Corps of Engineers (1980).

Mustran Cell

The *Mustran cell* was developed by personnel of the University of Texas under sponsorship of the Center for Highway Research (Barker and Reese, 1969), primarily for measuring strains in drilled shafts. The name is derived from *mu*ltiplying *s*train *tran*sducer. Strain gages are bonded to a steel rod (Figure 13.31) with square cross section. The rod has a large flange on each end and is surrounded by a length of rubber hose. The design attempts to match the stiffness of the cell to the stiffness of the displaced concrete, thereby ensuring conformance. The relationship depends on the size of the column and the diameter of the rubber hose. The plastic tube shown in Figure 13.31 is used for maintaining a nitrogen pressure, to minimize the possibility of false readings owing to moisture intrusion. The cell is pressure tested and filled with dessicant prior to installation.

Plastic Encased Gages

A *plastic encased gage* consists of a standard wire-type resistance strain gage, to which lead wires are attached, and hermetically sealed between thin plastic plates. The plastic is either given an irregular surface or coated with a coarse grit to promote bonding to the concrete. Quarter bridge gages are the most common, with either two or three lead wires, but gages molded into rosette forms are available for monitoring two-dimensional strains. The gage works well under dynamic loading conditions (Hirsch et al., 1970) but is suitable for short-term measurements only. Attempts by the author to use the version shown in Figure 13.32 for long-term tests in precast concrete piles showed severe gage drift, presumably caused by creep in the polyester, and usage appears to be limited to tests of 1 or 2 days duration.

Eaton Corporation Gage

The embedment strain gage manufactured by Eaton Corporation (formerly manufactured by Ailtech) consists of a thin resistance element placed within a tube in a similar way to the surface-mounted gage shown in Figure 13.24, and wired as a quarter bridge three-wire system (Figure 13.33). The annular space between the tube and resistance element is filled with magnesium oxide powder to ensure frictional contact with the concrete. The gage is very delicate and must be precast in a briquette prior to installation.

13.4.4. Miscellaneous Embedment Strain Gages

Various *electrical crack gages* (Section 12.3.2) can be packaged for measurement of strain in concrete. The *soil strain gage* (Section 12.6.5) can also be

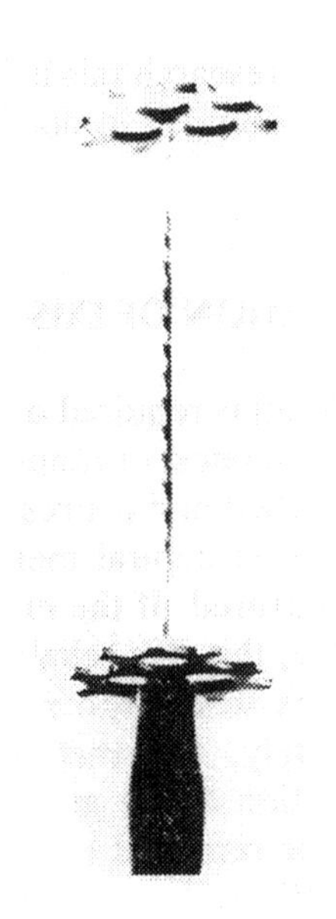

Figure 13.33. Strain gage for embedment in concrete (courtesy of Eaton Corporation, Los Angeles, CA).

used. The *sliding micrometer* (Section 12.5.4) is reported to be capable of strain measurements in concrete to an accuracy of ± 2 microstrain.

13.4.5. Guidelines on Use of Embedment Strain Gages

Criteria for the selection of strain gage length, cable, connectors, locations, and initial readings are all similar to the criteria given in Sections 13.3.5, 13.3.6, and 13.3.8 for surface-mounted strain gages. Additional guidelines for the use of embedment strain gages are given in the following sections.

13.4.6. Installation of Embedment Strain Gages

General guidelines on the installation of instruments are given in Chapter 17. The following additional guidelines apply to the installation of embedment strain gages.

Embedment strain gages are usually installed by tying them with soft iron wire to nearby reinforcing bars. In general, the tie wires should be aligned at 90 degrees to the gage axis, so that any movement of reinforcing bars during concrete placement will not cause a direct pull on the gage and resultant damage. Tie wires on vibrating wire gages fitted with end flanges should be wrapped around the protective tube, leaving the flanges free. Care should be taken to keep concrete vibrators away from gages, and concrete should not be poured directly on to a gage. When the gage is delicate or the risk of damage large, gages can be precast in briquettes of concrete identical to that used in the structure, and if necessary the larger aggregate can be removed. Briquettes **must** be cast not more than 24–48 hours ahead of the main concrete pour to ensure conformance.

The need for protection to gage cable depends on the cable type, and if heavy duty or armored cable is used, additional protection is usually unnecessary. Cables are normally routed alongside reinforcing steel to the exit point and tied at frequent intervals with plastic cable ties. A blockout box can often be attached to the inside of the form at the exit point, and a coil of cable stored within the box. After the form is stripped, the cable can be uncoiled and either terminated or extended to its terminal location.

13.4.7. Relationship Between Strain and Stress*

Relationships between strain and stress for surface measurements are discussed in Section 13.3.9 and apply also to measurements made **within** concrete.

An additional problem for embedment strain gages is the *inclusion effect*, whereby the presence of the gage may distort the strain field so that the measured strain is significantly different from the strain that would occur if the gage were not present; that is, there will be poor conformance.

The influence of the gage on measured strain depends on both the stiffness of the gage relative to the concrete and on geometry. If the stiffness of the gage is equal to that of the concrete, conformance requirements will be satisfied and the strain measurements will be correct. If the stiffnesses do not match, the measurement error depends on the geometry of the gage and structural member, and the gage will be more accurate if its stiffness is less than that of the concrete. Loh (1952) indicates that when the stiffness of the gage is less than that of the concrete, the error will not be excessive provided that the length of the gage is greater than five times its diameter. This conclusion applies primarily to measurements of compressive stress, where bond between the concrete and gage is not an issue, and to the case where the mass of the surrounding concrete is large relative to the size of the gage.

*Written with the assistance of Stanley L. Paul, Associate Professor, Department of Civil Engineering, University of Illinois at Urbana–Champaign, IL.

The use of *sister bars,* which are generally very long relative to their diameter, is based on a somewhat different approach. The bar is long enough on each side of the gaged portion at the midpoint, so that adequate bond is created between the bar and concrete, and therefore at the gaged portion the strain in the bar and concrete are equal. Also, the bar should be small relative to the size of the structural member, so that the response of the member is not influenced by the presence of the bar. To satisfy bond length requirements, the bond length at each end of the bar shown in Figure 13.29 should be equal to or greater than about 50 times the diameter of the bar. For such a long bar, the inclusion effect is small, even when the gage is stiff relative to the concrete. Threaded connections should not be included in the bar, as the threads tend to create a zone of the bar with lower stiffness than the unthreaded portion. The central portion of the bar around the strain gage should be debonded from the concrete for a minimum length of 6 in. (150 mm) to ensure an adequate gage length. This debonded length should not be included in the length calculated for bond requirements, and hooks can be provided at each end of the bar to assist with developing full bond length. If the cross section at the gage is different from that elsewhere, strain will not be uniform along the bar and measurements will be misleading. A goal in the design of sister bars should therefore be to create a uniform cross section throughout the debonded length and the length required for bond. This goal is satisfied by the configuration shown in Figure 13.29 and essentially also by the sister bar described by DiBiagio (1983). The length of the cylindrical part of the gage shown in Figure 13.29 results from the need to avoid damaging the transducer, thermistor, electrical coil, and cable splice while welding the cylinder to the reinforcing bar. The diameter of the sister bar does not need to be the same as the reinforcing steel in the structural member: it should be the minimum required for gage mounting.

When the size of the structural member is large relative to the size of a sister bar, the author believes that sister bars of the type described above should be given first consideration when selecting an embedment strain gage. They are robust, reliable, and easy to install. They should preferably have vibrating wire transducers. Despite this recommendation, and despite the belief that the inclusion effect is small, there is a need to document the accuracy of this type of gage, and practitioners who are in a position to research this issue and to publish the results are encouraged to do so.

13.5. DETERMINATION OF EXISTING STRESS

Where measurement is required of existing stress in in-place steel struts, piles, or tunnel supports, strain gages can be attached and a stress relief technique used, whereby the structural member is unloaded and the strain measured. If the member can be unloaded completely, this is a reliable method.

For those cases where the member cannot be unloaded completely, two methods are possible. First, a small portion of the member including the strain gages can be removed by cutting around the gages to stress relieve the material and the gages. However, fabrication stresses in the steel member and stresses associated with the cutting operations may complicate the results considerably, and the results may be misleading. This point is clearly illustrated by DiBiagio (1977), who describes laboratory tests to examine the effect of cutting and welding. Vibrating wire strain gages were mounted on each face of an unloaded steel plate that had previously been burned out of a larger plate. The plate was sequentially reduced in size by cutting with a metal saw, with an oxygen-acetylene torch, and again with a metal saw. Measured apparent stress changes ranged from a tensile stress of 2600 lb/in.^{2} (18 MPa) to a compressive stress of 11,700 lb/in.^{2} (80 MPa) with respect to the starting conditions. The second method for use on steel is the *hole-drilling strain gage method,* also referred to as the *blind hole drilling method* (Measurements Group, 1985). A rosette of electrical strain gages is bonded to the structural surface and a short blind hole drilled in the center of the rosette, using a milling guide (Figure 13.34). Strains created by the drilling are analyzed to determine existing stress. However, it seems reasonable to assume that this method may also suffer from inaccuracies caused by fabrication stresses and stresses associated with the drilling operation.

Attempts have been made to determine existing stress in concrete by using a stress relief *overcoring* procedure. The procedure was developed for determination of in situ stress in rock and entails setting a borehole deformation gage or strain cell (Chapter 11) within a small-diameter borehole and overcoring the gage with a larger-diameter borehole. Gage

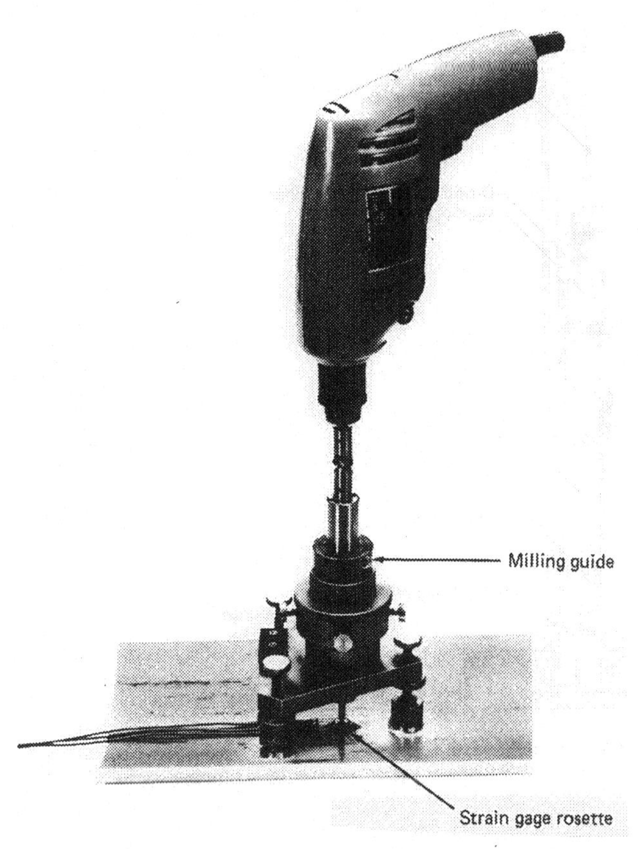

Figure 13.34. ***Hole-drilling strain gage method*** **for determination of existing stress (courtesy of Measurements Group, Inc., Raleigh, NC).**

measurements provide strain relief data, which are converted to stress by an analysis using an assumed or measured modulus. The U.S. Bureau of Mines overcoring procedure is described by Hooker and Bickel (1974) and Hooker et al. (1974). Although the overcoring procedure is used satisfactorily in rock, it has questionable value for measurements in concrete. While drilling to relieve stresses, cooling water is used on the bit, and strains resulting from concrete swell usually mask out strains resulting from stress relief. This problem does not usually arise in rock, because the rock will generally be below the water table. The method can, however, be used to determine approximate relative stress magnitudes in concrete. If a part of the same concrete is under a known stress, for example, zero stress, an approximate absolute stress determination can be made by making several overcoring tests also at this location and using the results as a "calibration" to account for strain resulting from swell at other locations.

13.6. CONCRETE STRESS CELLS

Problems of converting strain to stress in concrete have been discussed earlier in this chapter. An alternative approach is to measure stress directly, by using *concrete stress cells,* which are similar to the earth pressure cells described in Chapter 10.

Carlson (1984) indicates that the requirements for a concrete stress cell include the following:

1. It must not change appreciably the existing stress.
2. It must be embedded so as to be in close and intimate contact with the concrete.
3. Its composite modulus of elasticity must be at least one-half that of the concrete.

Concrete stress cells are subjected to a condition that does not exist for earth pressure cells. As concrete cures, its temperaure increases, and the temperature of the stress cell is therefore increased. The possibility therefore exists for the cell to expand, pushing the weak concrete away from the cell and, when the concrete takes its initial set and cools, for the cell to cool, contract, and uncouple itself from the surrounding concrete. In this condition the cell would not be responsive to subsequent stress changes in the concrete. The two approaches adopted for solving this problem are described in Sections 13.6.1 and 13.6.2.

Concrete stress cells have one notable weakness: they cannot be used to measure tensile stress. Carlson (1984) recommends that, whenever tensile stress is anticipated, an embedment strain gage should be installed alongside each stress cell. When the stress changes from compression to tension, the stress cell provides a way of knowing when the stress is zero, an all important reference point for interpretation of strain measurements.

13.6.1. Carlson Stress Meter

The *Carlson stress meter,* shown in Figure 13.35, consists of a mercury-filled circular cell with a Carlson elastic wire strain meter mounted on one face. The instrument has a composite elastic modulus of at least one-half that of concrete.

The problem of uncoupling between the meter and concrete during curing is overcome by minimizing the thickness of the mercury film. Carl-

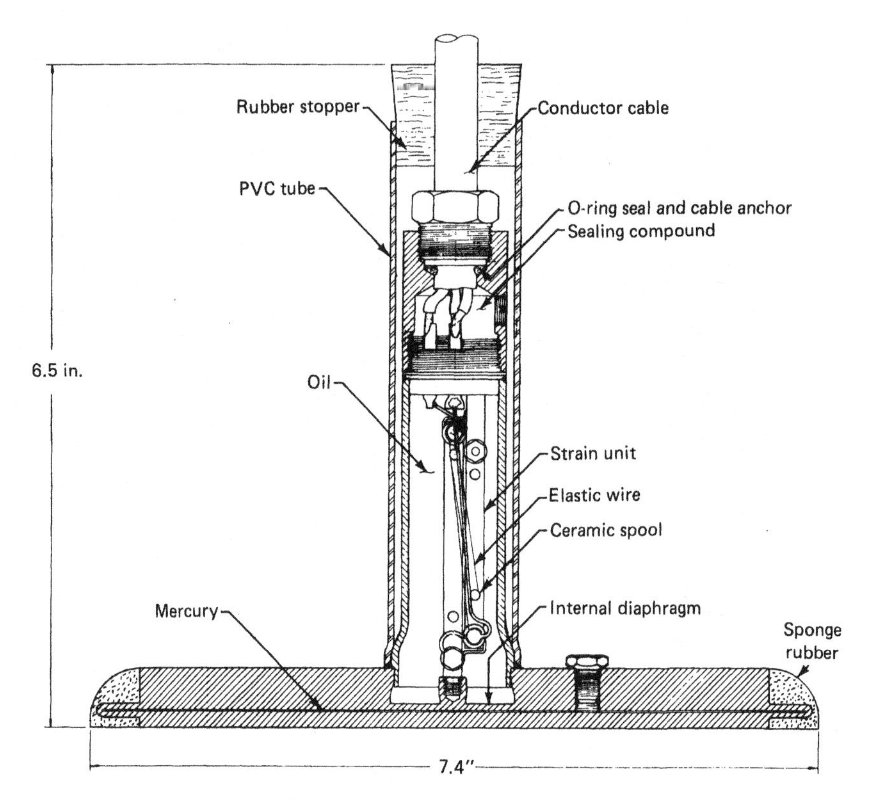

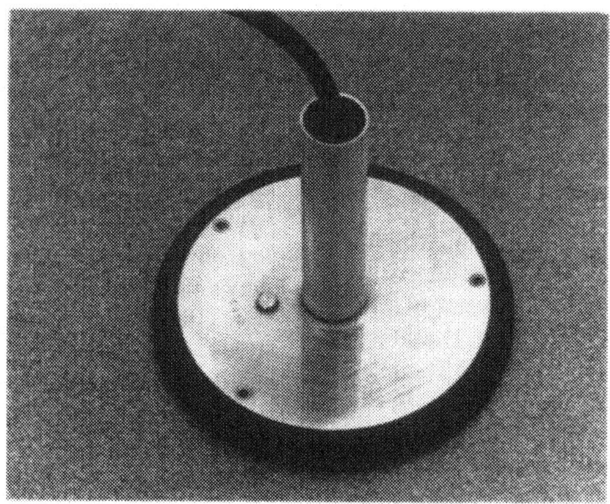

Figure 13.35. Carlson concrete stress meter (courtesy of Carlson Instruments, Campbell, CA).

son (1984) indicates that the mercury film should not be more than 0.011 in. (0.28 mm) thick and that even the finest machining is not adequate for preventing tiny grooves into which the mercury can squeeze under pressure. The surface tension of mercury is so great that it tends to span over even the slightest groove or imperfection, and the inner steel surfaces should be polished to a tolerance of 6 $\times 10^{-6}$ in. (150 $\times 10^{-6}$ mm) RMS. After assembly, each meter should be tested for compressibility under a pneumatic pressure, using an interferometer to measure compressibility at the low end of the range. Pirtz and Carlson (1963) describe comprehensive laboratory tests indicating that the meters, used individually, are suitable for use in concrete containing 6 in. (150 mm) aggregate in applications where accuracy of 10–15% is satisfactory. Carlson (1978) describes additional laboratory tests, concluding that the worst error registered during the tests was 12% of the major principal stress, and this only happened when the load was applied more quickly than would be the case in a field structure.

The standard installation procedure for measuring stress in a horizontal plane involves installing the meter on the surface of previously cured concrete and later covering with fresh concrete. Great care must be taken during installation to ensure good contact between the meter and surrounding concrete, using a technique described by Carlson (1975). The technique includes preparing the surface, placing a mortar pad containing steel filings to increase the modulus, painting a film of grout on the mortar pad, pressing the meter into place with a rotary motion to squeeze the mortar outward, weighting the meter temporarily with a metal tripod, placing the overlying concrete, and removing the metal tripod.

An alternative installation procedure (Green, 1985) was used at the Mount Baker Ridge Tunnel in Seattle, where restricted access and potential damage precluded use of the standard installation procedure. Meters were cast, with the active face vertical, within a 12 in. diameter by 12 in. long (300 mm by 300 mm) briquette, cast 24–48 hours ahead of the main concrete pour. Laboratory tests on the briquette, in a concrete testing machine, indicated an almost 1 : 1 ratio between applied and measured stress, and field data appear very reasonable. The small size of the briquette apparently allowed temperature to dissipate rapidly enough so that the meters did not become uncoupled from the concrete. The young age of the concrete in the briquette at the time of the main concrete pour is vital to the success of this procedure.

13.6.2. Glötzl Cell

The Glötzl cell, shown in Figure 13.36, is similar to the hydraulic type of earth pressure cell, with pneumatic transducer, described in Chapter 10. For measurements greater than 300 lb/in.^{2} (2 MPa), the transducer is pressurized with oil rather than gas. The cell is provided with a post-stressing tube, long enough to protrude from the concrete pour. If a length longer than 30 ft (10 m) is required, a check valve is inserted in the post-stressing tube near the cell. After the concrete has set, the tube is pinched repeatedly until the pressure measurement indicates that the cell has regained contact with the concrete. Thereafter, provided that the thickness of the mercury film is restricted to approximately the same as in the Carlson stress meter, and the composite modulus of elasticity of the cell is at least one-half that of the concrete, the cell will function correctly.

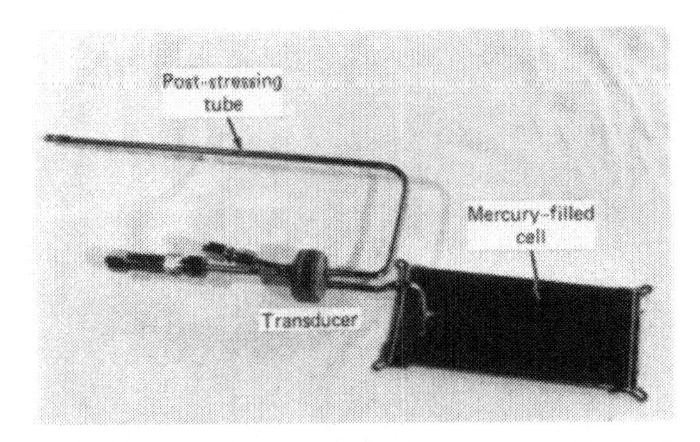

Figure 13.36. Glötzl concrete stress cell (courtesy of Glötzl GmbH, Karlsruhe, West Germany, and Geo Group, Inc., Wheaton, MD).

The cell is installed by wiring with soft iron wire to nearby reinforcing steel, with care to avoid entrapment of air as concrete is placed. For measurements in concrete containing aggregate larger than 1 in. (25 mm), it is advisable to surround the cell with a special concrete mix from which the coarser particles of aggregate have been removed. Alternatively, the cell can be installed in a briquette of concrete as described previously for the Carlson stress meter, ensuring that the briquette is cast within 24–48 hours of the main concrete pour.

13.6.3. Other Concrete Stress Cells

A post-stressing tube arrangement, similar to the Glötzl cell, has been adopted by manufacturers of concrete stress cells with vibrating wire transducers.

Interfels manufactures a concrete stress cell with post-stressing tube and a hydraulic/mechanical transducer, and changes in cell pressure create movement within a measuring unit that is monitored with a dial indicator.

CHAPTER 14

MEASUREMENT OF TEMPERATURE*

14.1. APPLICATIONS

Temperature measurements in geotechnical engineering fall into three general categories: first, where temperature itself is a primary parameter of interest; second, where temperature change causes a real deformation or stress change in the ground or in a structure; and third, where a transducer itself is sensitive to temperature change. Examples in each category are given in Sections 14.1.1–14.1.3.

14.1.1. Temperature as a Primary Parameter of Interest

Temperature measurements are often made while supporting excavations by ground freezing, while limiting the temperature of concrete during curing, during construction of projects in permafrost, during frost heave and frost penetration studies, and while studying ground response to high temperatures during studies for underground disposal of nuclear waste.

Birman et al. (1971) describe thermal monitoring in an earth dam to detect leakage and to monitor changes in the leakage pattern. Highways are frequently damaged during the "mud season," in spring when a subsurface frozen layer remains although the surface is unfrozen. Traffic speeds are often restricted until complete thaw occurs, using measurements of temperature to indicate the extent of freezing.

Where exposed rock cuts or tunnels are subjected to low temperatures, freezing at the surface can cause ice dams to form, holding unfrozen water in the rock mass with resulting buildup of joint water pressure and decrease in stability. Temperature measurements will show the depth and extent of the freezing problem. Finally, ambient temperature should be measured when temperature change may cause changes in other measured parameters, so that cause–effect relationships can be established.

14.1.2. Temperature Change Causing Real Deformation or Stress Change

A temperature increase causes most materials to expand and, if expansion is resisted, temperature increase will cause a compressive stress increase. Thus, steel struts across an open cut excavation can be subjected to substantial compressive stresses when heated by the sun: the greater the end restraint the greater is the stress. The same phenomenon is illustrated by thermal expansion of railroad tracks.

The coefficient of linear expansion is the change in length, per unit of length, for a temperature change of 1 degree. A bar, free to move, will increase in length with an increase in temperature and will decrease in length with a decrease in tempera-

*Written with the assistance of Ronald T. Atkins, Chief, Technical Services Division, U.S. Army Cold Regions Research and Engineering Laboratory, Hanover, NH, and J. Barrie Sellers, President, Geokon, Inc., Lebanon, NH.

Table 14.1. Linear Coefficients of Thermal Expansion

Material	Linear Coefficient of Thermal Expansion × 10^{-6} per Degree, at 20°C[a]	
	°C	°F
METALS AND ALLOYS		
Aluminum, wrought	23.1	12.8
Brass	18.8	10.4
Copper	16	9
Invar	1.4	0.8
Iron, cast, gray	10.6	5.9
Iron, wrought	12.0	6.7
Steel, mild	11.7	6.5
Steel, stainless, 18-8	17.8	9.9
Super-invar	0.4	0.2
STONE AND MASONRY		
Brick masonry	6.1	3.4
Concrete	10–13	5.5–7.2
Granite	8.0	4.4
Limestone	7.6	4.2
Marble	8.1	4.5
Sandstone	9.7	5.4
Slate	8.0	4.4

[a]Coefficients of many materials increase very substantially with increasing temperature.

Source: After American Institute of Steel Construction.

ture. The change in length will be $\varepsilon \Delta T l$, where:

ε = linear coefficient of thermal expansion,
ΔT = change in temperature,
l = length.

If the ends of the bar are fixed, a change in temperature will cause a change in the unit stress of $E\varepsilon\Delta T$, where:

$$E = \text{modulus of elasticity.}$$

As an example of stress magnitude when end restraint is total, a 1°C temperature increase causes a compressive stress increase of 340 lb/in.^{2} (2.3 MPa) in mild steel. Linear coefficients of thermal expansion are given in Table 14.1.

Sellers et al. (1972) give an example of temperature-induced structural distress in a 60 ft (18 m) wide mine opening in competent marlstone. During the winter the roof rock contracted, loosened, and sagged downward, and several roof falls occurred. No such falls occurred during summer months.

14.1.3. Transducer Sensitive to Temperature Change

The third category for temperature measurements, where a transducer itself is sensitive to temperature change, can be illustrated by hydraulic systems such as earth pressure cells and hydraulic load cells. Temperature changes cause expansion or contraction of the liquid in the cell—thus an indicated pressure or load change—and temperature corrections must be applied if these changes are significant. Similarly, strain gages and extensometers will, if their thermal expansion and contraction characteristics are not matched with the monitored structure, show extraneous sensitivity to temperature.

Three approaches are possible for taking account of temperature sensitivity: matching thermal coefficients of instrument and structure, use of low coefficient of expansion materials such as invar or super-invar, and application of temperature corrections. The first two approaches are often only partially successful, owing to design and/or cost limitations. The third approach requires measurement of temperature and use of temperature sensitivity relationships either provided by the instrument manufacturer or developed by users for their site-specific instrument–structure interaction. It also requires ensuring that the temperature transducer is in good thermal contact with the instrument to be corrected.

14.2. MERCURY THERMOMETER

The familiar mercury thermometer is useful for spot measurements at accessible locations. However, mercury thermometers are fragile, are not suitable for remote readout, and are limited to temperatures above about −30°C. They have a much larger thermal mass than most other devices for measurement of temperature. They are therefore a poor choice where measurements of rapidly changing temperature are required or where their use might alter the thermal characteristics of the material to be monitored. Maximum and minimum thermometers are available for indication of extreme temperatures.

14.3. BIMETAL THERMOMETER

Bimetal thermometers use a bimetallic element in which two strips of different metals are bonded together back to back. The different coefficients of

Figure 14.1. Bimetal thermometer with clock-driven recorder (courtesy of Omega Engineering, Inc., Stamford, CT).

thermal expansion of the two metals cause the strip to bend when cooled or heated. The strip is usually wound into a spiral, so that temperature changes cause the spiral to wind or unwind and twist a pointer over a circular scale. These devices are rugged, but accuracy is usually low if the temperature range is wide. However, narrow range bimetal thermometers with ½–1% full-scale accuracy are available. Figure 14.1 shows a clock-driven 7 day recorder, suitable for ambient temperature measurements. Accuracy is 2% full scale.

Johnston (1966) describes a self-contained ground temperature recorder in which a bimetallic strip moves a pointer on pressure-sensitive recorder paper. The device is battery powered and has a range of 11°C and an accuracy of about ±0.3°C. The instrument is 3.75 in. (95 mm) in diameter by 11 in. (280 mm) long and is designed for installation in a borehole for recording ground temperature over a 1 year period without attention or servicing.

14.4. THERMISTOR

The name is derived from *therm*ally sensitive res*istor*. A thermistor is composed of semiconductor material that changes its resistance very markedly with temperature. Lead wires are used to connect the thermistor to a measuring instrument. Readout may be in resistance units (ohms), requiring the use of commercially available digital multimeters and resistance–temperature tables, or directly in tem-

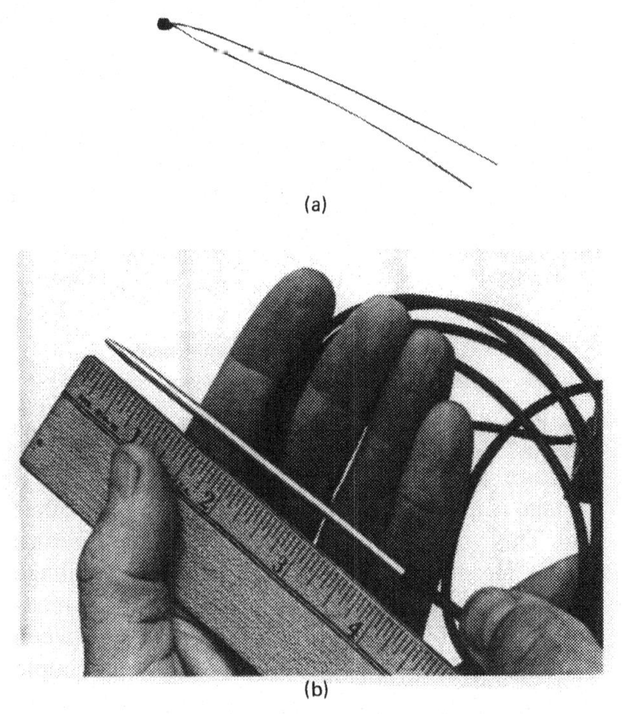

Figure 14.2. Thermistors: (a) suitable for attachment to another instrument for monitoring its temperature so that a temperature correction can be applied and (b) for use as a portable probe or for permanent embedment (courtesy of Atkins Technical, Inc., Gainesville, FL).

perature units by using a **matched** thermistor and readout instrument. Any device for reading thermistors must apply only a very small current, since the temperature of the thermistor will be changed significantly when as little as 0.001 watt is applied. The old type of volt–ohm–milliammeter (VOM) causes too much self-heating and should not be used.

Thermistors are most often used in a glass bead configuration, although they are available as disks, washers, rods, and in sizes as small as pin heads. Two configurations are shown in Figure 14.2. Thermistors are capable of highly accurate temperature measurements. For this reason they are often used where accuracy of ±0.1°C or better is required over a wide temperature range. Robertson et al. (1966) present comprehensive guidance for use of thermistors.

14.5. THERMOCOUPLE

A *thermocouple* is composed of two wires of dissimilar metals, with one end of each wire joined

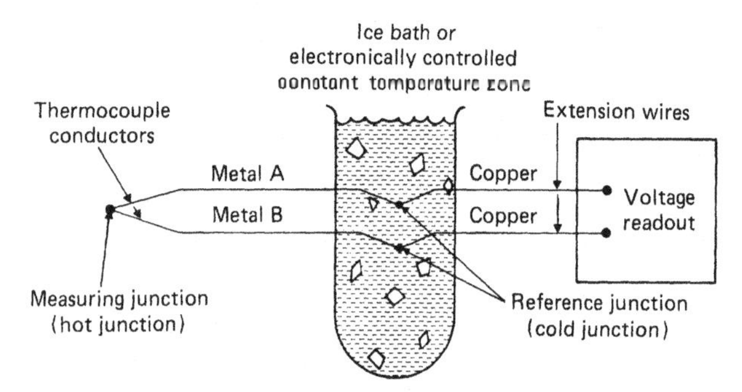

Figure 14.3. Schematic of thermocouple.

together to form a measuring junction. At any temperature above absolute zero (−273°C), a small voltage is generated between the wires at the other end. This voltage is proportional to the temperature of the measuring junction. However, the voltage cannot be read directly, since connecting any readout device will create two more contacts between dissimilar metals and thus unwanted thermocouple connections. To solve this problem, thermocouples must be "referenced" as shown in Figure 14.3, with the readout device connected to similar metals so that no thermocouple is formed. Modern thermocouple readout devices have internal electronic reference junctions, allowing direct connection to the thermocouple leads. They must be matched to a specific type of thermocouple: for example, an iron–constantan thermocouple cannot be read with a readout device designed for a copper–constantan thermocouple.

Thermocouples are widely used in applications requiring modest accuracy (±0.5°C over a wide temperature range) and in large installations requiring data logging equipment. Readout instruments read directly in temperature units and are available as battery operated units for field applications. Figure 14.4 shows an example.

Thermocouple manufacturers should be consulted for guidance when selecting thermocouple components. Factors that must be considered include type of thermocouple, proper installation for the wires, readout equipment, temperature range, length of service required, lead length, and the ambient temperature to which readout equipment will be subjected in the field. Copper–constantan (Type T) thermocouples are the most frequent choice for geotechnical applications. Properly selected and installed systems are capable of good long-term performance, meeting specification limits for 5 years and longer. Other sources that provide comprehen-

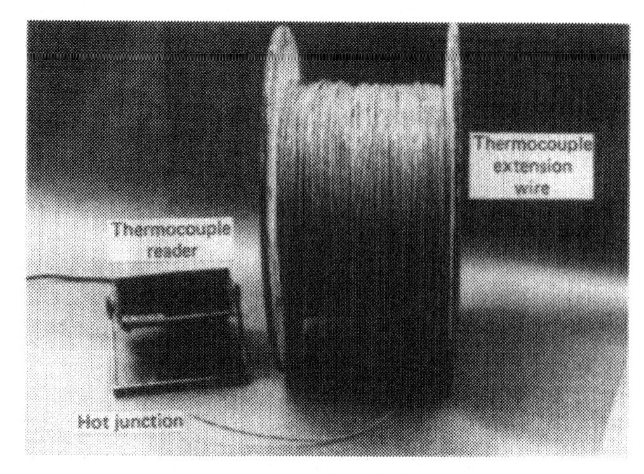

Figure 14.4. Thermocouple (courtesy of Conax Buffalo Corporation, Buffalo, NY).

sive guidance on the use of thermocouples are ASTM (1974) and ISA (1974).

14.6. RESISTANCE TEMPERATURE DEVICE (RTD)

Resistance temperature devices depend on the principle that change in electrical resistance of a wire is proportional to temperature change. The wire is usually mounted on a postage stamp-sized backing or wound on a small-diameter coil. Wire material is usually platinum, copper, nickel, or a copper–nickel alloy. Platinum has the greatest accuracy, linearity, stability, and repeatability but copper–nickel alloys have a much greater sensitivity. Readout devices normally contain a Wheatstone bridge circuit, since for a 100 ohm RTD a 10°C temperature change normally creates a resistance change of only 2 or 3 ohms. Figure 14.5 shows an example of an RTD.

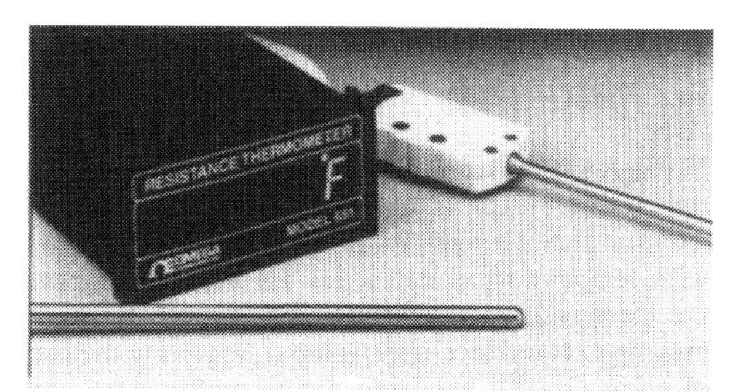

Figure 14.5. Resistance temperature device (courtesy of Omega Engineering, Inc., Stamford, CT).

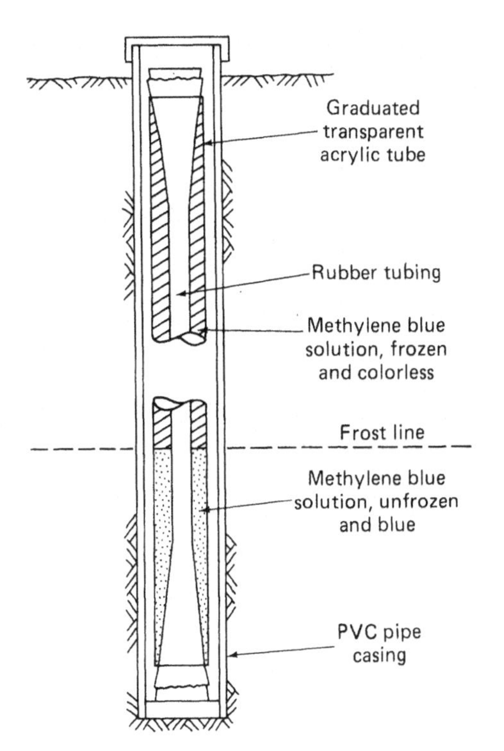

Figure 14.6. Schematic of Gandahl frost gage, used to measure depth to frost line.

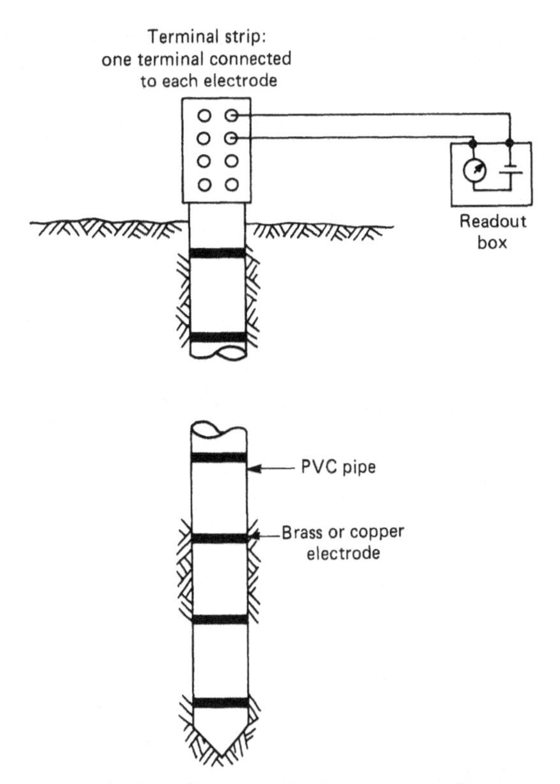

Figure 14.7. Schematic of Bergau frost gage, used to measure depth to frost line.

RTDs have not been used widely for measurement of temperature in geotechnical applications. However, their excellent stability and accuracy makes them well suited to long-term installations (15–20 years) where high accuracy is a requirement.

14.7. FROST GAGES

Two gages are described by Gandahl and Bergau (1957) for measuring the depth to the frost line. The Gandahl gage, Figure 14.6, consists of a transparent acrylic tube containing distilled water colored with methylene blue. A PVC pipe casing is installed in the ground, usually to a depth of between 4 and 8 ft (1.2–2.4 m) and the acrylic tube is left in place within the PVC pipe. To read the gage, the acrylic tube is withdrawn. Unfrozen methylene blue solution is blue but becomes colorless when frozen; thus, the depth to the frost line can be determined from the color transition, using graduations on the acrylic tube.

Although a simple and inexpensive device, the Gandahl gage will be in error if the freezing point of the groundwater is depressed to below 0°C, which is generally the case in highway applications where salt has been used on the pavement. The Bergau gage, shown in Figure 14.7, overcomes this problem. The electrical conductivity of frozen soil is much lower than that of unfrozen soil. Therefore, by attaching electrodes to a nonconductive pipe and installing the pipe in the ground, conductivity measurements between adjacent electrodes will indicate the depth to the frost line. The electrode spacing depends on the required accuracy. The gage must be read using alternating current, since direct current causes polarization of ions in the soil, leading to erroneous readings. The PVC pipe can be either driven into the ground while the soil is unfrozen or installed in an augured hole.

Atkins (1979) describes the design and evaluation of two resistivity gages for determining frost penetration and indicates their superiority to temperature transducers when salt has been used to depress the freezing point or when measurements are required during spring thaw. During spring thaw, subsurface temperatures often become nearly isothermal at 0.0°C, making it difficult to establish the frost line by measuring temperature. Atkins recommends use of both temperature and resistivity gages when

Table 14.2. Comparison Among Transducers for Remote Measurement of Temperature in Geotechnical Environments

Feature	Thermistor	Thermocouple	RTD
Readout	Digital ohmmeter or multimeter	Thermocouple reader	Wheatstone bridge with millivolt scale
Sensitivity	Very high	Low	Moderate
Linearity	Very poor	Fair	Fair
Accuracy	High	Moderate	Very high (but may be reduced by lead wire effects)
Stability	Excellent	Good	Excellent
Type of lead wire	Two-conductor	Special (bimetal)	Three-conductor
Repairability of lead wire	Straightforward	Less straightforward (can cause errors)	Straightforward
Temperature range	Wide	Wide	Wide
Rapidity of response	Rapid	Rapid	Rapid
Applicability for instrument temperature corrections	Preferred	Possible	Possible
Suitability for automatic data acquisition	Fair	Excellent	Good

determining maximum frost depth, since one complements and increases the confidence level in the other.

Other frost gages are described by Mackay (1973) and Wilen and MacConnell (1973).

14.8. OTHER TRANSDUCERS FOR MEASUREMENT OF TEMPERATURE

Vibrating wire transducers are available for measurement of temperature and may be the preferred type if other vibrating wire instruments are being used and an indicator is available, or if a frequency signal is required. However, they are more expensive than thermistors, thermocouples, and RTDs and appear to offer no other inherent advantages.

Some pressure or strain measuring instruments based on the vibrating wire principle use the resistance of the plucking coil as a measure of temperature. The Carlson type of unbonded resistance strain gage (Chapter 8) can also be used to measure temperature.

14.9. COMPARISON AMONG TRANSDUCERS FOR REMOTE MEASUREMENTS

Apart from vibrating wire and unbonded resistance strain gage transducers, the choice among transducers for remote or continuous measurement of temperature is restricted to thermistors, thermocouples, and RTDs. The three are compared in Sections 14.9.1–14.9.6, and comparisons are summarized in Table 14.2. All three are suitable for use in geotechnical environments, the choice depending on the application.

14.9.1. Readout

RTDs and thermistors provide a direct measurement of absolute temperature, whereas thermocouples provide a measure of the temperature differential between a measuring junction and a reference junction. In practice, however, direct temperature readout devices can be purchased for all three transducers. Thermistors can be read out using commonly available digital multimeters set to measure resistance. Battery operated 4½ digit multimeters are normally used with low readout currents (a few microamperes) when set to measure resistance. Battery operated thermocouple readout units are available with sufficient ruggedness for field use. RTDs are read out using devices containing a Wheatstone bridge circuit and with a millivolt scale.

14.9.2. Sensitivity, Linearity, Accuracy, and Stability

Typical sensitivities are: thermistors, 200 ohms/°C; thermocouples, 0.04 mV/°C; RTDs, 0.2 ohm/°C.

Thermistors have by far the greatest output signal and sensitivity but the worst linearity.

RTD transducers have the greatest accuracy, but system accuracy may be reduced substantially by lead wire effects. The overall system accuracy of thermistors and RTDs is similar and in the region of ±0.1°C. Thermocouple accuracies of about ±1°C are normal, but ±0.5°C accuracies are possible by using "special limits of error" wire.

Thermistors and RTDs have excellent long-term stability. Some thermistor manufacturers claim that thermocouples exhibit drift with time, but available evidence suggests that this is likely only where they are in an oxidizing environment and unprotected. If thermocouples are protected by heat-shrink polyolefin so that no metal oxidization can occur, they can be expected to remain within their accuracy limits for at least 5 years in most geotechnical applications.

14.9.3. Lead Wires

The high sensitivity and high nominal resistance of thermistors allows them to be used with two-conductor lead wires, since ambient temperature changes cause negligible changes in contact or transmission line resistance. For long lead wires where high accuracy is required, the resistance of the wires should be subtracted from the resistance measurement before converting to temperature. RTDs use a three-conductor cable, for the three-wire quarter bridge configuration. Thermocouples use special thermocouple wire between the transducer and the reference junction location.

Thermistors and RTDs may be procured preencapsulated in epoxy resin with lead wires attached. All lead wires must be suitable for outdoor use and should be selected in accordance with guidelines given in Chapter 8 for selection of cable.

Field repairs are possible to the lead wires for all three transducers. For thermistors and RTDs the leads may be repaired by using commercially available splice kits with crimp connectors, as described in Chapter 8. Field repair of thermocouple lead wires is also possible, though less straightforward because errors can be caused. Care must be taken to ensure that the two thermocouple materials are not interchanged at the splice and that close metal-to-metal contact is established. Each conductor should be repaired by gently twisting the two parts together for about 1 inch (25 mm), followed by the crimp procedure described in Chapter 8. Although a high-quality repair will often cause no error in thermocouple data, the addition of two junctions within the circuit creates the possibility of errors, and splices should be avoided whenever possible.

14.9.4. Temperature Range and Rapidity of Response

All three devices are suitable for the range of temperatures normally encountered and rapidity of response required in geotechnical environments.

14.9.5. Applicability for Instrument Temperature Corrections

When a transducer is sensitive to temperature change, and data are to be corrected in accordance with measured temperature, a temperature transducer is required at the instrument location. Thermistors are usually preferred, partly because of their small size and partly because two conductors in the multiconductor cable attached to the instrument can readily be used for the thermistor. Good thermal contact between the instrument and thermistor must be ensured.

14.9.6. Suitability for Automatic Data Acquisition

If automatic data acquisition equipment is to be used, thermocouples are the instruments of choice. Data loggers can readily be used with thermocouples since the necessary reference junctions are available from many commercial sources and, because the calibration is common to all thermocouples of the same type, it can be programmed into the data logger so that readout is in degrees.

It is less straightforward to read thermistors with an automatic data acquisition system because resistance changes are so very large. A few manufacturers of data loggers will supply a special input circuit designed to read thermistors directly, but the selection of compatible thermistors is limited, and readout may be in millivolts rather than in degrees. However, data can be logged automatically by using a data controller that has an autoranging digital voltmeter and that has programmed nonlinear scaling functions.

RTDs can be read with a data logger since their resistance change is relatively small. However, either each RTD must have its own bridge circuit or a single bridge circuit must be switched to each RTD, posing a switching problem for the data logger. Special RTD readout circuits are available from some data logger manufacturers, and these special circuits usually have readout in degrees.

14.10. INSTALLATION OF TRANSDUCERS FOR MEASUREMENT OF TEMPERATURE

When planning installation of transducers for measurement of temperature, care must be taken to ensure good thermal contact between the transducer and the material to be sensed.

When installing transducers in a borehole to measure subsurface ground temperature, the borehole must be backfilled completely, and the backfilling procedure should depend on the required measurement accuracy. When maximum accuracy is required, for example, when measuring the depth of frost penetration or when measurements are used to make decisions on limiting roadway traffic in the spring, great care must be taken to minimize disturbance to the thermal regime and to the pathway for moisture movement. In such cases an accuracy of about ± 0.1°C is required, and the borehole should be backfilled with material removed from the borehole. Either the material can be placed and tamped in layers, or it can be mixed with water and poured or tremied into the borehole. When a lesser accuracy is acceptable, the borehole can be backfilled with cement grout or other material that ensures complete backfilling.

When transducers are attached to a structural surface, thermal contact is sometimes obtained simply by taping the transducers to the surface. Some manufacturers supply special rugged versions that can be welded to a metal surface for more permanent contact. Good thermal contact can also be created by drilling a small hole in the structure and using a heat-conducting epoxy to cement the transducer in place. The same epoxy can also be used to cement transducers directly to a structural surface, but this arrangement is less rugged.

General guidelines for installation of cables are given in Chapter 17.

Part 4

General Guidelines on the Execution of Monitoring Programs

Part 4 is addressed to the same audience as Part 3: the people who are involved in the details of instrumentation use. Chapter 15 presents a recipe for reliability of performance monitoring, including both *instrument ingredients* and *people ingredients*. Guidelines are given in Chapters 16–18 for calibration, maintenance and installation of instruments, and for collection, processing, presentation, interpretation and reporting of instrumentation data. These general guidelines are applicable to all instrumentation and supplement the more instrument-specific guidelines given in Part 3.

CHAPTER 15

A RECIPE FOR RELIABILITY OF PERFORMANCE MONITORING

The overriding desirable feature for performance monitoring is **reliability.** In suggesting a recipe for reliability, the ingredients can be divided into two types: *instrument ingredients* and *people ingredients*.

15.1. INSTRUMENT INGREDIENTS IN A RECIPE FOR RELIABILITY

As a general goal, the three major ingredients described in Sections 15.1.1–15.1.3 should be sought when selecting reliable instruments.

15.1.1. Simplicity

Inherent in reliability is maximum simplicity. In general, transducers can be placed in the following order of decreasing simplicity and reliability: optical, mechanical, hydraulic, pneumatic, electrical.

15.1.2. Self-Verification

This term means that instrument readings can be verified in place, as in the following:

- Checking a fixed borehole extensometer fitted with electrical transducers by inserting a dial indicator at the head.
- Water pressure can be increased or decreased manually within open standpipe or twin-tube hydraulic piezometers and the system then allowed to regain equilibrium. Equal readings before and after the change provide a degree of self-verification. Rates of increase or decrease should be consistent with the permeability of the ground around the piezometer.
- Use of duplicate transducers. For example, a vibrating wire and a pneumatic transducer packaged within the same housing to create a piezometer with two independent methods of reading.
- When taking readings with an inclinometer, a degree of self-verification is obtained when *check-sums* remain reasonably constant for all depth intervals in a given data set.
- Addition of an arrangement to a fixed borehole extensometer whereby each rod can be disconnected temporarily from its anchor, so that a check can be made for free-sliding.

15.1.3. Durability in the Installed Environment

The transducer must have proven longevity to suit the application. Cables, tubes, or pipes that connect the transducer to its readout unit must be able to survive imposed pressure changes, deformation, water, sunlight, and chemical effects such as corrosion and electrolytic breakdown. Guidelines on maximizing the longevity of transducers and connecting linkages between transducers and readout units are given in Chapter 8. Chapter 8 also includes guidelines on lightning protection when electrical systems are used.

Above-ground field terminals must be able to survive adverse environmental conditions and physical abuse.

The three goals outlined above are ideals that cannot always be achieved. Simplicity is not always appropriate, but there is no excuse for unnecessary complexity. Self-verification is a rare luxury, but it is often possible to achieve a similar goal by using the procedures, described in Chapter 4, for ensuring reading correctness. Good instrument design can usually create adequate durability in the installed environment. In summary, we must often settle for a little less than the best recipe, but we should **try** for the best.

15.2. PEOPLE INGREDIENTS IN A RECIPE FOR RELIABILITY

As a general goal, people responsible for performance monitoring programs should provide five major ingredients, described in Sections 15.2.1–15.2.5.

15.2.1. Thorough Planning

The task of planning a monitoring program should be a logical and comprehensive engineering process that begins with defining the objective and ends with planning how the measurement data will be implemented. Twenty steps in the planning process are described in Chapter 4. All twenty steps should be taken with great care. Paramount among the steps is the requirement to prevent cost from dominating the selection of an instrument.

15.2.2. Initial Calibrations and Inspections

Instruments should be calibrated, inspected, and tested before shipment to the user. On receipt by the user, acceptance tests should be made to ensure correct functioning. Guidelines are given in Chapter 16.

15.2.3. Installation Care

No instrument will be reliable if it is not installed correctly. Guidelines for installation care are given in Chapter 17.

15.2.4. Regular Maintenance and Calibration

Readout units, field terminals and any accessible embedded components should be maintained and calibrated on a regular schedule. Guidelines are given in Chapter 16.

15.2.5. Care During Data Collection, Processing, and Interpretation

Careful data collection, processing, and interpretation require a commitment by experienced and motivated people. Guidelines are given in Chapter 18.

In contrast to the goals for instrument ingredients, these five goals for people ingredients can usually be achieved. If we remember that the primary *people* need is **a high degree of motivation,** the goals will be within our grasp.

CHAPTER 16

CALIBRATION AND MAINTENANCE OF INSTRUMENTS

An instrument reading is useful only if the correct calibration is known. Zero shifts and scale span changes of transducers can occur because of normal wear and tear, misuse, creep, moisture ingress, and corrosion. If these changes are not accounted for, the entire monitoring program can become worthless. To maximize their effectiveness, all instruments must be calibrated and maintained properly.

16.1. INSTRUMENT CALIBRATION

Calibration consists of applying known pressures, loads, displacements, or temperatures to an instrument, under controlled environmental conditions, and measuring the response.

Instrument calibrations are generally required at three stages: prior to shipment of instruments to the user (*factory calibrations*), when instruments are first received by the user (*acceptance tests*), and during service life. These are discussed in turn.

16.1.1. Factory Calibrations

Instruments should be calibrated at the factory before shipment to the user. Experience indicates that factory calibrations are often minimal and may be incomplete and insufficient. The responsibility for this shortcoming is not with manufacturers, but with users who opt for low-bid procurement procedures.

Chapter 5 indicates that the procurement specification should include required factory calibrations and quality assurance and provides guidelines on methods and documentation. The user may opt to **participate in** factory calibrations. This is strongly recommended whenever the procurement includes newly developed instruments, when calibration requirements differ from the manufacturer's standard procedures, or when users believe that their needs cannot be satisfied by the acceptance tests described in Section 16.1.2. The words *participate in* rather than *inspect* are used to convey the need for cooperation.

16.1.2. Acceptance Tests

Instruments often receive rough handling while in transit from the manufacturer to the user and must be checked by the user to ensure correct functioning before installation. Such checks are called *acceptance tests*. If an instrument does not work **perfectly** under laboratory conditions, it is unlikely to work at all when installed in the field (Cording et al., 1975).

As indicated in Chapter 5, the instruction manual should include a step-by-step acceptance test procedure. Whenever possible, acceptance tests should include a verification of calibration data provided by the manufacturer, by checking two or three points within the measurement range, with transducers and readout unit at the various temperatures anticipated in the field during service life. Tests at extreme anticipated temperatures are important and may reveal malfunctions that, if not corrected, would result in faulty data.

When comprehensive acceptance tests are not possible, simple tests should be performed to verify that instruments appear to be working correctly. These are referred to as *function checks*. Transducers should be connected to readout units and tilted, pressurized, squeezed, or pulled to induce changes of magnitude consistent with the calibrations supplied. Each electrical connector should be unmade and remade several times, and the cable immediately adjacent to each cable-end connector should be flexed carefully to disclose any malfunction. The zero readings should agree with the readings supplied by the manufacturer. All electrical transducers intended for burial should be immersed in water for as long as possible to check the waterproofing. Instruments such as dial indicators, depth micrometers, and mechanical strain gages are generally reliable but should be checked against reference standards. Piezometers should be checked by immersion under known depths of water, and inclinometers may be function-checked by placing in sections of inclinometer casing kept fixed at some constant, known angle in the laboratory.

Any instrument that fails an acceptance test or function check should be returned to the manufacturer for replacement or repair, with a description of failure characteristics. In addition to verifying calibrations and detecting faulty instruments, these tests and checks provide an opportunity for the user to learn how to operate the instruments correctly.

Because standard transit insurance often requires notice of damage within a relatively short period after delivery, acceptance tests and function checks should be made before this time has expired. Sufficient lead time must be allowed to permit adequate testing and for any needed repair or replacement of faulty instruments.

16.1.3. Calibrations During Service Life

Calibrations or function checks of readout units are required during service life. Calibrations of any embedded components provided with in-place calibration check features are also required. These calibrations are usually performed by personnel responsible for data collection and should be under the direct control of the owner or instrument specialists selected by the owner.

Portable readout units are especially vulnerable to changes in calibration, often resulting from mishandling and lack of regular maintenance. They can sometimes be checked and/or recalibrated by following the acceptance test procedure. When this is insufficient, calibrations can often be made at local commercial calibration houses, using equipment traceable to an accepted standard agency. Certain instruments such as inclinometers can only be function-checked by the user or at calibration houses and must be returned to the manufacturer for any needed complete calibration, adjustment, and repair.

When readout instruments are attached within a permanent terminal enclosure, they also require regular calibration. Calibrations can be made by use of a standard device located in or carried to the terminal enclosure, or by detaching the instruments for calibration elsewhere. For example, a "master" gage can be carried to the terminal enclosure for twin-tube hydraulic piezometers and connected to the terminal system when calibrations of Bourdon tube pressure gages are required. All readout instruments attached within a permanent terminal enclosure should be arranged for easy detachment for calibration and/or replacement, and shutoff valves should be provided at all pressure gage connections.

When an in situ calibration or function check is possible, this should be included in the regular calibration schedule. For example, some fixed borehole extensometers can be checked for free-sliding of the wire or rod by moving the instrument head outward and measuring wire or rod elongation. Some instruments have two different types of transducer. Permeability tests can be made in open standpipe and twin-tube hydraulic piezometers to examine correct functioning. Sometimes it is possible to remove the load from load cells temporarily by making *lift-off* tests, allowing a check on the calibration.

Sections 8.4.3 and 8.4.17 provide guidelines on field checks of electrical instruments by using a circuit tester and a volt–ohm–milliammeter.

Calibration frequency, of course, depends on the application and use environment, but as a general rule the user should arrange for regular calibrations on a frequent rather than infrequent schedule.

Many users have experienced the dilemma of discovering that changes in calibration have occurred, therefore being unsure of data correctness since the last calibration date. Frequent calibrations minimize this dilemma. Certain instruments merit calibration spot-checks on each day that readings are taken. For example, an inclinometer should be checked on each day that readings are taken, in a test casing, field casing, or test stand as described in Section 12.8.6, and a portable depth micrometer should be checked within a portable reference standard. Other instruments require calibration spot-checks on a weekly or monthly basis, and a complete calibration is recommended at intervals of between 3 months and 1 year. A sticker on each instrument should indicate the last and next calibration dates.

When instruments are used for monitoring long-term performance, procedures for calibrations during service life should be included in the operating and maintenance manual.

16.2. INSTRUMENT MAINTENANCE

Regular maintenance required during service life is usually performed by personnel responsible for data collection and should be under the direct control of the owner or instrument specialists selected by the owner. Personnel responsible for data collection should always be on the lookout for damage, potential for damage, and deterioration or malfunction and should initiate repair or replacement without delay. Detailed maintenance requirements vary with each instrument and should be stated in the manufacturer's instruction manual (Chapter 5). The manual should include a troubleshooting guide, cleaning, drying, lubricating, and disassembly instructions, and recommended maintenance frequency. If batteries are required, service and charging instructions should be given in the manual.

As indicated in Chapter 5, appropriate spare parts should be ordered when instruments are procured to replace accessible malfunctioning components during service life. Provision should also be made for spare or standby readout units for use in case of malfunction. Options are procurement of a spare, arrangements with the manufacturer for rental of a spare if needed, and a standby equipment lease. Maintenance personnel should ensure that adequate spare parts and readout units are on hand or readily available. When instruments are used for monitoring long-term performance, maintenance procedures should be included in the operating and maintenance manual.

The following general guidelines indicate the factors to be considered when planning maintenance details.

16.2.1. Readout Units

Readout units should be kept clean and dry, following the manufacturer's instructions. Particular attention should be given to connections, including O-rings. If the readout unit includes gold-plated electrical connectors, these must be kept clean to prevent wear of the plating and corrosion of the underlying metal. If the readout unit is fitted with a humidity indicator, it should be dried whenever indicated. If no indicator is fitted, a special effort must be made to dry the unit when appropriate. Lubrication instructions should also be followed, as some readout units require periodic lubrication, whereas others must not be lubricated.

Protective caps and plugs are important, and their integrity should be maintained. If the manufacturer has not provided protective caps and plugs for terminals on readout units, users are advised to add these to their readout units. Caps and plugs should preferably be anchored to the readout unit to avoid loss and to remind reading personnel that they must be used.

Insufficient care is often given to battery charging and maintenance. Nicad and lead–acid batteries require different charging schedules, and either can be damaged by use of incorrect procedures. Guidelines on battery charging and maintenance, field checks with a volt–ohm–milliammeter, operating spares, and handling and transporting readout units are given in Section 8.4.17.

16.2.2. Field Terminals

Regular inspections should be made of field terminals to be sure that they are clean, dry, and functioning, that protective plugs, caps, and covers are in place, and that instrument numbers are clearly visible. Terminal accessibility should be verified and the integrity of enclosures and barricades checked. The inspections should include locations where embedded components exit from the ground: these locations are often susceptible to damage.

16.2.3. Embedded Components

Various embedded and normally inaccessible components require maintenance. For example, twin-tube hydraulic piezometer tubes usually require occasional flushing to remove gas bubbles, and tubing connected to pneumatic transducers may occasionally require flushing with a dry gas to remove moisture. Probe extensometer access pipes and inclinometer casings may require occasional flushing to remove accumulated debris.

When embedded components include retrievable parts, these can be removed for maintenance and reinstalled. Examples include transducers for in-place inclinometers, and electrical piezometers either suspended in standpipes or packed-off with pneumatic packers.

CHAPTER 17

INSTALLATION OF INSTRUMENTS

Installation of instruments requires a special effort. In his foreword to this book, Ralph Peck states:

> Equipment that has an excellent record of performance can be rendered unreliable if a single essential but apparently minor requirement is overlooked during the installation. The best of instruction manuals cannot provide for every field condition that may affect the results. Therefore, even slavish attention to instructions cannot guarantee success. The installer must have a background in the fundamentals of geotechnics as well as knowledge of the intricacies of the device being installed. Sometimes the installer must consciously depart from the installation manual.

When planning for instrument installation, Murphy's law (*If anything can go wrong, it will*) should be foremost in mind. If any source of difficulty can be postulated, it should be taken care of at this stage.

Before installation of instruments can start, the monitoring program will have been planned and instruments will have been procured, following the guidelines given in Chapters 4 and 5. Planning includes selection of appropriate contractual arrangements for installing instruments, selection of instrument locations, preparation of detailed installation procedures, preparation of an installation schedule, and coordination of installation plans with other parties. These planning steps are described in detail in this chapter, followed by guidelines on field work and preparation of an installation report.

17.1. CONTRACTUAL ARRANGEMENTS FOR INSTALLING INSTRUMENTS

Five possible contractual arrangements for installing instruments are described in Chapter 6, required personnel qualifications are given, and a recommendation is made to avoid use of the low-bid method. Installation of instruments should preferably be under the direct control of the owner or of instrumentation specialists selected by the owner.

During the planning phase a designation will have been made of which tasks are within the capability of average construction contractors (*support work*) and which tasks require the skill of instrumentation specialists (*specialist work*). In the author's view, installation of instruments should be a team effort between support and specialist personnel, following the guidelines for cooperation given in Section 6.6.7. Drilling crews should not be relied on to install instruments in boreholes unless specialist personnel are intimately involved.

17.2. LOCATIONS OF INSTRUMENTS

Locations of instruments should be selected in accordance with the three-step approach given in Chapter 4, considering zones of particular concern, representative zones, and zones to serve as indices of comparative behavior. Locations should be shown on the contract plans, but the exact location of instruments should usually be determined in the field, when geologic details and construction procedures are defined more closely than during the design phase.

17.3. DETAILED INSTALLATION PROCEDURES

Planning for installation of instruments should include preparation of detailed written installation procedures. If instruments are to be installed by the owner's personnel, these procedures will be used directly. If they are to be installed by the construction contractor, an abbreviated version will be included in the specifications, retaining key items for enforcement by the owner's representative, and the procedures will be used when reviewing the contractor's submittal of proposed installation methods.

The instrument manufacturer's instruction manual will often be helpful when preparing detailed installation procedures, but many manuals provide insufficient guidelines. When site-specific constraints must be met, as is normally the case, the manufacturer cannot be expected to provide complete and definitive procedures, and the user must plan detailed procedures to suit the specific site geotechnical conditions. However, these procedures must be flexible enough to account for unexpected conditions that arise during installation, such as unexpected ground conditions or changes in the construction contractor's procedure or schedule.

The various factors to be considered when preparing detailed installation procedures are described in Sections 17.4–17.10. Factors that are applicable to most instruments are included, and specific factors relevant to one instrument category are included in Part 3.

Detailed installation procedures should be in step-by-step form and should include a complete "shopping list" of required materials and equipment, including spare parts for replacing components likely to be damaged during installation. An example of a detailed installation procedure is included in Appendix G.

17.4. INSTALLATION AT THE GROUND SURFACE

Methods for installing instruments at the ground surface depend on the type of instrument, and various specific guidelines are given in Part 3. Some general guidelines are added in this section, particularly relating to protection of exposed components to minimize the chance of damage.

Surface-installed instruments should usually be protected with robust cover plates, welded, bolted, or otherwise attached to the surface. Exposed tubes and cables are extremely vulnerable to damage and should usually be protected by conduit. Flexible conduit is convenient, as leads can often be snaked through the conduit prior to installation, thereby minimizing interruption to normal contruction activities. Flexible spiral metallic conduit is often used but cannot be bent round a radius less than about 6 in. (150 mm). If watertightness is important, plastic sheathed flexible metallic conduit can be used (ends of conduit must also be sealed), and where small radius bends are required, rubber hose may be a better choice. In some cases rigid PVC or steel pipe or flexible corrugated polyethylene pipe may be preferable. When considering the options, users should bear in mind the smallest allowable bend for the tubes or cables and the need for waterproofing. If there is any chance of water entering a supposedly sealed conduit, either at the ends or through a defect, the sealing feature may in fact cause ponding of water, and an unsealed conduit may be a better choice. Cushioning must be provided between the leads and conduit at the exit points, for example, by using garden hose or by packing the space with burlap or soft rags.

Instrument terminals must be protected from damage caused by construction activities, vandalism, and the environment. Strong metallic cover plates, boxes, or pipes, with locks if necessary, may be required as protection from construction activities and vandals.

17.5. INSTALLATION IN BOREHOLES

Detailed procedures for installation of instruments in boreholes should ensure conformance, should satisfy requirements for borehole diameter, length, alignment, and wall roughness, and should include methods of drilling, installation of downhole components, and borehole backfilling. These issues are discussed in turn.

17.5.1. Conformance

When either changes in the length of the borehole or local shear displacements are anticipated during instrument service life, downhole components must be selected so that conformance is maintained and so that components are not damaged as deformation occurs. For example, when inclinometer casings or probe extensometers are installed vertically in a soft clay foundation for monitoring deformations beneath an embankment, telescoping couplings will normally be required in the casing or access pipe. Alternatively, axially compressible corrugated plastic pipe can be installed around rigidly connected pipe, provided that the outer pipe is free to slide with respect to the inner pipe. When plastic tubes or electrical cables are installed in boreholes through soils that will undergo substantial vertical compression, tubes and cables can often be prespiraled prior to installation, by winding them around a pipe of suitable size. Spiral in a plastic tube can be made permanent by immersing the pipe and tube in hot water and allowing both to cool before removing the pipe (Handfelt et al., 1987). Of course, temperature limits for the plastic must not be exceeded, and tests should be made to determine the minimum required temperature and immersion time.

17.5.2. Borehole Requirements

Most instruments that are installed in boreholes require minimum and maximum allowable borehole diameters, and some have limitations on borehole alignment or wall roughness. Where the diameter is not critical, it should be ample, so that downhole components pass easily along the borehole during installation. Boreholes should generally be drilled a few feet longer than the instrument to ensure that debris falling or pushed to the end of the borehole will not obstruct the installation. Borehole requirements may include an enlargement or recess at the collar for the instrument head and additional small holes for mounting the head and protective arrangements.

17.5.3. Drilling Methods*

Various methods are used to drill boreholes for instruments. The drilling method selected depends on the soil or rock type to be drilled, type of instrument to be installed, availability of drilling equipment and trained personnel, requirements for borehole diameter, depth, inclination, straightness, and wall roughness, access for drilling equipment, and the need for sampling. Detailed discussion and comparison of drilling methods are beyond the scope of this book, but methods in common use are described briefly. Dimensions of drill rods, flush-joint casing, diamond coring bits, hollow-stem augers, and U.S. pipe are given in Appendix F.

*Written with the assistance of Charles O. Riggs, Manager of Research and Marketing, Central Mine Equipment Company, St. Louis, MO.

Wash Boring

The *wash boring* method (Figure 17.1) is used to advance driven casing through soil. Soil that enters the casing at the bottom during driving is removed by pumping water through a small-diameter wash pipe, with a washing or chopping bit attached to its lower end. Cuttings of soil are carried upward by the wash water within the annulus between the casing and wash pipe. Standard or heavy-duty threaded and coupled pipe is sometimes used as

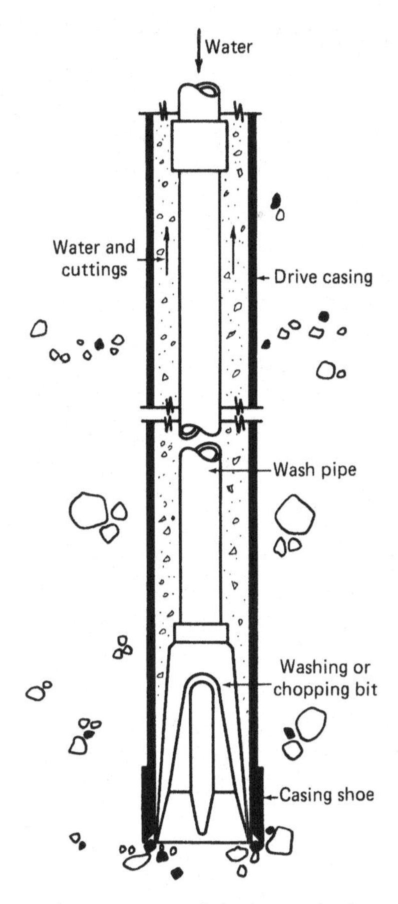

Figure 17.1. Wash boring method.

drive casing, but the exterior couplings can impede casing removal, particularly in granular soils. It should be used only when removal will present no problems and, when it is used, individual pipe ends should be reamed to ensure a flush inside diameter at the couplings. *Flush-coupled* casing, of the type sometimes used to case overburden for core drilling, is generally not suitable for supporting boreholes that are to be used for installation of instruments, because the inside diameter at a coupling is significantly less than the inside diameter of the casing. *Flush-joint* casing, another type often used during core drilling, is flush inside and outside and is the most suitable choice.

The wash boring method is frequently used for installing instruments in soil when there is a need to be certain of a clean stable borehole, as is often the case when installing piezometers, inclinometer casings, probe extensometers, and fixed borehole extensometers. The method is slow and therefore expensive but, in the view of the author, it provides a good defense against Murphy's law.

Auger Drilling with Hollow-Stem Augers

Auger drilling can be performed using either continuous flight solid-stem or hollow-stem augers, but solid-stem augers are not usually applicable for advancing boreholes for installation of instruments. As shown in Figure 17.2, a hollow-stem auger column has a continuously open axial stem that allows the borehole to be drilled and cased simultaneously. Steel flights are welded to the outside of the stem to form the auger, which is advanced by rotation and application of axial force to the top of the auger column. It is often necessary to plug the aperture at the auger head to prevent problems caused by soil entering the hollow stem. The pilot assembly shown in Figure 17.2 is sometimes used for this purpose and also assists in drilling. The pilot assembly is removed before inserting an instrument through the hollow stem. Alternatively, a flat expendable *knock-off* plate can be used on the bottom of the auger column.

Drilling with hollow-stem augers is used for installation of piezometers and inclinometer casings in soil and occasionally for installation of other instruments for measuring deformation. It is usually necessary to remove the augers from the ground without rotation, to prevent twisting of downhole instrument components.

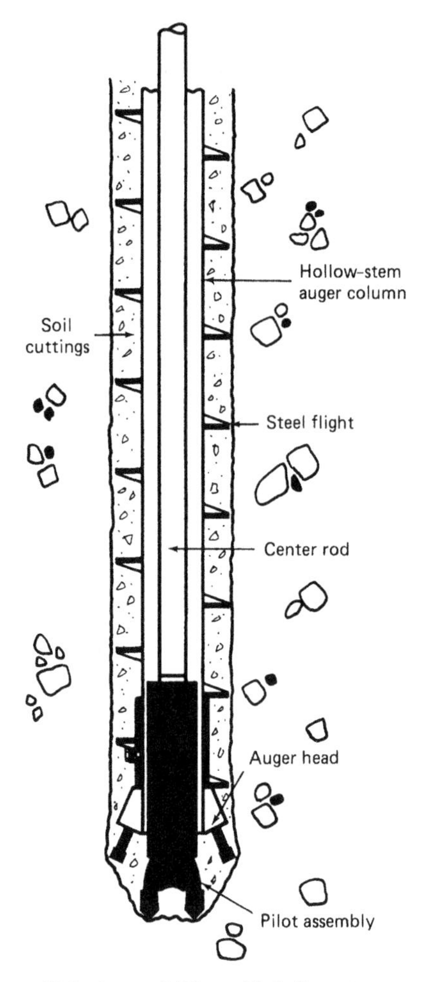

Figure 17.2. Auger drilling with hollow-stem augers.

Rotary Drilling

Rotary drilling (Figure 17.3), also called *hydraulic rotary drilling, straight rotary drilling,* or *mud rotary drilling,* can be used in soil or rock. No core is recovered while drilling but, when drilling in soil, samples can be recovered by intermittent sampling at the bottom of the borehole. A bit on the bottom of a column of drill rods is rotated and axially advanced. A drilling fluid is pumped downward through the drill rods, and cuttings of soil and rock are carried upward within the annulus between the drill rods and the wall of the borehole. A correctly "designed" drilling fluid will penetrate the wall of the borehole and form a thin "mud cake," thereby maintaining the level of fluid in the borehole to provide supporting pressure on the wall of the borehole. Bentonite (powdered sodium montmorillonite) is the most commonly used drilling fluid

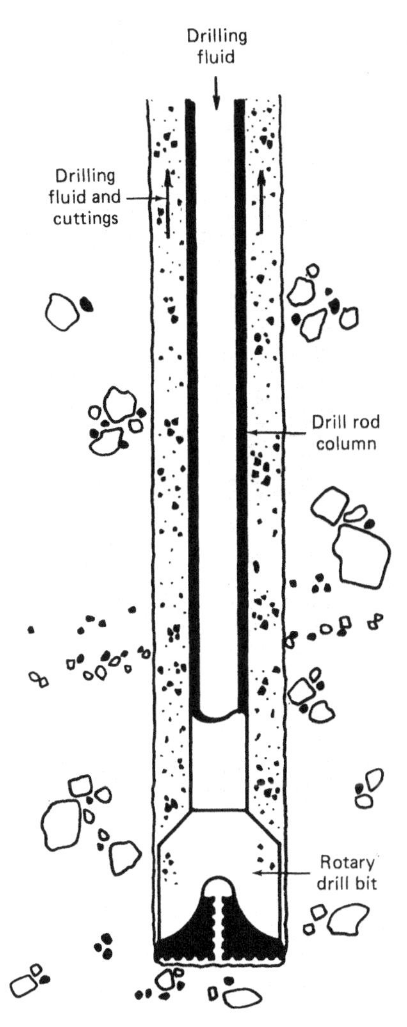

Figure 17.3. Rotary drilling.

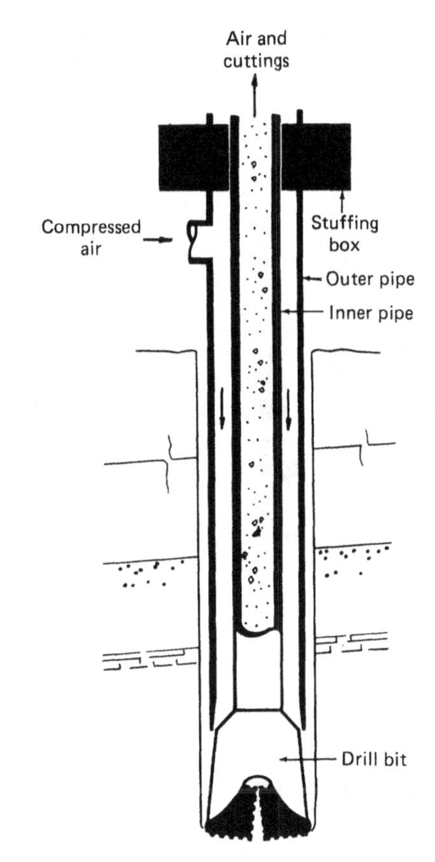

Figure 17.4. Drilling by double-tube reverse air circulation.

additive. The most commonly used bit is the *tricone* bit, fitted with three rotating toothed cones, but a *drag* bit with three or four blades is often used in soil.

Rotary drilling can be used for installation of instruments in soil and rock when the stability of the borehole is certain. If there are doubts about stability, casing or hollow-stem augers should be used instead of drilling fluid: the author has experiences of instruments becoming stuck in a fluid-filled borehole before they have reached the planned depth! Under some circumstances a flush-joint casing should be advanced with rotary procedures to ensure a stable borehole. Conventional bentonite drilling fluid should not be used in boreholes for piezometer installations, because the additives will permanently impede the functioning of the piezometer by severely decreasing the permeability of the surrounding soil. However, as described in Chapter 9, a drilling fluid containing a biodegradable organic polymer can be used.

A special type of rotary drilling, *double-tube reverse air circulation,* has some application for installation of instruments, and is shown in Figure 17.4. With this method an outer pipe is advanced as the cuttings are forced upward at high velocity within a small inner pipe.

Core Drilling

Core drilling can be used in soil or rock, and a core is recovered. A core barrel attached to a column of drill rods is rotated and advanced, and drilling fluid flushes and cools the bit on the end of the core barrel. The soil or rock at the advancing end of the core barrel is cut or abraded by the bit, cuttings are carried out of the borehole by the drilling fluid, and a cylinder of soil or rock (the *core*) enters and is protected by the core barrel. Drill casing is used when necessary to ensure that overburden soil or

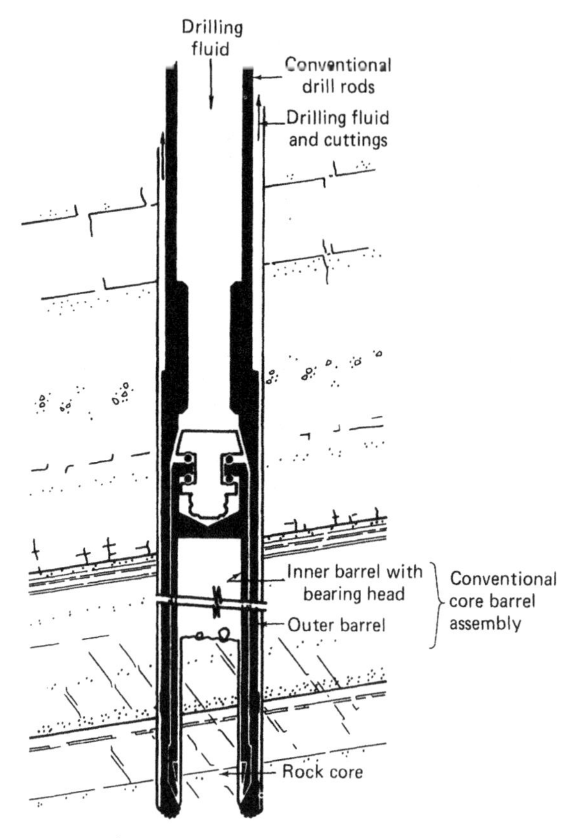

Figure 17.5. Conventional core drilling.

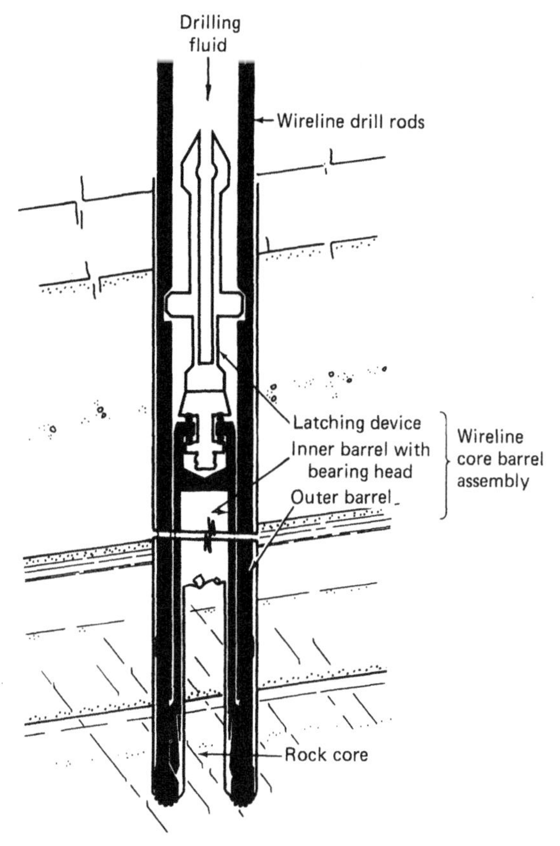

Figure 17.6. Wireline core drilling.

rock does not enter the borehole and interfere with coring, and it also provides a reliable conductor for the return of drilling fluid. Core drilling is performed either with *conventional* or *wireline* systems, as shown in Figures 17.5 and 17.6. When a conventional core barrel is used, the complete drill rod column must be withdrawn from the borehole to remove the increment of core from the core barrel. When a wireline core barrel is used, the inner barrel, containing the increment of core, is withdrawn from the borehole on a "wireline," which attaches to the latching device shown in Figure 17.6. The bit, outer barrel, and drill rods remain in the borehole. The hardest rocks, such as quartzite, are often cored most economically by using impregnated diamond bits, with small manufactured diamonds contained in the bit crown matrix. Softer rocks, such as limestone, are often cored using surface-set bits, with larger natural diamonds set at the surface of the crown matrix. Soft rock is often cored using carbide insert or polycrystalline (carbide and manufactured diamond) insert bits. The evaluation of rock cores is usually enhanced greatly if split inner tube core barrels are used.

Core drilling is the method of choice for installation of instruments in rock when there is a need to define subsurface stratigraphy and the locations of discontinuities: this is typically the case when installing instruments to measure deformation, stress change, and joint water pressure. Core drilling is not often used for installation of instruments in soil.

Rotary Percussion Drilling

Rotary percussion drilling (Figure 17.7) involves simultaneous rotation and percussion of a drill bit to fracture the material at the end of a borehole. No core is recovered. The most common drills are rotary-impact rock drills, called *drifters*, usually mounted on track carriers. Alternatively, down-the-hole rotary-impact hammers can be used. The borehole is flushed with air or water. Rotary percussion drilling is not a frequent choice for installation of instruments, because locations of discontinuities cannot be defined. Moreover, significant deviations in alignment often occur, so that the locations of downhole components are unknown. When these limitations are acceptable, the method can be used

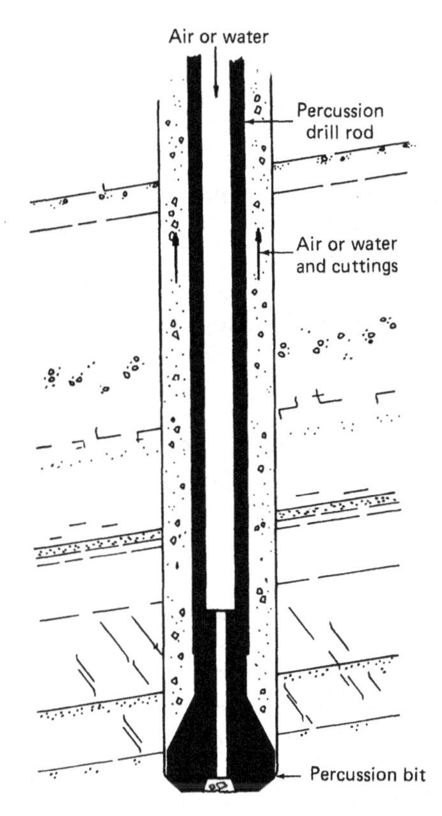

Figure 17.7. Rotary percussion drilling.

economically, particularly when drilling equipment is on the site for other purposes, such as drilling holes for blasting.

Controlling Borehole Alignment

When borehole alignment is critical, for example, when installing piezometers between closely spaced vertical drains after the drains have been installed or when installing fixed borehole extensometers or inclinometer casings above or alongside a future tunnel, care must be taken to set up the drill rig at the correct alignment and to use drilling procedures that minimize deviation from this alignment. When using the wash boring method to install drill casing in soft ground, an oversized casing shoe should be used and the casing advanced by spinning and washing, rather than by driving.

Directional drilling techniques can be used to maintain borehole alignment or to deviate deliberately from the initial alignment. They have been developed for use in production of oil and gas but are occasionally relevant for installation of instruments, for example, plumb lines and inverted pendulums. The technique involves inserting a deflection wedge at the bottom of the borehole. Many experienced and competent core drillers have never used directional drilling techniques.

Guidelines Applicable to All Drilling Methods

Whichever drilling method is selected, the detailed written installation procedures should include the drilling method and requirements for drilling equipment, sampling methods and frequency, maximum and minimum borehole diameters, casing or auger diameter, drilling fluid additives, drilling records, sample containers, and storage and submittal of samples.

During drilling, a conventional borehole log should be maintained, and any required samples should be taken.

After completion of drilling, the borehole should be thoroughly cleaned with air, water, or drilling fluid and the length and diameter checked. When water is used for drilling, many drillers will wash the borehole by inserting drill rods or a jet pipe to the bottom of the borehole and flushing water through the rods or pipe. However, better results are obtained by reversing the jetting pump and using the drill rods or jet pipe as the intake, with the lower end a few inches from the bottom of the borehole. The borehole is kept filled by pouring clean water in until all cloudy water is pumped out. This is referred to as the *reverse circulation method* and maximizes the velocity of outgoing water. When a borehole in rock is advanced by diamond coring methods, the hole should require minimal cleaning. Minimum diameter can be checked by passing a suitably sized "dummy" along the borehole and, where more complete data are required, a borehole caliper can be used. When borehole alignment, inclination, or straightness is critical, a borehole directional survey (Chapter 12) may be appropriate.

If grout is to be used while installing downhole components, a test should be made to ensure that grout will not leak from the borehole. An estimate of grout-tightness can sometimes be made from knowledge of stratigraphy and previous boring experience at the site. When using grout backfill in an unsupported borehole, the borehole should first be tested for grout-tightness by removing drill rods and conducting a permeability test, usually by running water into the borehole to maintain water level at the borehole collar. The author has used a maximum water inflow of 1 U.S. gallon per 10 feet (1.3 liters/meter) of borehole per 15 minutes as an acceptability criterion. If the inflow exceeds an acceptable value, the borehole must be sealed with bentonite or other sealing compound, or grouted and redrilled, and tested again. Such testing is not possible in boreholes containing drilling fluid or when drill casing is used throughout the borehole length, and a conservative estimate of grout-tightness must be made. Testing and sealing can of course be performed in any borehole length below the bottom of the drill casing.

17.5.4. Installation of Downhole Components

If significant time has elapsed since completion of drilling and checking for length, diameter, and grout-tightness, the checks should be repeated prior to inserting downhole components and the borehole again cleaned and sealed if necessary. The method of installing downhole components depends on the type of instrument and drilling method, but several general guidelines can be given.

When maximum tensile strength is required across socket couplings in plastic pipe or when plastic pipe is to be used for high-pressure applications, a primer should be used before the solvent cement is applied. The primer etches the surfaces and allows proper adhesion. Dimensions of steel and PVC pipe that are often used as downhole components are given in Appendix F.

Downward Boreholes

If the instrument is heavy, a ¼ in. (6 mm) manila rope or ⅛ in. (3 mm) airplane cable should be attached securely to the bottom and used for lowering the instrument, thereby avoiding excessive tension in the components and allowing for recovery if problems arise. A hand-operated winch is useful for controlling this safety line when instruments are heavy or when boreholes are deep. The safety line will normally be cut after it has served its purpose and left in place. If, on the other hand, the instrument is very light, it may be necessary to add weight before installation. Instruments such as inclinometer casings can often be installed in downward boreholes under neutral buoyancy by ensuring that the borehole is water-filled and filling the casing with water as it descends.

If a grout backfill is to be used, an estimate of grout density should be made, the volume and weight of downhole components determined, and any weight necessary for overcoming buoyancy should be added to the bottom of the instrument before installation. Typical grout density ranges from 70 to 100 lb/ft^3 (1.1–1.6 Mg/m^3).

Any required flexible tubes or rigid pipes for backfilling the borehole with grout or granular material should be inserted in the borehole as installation of downhole components proceeds. Typically, 1 in. (25 mm) diameter plastic tubes or pipes are required, but a smaller diameter can be used for thin grout mixes, and a larger diameter may be required for very thick mixes or for compressed bentonite pellets (Chapter 9). If rigid pipe is used, joints should be flush on both the inside and outside so that the pipe does not "hang up" on other downhole components. Schedule 80 rigid PVC pipe, with square threads or threads of the type shown in Figure 9.3, is usually the best choice: sizes are given in Appendix F. If tubes or pipes are to be left in place, they should be taped securely to other downhole components. If they are to be recovered, light taping at the bottom is necessary for tubes but usually unnecessary for pipes. If backfill is to be placed in stages, a single tremie tube or pipe can be pulled upward as backfilling progresses, or a separate length can be installed for each stage.

Many borehole instruments are shipped in short lengths, and installation time can often be reduced by preassembling components into longer lengths prior to installation. Certain instruments such as fixed borehole extensometers with spring-loaded anchors require insertion rods, and these must be inserted in the borehole as installation of downhole components proceeds. Other instruments such as inclinometer casings and multipoint fixed borehole extensometers require careful control of orientation as they are installed.

Near-Horizontal Boreholes

When installing in near-horizontal boreholes, insertion rods are often required. These can usually be rented from the instrument manufacturer, or lightweight pipe such as aluminum electrical conduit can be used. Fiberglass chimney sweep rods can also be considered for this purpose.

If a grout backfill has to be used, it is preferable to incline the borehole downward 5–10 degrees and to use a single grout tube with its outlet at the bottom of the borehole. If there is any doubt about the alignment of the borehole or if it must be as horizontal as possible, an alignment survey should be made prior to locating grout outlets and vents. For example, alignment surveys are essential when near-horizontal holes are drilled in rock by percussion methods and when a grout backfill will be used, because borehole alignment is always uncertain. A borehole directional survey instrument or full-profile liquid level gage can be used. Downward boreholes that hold water can be surveyed by traversing a diaphragm piezometer along the hole, and at several depths measuring water head and lead length. If the borehole will not hold water, it may not hold grout, and it should be sealed as described above. Grout outlets must be provided at the lowest point in the borehole and at each trough and a vent at the highest point and at each crest.

Upward Boreholes

When installing in upward boreholes, insertion rods are also required. A hydraulic anchor designed for use with fixed borehole extensometers can often be attached to the upper end of the instrument and used to hold the system in place after insertion rods are withdrawn. If a grout backfill is to be used, stage grouting is necessary, first to seal around the instrument head at the collar of the borehole and later to fill the remainder of the borehole in one or more stages, and grout outlets and vents must be installed for each stage.

17.5.5. Selection of Backfill Material

Alternative backfill materials for boreholes include grout, granular fills such as sand and pea gravel, and bentonite pellets. Grout backfill is most frequently used, is suitable when the borehole must be sealed against water flow, and can be used for all borehole alignments. Grout backfill is more likely than granular backfill to fill the borehole completely but cannot be used if grout would bleed into the surrounding ground. Where permeable strata overlie impermeable strata, grout backfilling is possible in the impermeable zone, and granular backfill can be placed in the permeable zone after grout has set. Granular backfill cannot readily be installed in upward boreholes and should not be used where its high permeability would create nonconformance by allowing flow of water along the borehole. Bentonite pellet backfill is used for installation of piezometers in boreholes. Alternative backfill materials are discussed in Sections 17.5.6–17.5.8.

17.5.6. Backfilling Boreholes with Grout*

When selecting a mix for grout backfill, the first task is to define the required engineering properties. As a goal, the grout should ensure conformance between the instrument and the surrounding soil or rock and should not alter the value of the parameter being measured. For example, grout in boreholes for piezometers should satisfy permeability criteria and, if compression is anticipated along the axis of the borehole, also compressibility criteria. When probe extensometers rely on grout to ensure conformance, the grout should satisfy criteria for compressibility and shear strength. Grout for fixed borehole extensometers in soft ground should not have significant compressive or tensile strength. Grout for inclinometer casings should satisfy criteria for maximum and minimum strength. For these reasons there is no universally suitable grout, and each installation must be considered individually.

While assembling material for this section of the book, by reviewing the contents of the technical literature and by discussing with engineers who have expertise in grouting, two overriding factors became evident. First, little is known on the subject of grouting around instruments in boreholes, presumably because this use of grout represents such a tiny proportion of overall grout use. The following are useful references, but none relates specifically to instrumentation: ACI (1981), Dames & Moore (1979), Deere (1982), Jefferis (1982), and Littlejohn (1982). Second, even if the "perfect" grout mix can be determined, it probably will not set as a uniform column throughout the borehole. In the view of the author, these realizations should color our reliance on grout where properties and uniformity are critical. For example, when a probe extensometer is to be installed to monitor substantial vertical compression, reliance on grout may be unwarranted, and conformance should be ensured by using a positive anchorage at each measuring point.

Some typical mixes will be given, but it is emphasized that they should not be used as a "cookbook." Trial mixes should be made for each application and judgment often made by visual observation and supplemented with simple tests such as pressing with a thumb or use of a Torvane. Where a more exact measurement of grout properties is required, laboratory tests of trial mixes will be needed, following the method described by ASTM (1980).

The properties of grout are often dependent on the sequence of adding ingredients, and the sequence should be standardized. As a general rule, liquids should be mixed first, followed by the finest through the coarsest materials. When using cement/bentonite grout, bentonite should be added to the

*Written with the assistance of Joseph P. Welsh, Vice President, GKN Hayward Baker, Inc., Odenton, MD.

water first, because if bentonite is added to a cement and water mix an ion exchange takes place and the expansion of the bentonite is reduced significantly. Properties are often also dependent on the chemical constituents of the mixing water, and water for trial mixes should be from the same source as the field mix.

The following are some examples of grout mixes used during installation of instruments in boreholes.

1. Water/cement/bentonite, 1/0.15/0.06 by weight (Easton, 1984). Bentonite and water mixed first. Used for probe extensometers that rely on grout to ensure conformance in soft clay. The 28 day unconfined compressive strength tested as approximately 500 lb/ft^2 (25 kPa). Minnitti (1985) tested a similar mix, concluding with a 28 day unconfined compressive strength of approximately 300 lb/ft^2 (15 kPa). Minnitti also tested a water/cement/bentonite mix, 1/0.08/0.11 by weight, giving a 28 day unconfined compressive strength of approximately 400 lb/ft^2 (20 kPa).
2. Bentonite/cement ratio of 0.10/1 by weight, with sufficient water to allow pumping. Recommended by AASHTO (1978) for backfilling around inclinometer casings.
3. Hydrated lime/cement ratio of 1.6/1 by weight, with sufficient water to allow pumping. Recommended by AASHTO (1978) for backfilling around inclinometer casings. However, the high pH in hydrated lime can cause corrosion of aluminum and should therefore not be used with aluminum inclinometer casing.
4. Bentonite/cement ratio of 0.15/1 by weight, with sufficient water to allow pumping. Recommended by Corps of Engineers (1971) for sealing above open standpipe piezometers.
5. Sand/cement ratio of 5/1 by weight. Recommended by USBR (1974) for sealing above open standpipe piezometers.
6. Fetzer (1982) used two mixes to seal above open standpipe piezometers. A 5 ft (1.5 m) seal was first made with a water/bentonite, 1/0.3 mix by weight, followed by a water/cement/fine sand, 1/1.9/2.7 mix by weight, with expansion agent and retarder.

Whenever quoting proportions of grout mixes, it is essential to indicate whether proportions are by weight or by volume: there is significant confusion in the technical literature because this is not always done. Useful conversion factors for preparation of grout mixes are given in Appendix H.

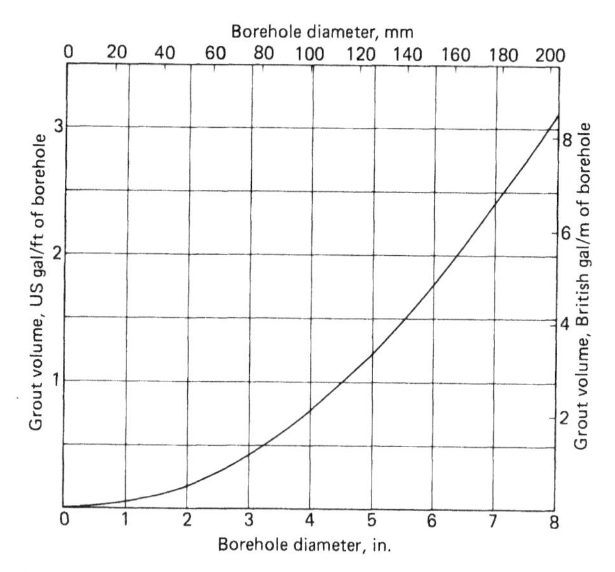

Figure 17.8. Required grout volumes for borehole installations (allowing for 20% waste).

Bentonite is available in a form specifically designed for use in salt water, but a much larger quantity of bentonite is required (Handfelt et al., 1987) and experience has not always been satisfactory. It may be more appropriate to mix conventional bentonite with fresh water when grouting boreholes in a salt-water environment or to use a chemical sealant (Driscoll, 1986; Senger and Perpich, 1983).

Sand is rarely used in grouts for backfilling boreholes, but additives such as expanders, fluidizers, and retarders* may be included, and on occasion chemical grouts may be appropriate, but in all cases the advice of grouting specialists should be obtained.

Various practical guidelines for backfilling boreholes with grout are given in this paragraph. Boreholes at all alignments can be backfilled with grout but, as discussed earlier in this chapter, multiple grout outlets and vents and stage grouting may be required. Prior to grouting, a check should be made to ensure that a sufficent quantity of each ingredient is readily available. Figure 17.8 indicates required grout volumes, including a 20% margin for waste, for various borehole diameters. Water should be circulated through the grout tube or pipe

*Some commercial sources are listed in Appendix D.

to verify that the circuit is open. When downhole components include a vertical pipe of adequate diameter, the borehole can be grouted from within the pipe by using one of the check valve arrangements shown in Figure 12.74 instead of an outside tremie pipe. These arrangements often allow use of a smaller-diameter borehole. In general, downhole vertical pipes and tubes should be filled with water or oil while grouting to avoid collapse owing to external pressure, and grout pressures should be controlled to avoid collapse. For very deep installations stage grouting may be required, in which case an outside tremie pipe must be used. When there is a possibility of grout entering instrument components, for example, the top of a vertical pipe when grouting outside the pipe, the possible grout entry point should be sealed. Grout should be mixed thoroughly by circulating through the grout pipe for sufficient time or by use of a colloidal mixer if available. A fine mesh such as window screen should be fitted over the suction intake during grouting to ensure that lines are not blocked by lumps resulting from incomplete mixing. When withdrawing any drill casing or augers during or after backfilling a borehole, rotation should not be allowed, otherwise downhole components may become spiraled.

When grouting under artesian conditions, one of the methods described in Chapter 9 for installing piezometers can be used. Alternatively, barium sulfate can be added to the grout to increase density, taking care to top up with grout while withdrawing casing.

17.5.7. Backfilling Boreholes with Sand or Pea Gravel

Use of sand or pea gravel backfill is limited to downward boreholes, and the borehole diameter should be large enough to discourage bridging, generally at least 2 in. (50 mm) larger than the outside diameter of the downhole components. Rounded grains are less likely to bridge than angular grains.

Two backfilling methods are possible. First, sand or pea gravel can be poured down the annular space. The material should be saturated with water before pouring, should be poured slowly to avoid bridging in the borehole, and should be tamped thoroughly in 2 ft (600 mm) maximum layers. A flush-coupled rod or pipe, such as fixed piston sampler actuating rods with a special tamping foot, is useful for tamping, measuring depths, and breaking bridges. When it is important to center the components in the borehole, the sounding hammer described in Chapter 9 should be used. If possible there should be no protruding couplings in downhole components. Internal vibrators supplied by some manufacturers of inclinometer casing often have insufficient power for proper compaction of backfill.

Second, sand or pea gravel can be tremied to the bottom of the borehole. A 1 in. (25 mm) Schedule 80 PVC pipe with flush threaded couplings is inserted within the borehole as downhole components are installed, and a 45 degree Y-branch is fitted to the pipe near its upper end. A water supply is connected to the branch, and water is circulated down the pipe until it spills out of the top of the borehole. Sand or pea gravel is poured slowly into the top of the pipe and washed to the bottom of the borehole, and the pipe is gradually raised as backfilling progresses.

The second method generally results in more complete backfilling than the first. For both methods, if drill casing or hollow-stem augers have been used, they should be withdrawn, without rotating, in small increments to avoid collapse of the borehole prior to backfilling. Experience with each installation will indicate the allowable increment, which may vary from 6 to 30 in. (150–750 mm). Withdrawal should usually precede backfilling, because if backfill is allowed to settle within the casing or augers it may "grab" downhole components and cause them to be lifted when casing or augers are withdrawn. Depth measurements should be made repeatedly while backfilling to verify correctness of the backfill level. It is helpful to mark the tamping rod or tremie pipe at a distance from its bottom equal to the length of casing or augers, so that the mark is alongside the top when the backfill is placed up to the bottom.

Figure 17.9 indicates the quantity of Ottawa sand required to fill various sizes of borehole.

17.5.8. Backfilling Boreholes with Bentonite Pellets

Backfilling boreholes with compressed bentonite pellets for installation of piezometers is described in Chapter 9. Figure 17.9 indicates the quantity of compressed bentonite pellets required to fill various sizes of borehole, assuming a unit weight of 74 lb/ft^3 (1.19 Mg/m^3). This unit weight is typical, but some commercial sources have unit weights up to 82 lb/ft^3 (1.31 Mg/m^3).

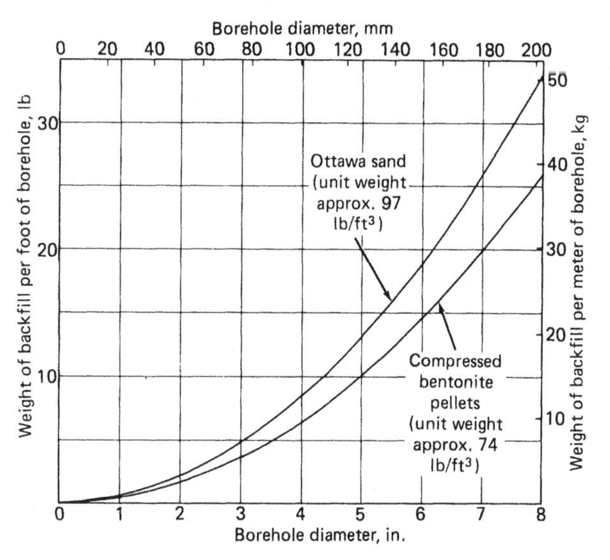

Figure 17.9. Required volumes of Ottawa sand and compressed bentonite pellets for borehole installations (no allowance for waste).

17.6. INSTALLATION IN FILL

When deformations are anticipated during instrument service life, instrument components must be selected so that conformance is maintained and so that they are not damaged as deformation occurs. For example, probe extensometers, inclinometer casings, and fixed embankment extensometers will normally require either axially compressible access pipe or telescoping couplings in rigid access pipe. Alternatively, a rigid access pipe can be surrounded by an axially compressible or telescoping pipe, provided that this outer pipe is free to slide with respect to the access pipe. As an illustration of damage that can be caused to pipes by deformation of fill, Figure 17.10 shows dramatic rupture of rigid plastic standpipes attached to piezometers, found in an exploratory shaft in an embankment dam. Plastic tubes and electrical cables also often require special measures to protect them from damage by localized ground deformation.

The following sections provide guidelines on installation of pipes, tubes, and cables. Factors relevant to a single instrument category and guidelines

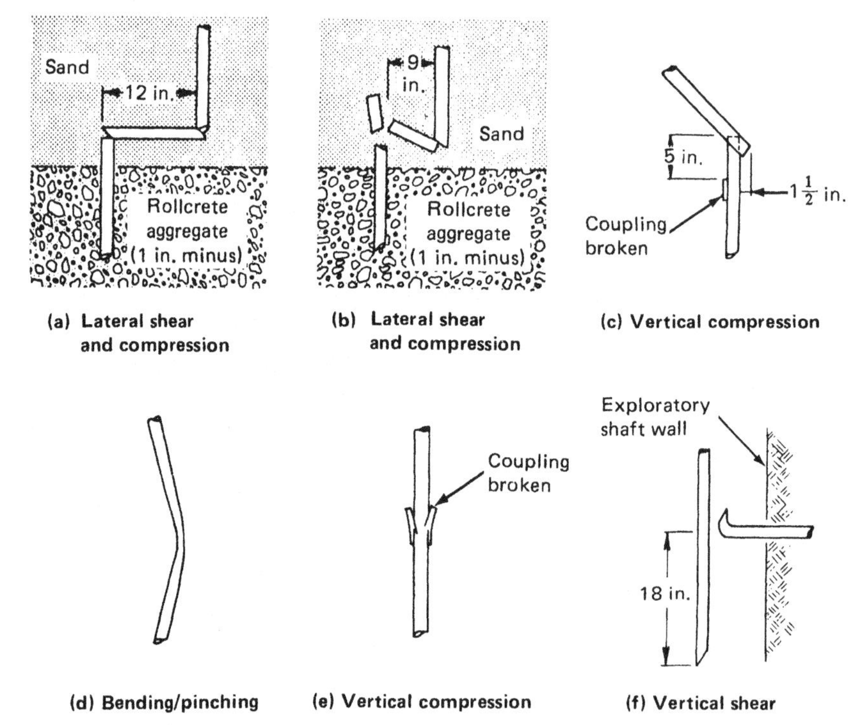

Figure 17.10. Rupture of piezometer standpipes found in an exploratory shaft in an embankment dam (after Mikkelsen and Wilson, 1983).

on installation of various individual instruments are included in Part 3. In general, the engineering properties of fill material around pipes, tubes, and cables should match, as closely as possible, the properties of the surrounding fill and should not provide a leakage zone for water. This need is of paramount importance in embankment dams. When pipes, tubes, and cables are installed in fill containing large-sized particles, finer material may be required around them.

Pipes, tubes, and cables should be checked for integrity and correct functioning before installation and again before backfilling. As discussed in Section 17.5.4, when maximum tensile strength is required across socket couplings in plastic pipe or when plastic pipe is to be used for high-pressure applications, a primer should be used before the solvent cement is applied. Before tubing is installed, it should be tested at a pressure well in excess of the anticipated pressure in the field. Constrictions in tubing have been found on several occasions and, unless prior experience or testing indicates that constrictions are not present, an appropriately sized ball bearing should be blown through the tubing. Tubing should be stored away from direct sunlight, and ends should be capped during storage to prevent the entry of dust and insects. All pipes, tubes, and cables that are installed in groups should be marked individually at approximately 20 ft (6 m) intervals so that they can be identified if they are damaged and repairs are attempted.

17.6.1. Horizontal Pipes

Telescoping couplings must be sealed to prevent intrusion of fill material. O-rings form the best seal or, alternatively, soft rubber tubing can be installed to span between coupling and pipe, with a hose clamp at each end. Various filler materials can be packed in the annular space between pipe and coupling, including pliable sealing compounds manufactured for waterproofing and insulating electrical connections, or grease-impregnated tapes.

Pipes are usually placed in a trench 1–2 ft (300–600 mm) wide by 2–3 ft (600 mm to 1 m) deep, bedded on and covered by approximately 6 in. (150 mm) of suitably graded fill material. The allowable maximum particle size in the material depends on the required engineering properties of the backfill, the strength of the pipe, the depth of cover, and the type of construction equipment that will operate over the backfilled trench and will normally be between 0.2 and 0.5 in. (5–12 mm). Horizontal pipes installed in rockfill are normally surrounded with a layer of finer material such as sand or pea gravel, and intermediate layers of coarser material may be required to prevent migration of particles. A deeper trench may be necessary if heavy compaction equipment will operate over the backfilled trench. The remainder of the trench is then filled with excavated material in 4–8 in. (100–200 mm) lifts and compacted with hand-controlled power-operated equipment.

An alternative installation method, whereby pipes are installed on the surface of previously compacted fill rather than within a trench, is included in Section 17.6.3, describing installation of horizontal tubes and cables.

When horizontal pipes are installed to allow passage of a probe, such as a probe extensometer, inclinometer, or full-profile liquid level gage, a method of traversing the probe must be devised. The simplest method, used where access is available to both ends of the pipe, involves leaving a traction wire within the pipe, clipping the probe to one end of the wire when readings are required, and pulling on the other end. When only one end of the pipe is accessible, three methods are possible. First, fiberglass chimney sweep rods can be used to push the probe up to about 200 ft (60 m), using a fitting on the probe for attachment of the rods. This method is suitable only if the pipe is reasonably straight and if there are no irregularities on the inside wall of the pipe, such as telescoping couplings. Second, an endless traction wire can be installed. The wire can pass around a pulley at the inaccessible end of the pipe and return along the same pipe, but this arrangement runs a severe risk of the probe becoming entangled with the return wire, such that it cannot be moved, and is not recommended. A better arrangement is to install a second pipe, parallel to and of smaller diameter than the main pipe, connecting the two pipes at the inaccessible end with two large-radius 90 degree steel elbows, such that the wire passes along both pipes. Plastic elbows are not suitable, because repeated passage of the traction wire may wear a slot in the inside wall, allowing soil particles to enter and block the pipe or causing the wire to become jammed. The third method uses a duct motor developed by post office engineers (Deadman and Slight, 1965; Penman and

Charles, 1973) for threading telephone wires along ducts. The duct motor is driven by compressed air and consists of a double-acting telescopic plunger with a valve mechanism causing it alternately to extend and contract as long as the air supply is maintained. The duct motor and probe are driven to the inaccessible end of the pipe and readings taken while withdrawing the probe.

17.6.2. Vertical Pipes

Telescoping couplings in vertical pipes must be sealed, as described above for horizontal pipes, to prevent intrusion of fill material. When settlement is anticipated, the couplings must be installed in an open position, and the sealing arrangements help to prevent closing during installation. When the instrument consists of a rigid access pipe surrounded by an axially compressible or telescoping pipe and when water leakage must be prevented, the annular space between the pipes should be filled with a slurry.

Two methods are available for extending a pipe upward as fill is placed: the *mounding* method and the *excavate and replace* method. In the mounding method, the pipe is installed to protrude above the fill level and protected with a hand-compacted mound of fill. Typical maximum height of the mound is 5 ft (1.5 m). In the excavate and replace method, the pipe is capped, filling and compaction continued over the pipe for about a 5 ft (1.5 m) depth, a local excavation made to locate the cap, a new pipe length added, and the excavation backfilled and compacted with hand-controlled power-operated equipment. The latter method usually results in a more straight and vertical extension than the former. Whichever method is used, great care is needed to avoid damaging the pipe, compaction around the pipe tends to be inferior, and interruption to normal filling operations is costly. Maximum particle size and engineering properties of fill adjacent to the pipe should follow the guidelines given above for horizontal pipes.

17.6.3. Horizontal Tubes and Cables

When planning installation of horizontal tubes and cables, special attention should be paid to routing, avoiding wherever possible traversing zones of large differential strain and providing slack at any such transitions. Tube and cable splices should be reserved for repair work and should not be included in the planned arrangements unless absolutely necessary. When splices are used, they should be made with great care as described in Chapter 8.

Horizontal tubes and cables are normally installed in trenches, bedded on and covered by suitably graded material. Trench size and backfilling arrangements are generally as described in Section 17.6.1 for horizontal pipes, but the allowable maximum particle size will usually be smaller than for pipes. Where armoring or conduit is not used, typical maximum size is 0.2 in. (5 mm). Particle size restrictions may also be needed in the first two lifts of backfill overlying the 6 in. (150 mm) covering layer. In rockfill, two or more layers of protection will usually be required, for example, sand and pea gravel, and gradations should follow filter design criteria. Individual tubes and cables should be separated by approximately 0.5 in. (12 mm) and, where several are installed, a wood or metal rake similar to a garden rake is helpful when controlling spacing. No tube or cable should be allowed to cross over another. Installation of tubing for twin-tube hydraulic piezometers, following the general procedures outlined previously, is described by Bartholomew et al. (1987) and USBR (1974).

Clements (1982) describes an alternative, which he refers to as the *surface installation method,* to installation of horizontal pipes, tubes, and cables in trenches. Clements describes in detail the disadvantages of trench installation, including high cost, lengthy time required, difficulty in excavating to a level bottom, difficulty in maintaining a steep-sided trench in granular fill, problems of maintaining the maximum particle size, and ponding of rainwater. The writer describes a "tramline" effect, whereby continuous lines of material are disturbed by backhoe teeth to a depth of up to 8 in. (200 mm) below the bottom of the trench. He states that there has been recent evidence to suggest that isolated instances of limited piping have occurred along instrumentation trenches within completed embankment dams and comments that although the tramline factor is not necessarily the cause of such happenings, it represents a weakness in the installation method that could result in piping effects.

The alternative recommended by Clements entails installation of pipes, tubes, and cables on the surface of previously compacted fill as shown in Figure 17.11. The first stage requires cutting a swath with a bulldozer and dumping a mound of loose parent material alongside the swath to provide protection from construction traffic. Figure 17.11

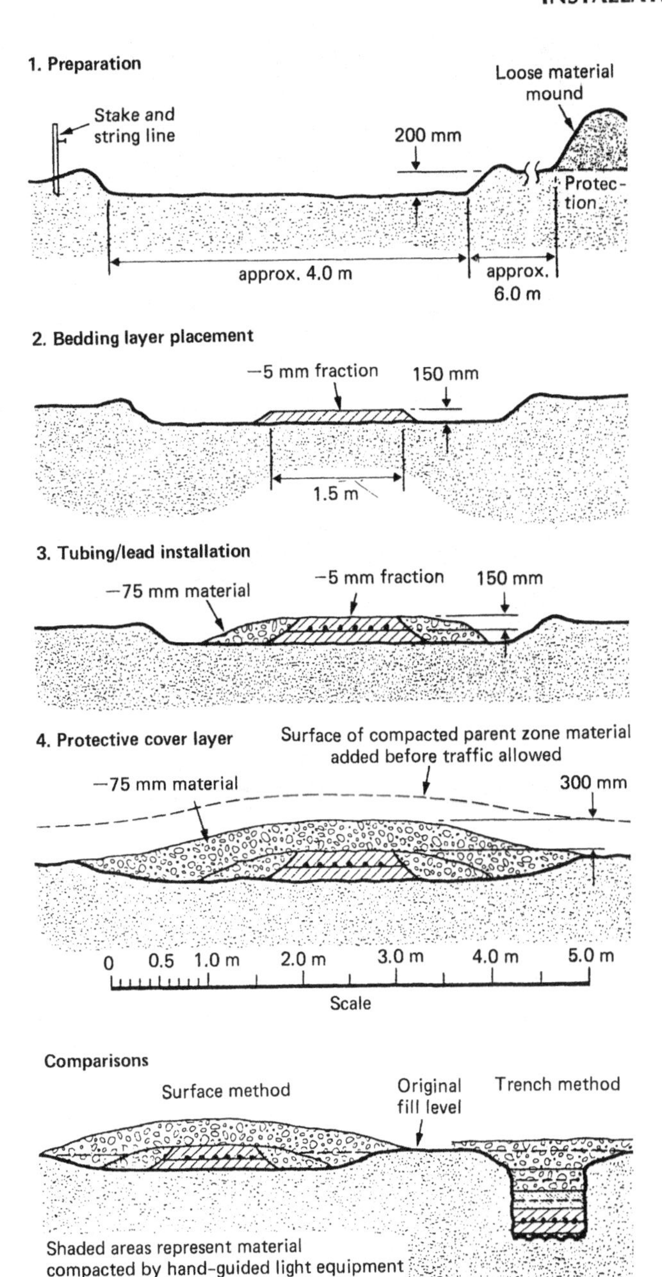

Figure 17.11. Stages in the *surface installation method* for tubes and cables (after Clements, 1982).

shows a swath 8 in. (200 mm) deep but, depending on the surface level control prior to cutting, the depth may vary. During the second stage, a bedding layer is placed, using a skid-mounted hopper approximately 8 ft (2.5 m) wide and 13 ft (4 m) long, with fixed top and side guide plates at the hopper discharge mouth to ensure a continuous and regular layer. The hopper is towed along the swath centerline by a bulldozer and filled using a front end loader. The layer is compacted using a self-propelled vibratory roller with a dynamic force of 5000 lb (22 kN), the first pass without vibration, followed by a specified number of vibrating passes. Instrument leads are then placed and covered using a skid-mounted guide and hopper. Before the initial protection layer is compacted, it is necessary to support the sides by placing parent zone material against it, and the ridge of protection material at the side of the swath provides a convenient source. Once placed, the material is hand-picked to remove rocks larger than 3 in. (75 mm) in diameter and to match the level of the uncompacted protection layer. The whole is then compacted with hand-guided vibratory plate compactors. This is the only stage at which heavy compaction equipment is not used. The final stage of the operation involves placing a 12 in. (300 mm) layer of up to 3 in. (75 mm) material over the swath area. Much of this work can be done by a bulldozer pushing loose material from the protective ridge at the side of the swath. Once spread, the material is hand-picked to remove oversize rocks. During this operation, care must be taken to prevent vehicle tracks from advancing over the instrument leads. Once the layer is leveled, compaction is accomplished by the self-propelled vibratory roller, the first pass being made without vibration. Normal construction and compaction techniques are restored for the next and subsequent lifts. It is also recommended that 12 in. (300 mm) of added cover is placed over the installation before loaded haulage equipment is allowed to cross it. The installation procedure is now complete and the area may be handed over to normal fill placement operations.

Clements indicates that the surface installation method is much quicker than the trench method, uses less personnel, is less disruptive to routine construction activities, results in better conformance between backfill and surrounding compacted fill, and does not allow ponding of rainwater. Clements also comments that, because the method is quicker, manual screening of protection material may create difficulties, and a small mechanical screening plant may be worthwhile.

Despite the advantages of the surface installation method, the author believes that it should be used only when there is good supervision of general fill placement: it is all too easy for mounds of fill to be pushed around during subsequent fill placement, and bulldozer drivers have a natural tendency to level out the surface for the next layer. The attractiveness of the surface installation method clearly depends on specific site factors, and the primary application appears to be for installation of instrument leads during construction of embankment dams.

Armoring or conduit is often the preferred method for protecting tubes and cables from damage. Use of conduit alleviates stringent criteria on maximum particle size of backfill and greatly reduces the likelihood of damage when the fill deforms. If no conduit is provided, the tubes or cables must accommodate all strains experienced by the fill, but if they are intentionally left slack within conduit, a greater length is available for accommodating local deformation. The benefit of this arrangement for accommodating tensile strains is self-evident, but less so for compressive strains. However, when electrical cables are installed without conduit and subjected to compressive strains, conductors can break, protrude through the cable insulation, and create a short circuit. Rigid PVC pipe is most frequently used as conduit when installing horizontal leads in fill, but corrugated flexible polyethylene pipe is an alternative in cases where it is adequately robust. Conduit should not be used where passage of water along the instrument leads must be prevented, but even in these cases individual leads can be threaded within minimum-sized plastic tubing where they pass through zones that will experience significant strain, and the annular space filled with petroleum jelly. This arrangement, for example, can be used for instrument leads that pass from the core of an embankment dam to the downstream filter, the tubing extending 3–10 ft (1–3 m) either side of the interface.

Many specifications require that instrument leads are installed in a meander from one side of the trench to the other to create slack, but this arrangement appears to be of marginal value. It cannot provide slack to accommodate tensile strain along the leads, because any such benefit assumes that leads can move laterally with respect to the fill surrounding them. Perhaps some benefit is gained when deformation occurs transverse to the leads, particularly during compaction, but a better solution appears to be selection of leads that can accommodate deformation and use of conduit where possible.

Special arrangements must be made for tubes and cables where they emerge from the ends of con-

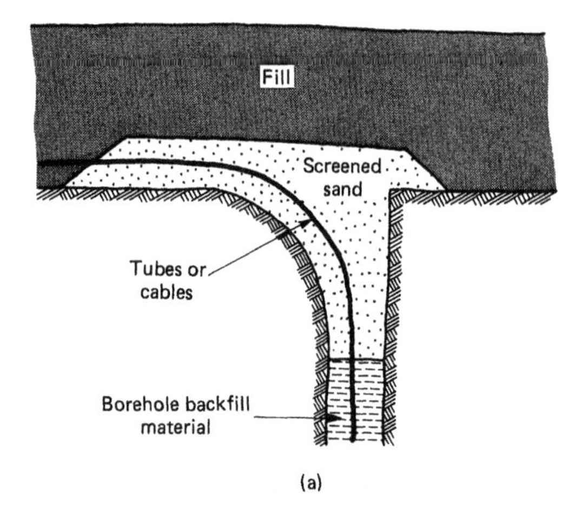

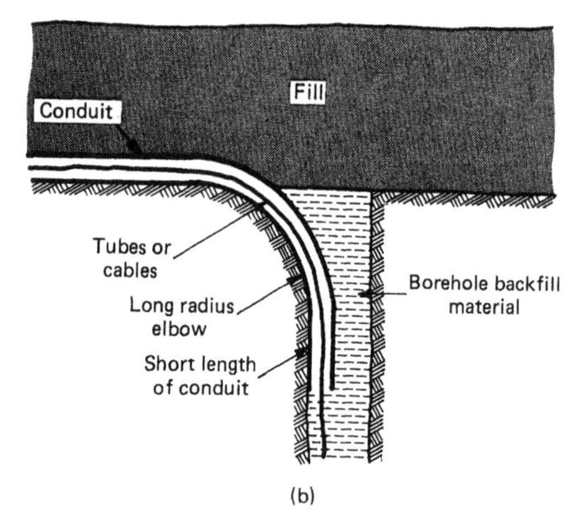

Figure 17.12. Typical arrangements for protecting tubes and cables where they emerge from a borehole: (a) without conduit and (b) with conduit.

duit or from boreholes. If there is any possibility of shear deformation or kinking at the ends of conduit, cushioning must be provided between the leads and conduit at the exit points. Rubber hose, such as automobile radiator hose or thick-walled garden hose, can be used, or the space can be packed with burlap or soft rags. Where leads emerge from vertical boreholes to pass horizontally within fill or at the bottom of fill, there is generally a concern for shearing at the top of the boreholes. If conduit is not used, the leads should be cushioned with screened sand and routed along a curve as shown in Figure 17.12a. If leads emerge from a vertical borehole and enter pipe conduit, the conduit should be anchored to the borehole by attaching a long-radius elbow and short length of pipe and inserting the short length down the borehole, as shown in Figure 17.12b.

17.6.4. Vertical Tubes and Cables

The guidelines on routing, splices, and identification tags given above for horizontal leads apply also to vertical leads. However, vertical runs of tubes and cables should be avoided wherever possible: they are a nuisance to construction, are liable to be damaged, and impede the quality of compaction. Vertical runs have sometimes been protected within a vertical pipe, leaving the pipe in place and without backfilling inside the pipe, but this practice runs the risk of damaging leads at the point of entry or exit and is usually unsatisfactory. When vertical runs cannot be avoided, they should be kept to a minimum. Protection options include surrounding the run with suitably graded material or installing within telescoping pipe fitted with large-radius 90 degree elbows at top and bottom.

17.7. INSTALLATION IN UNDERGROUND EXCAVATIONS

When instruments are installed within underground excavations, additional constraints often exist, and cooperation with the construction contractor is essential. The contractor's primary objectives will usually be to excavate soil or rock and to install support. Interruptions for installation of instruments can delay an entire construction cycle, and installation plans must be coordinated with the contractor well ahead of installation dates. Guidelines on coordination are given in Section 17.12.

Special attention must be given to the provision of any necessary platforms to support installation equipment and personnel and to the movement of installation equipment such as drill rigs. Whenever possible, any required boreholes should be drilled with equipment that is used for other construction activities. Special care must be taken to protect instruments from damage, following the guidelines given in Sections 17.4 and 17.8.

17.8. PROTECTION FROM DAMAGE

Various procedures for minimizing damage to surface and downhole components, to pipes, tubes, and cables installed in fill, and to instruments installed in underground excavations have been described in the previous sections. Some additional guidelines can be given to minimize the chance of damage.

When embedded components that terminate at the ground surface are subject to damage by construction activities, special precautions must be taken. For example, inclinometer casings and probe extensometers will often require protective barricades, marked clearly to warn operators of construction equipment. When vandalism is the overriding issue, terminals should if possible be buried and made unobtrusive, since a strong protective box often encourages a vandal to look for a stronger vandal. Burying of course necessitates a foolproof method of locating the terminal. It is hoped that readers of this book will not endure the embarrassment of the author, who on one occasion buried terminals to instruments installed in a city suburb prior to construction of a tunnel and located them with respect to nearby buildings. The project was delayed, urban renewal took place, and when the project was reactivated extensive digging was required before the terminals were found!

All vertical pipes should be provided with a cap to prevent intrusion of debris. Where construction activities might damage the tops of vertical pipes or where vandals might block pipes by dropping stones, a removable plug should be installed at an appropriate depth. A suitable plug is shown in Figure 17.13. The plug is attached to an insertion tool, consisting of two concentric pipes, the inner pipe connecting to the left-hand thread on the plug, the outer pipe terminating in a socket wrench to fit the hexagonal nut. Each pipe has a tee handle on the upper end. The plug and tool are inserted within

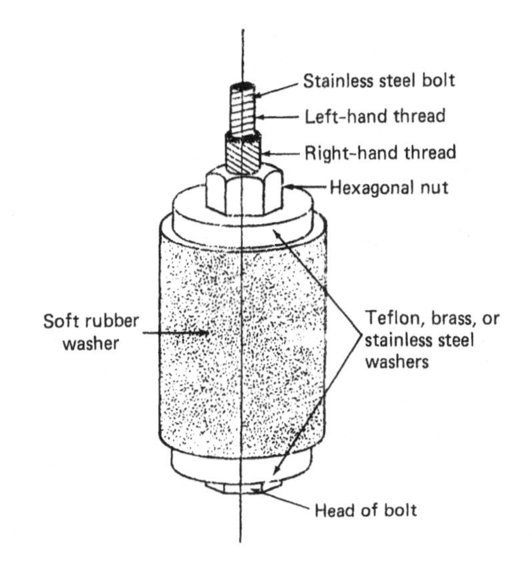

Figure 17.13. Removable plug for vertical pipe (courtesy of Syncrude Canada Ltd, Fort McMurray, Alberta, Canada).

the pipe, the plug held in place with the inner pipe and set by turning the outer pipe clockwise, the insertion tool removed, and the depth of the plug recorded. Plug removal follows the reverse procedure.

Protection from damage by the environment should consider all the factors listed in Table 4.4. Of particular relevance to terminals are corrosion, water, temperature extremes, ice, shock, dust, dirt, mud, chemical precipitates, humidity, and loss of accessibility. Piezometer tubes and electrical cables have occasionally been eaten by rodents, who apparently have a taste for plastic. When appropriate, protection from this hazard should be provided by use of metal conduit. All relevant environmental factors must be addressed on a case-by-case basis.

17.9. ACCEPTANCE TESTS

Acceptance tests should always be performed to ensure, to the extent practicable, that installations have been completed satisfactorily. For example, permeability tests can be made in twin-tube hydraulic piezometers, groove tracking tests and spiral survey measurements can be made in inclinometer casings, and rod sliding tests can be made in fixed borehole extensometers equipped with a disconnect between rod and anchor. Repeatability of readings can also provide acceptability criteria: a minimum of three sets of instrument readings can be made, repeatability examined, and an evaluation made of whether data are within expected tolerances.

17.10. INSTALLATION RECORDS

As discussed in Chapter 4, "installation record sheets" should be prepared during the planning phase. These records serve two purposes. First, when installation personnel are required to enter data in blank spaces on a field form, they are more likely to follow the installation procedure with care. Second, "as-built" data are required both for record purposes and for use during evaluation of data. A major task during evaluation of data is development of a relationship between measurements and causes, and details of installation may be one of those causes. For example, if installation personnel are unable to set a borehole extensometer anchor correctly, this fact may explain later measurements. If more or less backfill than normal is used in a borehole installation, this fact may be evidence that is helpful when evaluating data. If an instrument is installed in adverse weather conditions, less reliance on data may be justified.

Installation record sheets should include a record of appropriate items from the following list:

- Project name.
- Instrument type and number, including readout unit.
- Planned location in plan and elevation.
- Planned orientation.
- Planned lengths, widths, diameters, depths, and volumes of backfill.
- Personnel responsible for installation.
- Plant and equipment used, including diameter and depth of any drill casing used.
- Date and time of start and completion.
- Spaces for necessary measurements or readings required during installation to ensure that all previous steps have been followed correctly, including acceptance tests.
- A log of appropriate subsurface data.
- Type of backfill used.
- As-built location in plan and elevation.
- As-built orientation.
- As-built lengths, widths, diameters, depths, and volumes of backfill.
- Weather conditions.
- A space for notes, including problems encountered, delays, unusual features of the installation, and any events that may have a bearing on instrument behavior.

17.11. INSTALLATION SCHEDULE

A deadline should be determined before which each instrument must be installed. The deadlines will be based on the need to establish preconstruction conditions, construction activity, availability of installation personnel, instrument delivery dates, and other relevant factors.

Instruments should be installed as early as possible to provide vital initial readings for establishing reliable base conditions. Unfortunately, this requirement is often not fulfilled, and important baseline data are not obtained. Early installation will also provide information on faulty instruments so that they can be repaired or replaced before construction begins.

In cases where an instrument is required to monitor long-term performance but not required during construction, it may be possible and expedient to install the instrument in a borehole after construction is complete, thereby reducing interference to construction and the chance of damage to the instrument.

17.12. COORDINATION OF INSTALLATION PLANS

Detailed planning for installation should be coordinated among all parties involved, including the instrument supplier, owner's personnel, and the construction contractor. Arrangements must be made to ensure that instruments arrive on time.

When the owner's personnel are responsible for installation, a cooperative working relationship with the construction contractor is essential, and the owner's field personnel should make a special effort to establish a cooperative relationship. The best way of establishing such a relationship is for instrumentation personnel to initiate thorough communication with all levels of the contractor's personnel several weeks or months before the start of installation work. The instrumentation personnel should meet with the contractor's engineers and supervisors to explain what will be done, why it must be done, and what will be required of the contractor. They should discuss access for installation and prepare sketches of detailed installation arrangements, linked to the contractor's actual method of construction, to forewarn the contractor of the impact on normal construction work. They should provide lists of materials that will be required for support work and be willing to tailor their installation plans to create minimum interference to the contractor's work. In short, mutual respect can usually be established by thorough communication, dispelling the adversary relationship that sometimes arises between architect-engineers and construction contractors. Having established this respect, instrumentation personnel should maintain the contractor's respect by performing top-quality work, by being responsive to the effects of the program on the contractor, and by working **with** the contractor to minimize any adverse effects.

17.13. FIELD WORK

If the guidelines provided in this chapter have been followed, intensive planning will have been completed before field work is started. Decisions will have been made on contractual arrangements for installing instruments, locations of instruments, and installation schedules. Detailed installation procedures will have been planned, and they will have been coordinated with all parties involved.

Despite such intensive planning, surprises are the norm rather than the exception, and field personnel must make a special effort to complete each installation satisfactorily. Chapter 6 indicates that the basic requirements for field instrumentation personnel are reliability and patience, perseverence, a background in the fundamentals of geotechnical engineering, mechanical and electrical ability, attention to detail, and a high degree of motivation. If installation personnel possess these qualities, they should be well equipped to solve problems created by surprises.

Before starting installation work, field personnel should study and understand the detailed installation procedure. They should be aware that the instrument may not serve its purpose if a single essential but apparently minor requirement is overlooked during the installation. A complete check should be made to ensure that all listed materials and equipment are either on hand or readily available, and all components should be identified.

A few golden rules apply to field work. The first rules apply to care of hardware. All instruments and components should be stored and protected prior to installation. They should be handled carefully. Cables and tubes can readily be damaged by nicking, bending, or twisting, and they should always be removed from a spool by rotating the spool. Components should be kept as clean as possible, and ends of tubing and pipe, connectors, and threads should be covered with protective material or placed on clean above-ground surfaces. Adequate cleaning materials should be on hand. Second, when installing instruments in boreholes, listen to the advice of the driller. Experience with drilling can often contribute more to the success of a borehole installation than a university degree. Third, if problems arise while installing downhole components and if there is any possibility of collapse of the borehole or of premature setting of grout, act fast and discuss later: *if in doubt, pull it out*. Fourth, remember that Murphy's law was written especially for installation of geotechnical instruments.

17.14. INSTALLATION REPORT

On completion of installation, an installation report will usually be required to provide a convenient summary of information needed by personnel responsible for data collection, processing, presentation, and interpretation. The installation report should contain at least the following information:

- Plans and sections sufficient to show instrument numbers and locations.
- Appropriate surface and subsurface stratigraphic and geotechnical data.
- Descriptions of instruments and readout units, including manufacturer's literature and performance specifications (photographs are often helpful).
- Details of calibration procedures.
- Details of installation procedures (photographs are often helpful).
- Initial readings.
- A copy of each installation record sheet.

CHAPTER 18

COLLECTION, PROCESSING, PRESENTATION, INTERPRETATION, AND REPORTING OF INSTRUMENTATION DATA

Chapter 4 indicates that data collection, processing, presentation, interpretation, and reporting procedures should be planned before instrumentation work commences in the field. This chapter provides guidelines on planning and execution.

A detailed draft of procedures should be prepared during the planning phase and finalized after instruments have been installed, by which time responsible personnel are more familiar with operation of instruments and specific site considerations. Chapter 5 includes a listing of data collection, processing, presentation, and interpretation information that should be included in a comprehensive instruction manual, but users should recognize that, although manufacturers can provide some basic information, they are usually not familiar with specific site conditions. Users must therefore prepare their own procedures.

18.1. COLLECTION OF INSTRUMENTATION DATA

Responsibility for collection of instrumentation data will have been determined during the planning phase and should preferably be under the direct control of the owner or instrumentation specialists selected by the owner. The construction contractor may be responsible for support work, such as provision of access to instrumentation locations, and will sometimes be responsible for optical survey work associated with data collection.

18.1.1. Personnel Qualifications and Responsibilities

Data are normally collected by technicians or junior engineers, under the supervision of a more experienced geotechnical engineer. Qualifications of data collection personnel should be similar to those of installation personnel: reliability, mechanical and electrical ability, computational skill, attention to detail, a background in the fundamentals of geotechnical engineering, and a high degree of motivation.

Data collection personnel should work as a team, the division of responsibilities depending on their individual capabilities and on project requirements. Technicians and junior engineers should be encouraged to take an interest in the instrumentation program and be given as much responsibility as they can handle, thereby creating a sense of involvement and motivation and permitting the more experienced engineer to concentrate on aspects that require mature engineering capability.

Typical personnel responsibilities are given in Tables 18.1 and 18.2, so that data collection personnel can be aware of the full scope of their tasks. These responsibilities are listed as if data are to be collected manually. When automatic data acquisition systems are used, some of the tasks will be handled automatically.

18.1.2. Role of Automatic Data Acquisition Systems

Chapter 8 includes a description of automatic data acquisition and communication systems and a discussion of suitability for monitoring various transducers. This section is confined to a discussion of advantages and limitations of automatic systems.

Until a few years ago, almost all data were recorded by hand: the *long way*. The advent of automatic data acquisition systems has changed the state of the practice, for better and for worse, and users should be aware of the limitations as well as the advantages of these systems. No automatic system can replace engineering judgment. These remarks should not be taken as a vote against the use of automatic data acquisition systems, but as a plea for an honest appraisal of their suitability before they are selected. Advantages and limitations of automatic data acquisition systems are given in Table 18.3.

The guidelines in the remainder of Section 18.1 are written as if data are to be collected manually because, when data are collected automatically, the goals are the same. When planning to use automatic data acquisition systems, users are encouraged to study these guidelines and to ensure that an essential need is not forgotten.

18.1.3. Written Procedure for Data Collection

The experienced geotechnical engineer responsible for data collection should finalize the draft procedure that has been written during the planning phase. The detailed written procedure will be based on information provided in the manufacturer's instruction manual and on specific site conditions and will include the items in the following list. It will be seen that this list has many similarities to the responsibilities listed in Tables 18.1 and 18.2, but it serves a different purpose.

- Needs and procedures for coordinating with other parties for access and other support work.
- Step-by-step procedure for equipment set-up and turn-on.
- Step-by-step procedure for taking readings, with appropriate illustrations, including a list of equipment and tools required during reading and cautions pertaining to personnel and equipment. (It is often advisable to read each instrument more than once and to use judgment in selecting the most appropriate value. Use of the same data collection personnel for all measurements throughout a particular project is highly desirable. Interchange of readout units should be avoided whenever possible, because readings are often dependent on the particular transducer and readout unit combination.)
- Special procedures for taking initial readings.
- Guidelines on reading frequency.
- Any required instructions for reading two or more instruments concurrently, so that data can be correlated on a common time basis.
- Guidelines on handling readout units, as given in Section 8.4.17.
- Field data record.
- Sample completed field data record.
- Guidelines to enable data collection personnel to answer the question, *Is the instrument functioning correctly?* (Procedures for ensuring reading correctness are discussed in Chapter 4 and include visual observations, duplicate instruments, a backup system, regular in-place calibration checks, consistency between two or more measured parameters, and data repeatability.)
- Procedure for comparing the latest readings with the previous readings, so that any significant changes can be identified immediately. (If data are recorded automatically, a comparison step should be programmed into the system. When manual recording and field data sheets are used, the previous set of readings should be taken to the field, either by copying the previous field data sheets or by transcribing key data to a column on the new data sheets. Originals of previous data sheets should not be taken into

Table 18.1. Typical Responsibilities of the Experienced Geotechnical Engineer

Phase of Work	Responsibility
Before and during collection of field data	Coordinating with and ensuring the cooperation of all parties Training and supervising technicians or junior engineers and assisting them whenever they have questions or concerns
Before collection of field data	Familiarity with the planning of the monitoring program, as described in Chapter 4 Familiarity with the project construction procedure Familiarity with the calibration and maintenance requirements described in Chapter 16 Familiarity with details of instrument installation, including the installation record sheets and installation report, as described in Chapter 17 Familiarity with the general requirements for data collection, as described in this chapter Preparing the final version of detailed data collection procedures Establishing and updating data collection schedules Conveying, to technicians or junior engineers, hazard warning levels and other critical measurements that should be brought to the immediate attention of the experienced engineer
During collection of field data	Reviewing data Advising the owner, the design consultant, or the construction contractor of any condition requiring their attention Communicating with personnel responsible for data processing, presentation, and interpretation Recording all factors that may influence measured data[a] Making visual observations of behavior, for correlation with instrumentation data[a]

[a]If technicians or junior engineers have sufficient ability and motivation, they can handle the last two responsibilities, under the close supervision of the experienced geotechnical engineer.

Table 18.2. Typical Responsibilities of Technicians and Junior Engineers

Phase of Work	Responsibility
Before collection of field data	Understanding why the instruments are necessary, how they work, and what parameters are being measured Understanding the project construction procedure
During collection of field data	Consciousness of safety considerations Reading instruments and recording data on field data records Comparing the latest readings with the previous readings Ability to answer the question, *Is the instrument functioning correctly?* Alerting the experienced geotechnical engineer when magnitudes of data change exceed predetermined critical magnitudes Calibrations during service life and maintenance of readout units, field terminals, and embedded components, as described in Chapter 16 Being on the lookout for damage, potential for damage, and deterioration or malfunction of instruments and initiating preventative action or repair or replacement without delay Verifying that sufficient spare parts, including spare readout units, are always on hand or readily available

the field. When data are recorded manually in a field book, previous data are in the book.)

- Procedure for alerting the experienced geotechnical engineer when magnitudes of data change exceed predetermined critical values and for advising the owner, the design consultant, or the construction contractor of any condition requiring their attention.
- Procedure for recording all factors that may influence measured data, including construction details and progress, geology and other subsurface conditions, and environmental factors

Table 18.3. Advantages and Limitations of Automatic Data Acquisition Systems

Advantages	Limitations
Reduced personnel costs for reading instruments and analyzing data	Replacement of a knowledgeable observer by an item of hardware (there is a real possibility that visual observations will not be made, that other factors influencing measured data will not be recorded, and that causal information will therefore not be available)
More frequent readings	Possibility of generating an excess of data, encouraging a "file and forget" attitude, and therefore failing to take timely action in response to data
Retrieval of data from remote or inaccessible locations	Possibility of blind acceptance of data, which may or may not be correct (garbage in, garbage out)
Instantaneous transmittal of data over long distances, using telemetry	High initial cost and often high maintenance cost
Increased reading sensitivity and accuracy	Currently, often requires some custom designed components that are unproven
Increased flexibility in selecting required data	Complexity, requiring an initial "debugging" period
Measurement of rapid fluctuations, pulsations, and vibrations	Need for regular field checks and maintenance by specialized personnel
Recording errors are fewer and immediately recognizable	Need for backup manual recording arrangements
Electronic data storage is in a format suitable for direct computer analysis and printout	Need for a reliable and continuous source of power
	Susceptibility to damage caused by weather conditions and construction activity

such as temperature, rainfall, snow, sun, and shade.

- Procedure for making visual observations of behavior for correlation with instrumentation data. (Special care should be taken to continue making visual observations during periods when instruments are read infrequently, so that any signs of adverse behavior are identified and a more intensive reading schedule resumed.)
- Procedure for inspecting for damage, potential for damage, and deterioration or malfunction of instruments and for initiating preventative action or repair or replacement.
- Procedure for calibrations during service life and for maintenance of readout units, field terminals, and embedded components.
- Procedure for communicating with personnel responsible for data processing, presentation, and interpretation.

18.1.4. Field Data Records

When data are collected manually, field data records are required. Readings can be recorded either in a field book or on field data sheets.

Field books contain previous readings and therefore facilitate immediate comparisons. They are easier to handle and to keep dry than field data sheets, but loss of a field book can be a serious issue. The author prefers the use of field data sheets, specially prepared for each project and instrument.

Field data sheets should include project name and instrument type, and spaces are required for date, time, observer, readout unit number, instrument number, readings, remarks, data correctness checks, visual observations, and other causal data including weather, temperature, and construction activities. Data collection personnel are less likely to write down all these important factors when entering data on a blank page in a field book. The need for comparing readings immediately with the previous set of readings is handled by taking a copy of previous readings into the field or by transcribing key data to a column on the new data sheets. Special paper, available from suppliers of weatherproof field books, can be used to allow writing in wet conditions. One or more field data sheets will be used for each date, with later transcription of data to one calculation sheet for each instrument. Raw data should be copied and the copy and original stored in separate safe places to guard against loss.

CURIOUS GEORGE PROJECT
FIELD DATA SHEET FOR PNEUMATIC PIEZOMETERS

Date ______________________ Observer ______________________
Readout Unit No. ______________________ Air Temperature ______________________

Piez. Number	Time	Readings (lb/in.^{2})			Remarks
		Individual		Best	

Additional Remarks (Data correctness checks, visual observations, causal data, weather, construction activities. Continue on back of sheet if necessary.)

____

____

____

____

____

____

Figure 18.1. Typical field data sheet for pneumatic piezometers.

Figure 18.1 shows a typical field data sheet for a pneumatic piezometer, assuming that a copy of previous data will be taken to the field. Space is provided for four individual readings, so that data correctness can be judged on the basis of repeatability and a judgmental *best* reading entered. The *best* reading can be an average or can be a median value. The median value is obtained by deleting highest and lowest values until only one central reading remains, this being the median. If two readings remain (as is the case for an even number of readings), the average of these two is taken as the median value. A space is also provided for air temperature, because pneumatic readout units can be sensitive to temperature, and on occasion a column for air temperature may be appropriate. If there is a

CURIOUS GEORGE PROJECT
FIELD DATA SHEET FOR PROBE EXTENSOMETER

Extensometer No. ____________ Date ____________
Readout Unit No. ____________ Start Time ____________
Survey Tape No. ____________ Finish Time ____________
Observer ____________

Meas. Point No.	Tape Reading (ft)				Remarks
	Individual			Best	

Additional Remarks (Data correctness checks, visual observations, causal data, weather, construction activities. Continue on back of sheet if necessary.)

____

____

____

____

____

____

Figure 18.2. Typical field data sheet for a probe extensometer, with deepest measuring point in bedrock.

predominant cause of pore water pressure change, a column may be appropriate for recording those causal data. For example, if the piezometers are installed in soft ground beneath an embankment, a column may be appropriate for the current elevation of the top of the fill. To avoid unnecessary writing, it may be convenient to enter instrument numbers on the original of the field data sheet.

Figure 18.2 shows a typical field data sheet for a probe extensometer, for which a set of data are required rather than a single value, and one full sheet is needed for a single set of readings of one instrument. In this example the deepest measuring point is always in bedrock. If there is a possibility of settlement at the deepest measuring point, a space will be required on the sheet for collar elevation, requiring a surveyed elevation of the point on the instrument terminal at which survey tape readings are made. Again, it is assumed that a copy of previous data will be taken to the field. Space is provided for three individual readings, but the third set is necessary only if significant differences occur between the first two sets. The sheet should be designed for the largest number of measuring points at an individual instrument, in this case eight.

The U.S. Bureau of Reclamation (Bartholomew et al., 1987; Carpenter, 1984b) has pioneered use of hand-held field data entry devices to replace field data sheets. A bar-coded tag is attached to the field terminal for each instrument, containing the project name and instrument number. The tag is read using an optical reader pen attached to the data entry device. The device is programmed in a question and answer mode to prompt data collection personnel on what entry is expected, and previous readings are stored in the device so that new data can be compared immediately with previous data. When all field data are entered, the device is taken to a telephone or computer terminal where it transmits the data to a file at a central computer. Data processing personnel then access the file, add it to the project data base, and process the data by computer. Simpler devices are also available, without the bar-coded tag, such as are used by electric and gas meter readers.

18.1.5. Initial Readings

Most instrumentation data are referenced to initial data, and engineering judgments are usually based on changes rather than on absolute values. Correct initial readings are therefore essential.

Many instruments, including instruments installed in boreholes and many types of strain gage, take a few days to stabilize after installation. Instruments should therefore be installed as early as possible, before they are affected by construction activities, and initial readings taken after stabilization.

The following guidelines are given as a goal when establishing initial readings. The guidelines can usually be followed with instruments requiring a single reading, such as strain gages and piezometers. However, for instruments requiring multiple readings, such as probe extensometers and inclinometers, this goal may be too demanding and should be tempered by considerations of a reasonable level of effort.

Typically, a minimum of two readings will be taken immediately after installation, as part of the acceptance test (Chapter 17). Daily readings should then be taken until data are stable, at which time the formal initial readings should be made. Initial values should be based on a minimum of two readings, and repeatability between these readings should satisfy the expected tolerance. If the tolerance is exceeded, the transducer or readout unit may be faulty, installation may be faulty, real changes may be occurring, or stabilization may not be complete. The cause must be evaluated and the problem remedied. After the formal initial readings have been taken, daily readings should be continued for a few days to ensure that data are indeed stable.

In summary, the goal for the initial reading process includes four steps: acceptance tests, daily readings until stable, formal initial readings, and daily readings to verify stability.

Readings are often dependent on the particular transducer and readout unit combination, and interchange of readout units should therefore be avoided whenever possible. When a spare readout unit has been procured, for use if the primary unit malfunctions, or when there is a possibility of interchanging units, initial readings should be made with each unit. When an interchange occurs and this has not been done, formal "initial readings" must be taken with the replacement unit, and for data continuity it must be assumed that either there has been no change since the previous reliable reading or data have changed in accordance with an established trend.

To avoid the possibility of disputed measurements of change, the contractor and owner's representative should jointly make initial readings and agree on appropriate values.

18.1.6. Data Collection Frequency

The frequency of data collection should be related to construction activity, to the rate at which the readings are changing, and to the requirements of data interpretation. Too many readings overload the processing and interpretation capacity, whereas too few may cause important events to be missed and prevent timely actions from being taken. Good judgment in selecting an appropriate frequency is vital if these extremes are to be avoided.

When construction commences and approaches the instrumented location, readings should be taken frequently; for example, once a week, once a day, once a shift, or even more frequently in relation to construction activity (such as before and after each blast, during pile driving, or during the placement or removal of surcharge). It is often wise to increase the frequency of readings during heavy rainfalls. As construction activity moves away from the instrument location or ceases altogether, and when readings have stabilized and remained constant, the frequency may then be decreased.

Cording et al. (1975) indicate that data collection needs generally fall into two categories: those that are construction dependent and those that are time dependent. A preliminary schedule should be developed on this basis during the planning phase so that requirements can be estimated for access and personnel. This preliminary schedule should assume a maximum reading frequency, and once construction has started and procedures are known, a more realistic schedule can be developed. The schedule should be flexible and should be reviewed frequently, in light of observed trends and predicted behavior.

As indicated in Chapter 6, if the construction contractor is responsible for data collection under a unit price contract, there should not be a predetermination of the reading schedule and frequency. Frequency should be at the discretion of the owner's representative, necessitating a *data set* or *man hour* pay item unit.

18.2. PROCESSING AND PRESENTATION OF INSTRUMENTATION DATA

The first aim of data processing and presentation is to provide a rapid assessment of data in order to detect changes that require immediate action. The second aim is to summarize and present the data in order to show trends and to compare observed with predicted behavior so that any necessary action can be initiated.

Responsibility for processing and presentation of instrumentation data will have been determined during the planning phase and should preferably be under the direct control of the owner or instrumentation specialists selected by the owner. Personnel requirements for these tasks are frequently underestimated, resulting in the accumulation of unprocessed data and failure to take appropriate action. The time required for data processing and presentation is usually similar to, and may even exceed, the time required to collect data.

18.2.1. Personnel Qualifications

Processing and presentation should be under the supervision of the experienced geotechnical engineer who has been responsible for data collection. The engineer will be assisted by other engineers or by technicians, but only those who have computational skill and a background in the fundamentals of geotechnical engineering. Processing and presentation require significant judgments and should not be delegated to inexperienced personnel.

18.2.2. Role of Automatic Data Processing and Presentation

The advent of automatic data processing and plotting systems has made an enormous increase in the efficiency of data processing and presentation. Despite their obvious advantages they have their limitations, of which users should be fully aware. No automatic system can replace engineering judgment.

Advantages of Automatic Data Processing and Presentation

DiBiagio (1979) lists the following advantages of automatic data processing and presentation:

- Computers provide for rapid and accurate processing of data, providing the programs are available and thoroughly tested.
- Once data are stored in a form that is readily accessible by a computer, it is a relatively simple task to reprocess batches of data if changes in processing or computational procedures are desired.
- Computers can be used to generate summary tables or results for direct use in reports. Thus, the costly and time consuming task of hand typing and proofreading tables is avoided.
- Curves or other forms of graphical representation of data can be generated automatically by the computer. Very little effort is required, for example, to change scales or replot data. This is often necessary in reporting data obtained from instrumentation projects, in particular when they extend over a long period of time.
- Highly skilled personnel are not required to administer routine computer data processing tasks.
- Computers or microprocessors can be used to scale signals or to perform computations on input

data in real time if immediate access to results is a requirement.

- On projects involving large amounts of data [such as offshore structures] computers may be used advantageously to perform a running statistical analysis of recorded data, to produce separate records of compact statistical data. Analysis of reduced data of this kind is useful as a first step in planning subsequent data processing and analysis procedures to be used on the bulk of the recorded data.

Disadvantages of Automatic Data Processing and Presentation

Disadvantages of automatic data processing and presentation include the following:

- Replacement of a knowledgeable engineer by an item of hardware. There is a real possibility that engineering judgment will be given second place, that correlations will not be made with visual observations and with factors that influence measured data.
- Computer programs must be thoroughly tested before they are accepted. Whenever possible, self-checks should be built into computer programs.

The author believes that when automatic data processing procedures are used they should be designed so that the product is never inferior to the product of thorough and judgmental manual data processing. With this in mind, the remaining subsections of Section 18.2 are written as if data are processed manually. This should not be taken as a vote against the use of automatic data processing, but merely as a way to define a quality standard to be used as a model when developing detailed manual or automatic data processing for a particular project.

Figures 18.3–18.5 show examples of data plots prepared by automatic systems during construction of the Mt. Baker Ridge Tunnel in Seattle.

Bartholomew et al. (1987) provide guidelines on automatic processing of data for embankment dam instrumentation.

18.2.3. Written Procedure for Data Processing and Presentation

The geotechnical engineer responsible for data processing and presentation should finalize the draft procedure that has been written during the planning phase. The detailed written procedure will be based on information provided in the manufacturer's instruction manual and on specific site conditions and will include the following:

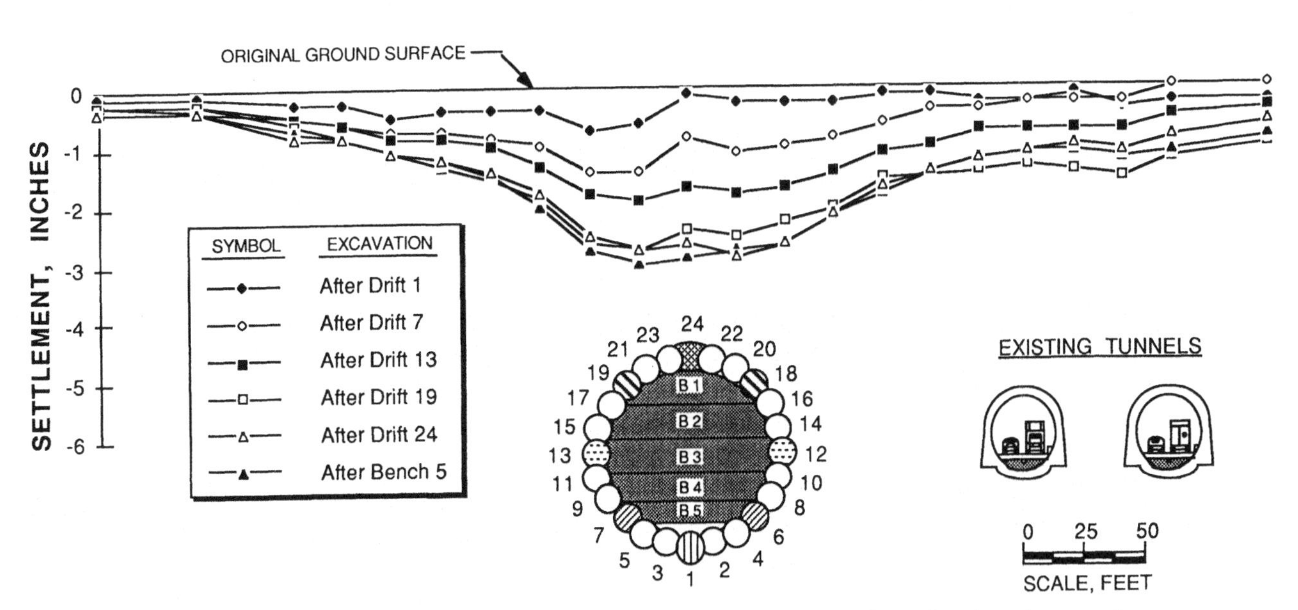

Figure 18.3. Typical plot of surface settlement data, prepared using Apple Macintosh 512k personal computer. Mt. Baker Ridge Tunnel, Seattle, WA (courtesy of Shannon & Wilson, Inc., Seattle, WA).

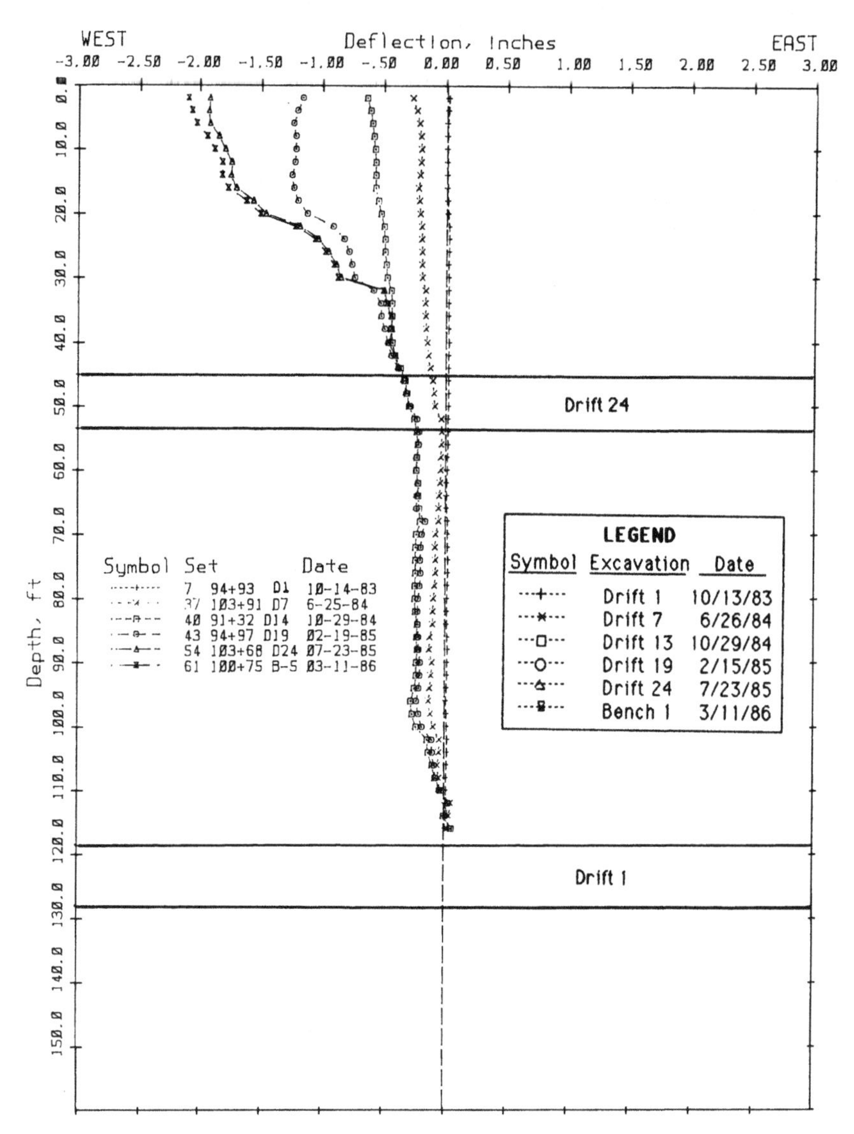

Figure 18.4. Typical plot of inclinometer data, prepared using Digital Equipment Corporation PDP 11/34 computer and Houston Instruments DMP 29 X-Y plotter. Mt. Baker Ridge Tunnel, Seattle, WA (courtesy of Shannon & Wilson, Inc., Seattle, WA).

- List of information that should be provided by data collection personnel, including initial readings, field data records, information on reading correctness, information on factors that may influence measured data, and visual observations of behavior.
- Factory calibrations, results of acceptance tests, and procedure for updating calibration data in accordance with calibrations made during service life.
- Use of automatic data processing procedures.
- Procedures for data screening.
- Data calculation sheet.
- Step-by-step calculation procedure.
- Sample data calculation.
- Methods of plotting data.
- Sample data plots.
- Procedure for advising the owner, the design consultant, or the construction contractor of any conditions requiring their attention.
- Procedure for communicating with personnel responsible for interpretation of data.

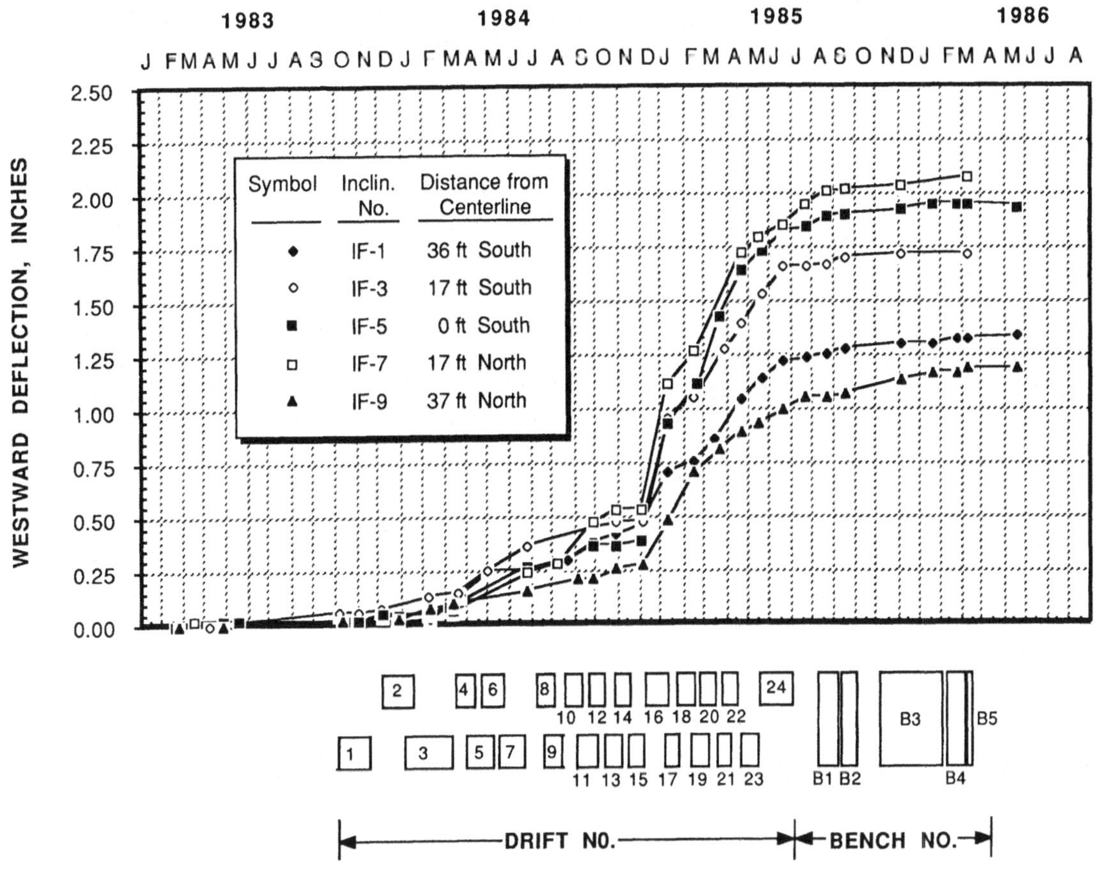

Figure 18.5. Typical plot of inclinometer data, prepared using Apple Macintosh 512k personal computer. Mt. Baker Ridge Tunnel, Seattle, WA (courtesy of Shannon & Wilson, Inc., Seattle, WA).

18.2.4. Screening of Data

Data are usually screened, in two steps, prior to data calculation and plotting.

The first data screening step is made in the field by data collection personnel when they examine reading correctness, by using one or more of the following procedures: visual observations, duplicate instruments, a backup system, regular in-place calibration checks, consistency between two or more measured parameters, and data repeatability. After this step, an assessment of data can be made in order to detect changes requiring immediate action.

The second data screening step is made in the office by data processing personnel and entails scrutinizing the field data, marking any obvious errors on the field data records, and reviewing the evaluations of data correctness made earlier in the field. When obvious errors are found, repeat readings will usually be required. On occasion, additional screening is possible by evaluating plots of raw data or computer-generated tabulations designed specifically for screening purposes, as discussed in Section 18.2.6.

18.2.5. Calculations

After data that appear false are discarded, raw readings are transcribed from field data records to calculation sheets for conversion to engineering units. Usually this task should be accomplished within 24 hours of data collection and should be followed immediately by updating routine plots of data versus time.

Calculation sheets should be made specially for each project and instrument type, to record project name, instrument type and number, date and time of readings, initials of persons making and checking

CURIOUS GEORGE PROJECT
CALCULATION SHEET FOR PNEUMATIC PIEZOMETER

Piezometer No. ____________________ Initial Piezometric Elev. ____________ ft.

(1) Date & Time	(2) Best Reading (lb/in.^{2})	(3) Piezo-metric Head (ft) (2) × 2.31	(4) Elev. of Piez. (ft)	(5) Piezo-metric Elev. (ft) (3) + (4)	Calcs. by,	Checked by	Remarks

Figure 18.6. Typical calculation sheet for a pneumatic piezometer.

calculations, readings transcribed from field data record, equations used for calculations (including any calibration or correction factors), and any remarks. Numbering the columns and using these numbers in defining each calculation step at the head of subsequent columns can be helpful.

Figure 18.6 shows a typical calculation sheet for a pneumatic piezometer. The initial piezometric elevation is included so that a comparison can be made with subsequent values. The first two columns are transcribed from the field data sheet (Figure 18.1). The third column is for a change of units from pressure to head and can include any instrument calibration factor. The fourth column is required so that head can be converted to elevation. The elevation must be either estimated or measured at a nearby settlement gage. The fifth column of data will be plotted against time. The remarks column will include key items transcribed from the field data sheet and may also include remarks on reading repeatability.

Figure 18.7 shows a calculation sheet for a probe extensometer, compatible with the field data sheet in Figure 18.2. In this case a full calculation sheet is required for each set of data. The survey tape constant is the distance between the sensing point in the probe and the zero on the survey tape. The elevation of the bottom measuring point is determined when initial readings are made and will not change, because in this example the deepest measuring point is in bedrock. Spaces (3), (5), and (6) are completed by transcription from the field data sheet. The collar is the point on the instrument terminal at which survey tape readings are made. The collar is settling, and data must be referenced to the fixed bottom measuring point by calculating as shown on the sheet. The calculated settlement for the bottom measuring point should of course be zero. Settlement will usually be plotted against time, and the plot for each measuring point will be labeled with the initial elevation of the measuring point. If there is no interest in elevation data and if the same survey tape is always used, the calculations can be simplified by omitting the survey tape constant and revising the calculation sheet.

CURIOUS GEORGE PROJECT
CALCULATION SHEET FOR PROBE EXTENSOMETER

Extensometer No. ____________ Date ____________ Survey Tape Const. ____________ ft (1)
Elev. at Bottom Measuring Point ____________ ft (2)
"Best" Tape Reading for Bottom Measuring Point ____________ ft (3)
Collar Elev. ____________ ft (4)[(1) + (2) + (3)] Calcs. by ____________ Checked by ____________

(5) Meas. Point No.	(6) Best Tape Reading (ft)	(7) Depth Below Collar (ft) (6)+(1)	(8) Meas. Point Elev. (ft) (4)−(7)	(9) Initial Meas. Point Elev. (ft)	(10) Settlement (ft) (9)−(8)	Remarks

Figure 18.7. Typical calculation sheet for a probe extensometer, with deepest measuring point in bedrock.

18.2.6. Plots of Data

After calculations have been made, plots of data should always be prepared. Some engineers and geologists struggle to interpret data that is presented on field data records or calculation sheets, but the human eye and brain are not able to do this efficiently. *Prepare a plot and all becomes clear!*

Various types of plot play a role in the data processing, presentation, and evaluation process. Some are described in the following subsections, and examples are given. The examples are intentionally simplistic, because their purpose is to show concepts rather than actual data.

Plots to Assist with Data Screening

As indicated in Section 18.2.4, data are screened both in the field and in the office. On occasion, data quality can be improved further by plotting raw readings or by generating computerized tabulations designed specifically for screening purposes and exercising engineering judgment. Two prerequisites are needed on these occasions: sufficient data for use as input to the screening process and the existence of a known truth about performance, for use as a basis when judging data. An example will illustrate the procedure.

A probe extensometer has been installed vertically in an alluvial deposit. The first stage of an embankment has been placed on the deposit, and probe extensometer readings are being used to evaluate the progress of consolidation settlements during a 1 month waiting period. The designers are analyzing vertical compression, within the deposit, in horizontal slices defined by known stratigraphic boundaries. Measuring points for the probe exten-

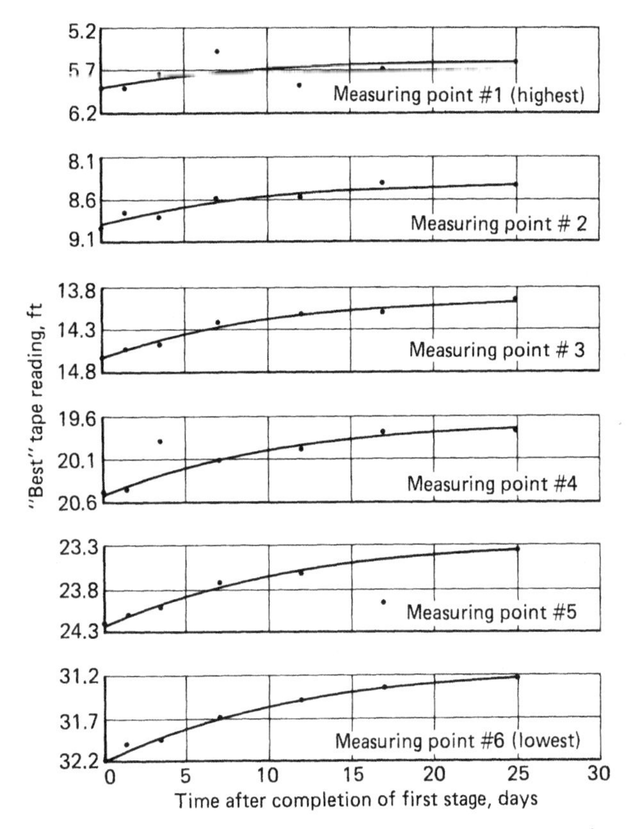

Figure 18.8. Plot of raw data for probe extensometer, to assist with screening.

someter have been installed at these boundaries. It would be possible to process data directly from raw field data by plotting vertical compression between each pair of measuring points versus time, but that procedure would not make use of the following known truth: *under constant applied loading conditions, a plot of settlement versus time should follow a logical pattern; the slope of the plot may change, but there should be no significant scatter on either side of a smooth plot.* This rule applies to all measuring points in the soil profile and allows an opportunity to improve data quality by correcting for random reading errors. Plots would first be made, for each measuring point, of tape reading versus time, and logical and smooth lines would then be drawn as best fits to the points, as shown in Figure 18.8. These smooth lines would then be used, in preference to the raw data, as a basis for analyzing vertical compression.

Routine Plots of Data Versus Time

Immediately after converting data from raw readings to engineering units, they should be added to time plots. These plots will be used by data collection and processing personnel to assist in an assessment of data quality and to show data trends and will be used as base material for data interpretation. They will be updated whenever new data are collected.

Plots to Assist with Predictions

Frequently, the routine plots of data versus time are adequate for predicting future trends. For example, when monitoring performance of an embankment on soft ground, plots of pore water pressure versus time may be adequate for predicting the length of time required for full consolidation to occur. On other occasions, for example, when deformation is monitored around a tunnel in rock, trends may not stand out clearly on plots of deformation versus time, and it is often helpful to plot velocity or even acceleration versus time. In many cases the velocity of deformation is much more important than the absolute magnitude, because an increasing velocity (acceleration) will generally indicate potential hazard.

Plots for Comparing Observed and Predicted Behavior

When comparison between observed and predicted behavior is part of data interpretation, the two sets of data should generally be plotted against time and on the same axes.

Plots for Comparison of Measurements and Observations

Comparisons between two sets of different data often form part of data interpretation. For example, when monitoring both settlement and pore water pressure in soft ground beneath an embankment, there should be consistency between the two sets of data, and data can be evaluated by plotting both sets on the same axes. When both a load cell and hydraulic jack are used to measure load during a pile load test or tieback proof test, one can be plotted against the other, or the ratio of the two can be plotted against the one assumed to be more correct. An example of such a plot, in which the load cell is assumed to be more correct, is given in Figure 18.9. When visual observations can be quantified, plots can be made to compare observations with measurements and used to assist with data interpretation.

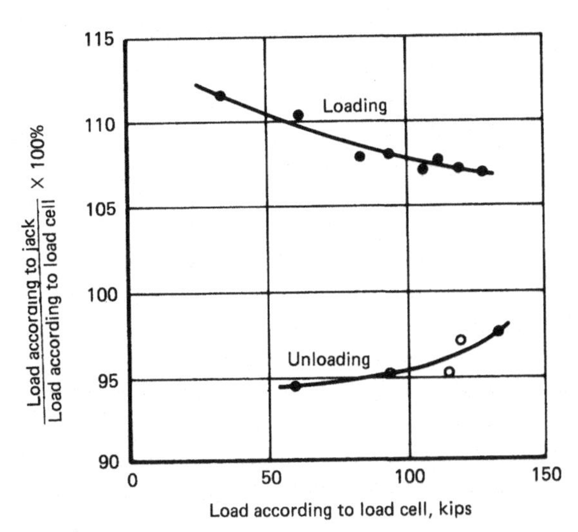

Figure 18.9. Comparison between load cell and hydraulic jack during proof test of tieback.

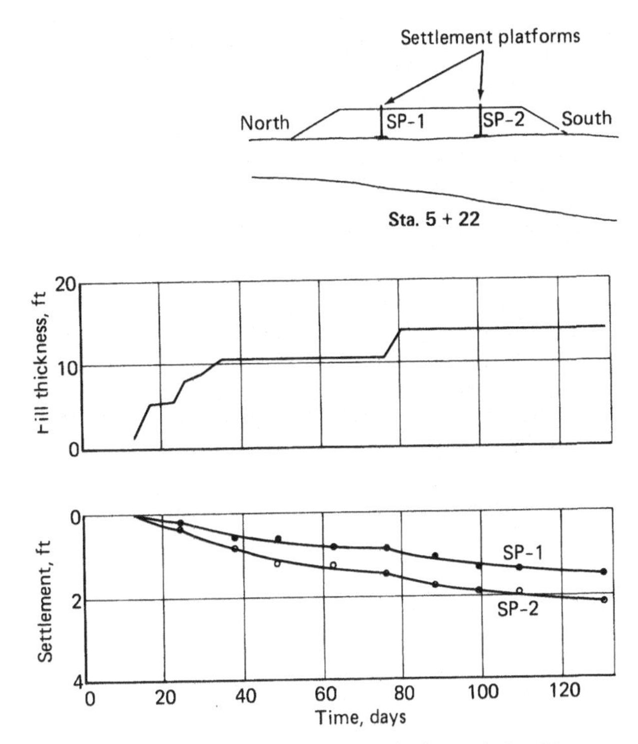

Figure 18.10. Plot to examine cause and effect relationships: settlement of soft ground below an embankment.

Plots to Examine Cause and Effect Relationships

Data interpretation nearly always includes a study of cause and effect relationships. Causal information includes construction details and progress, geology and other subsurface conditions, and environmental factors such as temperature, rainfall, snow, sun, and shade. Major causal information and visual observations should be added to the routine plots of data versus time, either by noting the occurrence of relevant events or by plotting causal data to the same time scale.

Figure 18.10 shows plots of fill thickness and settlement for an embankment on soft ground, indicating the relationship between cause and effect.

Summary Plots

Many of the plots described above may show too much detail for submittal to senior personnel or for use in a report, and often simplified summary plots are required for presentation of data to personnel who may not have the time or technical background to understand or digest all the measurements.

As an example of a summary plot, deformation can often be summarized by preparation of a contour plot on a plan of the deforming area, each contour indicating a different magnitude of deformation. Such plots are useful, for example, to show deformation of an embankment dam. Alternatively, vector plots can be prepared, showing arrows that indicate the magnitude and direction of deformation.

Cording et al. (1975) suggest that summary plots should not contain all the instruments or all the data. In most cases, they should consist only of a few selected instruments that show the significant trends most clearly and enough data points to show the trends and any significant fluctuations. The plots should be arranged to show which factors are influencing the data trends. In some cases, both the predicted behavior and the hazard warning levels can be included on the plots.

Guidelines on Plotting

Lambe (1970) discusses portrayal of data:

> The proper portrayal of data is now, and always has been, essential to good soil [and rock] mechanics. Important trends can be completely hidden in poorly portrayed data. Further, valuable experimental or theoretical data incorrectly portrayed can be worthless, or even worse, can be misleading. The engineer needs to recognize that he must aim his plots towards the intended use. He might display his data one way, if they are to be used by a researcher looking for a specific phenomenon, but display them quite differently if he is trying to show a general trend to a non-technical person.
>
> Several preliminary stages are usually needed to produce an outstanding plot. To make an outstand-

> ing plot, the engineer needs to know what is significant about the data, but to determine what is truly significant he needs a good plot! He thus must make a plot, study it for trends, mechanisms, etc., and after discovering what is the most important result to be gathered from the plot, finally prepare it to show very clearly the discovered result. The engineer is not likely to obtain outstanding plots if he does not fully understand his subject. He is usually doomed to failure if he depends on a laboratory or field technician aided by a draftsman to prepare his plots. The engineer is guaranteeing second-rate work if he so delegates this responsibility. Years ago, I made an unsuccessful attempt to establish rules for plotting data. This attempt failed, as one would expect, since the most important contribution to outstanding plots is the preparer's own understanding of his subject and his own creativity.

Good portrayal of data therefore requires a full understanding of the instrumentation program combined with imagination, ingenuity, sufficient time to digest the data, and some trial and error. Lambe (1970) and Hansmire (1978) provide guidelines on plotting, including the following:

- All work should be done in pencil. Use symbols to distinguish between different instruments or different times. Colored pencils should not be used, as the colors do not reproduce.
- Choose scales so that observations fill the space available, but do not use exaggerated scales that would magnify minor changes to make them appear alarmingly large.
- Draw plots slightly darker than the underlying grid so that when copied both are visible but the plots stand out. Make sure data points are visible.
- Plot elevations and depths on the vertical axis.
- Plots should be self-explanatory. Show project name, the type of instrument, the scale and units of measurement, and the time of measurement.
- Use sketches on the plot to show the locations of the instrument relative to the construction activity and the geology.
- Maintain consistency of scales in a given document so that plots can be compared.
- Prepare time plots showing the progress of construction as well as the measurements.
- Plot predicted behavior or the limits of safety on the same axes as the field readings.
- All plots should be initialed by the engineer responsible for their preparation, thereby ensuring a sense of responsibility.
- Whenever possible summary plots should be on a single sheet of report-sized paper.

Lambe (1970) presents Figure 18.11 to illustrate poor and good plots of the same data. The poor plot is on the left, the good plot on the right. Shortcomings of the poor plot include lines drawn too lightly, a poor selection of scales, use of a legend to denote the large and small footings rather than direct labeling, subsoil described in words rather than in a simple sketch, poor placement of numbers and titles on the axes, poor arrangement of title, and data points not shown.

18.3. INTERPRETATION OF INSTRUMENTATION DATA

Monitoring programs have failed because the data generated were never used. If there is a clear sense of purpose for a monitoring program, the method of data interpretation will be guided by that sense of purpose. Without a purpose there can be no interpretation.

Early data interpretation steps have already been described: screening of data in the field and office to examine reading correctness and to detect changes requiring immediate action. The essence of subsequent data interpretation steps is to correlate the instrument readings with other factors (cause and effect relationships) and to study the deviation of the readings from the predicted behavior. By its very nature, interpretation of data is a people-intensive activity, and no technique has yet been developed for automatic interpretation of data.

18.3.1. Personnel Qualifications

Responsibility for data interpretation will have been determined during the planning phase and should be under the direct control of the owner, the design consultant, or instrumentation specialists selected by the owner. Interpretation is a task for experienced geotechnical engineers. They should be familiar with the planning of the monitoring program, the project construction procedure, and details of instrument installation, data collection, processing, and presentation. They should have computational and analytical skills and should be

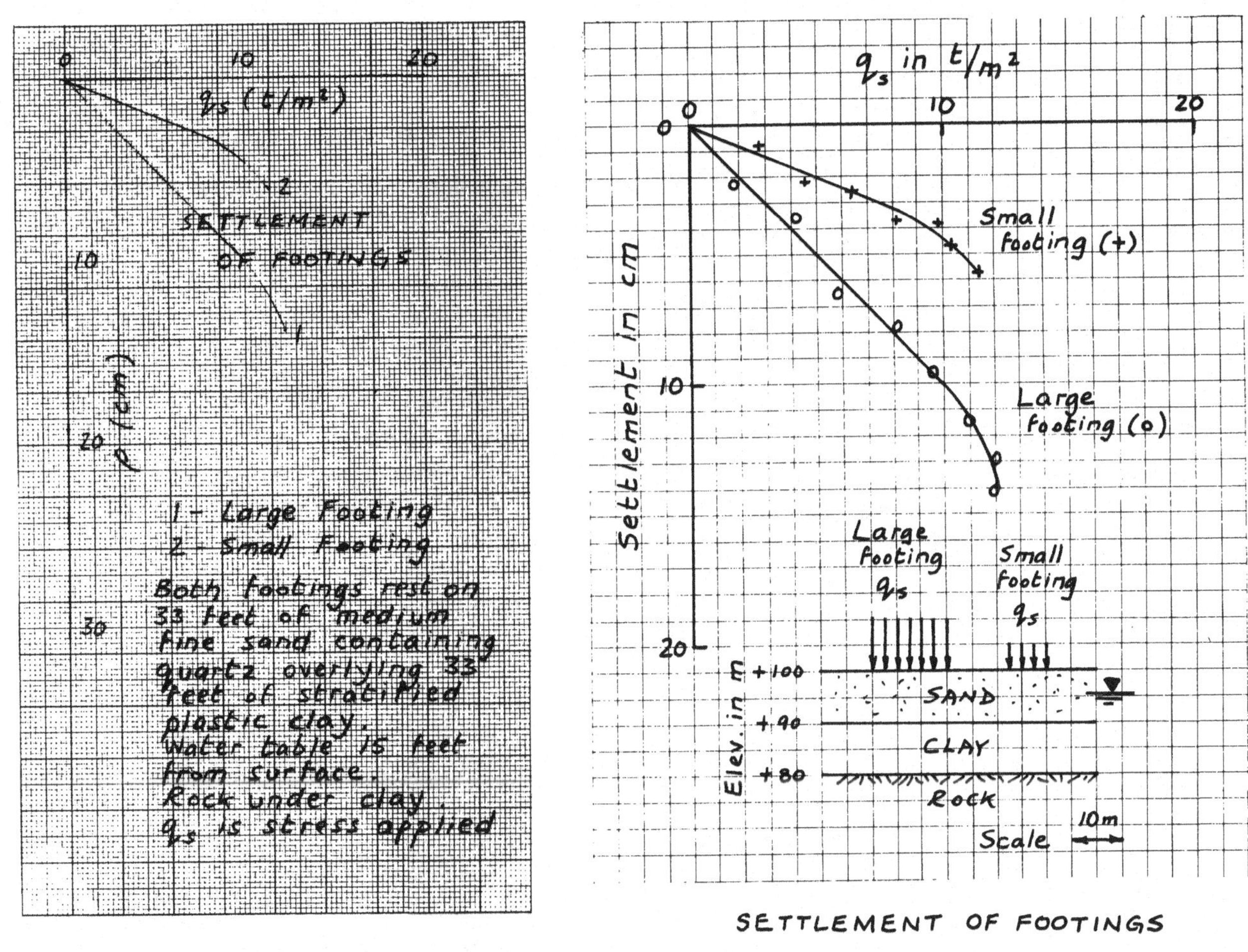

Figure 18.11. Poor and good portrayal of same data (after Lambe, 1970).

capable of exercising engineering judgment. In many cases, the same geotechnical engineer who has been responsible for data collection, processing, and presentation will play a lead role in data interpretation. If a lead role is inappropriate, this person must at least act in a coordinating and review capacity, because interpretation cannot be divorced from all preceding tasks.

18.3.2. General Guidelines on Interpretation of Data

Details of data interpretation are dependent on factors specific to each instrument and project, but a few general guideines can be given.

Preliminary Work

A draft procedure for data interpretation should havė been written during the planning phase, and this should be studied so that interpretation is tuned to the intended purpose of the instrumentation program.

Schedule

Interpretation should not be delayed until a large quantity of data has been collected and processed, because the tasks of data collection, processing, and interpretation should influence each other. The process should be iterative, allowing the needs of any one task to alter any other.

Interpretation and Reinterpretation

Interpretation and reinterpretation is an ongoing process. Initial interpretations will be tentative, dependent on collection of further data. Interpretations may change as a clearer understanding of real behavior is developed. Assessment of the performance of an individual instrument often requires a study of data over a significant time period.

Selection of Plots

Various types of plot that are useful when interpreting data have been described in Section 18.2.6. They include routine plots of data versus time, plots to assist with predictions, plots for comparing observed and predicted behavior, plots for comparison of measurements and observations, and plots to examine cause and effect relationships. Selection of plots must be made on a case-by-case basis, with the purpose of the instrumentation program clearly in mind.

Questionable Data

When faced with data that on first sight do not appear to be reasonable, there is a temptation to reject the data as false. However, such data may be real and may in fact carry an important message. A significant question to ask is: *Can I think of a hypothesis that is consistent with the data?* The resultant discussion, together with the use of appropriate procedures for ensuring reading correctness, will often lead to an assessment of data validity.

Details recorded on installation record sheets (Chapter 17) are often helpful when evaluating questionable data, because difficulties encountered during installation may cause abnormal data. Questionable data can also be evaluated by reviewing the studies of data correctness that have been made earlier, in the field during the data collection phase and in the office during the data processing phase.

Communication

An open communication channel should be maintained between design and construction personnel, so that discussions can be held between design engineers who raised the questions that caused instrumentation to be used and field engineers who provide the data. A special effort will often be required to keep this channel open, both because the two groups sometimes tend to avoid communication and because the contract for design personnel may have been terminated.

18.4. REPORTING OF CONCLUSIONS

After each set of data has been interpreted, conclusions must be reported in the form of an interim monitoring report and submitted to personnel responsible for implementation of data. In addition, a final report of the monitoring program is often required, and a technical paper may be prepared.

18.4.1. Interim Monitoring Reports

The conclusions from data interpretation should be communicated to all parties who have a role in data implementation. The initial communication may be verbal but should be confirmed in an interim monitoring report. Reporting will be on a regular schedule to allow timely implementation and will include the following:

- Updated summary plots.
- A brief commentary, drawing attention to all significant changes that have occurred in the measured parameters since the previous interim monitoring report, together with probable causes.
- Recommended action

It may be appropriate also to include copies of calculation sheets, but the reports should be kept as simple as possible to avoid delay in preparation and submittal.

18.4.2. Final Report of Monitoring Program

A formal report is often prepared to document key aspects of the monitoring program and to support any remedial actions. The report also forms a valuable bank of experience and should be distributed to the owner and design consultant so that any lessons learned may be incorporated into subsequent designs. If an installation report (Chapter 17) has been written, the final report can either reference that report or repeat appropriate content. The final report should contain at least the following information:

- Summary of the report.
- Introduction, including a brief description of the project and the reason for using geotechnical instrumentation.
- Any project design and construction information that is relevant to the monitoring program.
- Summary of the monitoring program planning phase.
- Description of instruments and readout units.*

* For a major monitoring program, brief material can be included in the body of the report and details in appendixes.

- Plans and sections sufficient to show instrument numbers and locations.
- Appropriate surface and subsurface stratigraphic and geotechnical data.*
- Instrument calibration and maintenance procedures.*
- Instrument installation procedures.*
- Procedures for data collection, processing, presentation, and interpretation.*
- Observed behavior, including summary plots and factors that influence measured data.
- Analysis of observed behavior, including comparisons between measurements and predictions, a discussion of significant changes and probable causes, and comparisons with published information.
- Conclusions, discussion, and recommendations, including a statement of any remedial actions taken.

Whenever possible, the report should also include an assessment of the monitoring program. This assessment should cover instrument performance, calibration and installation techniques, adequacy of data collection, processing, presentation, and interpretation procedures, and recommendations for specifying future monitoring programs. Inclusion of such an assessment may entail some self-criticism but will provide valuable input to planning future monitoring programs. We learn from our mistakes, and we very much welcome learning from the mistakes of others!

When instruments are used for monitoring long-term performance, the final report should form part of the operating and maintenance (O&M) manual. The O&M manual should also provide guidelines to assist O&M personnel in judging satisfactory performance based on the measurement data. An example of guidelines for O&M personnel responsible for embankment dams is given by Dunnicliff (1981).

On completion of the final report, all data should be filed in an orderly manner in one location.

18.4.3. Technical Publication

Well-documented case histories of monitoring programs are invaluable for advancing the state of the art of geotechnical design and construction. The fact that many geotechnical journals contain more research-oriented papers than case history papers generally reflects the nature of papers submitted for publication rather than the intent of journal publishers to emphasize research. Well-documented case histories are welcomed by journal publishers and readers, and engineers who are in a position to prepare these papers are strongly encouraged to do so. The content will usually be similar to the content listed above for the final report.

This plea for well-documented case histories should not be taken as a plea for papers describing every monitoring project. Before embarking on writing a paper, the writer should have positive answers to two questions: *Do I have something new to say?* and *Will my work help to fill a knowledge gap?*

Part 5

Examples of Instrumentation Applications

The goal of Part 5 is to describe the role of instrumentation in addressing geotechnical questions that may arise during the design, construction, or performance of typical civil engineering projects.

The chapters are divided into four sections. The first section indicates the general role of geotechnical instrumentation. The second section suggests the principal geotechnical questions that may arise and indicates the types of instruments that may be used to help provide answers to those questions. The third section provides an overview of what is typically *routine* monitoring and what is monitoring for *special applications*. The fourth section presents a tabular summary of selected case histories that illustrate effective use of geotechnical instrumentation. Case histories are summarized in chronological order, starting with the most recent.

The sequence of geotechnical questions in the second part of each chapter is intended to match the time sequence in which the question may be addressed during the design, construction, and performance process and does not indicate any rating of importance. Sketches are included to convey a possible layout of instruments for addressing some of the geotechnical questions. However, readers are cautioned that these layouts are intended merely to convey a general approach: as indicated in Chapter 4, the selection of instrument locations should reflect predicted behavior for each individual case and should be compatible with the method of analysis that will later be used when interpreting the data.

The format and content of this part of the book have been strongly influenced by DiBiagio and Myrvoll (1981), describing use of instrumentation for projects in or on soft clay.

These chapters are not intended to be exhaustive summaries, state-of-the-art papers, or "cookbooks." They are intended merely to open the minds of readers to the possible role of geotechnical instrumentation and to guide them toward implementation.

It is said in Chapter 4 that geotechnical instrumentation should not be used unless there is a valid reason that can be defended. Every instrument on a project should be selected and placed to assist with answering a specific question: if there is no question, there should be no instrumentation. If a reader uses this book by (1) studying the chapters in Part 5, (2) noting the types of instruments, (3) noting the sketched layouts, (4) studying Part 3, *Monitoring Methods,* for details of the instruments, and (5) proceeding with a monitoring program, that reader is misusing the book. Readers are strongly encouraged to follow the more logical procedure outlined in Chapter 4.

CHAPTER 19

BRACED EXCAVATIONS*

Several types of wall are used for supporting the sides of braced excavations, including steel sheet piles, soldier piles and lagging, auger piles, interlocking wood sheeting, and slurry walls. Lateral support can be provided either by *internal* or *external bracing*. Internal bracing consists of structural members of steel or wood, placed either as horizontal *cross-lot bracing* or as inclined or horizontal *rakers*. The most commonly used external bracing consists of *tieback anchors*.

Figures 19.1–19.3 show the contrast between how projects will usually appear on engineering drawings, and what may actually happen in the field: the *field* figures are exaggerated to emphasize the contrast. Figure 19.1 depicts a typical excavation in clay, supported by interlocking sheet piles and cross-lot bracing, for the case where the sheet piles are not driven to bedrock or underlying firm stratum. Figure 19.2 shows a typical excavation supported by soldier piles, lagging, and tieback anchors. Figure 19.3 shows a cross section through a typical excavation for a slurry wall, prior to concreting.

19.1. GENERAL ROLE OF INSTRUMENTATION

The design of braced excavations is based for the most part on empirical procedures and past experience. The consequences of poor performance can be severe and may on occasion be catastrophic. A monitoring program may not be required if the design is very conservative, if there is previous experience with design and construction of similar facilities under similar conditions, or if the consequences of poor performance will not be severe. However, under other circumstances a monitoring program will normally be required to demonstrate that the excavation is stable and that nearby structures are not affected adversely. Depending on the specific needs of each case, the monitoring program may apply to the wall and bracing, to the ground beneath or surrounding the excavation and/or to adjacent structures or utilities.

The choice between internal and external bracing has a major influence on the type and role of instrumentation for a braced excavation. It is possible to make regular visual inspections of internal bracing, but external bracing cannot be seen. Although confidence in the performance of an externally braced excavation is increased by conducting a proof test on every tieback anchor, if an anchor subsequently fails, the failure may be progressive and catastrophic. In general, therefore, instrumentation plays a role in three phases of external bracing that are not applicable to internal bracing: testing of *test anchors* during the design phase or at the start of construction, *performance* and *proof testing* of anchors during construction, and subsequent *monitoring* of selected representative anchors. The third phase may be omitted if a conservative design has been used.

*Written with the assistance of Mark X. Haley, Associate and Vice President, Steven R. Kraemer, Associate & Vice President, and Cetin Soydemir, Senior Engineer, Haley & Aldrich, Inc., Cambridge, MA.

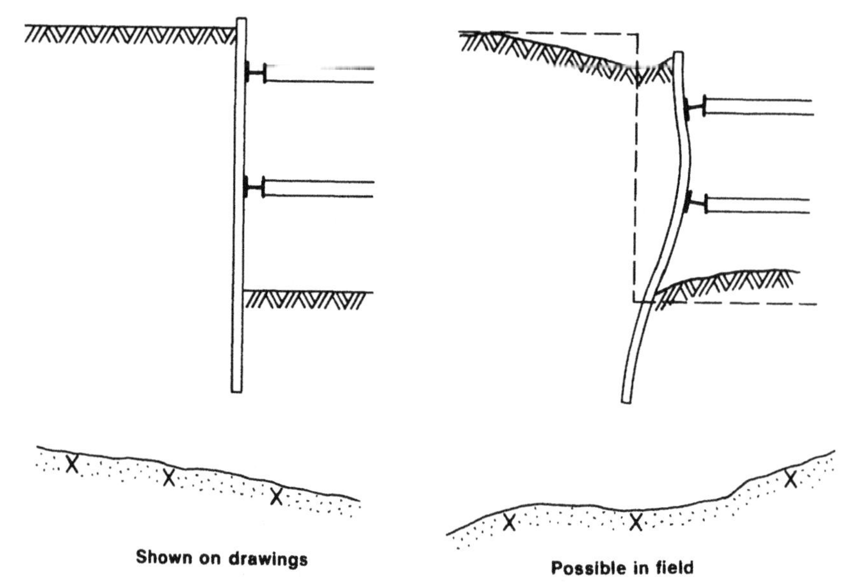

Figure 19.1. Open cut excavation supported by interlocking sheet piles and cross-lot bracing (after DiBiagio and Myrvoll, 1981).

19.2. PRINCIPAL GEOTECHNICAL QUESTIONS

The following principal questions are presented in the normal order of occurrence for a braced excavation. The order does not reflect a rating of importance.

19.2.1. What Are the Initial Site Conditions?

Initial site conditions are determined by use of conventional site investigation procedures, sometimes supplemented by in situ testing. However, performance monitoring instrumentation sometimes plays a role. For example, initial groundwater pressures and fluctuations must often be determined for design purposes. Piezometers can be installed well before the start of excavation to define the preconstruction groundwater pressure regime, including any perched or artesian water. If the instruments are installed sufficiently early, seasonal variations can be defined.

A *preconstruction conditions survey* will normally be made by representatives of the construction contractor's insurance company and some-

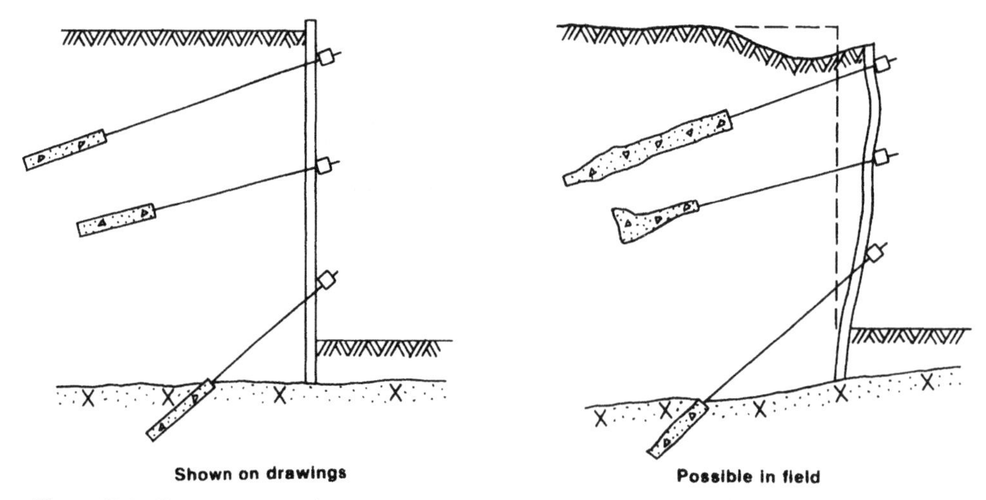

Figure 19.2. Open cut excavation supported by soldier piles, lagging, and tieback anchors (after DiBiagio and Myrvoll, 1981).

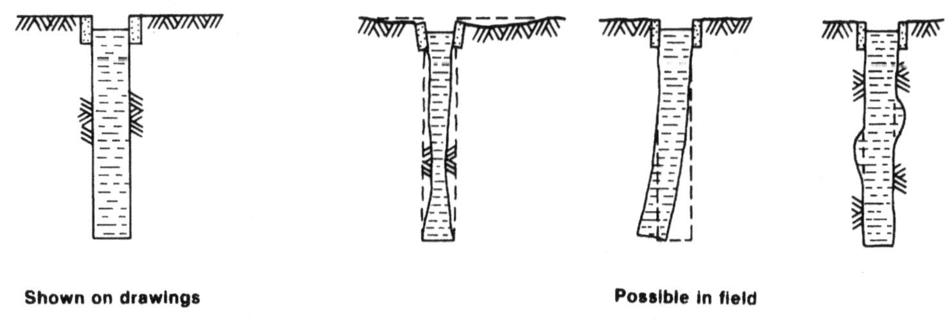

Figure 19.3. Excavated trench for slurry wall (after DiBiagio and Myrvoll, 1981).

times also by the owner. The purpose of the preconstruction survey is to observe and document the conditions of structures that may be influenced by the excavation, using photographs, written descriptions, and/or videotapes of existing defects in the structures. The preconstruction conditions survey is frequently supplemented by use of surveying methods to define the elevations of reference points on the structures and on the ground surface. The widths of any existing cracks should also be measured.

19.2.2. What Information Can Be Provided by Instrumenting a Full-Scale Test of the Complete Excavation and Support System?

The primary goals when designing support are to control ground movements and to establish a stable excavation. The designers may identify a combination of circumstances that lead them to opt for a full-scale test section, as a means of ensuring design adequacy and economy. The combination of the following four circumstances provides an example. First, the planned excavation is for a long subway or water tunnel. Second, subsurface conditions are judged to be relatively uniform along the length of the excavation. Third, there is little precedent for constructing a braced excavation in similar ground. Fourth, the designers believe that a full-scale test section can reduce the conservatism of the loading envelope and can therefore be economical.

Instrumentation and observation are essential components of a full-scale test section. The intensity of instrumentation is usually greater than for other applications, so that maximum information can be learned from the test. The types of instrumentation are discussed elsewhere in this chapter.

The test section should be long enough so that end effects are minimized. As a general rule, the length of the test section should be at least four times the width of the excavation, and it should be located at least three braces away from any nonuniformity in the cross section. However, this ideal must be tempered by cost considerations and by knowledge of site and project conditions.

19.2.3. What Is a Suitable Design for Tieback Anchors?

The load–movement relationship of a tieback anchor depends on the soil or rock properties and the size and shape of the grouted body of the anchor. The relationship is generally not governed by the properties of the tendon, and computations of allowable loads for anchors are therefore only rough approximations. When design uncertainties are unacceptable, *test anchors* can be installed and tested either during the design phase or at the start of construction. The tests include measurement of load and displacement at the anchor heads.

As indicated in Chapter 13, calibrated hydraulic jacks do not provide an accurate measurement of load, and a good-quality load cell should be used adjacent to the jack. Electrical resistance load cells are normally used, but mechanical, vibrating wire, and hydraulic cells are also suitable. The author favors use of the Proceq mechanical load cell (Figure 13.1) for this purpose: it is robust, has adequate accuracy, is read simply by observing a dial indicator, and does not require attachment of an electrical cable or remote readout unit.

Displacement is normally measured by attaching a long-range dial indicator to a survey tripod, with the dial indicator stem bearing on the anchor head. A universal joint at the connection between dial indicator and tripod eases the setting up task.

If the objective of test anchors includes determination of the pattern of load transfer in the grouted zone, strain gages and/or multiple telltales should be used, as described in Section 19.2.9.

19.2.4. What Is a Suitable Design for Slurry Trench Excavations?

If soil conditions are well defined and if the performance of slurry trench excavations has already been proved in similar soils, feasibility is not in doubt. However, in other cases, either the safe and economical construction method or even the feasibility of constructing a slurry trench may be in doubt, and a full-scale instrumented test may be warranted. Various aspects of a typical full-scale test are discussed in the following subsections.

Stability in Relation to Time

One objective of a full-scale test is to monitor lateral and vertical ground deformations in the vicinity of the trench in relation to time. The magnitude and rate of observed deformations will provide an indication of trench stability or instability.

Lateral deformations in the soil alongside the trench can be monitored by using inclinometer casings installed close to the walls of the trench. The alignment of boreholes for inclinometer casings should be controlled carefully, so that casings do not intercept the excavation. However, because neither the casing nor the trench will be exactly vertical, the top of an inclinometer casing should not be closer than about 18 in. (460 mm) to the wall of the trench, and a backup casing is advisable at about 4 ft (1.2 m) from the wall. Because inclinometers cannot provide data on lateral deformation of the wall of the trench itself, a comprehensive test includes monitoring of convergence across the trench, using one of the gages described in Section 12.11.2.

Vertical movements of the ground surface and settlement of the guide walls are normally monitored by surveying methods. Although not of primary importance, monitoring of subsurface vertical deformation may provide valuable input to the evaluation of the test and is usually justified. Probe extensometers will normally be used. Trends in measured pore water pressure in the surrounding soil can also serve as a guide to indicate how fast overall stability is deteriorating with time, and pore water pressures can be monitored with rapidly responding diaphragm piezometers. Figure 19.4 shows a possible layout of instrumentation for examining the stability of a slurry trench in relation to time. The piezometers remote from the test panel are required as reference piezometers, to monitor any variation in groundwater pressure that may result from other causes.

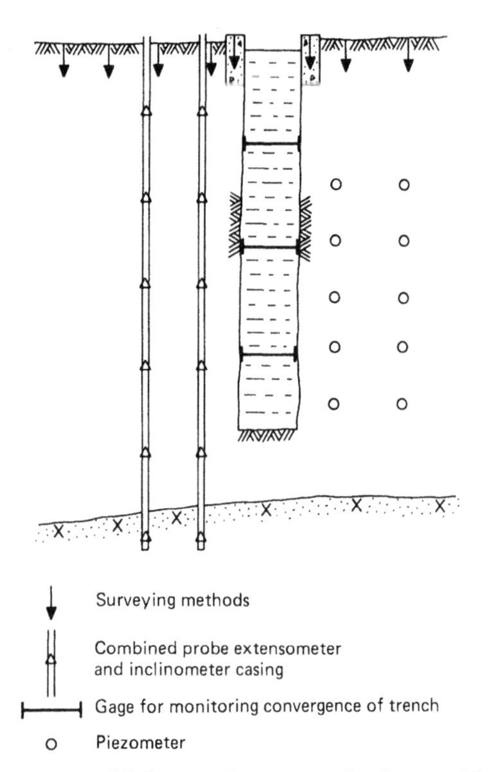

Figure 19.4. Possible layout of instrumentation for examining the stability of a slurry trench in relation to time (after DiBiagio and Myrvoll, 1981). Note: Use of surveying methods requires benchmark, remote from test panel. Piezometers also required, remote from test panel.

Stability in Relation to Panel Length

The limiting length to width ratio for which a single panel can be excavated safely can be established on the basis of full-scale test panels of varying length. Comprehensive measurements of the displacement field are needed to evaluate the restraint provided by the soil at the ends of a trench, using the instruments shown in Figure 19.4.

Stability in Relation to Slurry Level and Density

Trench stability is affected by the difference between the level of slurry and the groundwater level and by the density of slurry. The effects can be assessed by varying the slurry level and density while monitoring ground deformation and groundwater level in the vicinity of the panel, again using the instruments shown in Figure 19.4.

Shape and Alignment of Slurry Trench

The shape and alignment of the trench may not be as intended. It is impossible to make a visual examination of an excavated trench; therefore, any final inspection must rely on instrumentation. The width of the trench at any point can be determined with remote reading caliper tools, and the same instrument can be used to search for local failures or overbreak, particularly in fissured clay where blocks of material may fall out during or subsequent to excavation. Errors in vertical alignment of slurry trenches may have serious consequences, and verticality of the test trench can be measured using an inclinometer, by lowering inclinometer casing temporarily into the trench and pressing the casing against one side with small double-acting pneumatic jacks. Other measurement methods are described in Section 19.2.6.

Deformation of Trench During Concreting

Filling a deep slurry trench excavation with fresh concrete may cause a significant increase in loading on the bottom and sides of the trench and may lead to local instability. If the trench is excavated in very soft clay, large local deformations may occur, and there may even be a chance of a bearing capacity type failure of the bottom or sides of the excavation. When a test panel is constructed, lateral and vertical deformations and pore water pressures should be monitored closely during concreting. Instrumentation is again as shown in Figure 19.4, with the addition of small settlement platforms set on the bottom of the trench. A sleeve should be provided around the riser pipe for each settlement platform.

19.2.5. Is the Bracing Being Installed Correctly?

Field measurements play a role in verifying that internal and external bracing is being installed correctly.

Internal Bracing

Specifications for cross-lot bracing normally require that ground deformation is minimized by preloading the struts. Preloading involves loading the strut to some fraction of the design load by using hydraulic jacks and then wedging the strut between opposite walls of the excavation. Contractor performance can be monitored by observing hydraulic jack pressures, and a special effort must be made to ensure that the jack is calibrated and aligned correctly, as discussed in Chapter 13.

External Bracing

Specifications for external bracing normally require that each tieback anchor is proof tested to a value in excess of the working load used in design. A proof test either verifies that the carrying capacity is within acceptable limits or indicates that an anchor is defective.

The test normally involves loading the anchor incrementally to a load greater than the design load, holding for a creep test, and locking-off at a lesser load. Load and displacement at the anchor head are monitored throughout the test, using a load cell and dial indicator as described in Section 19.2.3 for test anchors. The load cell is positioned immediately above or below the jack and is removed after completion of the proof test.

In addition to proof testing each anchor, it is customary to conduct a performance test on a specified percentage of anchors. A performance test normally entails loading and unloading the anchor incrementally in cycles, increasing the load at the end of each cycle to the same maximum load as for the proof test, holding for a longer creep test, and locking-off as usual. Instrumentation is the same as for proof testing.

When a tieback anchor is proof tested and locked-off, the load may be changed in adjacent anchors that have already been stressed and locked-off. For example, in the case of a slurry wall or a tied-back bulkhead, locking-off each anchor at a predetermined load does not necessarily produce the desired condition, and one or more rounds of adjustments may be required. Measurements are normally made by *lift-off* testing, using a load cell adjacent to the hydraulic jack.

19.2.6. Is Construction of Slurry Trenches Adequately Controlled?

Observations and measurements are generally required for construction control of slurry trenches. Construction contractors will generally measure and control the level and properties of the slurry, including density, pH, viscosity, sand content, and filtration. They may also monitor verticality, but this is not common practice, and when verticality is critical to use or appearance, inspection personnel should initiate a measuring program. The author is aware of five methods for measuring verticality. First, a tiltmeter can be attached to the kelley bar, cable, or bucket. Second, a short length of pipe, with outside diameter slightly less than the width of the trench, can be lowered within the trench on the end of a wire, and the lateral position of the wire with respect to the guide walls can be monitored as the pipe is lowered. Third, if a pipe is used as an end-stop in the panel, it can first be traversed along the length of the panel: if it will not pass easily along the length, the trench is not likely to be vertical. Fourth, an ultrasonic monitoring instrument can be used.* Fifth, an inclinometer can be used, as described in Section 19.2.4, but this method is time-consuming and its use is normally limited to test trenches.

Vertical movements of the ground surface, guide walls, and adjacent structures should be monitored as panels are excavated, and rate of movement is a key factor in evaluating trench stability. Trench stability can also be evaluated by comparing the quantity of excavated material with the depth of the trench: if material is removed, but the depth does not increase, caving is occurring.

Sonic coring, as described in Chapter 24, can be used to test the integrity of concrete in a slurry trench.

19.2.7. Is the Excavation Stable and Are Nearby Structures Being Affected Adversely by Ground Movements?

Ground movements may be caused by lateral deformation of the supporting wall and also by bottom heave. The movements may lead to an unstable excavation and may also affect nearby structures or utilities adversely. Instrumentation can often be used to provide a forewarning of adverse behavior, thereby allowing remedial measures to be implemented before critical situations arise. A planned instrumentation program, which indicates that the construction will be watched carefully, can give reassurance to building owners and thus can expedite the legal approval and public acceptance of a project.

Whenever there is concern for instability, surface deformation should be monitored, using surveying methods. Monitoring should include settlement of the ground, the top of the supporting wall, and nearby structures. When tiebacks are used with soldier piles and lagging, settlement measurements should normally be made on the tops of all soldier piles, to monitor any downward movement of a pile owing to the downward component of anchor load. A *rod settlement gage* (Section 12.7.7) alongside each important structure will provide accurate settlement data without need for a survey crew. Horizontal deformation measurements on the top of the supporting wall, using surveying methods and a horizontal control station, provide routine data for stability control when it is critical to minimize ground movements.

Crack gages across existing cracks in structures provide important information when there is concern that excavation may cause additional damage.

Subsurface settlement alongside important underground utilities should also be monitored, using subsurface settlement points.

If the monitoring program consists only of deformation measurements on the surface and alongside utilities, and the measurements indicate adverse behavior, there will often be inadequate data for defining the cause of the problem and for developing an economical solution. A well-planned program for monitoring stability and for providing forewarning of adverse behavior should therefore include instrumentation that indicates **causes** of surface deformation. Inclinometers are primary tools for this purpose. When soldier piles and lagging are used, inclinometer casings can be attached to soldier piles, but the bottom should be deep enough to ensure base fixity, and this may require attaching steel pipes to the soldier piles and drilling through the pipes prior to installation of inclinometer casing. When a slurry wall is used, inclinometer casings can be installed within the trench prior to concreting, again ensuring that the bottom is deep enough for base fixity, possibly requiring drilling below the base of the trench. For all types of wall, inclinometer measurements are often supplemented by using convergence gages across the excavation. Fixed borehole extensometers are an alternative to inclinometers if external bracing is used. A simple plumb line can sometimes be useful for providing horizontal deformation data at low cost. Subsurface settlement measurements, using either probe extensometers or fixed borehole extensometers, can be used to provide additional causal data.

Figure 19.5 shows a possible layout of instrumentation for monitoring an internally braced excavation in clay when there is concern that a

*A commercial source is given in Appendix D.

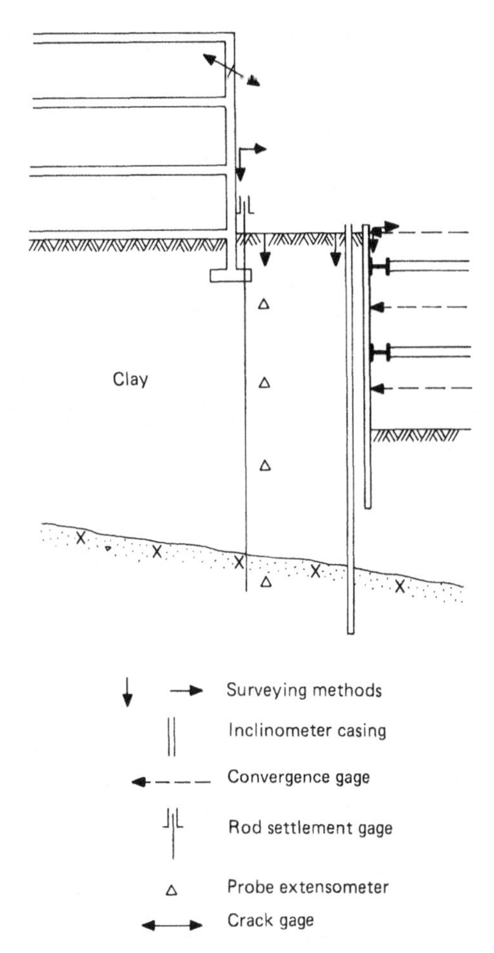

Figure 19.5. Possible layout of instrumentation to provide forewarning of adverse effect on a building alongside an internally braced excavation in clay. Note: Use of surveying methods requires benchmark and horizontal control station, remote from excavation. Piezometers also required if lowering of the groundwater table could cause problems.

building may be affected adversely by ground movements. This is only one example, and it must be stressed that many other configurations are possible, as indicated by the case histories described later in this chapter.

Adverse ground movements, of course, can also be caused by overload of individual braces, lowering of the groundwater table, or bottom heave, and these issues are addressed in subsequent sections.

Table 19.1 lists suitable instruments for monitoring whether the excavation is stable and whether nearby structures are being affected adversely by ground movements.

Table 19.1. Instruments Suitable for Monitoring Whether Excavation Is Stable and Whether Nearby Structures Are Being Affected Adversely by Ground Movements

Measurement	Suitable Instruments
Settlement of ground surface, structures, and top of supporting wall	Surveying methods Rod settlement gages
Horizontal deformation of ground surface, structures, and exposed part of supporting wall	Surveying methods Convergence gages Plumb lines
Change in width of cracks in structures and utilities	Crack gages
Subsurface horizontal deformation of ground	Inclinometers Fixed borehole extensometers In-place inclinometers
Subsurface settlement of ground and utilities	Subsurface settlement points Probe extensometers Fixed borehole extensometers
Load in internal bracing	Surface-mounted vibrating wire strain gages Surface-mounted mechanical strain gages Calibrated hydraulic jack and load cell (lift-off test)
Load in external bracing	Load cells Calibrated hydraulic jack and load cell (lift-off test) Surface-mounted vibrating wire strain gages
Groundwater pressure	Piezometers
Bottom heave	Magnet/reed switch gages Mechanical heave gages Fixed borehole extensometers Inclinometers

19.2.8. Is an Individual Brace Being Overloaded?

Monitoring of individual braces is not normally required if the design of the support system is very conservative, if there is previous experience with design and construction of similar facilities under similar conditions, or if the consequences of poor performance will not be severe. However, under other circumstances, loads developed in selected representative braces should be monitored.

Internal Bracing

For an internally braced excavation, strut loads should be monitored on as many struts as is practicable or economically justified, and reserve struts should be available for immediate installation in the event that measurements indicate that allowable limits have been reached. Surface-mounted vibrating wire strain gages are usually the preferred instruments, and mechanical gages are sometimes used as backup. Load cells are not favored for this application, because insertion of a load cell usually alters the method of strut installation, thereby creating nontypical loading conditions on the strut. Strut load can also be determined by temporary insertion of a calibrated hydraulic jack between the waler and a reaction member welded to the strut and by performing a lift-off test, but this method is cumbersome and accuracy is not great.

When using strain gages for determination of strut loads, a check on strain gage performance can be made during preloading. O'Rourke and Cording (1975) report good correlation between hydraulic jack and strain gage measurements soon after calibration of the jack, but the correlation became less good over a period of months without recalibration. A load cell in series with each hydraulic jack provides a better check on strain gage performance. Chapter 13 includes a discussion of temperature effects on strut loads and on strain gage readings and indicates that an error can occur if the strain gage and the structural member are at different temperatures. This problem can be acute when sunlight causes differential heating of a steel strut, and the guidelines given in Chapter 13 should be followed.

External Bracing

For an externally braced excavation, anchor loads subsequent to lock-off can be determined by lift-off testing. If several measurements are required on an anchor, at different times, it will usually be more economical to install a load cell. Any of the load cells discussed in Section 19.2.3 are adequate, and *telltale load cells* provide an economical method, provided that sufficient annular space is available and that access to the anchor head is maintained. Surface-mounted strain gages, preferably of the weldable vibrating wire type, are an alternative to load cells if *threadbar* (threaded solid bar) tendons are used.

19.2.9. What Is the Magnitude and Distribution of Load in the Support System?

A change in distribution of load with time can be an indicator of impending instability, but the support system will not normally be instrumented for this purpose. The question more normally implies an interest in improving the state of the art of design, rather than in performance monitoring to benefit the project under construction. The magnitude and distribution of load can be determined by making measurements in a test section of adequate length, as discussed in Section 19.2.2.

Internal Bracing

Determination of stress in internal bracing requires either lift-off testing or use of strain gages, as discussed in Section 19.2.8.

External Bracing

Determination of total load in external bracing requires either lift-off testing or use of load cells or strain gages, as discussed in Section 19.2.8.

Determination of the pattern of load transfer in the grouted zone requires measurement of strain, using strain gages and/or multiple telltales. Very great cooperation is essential between personnel responsible for instrumentation and personnel responsible for installation of the anchors. When an anchor instrumented for this purpose is installed, strain measurements will normally be made during the proof test to determine a relationship between measured strain and load and to determine the quality of strain data. An estimate of the combined modulus of the steel and grout is needed so that measured strains can be converted to stresses.

When detailed load transfer data are required for threadbar tendons, weldable vibrating wire strain gages are the first choice. Fuglevand et al. (1984) report that a manufacturer of threadbar tendons recommended against use of weldable strain gages, believing that the heat from welding could affect the metallurgical characteristics of the bar. However, gages were attached to test lengths of bar that were then pulled to failure, with no notable reduction in capacity. Weldable resistance strain gages are an alternative to vibrating wire gages but, as discussed in Chapter 13, they have many limitations. Damage to all types of strain gages and cables is not unusual while pulling drill casing and while grouting; there-

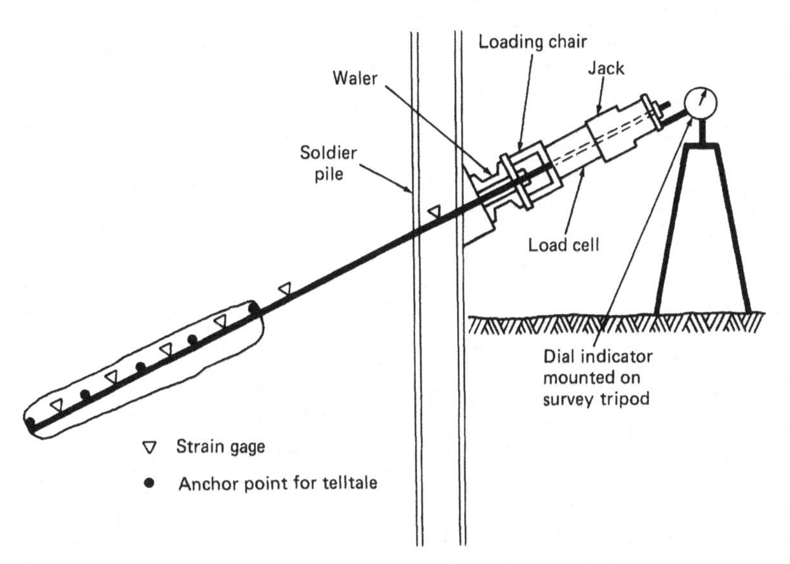

Figure 19.6. Setup during proof test, showing possible layout of instrumentation for a *threadbar* anchor, when detailed load transfer data are required.

fore, great care must be taken with protection arrangements, and significant redundancy is recommended. Whenever adequate annular space is available, multiple telltales are recommended as backup. Two configurations are possible: first, unstressed 0.25 in. (6 mm) diameter flush-coupled rods within oil-filled sleeves of PVC pipe; and second, the *slack-wire* type of fixed borehole extensometer system described in Chapter 12, with each wire housed within an oil-filled 0.25 in. (6 mm) outside diameter nylon tubing. The first alternative is simpler but requires more annular space than the slack-wire system.

Figure 19.6 shows a possible layout of instrumentation for a threadbar anchor when detailed load transfer data are required. The figure shows the setup during the proof test.

When detailed load transfer data are required for stranded cable tendons, attachment of strain gages is impracticable, and multiple telltales are the only available option. However, an additional difficulty arises with stranded cable tendons that does not arise with threadbar tendons. As the drill casing is pulled, it must normally be rotated, tending to cause a spiral in the tendons and binding of the telltale rods or wires within their sleeves, with significant reduction in accuracy of data. Provisions must be made for examining the extent of binding and, if possible, for gathering enough data so that frictional effects can be screened out. The slack-wire extensometer is a suitable choice because readings can be made at various wire tensions and adequate data obtained for screening. If unstressed rod telltales are used, they should be provided with a disconnect between the lower end of each rod and its anchor (Section 12.7.1), so that binding can be examined. If possible, both wire and rod telltales should be used on a stranded cable tendon, one as backup to the other. A telltale can be attached only to a single cable, and the assumption must therefore be made that the behavior of this single cable is representative of the behavior of all the cables that form the tendon: experience has shown that this appears to be a valid assumption.

Sheeting and Walers

Stresses in sheeting and walers can be determined by use of strain gages, but because stresses may vary widely from point to point, data may be misleading. Attempts to measure total stress acting on a sheet pile wall by attaching contact earth pressure cells to the sheeting have generally not been successful.

Slurry Walls

Stresses in slurry walls can be determined by use of strain gages, with a preference for the *sister bar* type of embedment gage with a vibrating wire transducer. Alternatively, strain measurements can be made and bending moments determined by measuring within sliding micrometer casings installed near the front and back of the wall. Earth pressure cells and piezometers can be installed at the interface between the slurry trench and the soil, using special transducer housings and hydraulic jack installation procedures, for subsequent monitoring of total pressure and pore water pressure on the completed wall (DiBiagio and Roti, 1972).

19.2.10. Is the Groundwater Table Being Lowered?

When dewatering has been implemented to predrain the ground around the excavation, its effectiveness may be monitored by using piezometers.

Prolonged lowering of the groundwater table may cause serious consolidation settlements over a large area when excavations are made in compressible soils. Lowering of the groundwater table can be diagnosed by installing piezometers in any permeable layers and, if economy permits, also with depth along several vertical sections behind the supporting wall. Whenever piezometers are installed in the vicinity of an excavation, reference piezometers should also be installed, remote from the excavation, to monitor any variation in groundwater pressure that may result from other causes.

19.2.11. Is Excessive Bottom Heave Occurring?

When excavating in soft clay, the possibility must be considered that the bottom may fail by heaving. Heave gages can be installed below the bottom of the excavation and time–heave plots used to indicate if and when critical soil movements start to occur. Measurements of horizontal deformation within inclinometer casings installed alongside and below the supporting wall can also be used as indicators of any instability caused by bottom heave.

Three types of heave gage are described in Chapter 12. First, there is a mechanical heave gage (Section 12.5.1) that is simple and inexpensive, but its accuracy is not great, its use requires a survey crew, and there is a risk of the borehole caving during excavation. Second, there is a magnet/reed switch gage (Section 12.5.7) that can be used to provide reliable and accurate data without interfering with excavation work. Third, there is a fixed borehole extensometer fitted with an electrical linear displacement transducer (Section 12.7.4) that is accurate but prone to damage and malfunction. For most applications the magnet/reed switch gage is the instrument of choice.

19.2.12. Is Long-Term Performance of the Bracing Satisfactory?

Internal bracing is either removed or incorporated into the final structure. The question is therefore limited to external bracing.

Tieback anchors are now a proven and accepted method for temporary support, and there is an increasing trend to install permanent tiebacks for long-term support. This trend leads to an increased role for instrumentation on critical projects where there is a need to demonstrate long-term satisfactory performance.

Long-term performance monitoring can consist of monitoring load in and deformation of selected anchors, or monitoring deformation on a larger scale, or both. Monitoring long-term deformation on a larger scale than at selected anchors normally requires surveying methods, with electronic distance measurements playing a primary role. Monitoring of selected anchors will generally follow the procedures outlined in Section 19.2.8 for examining whether an individual anchor is overloaded. However, two additional issues must be taken into account when monitoring anchors on a long-term basis: longevity of load cells and interruption to corrosion protection.

Longevity of Load Cells

Longevity can be maximized by following the guidelines for reliability, given in Chapter 15.

The arrangement shown in Figure 19.7 can be

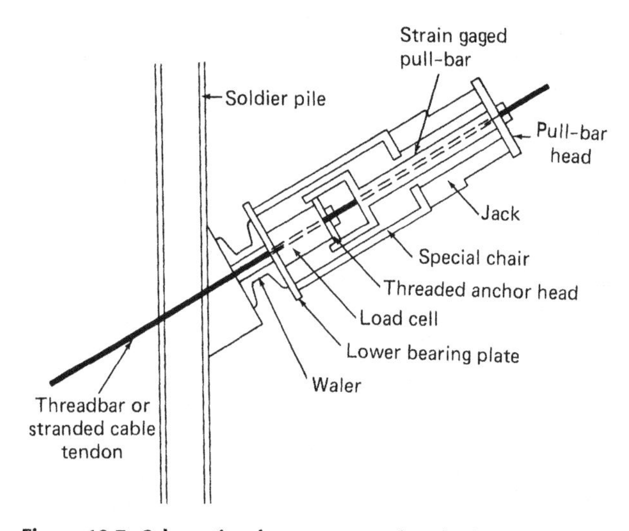

Figure 19.7. Schematic of arrangement for check-calibrating a load cell by *lift-off* testing.

adopted to check-calibrate a load cell by *lift-off* testing. The anchor head must have a threaded outside diameter or threaded central hole, for later connection to a *pull-bar*. The procedure involves placing a special chair over the load cell and anchor head, to bear on the lower bearing plate, threading a calibrated strain gaged pull-bar on to the anchor head, and inserting a center-hole hydraulic jack over the pull-bar to bear between the top of the chair and the pull-bar head. As the jack is actuated, load is transferred from the load cell to the pull-bar, a calibration of the load cell is made, and the zero reading of the load cell is determined after lift-off.

Nicholson Construction Company, Bridgeville, PA, has devised a method for removing a load cell without unloading the tieback, so that the cell can be checked and a new cell inserted if necessary. The arrangement is shown in Figure 19.8. A threaded anchor head is used, and the inside diameter of the load cell must be larger than the outside diameter of the anchor head/pull-bar arrangement. Prior to lock-off, split shims are inserted between the load cell and anchor head, with adequate total thickness for transferring load from anchor head to load cell. When preparing to remove the load cell for checking, a special chair is placed over the load cell and anchor head to bear on the lower bearing plate, with sufficient internal height to contain the load cell as described below. A pull-bar is threaded on to the anchor head, and a center-hole hydraulic jack is inserted over the pull-bar to bear between the top of the chair and the pull-bar head. The jack is actuated to lift the anchor head off the split shims, the split shims removed, the load cell lifted over the anchor head, and the split shims replaced together with additional split shims in the space previously

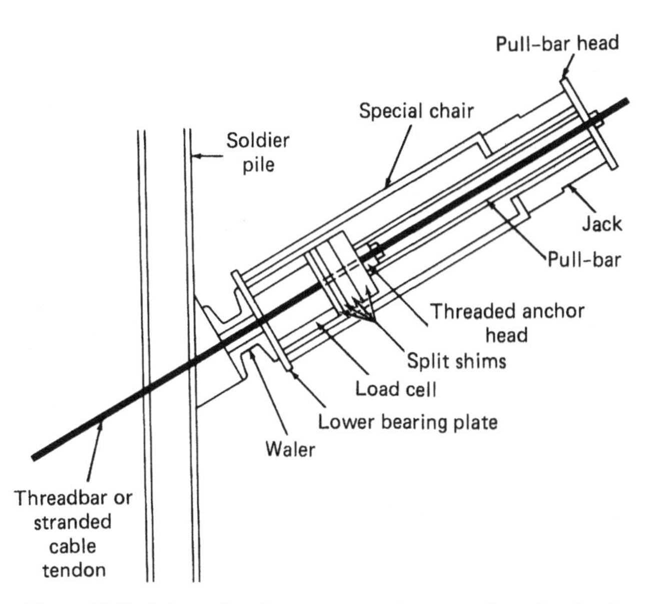

Figure 19.8. Schematic of arrangement for removing a load cell without unloading the tieback (after Nicholson Construction Company, Bridgeville, PA).

Table 19.2. Overview of Routine and Special Monitoring

Applications	Parameters
Routine monitoring: all types of ground support	Settlement of ground surface, structures, top of supporting wall, and utilities Horizontal deformation of ground surface, structures, and exposed part of supporting wall (when it is critical to minimize ground movements) Change in width of cracks in structures and utilities Convergence within excavation Groundwater pressure
Routine monitoring: external bracing	Load and deformation at anchor head during proof and performance testing of tieback anchors
Routine monitoring: internal bracing	Preload in bracing
Routine monitoring: slurry trench excavations	Level and properties of slurry Quantity of excavated material Verticality of trenches
Special applications: all types of ground support	Subsurface horizontal and vertical deformation around the excavation Bottom heave Stress in support system Deformation and support stress in full-scale test section of complete braced excavation and support system
Special applications: external bracing	Load and deformation at anchor head while loading test anchors during design phase or at the start of construction Long-term load and deformation in permanent tieback anchors
Special applications: slurry trench excavations	Deformation, groundwater pressure, level, and properties of slurry during full-scale test of slurry trench excavation Integrity of concrete

occupied by the load cell. The jack pressure is then released, and the pull-bar, jack, and special chair are removed. After checking, the load cell is replaced, using the reverse of the above procedure.

The second method of checking is more comprehensive than the straightforward lift-off test, because it allows for replacement of the load cell with a new cell in the event of serious malfunction. Both methods require use of special load cells designed for long-term use, and of course require that a sufficient length of tendon is left in place beyond the anchor head and that access is available for testing.

Interruption to Corrosion Protection

Instruments installed on a tieback anchor for monitoring long-term performance may interrupt the corrosion protection, and in general simple manual instruments create greater interruption than more complex remotely read instruments. Because simplicity leads to longevity, a dilemma is faced, and the designer must make a selection between two approaches.

First, simple instruments such as mechanically read telltale load cells can be used, or one of the above two methods for checking conventional load cells can be adopted, the interruption to corrosion protection accepted, and the monitored anchors regarded as *worst cases*. Second, a waterproof protective cover can be provided over the head assembly and filled with a rust-inhibiting protective grease, but this option requires use of inaccessible remotely read instruments, with reduced ability to check and replace them if necessary. Selection between the two approaches depends on likely consequences of interrupting the protection and on the availability of instruments that satisfy the criteria for long-term reliability, discussed in Chapter 15. For the particular case of monitoring load at the anchor head when threadbar tendons are installed, the best approach appears to be use of a pair of weldable vibrating wire strain gages on the bar (Figure 13.26), about 6 ft (2 m) from the nut. These gages can easily be incorporated into a double corrosion protection system, and are about the same size as a coupler in the tendon.

19.3. OVERVIEW OF ROUTINE AND SPECIAL APPLICATIONS

For those readers who seek general guidelines on what is typically *routine* monitoring and what is monitoring for *special applications*, Table 19.2 presents a broad generalization.

19.4. SELECTED CASE HISTORIES

Table 19.3 summarizes selected case histories of instrumented braced excavations.

Table 19.3. Summary of Selected Case Histories of Braced Excavations

Project	Principal Concerns	Type of Support and Ground	Measurement[a]	Instrumentation Discussed	Special Features	Reference
Harvard Square Subway Station, Cambridge, MA, U.S.A.	Stability of excavation Integrity of nearby structures Development of improved design procedures for slurry wall systems	Externally braced slurry walls in fill, sand, and glacial till	D	Surveying methods	Very comprehensive instrumentation	Rawnsley et al. (1985) Hansmire et al. (1984)
			D	Tiltmeters		
			D	Inclinometers		
			D	In-place inclinometers		
			D	Probe extensometers		
			D	Fixed borehole extensometers		
			G	Open standpipe piezometers		
			L	Load cells on external bracing		
			S	Surface-mounted and embedment strain gages in slurry wall		
			TS	Contact earth pressure cells at face of wall		
Good Samaritan Hospital, Cincinnati, OH, U.S.A.	Stability of excavation	Externally braced soldier piles in residual and unweathered shale and limestone	D	Surveying methods	Instrumentation planned for monitoring long-term performance	Anderson et al. (1984)
			D	Inclinometers		
			D	Fixed borehole extensometers		
			L	Load cells on external bracing		
Sewer culvert in Islais Creek Basin, San Francisco, CA, U.S.A.	Stability of excavation Bottom heave	Internally braced sheet pile excavation in very soft clay	D	Surveying methods	Large movements measured	Clough and Reed (1984)
			D	Inclinometers		
			G	Pneumatic piezometers		
			S	Vibrating wire strain gages on internal bracing		
Two excavations for buildings in Taipei, Taiwan	Stability of excavations	Internally braced slurry walls in silty clay and silty sand	D	Surveying methods		Moh and Song (1984)
			D	Inclinometers		
			D	Heave gages		
			G	Observation wells		
			G	Open standpipe piezometers		
			G	Pneumatic piezometers		
			S	Vibrating wire strain gages on internal bracing		

Table 19.3. Summary of Selected Case Histories of Braced Excavations (*Continued*)

Project	Principal Concerns	Type of Support and Ground	Measurement[a]	Instrumentation Discussed	Special Features	Reference
			S	Strain gages on reinforcement in slurry wall		
			TS	Contact earth pressure cells on slurry wall		
Six-story building, Westminster, London, England	Stability of excavation	Internally braced slurry walls in London clay	D	Inclinometers		Wood and Perrin (1984)
			D	Probe extensometers		
			G	Open standpipe piezometers		
			S	Vibrating wire strain gages on internal bracing and on reinforcement in slurry wall		
Charles Center Station, Baltimore Metro, Baltimore, MD, U.S.A.	Integrity of nearby structures	Internally braced slurry walls in silty sand, gravel, and residual soil	D	Surveying methods		Zeigler et al. (1984)
			D	Subsurface settlement points		
			D	Inclinometers		
			G	Observation wells		
			S	Vibrating wire strain gages on internal bracing		
Rio de Janeiro Underground, Brazil	Development of improved design procedures for slurry wall systems	Internally braced slurry walls in sandy and silty clay and sand	D	Inclinometers in soil and slurry wall	Inclinometer data used to estimate bending moments in slurry wall	Soares (1983)
			D	Probe extensometers for monitoring bottom heave		
			D	Subsurface settlement points		
			G	Pneumatic piezometers		
			S	Vibrating wire strain gages in concrete and on reinforcing steel in walls		
			S	Vibrating wire strain gages on bracing		
			TS	Contact earth pressure cells on slurry walls		

Table 19.3. Summary of Selected Case Histories of Braced Excavations (*Continued*)

Project	Principal Concerns	Type of Support and Ground	Measurement[a]	Instrumentation Discussed	Special Features	Reference
Underpass in Edmonton, Alberta, Canada	Loading on external bracing in winter conditions	Externally braced sheet piles in stiff fissured clay and till	L	Electrical resistance strain gage load cells on external bracing	Effect of temperature change on loading in external bracing	Morgenstern and Sego (1981)
University of Quebec, Montreal, Canada	Integrity of nearby structures	Externally braced slurry walls in glacial till	D D D L	Surveying methods Crack gages Inclinometers Hydraulic load cells on external bracing		Rosenberg et al. (1977)
Thorold Tunnel, St. Catharines, Ontario, Canada	Define cause of damage	Reinforced concrete highway tunnel in dolomite and limestone	D D D G S T	Surveying methods Convergence gages Inclinometers Piezometers Mechanical strain gages Thermocouples	Cracking of 6 ft (1.8 m) thick reinforced concrete side walls	Bowen and Hewson (1976)
Railway Tunnel, Oslo, Norway	Stability of excavation Development of improved design procedures	Internally braced sheet pile excavation in quick clay	D D G S TS	Surveying methods Inclinometers Open standpipe and vibrating wire piezometers Vibrating wire strain gages on bracing Contact earth pressure cells on sheeting	Excessive movement arrested by casting concrete beams across excavation below bottom	Karlsrud and Myrvoll (1976)
Washington Metro, Washington, DC, U.S.A.	Stability of excavation Development of improved design procedures Performance of strain gages	Two internally braced excavations	S	Vibrating wire strain gages on internal bracing		O'Rourke and Cording (1975)
Three excavations in Toronto, Canada	Improvement of design procedures	Externally braced excavations in sand, silt, and till	L	Load cells on external bracing		Trow (1974)
Embarcadero Subway Station, BARTD, San Francisco, CA, U.S.A.	Stability of excavation	Internally braced slurry walls in clay and sand	D D G S	Surveying methods Inclinometers Observation wells Vibrating wire strain gages on internal bracing		Armento (1973)

Table 19.3. Summary of Selected Case Histories of Braced Excavations (*Continued*)

Project	Principal Concerns	Type of Support and Ground	Measurement[a]	Instrumentation Discussed	Special Features	Reference
Bank of California Center, Seattle, WA, U.S.A.	Stability of excavation Verification of design methods	Soldier piles and external bracing in overconsolidated clay	D D D L	Surveying methods Inclinometers Heave gages (fixed borehole extensometers) Load cells on external bracing		Clough et al. (1972)
CNA Center and Standard Oil Building, Chicago, IL, U.S.A.	Stability of excavation	Externally braced slurry walls in silty clay Internally braced slurry walls in medium clay	D D	Surveying methods Inclinometers		Cunningham and Fernandez (1972)
Test Site at Studenterlunden, Oslo, Norway	Behavior of slurry trench excavation: research test	Slurry trench excavation in soft clay: full-scale test	D D D G	Surveying methods Inclinometers Slurry trench convergence gages Vibrating wire piezometers		DiBiagio and Myrvoll (1972)
Subway extension, South Cove, Boston, MA, U.S.A.	Integrity of adjacent building	Internally braced slurry walls in clay	D D G S	Surveying methods Inclinometers Open standpipe and vibrating wire piezometers Vibrating wire strain gages on internal bracing		Lambe et al. (1972)
Office Building, Ottawa, Canada	Stability of excavation Comparison between measured and predicted movements	Externally braced sheeting in fill, stiff clay, and glacial till	D D G L	Surveying methods Fixed borehole extensometers Open standpipe and vibrating wire piezometers Load cells on external bracing		McRostie et al. (1972)
Greenway Pollution Control Centre, London, Ontario, Canada	Stability of excavation	Internally braced sheeting in fine sand	D G S	Surveying methods Open standpipe piezometers Mechanical strain gages on internal bracing		Scott et al. (1972)

Table 19.3. Summary of Selected Case Histories of Braced Excavations (*Continued*)

Project	Principal Concerns	Type of Support and Ground	Measurement[a]	Instrumentation Discussed	Special Features	Reference
			TS	Contact earth pressure cells on sheeting		
Various	Various	Internally and externally braced excavations	D G L S TS	Deformation gages Piezometers Load cells Strain gages Contact earth pressure calls	State-of-the-art summary of measurements	Gould (1970)
Pierre Laclede Building, Clayton, MO, U.S.A.	Stability of excavation	Soldier piles and external bracing in silty clay and clayey silt	S S	Mechanical strain gages on soldier piles Mechanical and electrical resistance strain gages on external bracing		Mansur and Alizadeh (1970)
Various	Various	Internally and externally braced excavations	D S	Deformation gages Strain gages on internal bracing	State-of-the-art summary of measurements	Peck (1969b)
Harris Bank and Trust Company, Chicago, IL, U.S.A.	Load in internal bracing	Internally braced excavation	S	Mechanical strain gages on internal bracing	Measurements initiated by construction contractor to justify reduced number of struts	Peck (1969a)

[a]D: deformation L: load in bracing T: temperature
G: groundwater pressure S: strain in support TS: total stress between supporting wall and soil

CHAPTER 20

EMBANKMENTS ON SOFT GROUND*

This chapter deals with instrumentation of embankments that do not retain water. Guidelines for water-retaining embankments are given in Chapter 21.

In many cases, selection of soil parameters for the foundation soil is reliably conservative. The embankment is therefore designed with confidence that performance will be satisfactory, and "comfortable" factors of safety are used. In such cases, many projects will proceed without the use of instrumentation. However, some uncertainties always exist.

Where design uncertainties are great, factors of safety small, or the consequences of poor performance severe, a prudent designer will include a performance monitoring program in the design.

20.1. GENERAL ROLE OF INSTRUMENTATION

In spite of a long record of embankment construction throughout the history of civil engineering, embankments that are designed with a factor of safety greater than unity fail embarrassingly often. On the other hand, some test embankments that are designed to fail intentionally, never do. Thus, it is not surprising that instrumentation plays a significant role in design and construction of embankments on soft ground.

Figure 20.1 shows a typical embankment on soft ground and conveys the contrast between how the project will usually appear on engineering drawings and what may actually happen in the field. Primary mechanisms that control the behavior of embankments on soft ground are described in Chapter 2.

The most frequent uses of instrumentation for embankments on soft ground are to monitor the progress of consolidation and to determine whether the embankment is stable. For example, instrumentation will normally be used to indicate the progress of consolidation beneath a single-stage embankment, a surcharged embankment, or staged construction, so that construction schedules can be determined. When vertical drains are installed, instrumentation will normally be used to evaluate whether they are being effective in accelerating consolidation. If the calculated factor of safety is likely to approach unity, instrumentation will generally be installed to provide a warning of any instability, thereby allowing remedial measures to be implemented before critical situations arise.

If uncertainties in the selection of soil parameters are unacceptably great, or if construction feasibility is in doubt, it may be appropriate to construct a test embankment. Instrumentation data provide an essential role in evaluating the performance of such a test.

20.2. PRINCIPAL GEOTECHNICAL QUESTIONS

The following principal questions are presented in the normal order of occurrence for an embankment on soft ground. The order does not reflect a rating of importance.

*Written with the assistance of James R. Lambrechts, Senior Engineer, Haley & Aldrich, Inc., Cambridge, MA.

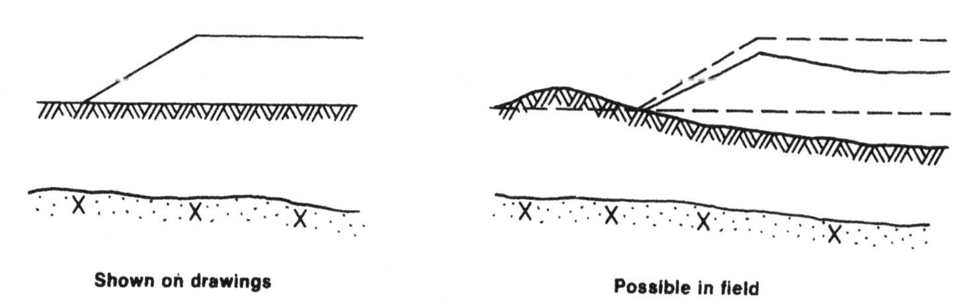

Figure 20.1. Embankment on soft ground (after DiBiagio and Myrvoll, 1981).

20.2.1. What Are the Initial Site Conditions?

Initial site conditions are determined by use of conventional site investigation procedures, often supplemented by in situ testing. However, performance monitoring instrumentation sometimes plays a role. For example, initial groundwater pressures and fluctuations must often be determined for design purposes. Piezometers can be installed well before the start of filling, to define the preconstruction groundwater pressure regime, including any perched or artesian water. If the piezometers are installed sufficiently early, seasonal variations can be defined. Settlement measurements may also be required to establish preconstruction settlement behavior.

20.2.2. What Information Can Be Provided by Instrumenting a Test Embankment?

Test embankments are sometimes constructed to resolve uncertainties in the selection of soil parameters, to examine alternative construction methods, or to demonstrate construction feasibility. The goal of instrumentation in these tests will generally be to provide an indication of incipient failure and/or to evaluate the progress of consolidation. In addition, instrumentation may be installed to permit a back-analysis for determining the engineering properties of the underlying soil. In all cases, the primary parameters of interest are vertical and horizontal deformations and pore water pressure.

Depending on the specifics of the case, instrumentation for a test embankment will include any or all of the monitoring methods discussed elsewhere in this chapter. The intensity of instrumentation will usually be greater than for a prototype embankment (one that is part of the constructed project), so that maximum information is gained from the test.

Figure 20.2 shows a possible layout of instrumentation for monitoring the progress of consolidation beneath a test embankment when vertical drains have been installed. If vertical drains have not been installed, a similar layout may be appropriate. This is only one example, and it must be stressed that many other configurations are possible, as indicated by the case histories described later in this chapter.

When the test embankment is designed to fail, a knowledge of the location of the failure plane is required. For this purpose, shear plane indicators usually provide adequate data, and it is not always necessary to install inclinometer casings.

20.2.3. What Is the Progress of Consolidation?

When an embankment is constructed on a soft compressible foundation, the added weight of the em-

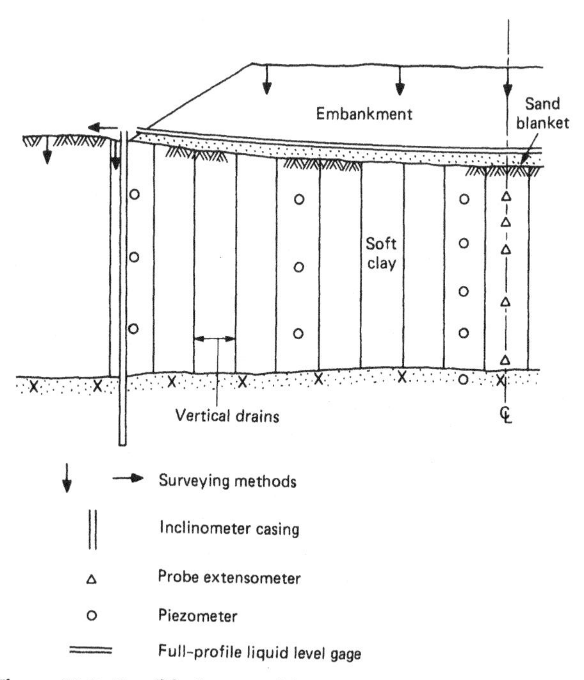

Figure 20.2. Possible layout of instrumentation for monitoring progress of consolidation beneath a test embankment when vertical drains have been installed. Note: Use of surveying methods requires benchmark and horizontal control station, remote from embankment. Piezometers also required, remote from embankment.

bankment causes consolidation in the foundation. The process of consolidation is described in Chapter 2. Construction of a facility on the top of the embankment, such as a building or a highway pavement, will normally be delayed until there is assurance that the facility will not be damaged by settlement that occurs during subsequent consolidation. If the predicted delay is unacceptable, the time required for consolidation can be reduced by *surcharging* or by installation of *vertical drains*.

Surcharging involves placement of fill material to a level higher than required for the embankment, so that consolidation proceeds more quickly. The excess material is removed when adequate consolidation has occurred.

Vertical drains are highly permeable columns of material installed at close intervals within the foundation soil. A continuous sand blanket is placed over the tops of the drains, so that the drains communicate with the blanket and allow pore water pressures to dissipate more rapidly, thus minimizing the time required for consolidation.

If the foundation might become unstable by placing the entire embankment rapidly, *staged construction* can be used. Staged construction involves placement of a first stage of fill material, waiting for consolidation and strength gain, and repeating the process until the embankment is complete. Staged construction can be used in conjunction with surcharging or vertical drains or can be used alone when sufficient time is available for consolidation.

When surcharging, vertical drains, or staged construction is employed, instrumentation will normally be used as a tool to indicate the progress of consolidation. Instrumentation data allow schedules to be determined for surcharge removal and for placement of stages during staged construction and also allow an evaluation of the effectiveness of vertical drains.

When evaluating the progress of consolidation beneath a single-stage embankment, a surcharged embankment, or staged construction, both settlement and pore water pressure measurements are normally made. Whenever piezometers are installed in the vicinity of the embankment, reference piezometers should also be installed, remote from the embankment, to monitor any variation in groundwater pressure that may result from other causes.

A predictive extrapolation can be made from a plot of pore water pressure versus time because the equilibrium piezometric level corresponding to full consolidation is known to be equal to the reference value remote from the embankment. Prediction from a plot of settlement versus time is less reliable, because the magnitude of ultimate settlement cannot be known with certainty. Ladd (1984) provides guidelines on the interpretation of field data when vertical drains are used.

Instruments for monitoring the progress of consolidation can be selected from the list given in Table 20.1. The list is not intended to be exclusive: it merely indicates instruments that have provided satisfactory data on projects with which the author has been involved. Selection among the various options will depend on the factors discussed in Chapters 9 and 12, including the feasibility of extending instrument pipes and leads up through the embankment without risk of damage and the need for redundancy. In all cases predictions should be made of maximum vertical foundation compression and horizontal spreading, and a special effort must be made to select instruments that are both capable of surviving the deformations and capable of providing reliable data as deformation occurs. Installation procedures must also be planned in recognition of the future deformations, following the guidelines given in Chapter 17.

The arrangement shown in Figure 20.2 for a test embankment is also a possible layout for monitoring the progress of consolidation beneath a prototype embankment.

When planning a program to monitor the performance of vertical drains, a decision must be made on whether to install instruments before or after installation of drains. Installing instruments well before drains ensures that baseline data are obtained but runs the risk of damage to instruments as drains are installed. The author suggests installing up to half of the instruments well before installing drains, and the remainder afterward.

20.2.4. Is the Embankment Stable?

If the possibility of lateral instability is minor, instrumentation will not be required during construction of an embankment on soft ground. However, in other cases a monitoring program is generally required to provide a forewarning of instability, thereby allowing remedial measures to be implemented before critical situations arise. Remedial measures may include a waiting period to allow the foundation material to increase in strength as pore water pressures dissipate and/or the construction of stabilizing berms or the removal of some fill.

Table 20.1. Instruments Suitable for Monitoring Progress of Consolidation

Measurement	Suitable Instruments
Vertical deformation of embankment surface and ground surface at and beyond toe of embankment	Surveying methods
Vertical deformation of original ground surface below embankment	Settlement platforms Buried plates Single-point liquid level gages Full-profile liquid level gages with pressure transducers and attached liquid-filled tubes[a] Horizontal inclinometer casings[a]
Vertical deformation and compression of subsurface	Spiral-foot settlement gages Probe extensometers with induction coil or magnet/reed switch transducers[b] Single-point liquid level gages
Groundwater pressure	Open standpipe "push-in" piezometers, as in Figure 9.23 Pneumatic and vibrating wire piezometers pushed in place below bottom of boreholes, as in Figures 9.31 and 9.32

[a]Casing or pipe should either have telescoping couplings or be surrounded with a corrugated plastic pipe.
[b]Requires careful selection of method for anchoring the measuring points to the soil, depending on the predicted vertical compression (see Chapter 12).

When a monitoring program is required to provide an indication of incipient failure, horizontal deformation measurements will normally provide the most direct data. An inclinometer is the primary tool, supplemented with surface deformation monitoring by surveying methods. However, measurements of pore water pressures may give a better first indication of failure conditions developing than can be derived from deformation data (Clausen et al., 1984). Rapidly responding diaphragm piezometers are required and, to be certain of having measurements in the zones of initial yielding, a relatively large number of piezometers may be required unless experience and proven methods of analysis can be used to locate the critical zones.

20.2.5. What Are the Fill Quantities?

Soft clay foundations may experience substantial settlement during embankment construction. When embankment fill is paid for on a volume basis and measured to the actual base of the fill, a determination must be made of the elevation of the base of the fill. Settlement platforms are often used, but full-profile liquid level gages are suitable alternatives.

20.3. OVERVIEW OF ROUTINE AND SPECIAL APPLICATIONS

For those readers who seek general guidelines on what is typically *routine* monitoring and what is monitoring for *special applications,* Table 20.2 presents a broad generalization.

20.4. SELECTED CASE HISTORIES

Table 20.3 summarizes selected case histories of instrumented embankments on soft ground. A study of these case histories reveals major similarities among many of the instrumentation programs, indicating that the general aspects of the state of the practice are well established. The general approach to instrumenting test and prototype embankments is similar, although the intensity of instrumentation is usually greater for test embankments.

Table 20.2. Overview of Routine and Special Monitoring

Application	Measurement
Routine monitoring	Vertical deformation of the embankment surface Vertical and horizontal deformations of the ground surface at and beyond the toe of the embankment Vertical deformation of the original ground surface below the embankment Groundwater pressure
Special applications	Subsurface vertical and horizontal deformations of the prototype Deformation and groundwater pressure during performance monitoring of a test embankment

Table 20.3. Summary of Selected Case Histories of Embankments on Soft Ground

Project	Principal Concerns	Type of Embankment	Measurement[a]	Instrumentation Discussed	Special Features	Reference
Airport at Chek Lap Kok, Hong Kong	Feasibility of construction Need to establish construction procedures Performance of wick and sand drains	Test, for airport	G D D D D D	Open standpipe and pneumatic piezometers (push-in type) Surveying methods Settlement platforms Probe extensometers Single-point and full-profile liquid level gages Inclinometers	Offshore reclamation Staged construction used Instrumentation designed to accommodate up to 30% vertical strain in foundation, large lateral deformation, over-water installation	Handfelt et al. (1987)
Geotextile-reinforced embankment, Bloomington Road, Aurora, Ontario, Canada	Performance of embankment during construction	Prototype, for highway	G D D S	Open standpipe piezometers (push-in type) Full-profile liquid level gages Inclinometers Strain gages on fabric (hydraulic and induction coil type)		Rowe et al. (1984a,b)
Highway B5, Brunsbüttel-Itzehoe, West Germany	Performance of embankment during construction: staged construction and surcharge used	Prototype, for highway	D D	Settlement platforms Combined probe extensometers and inclinometers		Thamm (1984)
Jourdan Road Terminal, New Orleans, LA, U.S.A.	Performance of stone columns and wick drains	Prototype, for wharf	G D D D	Pneumatic piezometers Settlement platforms Probe extensometers Inclinometers		Castelli et al. (1983)
Jourdan Road Terminal, New Orleans, LA, U.S.A.	Performance of stone columns in stabilizing clay deposit Magnitude of load carried by stone columns	Test, for wharf	G D D D TS	Pneumatic piezometers Probe extensometers Inclinometers Shear plane indicators Embedment earth pressure cells	Test embankment purposely failed by excavating adjacent ground Pneumatic piezometers fitted with two hydraulic tubes for flushing gas from cavity between diaphragm and filter tip	Munfakh et al. (1983)

Table 20.3. Summary of Selected Case Histories of Embankments on Soft Ground (*Continued*)

Project	Principal Concerns	Type of Embankment	Measurement[a]	Instrumentation Discussed	Special Features	Reference
Site near Rio de Janeiro, Brazil	Progress of consolidation Stability Improvement of design procedures	Test	G D D D D D	Twin-tube hydraulic piezometers Surveying methods Settlement platforms Inclinometers Probe extensometers (installed horizontally) Fixed embankment extensometers (installed horizontally)	Failure of test embankment	Ramalho-Ortigão et al. (1983)
Site near Szolnok, Hungary	Progress of consolidation	Prototype, for highway	D	Buried plates		Rózsa and Vidacs (1983)
Queenborough Bypass, Isle of Sheppey, Kent, England	Progress of consolidation Stability: staged construction controlled by measurements Performance of wick and sandwick drains	Prototype, for highway	G D	Twin-tube hydraulic piezometers (push-in type) Probe extensometers	Vertical drains installed after first stage measurements indicated need to accelerate consolidation	Nicholson and Jardine (1981)
River Street Viaduct, Tacoma, WA, U.S.A.	Progress of consolidation Stability: staged construction controlled by measurements	Prototype, for highway	G D D	Pneumatic piezometers (push-in type) Settlement platforms Inclinometers	Embankment constructed alongside railroad tracks and water main	Mikkelsen and Bestwick (1976)
I95, Saugus, MA, U.S.A.	Pore water pressures, deformation, and stability, as additional fill was placed	Test, for highway	G D D	Piezometers Settlement platforms Inclinometers	Ten teams of geotechnical engineers predicted behavior (before fill placement) Predictions and actual behavior are reported Failure of test embankment	MIT (1975)
Various	Various (numerous case histories)	Test, for highways, levees, research, dams	G D TS	Numerous types of instruments for measurement of groundwater pressure, deformation, and total stress	Discussion of role of test embankments	Bishop and Green (1974)

Table 20.3. Summary of Selected Case Histories of Embankments on Soft Ground (*Continued*)

Project	Principal Concerns	Type of Embankment	Measurement[a]	Instrumentation Discussed	Special Features	Reference
Thames Oil Refinery, Canvey Island, Essex, England	Need to establish design criteria	Test, for oil storage tanks	G D D D D	Open standpipe piezometers Surveying methods Probe extensometers Single-point and full-profile liquid level gages Inclinometers		George and Parry (1974)
Flood Defense Banks, River Thames, England	Progress of consolidation Stability Comparison with predictions	Prototype, for flood defense	G D D D	Pneumatic and electrical resistance piezometers Surveying methods Probe extensometers Inclinometers	Pneumatic piezometers in organic deposits fitted with two hydraulic tubes for flushing gas from cavity between diaphragm and filter tip	Marsland (1974b)
M5 Motorway, Edithmead and Huntworth, England	Progress of consolidation: surcharge used Stability	Prototype, for highway	G D D D D D	Twin-tube hydraulic piezometers Surveying methods Settlement platforms Probe extensometers (installed horizontally and vertically) Full-profile liquid level gages Inclinometers		McKenna and Roy (1974)
Embankments at Saint-Alban, Quebec, Canada	Need to develop methods for predicting magnitude and rate of settlement for high embankments	Test, for research	G D D D D	Pneumatic and vibrating wire piezometers (push-in type) Surveying methods Settlement platforms Probe extensometers Inclinometers	Three test embankments, with same height but different side slopes	Tavenas et al. (1974)
Hampton Roads Crossing, VA, U.S.A.	Progress of consolidation: jetted sand drains installed Stability Measurement of fill quantity	Prototype, for highway	G D D	Open standpipe and pneumatic piezometers Surveying methods Settlement platforms	Up to 12 ft (3.7 m) of settlement was measured	Kuesel et al. (1973)

Table 20.3. Summary of Selected Case Histories of Embankments on Soft Ground (*Continued*)

Project	Principal Concerns	Type of Embankment	Measurement[a]	Instrumentation Discussed	Special Features	Reference
			D	Subsurface settlement points		
			D	Inclinometers		
I295, Portland, ME, U.S.A.	Performance of driven, augered, and jetted drains	Test, for highway	G	Vibrating wire piezometers		Aldrich and Johnson (1972)
			D	Settlement platforms		
			D	Subsurface settlement points		
Gloucester Test Fill, Ottawa, Canada	Comparison with predictions	Test, for research	G	Open standpipe piezometers		Bozozuk and Leonards (1972)
			D	Probe extensometers		
			D	Full-profile liquid level gages		
			TS	Embedment earth pressure cells		
Bangkok-Siracha Highway, Thailand	Bearing capacity of soft clay Need for stabilizing berm Performance of sand drains	Test, for highway	D	Settlement platforms		Eide and Holmberg (1972)
			D	Shear plane indicators		
Embankments at Skå-Edeby, Stockholm, Sweden	Performance of sand drains in varved clay Comparison with predictions	Test, for research	G	Hydraulic piezometers	Summary of 14 years of observations	Holtz and Broms (1972)
			D	Subsurface settlement points		
I95, Portsmouth, NH, U.S.A.	Need to establish design criteria	Test, for highway	G	Open standpipe and vibrating wire piezometers		Ladd (1972)
			D	Surveying methods		
			D	Settlement platforms		
			D	Subsurface settlement points		
			D	Inclinometers		
I95, Portsmouth, NH, U.S.A.	Progress of consolidation: sand drains and surcharge installed Stability: staged construction and stabilizing berms used	Prototype, for highway	G	Observation wells		Ladd et al. (1972)
			G	Open standpipe piezometers (push-in type)		
			G	Pneumatic and twin-tube hydraulic piezometers		
			D	Settlement platforms		

Table 20.3. Summary of Selected Case Histories of Embankments on Soft Ground (*Continued*)

Project	Principal Concerns	Type of Embankment	Measurement[a]	Instrumentation Discussed	Special Features	Reference
			D	Subsurface settlement points		
			D	Inclinometers		
Watergate Complex, Emeryville, CA, U.S.A.	Progress of consolidation: jetted sand drains installed	Prototype, for apartment and office complex	G	Pneumatic piezometers		Margason and Arango (1972)
King's Lynn Bypass, England	Need to establish design criteria: augered sand drains installed	Test, for highway	G	Twin-tube hydraulic piezometers	Failure of test embankment	Wilkes (1972)
			D	Surveying methods		
			D	Settlement platforms		
			D	Probe extensometers		
			D	Liquid level gages		
			D	Inclinometers		
			D	Shear plane indicators		
I95, Saugus, MA, U.S.A.	Need to develop methods for predicting deformation and stability	Test, for highway	G	Observation wells		Wolfskill and Soydemir (1971)
			G	Twin-tube hydraulic and vibrating wire piezometers		
			D	Settlement platforms		
			D	Subsurface settlement points		
			D	Inclinometers		
			TS	Embedment earth pressure cells		
Atchafalaya Levees, Mississippi River, LA, U.S.A.	Feasibility of using stabilizing berms	Test, for levee	G	Open standpipe, twin-tube hydraulic and pneumatic piezometers		Kaufman and Weaver (1967)
			D	Surveying methods		
			D	Settlement platforms		
			D	Subsurface settlement points		
			D	Inclinometers		
			D	Shear plane indicators		

[a]D: deformation S: strain in structural member
G: groundwater pressure TS: total stress in soil

CHAPTER 21

EMBANKMENT DAMS*

Embankment dams generally include earthfill dams of various types, rockfill dams with impervious earth cores, and rockfill dams with upstream concrete or asphaltic concrete facings. Primary mechanisms that control the behavior of embankment dams are described in Chapter 2.

21.1. GENERAL ROLE OF INSTRUMENTATION

The main purpose of instrumentation installed within an embankment dam is to study whether or not the dam is behaving according to design predictions. This general statement can be subdivided into two categories: first, the study of special problems at individual sites that are related to special foundation conditions or uncommon design features and second, the study of behavior when there are no such special problems.

21.1.1. Dams with Special Foundation Conditions or Uncommon Design Features

When instrumentation is used for this purpose, it helps to determine whether design assumptions for the special conditions and features are being realized during construction and operation. The design of the monitoring program is tuned directly for the special conditions and features and, because the designers of the dam know the weaknesses of the particular site and the sensitive features of the design, they play a leading role in the choice of type and location of instruments.

*Written with the assistance of Arthur D. M. Penman, Consulting Geotechnical Engineer, Harpenden, England, and James L. Sherard, Consulting Engineer, San Diego, CA.

21.1.2. Dams Without Special Foundation Conditions or Uncommon Design Features

The practice of placing instruments routinely in embankment dams where there is no special problem to be studied, for example, a conventional dam of moderate height on a good foundation, has seen several ups and downs in the last 30 years. In the 1960s, many important dams were constructed with essentially no internal instruments. Each failure or major problem tends to be widely publicized, and dam engineers feel pressure to install instruments routinely for their own protection, even if they do not believe it absolutely necessary. The situation is similar to the dilemma of medical doctors who, fearing lawsuits if they do otherwise, prescribe suites of laboratory tests that may not be justified on medical grounds alone. At the present time, the failure of the Teton Dam is so recent that most engineers would hesitate to eliminate instruments in an important dam, for fear of criticism.

Most older dams of less than 50 ft (15 m) height have no instruments other than V-notch weirs or other means of measuring leakage. There are very many up to 300 ft (90 m) built during 1960–1975 without additional instruments. Although this is often not considered by inspecting engineers to be a shortcoming, the importance of many higher dams built during the last two decades has resulted in significant use of instrumentation.

Peck (1985) comments:

> Instrumentation, vital for obtaining quantitative answers to significant questions, is too often misused, especially in earth and rockfill dams. In some countries regulations concerning the safety of dams demand the incorporation of inclinometers, settlement indicators, and piezometers in the cores of virtually all new dams, but for what purpose? Not for research, because the patterns of deformation and pore pressure development for ordinary geometrics and materials are now well known and can be predicted by calculation. Only under unusual circumstances can it be said that design assumptions in these regards require verification. Yet, installation of instruments, even under the best of circumstances, introduces inhomogeneities into the cores, and occasionally is the direct cause of such local defects as sinkholes. The potential weakness introduced by an installation should be balanced against the potential benefit from the observations. In contrast to those located in cores, piezometers in foundation materials near the downstream toes detect upward seepage pressures that cannot be predicted reliably, and can thus give timely warning if measures are needed to ensure safety. There is a danger that instrumentation may be discredited because of indiscriminate use.

Of all the internal instruments in embankment dams, piezometers are by far the most common, and an installation consisting only of piezometers can serve two purposes. First piezometers in a clay core indicate rates of dissipation of pore water pressure and the approach of equilibrium conditions after impounding. Second, piezometers may be appropriate where it is desired to avoid criticism when designing a dam of conventional design and moderate height on a good foundation

21.1.3. Instruments that Cause Problems

Instruments can be a direct cause of problems. Peck refers above to instruments causing local defects such as sinkholes, and these cases are generally where inferior compaction has resulted from carrying tubes, cables, or pipes upward as filling proceeds. The solution to this problem is to avoid vertical risers (except within filters as described in Section 21.4), following the guidelines given in Section 21.4 and Chapter 17.

There is another category of problems, which can be classified as *rogue instruments*. In this category the dam is performing perfectly well but concern arises because a malfunctioning instrument indicates unexpected or alarming measurements. Often it is impossible to examine the health of the instrument directly and, after intensive study and worry, a final conclusion is reached that the instrument is not functioning correctly.

There have been a sufficient number of cases in which instruments have been the cause of problems, so that the designer of an embankment dam is likely to be very wary of geotechnical instrumentation. The solution to this dilemma has two parts. First, instruments should be used only when there is a valid reason that can be defended. Second, the overriding feature when selecting instruments should be *reliability*. A recipe for reliability is given in Chapter 15.

21.1.4. Summary to the General Role of Instrumentation

Despite the reservations given above, instrumentation **can** indeed be used to assist in obtaining quantitative answers to significant questions during the design, construction, and performance phases, and guidelines are given in the remainder of this chapter.

The general role of instrumentation for embankment dams is described by ICOLD (1969). The International Commission on Large Dams, Committee on Monitoring Dams and their Foundations, is currently assembling a series of national state-of-the-art reports presenting the instrumentation practices in various countries, which are intended to supersede ICOLD (1969). USCOLD (1986) is the U.S. contribution.

21.2. PRINCIPAL GEOTECHNICAL QUESTIONS

The following principal questions are presented in the normal order of occurrence for an embankment dam. The order does not reflect a rating of importance.

21.2.1. What Are the Initial Site Conditions?

Initial site conditions are determined by use of conventional site investigation procedures, sometimes supplemented by in situ testing. However, piezometers also play a role.

It is usually of value to install piezometers, preferably of the open standpipe type, in the exploratory boreholes made during the investigation of a dam site, rather than simply backfilling the boreholes. Regular readings can reveal much about groundwater pressure conditions on the site before dam construction. They are particularly useful in the abutments, where knowledge of groundwater pressures and their seasonal variations is vital to an understanding of abutment permeability and the need for grouting in rock abutments. Selected piezometers can later be converted, as described in Chapter 9, to be read remotely and form part of the long-term monitoring system.

21.2.2. Is Performance Satisfactory During Construction?

Dams with Special Foundation Conditions or Uncommon Design Features

If the site includes special foundation conditions or the design of the dam has uncommon features, instrumentation is a valuable tool to assist with construction control.

For example, soft foundation conditions may lead to the use of instrumentation for monitoring foundation stability and the progress of consolidation, as described for embankments on soft ground in Chapter 20. A similar need may arise for monitoring within the embankment itself, when there is concern that pore water pressures may become excessive during construction.

As another example, if there is concern that tensile strains may cause adverse behavior, such as in high dams with irregular foundations or steep abutments, instrumentation can be used to provide performance data and indicate the need for remedial action. Figure 21.1 illustrates such an application.

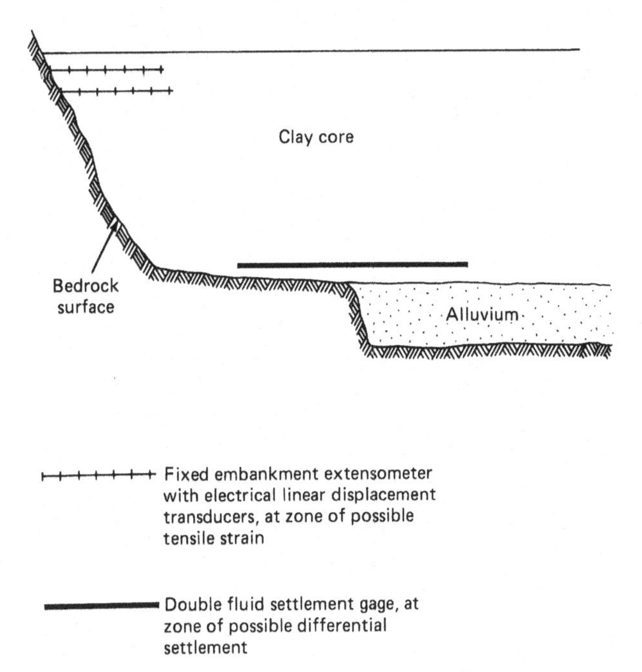

Figure 21.1. Possible layout of instrumentation for monitoring zones of potential tensile strain and differential settlement.

It is not reasonable to make a complete list of recommended instruments for monitoring such special conditions or uncommon features, because the selection depends on the details of each case. The primary monitored parameter will generally be either deformation or pore water pressure, but on occasion measurements of total stress may be included.

Dams Without Special Foundation Conditions or Uncommon Design Features

When the dam does not have special foundation conditions or uncommon design features, instrumentation can be used to determine whether or not the dam as a whole is behaving according to design predictions.

As an example, Figure 21.2 shows a possible arrangement of instrumentation for monitoring deformation of a dam that has been designed in accordance with acceptable deformation criteria. If measured deformations exceed acceptable values, construction procedures can be amended, for example, by building berms, flattening slopes, or slowing down the rate of construction.

One proven combination of instruments shown in Figure 21.2 is a probe extensometer with a current-displacement induction coil transducer (sometimes called a *horizontal plate gage*) and an overflow-type full-profile liquid level gage, both of which are described in Chapter 12. Figure 21.2 shows the system installed within a rockfill shell, but it can also be installed within downstream shells of other materials, provided that drainage is provided downstream of the core. The combined version for use in embankment dams is described by Penman and Charles (1982), and a precision of ±0.02 in. (±0.5 mm) is possible for both horizontal and vertical deformation measurements. The pipes are placed at the same inclination as lifts of fill, thereby minimizing interference to construction, and no delicate components are embedded. When used, such an ar-

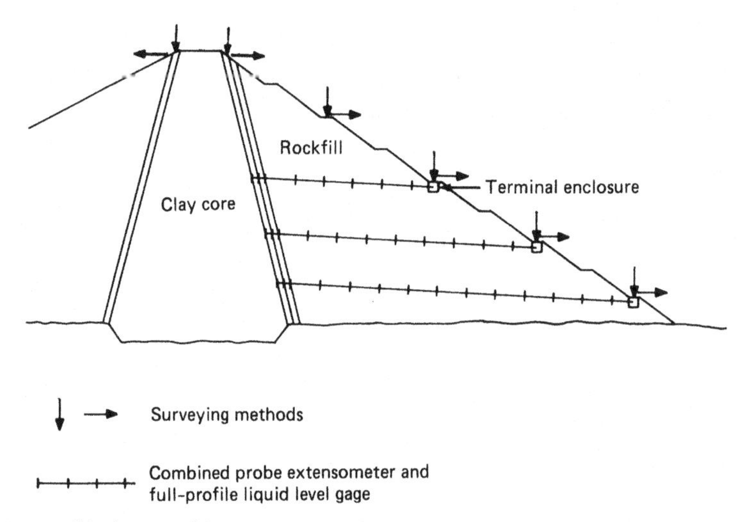

Figure 21.2. Possible layout of instrumentation for monitoring vertical and horizontal deformation during construction. Note: Use of surveying methods requires benchmarks and horizontal control stations, remote from dam.

rangement is typically installed at two cross sections, one at or near the maximum height, and permits comparison with a two-dimensional analysis.

It should be noted that the instruments shown in Figure 21.2 do not require extending leads or pipes upward through the fill. They are therefore preferable to use of vertically installed inclinometers and probe extensometers, both of which cause interruption to construction and often inferior compaction around the pipes. As indicated in Section 21.1, this practice can cause significant problems. Despite these comments, inclinometers are being used with increasing frequency, and the author recommends against this trend. However, when horizontal deformation data cannot be obtained by alternative instruments, such as for the special problem of potential shear sliding on weak horizontal clay or claystone beds in the foundation, inclinometers are the instruments of choice but, as discussed later, they should be installed only within filter zones.

On occasion, designers opt for monitoring horizontal and vertical deformations in a plane parallel to the axis of the dam, so that analyses can be made in two perpendicular planes. However, the types of instruments shown in Figure 21.2 can be read only from galleries in the abutments, and galleries are not normally justified for instrumentation alone. When such an instrumentation program is warranted, settlement measurements should be made with liquid level gages, with a preference for either the single-point overflow version with duplicate weirs or the double fluid settlement gage. Measurements of horizontal deformation can be made with inclinometers, but this practice is discouraged. Fixed embankment extensometers with electrical linear displacement transducers can also be used for measuring horizontal deformation, but these are expensive and much more prone to malfunction than the instruments shown in Figure 21.2.

21.2.3. Is Performance Satisfactory During First Filling?

Instrumentation data during first filling of the reservoir provide verification that the performance of the dam is within acceptable limits or that an unexpected event is occurring.

Regular visual observations and monitoring of leakage that emerges downstream are essential to assessing the behavior of a dam during first filling. The first indication of a potential problem is often given by an observed change of leakage rate, and monitoring the solids content in the leakage water can provide important information. Effectiveness of drains, relief wells, and grout curtains can be evaluated by measuring pore water pressures.

Equipment to measure leakage quantities formed some of the earliest instrumentation installed at dam sites. Where underdrains or relief wells are connected to manholes, flows are often measured with V-notch weirs or, when rates are low, by diverting the flow temporarily into containers of known volume. More sophisticated equipment in-

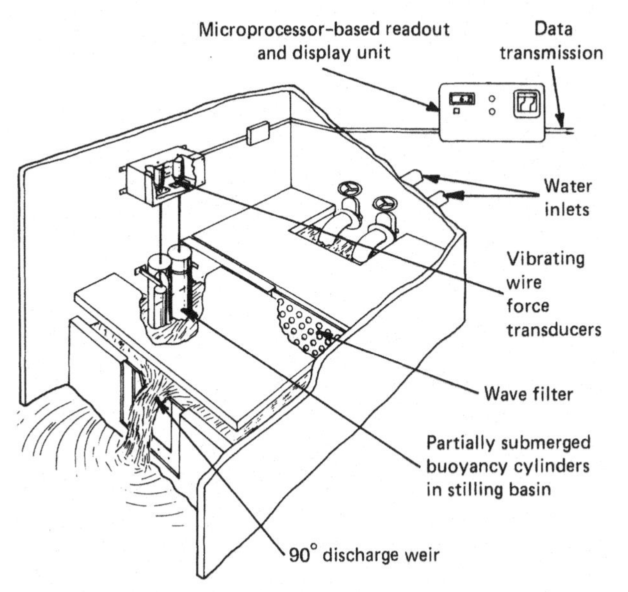

Figure 21.3. Remote-reading weir for monitoring leakage quantities (courtesy of Geonor A/S, Oslo, Norway).

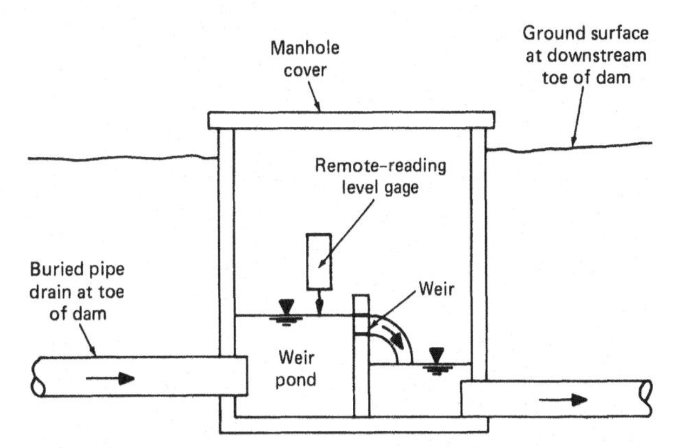

Figure 21.4. Arrangement for monitoring water flows in a buried pipe drain near the toe of a dam.

cludes remote-reading level sensors to record water levels and hence flows over weirs. Figure 21.3 shows a useful version that uses vibrating wire transducers to measure buoyancy of a partially submerged cylinder in the stilling basin. Duplicate cylinders and transducers are included to provide redundancy. However, DiBiagio and Myrvoll (1985) describe false measurements with this device, caused both by blockage of water flow and by freezing, and recommend heating of the weir station, temperature monitoring of both water and air, and electrical level switches in the stilling basin. Bartholomew et al. (1987) describe various methods for monitoring the quantity and quality of leakage water.

Figure 21.4 illustrates an arrangement for monitoring water flows in a buried pipe drain near the toe of a dam, and the addition of the remote-reading system greatly reduces the reading effort. At Svartevann Dam (DiBiagio et al., 1982), a small seepage barrier was built on the impervious bedrock directly under the crest of the sloping core so as to catch core leakage but reduce the contribution from rainfall. The arrangement, shown in Figure 21.5, in-

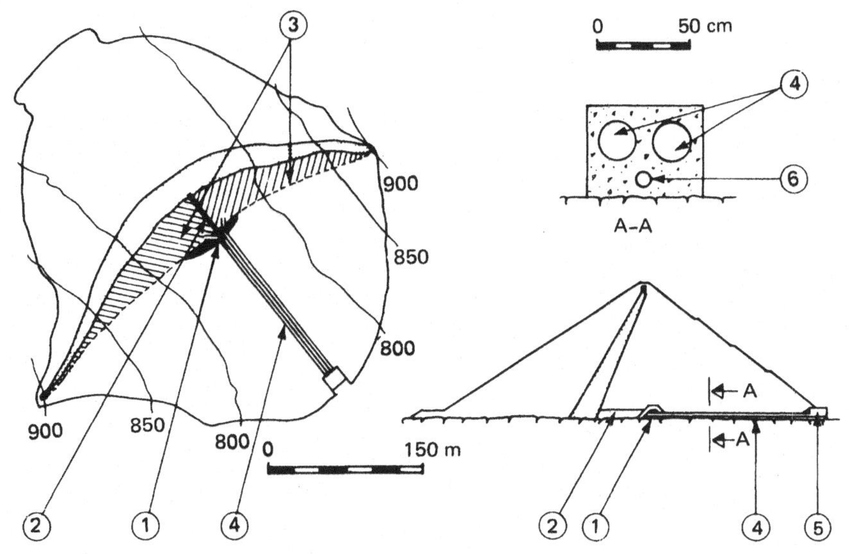

Figure 21.5. Arrangement for measurement of leakage water at Svartevann Dam: 1 seepage barrier, 2 partition wall, 3 catchment area, 4 pipelines encased in concrete, 5 instrument house with weir station, 6 conduit for instrument cables (after DiBiagio and Myrvoll, 1985).

cluded a longitudinal concrete partition wall between the core and the seepage barrier to separate right- and left-side leakage flows. Pipes were laid from the seepage barrier to a terminal enclosure at the downstream toe, and leakage was measured with the system shown in Figure 21.3.

When interpreting measurement data from downstream weirs, one must remember that the weirs also capture rain runoff. DiBiagio and Myrvoll (1985) report on the development of a precipitation gage suitable for remote measurements at dam sites.

When internal instrumentation has been installed to monitor performance during construction or drawdown or to monitor long-term performance, the same instrumentation will be monitored during first filling to determine whether or not the dam is behaving according to design predictions. Apart from instrumentation to monitor leakage, internal instruments installed specifically for monitoring performance during first filling are generally limited to piezometers for determining effectiveness of drains, relief wells, and grout curtains.

For a rockfill dam with upstream concrete facing, additional instrumentation is typically used to monitor deformation at perimeter joints in the facing. An electrical crack gage with a variable reluctance transducer is typically used.

21.2.4. Is Performance Satisfactory During Drawdown?

Rapid drawdown may cause instability because of the removal of stabilizing water forces. For dams with upstream slopes of impervious soil, it is difficult to predict pore water pressures following rapid reservoir drawdown. For any such dam retaining a reservoir that will be subject to drawdown at periodic intervals, a piezometer installation is desirable. This applies particularly to upper reservoirs for pumped storage hydro projects and to reservoirs that will be emptied partially or completely at periodic intervals for any purpose. When evaluating the need for piezometer installations, one must consider the rate of drawdown in light of the properties of the upstream shell material. With a clay, "rapid" may be a drawdown taking several weeks.

Penman (1971) gives examples of pore water pressures measured during drawdown.

21.2.5. Is Long-Term Performance Satisfactory?

A perspective on the use of instrumentation for monitoring long-term performance has been given in Section 21.1.

The subject of long-term performance monitoring of embankment dams merits a main section in this chapter and is included as Section 21.3.

21.2.6. Can the State of the Art Be Improved?

Instrumentation has provided and will continue to provide basic data for improvement of design practices, and without the knowledge gained from instruments during the past 20 years we would not be able to design and build embankment dams as we do today. Past measurements have strengthened our confidence in building dams with increasing height at sites with steep abutments and narrow valleys. Past measurements have also shown that well-designed dams are safe despite the important transfer of stress that is caued by differential settlements. Measurements with piezometers have revealed construction pore water pressures and provided essential data for the design and construction control of embankment dams with "wet" cores. As more instrumentation data become available, future design improvements will be made with confidence.

In recent years, the instrumentation of concrete face dams has contributed significantly to the data base for the design of future dams of greater height. A typical layout for a dam exceeding 400 ft (120 m) in height is shown in Figure 21.6. The joint meters are used to measure joint opening, offset, and parallel movement below normal minimum water level. The joint pins are above minimum water level to measure joint opening and are located where opening is anticipated. For lower dams, but above about 200 ft (60 m) in height, a typical program would consist of surface settlement points on the crest, two levels of three liquid level gages at the maximum section, and a leakage weir. Ninety percent of the face movements typically take place during first filling, and crest settlement is essentially complete in 2 years.

The profession has no reliable method for predicting total stresses between earth and concrete dams at connections, except with a soft clay core. The same difficulty exists for total stresses on the surfaces of concrete conduits passing through the

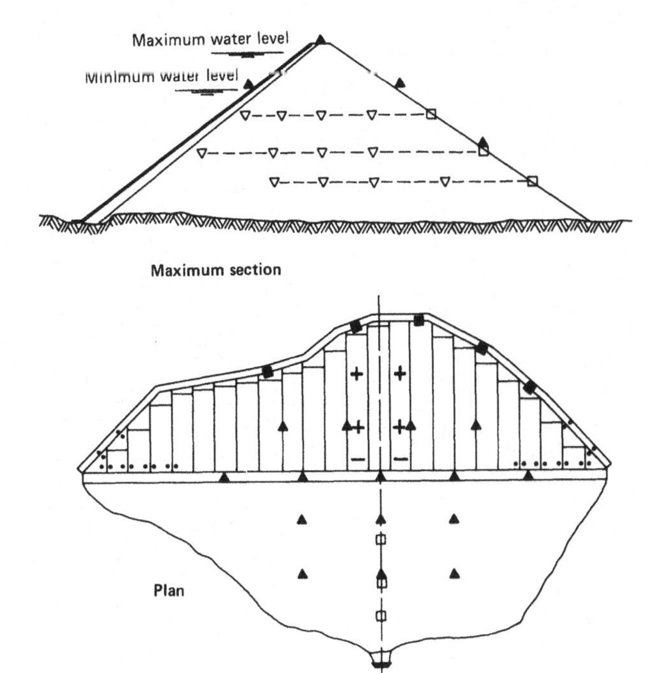

Figure 21.6. Typical layout of instrumentation to contribute to the data base for the design of future concrete face rockfill dams (after Cooke, 1986). Note: Use of surveying methods requires benchmark, remote from dam. For dams higher than about 450 ft (140 m), typical layout includes inclinometer casing on or below face slab.

bases of dams. Contact earth pressure cells are being used to provide case history data, and piezometers are usually installed nearby so that effective stresses can be determined.

21.3. LONG-TERM PERFORMANCE MONITORING OF EMBANKMENT DAMS

Once an embankment dam has reached full height, most failures are the result of internal erosion, overtopping, or other unexpected event, and a designer should be reluctant to depend on instrumentation for an advance warning of adverse behavior.

The need for long-term instrumentation depends greatly on the basic conservatism and foundation conditions of the dam. For example, a concrete face rockfill dam on a sound rock foundation forms one extreme, and an argument can be supported that instrument readings are only a formality, that the dam cannot fail except by severe overtopping. A dam on a horizontally bedded shale is an example of the other extreme, where measurements during construction show shear movement in the shale, and where calculations indicate a marginal safety factor against shear movement. In this case measurements of deformation and pore water pressure may be justified for several decades.

Table 21.1 gives a broad generalization, in priority order, of measurements that may be made for monitoring long-term performance. However, the priority must be tempered by the perspective given in the previous paragraph and in Section 21.1. The table also indicates recommended instruments and additional instruments for use in special cases: justifications for these recommendations are given later in this section.

Monitoring emphasis should be on regular visual observations and leakage quantities. Observations and instrumentation complement each other. When reviewing the performance of an existing dam on the basis of visual observations only, one may miss important behavioral trends. However, if judgments are made using instrumentation records without visual observation, incorrect conclusions may be drawn if instruments are not at the critical location or if they are not read at the critical time.

Instrumentation to monitor seismic events is now routine for important dams in areas of potentially strong earthquakes.

The need for additional external or internal instrumentation depends on the individual characteristics of the dam. The selection of instrument quantities and types must consider both technical factors and the capability of the owner to read instruments and interpret the data and will often be a compromise between criteria put forward by the designer, the owner, and regulating agencies.

If additional instrumentation is used, piezometers should be the first choice, and deformation monitoring should be the second choice. If instrumentation has been installed to monitor deformations during construction, instrument readings are often continued routinely in the long term. Measurements of total stress within the embankment are rarely reliable enough for long-term performance monitoring, but contact earth pressure cells can be used to monitor total stresses between earth

Table 21.1. Measurements and Instruments for Long-Term Performance Monitoring

Measurement, in Priority Order	Recommended Instruments	Additional Instruments for Special Cases
Condition of entire structure	Visual observations	
Leakage emerging downstream	Leakage weirs Precipitation gage	
Performance of relief wells	Leakage weirs Open standpipe piezometers	
Seismic events	Strong motion accelerographs Microseismographs	
Pore water pressure within the embankment	Open standpipe piezometers[a] Twin-tube hydraulic piezometers	Vibrating wire piezometers Pneumatic piezometers
Vertical movement of the embankment surface	Optical leveling Trigonometric leveling Satellite-based system Benchmarks	
Lateral movement of the embankment surface	Electronic distance measurements Triangulation Satellite-based system Horizontal control stations	
Vertical deformation within the embankment	Single-point and full-profile liquid level gages, overflow type Double fluid settlement gages Horizontal inclinometers Benchmarks	Probe extensometers, installed vertically[a]
Lateral deformation within the embankment	Probe extensometers with multiple induction coil or magnet/reed switch transducers, connected by rods and installed horizontally Horizontal control stations	Fixed embankment extensometers with vibrating wire transducers, or induction coil transducers with frequency output Inclinometers[a]
Total stress at contact between the embankment and a structure	Contact earth pressure cells	

[a] If carried up through fill, there will be significant interruption to construction and the probability of inferior compaction.

and concrete dams at connections and on the surfaces of concrete conduits passing through the bases of dams.

Instrumentation installed for monitoring long-term behavior during the life of a dam will in general impose selection criteria different from short-term instrumentation, and a dam designer is advised to view each of the two needs separately. Although in both cases the overriding desirable feature is reliability, instruments for short-term monitoring may be called on to provide extensive data in a short time period and, provided appropriate contractual arrangements have been made, will be read and maintained by competent personnel. This is generally not the case for long-term instruments, when durability, simplicity, and ease of reading are primary criteria.

Recommendations for monitoring various parameters during service life of an embankment dam are given in Sections 21.3.1–21.3.10. Recommendations for monitoring the reservoir rim are given by USCOLD (1979).

21.3.1. Regular Visual Observations

In many countries, requirements for regular visual observation are laid down by regulation. A detailed discussion of regular visual observations of embankment dams is beyond the scope of this book. They are mentioned here to remind readers that regular visual observations are an essential aspect of a program for monitoring long-term performance. In fact, visual observations by reservoir staff, trained to look for seeps, boils, shallow sloughing, cracks, or any other signs of distress, to log their observations, and to contact the responsible engineer when appropriate, is the primary approach to monitoring long-term performance of embankment dams. The visual observations should include the spillway and the abutments. If visual observations indicate a potential problem, it may be necessary to initiate a quantitative monitoring program to define the problem and assist in selecting a solution.

21.3.2. Monitoring of Leakage

Recommendations for monitoring leakage are given in Section 21.2.3.

Long-term performance of relief wells deserves special attention. Their efficiency can be reduced by plugging of filter packs or by accumulation of carbonates or bacteria, and it is often worthwhile to install open standpipe piezometers midway between selected wells and also directly adjacent to some wells. If indicated pressures become excessive, wells should be cleaned out or replaced.

21.3.3. Monitoring of Seismic Events*

The outdated concept that seismic instrumentation of embankment dams and reservoir sites is only a research tool has given way to the modern concept that seismic instrumentation is necessary for moderate to high dams in seismic areas. It is also desirable in traditionally nonseismic areas. This modern concept was graphically illustrated in the 1975 earthquake that subjected Oroville Dam, previously considered to be located in a nonseismic area, to significant levels of seismic shaking (Department of Water Resources, 1979).

Two classes of seismic instrumentation are used for monitoring dams and reservoirs: *strong motion monitoring* on and near the dam itself and *reservoir seismicity networks* in the area of the project site. USCOLD (1985) presents a compilation of U.S. dams with instruments for monitoring strong motion and reservoir seismicity.

*Written with the assistance of Marshall L. Silver, Marshall Silver & Associates Ltd., Chicago, IL.

Strong Motion Monitoring

Strong motion monitoring is used to measure the response of the dam to ground shaking. Although measurements can be used to validate seismic design assumptions, the most important benefit is to guide decisions on inspection and repair after the dam has been subjected to a seismic event. For example, if an earthquake has caused structural damage to the dam or to its appurtenant structures, the following questions must be answered immediately:

1. Was the earthquake larger or smaller than the design earthquake?
2. What will be the performance of the dam in the event of a larger shock?
3. What repair or strengthening is required?

If no *obvious* earthquake damage has occurred, a decision must be made on the extent of elaborate and expensive inspection operations, and data provided by strong motion monitoring equipment provide vital input to answering questions and reaching decisions. Deformation data provided by static instrumentation are also valuable, because seismic events often cause regional tectonic movements and also deformation within cohesionless soil deposits. On special occasions the pulses of seismically induced excess pore water pressures can be monitored during ground shaking, requiring the installation of piezometers with high dynamic response (Banister et al., 1976; Ishihara, 1981).

Instruments for strong motion monitoring are normally called *strong motion accelerographs* or *seismographs*. The key element of the instrument is an accelerometer, which can be thought of as a mass suspended in a case. The case is itself securely anchored to the dam. During an earthquake, relative movement between the mass and the case provides an electrical signal proportional to either the acceleration or velocity of the ground motion. An accelerograph also contains signal amplifiers, a recording device (paper, photographic film, or magnetic tape), a rechargeable battery power supply, a highly accurate clock, and a seismic trigger set to turn on the instrument when a preset level of ground shaking is exceeded.

Strong motion accelerographs should, as a minimum, be located at the base and crest of the dam, and additional instruments may be located on the downstream slope. If possible, an additional instru-

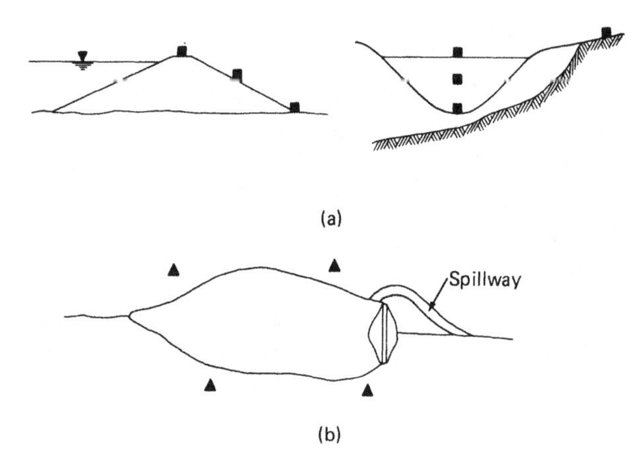

Figure 21.7. Recommended locations for (a) strong motion accelerographs and (b) microseismographs.

ment should be placed on a rock outcrop as close as possible to the dam to record bedrock motion. A recommended layout is shown in Figure 21.7a. Installations should be made as soon as possible after embankment construction.

Details of instrument selection, installation, and data evaluation are given by Bolt and Hudson (1975) and the Corps of Engineers (1981). Six important details are summarized below. First, because mains power will often fail during the seismic event being recorded, all instruments must be battery operated. Second, lightning protection must be provided. Third, it is important to ensure that any structures constructed to house the instruments do not influence dynamic measurements. Fourth, time history records of induced ground motions must be available at the site immediately after the earthquake, and little useful information results if the seismic record must be sent to the instrument manufacturer to be read. Fifth, regular maintenance of the instruments is vital. Sixth, all instruments and transmission devices must be secured against damage from earthquake shaking.

Reservoir Seismicity Networks

Seismicity networks are used to measure local small earthquakes in the area of the project site. Measured data provide information on the frequency of local earthquakes, the location and depth of seismic activity, and the magnitude and mechanisms of ground shaking.

Instruments for reservoir seismicity networks are normally called *microseismographs*. They should be located around the reservoir rim at intervals of from 3 to 20 miles (5–30 km), as shown in Figure 21.7b. Wherever possible, they should be located on rock in shallow pits away from noise from quarries, streams, and spillways. Microseismographs should be installed as soon as the project site has been selected, using the advice of a seismologist familiar with the seismic characteristics of the region.

21.3.4. Monitoring of Pore Water Pressure

Instruments for measurement of pore water pressure are described and evaluated in Chapter 9.

There is general agreement that open standpipe piezometers are the most reliable instruments for measurement of pore water pressure. When used for monitoring long-term performance of an embankment dam, they can often be installed in the foundation under the downstream slope, and at the downstream toe, after completion of the dam. They are generally unsuitable for installation during construction, because they interfere with fill placement. If open standpipe piezometers are not acceptable, the choice for long-term monitoring is among twin-tube hydraulic, pneumatic, and vibrating wire piezometers.

21.3.5. Diversity of Views on Selection of Remote-Reading Piezometer Type

During recent years several engineers have expressed their views in the technical literature on the advantages and limitations of twin-tube hydraulic, pneumatic, and vibrating wire piezometers for long-term monitoring of pore water pressure in embankment dams. There is no clear unanimity of view. The key points in each view are summarized in this section and followed by the author's suggestions for good practice.

Sherard (1981) generally favors the use of vibrating wire piezometers. He argues that although twin-tube hydraulic piezometers are good instruments, diaphragm piezometers are easier to use and require less skill and care in their maintenance. Sherard points out that some models of pneumatic piezometer have not given uniformly good service. When diaphragm piezometers are used with saturated high air entry filters and saturated interior cavity to measure pore water pressure, the vibrating wire piezometer has important advantages over the pneumatic piezometer because the initial saturation and air tightness of the piezometer can be tested before installation. The test is made by removing the assembled and saturated piezometer from its container of water and taking periodic readings while evaporation of the water occurs in the pores of the filter. Sherard concludes that, although the vibrating wire piezometer and the better types of pneumatic piezometer are satisfactory instruments, the vibrating wire piezometer has substantial advantages and will probably be used more widely in the future.

Green (1982), in his discussion of Sherard's paper, presents some points in favor of twin-tube hydraulic piezometers and favors pneumatic over vibrating wire versions. Green points out that the twin-tube hydraulic piezometer is the simplest, but the pneumatic piezometer is also mechanically simple and can be made entirely of corrosion-resistant plastics. In contrast, the vibrating wire piezometer is a complicated electromechanical device in which the calibrated part of the measuring system is buried and is inherently susceptible to corrosion, zero drift, calibration slope changes, and grounding problems owing to water ingress to the electrical system via the buried leads. Green argues that, although these problems can be minimized by good design, defects are all too often present in any instrument owing to the manufacturer's lack of awareness, inadequately detailed design, or cost-cutting production methods forced on manufacturers by the user's low-bid procurement procedures. Green states that the order of increasing complexity is: twin-tube hydraulic, pneumatic, vibrating wire. Provided a pneumatic piezometer is designed to operate at subatmospheric pressure, its initial saturation and air tightness can be checked before installation as described by Sherard. Green concludes that the simpler pneumatic piezometer appears to have much more to offer the geotechnical profession and the dam designer than does the vibrating wire piezometer.

Mikkelsen (1982), in his discussion of Sherard's paper, also favors pneumatic over vibrating wire piezometers, citing many factors similar to those given by Green. Mikkelsen indicates that if leads over 1000 m (3300 ft) are necessary, pneumatic piezometers start to become impracticable, whereas cables over 4000 m (13,000 ft) long have been used on vibrating wire piezometers with success. Pneumatic piezometers are not subject to damage by lightning strikes, can be read automatically using high-precision electronic pressure gages and microprocessor controlled pneumatic flow systems, and are less expensive than vibrating wire piezometers. Mikkelsen also states that because the pneumatic piezometer has a simple valve that is either open or closed, there is little or no concern for stability of calibration. It is nearly an ideal gage in that when the gage functions, the reading will tend to be correct.

Sherard (1982) responds to the discussions by Green and Mikkelsen. He states that many years of experience have shown that currently available vibrating wire piezometers have not had significant problems because of zero drift, corrosion, calibration, slope changes, and water leakage. This generally good performance record has convinced a large fraction of engineers designing dams that the vibrating wire piezometer is a basically good instrument.

Londe (1982) states that vibrating wire instruments are more reliable and more durable than mechanical or hydraulic instruments. In Londe's view, vibrating wire instruments are not subject to rust, leaks, or wear and, whenever long-term readings are required, they are preferable to other types. They have the merit of accuracy, extreme reliability, and ability to transmit readings without alteration over long distances.

DiBiagio and Myrvoll (1985) describe a very robust vibrating wire piezometer, shown in Figures 9.33 and 9.34, that has been designed specifically for embankment dams and report on good performance.

Dunnicliff (1981) gives recommendations for selection among twin-tube hydraulic, pneumatic, and vibrating wire piezometers and contends that twin-tube hydraulic piezometers have the best chance of longevity. Houlsby (1982) endorses this view. Suggestions by both are included in the recommendations given in the following section. Kleiner and Logani (1982), in their discussion of Dunnicliff's paper, recommend use of open standpipe piezometers wherever possible and indicate that both pneumatic and vibrating wire piezometers can provide reliable data over long periods of time, but that a substantial number of instruments can fail.

Penman (1986) favors hydraulic piezometers. He believes that the primary application for vibrating wire piezometers is to provide rapid response and it seems unreasonable to expect the thin diaphragm of a pneumatic piezometer to last for the life of a dam. Penman comments that the de-airing process seems to have given twin-tube hydraulic piezometers a bad name. In the early days with polyethylene (polythene) tubing, air passed through the walls, but with polyethylene coated nylon tubing and Bishop piezometer tips, de-airing needs are infrequent. With the recent development of automatic de-airing, de-airing is no longer an arduous procedure. Twin-tube hydraulic piezometer tubing will accommodate much more stretch than electrical cable.

21.3.6. Recommended Piezometer Types for Long-Term Monitoring

The author's suggestions for good practice are given in the following subsections.

Open Standpipe Piezometers

Open standpipe piezometers are the most reliable instruments for measurement of pore water pressure. In many countries they are commonly used for monitoring seepage pressures in the foundation under the downstream slope and at the downstream toe. Whenever possible, piezometers under the downstream toe should be installed in boreholes after completion of the dam so that there is no interference to fill placement.

When open standpipe piezometers are used, they should have low air entry filters. Standpipes should be of thick wall PVC or ABS plastic pipe, rather than steel pipe or plastic tubing. Standpipe inside diameter should be adequate for self-de-airing (minimum of 0.3 in., 8 mm) and, unless considerations of response time dictate otherwise, inside diameters of 0.75 or 1.0 in. (19 or 25 mm) are preferred so that strength is adequate and insertion of the reading device presents no problems. Standpipes should always be protected by an outer telescoping pipe to avoid ruptures of the types shown in Figure 17.10. When it is necessary to prevent leakage of water along the annulus between the pipes, the annulus should be filled with a slurry.

Friction problems and breakage of electrical dipmeter cables have been common features in very deep installations. When a standpipe is deeper than about 300 ft (90 m), the deflections of the standpipe owing to its own weight, within the protective pipe, can cause the dipmeter to be trapped by friction so that the effort required to extract it causes cable breakage (Wilson, 1982). This difficulty can be overcome by reading with the purge bubble device, leaving a small-diameter plastic tube within each standpipe. Because the tube cannot be pushed down a long standpipe after installation, it must be inserted as the standpipe is assembled.

If access to the top of the standpipes is difficult in the long term, readings can be made from a remote location by using one of the methods described in Chapter 9.

If open standpipe piezometers are installed in boreholes after completion of the dam, conformance between the instrument and fill should be ensured by selection of appropriate standpipes and backfill materials, and the potential for hydraulic fracturing should be recognized. The concern over hydraulic fracturing caused by drilling fluid is debatable. Sherard (1986) indicates that hydraulic fracturing probably occurs to some degree in most embankment dams, and usually no erosion is caused. It appears that the only certain method of preventing hydraulic fracturing during drilling is by using augers, either to advance the borehole directly or to clean soil from within driven casing. Hydraulic fracturing while grouting the borehole above a piezometer can be avoided by stage grouting.

Twin-Tube Hydraulic, Pneumatic, and Vibrating Wire Piezometers

If open standpipe piezometers are not acceptable, the choice is among twin-tube hydraulic, pneumatic, and vibrating wire piezometers. Based on his own experience and the diverse views summarized above, the author puts forward the following suggestions. There appear to be three primary questions that need to be answered when selecting among the three types of piezometer for long-term measurements. These are discussed in the following subsections.

First Question

Can I accept not knowing whether the measured pressure corresponds to the pore water pressure, pore gas pressure, or something in between?

As indicated in Section 9.13, the compacted fill in an embankment dam may remain unsaturated for a prolonged period after the reservoir is filled, and in fact the fill may never become permanently saturated by reservoir water. Increase of water pressure causes air to go into solution, and the air is then removed only when there is enough flow through the fill to bring in a supply of less saturated water. The pressure and time required to obtain saturation depend on the soil type, degree of compaction, and degree of initial saturation. Pore gas pressure may therefore remain significantly higher than pore water pressure for a substantial length of time, perhaps permanently. Despite use of saturated high air entry filters, they may not remain saturated during this time because gas may enter the filters by diffusion, and flushing of the filter and cavity may be necessary to ensure that pore water pressure continues to be measured. Twin-tube hydraulic piezometers allow for flushing with de-aired liquid, but this is not possible with conventional diaphragm piezometers. Although it is possible to add flushing tubes to diaphragm piezometers, their addition dictates that the instruments must be installed, terminated, operated, and maintained as required for twin-tube hydraulic piezometers.

If the answer to this first question is **no**, twin-tube hydraulic piezometers are required. If the answer is **yes**, all three types are suitable.

Second Question

To what extent are skilled personnel available to maintain instruments during the operating life of the dam?

Maintenance of all types of piezometer requires regular checking of all components in the terminal enclosure. Twin-tube hydraulic piezometers may also require occasional flushing of the system with de-aired liquid as described in Appendix E. Some users will not consider using a twin-tube hydraulic piezometer system because they claim that the maintenance burden is unacceptable. However, the author believes that the maintenance burden has been exaggerated by improper selection of components, and flushing is seldom necessary in a properly designed and installed system, except when subatmospheric pore water pressures are being read. A properly designed system entails attention to the many details discussed in Appendix E.

Maintenance of pneumatic and vibrating wire piezometers also requires regular checking and calibration of readout units, which can if necessary be done at an appropriate facility away from the site. The need for skilled maintenance is therefore applicable to all piezometers, but the availability of skilled personnel (local, visiting on a regular basis, and/or at off-site facility) is likely to impact on the selection among the three types.

Third Question

Do connecting tubes or cables need to be longer than about 2000 ft (600 m)?

If **yes**, twin-tube hydraulic and pneumatic piezometer readings are sluggish and questionable. The length of tubing can sometimes be minimized by constructing terminal houses within drainage or inspection galleries or by use of almost vertical runs of tubing in filters as discussed in Section 21.4. Penman (1986) indicates that the sluggishness of twin-tube hydraulic piezometers ceases to be a problem if an automatic reading and de-airing system is used. The same argument can be advanced for automatically read pneumatic piezometers of the *normally closed* type. Thus, for very long connecting tubes or cables, the options are vibrating wire piezometers or the remaining two types with automatic reading equipment.

Summary of Recommendations for Selection Among Twin-Tube Hydraulic, Pneumatic, and Vibrating Wire Piezometers

Answers to the above three questions may dictate selection among the three piezometer types for long-term measurements. The selection of filter type, shape, size, and saturation procedure should follow the guidelines given in Chapter 9.

Sherard (1981) states that the likely useful life of any of the three types of piezometer is more related to the conservatism of the design and construction details of the specific piezometer, connecting lines, and installation care, and to the conditions in which it is placed, than it is related to differences among the three types. The author totally agrees with this view.

The author also agrees strongly with Green (1982) that cost-cutting production methods, forced on the manufacturers by use of the user's low-bid procurement procedures, tend to reduce conservatism of the design and construction details. Perhaps this last observation helps to explain the diversity of views presented in the previous section: the good experiences with vibrating wire piezometers that have been reported by Londe, Sherard, and DiBiagio and Myrvoll are generally with Telemac and Geonor instruments in situations where lowest cost does not dominate the choice of instruments.

Although some manufactured versions have a better success record than others, there appears to be no certain and intrinsic longevity difference between pneumatic and vibrating wire piezometers. The inaccessible parts of both are self-evidently more complex than the twin-tube hydraulic piezometer.

General requirements for maximizing reliability of performance monitoring are given in Chapter 15, including a listing of three major requirements relating to the design of the instruments themselves: simplicity, self-verification, and durability in the installed environment. In the view of the author, these three requirements are satisfied by twin-tube hydraulic piezometers to a greater extent than by the other two types. Open standpipe piezometers installed in the foundation can be converted to remotely read twin-tube hydraulic piezometers by using the arrangement shown in Figure 9.10. **If engineers are truly committed to long-term readings that are as verifiable as possible, twin-tube hydraulic piezometers should be their first choice.**

Those who disagree with this preference for twin-tube hydraulic piezometers and favor use of pneumatic or vibrating wire piezometers for **long-term** measurements are encouraged to make a close study of recommendations by DiBiagio and Myrvoll (1985), Green (1982), Mikkelsen (1982), and Sherard (1981, 1982). They are also encouraged to consider very seriously the implications of low-bid procurement procedures and to favor heavy-duty instruments that are designed for long-term survivability, for example, the Geonor vibrating wire piezometer shown in Figure 9.34 or the heavy-duty version manufactured by Geokon. They should review published opinions on the likelihood of zero drift of vibrating wire piezometers: reference citations are given in Chapter 8. Chapter 8 also reports on two unpublished case histories, involving measurements with vibrating wire piezometers supplied by two different manufacturers and installed in two embankment dams. During first filling of each reservoir, one piezometer indicated a high piezometric level that caused concern, and filling was stopped. The piezometer reading continued to rise and, when the indicated piezometric level rose above pool level, the measurements were discounted and filling continued. These cases of a *rogue instrument* reinforce the author's preference for twin-tube hydraulic piezometers. They also emphasize the view expressed by LeFrancois (1986) (Section 8.4.9) that evaluations of data should not be made on the basis of readings from a single instrument.

There is a need for comparative evaluations on long-term performance of the three types of remote-reading piezometer, with care to avoid incorrect general conclusions because of commercial bias, unusual project conditions, improper installation, improper selection of components, or improper operation. Engineers who are in a position to pursue such practical comparisons are strongly encouraged to do so and to report their findings to the profession.

21.3.7. Recommended Instruments for Monitoring Long-Term Settlement

Measuring points on the crest and downstream face should be the first choice for long-term measurements of surface settlement. Measurements are normally made by optical or trigonometric leveling, but the satellite-based Navstar Global Positioning System is likely to be used with increasing frequency in the future.

If subsurface settlement measurements are required, the choice is among liquid level gages, horizontal inclinometers, and probe extensometers, and selection depends on the specific needs and details of each project. Liquid level gages and horizontal inclinometers are greatly preferable to vertically installed probe extensometers, because they cause much less interference to construction. Of the many available types of liquid level gage, the most suitable choices are the following:

1. Single-point overflow gages with duplicate weirs.
2. Full-profile overflow gages, often combined with probe extensometers installed horizontally as shown in Figure 21.2.
3. Double fluid settlement gages.

Details of all instrument types are given in Chapter 12.

A reliable benchmark is essential, beyond the influence of the dam and fluctuating reservoir. When benchmarks are located close to the dam or on steep slopes of native overburden soil, their uncertain reliability can invalidate the monitoring objective.

21.3.8. Recommended Instruments for Monitoring Long-Term Lateral Deformation

If lateral deformation measurements are required, measuring points on the crest and downstream face should be the first choice. Data are normally obtained by trilateration with electronic distance measuring equipment, but triangulation is also used and the satellite-based Navstar Global Positioning System is likely to be used more widely in the future. Stable horizontal control stations are critical and should be located as described in Section 21.3.7 for benchmarks.

If subsurface measurements of lateral deformation are required, the choice is among probe extensometers, fixed embankment extensometers, and inclinometers, and selection depends on the specific needs and details of each project. For measurements in the downstream shell, the best choice is usually a probe extensometer with a current-displacement induction coil transducer (sometimes called a *horizontal plate gage*) installed horizontally as shown in Figure 21.2. Precision is maximized by using the version with a primary induction coil near each steel plate and with interconnecting rods. Fixed embankment extensometers with electrical linear displacement transducers can be used to determine the pattern of horizontal strain in the core or elsewhere. Preferred transducers for long-term measurements are vibrating wire transducers or induction coil transducers with frequency output. Inclinometers can be installed to monitor lateral deformation throughout the length of a vertical casing but, unless the casing is extended upward within a filter as described in Section 21.4, this practice is discouraged. As discussed in Section 12.8.2, accuracy of inclinometer data in inclined casings is poor (e.g., casings within a filter alongside a sloping core).

Details of available instruments are given in Chapter 12.

21.3.9. Additional Approaches to Long-Term Monitoring

Automatic Monitoring

Recent developments in automatic data acquisition systems (ADASs), power supplies, and communication systems are having a major impact on long-term monitoring of embankment dams. These developments are described in Chapter 8, including a discussion of the applicability of ADASs for monitoring various transducers. Advantages and limitations of ADASs and automatic data processing and presentation systems are discussed in Chapter 18. ICOLD (1982) provides guidelines and examples for automatic monitoring, and the USCOLD Committee on Measurements is currently preparing a summary of U.S. experience (USCOLD, 1988).

One caution needs to be given. Adoption of an automated monitoring program is not usually a justifiable reason for selecting inaccessible electrical transducers. For example, the decision to read automatically should not in itself cause vibrating wire piezometers to be selected in favor of open standpipe or twin-tube hydraulic piezometers, because these last two types can also be read automatically by use of accessible hydraulic/electrical transducers. The selection of an inaccessible transducer should be based on inherent longevity criteria, not on the fact that it provides an electrical signal at the source.

Streaming Potential

Streaming potential (also called *self-potential*) appears to be a promising new method for long-term detection of seepage paths (Bartholomew et al., 1987; USCOLD, 1986). When water is forced to flow under laminar conditions through porous earth materials, it generates an electrical current. The current is monitored by embedding electrodes within the dam. However, there appears to be little practical experience to date.

Acoustic Emission Monitoring

Acoustic emission monitoring is being considered for detection of seepage characteristics (Chapter 12) but, as for streaming potential, there appears to be little practical experience to date.

Thermal Monitoring

Attempts have been made to monitor leakage through embankment dams by thermal monitoring, also called *thermotic surveys* (Bartholomew et al., 1987; Birman et al., 1971; USCOLD, 1986). The principle is based on embedding thermistors in the fill, drawing isotherms, and deducing flow patterns. The key to success of this approach appears to rest on having a significant temperature difference between the water in the reservoir and in the embankment fill.

21.3.10. Guidelines for Operating and Maintenance Personnel

Most major public agencies that are responsible for long-term performance of embankment dams have guidelines for emergency preparedness planning. These guidelines generally include regular visual inspections and the use of instrumentation for safety evaluation. For example, the U.S. Army Corps of Engineers has prepared several *Engineering Regulations* that are listed by the Corps of Engineers (1984). The Corps of Engineers (1985) provides additional guidelines. The U.S. Bureau of Reclamation has a similar list (Misterek, 1986), and the Australian National Committee on Large Dams has prepared guidelines (ANCOLD, 1983).

Dunnicliff (1981) attempts to answer the question:

> When instrumentation is installed on and in an embankment dam for the purpose of monitoring long-term performance, what guidelines should be given to operating and maintenance (O&M) personnel to assist them in judging satisfactory performance based on the measurement data?

The suggested guidelines relate to measured pore water pressures, settlements, lateral deformations, and seepage flows and to regular visual observations of the condition of exposed parts of the embankment and spillway.

An operating and maintenance manual should be prepared for the guidance of O&M personnel at all dams. If instrumentation has been installed for long-term performance monitoring, the manual should include guidelines on regular calibration and maintenance procedures (Chapter 16) and on procedures for collection, processing, presentation, interpretation, and reporting of instrumentation data (Chapter 18). The instrument installation report (Chapter 17) should be made available to O&M personnel.

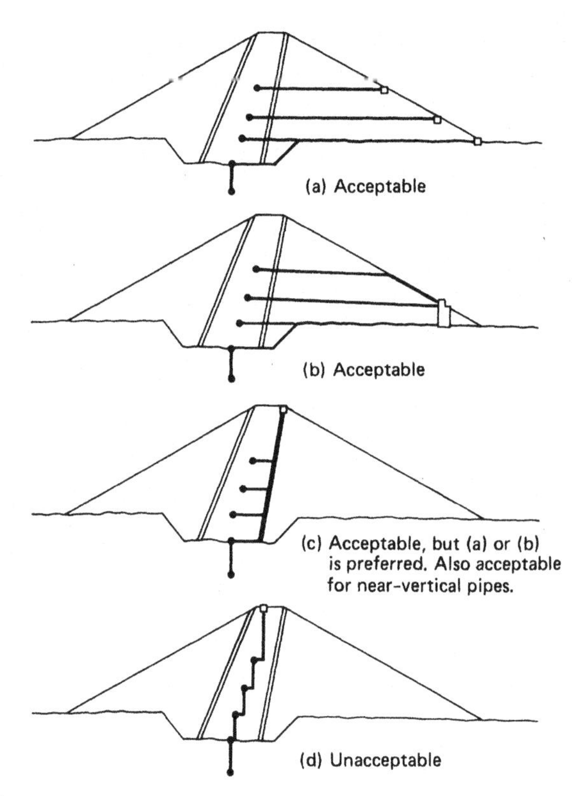

Figure 21.8. Examples of acceptable and unacceptable routings for instrumentation tubes and cables.

21.4. GENERAL GUIDELINES ON THE EXECUTION OF MONITORING PROGRAMS FOR EMBANKMENT DAMS

General guidelines on the execution of monitoring programs are given in Part 4 of this book. The guidelines include instrument calibration, maintenance and installation, and collection, processing, presentation, interpretation, and reporting of instrumentation data. This section provides general guidelines specific to execution of monitoring programs for embankment dams. Most of the guidelines relate to installation of instruments.

21.4.1. Installation of Pipes, Tubes, and Cables

Methods of installing pipes, tubes, and cables are described in Chapter 17. The following subsections elaborate on some important issues for embankment dams.

Comparison Between Vertical and Horizontal Runs

In embankment dams, vertical and near-vertical runs of pipes, tubes, and cables should be avoided wherever possible. There is one exception to this rule: installations can be extended upward through a filter as discussed later in this section. If installed other than in a filter, they impede the quality of compaction, and may cause significant problems with the performance of the dam.

It is preferable, wherever possible, to select instruments that do not require vertical runs. For example remote-reading piezometers can often be selected in preference to open standpipe piezometers. Liquid level gages can often be selected in preference to vertically installed probe extensometers, and horizontally installed probe extensometers can often be selected in preference to inclinometers. Pipes, tubes, and cables for these preferable instruments should be routed along the fill at the slope adopted for general placement purposes and terminated in an enclosure buried in the fill near the downstream face. Figure 21.2 shows an acceptable routing for horizontal pipes. Figure 21.8 shows acceptable and unacceptable routings for tubes and cables.

The arrangement shown in Figure 21.8d has been

used in several dams. Tubes or cables have been bundled together, and the vertical parts of the step have been protected by enclosing them within a temporary protective steel pipe, often about 3 ft (1 m) in diameter, jacked up vertically as filling proceeds. There is ample evidence that this method causes inadequate compaction around the bundle, damage to instrument leads, and a potential water pathway. Adverse behavior of major dams has been caused as a result of this practice (e.g., Harder, 1982; Mikkelsen and Wilson, 1983), but it is still considered acceptable by many engineers. The practice is **strongly** discouraged. However, if the pipes, tubes, or cables are extended upward through a filter as shown in Figure 21.8c, the result is much more satisfactory. Electrical cables were carried up the downstream filter of Srinagarind Dam in Thailand (Champa and Mahatharadol, 1982) and pneumatic tubes in the downstream filter of Wadaslintang Dam in Indonesia, using the jacked-up protective steel pipe arrangement described above. Filter material was added inside the pipe as it was jacked up, keeping the internal level close to the bottom end of the pipe to avoid creating tension in the leads as the pipe was jacked up.

For both horizontal and vertical runs, the engineering properties of fill material around pipes, tubes, and cables should match, as closely as possible, the properties of the surrounding fill. When installed in "impervious" fill, the installations should not provide a leakage zone for water.

Installation of Vertical Runs

When it is necessary to extend vertical or near-vertical runs of pipes, tubes, or cables upward through the fill, they should be restricted to installation within a filter as described above. Either the jacked-up protective steel pipe procedure, or alternatively, the *mounding* method or the *excavate and replace* method can be used, as described in Chapter 17.

Installation of Horizontal Runs

Mikkelsen (1986b) describes an innovative method for installing a horizontal pipe in rockfill during construction of a concrete face rockfill dam. The pipe was plastic inclinometer casing, for use with a horizontal inclinometer, and was installed on 6 in. (13 mm) of compacted sand as shown in Figure 21.9. Mikkelsen reports on several advantages of this method of installation when compared with a trenched-in method: maximized use of fill production and compaction equipment; no disturbance owing to trenching; improved initial grade and line; much faster installation; better access for compactors; less potential for casing damage; and better compaction.

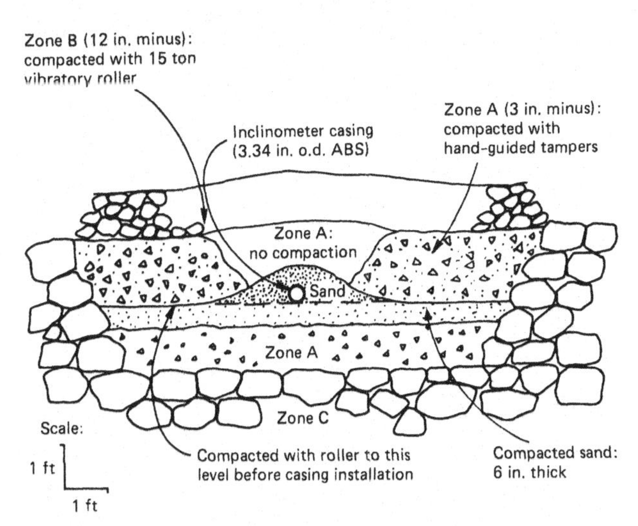

Figure 21.9. Installation of horizontal inclinometer casing in concrete face rockfill dam (after Mikkelsen, 1986b).

Mikkelsen and Wilson (1983) maintain that when trenches are excavated for instrumentation tubes or cables, they should not extend in the upstream–downstream direction through more than one-half, or preferably not more than one-third, of the width of the core. It is worthwhile to consider three likely reasons for this prohibition.

First, some engineers are concerned that instrument leads may be broken where they cross zones of different compressibility. This issue is discussed in the following subsection and appears not to be a major concern. Second, leakage can develop along the route if instrument leads deteriorate and leave an open hole. This issue can be minimized by selection of the types of tubes and cables that are recommended in Chapter 8 and by appropriate design of the downstream filter. Third, leakage can develop along the route if fill around the leads is poorly placed, such that it does not develop a total pressure greater than the pressure from reservoir water at that elevation. If great care is taken with fill placement, there should be no potential for leakage. The guidelines given in Chapter 17 should be followed, using the trenching and backfilling method. In soft clay cores, excavated material should be recompacted into the trench at placement water content and with at least as much compactive effort as used for the main fill. In more sandy materials it may be useful to add some bentonite to the excavated material to improve its plasticity and reduce its permeability. The "bentonite cutoffs" described by USBR (1974) are not necessary. Alternatively, the *surface installation method* described by Clements (1982) and in Chapter 17 can be used, provided that there is good supervision of general fill placement.

If a leak occurs through the core, of a type that can be imagined as the most extreme result of the actions of the above second and third points, the leak will be controlled and sealed safely by a good downstream filter (Sherard, 1985). For most dams it is necessary for the designer to assume that a small concentrated leak may develop through the core from other causes, and Sherard et al. (1984) give criteria for the design of appropriate filters. Others argue that leaks should be prevented to the extent possible, regardless of the filter design provided, because no filter design accounts for placement errors during construction.

In summary, if instrumentation of an embankment dam can be achieved without running tubes and cables through the core, this is the preferred method. However, if instruments are considered essential and if there are substantial economic advantages to running leads completely through a core, the practice is acceptable provided that great care is taken while placing surrounding fill material.

Installation of Tubes and Cables Across Zones of Different Compressibility.

The question of running piezometer tubes and cables through the boundaries between embankment zones is widely debated, particularly through the boundary between the downstream face of a central core and the downstream filter. Dogmatic guidelines cannot be given, but it appears to the author that the practice is acceptable.

Shear movements have been measured across zones of different compressibility in several dams, for example, between a vertical core and downstream filter, using double fluid settlement gages. Wilson (1982) reports on such measurements in Angostura Dam, a 500 ft (150 m) high rockfill dam with a central vertical clay core, showing that shear movements were not abrupt. This and other experiences convinced Wilson that, in general, it is satisfactory to route tubes and cables through the boundary between a vertical core and downstream filter, provided that they can accommodate minor tensile and compressive strains. Penman (1986) shares this view and comments that plastic tubing of up to 0.25 in. (6 mm) diameter survives the deformations. Suitable tubing and cable is recommended in Chapter 8. When using electrical cables, the water-blocked type should be selected to avoid water passing between the jacket and conductors. Because most electrical cables have limited ability to tolerate axial strain, it may be worthwhile to reduce concentrations of strain by threading individual cables within minimum-sized plastic tubing where they pass through zones that will experience significant strain and to fill the annular space with petroleum jelly. The tubing should extend at least 10 ft (3 m) on either side of the interface to help distribute the strain.

If designers predict more than normal differential movement between boundaries, an alternative routing will be more conservative. For example, it seems reasonable to expect a greater tendency for the face of an upstream sloping core to "slip" with respect to the first downstream filter layer than is the case for a vertical core, and more tendency near the top of a dam than near the bottom. For high dams with steep rock abutments, the core may slip abruptly downward with respect to the abutments as the core settles, and leads passing through the interface may be sheared. However, it is not usual to route instrument leads into abutments unless drainage galleries are being used for terminal arrangements. At New Don Pedro Dam (Taylor, 1968), where substantial differential movements between zones were anticipated, all tubes and cables were carried parallel to the axis of the dam, entirely within each zone either to the abutments or to a central backfilled shaft within the core. Today, a better practice would be to extend the leads upward in a filter, as described earlier.

21.4.2. Completion of Terminal Enclosure

On some projects, instruments are installed and leads are left for a significant time as coils on the fill. Many instruments, such as twin-tube hydraulic piezometers and liquid level settlement gages, cannot be read until terminal arrangements are completed, and if coils are left on the fill no initial readings are taken and early data are not available. Without initial readings, no calculations can be made of total changes. To avoid this shortcoming, terminal enclosures must be completed in a timely manner so that initial instrument readings can be made immediately after instruments are installed. Other instruments, such as vibrating wire and pneumatic piezometers, can be read while leads are coiled, and this issue does not arise.

21.4.3. Interpretation of Settlement Measurements in the Core

When measurements of settlement are made within the central core of an embankment dam, the data are often used to compute the modulus of compression. However, the engineer rarely knows whether the "compression" measured between two points is vertical compression (decrease in volume) or if it may primarily be due to shear deformation at constant volume as the sides of the core spread outward. In any given case both actions occur. If the shear deformation is ignored, interpretations of data showing that the central core of a rockfill dam is more compressible than the shells may be incorrect. When intending to use settlement data for this purpose, one must include additional measurements to resolve the uncertainty. The extent of horizontal spreading of the downstream face of the core can be measured by using the arrangement shown in Figure 21.2. Double fluid settlement gages can be used to measure settlement in both the core and shell.

21.4.4. Data Collection Frequency

Chapter 18 gives general guidelines for the frequency of data collection, and suggested schedules for embankment dams are given by Bartholomew et al. (1987) and USCOLD (1986).

When instruments have been installed in an embankment dam to monitor performance during construction, first filling, or drawdown, a decision must be made on whether to continue readings in the long term. The decision should be based on long-term needs, **not** on the fact that instruments are **there.** The need for monitoring a typical dam, after the reservoir has been filled for 2 or 3 years, decreases rapidly with each passing year. In particular, during the first few years the settlements and internal strains usually come to equilibrium, after which there is little interest in continued monitoring of internal deformation.

When long-term instruments have been read intensively for a period after the dam is in operation, and when long-term reading costs must be reduced, data collection frequency will be reviewed. A good approach is to select key instruments, based either on their importance or on previous behavior, and to continue to read these at an unchanged frequency. Others will then be read at a greatly reduced frequency. For instruments requiring multiple readings for a single data set, such as probe extensometers or inclinometers, frequent readings can be limited to the zone of greatest importance or of greatest previous change.

21.5. SELECTED CASE HISTORIES

Table 21.2 summarizes selected case histories of instrumented embankment dams. Several of the case histories are selected from the *Transactions of the Fifteenth International Congress on Large Dams* (ICOLD), held in Lausanne in 1985: there are many more good case histories in Volume 1 of these *Transactions*. A summary is given by Post (1985) and a brief overview by Water Power & Dam Construction (1985). Post also lists other recent occasions when monitoring of dams and their foundations has been the subject of earlier ICOLD Congresses.

Table 21.2. Summary of Selected Case Histories of Embankment Dams

Project	Principal Concerns	Type of Dam	Measurement[a]	Instrumentation Discussed[b]	Special Features	Reference
Fifteen dams in Canada	Post-construction deformations	Various rockfill dams, 30–500 ft (10–170 m) high	D	Surveying methods	Summary of post-construction settlement and horizontal deformation data	Dascal (1987)
Shiroro Dam, Nigeria	Performance during construction and operation	410 ft (125 m) high concrete face rockfill	LE G D D D D D T	Leakage weirs Pneumatic and vibrating wire piezometers Surveying methods Probe extensometers (H) Full-profile liquid level gages Crack gages Inclinometers (I,V) Thermistors	Inclined inclinometer casings installed below concrete face.	Bodtman and Wyatt (1985)
Various dams in Norway	Various	Various	LE G D D D D TS	Leakage weirs Piezometers Probe extensometers (V) Liquid level gages Fixed embankment extensometers Inclinometers (H,I,V) Embedment earth pressure cells	Summary of types and quantities of instruments in Norwegian dams	DiBiagio and Kjaernsli (1985)
Salvajina Dam, Colombia	Performance during first filling and operation	485 ft (150 m) high concrete face gravel/rockfill	LE G D D D S TS SE	Leakage weirs Pneumatic piezometers Surveying methods Liquid level gages Crack gages Embedment strain gages Embedment earth pressure cells Strong motion accelerographs		Hacelas et al. (1985)
Monasavu Dam, Fiji	Performance during construction and operation	280 ft (85 m) high rockfill with central core	LE G D	Leakage weirs Twin-tube hydraulic piezometers Surveying methods	Very wet residual clay core	Knight et al. (1985)

Table 21.2. Summary of Selected Case Histories of Embankment Dams (*Continued*)

Project	Principal Concerns	Type of Dam	Measurement[a]	Instrumentation Discussed[b]	Special Features	Reference
			D	Liquid level gages		
			D	Fixed embankment extensometers		
			TS	Embedment earth pressure cells		
			SE	Strong motion accelerographs		
Foz do Areia Dam, Brazil	Deformations during and after construction	525 ft (160 m) high concrete face rockfill	LE	Leakage weirs		Lagos Marques et al. (1985)
			D	Surveying methods		
			D	Probe extensometers (H)		
			D	Liquid level gages		
			D	Crack gages		
			S	Embedment strain gages		
Khao Laem Dam, Thailand	Performance during construction and operation	430 ft (130 m) high concrete face rockfill	LE	Leakage weirs	Inclinometer casings attached to concrete face	Mahasandana and Mahatharadol (1985)
			G	Observation wells		
			D	Surveying methods		
			D	Settlement platforms		
			D	Liquid level gages		
			D	Crack gages		
			D	Inclinometers (I)		
			S	Embedment strain gages		
			TS	Contact earth pressure cells		
Elandsjagt Dam, South Africa	Performance during construction and operation	250 ft (75 m) high earth and rockfill	LE	Leakage weirs	Very rapid first filling	Melvill (1985)
			G	Twin-tube hydraulic and vibrating wire piezometers		
			D	Probe extensometers (H,V)		
			D	Liquid level gages		
			TS	Contact and embedment earth pressure cells		
Meggett Dam, Scotland	Performance during construction and operation	180 ft (55 m) high dam with central asphaltic core and gravel shells	D	Surveying methods	Comparison between observed and predicted deformations	Penman and Charles (1985)
			D	Probe extensometers (H)		
			D	Liquid level gages		

Table 21.2. Summary of Selected Case Histories of Embankment Dams (*Continued*)

Project	Principal Concerns	Type of Dam	Measurement[a]	Instrumentation Discussed[b]	Special Features	Reference
Nurek Dam, USSR	Performance during construction and operation	980 ft (300 m) high zoned earthfill	G D D TS	Piezometers Fixed embankment extensometers Inclinometers Contact and embedment earth pressure cells	Height of dam (world's highest)	Sokolov et al. (1985)
W. A. C. Bennett Dam (Portage Mountain Dam), BC, Canada	Performance during construction and operation	600 ft (185 m) high zoned earthfill	LE G D D D S TS	Leakage weirs Open standpipe, twin-tube hydraulic, pneumatic, and vibrating wire piezometers Surveying methods Probe extensometers (V) Inclinometers (V) Embedment strain gages Embedment earth pressure cells	16 years of performance records	Taylor et al. (1985) Taylor and Chow (1976)
Bath County Hydroelectric Pumped-Storage Project, Virginia, U.S.A.	Performance during construction and operation	460 ft (140 m) and 150 ft (45 m) high zoned earth and rockfill	G G	Pneumatic piezometers Dual pneumatic and vibrating wire piezometers	Dual piezometers	Kleiner (1983)
Svartevann Dam, Norway	Performance during construction and operation: height of dam substantially exceeded other dams in Norway	425 ft (130 m) high rockfill with central earth core	LE G D D D D TS	Leakage weirs Vibrating wire piezometers Surveying methods Probe extensometers (I,H) Fixed embankment extensometers Inclinometers (I,H) Embedment earth pressure cells		DiBiagio et al. (1982)
Cethana Dam, Tasmania, Australia	Performance during construction and operation	360 ft (110 m) high concrete face rockfill	LE D D D	Leakage weirs Surveying methods Liquid level gages Crack gages	Inclinometer casings attached to concrete face	Fitzpatrick et al. (1982)

Table 21.2. Summary of Selected Case Histories of Embankment Dams (*Continued*)

Project	Principal Concerns	Type of Dam	Measurement[a]	Instrumentation Discussed[b]	Special Features	Reference
			D	Inclinometers (I)		
			S	Embedment strain gages		
Various dams in Britain	Long-term performance	Various earthfill dams	G	Twin-tube hydraulic piezometers	Emphasis on avoiding vertical pipes and tubes	Penman and Kennard (1981)
			D	Probe extensometers (H)		
			D	Liquid level gages		
			D	Fixed embankment extensometers		
			TS	Embedment earth pressure cells		
Various dams in western Canada	Various	Various earthfill dams	LE	Leakage weirs	Discussion of needs for limited and comprehensive monitoring programs	Peters and Long (1981)
			G	Piezometers		
			D	Surveying methods		
			D	Probe extensometers (V)		
			D	Inclinometers (V)		
El Infiernillo Dam, Mexico	Long-term performance	480 ft (145 m) high rockfill with central earth core	G	Open standpipe piezometers	18 years of performance records	Rosati and Esquivel (1981)
			D	Surveying methods		
			D	Probe extensometers (V)		
			D	Fixed embankment extensometers		
			D	Inclinometers (V)		
			SE	Strong motion accelerographs		
Tarbela Dam, Pakistan	Performance during construction and operation	460 ft (140 m) high zoned earth and rockfill	LE	Leakage weirs and flumes		Szalay and Marino (1981)
			G	Open standpipe, twin-tube hydraulic, pneumatic, and vibrating wire piezometers		
			D	Surveying methods		
			D	Single-point and full-profile liquid level gages		
			D	Fixed embankment extensometers		

Table 21.2. Summary of Selected Case Histories of Embankment Dams (*Continued*)

Project	Principal Concerns	Type of Dam	Measurement[a]	Instrumentation Discussed[b]	Special Features	Reference
			D	Tiltmeters		
			TS	Embedment earth pressure cells		
			T	Thermometers		
			SE	Strong motion accelerographs		
Mica Dam, BC, Canada	Cracking Control of seepage	800 ft (245 m) high zoned earthfill	LE	Leakage weirs	Height of dam	Khilnani and Webster (1976) Webster (1970)
			G	Observation wells		
			G	Piezometers		
			D	Surveying methods		
			D	Probe extensometers (V)		
			D	Fixed embankment extensometers		
			D	Inclinometers (V)		
			TS	Embedment earth pressure cells		
Ilha Solteira Project, Brazil	Performance during construction and operation	230 ft (70 m) high earth and rockfill	G	Open standpipe, pneumatic, and vibrating wire piezometers	Water loss, caves, and cracks resulting from inferior compaction around vertical pipes	Vargas and Hsu (1975)
			D	Probe extensometers (V)		
			D	Inclinometers (V)		
			TS	Embedment earth pressure cells		
Oroville Dam, California, U.S.A.	Performance during construction and operation Comparison between measured and predicted data	770 ft (235 m) high zoned earthfill	LE	Leakage weirs		O'Rourke (1974) Kulhawy and Duncan (1972)
			G	Twin-tube hydraulic piezometers		
			D	Probe extensometers (V)		
			D	Liquid level gages		
			D	Fixed embankment extensometers		
			S	Embedment strain gages		
			TS	Embedment and contact earth pressure cells		
			SE	Strong motion accelerographs		

Table 21.2. Summary of Selected Case Histories of Embankment Dams (*Continued*)

Project	Principal Concerns	Type of Dam	Measurement[a]	Instrumentation Discussed[b]	Special Features	Reference
Various	Various	Various earth and rockfill dams	D	Surveying methods		Wilson (1973)
			D	Probe extensometers (V)		
			D	Liquid level gages		
			D	Fixed embankment extensometers		
			D	Inclinometers (V)		
Briones Dam, California, U.S.A.	Performance during construction and operation	280 ft (85 m) high zoned earthfill	LE	Leakage weirs	Sloping clay core placed directly on upstream face	Anton and Dayton (1972)
			G	Open standpipe, twin-tube hydraulic, and pneumatic piezometers		
			D	Surveying methods		
			D	Probe extensometers (V)		
Galisteo Dam, New Mexico, U.S.A.	Large lateral deformations during construction	160 ft (50 m) high zoned earthfill	D	Surveying methods		Catanach and McDaniel (1972)
			D	Inclinometers (V)		
			D	Tiltmeters		
New Don Pedro Dam, California, U.S.A.	Performance during construction and operation	585 ft (180 m) high zoned earth and rockfill	G	Pneumatic piezometers	No pipes, tubes, or cables crossing interfaces between zones	Taylor (1968)
			D	Surveying methods		
			D	Probe extensometers (V)		
			D	Liquid level gages		
			D	Fixed embankment extensometers		
			D	Inclinometers (V)		
Mangla Dam, Pakistan	Performance during construction and operation	450 ft (135 m) high zoned earthfill	G	Open standpipe, twin-tube hydraulic, and vibrating wire piezometers		Binnie et al. (1967)
			D	Probe extensometers (V)		
			D	Inclinometers (V)		
			D	Plumb lines		
			D	Inverted pendulums		
			TS	Embedment earth pressure cells		

[a]D: deformation　LE: leakage　SE: seismic event monitoring
G: groundwater pressure　S: strain in structural members　T: temperature
TS: total stress in soil or at contact with structure.

[b]H: horizontal; I: inclined; V: vertical.

CHAPTER 22

EXCAVATED AND NATURAL SLOPES*

22.1. GENERAL ROLE OF INSTRUMENTATION

Analysis of slope stability is the principal geotechnical design task for temporary or permanent excavated slopes. Factors influencing stability include stratigraphy, groundwater levels, seepage gradients, strength of the soil or rock mass, geometry, and driving forces.

Stability of slopes in soil is controlled by the ratio between the available shearing resistance along a potential failure surface and the shear stress on the surface. Circular or wedge-shaped surfaces are often used in analyses that attempt to model actual conditions. Available strength includes cohesion and frictional components; for long-term considerations, the contribution of cohesion is often reduced significantly.

Stability of slopes in rock is usually controlled by the presence of discontinuities in the rock mass and the presence of water under pressure in these discontinuities. Failures most frequently occur as a result of sliding or separation along discontinuities.

Primary mechanisms that control the behavior of excavated slopes are outlined in Chapter 2. It is imperative that, prior to planning an instrumentation program for a slope in soil or rock, an engineer first develop one or more working hypotheses for a potential behavior mechanism. The hypotheses must be based on a comprehensive knowledge of the locations and properties of discontinuities.

*Written with the assistance of Robin B. Dill, Senior Engineer, and Douglas G. Gifford, Associate and Vice President, Haley & Aldrich, Inc., Cambridge, MA.

Figures 22.1 and 22.2 show typical excavated slopes in clay and in rock. The figures convey the contrast between how the project will usually appear on engineering drawings and what may actually happen in the field. A discussion of the role of instrumentation for slopes in soil is given by Wilson and Mikkelsen (1978) and for slopes in rock by Franklin and Denton (1973), Hoek and Bray (1981), and Patton (1983).

Instrumentation can be used to define the groundwater regime prior to excavating a slope. Results of measurements during excavation can be used as a basis for modification of the designed slope angle. Measurements of ground movement and groundwater pressure can assist in documenting whether or not performance during and after excavation is in accordance with predicted behavior. Measurements can also be used to document whether short- and long-term surface and/or subsurface drainage measures are performing effectively. If evidence of instability appears during or after construction, instrumentation plays a role in defining the characteristics of the instability, thus permitting selection of an appropriate remedy.

Instrumentation programs for natural slopes are essentially the same as for excavated slopes.

22.2. PRINCIPAL GEOTECHNICAL QUESTIONS

The following principal questions are presented in the normal order of occurrence for an excavated or natural slope. The order does not reflect a rating of importance.

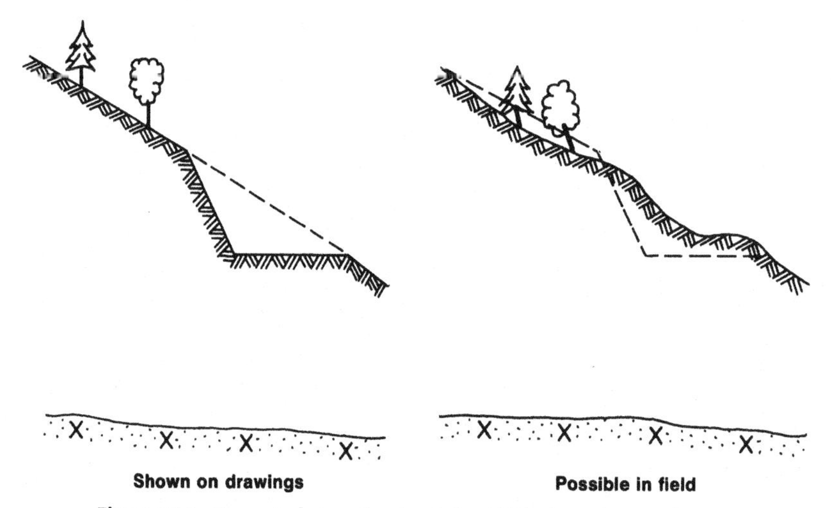

Figure 22.1. Excavated slope in clay (after DiBiagio and Myrvoll, 1981).

22.2.1. What Are the Initial Site Conditions?

Initial site conditions are determined by use of conventional site investigation procedures, sometimes supplemented by in situ testing. Special attention should be given to defining possible failure mechanisms. If potential failure zones are identified, the need for reinforcement or other methods of slope stabilization can then be addressed. For slopes in rock, comprehensive structural geologic mapping will indicate critical discontinuities. Particular attention should be given to persistent, adversely oriented joint sets, to possible low strength zones, and to continuous features such as faults and shears at the top of the slope that could allow release of potentially unstable blocks and wedges.

Performance monitoring instrumentation sometimes plays a role in defining initial site conditions. For example, groundwater pressures can have a large impact on the stability of slopes. Piezometers can be installed well before the start of excavation, to define the preconstruction groundwater pressure regime, including any perched or artesian water. If piezometers are installed sufficiently early, seasonal variations can be defined.

If there is evidence of instability prior to excavation of the slope, such as an old landslide, the available data should be analyzed to identify potential failure mechanisms. An instrumentation program can then be planned to test hypotheses and to determine whether adverse conditions are present. Methods of instrumentation for this purpose are described in the following sections.

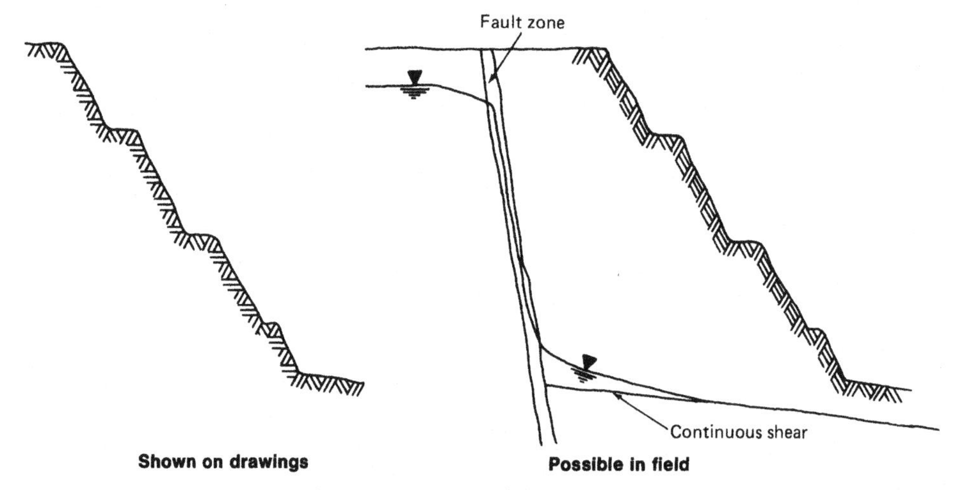

Figure 22.2. Excavated slope in rock.

22.2.2. Is the Slope Stable During Excavation?

A program to monitor stability during excavation is not usually required if the design is very conservative, if there is previous experience with design and construction of similar facilities under similar conditions, or if the consequences of poor performance will not be severe. However, under other circumstances a monitoring program will normally be required to demonstrate that the excavation is stable and that nearby structures are not affected adversely.

Deformation and groundwater pressure are the primary parameters that assist in the evaluation of stability during excavation. Deformation measurements are usually the primary interest but, because high groundwater pressures can cause deformation, groundwater pressure measurements are also often needed so that cause and effect relationships can be established.

Surface Monitoring of Deformation

Deformation monitoring will often be limited to surface measurements. Vertical and horizontal deformations are normally monitored by surveying methods, with electronic distance measurement playing a significant role. Surface measurements should extend beyond the uppermost limit of any possible movement zone to an area that is known to be stable, so that possible surface strain in advance of cracking can be monitored. Any toe heave should also be monitored. Tension cracks at the crest of the slope may be the first sign of instability. If cracks appear at the crest of the slope or elsewhere, their widths and vertical offsets should be monitored.

Crack measurements give clues to the behavior of the entire slope, and often the direction of movement may be inferred from the pattern of cracking, particularly by the matching of the irregular edges of the cracks.

Acoustic emission techniques can sometimes be used by experienced personnel over a wide area in shallow drillholes to determine deformation trends and locations.

For slopes in rock, monitoring the tilt of critical blocks can provide an assessment of stability if the deformation has a rotational component. Tiltmeters with electrolytic level transducers provide the most precise data, and the high precision allows trends to be determined in a minimum time period. Multipoint liquid level gages have been installed on benches of large excavations in rock where there is concern for a wedge failure (e.g., Stepanov, 1983): the instruments are used to monitor vertical deformation and are intended to provide a forewarning of any instability.

Subsurface Monitoring of Deformation

Subsurface deformation measurements will be required if sliding occurs and if the depth of sliding is not readily apparent from surface measurements and visual observations. Measurements of subsurface horizontal deformation are more important than measurements of subsurface vertical deformation.

For slopes in soil, inclinometers are the instruments of choice, although shear plane indicators can be used for crude measurements, and slope extensometers may be preferred if deformation is predicted to occur along thin shear zones.

Critical movements of slopes in rock are often smaller than critical movements of slopes in soil, and therefore the required accuracy of deformation measurements is generally greater. Fixed borehole extensometers, installed from the face of the slope following excavation of a rock bench, may therefore be selected for monitoring subsurface horizontal deformation of slopes in rock in preference to inclinometers.

Multiple deflectometers and in-place inclinometers can provide real-time monitoring of subsurface deformation, and these instruments can be connected to alarms if required.

Monitoring of Groundwater Pressure

Open standpipe piezometers are normally selected for slopes in soil, but diaphragm piezometers are appropriate if more rapid response is required. If any sliding is occurring, pore water pressures at or near the sliding surface must be measured to enable an effective stress analysis to be performed.

For slopes in rock, the heterogeneous nature of most rock masses results in a need for comprehensive monitoring of joint water pressure along, above, and below each possible failure plane. Patton (1983) presents a strong case for use of the movable probe type of multipoint piezometer, because it (1) provides a large number of measuring points, (2) does not create nonconformance, (3) allows the transducer to be calibrated at any time, (4) provides redundancy in field data, and (5) reduces the problems of creating multiple seals when more than one conventional piezometer is installed in a single borehole. Patton comments:

> In my opinion most existing field piezometer installations for [rock] slope stability investigations are deficient in the number of piezometers (probably by a factor of 5 to 10) unless the geology and hydrology of the slope are very simple. However, simple geology and hydrology cannot be demonstrated without a significant number of drillholes to document the geologic and piezometric data.

The Westbay Instruments Ltd. combined piezometer–inclinometer system provides a comprehensive profile of both joint water pressure and horizontal deformation in a single borehole.

If stability has been increased by improving groundwater drainage before the start of excavation, effectiveness of the drainage will usually be monitored by measuring groundwater pressure.

Predicting Stability of Completed Slope

Measurements during excavation can be used to assess the stability of the completed slope. Back-computations are made to examine the validity of parameters used in the original design. On the basis of these back-computations, new design calculations are made to predict the performance of the final slope, and slope angles are modified if necessary. If this approach is taken, clearly the calculations must be made in a timely manner, and a microcomputer will normally be used.

Table 22.1. Instruments Suitable for Examining Slope Stability During Excavation

Measurement	Suitable Instruments
Surface deformation	Surveying methods Crack gages Tiltmeters Multipoint liquid level gages
Subsurface deformation	Inclinometers Fixed borehole extensometers Slope extensometers Shear plane indicators Multiple deflectometers In-place inclinometers Combined piezometer–inclinometer system Acoustic emission monitoring
Groundwater pressure	Single piezometers Multipoint piezometers Combined piezometer–inclinometer system

Summary of Possible Instrumentation

In summary, possible instrumentation for examining slope stability during excavation can be selected from the list given in Table 22.1.

To the degree possible, key instruments should be installed and initial measurements taken before excavation starts. Additional instruments can be installed as excavation progresses.

22.2.3. How Much Ground Is Moving?

If there is evidence of instability during excavation or after completion of the slope, its characteristics must be defined so that any necessary remedial measures may be taken. The question *how much ground is moving?* can be answered by use of instrumentation. The question *why is the ground moving?* will not be answered by instrumentation alone: the answer of course also requires a complete geotechnical investigation and analysis.

Methods of instrumentation have been described in Section 22.2.2. Possible layouts for determining how much ground is moving in excavated slopes are shown in Figures 22.3 and 22.4. It must be stressed

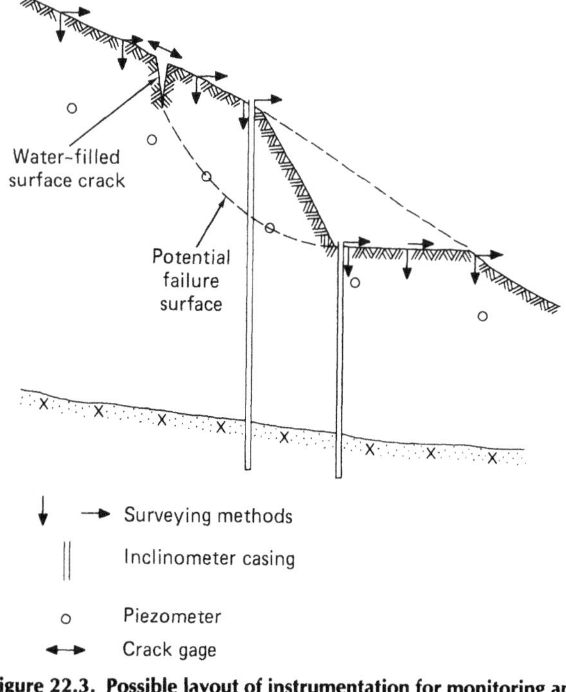

Figure 22.3. Possible layout of instrumentation for monitoring an excavated slope in soil when there is evidence of instability (after DiBiagio and Myrvoll, 1981). Note: Use of surveying methods requires benchmark and horizontal control station, remote from slope.

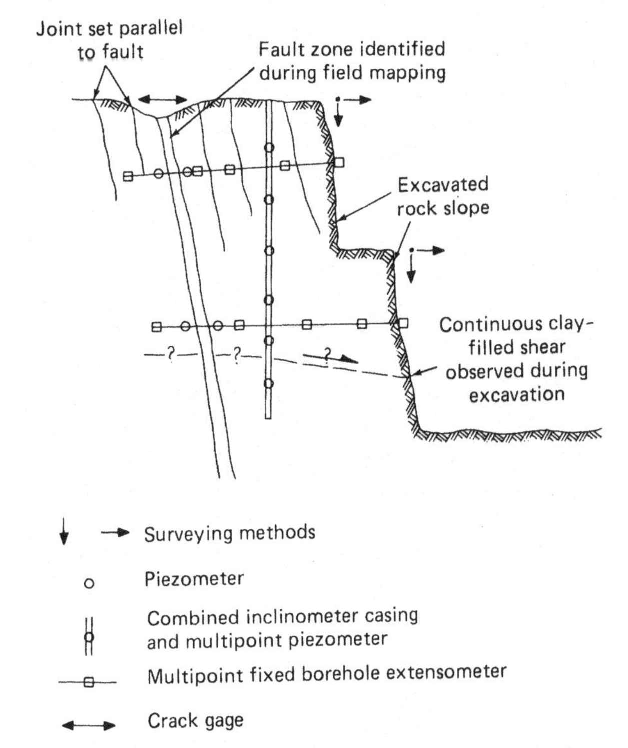

Figure 22.4. Possible layout of instrumentation for monitoring an excavated slope in rock when there is evidence of instability. Note: Use of surveying methods requires benchmark and horizontal control station, remote from slope.

that these layouts are only examples, and many other configurations are possible. Other configurations are indicated by the case histories described later in this chapter.

22.2.4. Is the Slope Stable in the Long Term?

The question applies primarily to excavated slopes that have been unstable during excavation. However, there are cases where the question applies also to excavated slopes with no history of instability, for example, when construction is planned near the toe. The question may also apply to natural slopes.

In general, a choice must be made among:

1. doing nothing and accepting the consequences of slope failure;
2. monitoring to provide a forewarning of instability, so that remedial measures can be implemented before critical situations arise; and
3. stabilizing the slope, perhaps including a monitoring program to verify that stability has been achieved.

The choice will be based on many factors, including the consequences of failure and the economics of stabilizing. When planning a monitoring program for either of the second two options, a two-step approach is recommended: first, measurements of deformation on the surface of the entire slope to indicate the existence of any instability and second, additional measurements at any location shown to be unstable. Measurement methods are similar to those discussed in Section 22.2.2, and Chapter 15 provides guidelines on maximizing reliability of long-term instrumentation.

For the first step, regular visual observations play an important role. Electronic distance measurement is the primary method for monitoring any deformation on the surface of the entire slope, and crack gages are used to monitor any deformation at existing cracks. If necessary, crack gages can be connected to alarms, set to trigger after a predetermined deformation has occurred. The crack gage shown in Figure 12.14 is particularly useful. Acoustic emission techniques and tilt monitoring can sometimes be helpful when assessing stability. If stability has been increased by improving groundwater drainage, long-term effectiveness of the drainage will usually be monitored as part of the first step, by measuring groundwater pressure.

If data gathered during the first step indicate an unstable area, the same surface observations and measurements will be concentrated in that area during a second step, to define the instability more closely. The depth of sliding may become readily apparent, and it may therefore be possible to plan and implement remedial measures without the need for subsurface instrumentation. When the depth and thickness of the zone of sliding are not apparent, measurements of subsurface horizontal deformation and groundwater pressure will normally be required.

Two additional factors may cause instability in the long term and may merit a monitoring program. First, the wedging action of ice can cause serious instability, and freezing at the surface can cause ice dams to form, which hold unfrozen groundwater inside the ground. The buildup of internal water pressure, unless relieved by drainage, can lead to slope failure. The same effect is observed when sunlight acts on the upper slopes of deep valleys, melting the ice and snow at upper levels, which then percolate downward and exert pressure against the still frozen ground in the lower, more shaded regions. To combat this effect, drainage of the slope may be required, and drains must be prevented from freezing. Temperature measurements will show the depth and extent of any freezing problems.

Second, loss of tension in rockbolts can cause instability. If rockbolts have been installed for slope reinforcement, the opportunity arises to measure load in the rockbolts, and these measurements may be useful for verifying assumed design loads. However, recognizing that the purpose of rockbolts is to restrain deformation, it is usually more effective to monitor deformation of the rock by using fixed borehole extensometers rather than to monitor stress in rockbolts. A typical approach is use of single-point fixed borehole extensometers, with rods and expansion shell anchors, anchored deeper than the rockbolts. If measurements of load in the rockbolts are also required for more comprehensive monitoring, load cells or strain gages can be used on end-anchored rockbolts but strain gages are the only option for use on fully grouted rockbolts.

Table 22.2. Overview of Routine and Special Monitoring

Application	Measurement
Routine monitoring	Surface deformation Groundwater pressure
Special applications	Subsurface deformation Load in rockbolts Temperature

22.3. OVERVIEW OF ROUTINE AND SPECIAL APPLICATIONS

For those readers who seek general guidelines on what is typically *routine* monitoring and what is monitoring for *special applications,* Table 22.2 presents a broad generalization.

In practice, the most frequent use of instrumentation for excavated and natural slopes is to investigate the characteristics of a slope that is observed to be moving.

22.4. SELECTED CASE HISTORIES

Table 22.3 summarizes selected case histories of instrumented excavated and natural slopes.

Table 22.3. Summary of Selected Case Histories of Excavated and Natural Slopes

Project	Principal Concerns	Type of Slope	Measurement[a]	Instrumentation Discussed	Special Features	Reference
New Liskeard College of Agriculture and Technology, New Liskeard, Ontario, Canada	Stability of slope after installation of horizontal drains	Varved clay	G	Vibrating wire piezometers		Lau and Kenney (1984)
Riverbank at Grand Coulee Dam, Washington, U.S.A.	Stability of slope during large fluctuations in river level	Granular alluvium, varved silts, and clays	G	Piezometers	Shafts and radial drains installed to reduce groundwater levels	Von Thun (1984)
Riverbank on Big Sunflower River, Clarksdale, MS, U.S.A.	Stability of slope, after stabilization with gravel trenches	Highly plastic clay	D	Inclinometers		Wardlaw et al. (1984)
Two sites in southern England	Rate of landside movement Effectiveness of trench drain remedy	Colluvium and clay	D D D G	Surveying methods Shear plane indicators Inclinometers Open standpipe piezometers		Barton and Coles (1983)

Table 22.3. Summary of Selected Case Histories of Excavated and Natural Slopes (*Continued*)

Project	Principal Concerns	Type of Slope	Measurement[a]	Instrumentation Discussed	Special Features	Reference
Vaiont Slide, Italy Pillar Mountain Slide, Alaska Downie Slide, Canada	Cause, physical characteristics, activity status of slides Planning remedial measures	Rock	D	Surveying methods		Patton (1983)
			D	Crack gages		
			D	Inclinometers		
			D	In-place inclinometers		
			D	Fixed borehole extensometers		
			D	Acoustic emission monitoring		
			G	Open standpipe piezometers		
			G,D	Combined multipoint piezometers and inclinometers		
Various hydroelectric power plants, USSR	Provide forewarning of failure	Rock	D	Tiltmeters	Use of multipoint liquid level gages to provide forewarning of failure	Stepanov (1983)
			D	Fixed borehole extensometers		
			D	Multipoint liquid level gages		
Downie Slide, Revelstoke, BC, Canada	Physical characteristics and activity status of slide	Rock	D	Surveying methods		Piteau et al. (1978)
			D	Inclinometers		
			D	Acoustic emission monitoring		
Saxon pit, Whittlesey, England	Stability of brickpit	Overconsolidated clay	D	Surveying methods		Burland et al. (1977)
			D	Mechanical crack gages		
			D	Subsurface settlement points		
			D	Inclinometers		
			D	Probe extensometers		
			D	Fixed borehole extensometers		
			G	Open standpipe piezometers		
Mine in western Canada Reservoir area of a dam	Stabilility of slopes	Rock	D	Electrical crack gages	Data transmitted to a remote location, using telemetry system	Weir-Jones and Bumala (1975)
			D	Tiltmeters		
			D	Inclinometers		
			D	In-place inclinometers		
			D	Fixed borehole extensometers		
			D	Acoustic emission monitoring		
			G	Vibrating wire and pneumatic piezometers		

Table 22.3. Summary of Selected Case Histories of Excavated and Natural Slopes (*Continued*)

Project	Principal Concerns	Type of Slope	Measurement[a]	Instrumentation Discussed	Special Features	Reference
Five open pit mines	Stability of slopes	Rock	D	Surveying methods		Brawner (1974)
			D	Mechanical crack gages		
			D	Inclinometers		
			D	Fixed borehole extensometers		
			G	Piezometers		
Cabin Creek Hydroelectric Project, Colorado, U.S.A.	Stability of slope	Rock	D	Fixed borehole extensometers		Dutro (1974)
Eleven sites in Canada	Stability of natural slopes	Champlain Sea clays	D	Inclinometers	Analyses of failures in natural slopes	Lo and Lee (1974)
			G	Piezometers		
Fountain Slide, Oregon, U.S.A.	Stability of slope	Rock	D	Inclinometers		Muñoz (1974)
			G	Pneumatic piezometers		
I40, Tennessee, U.S.A. I26, North Carolina, U.S.A.	Stability of slopes	Rock	D	Surveying methods		Tice and Sams (1974)
			D	Inclinometers		
Minneapolis Freeway, Minneapolis, MN, U.S.A.	Cause and physical characteristics of slide Planning remedial measures	Rock	D	Inclinometers	Slope failure along bentonite seam	Wilson (1974)
			G	Open standpipe piezometers		
Seattle Freeway, Seattle, WA, U.S.A.	Stability of eight slopes	Overconsolidated clay	D	Surveying methods		Palladino and Peck (1972)
			D	Inclinometers		
			G	Observation wells		
			G	Piezometers		
Chuquicamata Mine, Chile	Prediction of failure	Rock	D	Crack gages		Kennedy et al. (1971)
			D	Acoustic emission monitoring		
Potrero Hill Slide, San Francisco, CA, U.S.A.	Stability of slope	Rock	D	Surveying methods	Tunnel below toe of slope	Smith and Forsyth (1971)
			D	Inclinometers		
			D	Fixed borehole extensometers		
			D	Acoustic emission monitoring		
Tripp-Veteran Pitt, Ruth, NV, U.S.A.	Stability of slope	Rock	D	Mechanical crack gages		Stateham and Vanderpool (1971)
			D	Acoustic emission monitoring		

Table 22.3. Summary of Selected Case Histories of Excavated and Natural Slopes (*Continued*)

Project	Principal Concerns	Type of Slope	Measurement[a]	Instrumentation Discussed	Special Features	Reference
Eight slope failures	Various	Soil and rock	D D G	Surveying methods Inclinometers Open standpipe piezometers		Wilson (1970)
Steel Plant Expansion, Weirton, WV, U.S.A.	Cause and physical characteristics of slide Planning remedial measures	Colluvium	D D G S	Surveying methods Inclinometers Open standpipe piezometers Strain gages on tension ties	Slope stabilized with flexible sheet pile wall and tension ties	D'Appolonia et al. (1967)
Three slides alongside railroads in Japan	Prediction of failure	Soil and rock	D D	Surveying methods Crack gages	Recommendations for forecasting failure	Saito (1965)
Portugese Bend Landslide, Palos Verdes, CA, U.S.A.	Stability of slope	Rock	D D D	Surveying methods Crack gages Tiltmeters	Unsuccessful attempt to control sliding by installing caissons across sliding surface	Merriam (1960)

[a]D: deformation; G: groundwater pressure; S: strain in structural member.

CHAPTER 23

UNDERGROUND EXCAVATIONS*

Underground excavations include soft ground and rock tunnels, large caverns such as underground powerhouses, subway stations, underground repositories for nuclear waste, and mines.

Figures 23.1 and 23.2 show a typical soft ground tunnel and a typical tunnel in rock supported by rockbolts. The figures convey the contrast between how the projects will usually appear on engineering drawings, and what may actually happen in the field. The *field* figures are exaggerated to emphasize the contrast. Primary mechanisms that control the behavior of underground excavations are described in Chapter 2.

23.1. GENERAL ROLE OF INSTRUMENTATION

Analysis of stability is a principal geotechnical design task for underground excavations. Factors influencing stability include stratigraphy, groundwater levels, strength of the soil or rock mass, geometry, excavation method, type of support, sequence of excavation and support, and method of support installation.

Stability of underground excavations in rock is usually controlled by the presence and orientation of discontinuities in the rock mass and the presence of water under pressure in these discontinuities. Failures most frequently occur as a result of sliding or separation along discontinuities.

Stability of soft ground tunnels in soft clay is usually controlled by the undrained shear strength of the soil in relation to the total overburden stress at the depth of the tunnel. In silts and fine sands, stability is usually controlled by the effectiveness of construction dewatering or compressed air in controlling the inflow of groundwater. In desiccated hard clays stability is usually controlled by the presence of fissures and slickensides.

The consequence of poor performance of an underground excavation can be severe and may on occasion be catastrophic. A monitoring program may not be required if the design is very conservative, if there is previous experience with design and construction of similar facilities under similar conditions, or if the consequences of poor performance will not be severe. However, under other circumstances a monitoring program will normally be required to demonstrate that the excavation is stable and that nearby structures are not affected adversely.

As with all monitoring programs, planning of a monitoring program for an underground excavation should follow the systematic approach described in Chapter 4. Of particular importance is development of one or more working hypotheses for a mechanism that controls behavior, based on a comprehensive knowledge of the properties of the soil or intact rock material, and/or the locations, frequency, orientation, and properties of discontinuities. To

*Written with the assistance of Bruce E. Beverly, Associate & Vice President, Haley & Aldrich, Inc., Cambridge, MA.

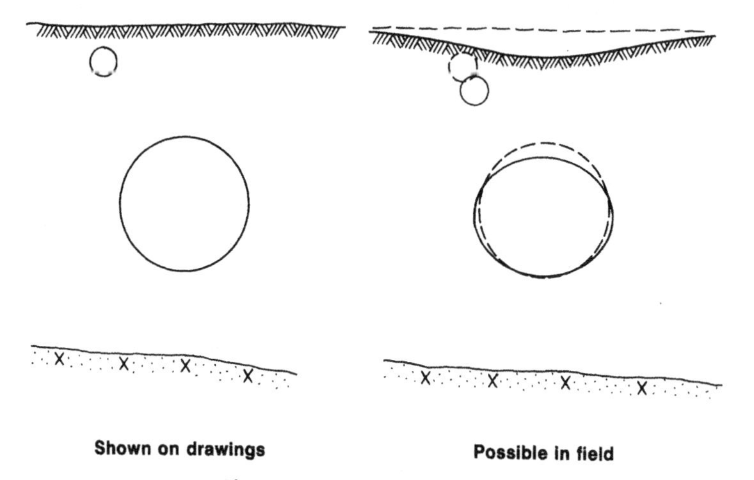

Figure 23.1. Soft ground tunnel.

the extent possible, planning of the program should also be based on construction experience with similar excavations in the same geologic formation, the complexity of the excavation geometry, the amount of rock or soil overburden, the location of the project with respect to structures and utilities, and the method of contracting for construction work.

Full-scale instrumented test sections can be very beneficial to underground excavation projects by confirming economical and safe construction procedures, and potentially large cost savings can be realized. For example, the understanding of the nature of earth pressures on tunnel liners has increased as a result of data obtained from instrumentation of soft ground tunnels during the past decade. Notable among these are U.S. subway tunnel projects in San Francisco, Washington, and Baltimore. Updating of the state of the art will continually occur from the accumulation of knowledge gained from these and similar future studies, thereby permitting better design and more efficient construction.

A program of instrumentation and observation can play a major role in evaluating stability of an underground excavation. When an excavation is **not** stable, underground safety and integrity of nearby structures become a major concern. When an excavation is marginally stable, early detection of instability is possible through use of instrumentation and observations such that supplemental support work can be implemented on a timely basis.

When the underground excavation encounters a known or unexpected major geologic feature such as a fault, shear zone, or a highly jointed or weathered rock zone, instrumentation can be used to monitor behavior, thereby confirming that the implemented construction methods address the major

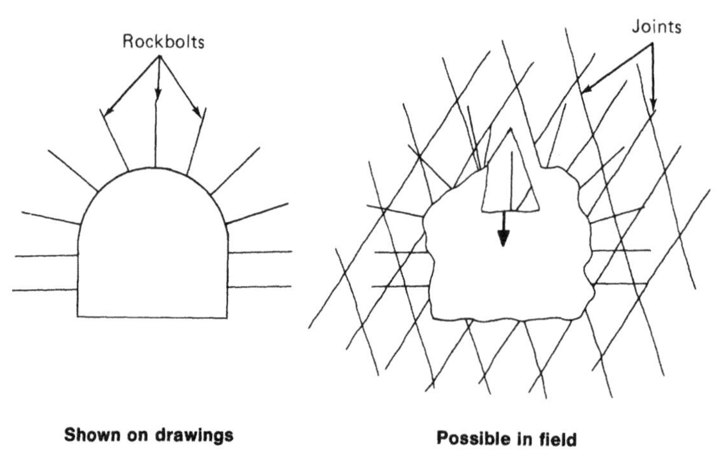

Figure 23.2. Tunnel in rock, supported by rockbolts.

geologic feature such that work proceeds safely and economically. Instrumentation can play an important and critical role where two or more excavations intersect, by confirming that the chosen excavation and support methods and sequence result in stable openings.

Where pillars in mines or other underground excavations are designed to carry high loads, and where integrity of pillars is crucial to stability, instrumentation can be used to monitor loads and deformations, thereby allowing a rational judgment of stability.

An instrumentation program can provide performance documentation for use in evaluating claims from the construction contractor associated with support requirements or "differing site conditions." A program can also be planned to obtain facts for use in legal proceedings if third parties claim that construction work has caused damage to their property. The public can obtain reassurance from a planned instrumentation program, which indicates that the construction will be carefully watched, and this can expedite the legal approval and public acceptance of a project.

When new construction techniques are conceived, their implementation is difficult to justify without monitoring the results to permit an adequate assessment. As an example, instrumentation was used to evaluate the effectiveness of compaction grouting to reduce building settlement caused by tunneling for the Baltimore subway. This gave confidence in the method, which permitted elimination of costly underpinning of many buildings in later parts of the tunnel construction.

More detailed discussions on the role of instrumentation for monitoring underground excavations are given by ASCE (1984), Cording et al. (1975), Dunnicliff et al. (1981), FHWA (1980), Lane (1975), and Schmidt (1976). A keyworded bibliography of about 300 publications describing instrumentation methods and applications for underground excavations has been prepared by FHWA (1980). Keyword groups include ground type, support method, excavation method, reason for monitoring, and monitored parameters.

23.2. PRINCIPAL GEOTECHNICAL QUESTIONS

The following principal questions are presented in the normal order of occurrence for an underground excavation. The order does not reflect a rating of importance.

23.2.1. What Are the Initial Site Conditions?

Information on anticipated subsurface conditions is determined by use of conventional site investigation procedures, sometimes supplemented by in situ testing. However, performance monitoring instrumentation sometimes plays a role. For example, piezometers play a role in defining the groundwater regime, including any perched or artesian water. If they are installed sufficiently early, seasonal variations can be defined. If highly permeable zones that may be connected to a source of groundwater recharge are anticipated, full-scale pumping tests may be made to assess dewatering requirements and methods, and these tests will include use of piezometers. Bonded resistance strain gage piezometer systems, with multichannel recorders, are available for this purpose on a rental basis from several manufacturers of geotechnical instruments.

When a shallow underground excavation is to be made near existing structures, a *preconstruction conditions survey* will normally be undertaken by representatives of the construction contractor's insurance company and sometimes also by the owner. The purpose of the preconstruction survey is to observe and document the conditions of structures that may be influenced by the excavation, using photographs, written descriptions, and/or videotapes of existing defects in the structures. The preconstruction conditions survey is frequently supplemented by use of surveying methods to define the elevations of reference points on the structures and on the ground surface. The widths of any existing cracks may also be measured.

Pilot tunnels or exploratory adits are sometimes excavated prior to construction of full-scale rock tunnels and large caverns such as underground powerhouses and subway stations. The primary purposes of these exploratory excavations are to study geologic conditions at excavation depth and to assess various excavation and support methods. In situ tests are sometimes made in holes drilled from the ground surface and also within the exploratory excavations to determine ground permeability and grouting requirements, rock mass deformation properties, and in situ stress conditions.

23.2.2. If an Exploratory Excavation Is Made, Can Field Measurements Assist with the Design of the Prototype?

Although the primary purposes of exploratory excavations are to study geologic conditions at excavation depth and to assess various excavation and support methods, their construction provides an opportunity to use geotechnical field instrumentation. For example, when squeezing ground conditions are anticipated, measurements of ground deformations can provide useful information for design of the prototype. However, attempts to extrapolate ground behavior from instrumentation measurements within exploratory excavations are often hampered severely by scale effects. Extrapolations to the scale of the prototype may be unwarranted unless the size of the exploratory excavation is large compared with the spacing between discontinuities. If the size is large enough, measurements of ground deformation and support stress may provide valuable input to the design of the prototype, although local geologic factors and construction details have a large influence on measurements. Extrapolation of the data must always be tempered by engineering judgment and experience.

When squeezing ground conditions are monitored in an exploratory excavation, instrumentation normally consists of surveying methods, convergence gages, and single-point fixed borehole extensometers with rods and mechanical readout. When support stresses are monitored, strain gages and/or load cells are used, of the types recommended below for full-scale test sections.

23.2.3. What Information Can Be Provided by Instrumenting a Full-Scale Test Section?

The design of underground excavations usually requires the selection of the most economical excavation and support method. The primary goals when designing support are to control ground movements and to establish a stable excavation. Ground movements and support stress are of course interdependent, and both are affected by the method and sequence of excavation and support.

The designers may identify a combination of circumstances that lead them to opt for a full-scale test section, as a means of ensuring design adequacy and economy. The combination of the following four circumstances provides an example. First, there is little precedent for excavations under similar conditions. Second, new excavation or support methods are to be used. Third, it appears that cost savings or increased safety may result from increasing site-specific knowledge of the relationships between excavation and support method, sequence of excavation and support, and ground movements and support stresses. Fourth, scale effects severely hamper the use of exploratory excavations for developing design parameters for the prototype.

Most full-scale test sections are initiated by the designers of the excavation, but Fox (1980) describes three cases in which the construction contractor initiated a full-scale instrumented test in a soft ground tunnel. First, the contractor needed to determine how much reinforcement was needed to a light steel liner plate so that the shield could be advanced by shoving against the liner. Second, there was a need to assess whether compressed air could be taken off when all liner plates were installed, so that the final concrete liner could be placed in free air rather than in compressed air. Third, the contractor needed to determine at what distance along the tunnel the shield thrust was dissipated in ground friction, so that a jacking frame could be removed from a shaft to improve access.

If both an exploratory excavation and a full-scale test section have been selected, the test section will normally be constructed by enlarging the exploratory excavation. In other cases the test section will be excavated and supported under the same construction contract as the prototype, as early as possible during the construction period. Ground conditions at a test section should be representative of a significant length of the underground excavation. Ideally, the excavation should be made to the same size as the prototype and the length should be adequate so that end effects are minimized, typically requiring a minimum length of about four times the width of the excavation. However, this ideal must be tempered by cost considerations and by knowledge of site and project conditions. The methods and sequence of excavation and support should match, as closely as possible, the likely methods and sequence for the prototype.

The primary parameter monitored in a test section will generally be deformation. Often support stresses are also monitored, and measurements of groundwater pressure may be included. Suitable instruments are listed in Table 23.1, and more details are given in the following subsections.

Deformation of Ground Around Soft Ground Tunnels

Inclinometers and probe extensometers can be installed from the ground surface alongside a soft ground tunnel and sometimes are also installed on the centerline in front of the heading. A probe ex-

Table 23.1. Suitable Instruments for Monitoring a Full-Scale Test Section

Measurement	Suitable Instruments
Deformation of ground surface and structures	Surveying methods Rod settlement gages
Settlement of utilities	Subsurface settlement points
Change in width of cracks in structures	Crack gages
Deformation within excavation	Surveying methods Convergence gages
Deformation of ground around soft ground tunnels	Inclinometers Probe extensometers Fixed borehole extensometers
Deformation of ground around underground excavations in rock	Fixed borehole extensometers Acoustic emission monitoring
Stress in or loading on soft ground tunnel liners	Strain gages Load cells Concrete stress cells Contact earth pressure cells
Stress in or loading on supports for underground excavations in rock	Strain gages Load cells Concrete stress cells
Stress change in rock	Soft inclusion gages Rigid inclusion gages
Groundwater pressure	Piezometers

tensometer and inclinometer casing can be combined in a single borehole. Fixed borehole extensometers, installed from the ground surface or from within the tunnel can provide supplementary deformation data.

Deformation of Ground Around Underground Excavations in Rock

Multipoint fixed borehole extensometers are the usual instruments of choice for monitoring deformations around underground rock excavations. They will normally be installed above the crown and also at appropriate angles from within the test section to monitor rock deformation along critical discontinuities. If the overburden is less than about 200 ft (60 m) deep, extensometers above the crown can be installed from the ground surface in vertical or angled boreholes prior to excavation, using groutable anchors and mechanical transducers. Use of groutable anchors creates a simple installation procedure and avoids possible communication of groundwater into the excavation. Borehole directional surveys will normally be required to determine anchor locations with respect to the excavation, and a colored plastic tube should be attached to the deepest anchor to protrude into the path of the excavation to indicate the location of the borehole.

When the excavation is too deep for installation of extensometers from the ground surface, extensometers can be installed in boreholes drilled upward from the excavation as soon as access becomes available. If a pilot tunnel has been driven at the crown of the future excavation, these extensometers can be installed from the pilot tunnel before production downward excavation starts, thereby avoiding interruption to production excavation. Extensometers installed from within the excavation should initially have mechanical transducers, to minimize the potential for damage while excavation continues nearby. Later they can be converted for reading with electrical transducers, either when access difficulties arise or when a remote reading arrangement is preferred.

Stress in Soft Ground Tunnel Liners

When stress data are required in soft ground tunnel liners, strain gages and/or load cells can be used. Circumferential stresses within steel and precast concrete liners can be determined by using either vibrating wire or electrical resistance strain gages, with a preference for the vibrating wire type. Weldable gages are usually preferred on steel liners, but bonded resistance gages may be appropriate if available space is too limited for weldable gages. The type of vibrating wire gage with two wires at different distances from the liner surface can be used to monitor bending strains by attaching the gage to the inside face only. Embedment gages with vibrating wire transducers are usually preferred in precast concrete liners. Measurements in cast-in-place or shotcrete liners normally require use of concrete stress cells, because if strain gages are used the strain data cannot meaningfully be converted to stress.

Load cells have been interposed between segments of precast concrete liner to measure circumferential loading (e.g., Tattersall et al., 1955), thereby overcoming the difficulties of converting strain in concrete to stress, but care must be taken to prevent the creation of a nontypical liner ring that would result in misleading data. When the total load and distribution of load imposed on the outside of the liner is required, circumferential strain gages can be placed at sufficiently close intervals so that imposed loads can be derived from stresses in the liner (Beloff et al., 1979). Alternatively, contact earth pressure cells with sensitive faces flush with the outside of the liner can be used, but data are very much affected by local contact irregularities, and a large number of cells are required for meaningful measurements. If there is a need to separate water loading from effective soil loading, piezometers can be installed in shallow holes, drilled through grout ports in the lining.

When the shield is advanced by shoving against the previously installed lining, the shoving jacks impose longitudinal stresses in the lining, and these stresses are often more critical than circumferential stresses caused by soil loading. Determination of longitudinal stress requires gages with adequate dynamic response, vibrating wire strain gages may not be suitable, and resistance gages may be required. Options are weldable or bonded gages on steel liners and embedment gages in concrete liners, monitored using a multichannel automatic data acquisition system with appropriate dynamic response. Correlations between longitudinal stress and jacking thrust can be made by including electrical resistance strain gage pressure transducers in the hydraulic lines for the jacks or by placing an electrical resistance strain gage load cell in series with each jack.

Stress in Supports for Underground Excavations in Rock

When stress data for steel support elements are required in underground rock excavations, strain gages and/or load cells can be used. As for soft ground tunnels, measurements in cast-in-place or shotcrete liners normally require use of concrete stress cells. However, although stress data can be useful, they are essentially *point* measurements, subject to variability in geologic or support characteristics, and may therefore not represent average or typical conditions on a larger scale. A large number of measurement points will usually be required before confidence can be placed in the data. In contrast, many deformation measuring devices respond to movements within a large and representative zone; data provided by a single instrument can therefore be meaningful, and deformation measurements are generally the most reliable and least ambiguous.

When load data are required for rockbolts, load cells can be used on end-anchored bolts, but fully grouted bolts require the use of strain gages. Weldable strain gages are normally used, with a preference for the vibrating wire rather than the resistance type.

Stresses in steel sets are normally calculated from strain data, using vibrating wire strain gages, either the arc welded or spot welded type. Load cells have been interposed between adjacent segments of a steel set, but care must be taken to prevent the creation of a nontypical set that would result in misleading data.

Figure 23.3 shows a possible layout of instrumentation in a shallow full-scale test section in rock. This is only one example, and it must be emphasized that many other configurations are possible. Other configurations are indicated by the case histories described later in this chapter.

23.2.4. Is the Excavation Stable?

The general role of instrumentation for stability monitoring has been described in Section 23.1. The types of instruments used are similar to those used in full-scale test sections, as listed in Table 23.1.

Routine Data for Stability Assessment

Measurements of crown settlement and wall-to-wall convergence, using surveying methods and convergence gages within the excavation, provide routine data for stability assessment. As an example, ovalling of the primary lining is a concern when tunneling in soft clay. These linings are generally thin and deform until all the earth load is carried in lining thrust. It is important to know when ovalling has stabilized, such that any secondary lining can be constructed and/or if continued ovalling is leading to overstress and failure of the lining.

When the underground excavation is shallow and the consequences of ground loss are significant, measurements of ground surface settlement will often be made above the excavation at several cross sections. Measurements of unacceptable settlement will indicate the need to alter the excavation or support method and/or sequence.

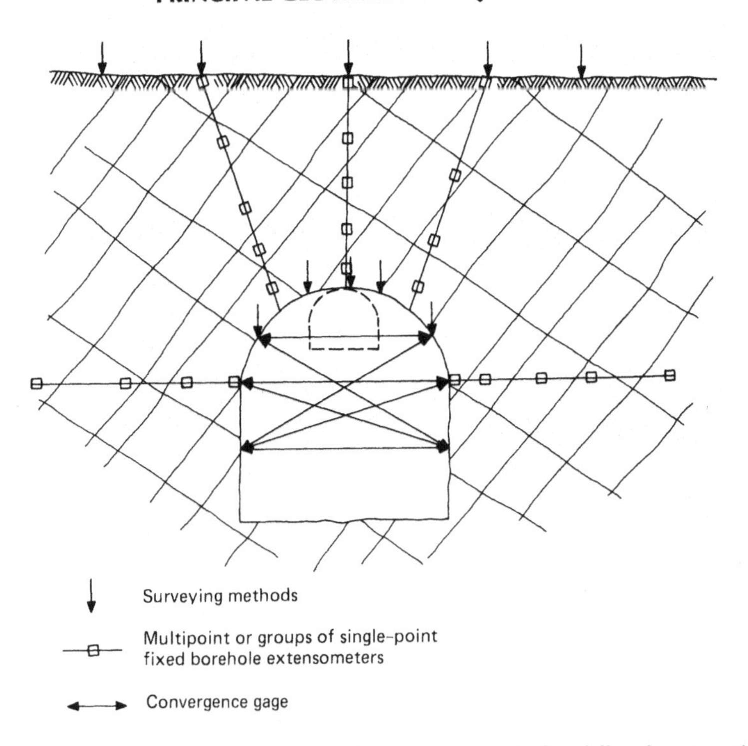

Figure 23.3. Possible layout of instrumentation for monitoring a shallow full-scale test section in rock. Note: Use of surveying methods requires benchmark, remote from excavation. Strain gages and/or load cells may be installed on steel supports, or concrete stress cells in shotcrete support.

Adverse Effects on Utilities and Buildings

Where there is a concern that ground movements may cause an adverse effect on nearby utilities, pavements, and buildings, additional measurements should be made. Instrumentation can provide a forewarning of adverse effects, thereby allowing remedial measures to be implemented before critical situations arise.

Settlement of important utilities can be measured using subsurface settlement points. A rod settlement gage alongside each important structure will provide accurate settlement data without need for a survey crew.

If the monitoring program consists only of surface and shallow subsurface settlement measurements and the measurements indicate adverse behavior, there will often be inadequate data for defining the cause of the problem and for developing an economical solution. A well-planned monitoring program for providing forewarning of adverse effects will therefore include instrumentation that indicates comprehensive causal data. Instruments for this purpose include inclinometers, probe extensometers, fixed borehole extensometers, crack gages, and piezometers. Acoustic emission monitoring can sometimes be used in shallow boreholes to provide an indication of deformation trends and locations.

Stability of Individual Rock Blocks

Concern for stability of individual rock blocks in underground rock excavations or of large zones bounded by adverse discontinuities is addressed using fixed borehole extensometers, convergence gages, and crack gages. For example, during construction of DuPont Circle Station in Washington, DC, in poor-quality rock, fixed borehole extensometer measurements provided essential input to daily decisions on excavation and support methods.

Areas of Special Concern

Instrumentation may be used effectively to monitor stability in areas of special concern, such as the intersections of the excavation and a major geologic feature or at locations where two or more openings intersect or cross in close proximity.

New Austrian Tunneling Method

Instrumentation plays an important role in the *New Austrian Tunneling Method (NATM)*. The method takes maximum advantage of the capacity of the rock to support itself. As the excavation is made, forces in the surrounding rock readjust and are carefully and deliberately controlled by selecting appropriate support (Rabcewicz and Golser, 1973). Support requirements are a minimum when the ground has reacted to the point where loosening of rock is about to occur. Various categories of ground conditions are established prior to construction, together with commensurate types and amounts of support required to stabilize that category of ground. The actual support is jointly decided on at the heading by qualified representatives of the owner and construction contractor, and contracts provide for appropriate payment in accordance with these decisions.

Visual observations and convergence measurements provide primary input to the decisions on actual support. Additional input is often gained by use of fixed borehole extensometers, to provide more comprehensive data on ground movement with distance from the heading.

Comparison Between Measurements and Predictions

In large underground excavations in rock, such as underground powerhouses, measurements are often made to assess stability, by comparing actual rock deformations with predicted values. Instrumentation generally includes multipoint fixed borehole extensometers and convergence gages.

Stability of Underground Mines

Concern for overload of mine pillars is often addressed by monitoring stress change with rigid or soft inclusion gages, supplemented with deformation measurements using fixed borehole extensometers, convergence gages, and crack gages. Bauer (1985) presents a thorough summary of methods in current use for monitoring the stability of underground mines.

Overload of Individual Supports

Concern for overload of individual supports can be addressed by using the types of strain gage and/or load cell described in Section 23.2.3. However, stress data are essentially *point* measurements and, recognizing that the purpose of support is to restrain deformation, it is often more effective and economical to monitor deformation rather than support stress.

Effect of Dewatering

When dewatering has been implemented to predrain the ground around the excavation, its effectiveness may be monitored by using piezometers.

Effectiveness of Innovative Construction Techniques

When an innovative construction technique is being used, instrumentation normally plays a role in monitoring the effectiveness of the technique. Selection of measured parameters and instruments of course depends on the construction technique. As an example, Figure 23.4 shows a possible layout of instrumentation for monitoring the effectiveness of chemical grouting in soft ground, to increase face stability and stand-up time and to decrease permeability.

23.2.5. What Is the Nature and Source of Undesirable Ground Movement?

If there is evidence of instability during excavation, the depth of sliding may be readily apparent from visual observations, and the routine stability monitoring described in Section 23.2.4 is often sufficient for obtaining the rate of movement. When the ex-

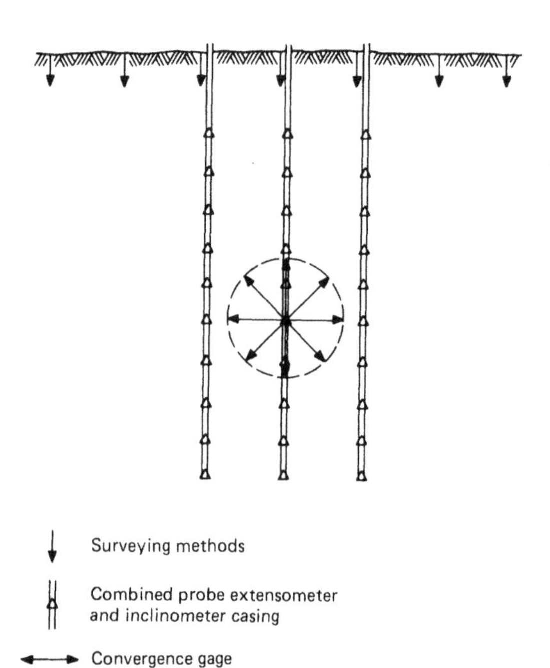

Figure 23.4. Possible layout of instrumentation for monitoring effectiveness of chemical grouting in soft ground (after Clough et al., 1977, 1979). Note: Use of surveying methods requires benchmark, remote from tunnel.

tent of the moving zone is not readily apparent, subsurface deformation measurements should be made so that an appropriate remedy can be planned. The question *how much ground is moving?* can be answered by the use of instrumentation, but the question *why is the ground moving?* will not be answered by instrumentation alone: the answer of course also requires a complete geotechnical investigation and analysis.

For soft ground tunnels, surveying methods, convergence gages, and inclinometers are the primary monitoring tools, supplemented if necessary by probe extensometers and fixed borehole extensometers. Deformation around the tunnel face is three dimensional, and measurements may include the use of inclinometer casings on the tunnel centerline ahead of the face.

For underground excavations in rock, surveying methods, convergence gages, and fixed borehole extensometers are the primary monitoring tools, supplemented as necessary by crack gages and acoustic emission monitoring.

If groundwater pressures are a possible cause of instability, they should be measured with piezometers.

Table 23.2. Overview of Routine and Special Monitoring

Application	Measurement
Routine monitoring	Surface settlement Settlement of structures and utilities Change in width of cracks in structures Convergence within excavation Groundwater pressure
Special applications: design phase	Groundwater pressure during full-scale pumping test Deformation of ground and support, within exploratory excavations Stress in or loading on support, within exploratory excavations Deformation of ground and support, within full-scale test section Stress in or loading on support, within full-scale test section
Special applications: construction phase	Deformation of ground and support, within excavation Stress in or loading on support, within excavation

23.3. OVERVIEW OF ROUTINE AND SPECIAL APPLICATIONS

For those readers who seek general guidelines on what is typically *routine* monitoring and what is monitoring for *special applications,* Table 23.2 presents a broad generalization.

In practice, the most frequent use of instrumentation for underground excavation projects is to determine whether stability of the excavation has been achieved.

23.4. SELECTED CASE HISTORIES

Approximately 300 case histories of instrumented underground excavations are included in a keyworded bibliography by FHWA (1980). Table 23.3 summarizes selected case histories from this and other sources.

Table 23.3. Summary of Selected Case Histories of Underground Excavations

Project	Principal Concerns	Type of Excavation	Measurement[a]	Instrumentation Discussed	Special Features	Reference
Various	Stability	Underground mines	D	Convergence gages		Bauer (1985)
			D	Fixed borehole extensometers		
			SCR	Soft and rigid inclusion stress gages		
Lafayette Park, Washington Metro, Washington, DC, U.S.A.	Improvement of design and construction procedures	Subway tunnel in sand, gravel, and clay	G	Open standpipe piezometers	Full-scale test section	Hansmire and Cording (1985, 1972)
			D	Surveying methods		
			D	Fixed borehole extensometers		
			D	Inclinometers		
Stillwater Tunnel, Utah, U.S.A.	Binding of tunnel boring machine	9.6 ft (2.9 m) diameter tunnel in highly fractured and sheared shales	D	Convergence gages	Deformations measured ahead of heading	Robinson et al. (1985)
			D	Fixed borehole extensometers		
			D	Horizontal inclinometers		
			D	Deflectometers		
			S	Mechanical and vibrating wire strain gages		
Milchbunk Tunnel, Zürich, Switzerland	Effectiveness of ground freezing in limiting ground movements	Highway tunnel in moraine	D	Surveying methods	Measurements led to modification of freezing procedure	Mettier (1983)
			D	Convergence gages		
			D	Fixed borehole extensometers		
			D	Inclinometers		
			T	Temperature sensors		
Demonstration storage cavern for crude oil, Kikuma, Japan	Stability of cavern Hydraulic behavior of rock	Storage cavern for crude oil in granite	G	Electrical resistance piezometers		Miyashita et al. (1983)
			D	Convergence gages		
			D	Fixed borehole extensometers		
			S	Strain gages on rockbolts		
Armco No. 7 Mine, Montcoal, WV, U.S.A.	Stability and subsidence	Longwall coal mine in sandstone, shale, and coal	D	Surveying methods		O'Connor (1983)
			D	Convergence gages		
			D	Fixed borehole extensometers		

Table 23.3. Summary of Selected Case Histories of Underground Excavations (*Continued*)

Project	Principal Concerns	Type of Excavation	Measurement[a]	Instrumentation Discussed	Special Features	Reference
			D	Combined inclinometers and probe extensometers		
			SCR	Soft and rigid inclusion stress gages		
Research mine, Kiruna, Sweden	Stability	Open stope mine	D	Convergence gages	Automatic data acquisition system used	Stillborg et al. (1983)
			D	Fixed borehole extensometers		
			D	Probe extensometers		
			SCR	Soft inclusion stress gages		
Kielder Water Scheme, Northumbria, England	Improvement of design and construction procedures for water tunnels in mudstone	11 ft (3.3 m) diameter experimental tunnel in mudstone	D	Convergence gages	Various support systems used: e.g., steel ribs, steel liner, rockbolts	Ward et al. (1983, 1976)
			D	Probe extensometers		
			D	Fixed borehole extensometers		
			L	Telltale load cells		
			S	Mechanical and vibrating wire strain gages		
Bolton Hill Tunnels, Baltimore, MD, U.S.A.	Effectiveness of compaction grouting in limiting ground movements	Subway tunnel in dense sand	D	Surveying methods	Full-scale test section	Baker et al. (1981)
			D	Subsurface settlement points		
			D	Combined inclinometers and probe extensometers		
Massachusetts Bay Transportation Authority, Davis Square, to Porter Square, Somerville, MA, U.S.A.	Stability	Subway tunnel in clay, glacial till, and argillite	D	Surveying methods	Mixed face excavation	Beloff et al. (1981)
			D	Subsurface settlement points	Floating crown bars used for primary support	
			D	Combined inclinometers and probe extensometers		
			S	Vibrating wire strain gages		
Various	Various	Various	G	Piezometers	Emphasis given to purpose and benefits of monitoring programs	Dunnicliff et al. (1981)
			D	Various deformation gages		
			L	Load cells		
			S	Strain gages		

Table 23.3. Summary of Selected Case Histories of Underground Excavations (*Continued*)

Project	Principal Concerns	Type of Excavation	Measurement[a]	Instrumentation Discussed	Special Features	Reference
Lexington Market Tunnels, Baltimore, MD, U.S.A.	Demonstrate suitability of precast concrete liners for soft ground tunnels	Twin subway tunnels in soft ground, one lined with precast concrete, one with steel	G	Open standpipe piezometers	Performance and cost of two systems compared	Wightman et al. (1980)
			D	Surveying methods		
			D	Convergence gages		
			D	Crack gages		
			D	Subsurface settlement points		
			D	Fixed borehole extensometers		
			D	Inclinometers		
			D	Tiltmeters		
			L	Pressure transducers in lines to shoving jacks		
			S	Embedment strain gages: sister bars		
Port Richmond Water Pollution Control Project, Staten Island, NY, U.S.A.	Justify reduction of liner thickness in future tunnels	10 ft (3 m) diameter tunnel in dense sand, organic silt, weathered serpentine, and schist	G	Pneumatic piezometers	Tunnel lined with 0.5 in. (13 mm) thick rolled steel liner plates Six full-scale test sections	Beloff et al. (1979)
			D	Surveying methods		
			D	Subsurface settlement points		
			D	Probe extensometers		
			D	Convergence gages		
			L	Pressure transducers in lines to shoving jacks		
			S	Electrical resistance strain gages		
Sewer tunnel in Mississauga, Ontario, Canada	Improvement of design procedures	14 ft (4.3 m) diameter tunnel in very dense till	G	Piezometers		DeLory et al. (1979)
			D	Convergence gages		
			D	Subsurface settlement points		
			D	Inclinometers		
			S	Electrical resistance strain gages		
			TS	Contact earth pressure cells		
Genesee River Interceptor Southwest, Rochester, NY, U.S.A.	Effect of high horizontal stresses	18.5 ft (5.6 m) diameter tunnel in highly stressed dolostone	D	Convergence gages		Guertin and Flanagan (1979)
			D	Fixed borehole extensometers		

Table 23.3. Summary of Selected Case Histories of Underground Excavations (*Continued*)

Project	Principal Concerns	Type of Excavation	Measurement[a]	Instrumentation Discussed	Special Features	Reference
			D	Inclinometers		
			D	In-place inclinometers		
East 63rd Street Tunnel, New York City Transit Authority, New York, NY, U.S.A.	Improvement of design and construction procedures	Tunnel in schist, 45 ft (13.7 m) high and wide	D	Surveying methods		Guertin and Plotkin (1979)
			D	Fixed borehole extensometers		
			S	Vibrating wire strain gages		
Various projects in Germany	Stability and settlement	Subway tunnels	D	Surveying methods	NATM used, resulting in minor settlements and no building damage	Müller and Spaun (1977)
			D	Convergence gages		
			D	Fixed borehole extensometers		
			D	Multiple deflectometers		
			TS	Contact earth pressure cells		
London Underground, Regents Park, London, England	Improvement of design and construction procedures	Subway tunnel in soft ground, with expanded concrete lining	G	Pneumatic piezometers		Barratt and Tyler (1976)
			D	Surveying methods		
			D	Convergence gages		
			D	Combined inclinometers and probe extensometers		
			L	Load cells		
			TS	Contact earth pressure cells		
Various	Various	Full-scale test sections	G	Piezometers	Emphasis on benefits of full-scale test sections	Lane (1975)
			D	Various deformation gages		
			L	Load cells		
			S	Strain gages		
Washington Metro, Washington, DC, U.S.A.	Improvement of design and construction procedures for subway tunnels	Subway tunnels in rock	D	Fixed borehole extensometers	Summary of rock deformations and strains in shotcrete linings	Mahar et al. (1972)
			S	Embedment strain gages		
Potomac Interceptor Sewer, Washington, DC, U.S.A.	Determine cause of excessive settlement Improve design and construction procedures	11 × 13 ft (3.6 × 4 m) tunnel in organic and clayey silts	D	Surveying methods		Rebull (1972)
			D	Crack gages		
			D	Convergence gages		
			D	Fixed borehole extensometers		
			TS	Contact earth pressure cells		

Table 23.3. Summary of Selected Case Histories of Underground Excavations (*Continued*)

Project	Principal Concerns	Type of Excavation	Measurement[a]	Instrumentation Discussed	Special Features	Reference
Anvil Points Oil Shale Research Center, Rifle, CO, U.S.A.	Improvement of design and construction procedures	Underground oil shale mine	D L SCR	Fixed borehole extensometers Load cells Soft and rigid inclusion stress gages	Full-scale test section	Sellers et al. (1972)
Morrow Point Power Plant, Colorado, U.S.A.	Stability	Underground powerhouse in high-quality metamorphic rock	D D D L	Surveying methods Fixed borehole extensometers Acoustic emission monitoring Rockbolt load cells	Monitoring revealed moving polyhedron of rock, requiring stabilization	Brown et al. (1971)
Various	Various	Soft ground tunnels	D D D L S	Surveying methods Convergence gages Inclinometers Load cells Strain gages	State-of-the-art report, including observational data and some design recommendations	Peck (1969b)
Garrison Dam, North Dakota, U.S.A.	Need to establish design criteria	36 ft (11 m) diameter test tunnel in clay-shale	G D D S TS	Piezometers Surveying methods Convergence gages Mechanical strain gages Contact earth pressure cells	Substantial cost savings for prototype tunnels	Burke (1957)

[a]D: deformation
G: groundwater pressure
L: load in support
S: strain in support
SCR: stress change in rock
T: temperature
TS: total stress between lining and soil.

CHAPTER 24

DRIVEN PILES*

24.1. GENERAL ROLE OF INSTRUMENTATION

The subsurface length of a driven pile cannot usually be inspected after driving; thus, its physical condition and alignment are unknown. Figure 24.1 shows a group of piles, driven to bedrock: the figure conveys the contrast between how the project will usually appear on engineering drawings and what may actually happen in the field. Primary mechanisms that control the behavior of driven piles are outlined in Chapter 2.

Subsurface geotechnical conditions are rarely known with certainty, and therefore the design of driven piles involves assumptions and uncertainties that are often addressed by conducting instrumented full-scale tests. Tests may examine the behavior of the pile under load applied to the pile butt (head) or under load caused by settlement of soil with respect to the pile.

Defects in piles can be created during driving, and inspection procedures are available for examining the condition and alignment after driving. Certain types of driven pile cause large displacements and changes of pore water pressure in the surrounding soil, and these may in turn have a detrimental effect on neighboring piles or on the stability of the site as a whole. Instrumentation can be used to quantify the consequences of pile driving and thus to assist in planning any necessary action.

*Written with the assistance of Douglas G. Gifford, Associate & Vice President, Haley & Aldrich, Inc., Cambridge, MA.

24.2. PRINCIPAL GEOTECHNICAL QUESTIONS

The following principal questions are presented in the normal order of occurrence for driven piles. The order does not reflect a rating of importance.

24.2.1. What Is the Load–Movement Relationship of the Pile? (Determination of Relationship by Static Testing)

The load–movement relationship of the pile and/or designed length of the pile may be uncertain during the design phase, and static load tests can be performed to resolve these uncertainties. Load tests may be conducted either during the design phase or at the beginning of the construction phase. A load test allows application of load to the pile butt so that the actual response to load can be verified experimentally. The loading mode may be axial compression or uplift, lateral, eccentric, moment, torsional, or a combination of these modes.

Static Load Tests Without Determination of Comprehensive Load Transfer Data: Loading in Axial Compression and Uplift

Procedures for static load tests in axial compression and uplift are described by ASTM (1981a) and ASTM (1983), respectively. Load at the pile butt should be measured using a load cell, equipped with a spherical bearing, because load determinations based on hydraulic jack pressures are often unreliable (Chapter 13). Vertical deformation at the pile

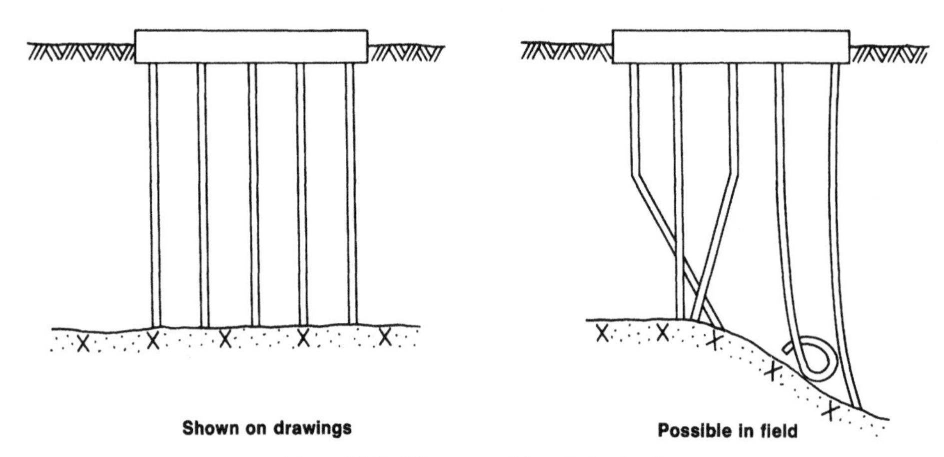

Figure 24.1. Pile group driven to bedrock.

butt is normally measured using dial indicators attached to a reference beam. A wire/mirror/scale arrangement is often used as backup. Similar measurements of horizontal deformation are recommended to indicate whether or not the applied load is axial and concentric. The reference beam should ideally be supported on ground that is not influenced by the test, but often this is not possible, and any movement of the beam should be measured throughout the test by surveying methods. Great care should be taken to minimize the effect of temperature changes on the reference beam.

Most conventional static load tests in axial compression include measurements of load and deformation at the pile butt only. However, such measurements are insufficient for evaluating load tests of relatively short duration on long end-bearing piles because much of the test load is carried by skin friction. They are also insufficient for concrete piles if they are tested while the concrete is "green," as is often the case when available time for testing is very limited. In general, therefore, tip movement should be measured, using a telltale and dial indicator, requiring an open hole along the length of the pile. Installation of telltales in closed-ended pipe piles is straightforward, but installation on other steel piles requires welding one or more pipes, channels, or plates to create an open hole along the length of the pile, through which telltales are later installed. Access for telltales in a concrete pile is created by casting a plastic pipe along the axis of the pile. The American Society for Testing and Materials (ASTM, 1981a) illustrates possible methods for installing telltales in or on steel H, pipe, and timber piles. A caution needs to be given about attachment of telltales (or any other instrumentation) to the outside of spiral-welded pipe piles: these piles tend to rotate as they are driven, and external instrumentation is not likely to survive. ASTM (1981a) also illustrates a *load transfer assembly* suitable for creating access to the tops of the telltales. When a load transfer assembly is used, its members should be welded together to ensure that no instability develops during load testing.

Evaluations of results of load tests in axial compression are severely hampered if tip load is not known, because total load cannot be separated into point resistance and skin friction. This statement leads to a recommendation for measurement of tip load and, although this is not standard procedure, measurements of tip load should be made if a financial payback can be identified. For steel piles, strain gages should be installed near the tip: some suggestions for gage type and methods for preventing damage are given in the following subsection. Multiple telltales can be used as backup but, as discussed in the following subsection, they are subject to several limitations. Measurements of tip load in concrete piles should be made by attaching a contact earth pressure cell to the pile tip: for the reasons given in the following subsection, strain measurements are not adequate. Figure 24.2 shows a possible layout of instrumentation during a static load test of a driven concrete pile, loaded in axial compression, for the case where comprehensive load transfer data are not required.

When driving a displacement pile through saturated clay, silt, or fine sand, temporary excess pore water pressures will be developed. If a pile is load tested while excess pore water pressures exist, the result is likely to be misleading; therefore, pressures should be allowed to dissipate before testing.

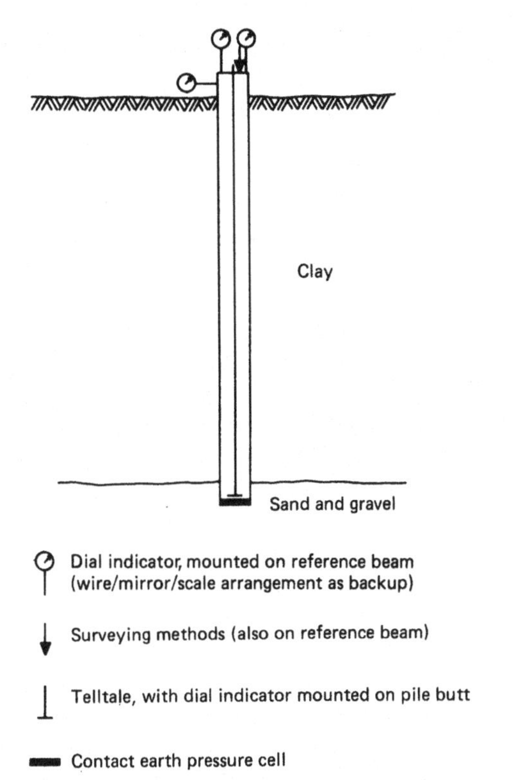

Figure 24.2. Possible layout of instrumentation during a static load test of a driven prestressed concrete pile, loaded in axial compression, for the case where comprehensive load transfer data are not required. Note: Loading arrangements, load transfer assembly, and reference beam are not shown. Also, a load cell is required. Use of surveying methods requires benchmark, remote from pile.

Normally, a waiting period of about a week is sufficient. However, if the test pile is selected from a group of piles, the excess pore water pressures may be significantly higher than for a single pile. Pore water pressures may also be significantly higher if the soil is excessively disturbed. In these cases, one or more piezometers should be installed so that effective stresses at the soil/pile contact can be evaluated.

Static Load Tests with Determination of Comprehensive Load Transfer Data: Loading in Axial Compression and Uplift

When comprehensive load transfer data are required along the lengths of driven piles during static load tests in axial compression or uplift, strain gages and/or multiple telltales are used.

Multiple telltales cannot be the primary measurement system because they can be installed only after pile driving and therefore cannot provide *residual stress* data. Residual stresses are stresses that remain in a pile after pile driving: without a knowledge of their magnitude, absolute load in a pile cannot be calculated from telltale data. As discussed in Chapter 13, multiple telltales are subject to additional limitations. First, load data are extremely sensitive to error in telltale data; therefore, duplicate telltales are preferred at each depth. The entire length of each telltale should be straight (the arrangement with an elbow, as shown in ASTM (1981a) for pipe piles, is suitable for monitoring tip settlement but introduces excessive errors when using multiple telltales for determination of strain), generally requiring use of a welded load transfer assembly as illustrated in ASTM (1981a). Second, the shape of the load transfer plot between telltale attachment points must be assumed; therefore, telltales at two depths are usually inadequate. Third, an extrapolation must be made for determination of tip load. When planning multiple telltale installations and interpreting data, one must pay careful attention to the guidelines given in Section 13.3.1.

For steel piles, strain gages provide the primary measurements, and multiple telltales can be used as backup. A possible layout is shown in Figure 24.3. Weldable vibrating wire gages are preferred on H-piles and also on pipe piles if they are installed on the outside. However, a method of installing the larger type of vibrating wire gage inside a closed-ended pipe pile is outlined in Chapter 13 and is an attractive option. Very great care must be taken to prevent damage to gages and cable during driving, and if there is likelihood of damage to the tip of the pile, gages should be located above the potentially damaged length. Gages on the outside of the pile should be protected with robust cover plates tapered toward the pile tip. Cables should usually be installed within steel angles or channels. An effective method involves welding channels to the pile, with 6 in. (150 mm) long windows for cable pull-through points. After pulling cables through, foam material (e.g., the two-part type used for home insulation) is injected through holes drilled in the channel at 2 ft (600 mm) intervals, to support the cable during driving. The windows are then closed with small lengths of channel, tack welded to the pile. It is often necessary to protect the cable, beyond the point of exit from the channel, with flexible metallic conduit. During driving, the cables and conduit can pass around a curved guide (a large-diameter pulley is suitable) mounted near the pile head, on the pile or hammer. A rope can be at-

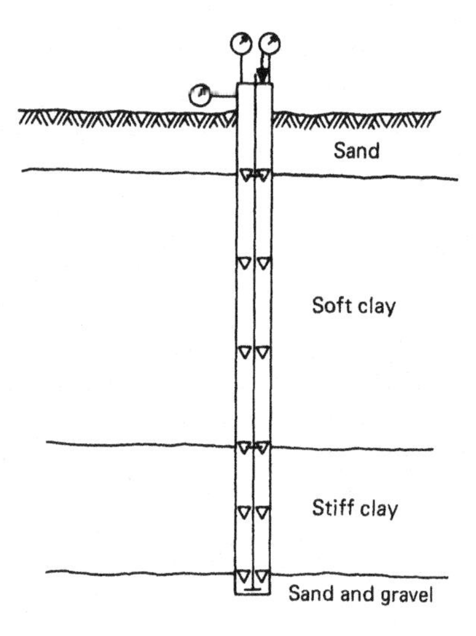

Dial indicator, mounted on reference beam (wire/mirror/scale arrangement as backup)

Surveying methods (also on reference beam)

Multiple telltales, with dial indicators mounted on pile butt. Duplicate telltales at each depth

▽ Surface-mounted strain gage

Figure 24.3. Possible layout of instrumentation during a static load test of a driven steel pile, loaded in axial compression, for the case where comprehensive load transfer data are required. Note: Loading arrangements, load transfer assembly, and reference beam are not shown. Also, a load cell is required. Use of surveying methods requires benchmark, remote from pile.

tached to the ends of the cables and held at the ground surface to prevent damage as the pile is driven. Special arrangements for cable protection must be made at any splices in the pile.

For concrete piles, strain data can be misleading, because they may be greatly influenced by cracks that develop in the concrete during driving. Ideally, several load cells should be installed along the length of the pile, but practical and economic considerations usually prevent this approach. When comprehensive load transfer data are required for concrete piles, both embedment strain gages and multiple telltales should be used, with full regard for their limitations. Conventional vibrating wire embedment strain gages are appropriate in small-diameter concrete piles, but in piles of 10 in. (250 mm) or greater diameter, the author favors use of the *sister bar* type of embedment strain gage, with a vibrating wire transducer. Very great care must be taken to prevent damage to gages and cables during driving, as discussed above for steel piles, and special care must also be taken in the casting yard. However, if the pile has a central hole for inspection or other purposes, or if a pipe can readily be cast on the axis of the pile, an alternative approach is possible for minimizing the damage potential. Weldable vibrating wire strain gages can be attached to a suitably sized steel pipe, angle, square tube, or plate and installed and grouted within the central hole after pile driving. A special effort must be made to ensure that the gages are subjected to the same strain as the concrete in the pile: the pipe used to form the central hole should be steel and not plastic, and the grout mix should be selected carefully. A mix of water and cement in a ratio of 1 : 0.3 by weight, with an expander additive, has been used successfully. This alternative approach does not allow determination of residual stresses but may nevertheless be the preferred approach in cases where there is a high potential for damaging cables and/or gages during pile driving. If a sliding micrometer is available, it can also be considered as an alternative (Carvalho and Kovari, 1983). Whenever possible, measurements in a concrete pile should include direct measurement of tip load, by installing a contact earth pressure cell at the tip.

As discussed in Chapter 13, special arrangements must be made for converting measured strain to stress in a prestressed concrete pile. Two methods are possible during an axial load test and are described below. Preferably, both methods should be used, to maximize confidence in the data. First, the top portion of the pile can be used as an *unconfined compression test specimen*. If possible, the length should be at least three times the diameter of the pile and should not include any length that may have been damaged during driving. Strain in the specimen is measured with strain gages, backed up with a pair of telltales if the specimen is long enough. Embedment strain gages can be used if the elevation to which the pile will be driven is known before driving; otherwise, surface-mounted gages can be used. The specimen can be entirely above the ground surface or within a sleeve just below the ground surface. Strain gages should be installed at the midheight of the specimen. If there is concern that gage readings may be influenced by damaged concrete or by end effects during loading, a second set of gages should be installed at or just below the lower end of the specimen. Because change in moisture content can cause significant strain, the specimen must be under the same moisture conditions as the length of the pile in which the other strain measurements are made. If that length is below the water table, the specimen should also be either below

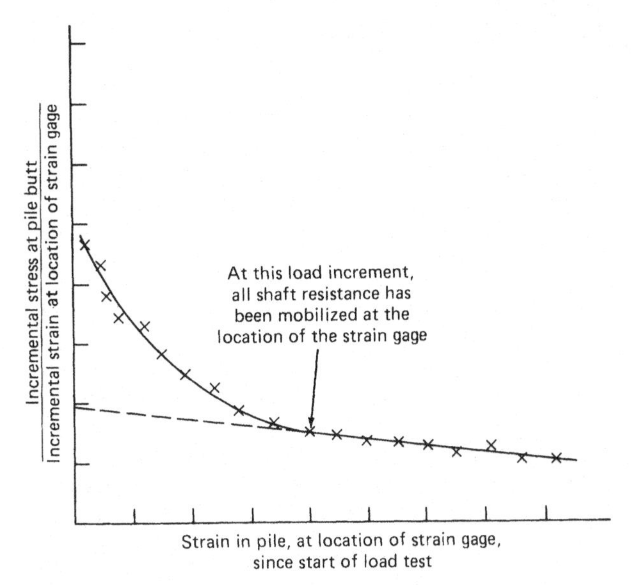

Figure 24.4. Typical plot for conversion of strain to stress in a prestressed concrete pile, using subsurface measurements of strain.

the water table or kept wet by artificial means. The relationship between strain and stress is then determined during the load test, using strain data from the specimen and load data from the load cell at the pile butt. The method provides data for every load increment.

The second method for strain/stress conversion makes use of load cell data and measurements of subsurface strain, determined from strain gages or multiple telltales in the pile, plotted as shown in Figure 24.4. During the load test, when all shaft resistance has been mobilized at the location of the strain gage, the pile acts as a column in compression and the plot becomes straight. The straight line is likely to be sloping, because the modulus of the pile generally decreases as the pile is loaded. The slope of the straight line part of the plot, together with the value of the intercept on the vertical axis, are used to determine the relationship between strain and stress at the location of the strain gage. Fellenius (1987) describes the method in detail.

Figure 24.5 shows a possible layout of instrumentation during a static load test of a driven prestressed concrete pile, loaded in axial compression, for the case where comprehensive load transfer data are required.

Static Load Tests: Lateral Loading Mode

The procedure for static lateral load tests is described by ASTM (1981b). Load measurements at the pile butt require use of a load cell or cable ten-

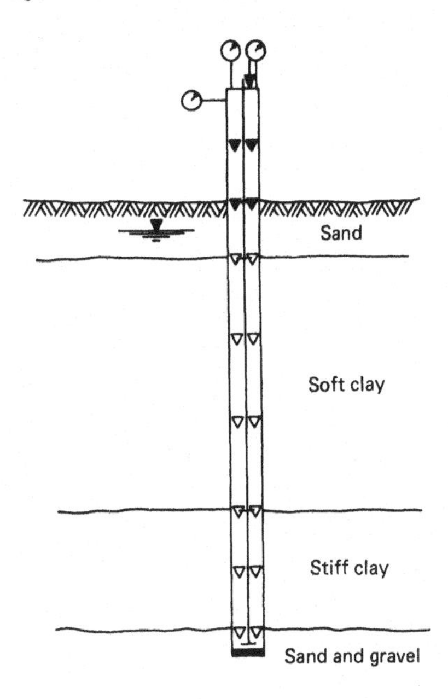

Figure 24.5. Possible layout of instrumentation during a static load test of a driven prestressed concrete pile, loaded in axial compression, for the case where comprehensive load transfer data are required. Note: Loading arrangements, load transfer assembly, and reference beam are not shown. Also, a load cell is required. Use of surveying methods requires benchmark, remote from pile. Top of pile should be kept wet.

sion meter. Deformation measurements at the pile butt require horizontal and vertical dial indicators and/or a wire/mirror/scale arrangement. Horizontal deformation measurements should be arranged to monitor whether or not the loading is causing torsion in the pile. Any movement of the reference beam should be measured throughout the test by surveying methods, and temperature effects should be minimized.

When bending data are required during a lateral load test, strain gages are installed on or near both faces of the pile in the plane of loading. Surface-mounted vibrating wire gages are normally preferred on steel piles, of the weldable type if installed on the outside and of the larger type with bolted end blocks if installed on the inside of a pipe pile. *Sister bars* with vibrating wire transducers are preferred

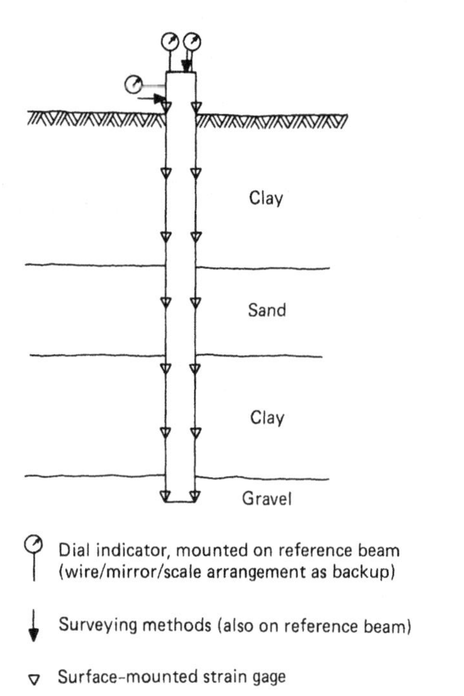

Figure 24.6. Possible layout of instrumentation during a static load test of a driven steel pile: lateral loading mode. Note: Loading arrangements and reference beam are not shown. Also, a load cell is required. Use of surveying methods requires benchmark, remote from pile.

in concrete piles of 10 in. (250 mm) or greater diameter. If the test is likely to develop significant lateral deformation at the butt (more than about 6 in., 150 mm), the opportunity exists to estimate bending stresses from inclinometer data as described in Chapter 12. However, this approach should only be used as backup. If it **is** used, inclinometer readings of maximum accuracy should be taken during each loading increment, and a tiltmeter should be attached to the butt so that more frequent readings of overall tilt can be made during the test. An inclinometer can also be used to determine the deformed shape of a pile at increments during a lateral load test.

When a lateral load test is conducted in cohesive soils such that deformation is affected by consolidation of soil on the compression side of the pile, piezometers in the soil will provide additional data for evaluation of the test.

Figure 24.6 shows a possible layout of instrumentation during a static lateral load test of a driven steel pile. This is only one example, and it must be emphasized that many other configurations are possible, as indicated by the case histories described later in this chapter.

Other Static Loading Modes

When moment, eccentric, torsional, or combined static loading modes are used, appropriate monitoring methods should be selected from those described for axial and lateral load tests. If loading is applied during moment or torsional tests by pulling on a steel cable, load will normally be monitored by using a cable tension meter. Torsional load tests may require the use of strain gages mounted horizontally to measure torsional strains.

Cyclic Loading Modes

Cyclic load tests require similar instrumentation to static tests, but the dynamic response of instrumentation must be sufficient to follow the loading sequence.

Remote Readings During Load Tests

During conventional load tests, instruments are normally read manually, but considerations of safety and/or convenience may lead to selection of instruments that can be read remotely and perhaps by an automatic data acquisition system. This option is particularly attractive when the number of instruments is large. DCDTs can be used instead of dial indicators to monitor deformation of the pile butt and telltales, and it is often worthwhile to use both DCDTs and dial indicators in parallel, to provide backup and to satisfy personnel who are concerned about trusting electrical devices.

24.2.2. What Is the Load–Movement Relationship of the Pile? (Determination of Relationship by Dynamic Testing)*

Static load tests are expensive and time consuming; therefore, the small number of piles that are tested may not be representative of the behavior of all piles at a site. As an alternative, the pile-driving system can be used as a loading system, or a load can be applied by use of a simple vibrator or hammer. These alternative methods allow dynamic load tests to be made on a large number of piles. Dynamic tests are much less expensive than static tests, primarily because a static loading system is not required. The recent development of on-site automatic data acquisition systems has caused a rapidly increasing application of analytic methods based on wave transmission.

*Written with the assistance of Allen G. Davis, Director, Testconsult CEBTP Ltd., Thelwall, England.

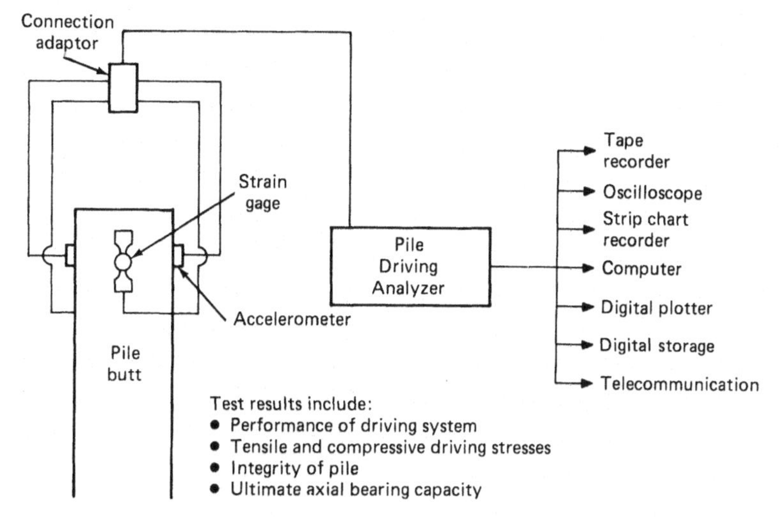

Figure 24.7. Schematic of Pile Driving Analyzer™ (courtesy of Pile Dynamics, Inc., Cleveland, OH).

Many successful correlations have been made between dynamic and static load tests. However, it is clear that much work still remains to be done to establish appropriate methods of dynamic testing and analysis, because pile performance is dependent on both strain rate and the magnitude of pore water pressure.

In the following paragraphs, three dynamic tests are described for determining axial load–movement relationships: *analysis of stress waves, steady-state vibration testing,* and *transient dynamic response testing.* The last two tests can also be used in the horizontal mode for determining lateral load–movement relationships.

Analysis of Stress Waves

The most commonly used and effective method for determining the axial load–movement relationships of driven piles by dynamic testing involves analysis of the stress wave during driving. The method, which has become known as the *Case Method,* after research programs conducted at Case Western Reserve University, requires measurement of strain and acceleration at the pile butt as the pile is driven, using a *Pile Driving Analyzer™*. By introducing these parameters into a pile–soil model based on a closed-form analysis of stress wave transmission through an elastic medium, an ultimate axial bearing capacity can be calculated for the pile under test.

Figure 24.7 shows a schematic of the test arrangement and indicates some of the test results.

The model requires a knowledge of the elastic properties of the pile, and a selection of soil damping constants must also be made, depending primarily on soil type. The recorded data may be processed by using an iterative analysis routine known as *CAPWAP*. This analysis, which is based on the wave equation theory as originally developed by Smith (1960), provides a refined prediction of pile behavior, including the load–movement relationships, and aids in evaluating the correctness of the soil damping constants. Rausche et al. (1985) describe the method and indicate correlations with static testing.

Axial Steady-State Vibration and Transient Dynamic Response Testing

Stain (1982) describes *steady-state vibration* and *transient dynamic response testing.*

Axial steady-state vibration testing involves attaching a vibrator to the pile butt to cause the butt to move up and down at the same frequency as the vibrator. A load cell is placed between the pile butt and the vibrator, and the velocity of movement is measured with an accelerometer, as shown in Figure 24.8. The measured axial stiffness of the pile butt allows calculation of the displacement of the butt under a given load (Davis and Dunn, 1974).

Axial steady-state vibration testing has largely been superseded by a development referred to as *transient dynamic response testing, TDR, shock testing,* or *impact testing.* As shown in Figure 24.9, the method involves placing a load cell centrally on the pile butt, striking the load cell with a hammer, and measuring velocity with a geophone.

Steady-state vibration and TDR tests were developed principally for drilled shafts, initially for concrete integrity determination. During their use in

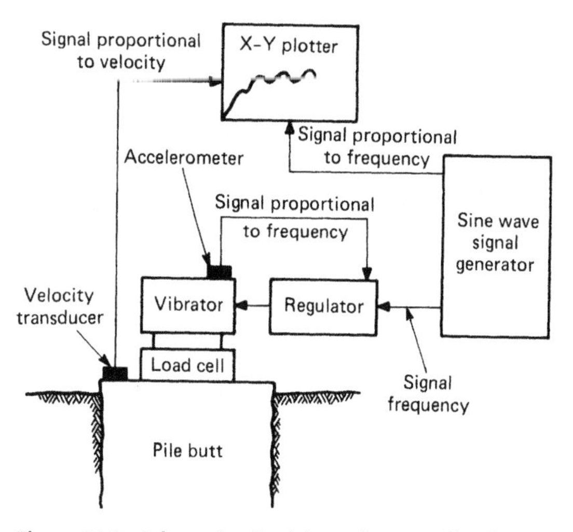

Figure 24.8. Schematic of axial steady-state vibration test.

the 1970s, it became apparent that the measurement of axial dynamic stiffness at the pile butt was a useful parameter in assessing performance under axial working load. Many correlations have been made between axial dynamic stiffness at the pile butt and the linear portion of the load–settlement curve as determined during static load testing in axial compression (Davis and Robertson, 1976). However, the tests cannot be used to estimate ultimate static axial capacity: to do so it is necessary to displace the tip of the pile, and neither method gives adequate input force.

Lateral Steady-State Vibration and Transient Dynamic Response Testing

S.A. Robertson (1979) describes a dynamic method for determining the lateral load–movement relationships of driven piles. The method is a development of the axial steady-state vibration test. It involves shaking the pile butt horizontally at controlled frequencies between 1 and 10 Hz and measuring the *lateral* velocity of the butt as a function of frequency. The lateral dynamic stiffness at the butt can be measured as in the axial steady-state vibration test, thus giving the displacement of the pile butt under a given load. If a vertical tube is built into the pile, a vertical profile of displacement for a given applied force can be obtained by measuring the pile velocity at different depths. The TDR test, used in the horizontal mode, has been demonstrated to give identical results to the lateral steady-state vibration test.

Comparison Among Dynamic Tests for Determining Load–Movement Relationships

When both the load–movement relationship and the ultimate axial bearing capacity of a driven pile are required, the only practical dynamic method to date is the Case method. The dependency on strain rate and magnitude of pore water pressure must be remembered when analyzing the results.

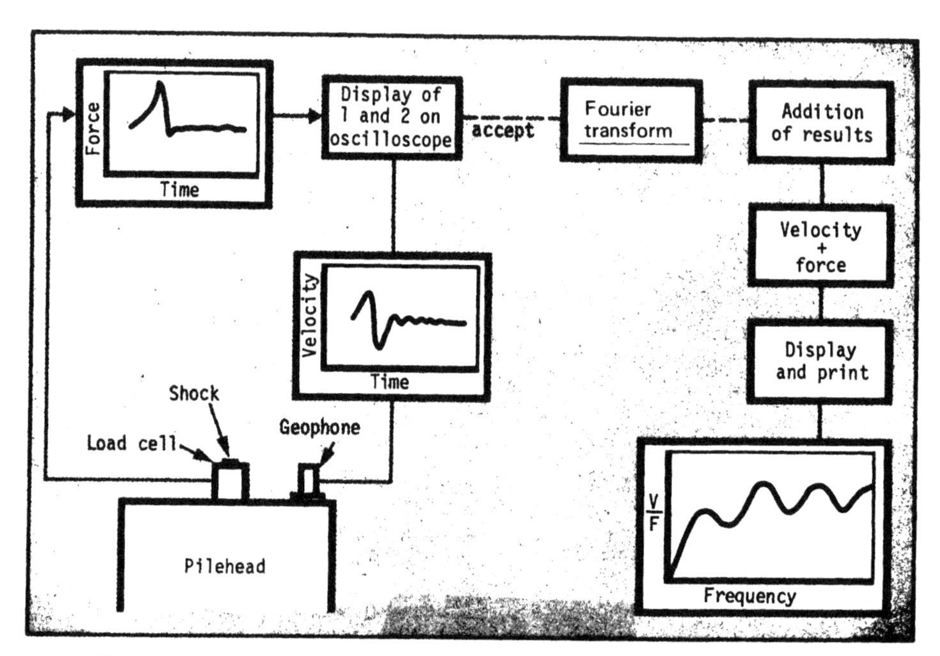

Figure 24.9. Schematic of axial transient dynamic response test (after Stain, 1982).

The Case method creates a higher strain level than the TDR test; therefore, it is generally preferred when only the axial load–movement relationships are required. However, the TDR test is a viable alternative, and it can be used rapidly and inexpensively to determine the axial or lateral load–movement relationships of all piles at a site. Although measurements are made at low strains only, results of the TDR test can be correlated with results of a static load test on each size and type of pile in a population. The TDR test has the great advantage of relatively low cost, and the test can be made without involving the pile-driving rig. The steady-state vibration test has largely been superseded by the TDR test.

24.2.3. Will There Be Significant Downdrag Loading?

When soil settles with respect to a pile, frictional forces between the soil and pile may cause a load increase in end bearing piles or increased settlement of friction piles. The loading is referred to as *downdrag* or *negative skin friction*. Possible causes of downdrag loading are described in Chapter 2.

Ongoing settlement prior to pile driving can be diagnosed by precise settlement surveys, and rod settlement gages are sometimes useful. Ongoing primary consolidation prior to pile driving can be diagnosed by using piezometers to measure groundwater pressure. The allowance for downdrag loading can be based on accepted design procedures but, when the allowance is substantial and a large quantity of piles are to be driven, a full-scale test to determine actual loading may be warranted. A full-scale test may also be warranted to prove the effectiveness of coating piles with a friction-reducing material such as bitumen. A full-scale test will normally entail driving several instrumented piles and placing a fill to cause the onset of downdrag loading. Primary parameters to be measured during the test are magnitude of load along the piles and relative settlement between the soil and piles.

For steel piles, adequate load data can be obtained by measuring strain, preferably by using vibrating wire strain gages, and very great care must be taken to prevent damage to gages and cables during driving. Despite the limitations discussed in Section 24.2.1, multiple telltales are recommended as backup. Figure 24.10 shows a possible layout of instrumentation during a full-scale test to determine downdrag loading on a steel pile. This is only one example, and it must be emphasized that many other configurations are possible: other configurations are indicated in the case histories described later in this chapter.

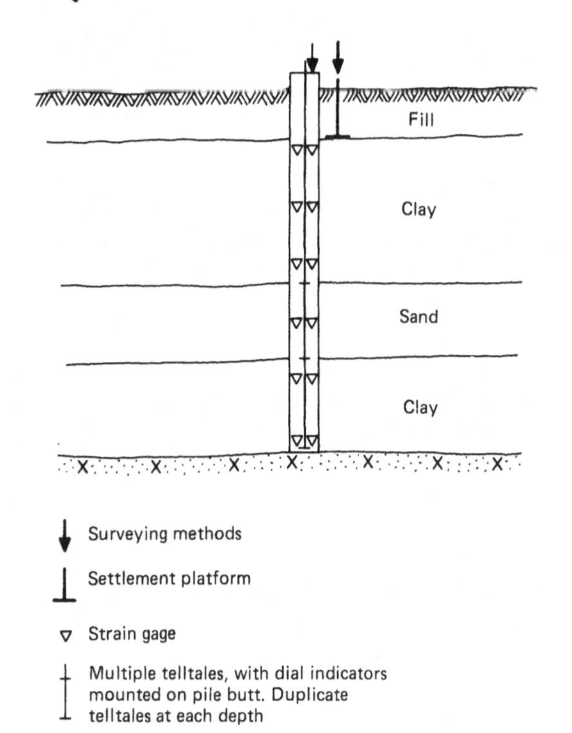

Figure 24.10. Possible layout of instrumentation during a full-scale test to determine downdrag loading on a steel pile. Note: Use of surveying methods requires benchmark, remote from pile.

For driven prestressed concrete piles, the methods described in Section 24.2.1 for converting strain to stress cannot be used, because no load is applied to the pile butt. Measurements of modulus made while load testing other similar piles also cannot be used, because a typical test to evaluate downdrag loading will be conducted for several weeks or months, during which time significant creep will usually occur. Load determinations in concrete piles must therefore include at least one **direct** measurement of load, by incorporating a load cell in a pile splice and/or by attaching a contact earth pressure cell to the pile tip. Because instruments for direct measurements of load are expensive, their number should be limited. If the soil profile is uniform, a single instrument deep down in the pile may be adequate, but in the general case measurements are preferable at two or three depths, to provide redundancy and to improve the accuracy of data evaluation. Load data should be supplemented by installing embedment strain gages, and multiple telltales should be used as backup, duplicating telltales at each depth as described earlier. Conventional vibrating wire embedment strain gages are appropriate in small-diameter concrete piles, but in piles of 10 in. (250 mm) or greater diameter, the author favors use of the *sister bar* type of embedment strain gage, with a vibrating wire transducer. The relationship between load and strain is determined, at points where load is measured directly, by relating load to strain measured near those points. The same relationship is applied to strains measured elsewhere.

Relative settlement between the soil and pile can be measured simply by installing a settlement platform at the original ground surface alongside each pile and measuring settlement of the platform with respect to the pile butt. However, if comprehensive data are required to define the relationship between relative settlement of soil and loading along the length of the pile, multiple telltales are needed in or on the pile, along with probe extensometers in the nearby soil.

For a full understanding of soil–pile interaction, it is necessary also to determine pore water pressure in the soil and effective stress at the soil/pile interface. Specially manufactured pneumatic piezometers and contact earth pressure cells have been used successfully for determining effective stress at the soil/pile interface (e.g., Clemente, 1979).

24.2.4. Has the Capacity of the Pile Been Reduced by Defects Created During Driving?

Many building codes impose allowable limits on the curvature of piles, because excessive bending reduces capacity and increases deformation of piles under long-term loading. Capacity is also reduced if piles are dog-legged. The curvature of piles can be determined by using an inclinometer. When used in pipe piles, the inclinometer can be fitted with long arms and guide wheels that ride on the inside wall of the pipe and can be lowered down the pile on orientation rods. When used in other types of pile, a steel or plastic guide pipe is required, attached to or embedded in the pile. For an approximate determination of curvature, the shear probe (Chapter 12) can be used within a guide pipe.

The application of the shear probe procedure can be extended to become part of the design/construct engineering package. A steel or plastic guide pipe, usually referred to as an *inspection pipe,* can be specified in or on all piles, and information on the condition of the pile below the ground surface can be obtained by using the shear probe as part of the routine inspection procedure. The specification includes acceptance/rejection criteria, a major design uncertainty is resolved, and the designer can allow a higher load per pile and therefore fewer piles than would be allowed without the inspection pipes.

Table 24.1. Types of Integrity Test

Category	Test
Tests carried out at pile butt	Analysis of stress waves (Case method) Echo test Axial transient dynamic response test Axial steady-state vibration test
Tests carried out within pipes in pile	Sonic coring Nuclear density probe, single-hole Nuclear density probe, cross-hole

The condition of a driven pile can also be determined by static load tests, as described above. However, they are too expensive for use as routine quality control tests.

The condition of a driven pile can also be determined by *integrity testing.** Integrity tests are listed in Table 24.1 and described in the following subsections. They are usually applied to concrete piles but some are also valid for wood and steel piles.

Analysis of Stress Waves

As indicated in Figure 24.7, the *Case method* can be used to determine integrity of driven piles.

Echo Test

The *echo test* (Stain, 1982), also referred to as a *surface reflection test* and *seismic test,* involves striking the pile butt with a hammer and measuring the time for the resulting compression wave to travel down the pile and back to the butt by reflection from the tip. The arrangement is shown in Figure 24.11. For a given quality of concrete, the approximate wave velocity is known; therefore, breaks or other irregularities will be indicated by a

*The text on integrity testing has been written with the assistance of Allen G. Davis, Director, Testconsult CEBTP Ltd., Thelwall, England.

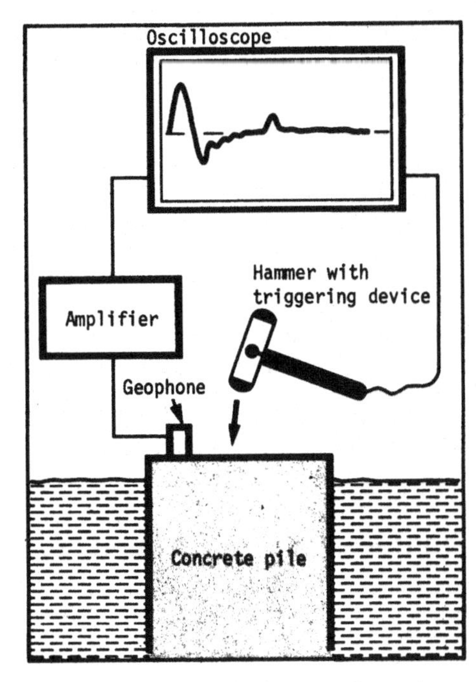

Figure 24.11. Schematic of echo test (after Stain, 1982).

short travel time. The method is rapid and inexpensive but cannot be used if there are splices in the pile. Damping of the signal causes a further limitation to the use of echo testing, particularly for long, slender piles or for piles driven into stiff soils. In stiff clays, for example, the pile tip can rarely be detected when the pile length/diameter ratio exceeds 30/1.

Axial TDR and Steady-State Vibration Tests

The *TDR* test has been used to test the integrity of driven piles of all types (Higgs and Robertson, 1979). Although having the same limitations as the echo test, measurements of axial dynamic stiffness can be used to compare piles within a population. The *axial steady-state vibration test* can also be used but has largely been superseded by the TDR test.

Sonic Coring

Sonic coring (Stain, 1982), also referred to as *sonic logging*, requires measuring the propagation time of a sonic signal between a transmitter and a receiver. Measurements can be made in a single hole on the axis of a pile, but the effective sampling area is small and significant defects may be missed. Reliable data can be obtained by cross-hole measurements between two metal pipes cast parallel to the

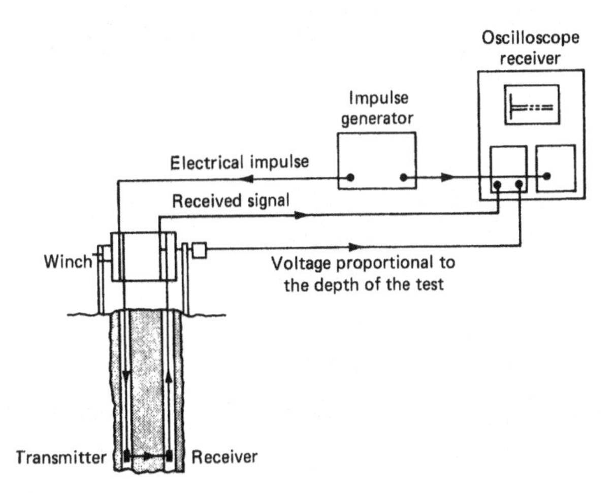

Figure 24.12. Schematic of sonic coring with cross-hole measurements (after Davis and Robertson, 1975).

axis of the pile during construction, as shown in Figure 24.12. Plastic pipes are not suitable, because the bond between concrete and plastic is sometimes imperfect, resulting in a loss of signal that can suggest that a fault is present in the pile. An emitter probe and receiver probe are lowered, one in each pipe, and measurements are made as the probes are raised in unison. Variations in sonic velocity indicate defects. Two, three, or four pipes are used, depending on the diameter of the pile. Sonic coring is not often used for testing driven piles because of the requirement for pipes in the piles: its primary application is for testing drilled shafts and slurry trench (diaphragm) walls.

Nuclear Density Probe

A nuclear density probe can be traversed along an embedded pipe to examine the integrity of concrete. Preiss and Caiserman (1975) describe use of a probe in a single pipe of plastic or metal, using the backscatter technique, which allows a relatively weak source to be used. Photons emitted from the source are partially absorbed by the material in the pile, partially scattered toward the detector, and counted. The count rate is indicative of the density of the material surrounding the probe. The arrangement is shown in Figure 24.13. The Preiss and Caiserman equipment allows a radial extent of inspection about the pipe of approximately 4 in. (100 mm), and measurements can be made within several individual pipes. Alternatively, cross-hole measurements can be made, using a transmission method, with the source and detector in different pipes. Cross-hole measurements require a stronger

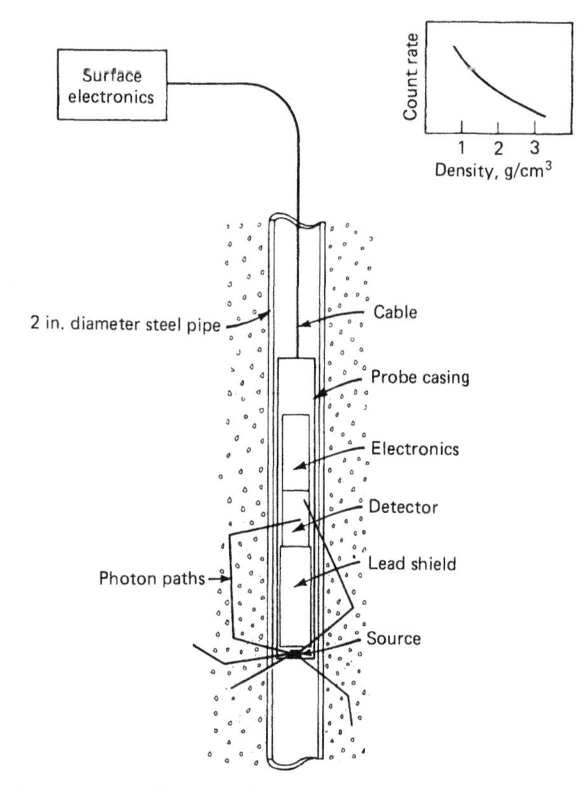

Figure 24.13. Schematic of nuclear density probe in single hole, using backscatter technique (after Preiss and Caiserman, 1975).

source than single-hole measurements and are limited to a cross-hole spacing of about 24 in. (600 mm).

Comparison Among Integrity Tests

The four tests carried out at the pile butt will be considered first. Driven precast concrete piles usually have a relatively high length/diameter ratio, and most integrity problems arise during handling and driving. These problems are best detected by monitoring during driving using the Case method. However, because of the need to instrument each pile butt during driving, this is a relatively expensive approach. As an alternative, integrity testing can be carried out at the pile butt after driving by using the echo or TDR method. As indicated earlier, the TDR method has largely superseded axial steady-state vibration testing. The TDR method gives more information than the echo method, including a value for axial dynamic stiffness and more detail of any defects within the top 6 ft (2 m).

In considering the three techniques carried out within pipes, these methods are more costly than those carried out at the pile butt. However, they are capable of locating **any** defect and giving a much greater definition of the type of defect. The sonic coring method is more able than the nuclear methods to locate small-scale defects such as fine cracks. In addition, the nuclear methods are sometimes restricted, because of problems inherent in the use of radioactive materials on construction sites.

24.2.5. Is Pile Driving Causing Excessive Lateral Deformation of the Ground?

When displacement piles are driven in soft clay, there will normally be a temporary increase of pore

Table 24.2 Overview of Routine and Special Monitoring

Application	Measurement
Routine monitoring during static axial load testing	Load and movement at pile butt Vertical movement of pile tip
Routine monitoring during static lateral load testing	Load and movement at pile butt
Special load testing applications	Load at pile tip during axial load testing Axial stress along length of pile during axial load testing Bending stress along length of pile during lateral load testing Moment, eccentric, torsional, or combined static loading modes: appropriate parameters as for static axial and lateral modes Load, velocity, strain, and/or acceleration as required for dynamic tests Deformation and stress during full-scale test to determine downdrag loading Excess pore water pressure
Other special applications	Integrity of pile As-driven alignment along length of pile Lateral deformation of ground and excess pore water pressure during pile driving Surface settlement and/or groundwater pressure for evaluation of downdrag loading

water pressure and a reduction of soil strength, perhaps leading to excessive lateral deformation and perhaps even a stability failure. Examples include pile driving on a slope that already has a low factor of safety and pile driving nearby an existing structure or nearby other piles. When this potential exists, an instrumentation program can be conducted to quantify displacements and pore water pressure changes, thus allowing adoption of a driving procedure or sequence that minimizes detrimental effects. Primary instruments used for this purpose include surveying methods, inclinometers or shear probes in the piles and/or soil, and piezometers.

24.3. OVERVIEW OF ROUTINE AND SPECIAL APPLICATIONS

For those readers who seek general guidelines on what is typically *routine* monitoring and what is monitoring for *special applications,* Table 24.2 presents a broad generalization.

In practice, the most frequent use of instrumentation for driven piles is to examine load–movement relationships during static axial load testing.

24.4. SELECTED CASE HISTORIES

Table 24.3 summarizes selected case histories of instrumented driven piles.

Table 24.3. Summary of Selected Case Histories of Driven Piles

Project	Principal Concerns	Type of Load Test	Type of Pile	Measurement[a]	Instrumentation Discussed	Special Features	Reference
Newport News Shipyard, Tidewater, VA, U.S.A.	Load–movement relationships	Static axial compression	Prestressed concrete	L S G	Load cell Strain gages Vibrating wire piezometers		Martin et al. (1987)
I75 near Tacoma, WA, U.S.A.	Load–movement relationships	Static axial compression	Steel H	D D,S L S	LVDTs Multiple telltales Load cell Strain gages		Rieke and Crowser (1986, 1987)
Various	Load–movement relationships	Static axial compression, uplift, lateral	Steel H and pipe	D D S	Telltales Inclinometers Strain gages	29 tests, conducted 1954–1984	AISI (1985)
Mar Pequeno Highway Bridge, São Paulo, Brazil	Load–movement relationships	Static axial compression	Large-diameter steel (650 mm, 26 in.) in silty clay and fine sand	D D S	Surveying methods Dial indicators Sliding micrometer	Load transfer data determined using sliding micrometer	Carvalho and Kovari (1983)
Various	Load–movement relationships	Static axial compression, uplift	Various	D D L S TS G	LVDTs Telltales Load cells Strain gages Contact earth pressure cells Piezometers	130 tests	Kulhawy et al. (1983)
Terminal 46, Port of Seattle, WA, U.S.A.	Stability of sloping ground during pile driving	—	Prestressed concrete in fill and loose silt	D G	Inclinometers Pneumatic piezometers	Prediction and use of hazard warning levels	Fellenius et al. (1982)
Group load test, Houston, TX, U.S.A.	Behavior of pile group in overconsolidated clay	Static axial compression: research test	Steel pipe Single and group of nine in overconsolidated clay	D,S D D S TS	Multiple telltales Subsurface settlement points Inclinometers Strain gages Contact earth pressure cells	Very comprehensive instrumentation	O'Neill et al. (1982a,b)

Table 24.3. Summary of Selected Case Histories of Driven Piles (*Continued*)

Project	Principal Concerns	Type of Load Test	Type of Pile	Measurement[a]	Instrumentation Discussed	Special Features	Reference
				G	Pneumatic piezometers in soil and pile		
Highway bridges between Berthierville and Yamachiche, Quebec, Canada	Load–movement relationships Performance of bridge pier foundations	Static axial compression, single and group	Timber, precast concrete, and steel pipe in soft sensitive clay	D D G	Surveying methods Subsurface settlement points Pneumatic piezometers in soil		Blanchet et al. (1980)
Test site at Hendon, England	Mechanism of load transfer from piles in stiff clay Behavior of piles at close spacing in stiff clay	Measurements made while jacking piles into clay	Tubular steel in overconsolidated clay	D D D L	LVDTs Subsurface settlement points Horizontal inclinometers with electrolytic level transducers Load cells within piles and at pile butts	Very comprehensive instrumentation	Cooke et al. (1980, 1979)
Bridge at Fredericton, New Brunswick, Canada	Load–movement relationships	Static axial compression, uplift	Steel H and pipe in granular fill and clayey silt	D D,S L	LVDTs Multiple telltales Load cell		Bozozuk et al. (1979)
Keehi Interchange, Honolulu, HI, U.S.A.	Downdrag loading	Downdrag	Prestressed concrete in soft underconsolidated clay, sand, and silty clay	D D,S D D D D S TS G G	DCDTs Multiple telltales Settlement platforms Subsurface settlement points Probe extensometers Inclinometer Strain gages Contact earth pressure cells Pneumatic piezometers in soil and pile Observation wells	Tests on bitumen-coated and uncoated piles Very comprehensive instrumentation	Clemente (1979)
Metropolitan Sewage Treatment Plant, Syracuse, NY, U.S.A.	Load–movement relationships	Static axial compression	Prestressed concrete in sand and silty clay	D D,S D L S	Dial indicators Multiple telltales Inclinometer Load cell Strain gages		Lacy (1979)
Fuel Reprocessing Facility, West Valley, NY, U.S.A.	Load–movement relationships	Static axial compression, cyclic axial and lateral	Steel H in glacial till	D D L S	Tiltmeters Inclinometer Load cells Strain gages	Load applied by automated closed-loop system	Lu et al. (1979)

Table 24.3. Summary of Selected Case Histories of Driven Piles (*Continued*)

Project	Principal Concerns	Type of Load Test	Type of Pile	Measurement[a]	Instrumentation Discussed	Special Features	Reference
Test Site at Brent, England	Load–movement relationships	Static and cyclic lateral	Steel pipe in weathered overconsolidated clay	D D S	Tiltmeters Electrical linear displacement transducers Strain gages		Price (1979)
Test Site at Hendon, England	Behavior of soil around friction piles	Static axial compression: research test	Steel pipe in overconsolidated clay	D D D L	Tiltmeters Subsurface settlement points Inclinometer Load cells	Measurements also made while jacking pile into clay	Cooke and Price (1978)
Test Site at Lewisburg, PA, U.S.A.	Development of improved design procedures for laterally loaded pile groups	Static lateral: research test	Steel H Single and group of six in clay	D S	Inclinometer Strain gages		Kim and Brungraber (1976)
Test Site at Melbourne, Australia	Downdrag loading	Downdrag: research test	Steel pipe in sand, clay, and gravel	D,S D S G T	Multiple telltales Settlement platforms Strain gages Piezometers Thermocouples	Tests on bitumen-coated and uncoated piles	Walker and Darvall (1973)
Quebec Autoroute, Berthierville, Quebec, Canada	Downdrag loading Load–movement relationships	Downdrag and static axial compression: research test	Steel pipe in marine clay	D,S D L S G	Multiple telltales Subsurface settlement points Load cell Strain gages Open standpipe piezometers	Pile filled with concrete and load tested after 10 years of downdrag measurements	Bozozuk (1972a, 1981)
Test Site at Gothenburg, Sweden	Downdrag loading	Downdrag	Precast concrete in soft clay, silt, and sand	D L G	Probe extensometers Load cells Piezometers	Vibrating wire load cells installed at splices in piles	Fellenius (1972) (Fellenius and Broms (1969)
Arkansas River Project, Arkansas, U.S.A.	Load–movement relationships	Static axial compression, uplift, lateral	Steel H and pipe, concrete, and timber in sand	D,S S	Multiple telltales Strain gages	Various pile lengths, diameters, and driving procedures	Mansur and Hunter (1970)
Sörenga and Heröya, Norway	Downdrag loading	Downdrag	Steel pipe driven to rock	S	Telltales	Tests with bentonite slurry, cathodic protection, and bitumen coating	Bjerrum et al. (1969)
Test site in Norway	Downdrag loading	Downdrag	Steel pipe driven to rock	D,S D D D G	Multiple telltales Settlement platforms Subsurface settlement points Inclinometers Piezometers		Johannessen and Bjerrum (1965)
Old River Control Project, Louisiana, U.S.A.	Load–movement relationships	Static axial compression and uplift	Steel H and pipe in sand, silt, and clay	S	Multiple telltales		Mansur and Kaufman (1956)

[a]D: deformation L: load in pile T: temperature
G: groundwater pressure S: strain in pile TS: total stress between pile and soil

CHAPTER 25

DRILLED SHAFTS*

25.1. GENERAL ROLE OF INSTRUMENTATION

Drilled shafts (also referred to as *drilled piers, caissons, drilled piles, bored piles, drilled caissons,* and *cast-in-place piles*) are constructed by augering a hole into the ground, placing a reinforcing cage, and backfilling the hole with concrete. A temporary outer casing may be required. Water or drilling mud may be used during the construction process, and in this case the water or drilling mud is displaced by tremie concrete as the casing is withdrawn. Construction procedures are described by Reese (1978).

Figure 25.1 shows drilled shafts bearing on bedrock: the figure conveys the contrast between how the project will appear on engineering drawings and what may actually happen in the field. Primary mechanisms that control the behavior of drilled shafts are outlined in Chapter 2.

Many uncertainties exist during design, and instrumentation plays a role in determining the load–movement relationship by conducting load tests. Concrete integrity is often uncertain during construction, particularly when shafts are constructed in granular soils below the water table or in softer, squeezing clays, when concrete slump is inadequate, or when concrete placement practices are inferior. Instrumentation can be used to examine the integrity of the concrete.

*Written with the assistance of Fred H. Kulhawy, Professor, School of Civil and Environmental Engineering, Cornell University, Ithaca, NY.

25.2. PRINCIPAL GEOTECHNICAL QUESTIONS

The following principal questions are presented in the normal order of occurrence for a drilled shaft. The order does not reflect a rating of importance.

25.2.1. What Is the Load–Movement Relationship of the Drilled Shaft?

The load–movement relationship of a drilled shaft will be uncertain during the design phase, and one or more load tests may be conducted either during the design phase or at the beginning of the construction phase. A load test allows application of load to the butt (head) of a shaft, so that the actual response to load can be verified experimentally. The loading mode may be axial compression or uplift, lateral. eccentric, moment, torsional, or a combination of these modes. Guidelines for conducting load tests are currently being prepared by Cornell University under the sponsorship of the Electric Power Research Institute, with an estimated publication date during 1988. This will be a comprehensive and important reference document.

Application of sufficient load to large drilled shafts can be expensive; therefore, the number of load tests is limited, and instrumentation programs

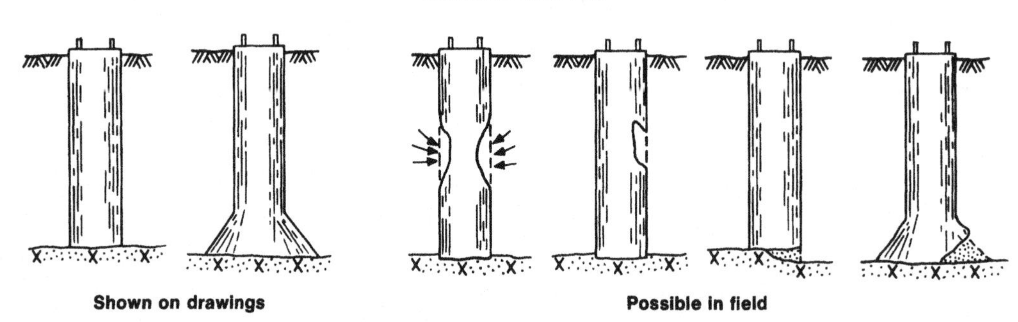

Figure 25.1. Drilled shafts (after DiBiagio and Myrvoll, 1981).

should be planned to provide comprehensive performance data. The cost of instrumentation is usually a minor part of the total cost.

Static Load Tests: Loading in Axial Compression and Uplift

Kulhawy et al. (1983) summarize 410 published and unpublished static load tests in axial compression or uplift, indicating that most such tests are conducted while measuring load and deformation at the shaft butt only. Such limited instrumentation does not provide comprehensive performance data, and subsurface measurements should also be made.

Measurement methods at the shaft butt are similar to those described for driven piles in Section 24.2.1. Vertical deformation is measured using dial indicators, often with a wire/mirror/scale arrangement as backup. Measurements of horizontal deformation, using similar methods, indicate whether or not the applied load is axial and concentric. As for driven piles, any movement of the reference beam should be measured throughout the test by surveying methods, and great care should be taken to minimize the effect of temperature changes on the reference beam. Load measurements require the use of a load cell.

Measurements of load in the shaft and at the tip should always be made when load testing large drilled shafts that have an anticipated capacity in excess of about 200 tons (1.8 MN). When conducting a load test to provide a basis for an efficient design, one cannot interpret the overall performance of the system completely unless load transfer data are available. When a load test is conducted during the construction period, as a proof test, the additional cost of subsurface instrumentation is minor. Load in the shaft should be measured using embedment strain gages, with a preference for the *sister bar* type with vibrating wire transducers. If a sliding micrometer is available, it can be used as an alternative to strain gages. Multiple telltales can be used as backup to strain gages, and the longest telltale provides data for tip settlement. Figure 12.106 shows one of a group of multiple telltales used during a recent static load test with which the author was involved. When planning multiple telltale installations and interpreting data, one must pay careful attention to the guidelines given in Section 13.3.1. Because the accuracy of loads that are calculated from telltale data is highly dependent on the quality of raw data, duplicate telltales should be installed at each depth. Data for conversion of strain to stress are provided as described in Chapter 24 for driven piles: by gaging an *unconfined compression test specimen* within a sleeve or above the ground surface, supplemented by use of strain data after all shaft resistance has been mobilized at a particular depth. However, for a drilled shaft it is not normally reasonable to leave a length of shaft above the ground surface, equal to three shaft diameters, and end effects in the *specimen* must be accounted for by averaging between several strain gages installed at one or more cross sections.

An attempt can be made to determine load at the tip from either strain gage or multiple telltale data. However, both are subject to inaccuracies in converting strain to stress, and use of multiple telltales involves an extrapolation. It is recommended that a hard data point be obtained by installing a contact earth pressure cell at the tip. Suitable cells are described in Chapter 10.

Figure 25.2 shows a possible layout of instrumentation during a static load test, loaded in axial compression. This is only one example, and it must be emphasized that many other configurations are possible, as indicated by the case histories described later in this chapter. A similar layout is applicable to a static load test, loaded in axial uplift, except that the contact earth pressure cell would be

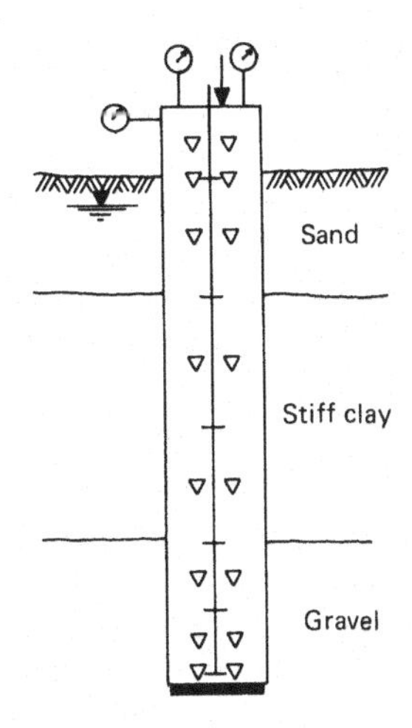

Multiple telltales, with dial indicators mounted on shaft butt. Duplicate telltales at each depth

Dial indicator, mounted on reference beam (wire/mirror/scale arrangement as backup)

Embedment strain gage

Contact earth pressure cell

Surveying methods (also on reference beam)

Figure 25.2. Possible layout of instrumentation during static load test: loading in axial compression. Note: Loading arrangements, load transfer assembly, and reference beam are not shown. Also, a load cell is required. Use of surveying methods requires benchmark, remote from shaft. Top of shaft should be kept wet.

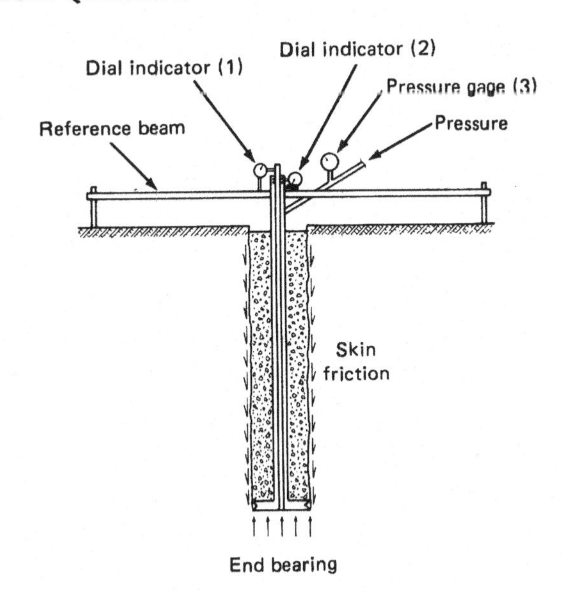

Figure 25.3. Arrangement for load testing a drilled shaft by applying load at the bottom (after Osterberg, 1984). Reprinted with the permission of the ADSC: an international association of foundation drilling contractors.

omitted. However, it should be recognized that determinations of stress from strain data during an uplift test are subject to significant errors if the concrete cracks. For this reason some engineers have installed surface-mounted electrical resistance strain gages on the reinforcing cage in preference to installing *sister bars* or other types of embedment strain gage. The author prefers use of *sister bars,* because well-designed sister bars will give the same data as gages on the reinforcing cage, and they are not subject to errors caused by installing gages in a field environment.

As an alternative to a hydraulic jack for load application, Horvath (1985) used three flatjacks, positioned in series with intervening bearing plates, thereby ensuring that the applied load was axial and concentric. The composite system was calibrated prior to use in the field. During a load test, one flatjack was inflated for load application and also used for load measurement. When the maximum aperture of the jack was reached, its valves were closed, the jack subsequently used as a passive load cell, and another flatjack activated for load application. Although this procedure is not accepted as a standard, it does present an interesting alternative.

Osterberg (1984) describes an innovative method for determining the load–movement relationship of a drilled shaft under static axial load by applying load at the bottom of the shaft. As shown in Figure 25.3, a special flatjack is placed at the bottom of the shaft before the concrete is poured. After the concrete has cured, the flatjack is pressurized internally, creating an upward force on the shaft and an equal but opposite downward force in end bearing. Attached to the bottom of the device is a steel pipe that extends upward within a steel sleeve. As pressure is applied, the inner pipe moves downward, and the movement is measured by using dial indicator (1). Dial indicator (2) is used to measure the upward movement of the shaft, and a pressure gage (3) is used to measure the internal pressure in the flatjack: this pressure is related to the total load during a separate calibration. Failure will occur either in skin friction or in end bearing, and the test is finished. Osterberg indicates that if the same failure load is applied as a downward load at the top of the shaft, the factor of safety against failure will be greater than 2 since, after allowing for the weight of the shaft, the skin friction and end bearing now both act upward to resist the downward load. When the test is completed, the flatjack is left in the hole.

Clearly, this test is considerably less expensive than a conventional load test, but additional comparative tests will be needed before its general applicability can be accepted. During the test, the major transfer of load occurs in the lower length of the shaft, the opposite of a foundation loaded in compression; thus, the load–movement pattern is different. This different pattern of load transfer distribution will be dependent on variation in soil type and on the state of in situ stress and requires further examination. Additional testing is planned.

Static Load Tests: Lateral Loading Mode

Measurement methods at the shaft butt include the use of a load cell or cable tension meter and horizontal and vertical dial indicators. A wire/mirror/scale arrangement is often used as backup. Horizontal deformation measurements should be arranged to monitor whether or not the loading is causing torsion in the shaft. Any movement of the reference beam should be measured throughout the test by surveying methods, and temperature effects should be minimized.

When subjected to lateral loads, the behavior of a drilled shaft is largely controlled by the flexural stiffness of the shaft in relation to the stiffness of the surrounding ground. When the shaft is rigid or short, it is often assumed to be infinitely stiff and to rotate as a rigid body, and failure usually occurs when the supporting capacity of the surrounding ground is reached. On the other hand, a flexible or long shaft exhibits elastic behavior; failure is usually structural and occurs when a plastic hinge forms at the point of maximum bending moment. Design of flexible or long shafts usually requires iterative techniques because the ground response depends on the magnitude of deformation. The designer of drilled shafts subjected to lateral loading must estimate the relative flexibility of the ground and shaft before deciding whether to use rigid or elastic design procedures. The information available to assist with this decision is very limited, and until a sufficient bank of data has been accumulated, load tests should include, wherever possible, determinations of bending stress in the shaft. Embedment strain gages can be used, again with a preference for the *sister bar* type with vibrating wire transducers, and again with limitations if concrete cracks. If a sliding micrometer is available, it can be used as an alternative to strain gages. A well-conducted test could develop up to 6 in. (150 mm) or more of lateral deformation at the butt; therefore, the opportunity exists to estimate bending stresses from inclinometer data as described in Chapter 12. However, this approach should only be used as backup. If it is used, inclinometer readings of maximum accuracy should be taken during each loading increment.

The understanding of the behavior under lateral load can be improved if the inclination of the butt of the shaft is measured, using a tiltmeter.

When a test with a lateral loading mode is conducted to determine how the soil reaction varies with depth, contact earth pressure cells can sometimes be installed at the interface between the shaft and the ground. However, this is possible only when constructing the shaft in an uncased hole. If casing is used, the only feasible approach is installation of an in situ testing device, such as a pressuremeter or dilatometer, in the ground alongside the shaft.

When a test is conducted in cohesive soils, such that deformation is affected by consolidation of soil on the compression side of the shaft, piezometers in the soil will provide additional data for evaluation of the test.

Figure 25.4 shows a possible layout of instrumentation during a static load test with a lateral loading mode.

Other Static Loading Modes

When moment, eccentric, torsional, or combined static loading modes are used, appropriate monitoring methods should be selected from those described for axial and lateral load tests. If loading is applied during moment or torsional tests by pulling on a steel cable, load will normally be monitored by using a cable tension meter. Torsional load tests may require use of embedment strain gages mounted horizontally to measure torsional strains.

Cyclic Loading Modes

Cyclic load tests require similar instrumentation to static tests, but the dynamic response of instrumentation must be sufficient to follow the loading sequence.

Remote Readings During Load Tests

During conventional load tests, instruments are normally read manually, but considerations of safety and/or convenience may lead to selection of

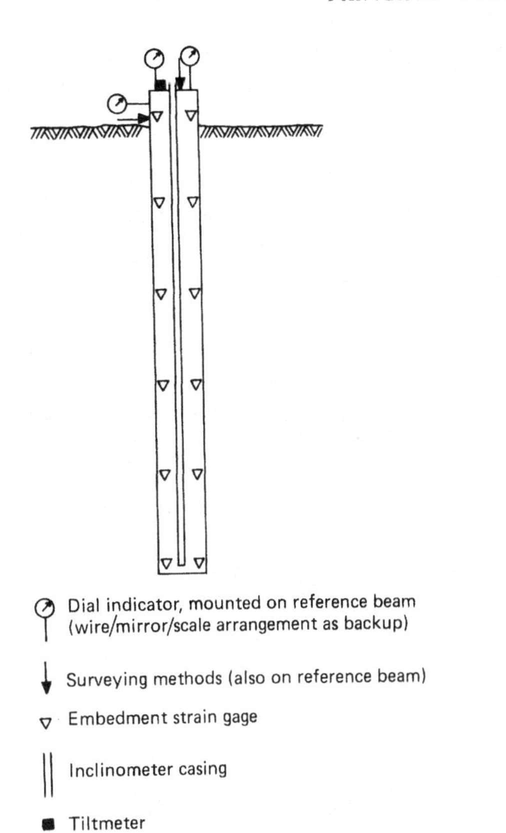

Figure 25.4. Possible layout of instrumentation during static load test: lateral loading mode. Note: Loading arrangements and reference beam are not shown. Also, a load cell or cable tension meter is required. Use of surveying methods requires benchmark, remote from shaft.

instruments that can be read remotely and perhaps by an automatic data acquisition system. This option is particularly attractive when the number of instruments is large. DCDTs can be used instead of dial indicators to monitor deformation of the shaft butt and telltales, and it is often worthwhile to use both DCDTs and dial indicators in parallel, to provide backup and to satisfy personnel who are concerned about trusting electrical devices.

*Dynamic Tests for Determining Load–Movement Relationships**

As discussed in Chapter 24 for driven piles, static load tests are expensive and time consuming, and less expensive dynamic tests are available. The *Case method* is not appropriate for drilled shafts, because no pile-driving hammer is used. *Transient dynamic response testing (TDR)* and *steady-state vibration testing*, described for driven piles in Chapter 24, are suitable for the determination of static axial and lateral load–movement relationships of drilled shafts. However, as for driven piles, the tests cannot be used to estimate ultimate static axial capacity, as neither method gives adequate input force.

An additional test, referred to as the *falling weight test*, has been developed specifically for determining the static axial load–movement relationships of drilled shafts and extends the application of stress wave methods to cases where no pile hammer is used. The test arrangement is identical to the arrangement for the *Case pile-driving analyzer*, shown in Figure 24.7: the pile hammer is replaced by a falling weight. Stress waves are generated by a mass of up to 4500 lb (2000 kg) dropped from varying heights on to a striker plate fixed to the butt of the shaft. The force transmitted to the butt is determined from strain measurements, using strain gages attached to the shaft near the butt. Simultaneously, the deformation of the butt is determined either by attaching an accelerometer to the butt or by use of a special laser theodolite. Static axial capacity is determined by using a similar analysis to the stress wave method. The major drawback of the falling weight method is the requirement to use strain measurements in the concrete near the butt of the shaft for a determination of force transmitted to the butt. An assumption must be made for the modulus of the concrete, and the concrete near the butt may not be representative of the total shaft. For determination of ultimate static axial capacity, it is necessary to displace the tip of the shaft during the test. This requirement imposes limitations in the case of large-diameter drilled shafts, because dynamic forces of about 4000 kips (18 MN) or more are required. Application of such large dynamic forces will often damage the concrete at the shaft butt, and it is usually preferable to lengthen the duration of the impact between the falling weight and the shaft butt by placing cushion material on the shaft.

In summary, the application of dynamic methods for the determination of load–movement relationships of drilled shafts has not progressed as rapidly as for precast concrete driven piles. Much of the experience at lower displacements (up to working loads) has been obtained on drilled shafts using TDR and vibration techniques. Extension of these methods, incorporating accelerometers in the body of the shaft, will undoubtedly increase their value.

*Written with the assistance of Allen G. Davis, Director, Testconsult CEBTP Ltd., Thelwall, England.

The falling weight test is an advance in the right direction. It has not reached the same status as the pile-driving analyzer for driven piles, principally because of the problem of using the concrete near the butt of the shaft for determination of force transmitted to the butt.

25.2.2. Will There Be Significant Downdrag Loading?

The causes and effects of downdrag loading on drilled shafts are described in Chapter 2 and the loading may be significant when shafts are constructed through soft cohesive soil and founded on firm soil. Settlement measurements, groundwater pressure measurements, and/or a full-scale test may be appropriate, as described for driven piles in Chapter 24.

25.2.3. What Is the Integrity of the Concrete?

Reese (1978) lists construction problems that can lead to a drilled shaft of inferior quality. These include improper excavation that allows collapse of the hole during drilling, inadequate cleaning at the base, poor concrete, improper placement of concrete, and improper pulling of casing. In all these cases, the integrity of the concrete may be questionable, and concerns are increased if the contractor is inexperienced or careless.

Continuous diamond core drilling is sometimes used to obtain samples for assessing the quality of the concrete and detecting flaws. However, this method is expensive and time consuming, and borehole direction surveys are often required to avoid misleading data in long shafts because of alignment problems.

An alternative procedure for detecting flaws is to survey the shaft by using one of the integrity testing procedures* described in Chapter 24 for driven piles. Arias (1977) and Weltman (1977) describe and compare the various procedures for integrity testing of drilled shafts by nondestructive methods. All methods that rely on measurements at the shaft butt gather maximum information from the upper part of the shaft and are therefore of limited value. In particular, for shafts cast in stiff soils, damping out of the response can result in very limited or no information on the integrity of concrete near the bottom of the shaft. If sensors are placed in the shaft during construction, additional information can be obtained, but at extra cost and inconvenience. The *echo test* cannot differentiate between bulbs formed by overbreak of the soil around the shaft and necking in the shaft.

As discussed for driven piles in Chapter 24, *sonic coring* and *single-hole* and *cross-hole nuclear density measurements* can be carried out within pipes. Sonic coring is the preferred method for a complete evaluation of concrete integrity in drilled shafts. The method requires three or four embedded metal pipes along the length of the shaft. When four pipes are used, measurements are made across the four sides and two diagonals. When one of the nuclear density methods is employed, at least four pipes should be used, but the methods are subject to the limitations given in Chapter 24.

*The text on integrity testing has been written with the assistance of Allen G. Davis, Director, Testconsult CEBTP Ltd., Thelwall, England.

Table 25.1. Overview of Routine and Special Monitoring

Application	Measurement
Routine monitoring during static axial load testing	Load and movement at shaft butt Load in shaft and at tip
Routine monitoring during static lateral load testing	Load and movement at shaft butt Bending stress in shaft
Special load testing applications	Earth pressure at contact with shaft during static lateral load testing Moment, eccentric, torsional, or combined static loading modes: appropriate parameters as for static axial and lateral modes Load, velocity, strain, and/or deformation, as required for dynamic tests Deformation and stress during full-scale test to determine downdrag loading
Other special applications	Groundwater pressure Integrity of concrete Surface settlement and/or groundwater pressure for evaluation of downdrag loading

25.3. OVERVIEW OF ROUTINE AND SPECIAL APPLICATIONS

For those readers who seek general guidelines on what is typically *routine* monitoring and what is monitoring for *special applications,* Table 25.1 presents a broad generalization.

In practice, the most frequent use of instrumentation for drilled shafts is to examine load–movement relationships during static axial load testing.

25.4. SELECTED CASE HISTORIES

Table 25.2 summarizes selected case histories of instrumented drilled shafts.

Table 25.2. Summary of Selected Case Histories of Drilled Shafts

Project	Principal Concerns	Type of Load Test	Ground Conditions	Measurement[a]	Instrumentation Discussed	Special Features	Reference
First International Plaza, San Antonio, TX, U.S.A.	Performance of foundation: comparison with predictions	—	Clay and clay shale	S	Embedment strain gages (Carlson type)		Briaud et al. (1985)
Test site at Burlington, Ontario, Canada	Load–movement relationships	Static axial compression	Weak mudstone	L S S	Flatjacks Multiple telltales Embedment strain gages (vibrating wire type)	Flatjacks used for load application	Horvath (1985)
Two sites in Denver, CO, U.S.A.	Integrity of concrete	—	Sand and clay-shale overlying bedrock Fill, sand, and gravel overlying bedrock Shafts socketed into bedrock at both sites	V	Modified version of sonic coring method with surface receivers		Olson and Thompson (1985)
Two sites in Texas, U.S.A.	Integrity of concrete	—	Various	V	Modified version of sonic coring method with surface and embedded receivers		Harrell and Stokoe (1984)
Power plant in South America	Integrity of concrete	—	Marine sediments, residual soil, weathered rock, overlying bedrock Shafts socketed into bedrock	V	Sonic coring with one and three embedded pipes		Kissenpfennig et al. (1984)
Various	Load–movement relationships	Static axial compression and uplift	Various	D D D D D L	Surveying methods Dial indicators Wire/mirror/scale LVDTs Telltales Load cells	410 tests	Kulhawy et al. (1983)
Two sites in southern Florida, U.S.A.	Load–movement relationships	Static axial compression	Fill and sands overlying weak porous limestone Shafts socketed into limestone	S S	Multiple telltales Embedment strain gages (Mustran type)		Gupton et al. (1982)

Table 25.2. Summary of Selected Case Histories of Drilled Shafts (*Continued*)

Project	Principal Concerns	Type of Load Test	Ground Conditions	Measurement[a]	Instrumentation Discussed	Special Features	Reference
Various	Integrity of concrete	—	Various	V	Steady-state vibration and transient dynamic response testing		Robertson (1982)
Test sites in Texas, U.S.A.	Load–movement relationships under lateral loading	Static lateral	Various	D L TS	Tiltmeters Cable tension meter Contact earth pressure cells		Bierschwale et al. (1981)
Sixteen-story building in North London, England	Magnitude of load taken by raft/clay interface and by shafts	—	London clay	D TS	Probe extensometers Contact earth pressure cells	Shafts capped with reinforced concrete raft	Cooke et al. (1981)
Eight-story building at Ponce, Puerto Rico	Load–movement relationships	Static axial compression	Sand and silty sand	S	Embedment strain gages (Mustran type)		Farr and Aurora (1981)
Sammis Plant, Steubenville, OH, U.S.A.	Load–movement relationships	Static axial compression, uplift, lateral	Sand and gravel overlying bedrock Shafts socketed into bedrock	S	Embedment strain gages (sister bars with weldable electrical resistance strain gage transducers)		Newman et al. (1981)
Power station in southeast Oklahoma, U.S.A.	Load–movement relationships	Static axial compression	Clay–shale	S	Embedment strain gages (Mustran type)		Goeke and Hustad (1979)
Various sites in Texas, U.S.A.	Load–movement relationships	Static axial compression	Various	S S CS TS	Multiple telltales Embedment strain gages (Mustran type) Concrete stress cells Contact earth pressure cells		Reese et al. (1976)
Test site in Houston, TX, U.S.A.	Load–movement relationships under lateral loading	Static lateral	Stiff and very stiff clay	L S	Load cell Strain gages	Strain gages bonded to steel pipe at axis of shaft	Reese and Welch (1975)
Westminster Bank Tower, London, England	Magnitude of load taken by raft/clay interface and by shafts	—	London clay	D L TS	Probe extensometers Photoelastic load cells Contact earth pressure cells	Shafts capped with reinforced concrete raft	Haws et al. (1974)
Test site in Wembley, London, England	Load–movement relationships	Static axial compression	Various	TS	Contact earth pressure cells		Whitaker and Cooke (1966)

[a]CS: stress in concrete L: load in shaft TS: total stress between shaft and soil
D: deformation S: strain in shaft V: velocity, load, strain, and/or deformation as required for dynamic tests.

Part 6

The Key to Success

Part 6 consists of Chapter 26. It is designed as a closing chapter to summarize the key issues for maximizing the success of geotechnical instrumentation for monitoring field performance.

CHAPTER 26

THE KEY TO SUCCESS: THE CHAIN WITH 25 LINKS

Full benefit can be achieved from geotechnical instrumentation programs only if every step in the planning and execution process is taken with great care. The analogy can be drawn to a chain with many potential weak links: this chain breaks down with greater facility and frequency than in most other geotechnical engineering endeavors. Shortcomings in instrumentation programs can usually be attributed to a weakness in one or more links. In this chapter, the major links are defined in chronological order, and guidance is given for maximizing the strength of each. There are 25 links.

The first 17 links in the chain are forged during the planning phase, and more detail is given in Part 2 of this book. The last eight links in the chain are forged during the execution phase, and more detail is given in Part 4.

1st Link. Predict Mechanisms that Control Behavior

One or more working hypotheses should be developed for mechanisms that are likely to control behavior.

2nd Link. Define the Geotechnical Questions that Need to be Answered

Every instrument on a project should be selected and placed to assist in answering a specific question: **if there is no question, there should be no instrumentation.** Before addressing measurement methods themselves, one should make a listing of geotechnical questions that are likely to arise during the design, construction, or operation phases.

3rd Link. Define the Purpose of the Instrumentation

Instrumentation can provide benefits during the design, construction, or operation phase of a project but should not be used unless there is a valid reason that can be defended.

4th Link. Select the Parameters to be Monitored

The question *which parameters are most significant?* should be answered. It is often found that deformation measurements are the simplest, most reliable, and least ambiguous.

5th Link. Predict Magnitudes of Change

Predictions should be made to establish the required range and accuracy of each instrument. Whenever measurements are made for construction control or safety purposes, a predetermination of *hazard warning levels* should be made.

6th Link. Devise Remedial Action

Inherent in the use of instrumentation for construction purposes is the absolute necessity for deciding, in advance, a positive means for solving any problem that may be disclosed by the results of the observations. If the observations should demonstrate that remedial action is needed, that action must be based on appropriate, previously anticipated plans. An open communication channel should be maintained between design and construction personnel, so that remedial action can be discussed at any time.

7th Link. Assign Tasks for Design, Construction, and Operation Phases

When tasks are assigned for monitoring, the party with the greatest vested interest in the data should be given direct line responsibility for producing it accurately. Reliability and patience, perseverance, a background in the fundamentals of geotechnical engineering, mechanical and electrical ability, attention to detail, and a high degree of motivation are the basic requirements for qualities needed in instrumentation personnel.

8th Link. Select Instruments

When instruments are selected, the overriding desirable feature is **reliability.** Inherent in reliability is maximum simplicity and, in general, transducers can be placed in the following order of decreasing simplicity and reliability: optical, mechanical, hydraulic, pneumatic, electrical. Lowest cost of an instrument should never be allowed to dominate the selection, and the least expensive instrument is not likely to result in minimum overall cost.

The state of the art in hardware design is far ahead of the state of the art in user technology. It is the responsibility of users to develop an adequate level of understanding of the instruments that they select. Users will often benefit from discussing their application with geotechnical engineers or geologists on the manufacturer's staff before selecting instruments, conveying as much as possible about the application and seeking out any limitations of the proposed instruments.

9th Link. Select Instrument Locations

The selection of instrument locations should reflect predicted behavior and should be compatible with the method of analysis that will later be used when interpreting the data.

A practical approach to selecting instrument locations involves three steps. First, zones of particular concern are identified, and appropriate instrumentation is located. Second, a selection is made of zones, normally cross sections, where predicted behavior is considered representative of behavior as a whole. These cross sections are then regarded as *primary instrumented sections,* and instruments are located to provide comprehensive performance data. Third, because the selection of representative zones may be incorrect, simple instrumentation should be installed at a number of *secondary instrumented sections* to serve as indices of comparative behavior.

When selecting locations, survivability of instruments should be considered, and additional quantities should be selected to replace instruments that may become inoperative.

10th Link. Plan Recording of Factors that May Influence Measured Data

Measurements by themselves are rarely sufficient to provide useful conclusions. The use of instrumentation normally involves relating measurements to causes, and therefore complete records and diaries must be maintained of all factors that might cause changes in the measured parameters.

11th Link. Establish Procedures for Ensuring Reading Correctness

Personnel responsible for instrumentation must be able to answer the question: *Is the instrument functioning correctly?* Input to the answer can often be provided by visual observations, by duplicate or backup instruments, or by studies of data consistency and repeatability.

12th Link. Prepare Budget

A budget should be prepared to ensure that sufficient funds are available for all monitoring tasks.

13th Link. Write Instrument Procurement Specifications

Procurement of other than the most simple geotechnical instruments should not be considered as a routine construction procurement item because, if valid measurements are to be made, extreme attention must be paid to quality and details. The cost is usually minor.

The *low-bid* method should never be used unless regulations allow for no alternative, and one of the following two methods is recommended:

- The owner or design consultant procures the instruments directly, negotiating prices with suppliers.
- The owner enters an estimate of procurement cost in the construction contract bid schedule and subsequently selects appropriate instruments for procurement by the contractor. Price is negotiated between the owner and suppliers of instruments, and the construction contractor is reimbursed at actual cost plus a handling fee.

In cases where neither of these methods can be used and the *low-bid* method with an *or equal* provision is unavoidable, a clear, concise, complete, and correct specification must be written. The specification should cover all salient features to guard against supply of an undesirable substitution, following the guidelines given in Chapter 5.

While writing instrument procurement specifications, one should determine the requirements for *factory calibrations,* and *acceptance tests* should be planned to ensure correct functioning when instruments are first received by the user.

14th Link. Plan Installation

Installation procedures and schedules should be planned well in advance of scheduled installation dates. Written step-by-step procedures should be prepared, including a detailed listing of required materials and tools and an *installation record sheet,* on which as-built installation details will be documented.

Procedures should ensure that the presence of the instruments do not alter the very quantities that instruments are intended to measure. When the owner or design consultant is responsible for installing instruments, a special effort must be made during the planning stage to establish a cooperative working relationship with the construction contractor.

15th Link. Plan Regular Calibration and Maintenance

Regular calibration and maintenance of readout units are required during service life. Planning should include procedures and schedules for regular maintenance of field terminals and any embedded components for which access is available.

16th Link. Plan Data Collection, Processing, Presentation, Interpretation, Reporting, and Implementation

Many consulting engineering firms have files filled with large quantities of partially processed and undigested data because sufficient time or funds were not available for tasks after completion of instrument installation.

Written procedures for data collection, processing, presentation, and interpretation should be prepared before instrumentation work commences in the field. A verification should be made to ensure that personnel responsible for interpretation of instrumentation data have contractual authority to initiate remedial action, that communication channels between design and construction personnel are open, and that arrangements have been made to forewarn all parties of the planned remedial actions.

17th Link. Write Contractual Arrangements for Field Instrumentation Services

Field services include instrument installation, regular calibration and maintenance, and data collection, processing, presentation, interpretation, and reporting. Contractual arrangements for the selection of personnel to provide these services may govern success or failure of a monitoring program.

Geotechnical instrumentation field work should not be considered a routine construction item, because successful measurements require extreme dedication to detail throughout all phases of the work. The *low-bid* method should never be used unless regulations allow for no alternative, and one of the following two methods is recommended:

- The owner or design consultant performs field work that requires specialist skill, if necessary retaining the services of a consulting firm specializing in instrumentation. Supporting work is performed by the construction contractor.
- The owner enters an estimate of specialist field service costs in the construction contract bid schedule. Subsequently, the owner and construction contractor select an appropriate specialist consulting firm, which is retained as a subcontractor by the construction contractor

to perform field work that requires specialist skill. Charges for specialist work are negotiated between the owner and consulting firm, and the construction contractor is reimbursed at actual cost plus a handling fee. Supporting work is performed by the construction contractor.

One of these two methods is essential for data processing, presentation, and interpretation. In cases where regulations do not allow either of these methods for installation, calibration, maintenance, and data collection, and where the *low-bid* method is unavoidable, a clear, concise, complete, and correct specification should be written to maximize the quality of field services, following the guidelines given in Chapter 6.

18th Link. Procure Instruments

Instruments should be calibrated, inspected, and tested before shipment to the user. On receipt by the user, *acceptance tests* should be made to ensure correct functioning.

19th Link. Install Instruments

Before starting installation work, field personnel should study and understand the written step-by-step installation procedure. They should be aware that the instrument may not serve its purpose if a single essential but apparently minor requirement is overlooked during the installation.

When installing instruments in boreholes, one should respect the driller's advice. If problems arise while installing downhole components, and there is any possibility of borehole collapse or premature grout setting, act fast and discuss later: *if in doubt, pull it out*.

20th Link. Calibrate and Maintain Instruments on a Regular Schedule

Readout units should be calibrated and maintained on a regular schedule. Field terminals should be maintained, and maintenance should include any embedded instrument components for which access is available.

21st Link. Collect Data

Special care should be taken when making initial readings, because most data are referenced to these readings, and engineering judgments are based on changes rather than on absolute values.

Data collection personnel should take the first step in determining whether the instrument is functioning correctly, by comparing the latest readings with the previous readings. Any significant changes can then be identified immediately, and if *hazard warning levels* have been reached, supervisory personnel should be informed. Data collection personnel should record factors that may influence measured data and should be on the lookout for potential for damage, deterioration, or malfunction of instruments.

22nd Link. Process and Present Data

The first aim of data processing and presentation is to provide a rapid assessment of data in order to detect changes requiring immediate action, and this should usually be the responsibility of data collection personnel. The second aim is to summarize and present the data in order to show trends and to compare observed with predicted behavior for determination of the appropriate action to be taken. Specially prepared data forms will usually be required, and data should always be plotted, normally versus time. Plots of predicted behavior and causal data are often included on the same axes.

23rd Link. Interpret Data

Monitoring programs have failed because the data generated were never used. If there is a clear sense of purpose for a monitoring program, the method of data interpretation will be guided by that sense of purpose. Without a purpose there can be no interpretation.

Communication channels between design and field personnel should remain open, so that discussions can be held between design engineers who framed the questions that need to be answered and field engineers who provide the data.

Early data interpretation steps have already been taken: evaluation of data to determine reading correctness and also to detect changes requiring immediate action. The essence of subsequent data interpretation steps is to correlate the instrument readings with other factors (cause and effect relationships) and to study the deviation of the readings from the predicted behavior. When faced with data that on first sight do not appear to be reasonable, there is a temptation to reject the data as false.

However, such data may be real and may in fact carry an important message. A significant question to ask is: *Can I think of a hypothesis that is consistent with the data?* The resultant discussion, together with the use of appropriate procedures for ensuring reading correctness, will often lead to an assessment of data validity.

24th Link. Report Conclusions

After each set of data has been interpreted, conclusions should be reported in the form of an *interim monitoring report* and submitted to personnel responsible for implementation of data. The initial communication may be verbal but should be confirmed in writing. The report should include updated summary plots, a brief commentary that draws attention to all significant changes that have occurred in the measured parameters since the previous interim monitoring report, probable causes of these changes, and recommended action.

A *final report* is often prepared to document key aspects of the monitoring program and to support any remedial actions. The report also forms a valuable bank of experience and should be distributed to the owner and design consultant so that any lessons may be incorporated into subsequent designs.

If important knowledge gaps have been filled, the conclusions should be disseminated to the profession in a technical publication.

25th Link. Implement Data

When instrumentation is used to provide input to the initial design of a facility or for the design of remedial treatment, data will be used directly in design. When instrumentation is used during construction, remedial action should have been planned and implementation should follow these plans.

Summary to the Key to Success

Peck (1972) states, "We need to carry out a vast amount of observational work, but what we do should be done for a purpose and be done well."

Part 7

Appendixes

A. Checklist for Planning Steps
B. Checklist for Content of Specifications for Procurement of Instruments
C. Checklist for Content of Specifications for Field Instrumentation Services
D. Commercially Available Geotechnical Instruments
E. Details of Twin-Tube Hydraulic Piezometer System
F. Dimensions of Drill Rods, Flush-Joint Casing, Diamond Coring Bits, Hollow-Stem Augers, and U.S. Pipe
G. Example of Installation Procedure, with Materials and Equipment List
H. Conversion Factors

APPENDIX A

CHECKLIST FOR PLANNING STEPS

Systematic planning of a monitoring program should proceed through the steps described in Chapter 4. The steps are summarized in this appendix in checklist form.

1. **Define the Project Conditions**
   - (a) Project type
   - (b) Project layout
   - (c) Subsurface stratigraphy and engineering properties
   - (d) Groundwater conditions
   - (e) Status of nearby structures or other facilities
   - (f) Environmental conditions
   - (g) Planned construction method
   - (h) Knowledge of crisis situation
2. **Predict Mechanisms that Control Behavior**
3. **Define the Geotechnical Questions that Need to Be Answered**
4. **Define the Purpose of the Instrumentation**
   - (a) Benefits during design
     - definition of initial site conditions
     - proof testing
     - fact-finding in crisis situations
   - (b) Benefits during construction
     - safety
     - observational method
     - construction control
     - providing legal protection
     - measurement of fill quantities
     - enhancing public relations
     - advancing the state of the art
   - (c) Verifying satisfactory performance after construction is complete
5. **Select the Parameters to Be Monitored**
   - (a) Pore water pressure or joint water pressure
   - (b) Total stress within soil mass
   - (c) Total stress at contact with structure or rock
   - (d) Stress within rock mass
   - (e) Vertical deformation
   - (f) Horizontal deformation
   - (g) Tilt
   - (h) Strain in soil or rock
   - (i) Load or strain in structural members
   - (j) Temperature
6. **Predict Magnitudes of Change**
   - (a) Predict maximum value, thus instrument range
   - (b) Predict minimum value, thus instrument sensitivity or accuracy
   - (c) Determine hazard warning levels
7. **Devise Remedial Action**
   - (a) Devise action for each hazard warning level, ensuring that labor and materials will be available
   - (b) Determine who will have contractual authority for initiating remedial action

(c) Ensure that communication channel is open between design and construction personnel
(d) Determine how all parties will be forewarned of planned remedial actions

8. **Assign Tasks for Design, Construction, and Operation Phases**
(a) Complete Table 4.2
(b) Assign supervisory responsibility for tasks by instrumentation specialist
(c) Plan liaison and reporting channels
(d) Plan who has overall responsibility and contractual authority for implementation

9. **Select Instruments**
(a) Plan for high **reliability:**
- study suggested recipe for reliability in Chapter 15
- maximum simplicity
- don't allow lowest cost to dominate selection
- maximum durability in installed environment
- minimum sensitivity to climatic conditions
- good past performance record
- consider transducer, readout unit, and communication system separately
- is reading necessarily correct?
- can calibration be verified after installation?

(b) Discuss application with manufacturer
(c) Recognize any limitations in skill or quantity of available personnel
(d) Consider both construction and long-term needs and conditions
(e) Ensure good conformance
(f) Ensure minimum interference to construction and minimum access difficulties
(g) Determine need for automatic data acquisition system
(h) Plan readout type and arrangements, consistent with required reading frequency
(i) Plan need for spare parts and standby readout units
(j) Evaluate adequacy of lead time
(k) Evaluate adequacy of time available for installation
(l) Question whether the selected instrument will achieve the objective

10. **Select Instrument Locations**
(a) Identify zones of primary concern
(b) Select primary instrumented sections
(c) Select secondary instrumented sections
(d) Plan quantities to account for less than 100% survival
(e) Arrange locations to provide early data
(f) Arrange locations to provide cross-checks
(g) Avoid nonconformance or weakness at clusters

11. **Plan Recording of Factors that May Influence Measured Data**
(a) Construction details
(b) Construction progress
(c) Visual observations of expected and unusual behavior
(d) Geology and other subsurface conditions
(e) Environmental factors

12. **Establish Procedures for Ensuring Reading Correctness**
(a) Visual observations
(b) Duplicate instruments
(c) Backup system
(d) Study of consistency
(e) Study of repeatability
(f) Regular in-place checks

13. **List the Specific Purpose of Each Instrument**

14. **Prepare Budget**
Include costs, being particularly careful to make a realistic estimate of project duration, for
(a) Planning monitoring program
(b) Making detailed instrument designs
(c) Procuring instruments
(d) Making factory calibrations
(e) Installing instruments
(f) Maintaining and calibrating instruments on a regular schedule
(g) Establishing and updating data collection schedule
(h) Collecting data

(i) Processing and presenting data
(j) Interpreting and reporting data
(k) Deciding on implementation of results

15. Write Instrument Procurement Specifications

(a) Assign responsibility for procurement
- construction contractor
- owner
- design consultant
- instrument suppliers acting as assigned subcontractors

(b) Select specifying method
- descriptive specification, with brand name and model number
- descriptive specification, without brand name and model number
- performance specification

(c) Select basis for determining price
- negotiation
- bid

(d) Write specifications
(e) Plan factory calibrations
(f) Plan acceptance tests when instruments are first received by user, and determine responsibility

16. Plan Installation

(a) Prepare step-by-step installation procedure well in advance of scheduled installation dates, including list of required materials and tools
(b) Prepare installation record sheets
(c) Plan staff training
(d) Coordinate plans with contractor
(e) Plan access needs
(f) Plan protection from damage and vandalism
(g) Plan installation schedule

17. Plan Regular Calibration and Maintenance

(a) Plan calibrations during service life
- readout units
- embedded components

(b) Plan maintenance
- readout units
- field terminals
- embedded components

18. Plan Data Collection, Processing, Presentation, Interpretation, Reporting, and Implementation

(a) Plan data collection
- prepare preliminary detailed procedures for collection of initial and subsequent data
- prepare field data sheets
- plan staff training
- plan data collection schedule
- plan access needs

(b) Plan data processing and presentation
- determine need for automatic data processing
- prepare preliminary detailed procedures for data processing and presentation
- prepare calculation sheets
- plan data plot format
- plan staff training

(c) Plan data interpretation
- prepare preliminary detailed procedures for data interpretation

(d) Plan reporting of conclusions
- define reporting requirements, contents, frequency

(e) Plan implementation
- verify that all Step 7 items are in place

19. Write Contractual Arrangements for Field Instrumentation Services

(a) Select field service contract method (complete Table A.1)
(b) Write detailed specifications

20. Update Budget

Include costs for all tasks listed in Step 14

Table A.1. Allocation of Responsibilities for Field Instrumentation Services

Services Provided by	Installation	Regular Calibration and Maintenance	Data Collection, Processing and Presentation	Data Interpretation and Reporting
Owner's personnel				
Construction contractor, using bid items in construction contract, without prequalification	Only very simple instrumentation	Only very simple instrumentation	Only very simple instrumentation	Not suitable
Construction contractor, using bid items in construction contract, with prequalification				Not suitable
Instrumentation specialist selected by and contracting with owner				
Instrumentation specialist selected by owner and contractor, and contracting with contractor as an assigned subcontractor				

APPENDIX B

CHECKLIST FOR CONTENT OF SPECIFICATIONS FOR PROCUREMENT OF INSTRUMENTS

Specifications for procurement of instrumentation are discussed in Chapter 5, and recommended content is given in Section 5.4. The content is summarized in this appendix in checklist form. As indicated in Section 5.4, it is not intended that all specifications should address all items in the checklist. Greatest detail is required when the low-bid procedure with an *or equal* provision is used.

The headings in this checklist correspond to the subheadings in Section 5.4.

Part 1. General (Wording Applicable to All Instruments)

5.4.1. Division of Responsibilities
- Procurement
- Factory calibration and quality assurance
- Review of proposed instruments
- Acceptance tests
- Summary statement of work included

5.4.2. Submittals
- Experience lists
- Requests for review of proposed instruments
- Calibration certificates
- Quality assurance checklists
- Warranties
- Instruction manuals
- Shipping documents
- Instrument samples
- Schedules for all submittals

5.4.3. Operating Environment
- Definition of environment
- Ground deformation

5.4.4. Experience
- Projects
- Users

5.4.5. General Material Requirements
- Factors in Table 5.4
- Longevity

5.4.6. Review of Proposed Instruments
- Requirements for review
- *Or equal* specification
- Submittals

5.4.7. Factory Calibration and Quality Assurance
- Factory calibrations
- Calibration certificates
- Unique identification
- Preshipment inspections
- Quality assurance checklists

5.4.8. Warranty

5.4.9. Instruction Manual

5.4.10. Shipment and Delivery

- Delivery dates
- Partial shipment schedule
- Method of shipment
- Prepayment of freight charges
- Insurance in transit
- Submittal of shipping and insurance documents

Part 2. Instrument Details (Separate Subheading for Each Instrument)

5.4.11. Instrument Operating Principles

5.4.12. Component Specifications

- Descriptive or performance specification
- Range
- Acceptable uncertainty
- Conformance requirements
- Method of verifying reading correctness
- Detail design of nonstandard components
- Verification that all required components have been included

5.4.13. Compatibility with Other Instruments

5.4.14. Physical Size Limitations

5.4.15. Submittal of Samples

- Requirements
- Characteristics to be tested

5.4.16. Installation Tools and Materials

5.4.17. Spare Parts

- Damage during installation
- Malfunction during operation
- Readout units
- Accessible terminals

Part 3. Measurement and Payment (Section 5.4.18)

- Measurement method
- Payment method
- Adequate definition of quantity and dimensions for payment purposes
- Pay item for every specified requirement
- Assigned subcontracts:
  - Definition of procedure
  - Markup
  - Adjustment of allowance

Reviews of Specification (Section 5.4)

- By experienced specification writer
- By instrumentation specialist
- By personnel familiar with other sections of the specifications

APPENDIX C

CHECKLIST FOR CONTENT OF SPECIFICATIONS FOR FIELD INSTRUMENTATION SERVICES

Specifications for field instrumentation services are discussed in Chapter 6, and recommended content is given in Section 6.6. The content is summarized in this appendix in checklist form. As indicated in Section 6.6, it is not intended that all specifications should address all the items in the checklist but, if the low-bid procedure has been selected for any task, extra care must be taken to write a comprehensive specification and all relevant requirements should be specified.

The headings in this checklist correspond to the subheadings in Section 6.6.

Part 1. General

6.6.1. Purpose of Instrumentation Program

- Overall purpose
- Parameters monitored by each instrument
- Use of data (brief statement)

6.6.2. Division of Responsibilities

- Procurement
- Factory calibration
- Acceptance tests
- Installation
- Regular maintenance
- Regular calibration
- Establishing and updating reading schedule
- Data collection
- Data processing and presentation
- Data interpretation
- Reporting of results
- Deciding on implementation of results
- Summary statement of contractor responsibilities: work included

6.6.3. Specification Method

- Definition and listing of support work
- Definition and listing of specialist work
- Description of specification method for:
  - Installation
  - Regular calibration and maintenance
  - Data collection, processing, and presentation
  - Data interpretation and reporting
- Definition and use of contingency allowance or assigned subcontract (brief statement)

6.6.4. Related Work Specified Elsewhere

- Listing
- Section numbers

6.6.5. Qualifications of Specialist Field Instrumentation Personnel
- General requirements
- Requirements for firm
- Requirements for individuals
- Schedule for qualification statements

6.6.6. Submittals During Construction
- Detailed requirements
- Schedule

6.6.7. Cooperation Between Construction Contractor and Owner's Specialist Field Instrumentation Personnel

Part 2. Products

6.6.8. Procurement of Instruments
- If procured by others:
  - Listing
  - Operating principles
  - Sizes
  - Factory calibration and quality assurance
  - Acceptance tests by owner
  - Warranties
  - Shipment and delivery
  - Schedules
  - Insurances
  - Handing over
  - Checking
  - Responsibility
- If procured by construction contractor:
  - Chapter 5 and Appendix B
  - Tabular listing of quantities
- Storage space

6.6.9. Support Work (Relating to Products)
- Detailed specification for each biddable and incidental work item (unless specified later)
- Description of each nonbiddable work item, paid for under a contingency allowance (unless specified later)
- Schedule requirements

Part 3. Execution

6.6.9. Support Work (Relating to Execution)
- Detailed specification for each biddable and incidental work item (unless specified later)
- Description of each nonbiddable work item, paid for under a contingency allowance (unless specified later)
- Schedule requirements

6.6.10. Locations of Instruments
- Approximate locations on drawings
- Owner's representative will finalize
- Record survey, with required accuracy

6.6.11. Installation of Instruments
- Details on drawings
- Step-by-step procedures
- General and instrument-specific drilling requirements
- Submission for review
- Log of installation activities
- Schedules
- Work restrictions
- Drilling specification
- Grouting specification
- Acceptance tests

6.6.12. Regular Maintenance and Calibration
- Procedures
- Schedules

6.6.13. Data Collection, Processing, and Presentation
- Procedures
- Format and schedule for data submission
- Frequency
- Start and termination schedule
- Joint initial readings

6.6.14. Availability of Data
- Availability of raw or interpreted data
- Schedule
- Contractor's responsibility for safety

6.6.15. Implementation of Data
- Use of data
- Action to be taken:
  - Predetermined remedial actions
  - Hazard warning levels
  - Contractual responsibility
  - Method of forewarning
- Consistency with non-instrumentation sections verified

6.6.16. Delay to Construction
- Delay time estimates
- Payment

6.6.17. Damage to Instrumentation
- Protection requirements
- Responsibility of contractor
- Timeliness for repair or replacement
- Work stoppage
- Financial penalty

6.6.18. Disposition of Instruments
- Ownership of portable instruments
- Salvage and ownership of recoverable components
- Restoration of surface and subsurface
- Overhaul of long-term instruments

Part 4. Measurement and Payment (Section 6.6.19)

- Every work item:
  Measurement method
  Payment method
  Work included and excluded
- Furnish and install work items:
  Definition of major component prices
- Data-set work items:
  Definition of unit
- Contingency allowance item for nonbiddable support work:
  Applicable work items
  Payment rates for labor
  Payment rates for plant and equipment
  Payment rates for materials
  Markup
- Assigned subcontract for specialist work:
  Applicable work items
  Definition of procedure
  Markup
  Adjustment of allowance

Reviews of Specification (Section 6.6)

- By experienced specification writer
- By instrumentation specialist
- By personnel familiar with other sections of the specifications

APPENDIX D

COMMERCIALLY AVAILABLE GEOTECHNICAL INSTRUMENTS

Instruments are grouped in the categories that have been defined in Part 3. The numbers alongside each category refer to the alphabetical listing of manufacturers given at the end of this appendix.

Certain manuacturers' names may not be listed alongside certain instrument categories. Any such omissions result solely from the author's lack of knowledge and do not imply any inadequacy in those omitted instruments.

Table D.1. Instruments for Measuring Groundwater Pressure

Instruments	Manufacturers
Observation wells	1, 2
Open standpipe piezometers	20, 36, 48, 49, 53, 54, 73, 95, 100, 102, 107, 108, 109, 110, 112
Twin-tube hydraulic piezometers	48, 53, 62, 73, 91, 95, 107, 109
Pneumatic piezometers	10, 49, 53, 55, 57, 107, 108, 109, 121, 124
Vibrating wire piezometers	46, 48, 51, 67, 82, 96, 109, 116, 120
Bonded electrical resistance piezometers	15, 53, 54, 68, 72, 91, 102, 107, 108, 121, 122
Unbonded electrical resistance piezometers	24, 62, 108
Multipoint piezometers	111, 112, 120, 124, 129

Table D.2. Instruments for Measuring Total Stress in Soil

Instruments	Manufacturers
Diaphragm earth pressure cells with vibrating wire transducer	41, 48, 116, 120
Diaphragm earth pressure cells with bonded electrical resistance strain gage transducer	22, 23, 62, 72
Hydraulic earth pressure cells with pneumatic transducer	49, 55, 57, 107, 108, 109, 124
Hydraulic earth pressure cells with vibrating wire transducer	46, 67, 82, 96, 109, 120
Hydraulic earth pressure cells with bonded electrical resistance strain gage transducer	54, 72, 91, 121
Hydraulic earth pressure cells with unbonded electrical resistance strain gage transducer	24, 62, 108

Table D.3. Instruments for Measuring Stress Change in Rock

Instruments	Manufacturers
Borehole pressure cells	46, 55, 104, 108
USBM borehole deformation gages	46, 67, 100, 101
Yoke borehole deformation gage	29
Biaxial strain cells	25, 100
Triaxial strain cells	25, 77, 80, 86
Vibrating wire stressmeters	46, 67, 82
Photoelastic stressmeters	117

Table D.4. Instruments for Measuring Deformation

Instruments	Manufacturers
Surveying instruments	70, 76, 132, 133
Electronic distance measuring instruments	4, 26, 70, 76, 87, 127, 132, 133
Photogrammetric instruments	70, 132, 133
Mechanical crack gages	1 (see also under *Surface-mounted mechanical strain gages,* in Table D.5)
Electrical crack gages	57, 100, 102, 107 (see also under *Surface-mounted vibrating wire* and *electrical resistance strain gages,* in Table D.5)
Convergence gages	46, 53, 66, 67, 70, 82, 100, 102, 108, 109, 110, 120
Mechanical tiltmeters	96, 109
Tiltmeters with accelerometer transducer	107, 108, 109, 111, 118, 121, 122
Tiltmeters with vibrating wire transducer	82, 96, 120
Tiltmeters with electrolytic level transducer and geodetic sensitivity	11, 40, 114, 126
Probe extensometers: mechanical heave gage	1
Probe extensometers: crossarm gage	1, 72, 96, 108, 109
Probe extensometers: mechanical probe within inclinometer casing	53, 66, 108
Probe extensometers: sliding micrometer	111
Probe extensometers: gages with induction coil transducer	62, 65, 66, 82, 96, 107, 108, 109, 120
Probe extensometers: magnet/reed switch gage	53, 109
Probe extensometers: magnetostrictive gage	46, 67
Settlement platforms	1
Fixed embankment extensometers: mechanical gage with tensioned wires	1 (also most manufacturers of fixed borehole extensometers)
Fixed embankment extensometers: gage with electrical linear displacement transducers	Most manufacturers of fixed borehole extensometers
Soil strain gages	18
Fixed borehole extensometers	1, 6, 46, 48, 49, 53, 55, 62, 66, 67, 72, 82, 99, 100, 102, 107, 108, 109, 110, 111, 118, 120, 122
Borros anchors	20, 48, 108
Spiral-foot gages	1, 100
Inclinometers	29, 48, 49, 53, 55, 66, 68, 72, 82, 102, 107, 108, 109, 110, 111, 120, 121, 122
Shear plane indicators	1, 17, 50, 53, 66, 108, 109
Plumb lines and inverted pendulums	62, 66, 72, 96, 109, 120
Remote monitoring system for plumb lines	62, 96, 113
In-place inclinometers and multiple deflectometers	66, 107, 108, 120, 121, 122
Portable borehole deflectometers	66, 108, 111
Liquid level gages	1, 20, 46, 49, 53, 54, 55, 57, 65, 66, 67, 88, 96, 107, 108, 109, 120, 124
Telltales	1
Convergence gages for slurry trenches	1
Time domain reflectometry	58, 120
Acoustic emission monitoring	3, 32, 66, 88, 94, 108, 131

Table D.5. Instruments for Measuring Load and Strain in Structural Members

Instruments	Manufacturers
Mechanical load cells	66, 97
Hydraulic load cells	55, 66, 102, 109, 118, 124
Electrical resistance load cells	19, 22, 29, 31, 45, 46, 60, 61, 62, 67, 72, 84, 102, 108, 109, 115, 118, 128
Vibrating wire load cells	41, 46, 48, 67, 82, 109, 116, 120
Photoelastic load cells	93, 117
Calibrated hydraulic jacks	31
Cable tension meters	22, 39, 100, 115
Surface-mounted mechanical strain gages	13, 33, 62, 66, 75, 84, 96, 100, 102, 103, 108, 110, 118
Surface-mounted vibrating wire strain gages	41, 46, 48, 67, 82, 108, 116, 120
Surface-mounted electrical resistance strain gages	19, 22, 23, 35, 59, 60, 72, 85, 123
Embedment strain gages: mechanical type	1
Embedment strain gages: vibrating wire types	41, 46, 48, 67, 82, 96, 116, 120
Embedment strain gages: electrical resistance types	19, 22, 24, 35, 62, 72, 107, 108, 118, 123
Concrete stress cells	24, 46, 55, 62, 66, 67, 72, 82, 96, 102, 109

Table D.6. Instruments for Measuring Temperature

Instruments	Manufacturers
Mercury thermometers	2
Bimetal thermometers	2
Thermistors	12, 27, 37, 89, 105
Thermocouples	12, 27, 37, 85, 89
Resistance temperature devices	12, 24, 27, 66, 89, 120
Vibrating wire transducers	46, 48, 67, 82, 120
Frost gages	1

Table D.7. Miscellaneous

Instruments	Manufacturers
Electrical cable	5, 16, 21, 38, 42, 130
Splices for electrical cable	9, 30, 42, 125
Automatic data acquisition systems for geotechnical instrumentation	22, 46, 47, 48, 52, 53, 55, 67, 72, 82, 108, 109, 120, 122, 124
Lightning protection devices	43, 44, 74, 81, 98
Borehole directional survey instruments	34, 56, 63, 90, 92
Ultrasonic monitoring instrument for measuring verticality of slurry trenches	71
Additives to water for hydraulic instruments	8, 79
Bentonite pellets	7, 46, 53, 67, 95, 100, 102, 108, 109, 124
Bentonite gravel	78
Additives to grout	28, 83, 106
Biodegradable drilling mud	69
Inflatable packers	14, 64, 102

NAMES AND ADDRESSES OF MANUFACTURERS AND NORTH AMERICAN SUPPLIERS*

1. Instrument can be fabricated by the user from locally available parts.
2. Instrument is available from numerous manufacturers.
   Listings may be found in:
   - Various trade journals
   - Local industrial buyers' guides
   - *Electronic Buyers' Guide* (McGraw-Hill Book Co., New York)
   - *Instruments and Control Systems Buyers' Guide* (Chilton Co., Radnor, PA)
   - *Directory of Instrument Society of America* (ISA, Research Triangle Park, NC)
   - *Thomas Register* (Thomas Publishing Co., New York)
3. Acoustic Emission Technology Corporation, 1812 J Tribute Road, Sacramento, CA 95815, U.S.A.
4. AGA Geotronics AB, Box 64, S-182 11 Danderyd, Sweden
5. Alpha Wire Corporation, 711 Lidgerwood Avenue, Elizabeth, NJ 07207, U.S.A.
6. Amberg Measuring Technique Ltd., Ausstellungsstrasse 88, P.O. Box 3141, CH-8031 Zürich, Switzerland (NAS: Roctest Ltd.)
7. American Colloid Company, Drilling Fluid Products Division, 5100 Suffield Court, Skokie, IL 60077, U.S.A.
8. American Cyanamid Company, 1 Cyanamid Plaza, Wayne, NJ 07470, U.S.A.
9. AMP, Inc., 449 Eisenhower Blvd., Harrisburg, PA 17105, U.S.A.
10. Apparatus Specialties Company, Box 122, Saddle River, NJ 07458, U.S.A.
11. Applied Geomechanics, Inc., 1336 Brommer Street, Santa Cruz, CA 95062, U.S.A.
12. Atkins Technical, Inc., 3401 S.W. 40th Blvd., Archer Interchange (I75), Industrial Area, Gainesville, FL 32608, U.S.A.
13. Avongard Products, U.S.A. Ltd., 2836 Osage, Waukegan, IL 60087, U.S.A.

*NAS = North American Supplier.

14. Baski Water Instruments, Inc., 1586 South Robb Way, Denver, CO 80226, U.S.A.
15. BAT Envitech AB, P.O. Box 27194, S-102 52, Stockholm, Sweden (NAS: BAT Envitech, Inc., P.O. Box 7826, Long Beach, CA 90807, U.S.A.; Soiltest, Inc.)
16. Belden Corporation, P.O. Box 1980, Richmond, IN 47374, U.S.A.
17. Bemek, Box 11063, S-95111 Luleå, Sweden
18. Bison Instruments, Inc., 5708 West 36th Street, Minneapolis, MN 55416, U.S.A.
19. BLH Electronics, 75 Shawmut Road, Canton, MA 02021, U.S.A.
20. Borros AB, Box 3063, S-17103, Solna, Sweden (NAS: Roctest Ltd.; Wykeham Farrance, Inc., 8000 Glenwood Ave., P.O. Drawer 30967, Raleigh, NC 27612, U.S.A.)
21. Brand-Rex Corporation, Electronic and Industrial Cable Division, Willimantic, CT 06226, U.S.A.
22. Brewer Engineering Laboratories, Teledyne Engineering Services, P.O. Box 288, Marion, MA 02738, U.S.A.
23. Cambridge Insitu, Little Eversden, Cambridge CB3 7HE, England
24. Carlson Instruments, 1190-C Dell Avenue, Campbell, CA 95008, U.S.A.
25. Chamber of Mines of South Africa, P.O. Box 91230, Auckland Park, 2006, Republic of South Africa
26. Com-Rad Electronic Equipment Ltd., 256 Ipswich Road, Trading Estate, Slough, Berkshire, England
27. Conax Buffalo Corporation, 2300 Walden Avenue, Buffalo, NY 14225, U.S.A.
28. Concrete Chemicals Company, 1705 Superior Building, Cleveland, OH 44114, U.S.A.
29. CSIRO, Division of Geomechanics, P.O. Box 54, Mount Waverley, 3149, Australia
30. Dow Corning Corporation, Midland, MI 48640, U.S.A.
31. Richard Dudgeon, Inc., 7A Market Street, Stamford, CT 06902, U.S.A.
32. Dunegan/Endevco, Rancho Viejo Road, San Juan Capistrano, CA 92675, U.S.A.
33. Eastlex Machine Corporation, 2170 Christian Road, Lexington, KY 40505, U.S.A.

34. Eastman Christensen Company, P.O. Box 14609, Houston, TX 77021, U.S.A.

35. Eaton Corporation, 5340 Alla Road, Los Angeles, CA 90066, U.S.A.

36. ELE International Ltd., Eastman Way, Hemel Hempstead, Hertfordshire, HP2 7HB, England (NAS: Engineering Laboratory Equipment, Inc., 2205 Lee Street, Evanston, IL 60202, U.S.A.)

37. Fenwal, Inc., Ashland, MA 01721, U.S.A.

38. Fluorocarbon Company, Samuel Moore Group, Dekoron Division, 1199 South Chillicothe Road, Aurora, OH 44202, U.S.A.

39. Fulmer Components Ltd., 231 Berwick Avenue, Slough, Berks, SL1 4QT, England

40. G + G Technics AG, Leimenweg 4, CH-4419 Lupsingen/BL, Switzerland

41. Gage Technique Ltd., P.O. Box 30, Trowbridge, Wilts BA14 8YD, England

42. General Electric Company, Wire and Cable Division, 1285 Boston Avenue, Bridgeport, CT 06486, U.S.A.

43. General Electric Company, Electronics Park, Syracuse, NY, 13201, U.S.A.

44. General Semiconductor Industries, Inc., 2001 West Tenth Place, Tempe, AZ 85281, U.S.A.

45. Geodesy, 832 High Street, East Kew, 3102, Australia

46. Geokon, Inc., 48 Spencer Street, Lebanon, NH 03766, U.S.A.

47. Geomation, Inc., 15000 W. 6th Avenue, Golden, CO 80401, U.S.A.

48. Geonor A/S, Grini Molle, P.O. Box 99 Roa, Oslo 7, Norway (NAS: Geonor USA, 1425 Broad Street, Clifton, NY 07013; Roctest Ltd.; Slope Indicator Company)

49. Geosistemas, S.A. de C.V., Aniceto Ortega No. 1306, 03100 Mexico, D.F., Mexico

50. Geosystems (AUST) Pty. Ltd., 3 Hocking Street, Coburg, 3058, Australia

51. Geotech AB, Klangfärgsg 10, 421-52 Västra Frolunda, Sweden (NAS: Roctest Ltd.)

52. Geotechnical Engineering and Mining Services (U.S.A.), Inc., A Subsidiary of Synergetics International, Inc., 6565 Odell Place, P.O. Box E, Boulder, CO 80306, U.S.A.

53. Geotechnical Instruments (U.K.) Ltd., Station House, Old Warwick Road, Leamington Spa, Warwickshire CV31 3NR, England

54. Geotechniques International, Inc., P.O. Box E, Middleton, MA 01949, U.S.A.

55. Glötzl GmbH, D-7512 Rheinstetten, 4-Fo, Karlsruhe, West Germany (NAS: Geo Group, Inc., 2209 Georgian Way, #12, Wheaton, MD 20902, U.S.A.)

56. Gyrodata, Inc., 1682 West Belt North, Houston, TX 77043, U.S.A.

57. Earl B. Hall, Inc., 1050 Northgate Drive, Suite 400, San Rafael, CA 94903, U.S.A.

58. Hewlett-Packard, 1820 Embarcadero Road, Palo Alto, CA 94303, U.S.A.

59. HITEC Products, Inc., P.O. Box 790, 87 Fitchburg Road, Ayer, MA 01432, U.S.A.

60. Hottinger Baldwin Measurements, Inc., 139 Newbury Street, Framingham, MA 01701, U.S.A.

61. Houston Scientific International, Inc., 4202 Directors Row, Houston, TX 77092, U.S.A.

62. Huggenberger AG, Tödistrasse 68, Postfach CH-8812 Horgen, Switzerland (NAS: Slope Indicator Company)

63. Humphrey, Inc., 9212 Balboa Avenue, San Diego, CA 92123, U.S.A.

64. Hydrophilic Industries, Inc., 5815 Meridian Avenue North, Puyallup, WA 98371, U.S.A.

65. Dr. Ing. Heinz Idel, Potthoffs Borde 15, 43 Essen, West Germany (NAS: Slope Indicator Company)

66. Interfels GmbH, Postfach 75, D-4444, Bad Bentheim 1, West Germany (NAS: Roctest Ltd.)

67. Irad Gage, Klein Drive, Salem, NH 03079, U.S.A.

68. ISMES S.p.A., Viale G. Cesare 29, 24100 Bergamo, Italy

69. Johnson Division of UOP, Inc., P.O. Box 43118, St. Paul, MN 55164, U.S.A.

70. Kern & Co. Ltd., CH-5001, Aarau, Switzerland (NAS: Kern Instruments, Inc., Geneva Road, Brewster, NY 10509, U.S.A.)

71. Koden Electronics Company Ltd., 2-10-45, Kami-Osaki, Shinagawa-ku, Tokyo, Japan

72. Kyowa Electronic Instruments Company Ltd., 5-1 Chofugaoka 3-chome, Chofu-shi, Tokyo, Japan (NAS: Kyowa Dengyo Corporation, 10 Reuten Drive, Closter, NJ 07624, U.S.A.)
73. Landtest Ltd., 43 Baywood Road, Rexdale, Ontario, M9V 3Y8, Canada
74. LEA, Inc., 12516 Lakeland Road, Santa Fe Springs, CA 90670, U.S.A.
75. Leigh Instruments Ltd., Engineering and Aerospace Division, 2680 Queensview Drive, Ottawa, Ontario, K2B 8J9, Canada
76. The Lietz Company, 9111 Barton, Box 2934, Overland Park, KS 66201, U.S.A.
77. LNEC, Laboratório Nacional de Engenharia Civil, 101, Av. do Brasil, P-1799, Lisbon, Portugal
78. L. V. Lomas Chemical Company Ltd./Ltée, 6365 Northwest Drive, Mississauga, Ontario L4V 1J8, Canada (NAS: L. V. Lomas Chemical Company Ltd./Ltée, 5500 Main Street, Suite 3-C, Williamsville, NY 14221, U.S.A.).
79. Lonza AG, 3930 Visp, Switzerland (NAS: Lonza, Inc., 22-10 Route 208, Fair Lawn, NJ 07410, U.S.A.)
80. Luleå University of Technology, Division of Rock Mechanics, S-951 87 Luleå, Sweden
81. Lumex Corporation, 540 North Court, Palatine, IL 60067, U.S.A.
82. Maihak AG, Semperstrasse 38, P.O. 60 17 09, D-2000 Hamburg 60, West Germany (NAS: Ampower Corporation, 1 Marine Plaza, North Bergen, NJ 07047, U.S.A.)
83. Master Builders, Division of Martin Marietta Corporation, 23700 Chagrin Blvd., Beachwood, OH 44122, U.S.A.
84. W. H. Mayes & Son (Windsor) Ltd., Vansittart Estate, Arthur Road, Windsor, Berkshire, SL4 1SD, England
85. Measurements Group, Inc., Micro-Measurements Division, P.O. Box 27777, Raleigh, NC 27611, U.S.A.
86. Mindata, 91A Orrong Crescent, Caulfield North, 3161, Australia (NAS: Geokon, Inc.)
87. Nippon Kogaku K. K., Fuji Bldg., 2-3, Marunouchi 3-chome, Chiyoda-ku, Tokyo 100, Japan
88. Walter Nold Company, Inc., 24 Birch Road, Natick, MA 01760, U.S.A.
89. Omega Engineering, Inc., One Omega Drive, Box 4047, Stamford, CT 06907, U.S.A.
90. Owl Technical Associates, Inc., 1111 Delaware Avenue, Longmont, CO 80501, U.S.A.
91. Oyo Corporation, 2-19 Daitakubo 2-chome, Urawa, Saitama 336, Japan (NAS: Oyo Corporation U.S.A., 7334 N. Gessner Road, Houston, TX 77040, U.S.A.)
92. Pajari Instruments Ltd., P.O. Box 820, Orillia, Ontario L3V 6K8, Canada
93. Perard Torque Tension Ltd., Claylands Avenue, Worksop, Nottinghamshire S81 7BQ, England (NAS: P. T. T. Mining Equipment, Inc., 3924 Youngfield Street, Wheat Ridge, CO 80033, U.S.A.)
94. Physical Acoustics Corporation, P.O. Box 3135, 743 Alexander Road, Princeton, NJ 08540, U.S.A.
95. Piezometer Research & Development, 939 Barnum Avenue, B341, Bridgeport, CT 06608, U.S.A.
96. Ing. Franco Pizzi, Via Ripoli 207E, 50126 Firenze, Italy (Successor to Officine Galileo, Firenze, Italy)
97. Proceq, SA, Riesbachstrasse 57, CH-8034, Zürich 8, Switzerland
98. RKS Industries, 208 Mount Herman Road, Suite 2, Scotts Valley, CA 95066, U.S.A.
99. Rock Instruments Ltd., Bell Lane, Uckfield, East Sussex, TN22 1QL, England (NAS: Irad Gage; Solinst Canada Ltd.)
100. Roctest Ltd., 665 Pine, St. Lambert (Montreal), Quebec J4P 2P4, Canada. Also Roctest, Inc., 7 Pond Street, Plattsburgh, NY 12901, U.S.A.
101. Rogers Arms & Machine Company, P.O. Box 2344, 1426 Ute Avenue, Grand Junction, CO 81501, U.S.A.
102. RS Technical Instruments Ltd., 18-1780 McLean Avenue, Port Coquitlam, British Columbia, V3C 4K9, Canada
103. Satec Systems, Inc., Grove City, PA 16127, U.S.A.
104. Serata Geomechanics, Inc., 1229 Eighth Street, Berkeley, CA 94710, U.S.A.

105. Sierracin/Western Thermistor, 354 Via Del Monte, Oceanside, CA 92504, U.S.A.

106. Sika Corporation, P.O. Box 297, Lyndhurst, NJ 07071, U.S.A.

107. SIS Geotecnica s.r.l., Via Del Santuario 39, 20096 Seggiano Di Pioltello, Milano, Italy

108. Slope Indicator Company, P.O. Box C-30316, Seattle, WA 98103, U.S.A. (also successor to Terrametrics, Golden, CO, U.S.A.)

109. Soil Instruments Ltd., Bell Lane, Uckfield, East Sussex, TN22 1QL, England (NAS: Irad Gage; Solinst Canada Ltd.)

110. Soiltest, Inc., 2205 Lee Street, Evanston, IL 60202, U.S.A.

111. Solexperts AG, P.O. Box 230, CH-8603, Schwerzenbach, Zürich, Switzerland

112. Solinst Canada Ltd., 2240 Industrial Street, Burlington, Ontario, L7P 1A5, Canada

113. Spectron Engineering, Inc., 800 West 9th Avenue, Denver, CO 80204, U.S.A.

114. Sperry Corporation, Sensing Systems, Aerospace & Marine Group, P.O. Box 21111, Phoenix, AZ 85036, U.S.A.

115. Strainsert Company, Union Hill Industrial Park, West Conshohocken, PA 19428, U.S.A.

116. Strainstall Ltd., Denmark Road, Cowes, Isle of Wight, PO31 7TB, England

117. Stress Engineering Services, Charlton Lane, Midsomer Norton, Bath, Avon, BA3 4BE, England

118. Structural Behavior Engineering Laboratories, Inc., P.O. Box 23167, Phoenix, AZ 85063, U.S.A.

119. Tektronix, Inc., P.O. Box 500, Beaverton, OR 97077, U.S.A.

120. Telemac, 2 Rue August Thomas, 92600 Asnières, France (NAS: Roctest Ltd.)

121. Terra Technology Corporation, 3860 148th Avenue, N.E., Redmond, WA 98052, U.S.A.

122. Terrascience Systems Ltd., 1574 West 2nd Avenue, Vancouver, British Columbia V6J 1H2, Canada

123. Texas Measurements, Inc., P.O. Box 2618, College Station, TX 77840, U.S.A.

124. Thor International, Inc., 13751 Lake City Way N.E., Seattle, WA 98125, U.S.A.

125. 3M Company, 3M Center 225-4N-05, St. Paul, MN 55144, U.S.A.

126. Tilt Measurement Ltd., Works Road, Letchworth, Herts SG6 1JU, England

127. Tokyo Optical Co. Ltd., 75-1 Hasunuma-cho, Itabashi-ku, Tokyo 174, Japan (NAS: Topcon Instrument Corporation of America, 65 West Century Road, Paramus, NJ 07652, U.S.A.)

128. Transducers, Inc., 14030 Bolsa Lane, Cerritos, CA 90701, U.S.A.

129. Westbay Instruments Ltd., 507 East Third Street, North Vancouver, British Columbia V7L 1G4, Canada

130. Westinghouse Electric Corporation, 25 Bridle Lane, Westborough, MA 01581, U.S.A.

131. Weston Geophysical Corporation, P.O. Box 550, Westborough, MA 01581, U.S.A.

132. Wild Heerbrugg Ltd., CH-9435 Heerbrugg, Switzerland (NAS: Wild Heerbrugg Instruments, Inc., 465 Smith Street, Farmingdale, NY 11735, U.S.A.)

133. Carl Zeiss, D-7082, Oberkochen, West Germany (NAS: Carl Zeiss, Inc., One Zeiss Drive, Thornwood, NY 10594, U.S.A.)

APPENDIX E

DETAILS OF TWIN-TUBE HYDRAULIC PIEZOMETER SYSTEM

The operating principle of the twin-tube hydraulic piezometer is described in Chapter 9, and advantages and limitations are given. The application for twin-tube hydraulic piezometers is almost exclusively limited to long-term monitoring of pore water pressures in embankment dams, and therefore longevity is a primary need. Systems have been developed in the United States and England, and each has been widely used in embankment dams throughout the world with very variable success. The system developed in England appears to have a better success record than the system developed in the United States, and is readily available from instrument manufacturers in England. Successful long-term use of twin-tube hydraulic piezometers entails close adherence to many proven details. Details are given in this appendix.

E.1. COMPONENTS

E.1.1. Piezometers

Piezometers are of two types, depending on whether they are to be installed in saturated or unsaturated soil. They are referred to here as *cylindrical piezometers* and *Bishop piezometers*, respectively. Cylindrical piezometers are typically installed below the groundwater table in boreholes, and Bishop piezometers are typically installed in compacted fill.

A Bishop piezometer is shown in Figures 9.12 and 9.13 and consists of a tapered ceramic filter retained with neoprene washers between two end caps. Two compression fittings are attached to the larger end cap. The filter should be a *high air entry filter* (Section 9.11) with a pore diameter of approximately 4×10^{-5} in. (1 micron), a permeability to water of about 3×10^{-6} cm/sec, and an air entry value of at least 15 lb/in.^{2} (100 kPa). A typical size of the filter is 4 in. (100 mm) long, with outside diameter tapering from 2 to 1.5 in. (50 to 38 mm). As indicated in Section E.1.3, the compression fittings connected to the top end cap should be of type 316 stainless steel. It is important that the diameter of the waterway through the compression fitting and top cap should not be less than the inside diameter of the tubing: a reduction of diameter at this point is thought to be the cause of blockage in some piezometers in Australia (Mitchell, 1985).

Available options for end cap material are brass, PVC, and stainless steel. Brass is incompatible with the stainless steel compression fittings, since galvanic corrosion may occur; therefore, it should not be used. Available options for the connecting rod material are also brass, PVC, and stainless steel. However, a connecting rod of PVC is too weak to give sufficient grip at the end caps for sealing on the ends of the filter, and a brass connecting rod is subject to corrosion. In summary, recommended materials are:

Compression fittings—type 316 stainless steel
End caps—PVC or type 316 stainless steel
Connecting rod—type 316 stainless steel
Washers—neoprene

Because the polyethylene sheath must be pared back (Section E.1.3) before attaching tubing to the compression fittings, the short lengths of bared nylon tubing above the compression fittings are a potential weakness if they are allowed to bend during installation. A worthwhile improvement to the standard procedure would be the addition of thick-wall heat shrink tubing over the compression fittings and the first few inches of tubing, shrunk in place after making the connections.

A cylindrical piezometer consists of a cylindrical ceramic filter between two circular plastic end caps, with two compression fittings attached to one end cap. The filter should be a *low air entry filter* (Section 9.11) with a pore diameter of approximately 0.002 in. (60 microns), a permeability to water of about 3×10^{-2} cm/sec, and an air entry value of about 0.7 lb/in.^{2} (5 kPa). The type of compression fittings and the size of the waterway through the fittings should be as described above for Bishop piezometers.

E.1.2. Tubing

Recommendations for tubing material and diameter are given in Section 8.2.3.

Unplasticized nylon 11 with a polyethylene (polythene) sheath is generally accepted as the tubing of choice. The absence of a plasticizer is very important, because this additive tends to form a layer of crystals inside the tubing and may ultimately cause blockage. Mitchell (1985) analyzed crystals that formed in plasticized tubing and found them to be N-butyl-benzene-sulfanamide.

The nylon should be of natural coloring and should not contain more than 0.15% of water-soluble material. The maximum water absorption by the nylon 11 should not be greater than 1.8% at a relative humidity of 100% (a lower-quality nylon 11 is available, with a water absorption of more than 10%) and the tubing should withstand a short-term bursting pressure of not less than 1200 ft head (350 m) of water. The tubing should be suitable for operation within the temperature range 0–50°C and should be capable of being bent at 0°C through 360 degrees around a 4 in. (100 mm) diameter rod without collapse or rupture. The tensile strength of the tubing should not be less than 7000 lb/in.^{2} (50 MPa) at 20°C and should be able to withstand 10% tensile strain at 20°C without rupture.

When the length of single tubing between piezometer and terminal enclosure is less than 600 ft (200 m), a nylon tubing with a 0.11 in. (2.8 mm) inside diameter and a 0.19 in. (4.8 mm) outside diameter is suitable. For longer lengths, these diameters should be either 0.15 in. (3.8 mm) and 0.25 in. (6.4 mm) or 0.2 in. (5.0 mm) and 0.3 in. (8.0 mm), respectively: both sizes have been used with success. The tolerance on the outside diameter should be $+0.002$ in. ($+0.05$ mm) and -0.008 in. (-0.2 mm), and the tolerance on the wall thickness should be ± 0.003 in. (± 0.08 mm). The radial thickness of the polyethylene sheath should be 0.04 ± 0.003 in. (1 ± 0.08 mm).

Unless absolutely impracticable, tubing should be in a continuous length from the piezometer tip to the terminal enclosure: any connections are potential sources of weakness and leakage. Tubing is available in continuous lengths of up to 2000 ft (600 m). It should be tested, before installation, at a pressure well in excess of the anticipated pressure in the field, and an appropriately sized ball bearing should be blown through to ensure that there are no constrictions: these have been found on several occasions. Prior to installation, tubing should be stored away from direct sunlight and the ends should be capped to prevent the entry of dust and insects.

In some cases, pairs of nylon 11 tubes have been sheathed within a coating of polyethylene. However, the polyethylene often does not completely fill between the two tubes, and to avoid seepage along this space, any connections should be encased in an epoxy resin to close off the ends of the passage. The electrical cable splice kits described in Section 8.4.17 are suitable.

E.1.3. Compression Fittings

Recommendations for compression fitting material and type are given in Section 8.2.3: type 316 stainless steel fittings with positive alignment of the compression sleeve. Brass compression fittings are often used on twin-tube hydraulic piezometers, and the author has no specific evidence that excessive corrosion has occurred. However, the author believes in recognizing that brass is subject to corrosion by dissolved gases in the liquid within the tubing and in water surrounding the fittings and thus in adopting a conservative approach.

Compression fittings should seal on the nylon tubing, **not** the polyethylene sheath. The sheath must therefore be pared back, using a cutting tool made for the purpose, and the protruding nylon tube inspected carefully to ensure that it has not been damaged.

E.1.4. Liquid in Tubing

Recommendations for liquid are given in Section 8.2.3: filtered distilled water, produced by steam distillation and de-aired to reduce the dissolved gas content to less than 1 ppm DO. QAC should be added as a wetting agent and bacterial inhibitor.

E.1.5. De-Airing Equipment

The correct functioning of a twin-tube hydraulic piezometer system depends on the absence of gas in the system. If liquid is not de-aired, gas will soon collect to create discontinuities in the liquid, and pressure readings will be incorrect. When this happens, fresh liquid can be flushed into the system to remove the gas, but new discontinuities will soon form. It is therefore necessary to de-air the liquid, and the greater the efficiency of de-airing, the less will be the required flushing frequency. The tubing and piezometer filters described earlier have been selected to minimize the entry of gas, either from unsaturated soil or from the atmosphere.

Some agencies responsible for embankment dams attempt to minimize the dissolved gas content in the liquid by using a simple air trap, consisting of a transparent cylinder in which air floats out of the water and is bled off periodically. Houlsby (1982) describes such a system. Houlsby (1983) reports that, with this system, flushing is needed at 3–6 month intervals, with additional attention to those piezometers showing marked differences in inlet and outlet gage readings. As indicated in Section 8.2.3, dissolved gas content should preferably be reduced to less than 1 part per million dissolved oxygen, either by using a water boiler made for the purpose or by using the patented Nold De-Aerator™. Both methods are described in Section 8.2.3.

E.1.6. Flushing Equipment

Various flushing systems are available from the manufacturers of twin-tube hydraulic piezometers, consisting of liquid containers, pumps, pressure gages, and associated pipework and valves. Examples, excluding the pumps, are shown in Figure E.1. Alternatively, the *Pressurized De-Aired Water Containment System,* manufactured by Walter Nold Company, Inc., can be used. The system consists of a portable tank that is filled, in a field laboratory or other on- or off-site facility, with de-aired liquid under pressure as shown in Figure E.2. The tank is then carried to the terminal enclosure, together with a regulator assembly, and used for flushing under controlled pressure. The arrangements in the terminal enclosure when this system is used are shown in Figure E.3.

As described in Section E.3, a vacuum is applied to the output line of each piezometer during flushing. Depending on site conditions, the vacuum may be created by a venturi-type ejector (e.g., Houlsby, 1982), driven by high pressure water or air, a hand pump, or an electric vacuum pump powered by mains supply or a portable generator. A trap is inserted at the output connection to contain returning liquid and is graduated for use in determining when the required volume of liquid has been replaced.

Figure E.1. Flushing equipment (courtesy of Geotechnical Instruments (U.K.) Ltd., Leamington Spa, England).

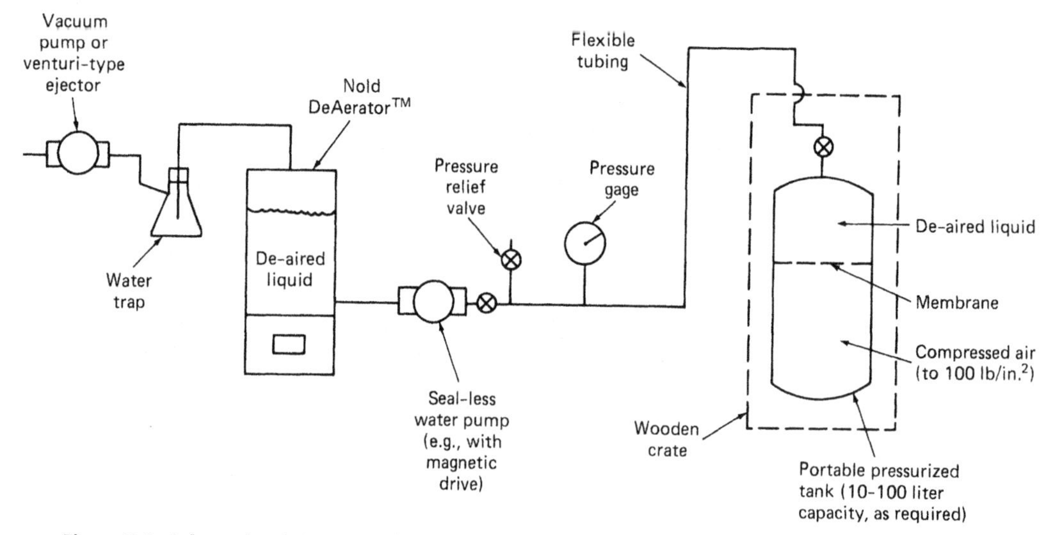

Figure E.2. Schematic of *Pressurized De-Aired Water Containment System* (courtesy of Walter Nold Company, Inc., Natick, MA).

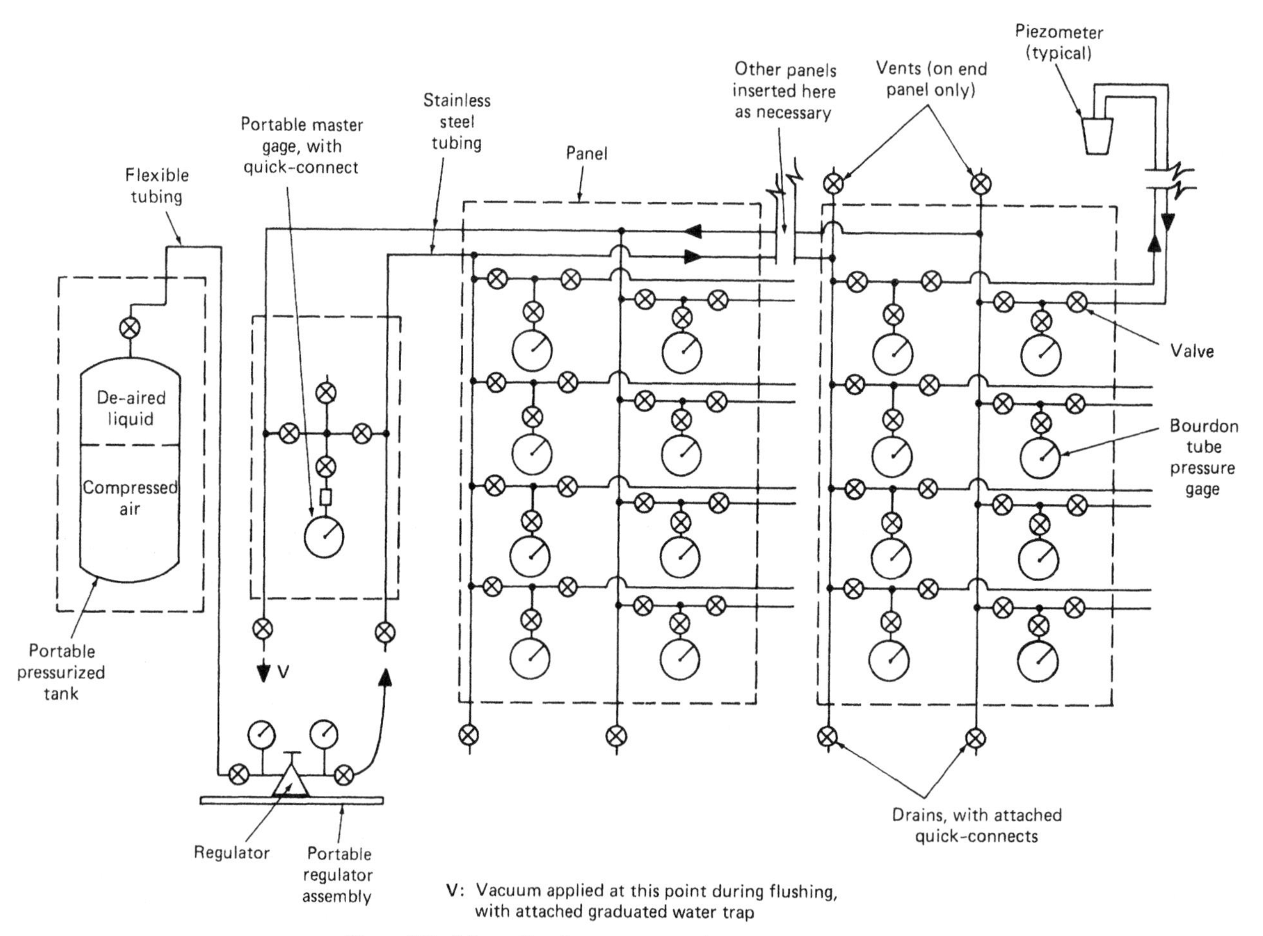

Figure E.3. Schematic of arrangements in terminal enclosure.

E.1.7. Pressure Measuring Equipment

Pressure measuring equipment can be Bourdon tube pressure gages, mercury manometers, or electrical transducers.

When opting for Bourdon tube pressure gages, the user must make a choice between inexpensive gages that will corrode and require replacement and gages that have a good chance of longevity. All readily available standard commercial gages in an affordable price range (less than about $500 U.S.) appear to have some aluminum components and are subject to corrosion. Some users in Australia favor a gage with a glass-filled nylon case, acetal copolymer movement, polycarbonate dial, nylon pointer, beryllium–copper Bourdon tube, and brass top connection, and Houlsby (1982) reports that these gages have virtually trouble-free internal workings. However, in the general case, any gage that allows air in contact with the internal parts may be subject to corrosion, and gages made from plastic and stainless steel and filled with transformer grade silicone oil are the most satisfactory choice when longevity is required. These gages are not standard items but are custom made by Walter Nold Company, Inc.* at a reasonable cost. Gages filled with glycerin are subject to excessive corrosion, because the glycerin contains water. A full-scale accuracy of ±1% is suitable.

Mercury manometers are sometimes the preferred choice, because of their durability and reliability, but mercury is a hazardous material and its use raises environmental concerns. Mercury manometers require greater reading skill than Bourdon tube pressure gages, and this skill may not always be available during the operating life of a dam. In fact, inexperienced personnel have on occasion blown mercury out of manometers while flushing the system, and it is not uncommon to find that mercury has entered the piezometer tubing. Electrical transducers with a digital display are available as an alternative to Bourdon tube pressure gages and manometers.

Figure E.3 shows the preferred layout of Bourdon tube gages, valves, and pipework in the terminal enclosure. Two gages are provided for each piezometer, so that no valve manipulation is needed when readings are taken, and relatively untrained personnel can therefore be used to take readings. A valve on the piezometer side of each gage allows isolation of the gage from the piezometer when checking and adjusting the gage against a *master* gage. This valve is normally left open at other times, and thus the pressure gages register direct hydraulic pressure from the tip and the reading operation consists merely of recording these pressures. The top connections to the gages minimize entrapment of gas, and the valve arrangement allows any malfunctioning gage to be removed and replaced easily.

The *master* gage is used for periodic checking of all active gages. Sometimes a gage with ±0.25% full-scale accuracy is permanently installed in the terminal enclosure, but such a gage is subject to deterioration. A better arrangement is to install a valve and quick-connect fitting for the gage and to store it in an indoor environment, attaching it to the terminal system only when periodic checks are to be made. There are two other available methods for checking active gages. First, a mercury/water manometer can be installed permanently in the terminal closure for this purpose, providing an inexpensive absolute standard that can be used both in the negative and positive range. However, use of mercury involves hazards and environmental concerns. Second, a dead weight tester can be used, but this is expensive and somewhat cumbersome and cannot be used to measure over the negative range.

*See Appendix D.

E.1.8. Valves and Fittings in Terminal Enclosure

Terminal panel valves should be *no-volume change* ball valves.

Two options are available for selection of valves and fittings: stainless steel and brass. Type 316 stainless steel creates a *minimum maintenance* system, but at slightly greater procurement cost than brass. Brass is subject to corrosion by dissolved gases in the internal liquid and by a humid atmosphere, the maintenance obligation must be accepted, and vigilance on the part of reading personnel must be assumed. The assumption of vigilance during the operating life of a dam will often be unwarranted, and stainless steel valves and fittings are the preferred choice.

E.1.9. Pipework in Terminal Enclosure

Copper tubing, nylon tubing, plastic pipe, and stainless steel tubing have all been used for terminal enclosure pipework. Copper tubing is subject to corrosion, and plastic pipe creates difficulties in connecting to valves and fittings. Nylon tubing can be used, but it is probably worthwhile to include a polyethylene sheath, to minimize the permeability to water (Section 8.2.3). However, the flexibility of plastic tubing creates a need for a large number of attachment points between the pipework and panel. The preferred material is ⅜ in. (10 mm) stainless steel tubing.

E.1.10. Equipment for Automatic Operation

When electrical transducers are used for pressure measurement, automatic reading is possible, thereby reducing reading effort and cost. The terminal panel is provided with no-volume change electric solenoid valves that are used to link with each piezometer in turn, and a scanner and printer are connected. Scanning can be manual, continuous, or on a programmed cycle, and data can be recorded on paper, punch, or magnetic tape. Data can also be logged and interfaced with a microcomputer so that, by using appropriate software, a rapid assessment of embankment dam performance can be made. Use of an automatic reading system overcomes the sluggishness that otherwise arises when using twin-tube hydraulic piezometers with tubes longer than about 2000 ft (600 m), because the system can be programmed to connect to a piezometer and wait for a prescribed time before recording pressure and linking with the next piezometer.

When electrical transducers are used, their zero readings and calibration must be checked at least once a year against known pressures, and they must be protected from severe temperature changes.

Semi-automatic flushing equipment is manufactured for circulating a predetermined volume of de-aired liquid through each piezometer at a selected pressure and vacuum. An automatic version of the Nold DeAerator™ is available for continuous preparation of de-aired liquid.

E.1.11. Terminal Enclosure

As discussed in Chapter 21, vertical runs of tubing should be avoided wherever possible. Figure 21.8 shows acceptable and unacceptable arrangements for terminal enclosures.

The terminal enclosures should be buried in the fill, and it is often convenient to provide access from a berm. They should have an external waterproof coating, a good drain under the floor, and should be protected from freezing, if necessary either by heating or by constructing below the frost line. Guidelines on terminal enclosure design are given by USBR (1974). However, in the author's experience a *walk-in* terminal enclosure is preferable to a pit, because a person who descends into a pit is likely to have an adverse psychological reaction, and care over reading and maintenance is likely to be reduced.

All components should be wall-mounted, because components placed on the floor are subject to damage by water, dirt, and personnel. If the inside walls of terminal enclosures are painted, the paint should be selected carefully because certain paint solvents dissolve plastic tubing.

E.2. INSTALLATION

E.2.1. Installation of Cylindrical Piezometers in Boreholes

Various installation methods are described in Sections 9.17 and 9.18.

E.2.2. Installation of Bishop Piezometers in Fill

General guidelines on the installation of piezometers in compacted clay fill are given in Sections 9.15 and 17.6. The following installation sequence is recommended for Bishop piezometers:

- Perform all necessary tests on components.
- Complete construction of the terminal enclosure, and fill all appropriate components in the enclosure with de-aired liquid.
- Prepare the fill as described in Sections 9.15 and 17.6.
- Install tubing from terminal enclosure to piezometer location.
- Saturate the piezometer, as described in Section 9.14.
- Fill tubing with de-aired liquid, by pumping from the terminal enclosure.
- Cut tubing to length and attach to piezometer under water in a container.
- Measure the elevation of the water in the container, using surveying methods, and verify that the measurement agrees with the pressure indicated in the terminal enclosure.
- Form a tapered hole in the fill and press the saturated filter into place.

- Place backfill material as described in Sections 9.15 and 17.6.
- Take sufficient readings to indicate the minimum pore water pressure (maximum suction) achieved, for use as the initial reading for that piezometer.

The exact sequence for filling appropriate components in the terminal enclosure with de-aired liquid depends on the layout of tubing, valves, and gages, and a sequence should be developed that ensures complete filling. A detailed procedure for use with their system is included in the hydraulic piezometer users' manual by Soil Instruments Ltd. (Soil Instruments Ltd., 1983). Usually, the first step is filling the lengths of terminal panel tubing that serve more than one piezometer (often referred to as the *busbars*), by applying suction at one point and de-aired liquid under a low pressure at another point. The system unique to each piezometer is then filled, following the flushing procedure described in Section E.3: calculation of required pressures and volumes, and circulation of de-aired liquid.

Procedures for installation of tubing from the terminal enclosure to the piezometer location are described in Section 17.6, including trenching, filling, splicing, labeling, and a method devised by Clements (1982) for the installation of tubing on the surface of previously compacted fill. Guidelines on routing are given in Chapter 21. De-aired liquid can withstand significant subatmospheric pressure without the formation of gas bubbles. When the dissolved gas content has been reduced to below 1 ppm dissolved oxygen, a practical limit of subatmospheric pressure appears to be about 20 ft (6 m), and tubes should be routed so that no part is above the internal pressure head level by more than this amount. Even within this limit there is a tendency for gas bubbles to form with time, and if possible tubes should be routed so that a positive pressure exists at all times. Tubing should be installed below the frost line.

The Bishop piezometer is installed in relatively stone-free material by pushing a special mandrel into the soil to form a hole 12 in. (300 mm) deep. This leaves an impression into which the piezometer should fit exactly, making close contact with the surrounding material. The piezometer is pressed into place with the aid of an extension handle that is later removed, and above the piezometer the hole is backfilled with the excavated material after removal of any large stones. This material should carefully be compacted by tamping. If the fill is too hard for pushing or driving the mandrel, an auger or chisel can be used to create a hole slightly smaller than the mandrel, and this hole can finally be shaped with the mandrel.

E.3. MAINTENANCE

Maintenance entails regular checking of all components in the terminal enclosure and occasional flushing of the system with de-aired liquid.

Bourdon tube pressure gages (and electrical transducers if used) should be checked on a regular schedule against the master gage, mercury/water manometer, or dead weight tester. Valves are closed to isolate the master gage and each gage in turn from the rest of the system, and pressure is applied to pressurize the gages throughout their range. A piston device is available from some manufacturers of geotechnical instruments for this purpose. Equipment for automatic operation of course requires regular skilled maintenance.

If the system is properly designed, by using the components described in Section E.1, flushing is seldom necessary unless subatmospheric pore water pressures are being read. Flushing frequency should be determined on a site-specific basis, dependent on reading changes caused by flushing, and with a properly designed system may be as infrequent as once every 5 years.

The flushing procedure is described by Little and Vail (1960). Refer to Figure E.4:

u = pore water pressure in the fill at the piezometer,
p = pressure on the pressure side of the flushing equipment,
v = pressure on the vacuum side of the flushing equipment,
h = difference in head between the piezometer and the pressure and vacuum gages of the flushing equipment,
f_{AB} and f_{BC} = friction losses in the piezometer tubing.

All pressures are measured with respect to atmospheric pressure. For equilibrium,

$$p = h + u + f_{AB}$$

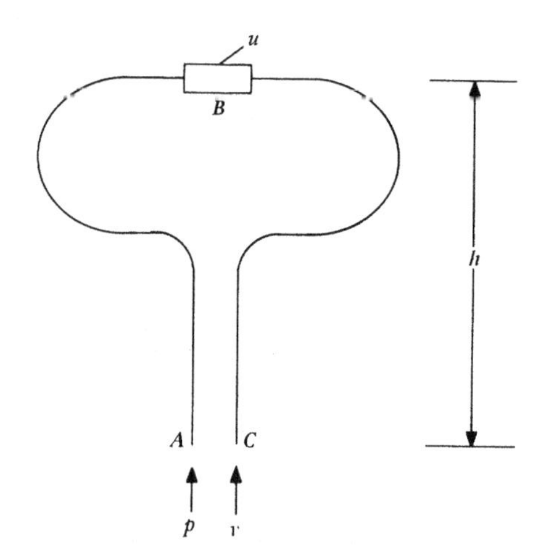

Figure E.4. Pressures during flushing (after Little and Vail, 1960).

and

$$v = h + u - f_{BC}.$$

Since $AB = BC$, and if no liquid flows in or out of the piezometer, the velocity will be the same throughout the system and

$$f_{AB} = f_{BC}.$$

Then

$$p = 2(h + u) - v.$$

In a piezometer installation h is known, v can be maintained at a known value, and u for each piezometer can be estimated using previous readings of pore water pressure. In this way, p can be calculated for each piezometer and de-airing carried out at the appropriate pressure. Examples of the calculation are given by Soil Instruments Ltd. (1983). Little and Vail (1960) advise that when a piezometer has accumulated a large amount of air, care should be taken to apply sufficient pressure to drive liquid at least as far as the piezometer before applying a vacuum to the other tube. If this is not done, air will be drawn in through the piezometer from the fill, because friction loss in the return tube carrying air is much less than in the one carrying liquid.

Vaughan (1974b) advises that it is desirable to maintain a small excess pressure at the piezometer tip for three reasons: first, to prevent air-saturated water being drawn from the soil; second, to avoid applying pressures that may rupture the soil; and third, to avoid a major drop in pressure in the return tube, which may cause air bubbles to come out of solution and create discontinuities in the liquid, and in some installations also cause cavitation. Figure E.5 shows the heads in the tubing during flushing. Vaughan comments that, to fulfill these requirements, it is often necessary to use a small differential pressure with a backpressure on the return tube. In these circumstances, the time for complete circulation may be long and automatic operation is helpful.

Prior to flushing, the volume of liquid required to fill the two tubes to each piezometer is calculated, and flushing is continued until at least this volume has emerged from the return tube. A check for leaks can be made by using tapered orifice flowmeters in the flushing panel system, one connected to the inlet and the other to the outlet tube. After flushing has been completed, appropriate valves closed or opened, and equilibrium has been established in the system, both pressure gages connected to each piezometer should show the same pressure. If the readings do not agree closely, it is probable that some air is present in the system, and further flushing is required.

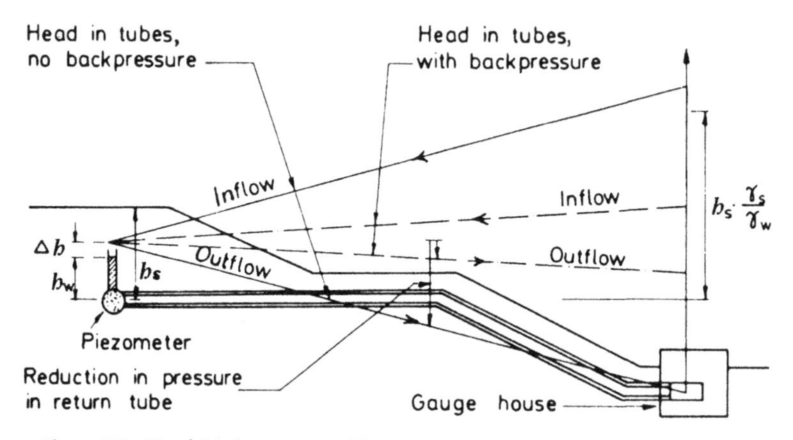

Figure E.5. Head in piezometer tubing during flushing (after Vaughan, 1974b).

APPENDIX F

DIMENSIONS OF DRILL RODS, FLUSH-JOINT CASING, DIAMOND CORING BITS, HOLLOW-STEM AUGERS, AND U.S. PIPE*

* Written with the assistance of Charles O. Riggs, Manager of Research and Marketing, Central Mine Equipment Company, St. Lous, MO.

Table F.1. Dimensions of Drill Rods

Size[a]	Outside Diameter of Rod		Inside Diameter of Rod		Inside Diameter of Coupling[b]	
	(in.)	(mm)	(in.)	(mm)	(in.)	(mm)
EX	1.312	33.3	0.844	21.4	0.437	11.1
AX	1.625	41.3	1.125	28.6	0.562	14.3
BX	1.906	48.4	1.406	35.7	0.625	15.9
NX	2.375	60.3	2.000	50.8	1.000	25.4
RW	1.094	27.8	0.719	18.3	0.406	10.3
EW	1.375	34.9	0.875	22.2	0.500	12.7
AW	1.750	44.4	1.219	31.0	0.625	15.9
BW	2.125	54.0	1.750	44.5	0.750	19.0
NW	2.625	66.7	2.250	57.2	1.375	34.9
HW	3.500	88.9	3.062	77.8	2.375	60.3
(L) AQWL	1.750	44.5	1.375	34.9	—	—
(C) AXWL	1.812	46.0	1.500	38.1	—	—
(L) BQWL	2.187	55.6	1.812	46.0	—	—
(C) BXWL	2.250	57.2	1.906	48.4	—	—
(L) NQWL	2.750	69.9	2.375	60.3	—	—
(C) NXWL	2.875	73.0	2.391	60.7	—	—
(L) HQWL	3.500	88.9	3.062	77.8	—	—
(C) HCWL	3.500	88.9	3.000	76.2	—	—
(L) PQWL	4.625[c]	117.5[c]	4.062	103.2	4.062	103.2
(C) CPWL	4.625	117.5	4.000	101.6	—	—

[a] X and W sizes are available from most manufacturers of drill rod. (L) indicates Longyear Company system. (C) indicates Christensen Dia-Min Tools, Inc. system.

[b] X and W series have a separate coupling. WL wireline series, except PQWL, are flush-joint, internally and externally, without a separate coupling.

[c] PQWL has a separate coupling, protruding outside the rod o.d., but internally flush. The outside diameter in table is o.d. of coupling. The outside diameter of rod is 4.500 in. (114.3 mm).

Source: Longyear Company and Christensen Dia-Min Tools, Inc.

Table F.2. Dimensions of Flush-Joint Casing

Size	Outside Diameter of Casing[a]		Inside Diameter of Casing[a]	
	(in.)	(mm)	(in.)	(mm)
RW	1.437	36.5	1.187	30.1
EW	1.812	46.0	1.500	38.1
AW	2.250	57.1	1.906	48.4
BW	2.875	73.0	2.375	60.3
NW	3.500	88.9	3.000	76.2
HW	4.500	114.3	4.000	101.6
PW	5.500	139.7	5.000	127.0
SW	6.625	168.3	6.000	152.4
UW	7.625	193.7	7.000	177.8
ZW	8.625	219.1	8.000	203.2

[a] No separate couplings. Connections are flush, internally and externally.

Source: Longyear Company and Christensen Dia-Min Tools, Inc.

Table F.3. Dimensions of Diamond Coring Bits

Size	Diameter of Core (in.)	Diameter of Core (mm)	Diameter of Borehole (in.)	Diameter of Borehole (mm)
EX, EXM	0.845	21.5	1.485	37.7
EWD3	0.835	21.2	1.485	37.7
AX	1.185	30.1	1.890	48.0
AWD4, AWD3	1.136	28.9	1.890	48.0
AWM	1.185	30.1	1.890	48.0
AQ Wireline, AV	1.065	27.1	1.890	48.0
BX	1.655	42.0	2.360	59.9
BWD4, BWD3	1.615	41.0	2.360	59.9
BXB Wireline, BWC3	1.432	36.4	2.360	59.9
BQ Wireline, BV	1.432	36.4	2.360	59.9
NX	2.155	54.7	2.980	75.7
NWD4, NWD3	2.060	52.3	2.980	75.7
NXB Wireline, NWC3	1.875	47.6	2.980	75.7
NQ Wireline, NV	1.875	47.6	2.980	75.7
HWD4	2.400	61.1	3.650	92.7
HXB Wireline, HWD3	2.400	61.1	3.650	92.7
HQ Wireline	2.500	63.5	3.790	96.3
CP, PQ Wireline	3.345	85.0	4.827	122.6

Source: Christensen Dia-Min Tools, Inc.

Table F.4. Dimensions of Typical Hollow-Stem Augers

Inside Diameter of Hollow-Stem		Outside Diameter of Flighting		Cutting Diameter of Auger Head	
(in.)	(mm)	(in.)	(mm)	(in.)	(mm)
2.25	57	5.625	143	6.25	159
2.75	70	6.125	156	6.75	171
3.25	83	6.625	168	7.25	184
3.75	95	7.125	181	7.75	197
4.25	108	7.625	194	8.25	210
6.25	159	9.625	244	10.25	260
8.25	210	11.625	295	12.5	318

Source: Central Mine Equipment Company.

Table F.5. Dimensions of U.S. Standard Weight Steel and PVC Pipe[a]

Nominal Diameter	Outside Diameter of Pipe		Inside Diameter of Pipe		Typical Outside Diameter of Steel Coupling[b]		Typical Outside Diameter of PVC Socket Coupling[b]	
(in.)	(in.)	(mm)	(in.)	(mm)	(in.)	(mm)	(in.)	(mm)
1/8[c]	0.405	10.3	0.269	6.8	0.563	14.3	—	—
1/4[c]	0.540	13.7	0.364	9.2	0.719	18.3	—	—
3/8[c]	0.675	17.1	0.493	12.5	0.875	22.2	—	—
1/2	0.840	21.3	0.622	15.8	1.063	27.0	1.125	28.6
3/4	1.050	26.7	0.824	20.9	1.313	33.4	1.312	33.3
1	1.315	33.4	1.049	26.6	1.576	40.0	1.625	41.3
1 1/4	1.660	42.2	1.380	35.1	1.900	48.3	2.000	50.8
1 1/2	1.900	48.3	1.610	40.9	2.200	55.9	2.250	57.2
2	2.375	60.3	2.067	52.5	2.750	69.9	2.750	69.9
2 1/2	2.875	73.0	2.469	62.7	3.250	82.6	3.500	88.9
3	3.500	88.9	3.068	77.9	4.000	101.6	4.000	101.6
3 1/2[c]	4.000	101.6	3.548	90.1	4.625	117.5	—	—
4	4.500	114.3	4.026	102.3	5.000	127.0	5.000	127.0
5[c]	5.563	141.3	5.047	128.2	6.296	159.9	—	—
6	6.625	168.3	6.065	154.1	7.390	187.7	7.687	195.2
8	8.625	219.1	7.981	202.7	9.625	244.5	9.937	252.4

[a] Standard weight pipe is also referred to as *Schedule 40* pipe. Steel pipe has threaded couplings. PVC pipe has cemented socket couplings (wall is generally too thin for threaded couplings).
[b] Outside diameters of couplings are not standard. Users should contact local pipe suppliers for diameters of couplings available from local sources.
[c] Not available in PVC.

Source: American Institute of Steel Construction, Nibco Inc., and Wheeling Machine Products Company.

Table F.6. Dimensions of U.S. Extra Strong Steel and PVC Pipe[a]

Nominal Diameter	Outside Diameter of Pipe		Inside Diameter of Pipe		Typical Outside Diameter of Steel Coupling[b]		Typical Outside Diameter of PVC Socket Coupling[b]	
(in.)	(in.)	(mm)	(in.)	(mm)	(in.)	(mm)	(in.)	(mm)
1/8[c]	0.405	10.3	0.215	5.5	0.563	14.3	—	—
1/4	0.540	13.7	0.302	7.7	0.719	18.3	1.000	25.4
3/8[c]	0.675	17.1	0.423	10.7	0.875	22.2	—	—
1/2	0.840	21.3	0.546	13.9	1.063	27.0	1.312	33.3
3/4	1.050	26.7	0.742	18.8	1.313	33.4	1.562	39.7
1	1.315	33.4	0.957	24.3	1.576	40.0	1.875	47.6
1 1/4	1.660	42.2	1.278	32.5	2.054	52.2	2.250	57.2
1 1/2	1.900	48.3	1.500	38.1	2.200	55.9	2.562	65.1
2	2.375	60.3	1.939	49.3	2.875	73.0	3.062	77.8
2 1/2	2.875	73.0	2.323	59.0	3.375	85.7	3.625	92.1
3	3.500	88.9	2.900	73.7	4.000	101.6	4.312	109.5
3 1/2[c]	4.000	101.6	3.364	85.4	4.625	117.5	—	—
4	4.500	114.3	3.826	97.2	5.200	132.1	5.500	139.7
5[c]	5.563	141.3	4.813	122.3	6.296	159.9	—	—
6	6.625	168.3	5.761	146.3	7.390	187.7	7.750	196.9
8	8.625	219.1	7.625	193.7	9.625	244.5	9.937	252.4

[a] Extra strong pipe is also referred to as *extra heavy* pipe and *Schedule 80* pipe. Steel pipe has threaded couplings. PVC pipe has either threaded or cemented socket couplings. PVC pipe can also be machined with threaded or cemented ends to form a connection that is flush on the inside and outside.
[b] Outside diameters of couplings are not standard. Users should contact local pipe suppliers for diameters of couplings available from local sources.
[c] Not available in PVC.

Source: American Institute of Steel Construction, Nibco Inc., and Wheeling Machine Products Company.

Table F.7. Dimensions of U.S. SDR 21, Class 200 PVC Belled End Pipe

Nominal Diameter (in.)	Outside Diameter of Pipe		Inside Diameter of Pipe		Typical Outside Diameter of Belled End[a]	
	(in.)	(mm)	(in.)	(mm)	(in.)	(mm)
½[b,c]	0.840	21.3	0.696	17.7	0.99	25
¾[c]	1.050	26.7	0.910	23.1	1.19	30
1[c]	1.315	33.4	1.169	29.7	1.46	37
1¼[c]	1.660	42.2	1.482	37.6	1.84	47
1½[c]	1.900	48.3	1.700	43.2	2.10	53
2	2.375	60.3	2.129	54.1	2.62	67
2½	2.875	73.0	2.581	65.6	3.17	81
3	3.500	88.9	3.146	79.9	3.86	98
4	4.500	114.3	4.046	102.8	4.96	126
5[c]	5.563	141.3	5.001	127.0	6.13	156
6	6.625	168.3	5.955	151.3	7.29	185
8	8.625	219.1	7.755	197.0	9.49	241

[a]Type and outside diameters of belled ends are not standard. Users should contact local pipe suppliers for types and diameters available from local sources.
[b]Not available in SDR 21. Sizes given are SDR 13.5, Class 315.
[c]May not be readily available with belled ends.
Source: Harvel Plastics, Inc.

Table F.8. Dimensions of U.S. SDR 26, Class 160 PVC Belled End Pipe

Nominal Diameter (in.)	Outside Diameter of Pipe		Inside Diameter of Pipe		Typical Outside Diameter of Belled End[a]	
	(in.)	(mm)	(in.)	(mm)	(in.)	(mm)
1[b]	1.315	33.4	1.175	29.8	1.46	37
1¼[b]	1.660	42.2	1.512	38.4	1.81	46
1½[b]	1.900	48.3	1.734	44.0	2.07	53
2	2.375	60.3	2.173	55.2	2.58	65
2½	2.875	73.0	2.635	66.9	3.12	79
3	3.500	88.9	3.210	81.5	3.79	96
4	4.500	114.3	4.134	105.0	4.87	124
5[b]	5.563	141.3	5.109	129.8	6.02	153
6	6.625	168.3	6.085	154.6	7.17	182
8	8.625	219.1	7.921	201.2	9.33	237

[a]Type and outside diameters of belled ends are not standard. Users should contact local pipe suppliers for types and diameters available from local sources.
[b]May not be readily available with belled ends.
Source: Harvel Plastics, Inc.

APPENDIX G

EXAMPLE OF INSTALLATION PROCEDURE, WITH MATERIALS AND EQUIPMENT LIST

The following example illustrates the level of detail required when preparing written installation procedures as described in Chapter 17.

The chosen example is installation of inclinometer casing in a vertical borehole through soft clay and into clayey sand, a few feet beyond the toe of a future embankment. Substantial vertical compression is predicted at the casing location; therefore, telescoping couplings have been selected. Because some lateral deformation may occur before vertical compression, couplings must allow for extension as well as compression. Previous borings in the clay have shown that borehole walls cannot be supported adequately by drilling mud, and drill casing will be required. The stratigraphy will not allow grout to bleed into the soil; thus, grout backfill can be used. The borehole is less than 100 ft (30 m) deep; therefore, buoyancy of the casing during grouting can be overcome by using a bottom weight, rather than by using the weight of the grout pipe. The "double-shutoff" arrangement (Figure 12.74) can therefore be used on the grout pipe.

It has been assumed that installations will be made by an instrumentation specialist working with a drilling contractor; therefore, no general contractor is involved. Plastic inclinometer casing will be used.

Some readers may argue that the level of detail in this example is not needed if *instrumentation specialists* are involved, and this of course is true if their experience includes previous identical installations. However, the author normally finds it worthwhile to write procedures in this detail, thereby ensuring that they are tailored to site-specific requirements and that all required materials and equipment are on hand.

Explanatory notes are in square brackets. In an actual installation procedure, these would be replaced by site specific requirements.

1. Complete appropriate parts of installation record sheet (Figures G.1 and G.2), including planned coordinates, ground surface elevation, two bottom cap dimensions as shown (F and length of perforated section), planned inclinometer casing groove orientation and top and bottom elevations, planned total length of inclinometer casing. Complete remainder of record as installation proceeds, and keep a time log.
2. Verify that all materials and equipment are readily available, as per Table G.1.
3. Prepare bottom casing section, with male quick-connect coupling. [Details depend on

Project Name ________________

Inclinometer Casing No. ________________

Planned Coordinates N ________________ E ________________

As-built Coordinates N ________________ E ________________

Start Date ______ Time ________________ Finish Date ______ Time ________________

Contractor ________________ Contractor's Engineer ________________

Driller ________________ Owner's Representative ________________

Type of Drill Rig ________________ Size of Drill Casing ________________

Ground Elevation ______ ft Borehole Diameter ________________

Planned Groove Orientation ________________

Planned Inclinometer Casing Bottom Elev. (A) ______ ft

Planned Inclinometer Casing Top Elev. (B) ______ ft

Planned Total Length of Inclinometer Casing (B − A) ______ ft

Length of Bottom Casing Section (C) ______ ft

Number of Standard Casing Sections = (B − A − C)/(section length + gap at coupling) = ______ (round up to integer)

Actual Total Length of Inclinometer Casing (C + D) ______ ft

Planned Depth to Bottom of Borehole ______ ft

Depth to Top of Clay ______ ft Depth to Bottom of Clay ______ ft

Actual Depth to Bottom of Borehole ______ ft

Calculated Grout Volume Required ______ gal

Inclinometer Probe Tracks Correctly: Yes/No

Length of Grout Pipe to Quick-Connect Shoulder (E) ______ ft

Stick-up of Grout Pipe ______ ft (Should be E + F − C − D)

Water Flows Through Grout Pipe: Yes/No

Grout Mix Used ________________

Total Volume of Grout Used to Fill Hole ______ gal

Grout Inside Inclinometer Casing: Yes/No

If Yes, Comments ________________

Actual Inclinometer Casing Top Elev. Before Cutoff (B′) ______ ft

As-built Inclinometer Casing Bottom Elev. (A′) ______ ft

Inclinometer Probe Tracks Correctly: Yes/No

Length of Inclinometer Casing Cut Off at Top ______ ft

As-built Inclinometer Casing Top Elev. After Cutoff ______ ft

As-built Groove Orientation at Casing Top ________________
(Also draw diagram showing groove orientation, and orientation reference for inclinometer probe)

Groove Spiral Survey Needed: Yes/No Acceptance Test Satisfactory: Yes/No

Components Used During Acceptance Test (identification numbers of probe, readout, cable)

____

NOTES ON BACK OF SHEET: (Notes on drilling, casing installation, grouting, problems encountered, delays, unusual features of the installation, or any other events that may have a bearing on inclinometer casing behavior.)

Figure G.1. Installation record sheet for inclinometer casing.

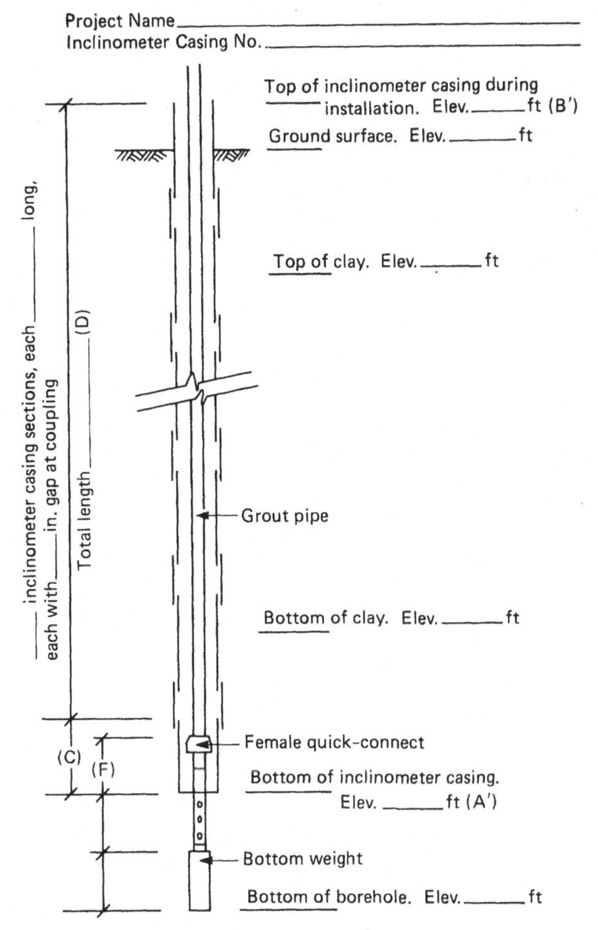

Figure G.2. Continuation of installation record sheet for inclinometer casing.

manufacturer. Use of the double-shutoff arrangement requires a special heavy-duty bottom cap assembly, which should be attached to the full wall thickness of the casing. Bottom cap connection must be fully grout-tight, often requiring filling casing grooves alongside cap with mastic. Tensile strength of connection must be sufficient to sustain total buoyancy force during grouting.] Allow solvent cement to dry overnight.

4. Estimate the buoyancy force that will be imposed on the casing when the annular space is filled with grout. [Use guidelines given in Chapter 17.] Assemble a weight, for attachment to the bottom cap, such that its buoyant weight is somewhat greater than the buoyancy force. Attach the weight to the perforated pipe on the bottom cap. Record length on record sheet. Attach a lowering line to the bottom cap.

5. Measure the length of the bottom casing section (C in Figure G.2). Determine the required initial gap between casing sections. [Base this on anticipated extension and compression.] Determine the number of standard casing sections required (see Figure G.1) and actual total length of inclinometer casing, and record. Lay out sufficient sections of casing for the depth of the borehole (estimated as ____ sections) on suitable aboveground clean surface such as sawhorses. If strong sunlight is anticipated, shade the casing. Inspect inside each casing section for groove burrs, and remove with casing brush. Measure groove spiral in each section [describe procedure, as in Chapter 12, if necessary], and mark magnitude and direction on outside of each section with waterproof marking pen. Reject any sections with excessive spiral. Sort casing sections and couplings into installation sequence to minimize cumulative spiral, and number in sequence.

6. Preassemble one coupling on to each casing section, to allow ____ in. [or ____ mm] extension and ____ in. [or ____ mm] compression at this connection. [These distances will depend on whether telescoping is possible at both or only one end of each coupling. If telescoping is possible at both ends, additional dimensions will be needed for Step 9. Connecting details depend on manufacturer, often requiring sealing mastic and tape to ensure grout-tightness. Care must be taken to avoid creating spiraled casing, and alternate couplings should usually be twisted left and right before fixing. Coupling sliding should be tight enough to hold in place during installation, but not too tight to inhibit later telescoping.] Do not attach a coupling to the upper end of the bottom casing section, as installation will be made with male end upward. [Often it is expedient to preassemble more than one casing section together, but generally 20 ft (6 m) is the maximum convenient length for handling.]

7. Plumb the drill rig carefully. Determine the required depth of borehole, allowing adequate depth for the perforated pipe on the bottom cap, weight, settling debris, base fixity, and stable length for inclinometer probe checks [if required]. Record on record

Table G.1. Materials and Equipment Required for Inclinometer Casing Installation[a]

Quantity	Description
____	Blank installation record sheets, on clipboard
____	Blank sheets for borehole log
____[b]	____ in. o.d. telescoping plastic inclinometer casing in ____ ft lengths
____[b]	____ in. o.d. telescoping inclinometer couplings, with necessary metal hardware
____	Top caps for inclinometer casing
____	Special heavy-duty bottom cap assemblies, with male quick-connect and perforated pipe: not attached to inclinometer casing
____ pint	Solvent cement (for bottom caps)
1	Pop-rivet gun (for bottom caps)
____[b]	Pop-rivets
1	Hand drill with two drill bits (for pop-rivets)
2	Wrenches or other tools for attaching coupling hardware
____ rolls	Sealing mastic [e.g., Scotch 2210 vinyl mastic rolls, 4 in. × 10 ft (100 mm × 3 m)]
____ rolls	Sealing tape [e.g., Scotch 33 vinyl plastic electrical tape, 1.5 in. × 44 ft (38 mm × 13 m)]
____	Bottom weights, ____ lb each
____ ft[b]	¼ in. (6 mm) manila rope for lowering line
1	Working surface for preassembly of casing (e.g., sawhorses), with shade cover if necessary
1	Inclinometer casing brush, with attached 20 ft (6 m) rope; also, means of attaching brush to bottom of grout pipe
1	Equipment for measuring groove spiral before installation
1	Builder's level for plumbing drill rig
1	Drill rig, with drilling and sampling tools [specify]
____ ft[b]	____ flush-joint drill casing
1	Clamping arrangement for withdrawing drill casing by attaching to outside
1	Cylindrical dummy, ____ in. o.d. for checking borehole depth
1	Grout mixer, suitable for mixing____/____ grout, with pump, hose and valve between pump and grout pipe, window screen
____ lb[b]	Materials for grout [Indicate mix required, and quantity of each ingredient. Allow at least 20% spare for waste. Use Figure 17.8 and add appropriate conversion factors in this table: see Appendix H, Section H.10.]
1	Calculator (for determining grout volumes)
1	Clean water supply, hose and valve [garden hose with gun-type spray fitting is ideal] for filling inclinometer casing with water during installation: alternatively, funnel and pail
2	Casing installation clamps
1	Inclinometer probe
____ ft	Graduated electrical cable for inclinometer
1	Inclinometer readout unit, with battery charger
1	Cable pulley and clamp
____ ft[b]	Grout pipe: 1 in. (25 mm) standard black steel pipe, threaded and coupled. Bottom thread connects to female thread on female quick-connect; top thread connects to grout hose
3	Female quick-connects, with o.d. to clear inner ends of metal hardware at couplings
____ rolls	Teflon tape (for grout pipe couplings)
____ sets	Top protective arrangements [e.g., length of 8 in. (200 mm) pipe with either screw cap or hinged cap with padlock, wooden barricade with flagging tape, and bright paint]
____	Paint for numbering installation
____	½ in. (13 mm) paint brushes
1 set	Equipment for surveying groove spiral after installation of casing
1	Pocket knife
1	Scissors
1	100 ft (30 m) measuring tape
1	Hacksaw and spare blades
1	Flat file
1	Pliers

Table G.1. Materials and Equipment Required for Inclinometer Casing Installation (*Continued*)

Quantity	Description
_____	Clean rags
_____	Paper towels
1	Field notebook
_____	Waterproof marking pens
1	Camera and film (for record photographs)

[a][Additional equipment needed for reading includes field data sheets and calculation sheets. If automatic recording or data processing will be used, the following may also be needed: casette tapes, recorder paper, computer program, special connectors and switches, cassette tape reader. Arrangements should also be made for a backup reading system in the event of malfunction.]

[b][Add quantity for spares to replace components damaged during installation and for possible deeper boreholes than expected.]

sheet. Advance the borehole, employing drilling procedures that minimize soil disturbance and using _____ drill casing [flush-joint casing is preferred, inside diameter should allow a minimum of 0.5 in. (13 mm) diametral clearance at sealed couplings] and take samples [depends on site-specific needs]. Maintain a conventional borehole log. Record depth to top and bottom of clay and actual borehole depth. Wash out to bottom, using the *reverse circulation method* [Chapter 17], until water runs clear. Verify that the borehole is fully open to the bottom, by lowering a dummy with diameter approximately 0.25 in. (6 mm) less than the inside diameter of the drill casing.

8. Verify that all grouting materials and equipment are at the site and that the grout pipe is clean inside. Calculate required grout volume, and enter on installation record sheet [Use Figure 17.8.]
9. Insert inclinometer casing into drill casing, male end upward, with bottom section at lower end, lowering on lowering line and filling inclinometer casing with water. Maintain groove orientation parallel and perpendicular to toe of future embankment, adding inclinometer casing sections as installation proceeds, sealing all couplings thoroughly, setting lower end of each coupling to allow _____ in. [or _____ mm] extension and _____ in. [or _____ mm] compression. Use the special casing clamps to support the inclinometer casing temporarily on top of the drill casing while new sections are added.
10. When the bottom weight reaches the bottom of the borehole, verify correct groove orientation, and adjust as necessary by raising off the bottom and working slowly around while pulling and lowering on the lowering line. Set back on bottom of borehole. The top of the inclinometer casing should be above the top of the drill casing. Verify correct inclinometer probe tracking by running the probe down and up both groove pairs, with inclinometer cable connected to readout unit and power on. If the probe will not pass along the casing, or if it returns with one or more wheels disengaged from the grooves, withdraw inclinometer casing and diagnose problem. Couple grout pipe to female quick-connect and insert within inclinometer casing, using Teflon tape on all threads. Keep track of pipe lengths, and record total length to the female quick-connect shoulder. Keep pipe water-filled: this is very important because mating the quick-connects with air in the grout pipe is likely to cause soil particles to enter the quick-connects and prevent later closure. Maintain tension on the lowering line so that the bottom weight does not penetrate below the borehole bottom. Mate quick-connect. Verify that distance from top of grout pipe to top of inclinometer casing checks out, using the installation record sheet (Figure G.1). Verify that the inclinometer casing is at the correct elevation and that the lowering line is tight. (If the lowering line is left slack, the inclinometer casing can sometimes descend and the grout outlet holes in the perforated pipe may become sealed.)
11. Pump water through grout pipe and observe return from drill casing, to check that circuit is open. If the installation can be completed in a continuous operation (e.g., without an

overnight), proceed with the following steps. If not, raise the grout pipe to close the check valves and tie grout pipe to drill rig and, on resuming work, lower the grout pipe and again check the circuit.

12. Verify that the drill casing can be raised. Do not rotate the drill casing because rotation may cause spiraling in the inclinometer casing. Seal the top of the inclinometer casing, around the grout pipe, using rags and tape to prevent grout entry. Mix grout [include required mix and mixing sequence], circulating through pump and back into mixing tank until there are no lumps. Pump grout into grout pipe, using window screen as a filter over the suction intake, until grout overflows from drill casing, keeping track of grout volume pumped. Cut lowering line. Verify that the seal at the top of the inclinometer casing is intact, and withdraw all drill casing, without rotating. Top up with grout. Record grout volume used, and compare with calculated volume required. Withdraw grout pipe, allowing quick-connects to close. Remove female quick-connect and wash thoroughly. Flush grout pipe with water. Insert open-ended grout pipe within inclinometer casing and wash out. There should be no grout in the wash water. If grout appears, keep flushing and watch grout level outside inclinometer casing. Use a stiff brush attached to grout pipe to clean grooves as necessary. If grout is entering inclinometer casing, withdraw inclinometer casing, flush with water, flush hole, reinstall drill casing, diagnose problem, and start again. If in doubt, act quickly and withdraw inclinometer casing. Record actual top and as-built bottom casing elevations on installation record sheet. Repeat inclinometer probe tracking test.
13. After grout has set, cut inclinometer casing to length with a horizontal cut, record as-built top elevation on installation record sheet, insert top cap, construct permanent protective arrangements. Leave space within protective arrangements for attachment of the cable pulley and clamp to the top of the inclinometer casing; ensure that the inclinometer casing is concentric within the protective arrangements. Paint the instrument number on the protective arrangements.
14. Record the as-built groove orientation at the casing top. If problems have been experienced during installation such that excessive groove spiral is suspected or if accurate directional deformation data are required, make a groove spiral survey.
15. Establish an orientation reference for the inclinometer probe, to be used for all readings, and record on installation record sheet. [For a biaxial inclinometer, generally accepted practice is to orient the A transducer so that it will register deformation in the principal plane of interest as a positive change].
16. Take a minimum of two sets of inclinometer readings, compare the readings, study the *check-sums,* and evaluate whether data are within expected tolerances. This constitutes the *acceptance test.* Record the identification numbers of components used during the acceptance test.

APPENDIX H

CONVERSION FACTORS

To convert from a unit in the first column to a unit in the second column, multiply by the factor in the third column. For example,

2 inches = 2 × 25.4 millimeters = 50.8 millimeters.

To convert from a unit in the second column to a unit in the first column, divide by the factor in the third column. For example,

$$2 \text{ millimeters} = \frac{2}{25.4} \text{ inches} = 0.0787 \text{ inches.}$$

H.1. LENGTH

To Convert From	To	Multiply By
inches	millimeters	25.4
inches	meters	0.0254
inches	microns	25400
feet	meters	0.3048
yards	meters	0.9144

H.2. AREA

To Convert From	To	Multiply By
square inches	square millimeters	645.2
square feet	square meters	0.0929
square yards	square meters	0.8361

H.3. VOLUME

To Convert From	To	Multiply By
cubic inches	cubic centimeters	16.39
cubic centimeters	milliliters	1.000
cubic feet	cubic meters	0.0283
cubic feet	liters	28.32
cubic feet	U.S. gallons	7.48
cubic yards	cubic meters	0.7646
U.S. gallons	cubic inches	231
U.S. gallons	liters	3.785
U.S. gallons	cubic meters	0.00378
British gallons	cubic inches	277
British gallons	liters	4.546

H.4. MASS

To Convert From	To	Multiply By
pounds-mass	kilograms	0.4536
short tons[a]	pounds-mass	2000
short tons	kilograms	907.2
metric tons[b]	kilograms	1000
metric tons	short tons	1.102
long tons[c]	pounds-mass	2240
long tons	short tons	1.120

[a] The short ton is currently used widely in the United States.
[b] The metric ton is sometimes spelled *tonne*. ASTM recommends that metric ton be restricted to commercial usage and that the term *tonne* be avoided altogether.
[c] The long ton is the Imperial avoirdupois ton, now largely superseded by the use of SI units.

H.5. FORCE

To Convert From	To	Multiply By
pounds-force	newtons	4.448
kilograms-force	newtons	9.807
short tons-force	newtons	8896
kips	pounds-force	1000
kips	newtons	4448
metric tons-force	kilograms-force	1000
metric tons-force	short tons-force	1.102
long tons-force	pounds-force	2240
long tons-force	short tons-force	1.120

H.6. PRESSURE AND STRESS

To Convert From	To	Multiply By
pascals	newtons per square meter	1.000
pounds per square inch	kilopascals	6.895
pounds per square inch	feet head of water[a]	2.307
pounds per square inch	kilograms per square centimeter	0.0703
pounds per square inch	atmospheres[b]	0.0680
pounds per square foot	kilopascals	0.0479
kilograms per square centimeter	kilopascals	98.07
kilograms per square centimeter	pounds per square inch	14.22
atmospheres[b]	kilopascals	101.3
bars	kilopascals	100
bars	pounds per square inch	14.50
short tons per square foot	kilopascals	95.76
metric tons per square meter	kilopascals	9.807
feet head of water[a]	inches head of mercury	0.8826

[a] At 4°C.
[b] At standard temperature and pressure.

H.7. DENSITY AND UNIT WEIGHT

To Convert From	To	Multiply By
pounds per cubic foot	kilograms per cubic meter	16.02

H.8. TEMPERATURE

Temp. in °F = (9/5 × Temp. in °C) + 32
Temp. in °C = 5/9(Temp. in °F − 32)

H.9. PLANE ANGLE

To Convert From	To	Multiply By
arc-minutes	arc-seconds	60
degrees	arc-minutes	60
degrees	radians	0.01745
radians	arc-seconds	2.063×10^5
gons[a]	radians	0.0157
gons	arc-seconds	3240

[a] 1 gon = 1 new degree = the 400th part of a circle.

H.10. USEFUL CONVERSION FACTORS FOR PREPARATION OF GROUT MIXES

1 U.S. gal = 231 in.^{3} = 3.79 liters
1 ft^3 = 7.48 U.S. gal
1 British gal = 277 in.^{3} = 4.55 liters
Density of water = 62.4 lb/ft^3 = 1000 g/cm^3
1 U.S. gal of water weighs approx. 8.3 lb
1 U.S. gal of cement weighs approx. 12.6 lb
1 U.S. gal of powdered or granular bentonite weighs approx. 9.0 lb
1 British gal of cement weighs approx. 6.9 kg
1 British gal of powdered or granular bentonite weighs approx. 4.9 lb
1 ft^3 of cement weighs approx. 94 lb
1 ft^3 of powdered or granular bentonite weighs approx. 67 lb
1 U.S. bag of cement usually weighs 94 lb
1 U.S. bag of powdered or granular bentonite usually weighs 50 lb or 100 lb
1 U.S. bag of hydrated lime usually weighs 50 lb
1/1 water/cement ratio by volume is equivalent to 0.67/1 by weight

H.11. PREFIXES FOR SI UNITS

The following prefixes are used in the SI (Système International: the international system of units) system:

Factor	Prefix	Symbol
10^9	giga	G
10^6	mega	M
10^3	kilo	k
10^2	hecto	h
10^1	deka	da
10^{-1}	deci	d
10^{-2}	centi	c
10^{-3}	milli	m
10^{-6}	micro	μ
10^{-9}	nano	n

REFERENCES

Abbreviations

AIME	American Institute of Mining, Metallurgical, and Petroleum Engineers
ASCE	American Society of Civil Engineers
ASTM	American Society for Testing and Materials
CSIRO	Commonwealth Scientific and Industrial Research Organization, Australia
ICOLD	International Congress on Large Dams
ISRM	International Society for Rock Mechanics
SM & FE	Soil Mechanics and Foundation Engineering
USCOLD	United States Committee on Large Dams

AASHTO (1984), "Standard Method for Measurements of Pore Pressures in Soils," American Association of State Highway and Transportation Officials, Des. T 252-84, Standard Specifications for Transportation Materials and Methods of Sampling and Testing, Part II, pp. 1047–1052.

AASHTO (1978), "Standard Method for Installing, Monitoring and Processing Data of the Traveling Type Slope Inclinometer," American Association of State Highway and Transportation Officials, Des. T 254-78, pp. 941–950.

Abramson, L. W., and G. E. Green (1985), "Reliability of Strain Gauges and Load Cells for Geotechnical Engineering Applications," Reliability of Geotechnical Instrumentation, *Trans. Res. Record,* No. 1004, pp. 13–19.

ACI (1981), *ACI Manual of Concrete Practice, Part 2—1981,* American Concrete Institute, Detroit, MI.

ACI (1982), "Prediction of Creep, Shrinkage, and Temperature Effects in Concrete Structures," Design for Creep and Shrinkage in Concrete Structures, Spec. Pub. SP-76, Committee 209, American Concrete Institute, Detroit, MI, pp. 193–300.

AISI (1985), *Steel Pile Load Test Data,* American Iron and Steel Institute, Washington, DC.

Alberro, J., and J. Z. Borbón (1985), "Testing of Earth Pressure Cells," *Behavior of Dams Built in Mexico (1974–1984),* Vol. 11, Comision Federal de Electricidad, Contribution to 15th ICOLD, Lausanne, pp. 2.1–2.15.

Aldrich, H. P., and E. G. Johnson (1972), "Embankment Test Sections to Evaluate Field Performance of Vertical Sand Drains for Interstate 295 in Portland, Maine," Soils and Bases: Characteristics, Classification, and Planning, *Highway Res. Record,* No. 405, pp. 60–74.

ANCOLD (1983), *Guidelines for Dam Instrumentation and Monitoring Systems,* Australian National Committee on Large Dams.

Anderson, T. C., M. E. Lockwood, and M. F. Nethero (1984), "Permanent Tieback Retention System," in *Proceedings of the International Conference on Case Histories in Geotechnical Enginering,* S. Prakash (Ed.), University of Missouri, Rolla, MO, published by University of Missouri, Rolla, MO, Vol. I, pp. 417–422.

Anton, W. F., and D. J. Dayton (1972), "Performance of Briones Dam," in *Proceedings of the ASCE, Specialty Conference on Performance of Earth and Earth Supported Structures,* Purdue University, Lafayette, IN, ASCE, New York, Vol. 1, Pt. 1, pp. 853–866.

Arias, R. P. (1977), "Evaluation of Drilled Shaft Integrity by Non-Destructive Methods," M.S. thesis, University of Texas, Austin, TX.

Armento, W. J. (1973), "Cofferdam for BARTD Embarcadero Subway Station," *J. Soil Mech. Found. Div. ASCE,* Vol. 99, No. SM10, Oct., pp. 727–744.

Arthur, K. (1970), *Transducer Measurements,* Measurement Concept Series, Tektronix, Inc., Beaverton, OR.

ASCE (1984), *Guidelines for Tunnel Lining Design,* T. D. O'Rourke (Ed.), Technical Committee on Tunnel Lining Design, Underground Technology Research Council, ASCE, New York.

ASCE (1987), *Consulting Engineering, A Guide for the Engagement of Engineering Services, Manual No. 45 (Revised),* Section III, ASCE, New York (Nov. draft).

ASTM (1974), *Manual on the Use of Thermocouples in Temperature Measurement,* ASTM STP 470A, ASTM, Philadelphia, PA.

ASTM (1980), "Standard Test Method for Compressive Strength of Hydraulic Cement Mortars (Using 2-in. or 50-mm Cube Specimens)," C 109-80, *Annual Book of Standards,* Book 13, Sec. 4, Vol. 4.01, ASTM, Philadelphia, PA.

ASTM (1981a), "Standard Method of Testing Piles Under Static Axial Compression Load," D 1143-81, *Annual Book of Standards,* Vol. 04.08, ASTM, Philadelphia, PA.

ASTM (1981b), "Standard Method of Testing Piles Under Lateral Loads," D 3966-81, *Annual Book of Standards,* Vol. 04.08, ASTM, Philadelphia, PA.

ASTM (1983), "Standard Method of Testing Individual Piles Under Static Axial Tensile Load," D 3689-83, *Annual Book of Standards,* Vol. 04.08, ASTM, Philadelphia, PA.

Atkins, R. T. (1979), "Determination of Frost Penetration by Soil Resistivity Measurements," U.S. Army Corps of Engineers, Cold Regions Research and Engineering Laboratory, Hanover, NH, Spec. Rep. 79-22.

Atkins, R. T. (1981), "Using Electronic Measurement Equipment in Winter," U.S. Army Corps of Engineers, Cold Regions Research and Engineering Laboratory, Hanover, NH, *Cold Regions Tech. Digest,* No. 81-1, July.

Audibert, J. M. E. (1985), personal communication, Sept. 18.

Baguelin, F., J.-F. Jézéquel, E. LeMee, and A. LeMehaute (1972), "Expansion of Cylindrical Probes in Cohesive Soils," *J. Soil Mech. Found. Div. ASCE,* Vol. 98, No. SM11, Nov., pp. 1129–1142.

Baquelin, F., J.-F. Jézéquel, and D. H. Shields (1977), *The Pressuremeter and Foundation Engineering,* Trans Tech Publications, Clausthal-Zellerfeld, F.R. Germany.

Bailey, D. J. (1980), "Land Movement Monitoring System," *Bull. Assoc. Eng. Geol.,* Vol. 17, No. 4, Fall, pp. 213–221.

Baker, C. (1978), "Surge Protection for Instrumentation," in Proceedings of the Tempcon Conference, London. Copy available from Measurement Technology Ltd., Power Court, Luton, Beds., LU1 3JJ, England or MTL, Inc., 7541 Gary Road, Manassas, VA 22110, U.S.A.

Baker, C. (1980), "Protecting Electronic Circuits from Lightning and Transients," Pulse, South Africa. Copy available from Measurement Technology Ltd., Power Court, Luton, Beds., LU1 3JJ, England or MTL, Inc., 7541 Gary Road, Manassas, VA 22110, U.S.A.

Baker, W. H., H. H. MacPherson, and E. J. Cording (1981), "Compaction Grouting to Limit Ground Movements: Instrumented Case History Evaluation of the Bolton Hill Tunnels," U.S. Department of Transportation, Urban Mass Transportation Administration, Rep. No. UMTA-MD-06-0036-81-1.

Ball, K. E. (1987), "An Overview of Off-Line Power," *Instrument Technol.,* Vol. 34, No. 1, Jan., pp. 29–31.

Banister, J. R., R. Pyke, D. M. Ellett, and L. Winters (1976), "In-Situ Pore Pressure Measurements at Rio Blanco," *J. Geotech. Eng. Div. ASCE,* Vol. 102, No. GT10, Oct., pp. 1073–1091.

Barker, W. R., and L. C. Reese (1969), "Instrumentation for Measurement of Axial Load in Drilled Shafts," Res. Rep. 89-6, Center for Highway Research, University of Texas, Austin, TX.

Barratt, D. A., and R. G. Tyler (1976), "Measurements of Ground Movement and Lining Behaviour on the London Underground at Regents Park," Lab. Rep. 684, Transportation and Road Research Laboratory, Crowthorne, England.

Bartholomew, C. L., B. C. Murray, and D. L. Goins (1987), *Embankment Dam Instrumentation Manual,* U.S. Department of the Interior, Bureau of Reclamation.

Barton, M. E., and B. J. Coles (1983), "Rates of Movement of Soil Slopes in Southern England Using Inclinometers and Surface Peg Surveying," in *Proceedings of the International Symposium on Field Measurements in Geomechanics,* Zürich, K. Kovari (Ed.), Balkema, Rotterdam, Vol. 1, pp. 609–618.

Base, G. D. (1955), "Further Notes on the Demec, a Demountable Mechanical Strain Gage for Concrete Structures," *Mag. Concrete Res.* (Cement and Concrete Assoc., London), Vol. 7, No. 19, Mar., pp. 35–38.

Bauer, E. R. (1985), "Ground Control Instrumentation. A Manual for the Mining Industry," U.S. Department of the Interior, Bureau of Mines, Inf. Circ. 9053.

Baur, P. S. (1987), "Radio Telemetry Challenges Wire in Remote Monitoring and Control," *Instrument Technol.,* Vol. 34, No. 1, Jan., pp. 9–14.

Bellier, J.-L., and P.-J. Debreuille (1977), "Three New Instruments for Measurements in Tunnels," in *Proceedings of the International Symposium on Field Measurements in Rock Mechanics,* Zürich, K. Kovari (Ed.), Balkema, Rotterdam, Vol. 1, pp. 351–360.

Beloff, W. R. (1986), personal communication, July 15.

Beloff, W. R., J. Dunnicliff, and W. R. Jaworski (1979), "Performance of a 10-foot Diameter Steel Tunnel Lining in Soft Ground," in *Proceedings of the 4th Rapid Excavation and Tunneling Conference,* Atlanta, A. C. Maevis and W. A. Hustrulid (Eds.), AIME, New York, Vol. 1, pp. 838–860.

Beloff, W. R., D. E. Puza, and F. M. Grynkewicz (1981), "Construction and Performance of a Mixed Face Tunnel Excavation," in *Proceedings of the 5th Rapid Excavation and Tunneling Conference,* San Francisco, R. L. Bullock and H. J. Jacoby (Eds.), AIME, New York, Vol. 2, pp. 1546–1562.

Bergdahl, U., and B. B. Broms (1967), "New Method of Measuring in Situ Settlements," *J. Soil Mech. Found. Div. ASCE,* Vol. 93, No. SM5, Sept., pp. 51–57.

Bierschwale, M. W., H. M. Coyle, and R. E. Bartoskewitz (1981), "Lateral Load Tests on Drilled Shafts Founded in Clay," in *Drilled Piers and Caissons,* M. W. O'Neill (Ed.), ASCE, New York, pp. 98–113.

Binnie & Partners (1979), *Geotechnical Manual for Slopes,* Chap. 10. Manual Prepared for Public Works Department, Government of Hong Kong.

Binnie, G. M., R. T. Gerrard, J. G. Eldridge, S. S. Kirmani, C. V. Davis, J. C. Dickinson, J. R. Gwyther, A. R. Thomas, A. L. Little, J. F. F. Clark, and B. T. Seddon (1967), "Mangla, Part 1, Engineering of Mangla," *Proceedings of the Institution of Civil Engineers,* London, Vol. 38, Nov., pp. 343–544.

Birman, J. H., A. B. Esmilla, and J. B. Indreland (1971), "Thermal Monitoring of Leakage Through Dams," *Bull. Geol. Soc. Am.,* Vol. 82, No. 8, Aug., pp. 2261–2284.

Bishop, A. W., and P. A. Green (1974), "The Development and Use of Trial Embankments," in *Proceedings of the Symposium on Field Instrumentation in Geotechnical Engineer-*

ing, British Geotechnical Society, Butterworths, London, pp. 13–37.

Bishop, A. W., M. Kennard, and A. D. M. Penman (1960), "Pore Pressure Observations at Selset Dam," in *Proceedings of the Conference on Pore Pressure and Suction in Soils*, Butterworths, London, pp. 91–102.

Bjerrum, L., I. J. Johannessen, and O. Eide (1969), "Reduction of Negative Skin Friction on Steel Piles to Rock," in *Proceedings of the 7th International Conference on SM & FE*, Mexico City, published by Mexicana de Mecanica de Suelos, Mexico, Vol. 2, pp. 27–34.

Blackwood, R. L. (1977), "An Instrument to Measure the Complete Stress Field in Soft Rock or Coal in a Single Operation," in *Proceedings of the International Symposium on Field Measurements in Rock Mechanics*, Zürich, K. Kovari (Ed.), Balkema, Rotterdam, Vol. 1, pp. 137–150.

Blanchet, R., F. Tavenas, and R. Garneau (1980), "Behaviour of Friction Piles in Soft Sensitive Clays," *Can. Geotech. J.*, Vol. 17, No. 2, May, pp. 203–224.

Blyth, F. G. H., and M. H. DeFreitas (1974), *A Geology for Engineers*, 6th ed., Edward Arnold, London.

Bodtman, W. L., and J. D. Wyatt (1985), "Design and Performance of Shiroro Rockfill Dam," in *Proceedings of the ASCE, Symposium on Concrete Face Rockfill Dams—Design, Construction and Performance*, Detroit, MI, J. B. Cooke and J. L. Sherard (Eds.), ASCE, New York, pp. 231–251.

Boissier, D., J. Gielly, R. Kastner, and J. C. Mangin (1978), "Détermination des Moments et des Pressions Exercées sur un Écran à Partir de Mesures Inclinométriques," *Can. Geotech. J.*, Vol. 15, No. 4, Nov., pp. 522–536.

Bolt, B. A., and D. E. Hudson (1975), "Seismic Instrumentation for Dams," *J. Geotech. Eng. Div. ASCE*, Vol. 101, No. GT11, Nov., pp. 1095–1104.

Bordes, J.-L. (1986), personal communication, Jan. 24.

Bordes, J.-L., and P.-J. Debreuille (1983), "Borehole Monitoring Instrumentation for Rock Mechanics," in *Proceedings of the International Symposium on Field Measurements in Geomechanics*, Zürich, K. Kovari (Ed.), Balkema, Rotterdam, Vol. 1, pp. 31–48.

Bordes, J.-L., and P.-J. Debreuille (1985), "Some Facts About Long-Term Reliability of Vibrating Wire Instruments," Reliability of Geotechnical Instrumentation, *Trans. Res. Record*, No. 1004, pp. 20–27.

Bouchard, H., and F. Moffitt (1987), *Surveying*, 8th ed., Harper & Row, New York.

Bowen, C. F. P., and F. I. Hewson (1976), "Rock Squeeze at Thorold Tunnel," *Can. Geotech. J.*, Vol. 13, No. 2, May, pp. 111–126.

Bozozuk, M. (1960), "Description and Installation of Piezometers for Measuring Pore Water Pressures in Clay Soils," Division of Building Research, National Research Council of Canada, Bldg. Res. Note No. 37.

Bozozuk, M. (1968), "The Spiral-Foot Settlement Gauge," *Can. Geotech. J.*, Vol. 5, No. 2, May, pp. 123–125.

Bozozuk, M. (1969), "A Fluid Settlement Gage," *Can. Geotech. J.*, Vol. 6, No. 3, Aug., pp. 362–364.

Bozozuk, M. (1970), "Field Instrumentation of Soil," in *Proceedings of the Conference on Design and Installation of Pile Foundations and Cellular Structures*, Lehigh University, Bethlehem, PA, Envo Publishing, Bethlehem, PA, pp. 145–157.

Bozozuk, M. (1972a), "Downdrag Measurements on 160-ft Floating Pipe Test Pile in Marine Clay," *Can. Geotech. J.*, Vol. 9, No. 2, May, pp. 127–136.

Bozozuk, M. (1972b), "The Gloucester Test Fill," Ph.D. thesis, Purdue University, Lafayette, IN.

Bozozuk, M. (1981), "Bearing Capacity of a Pile Preloaded by Downdrag," in *Proceedings of the 10th International Conference on SM & FE*, Stockholm, Balkema, Rotterdam, Vol. 2, pp. 631–636.

Bozozuk, M. (1984), "Measurement Problems Using Geotechnical Field Instrumentation," in *Proceedings of the Seminar on Recent Developments in Geotechnical Instrumentation and Monitoring*, Canadian Geotechnical Society and The Tunnelling Association of Canada, Toronto, Ontario.

Bozozuk, M., and B. H. Fellenius (1979), "The Bellow-Hose Settlement Gauge," *Can. Geotech. J.*, Vol. 16, No. 1, Feb., pp. 233–235.

Bozozuk, M., and G. A. Leonards (1972), "The Gloucester Test Fill," in *Proceedings of the ASCE, Specialty Conference on Performance of Earth and Earth-Supported Structures*, Purdue University, Lafayette, IN, ASCE, New York, Vol. 1, Pt. 1, pp. 299–317.

Bozozuk, M., G. H. Johnston, and J. J. Hamilton (1962), "Deep Bench Marks in Clay and Permafrost Areas," in *Field Testing of Soils*, ASTM STP 322, ASTM, Philadelphia, PA, pp. 265–279.

Bozozuk, M., M. C. Van Wijk, and B. H. Fellenius (1978), "Terrestrial Photogrammetry for Measuring Pile Movements," *Can. Geotech. J.*, Vol. 15, No. 4, Nov., pp. 596–599.

Bozozuk, M., G. H. Keenan, and P. E. Pheeney (1979), "Analysis of Load Tests on Instrumented Steel Test Piles in Compressible Silty Soil," in *Behavior of Deep Foundations*, ASTM STP 670, R. Lundgren (Ed.), ASTM, Philadelphia, PA, pp. 153–180.

Brand, E. W., and J. Premchitt (1982), "Response Characteristics of Cylindrical Piezometers," *Géotechnique*, Vol. 32, No. 3, pp. 203–216.

Brand, E. W., G. W. Borrie, and J. M. Shen (1983), "Field Measurements in Hong Kong Residual Soils," in *Proceedings of the International Symposium on Field Measurements in Geomechanics*, Zürich, K. Kovari (Ed.), Balkema, Rotterdam, Vol. 1, pp. 639–648.

Brawner, C. O. (1974), "Rock Mechanics in Open Pit Mining," *Advances in Rock Mechanics, Proceedings of the 3rd Congress of ISRM*, Denver, published by National Academy of Sciences, Washington, D.C., Vol. 1, Pt. A, pp. 755–773.

Brewer, G. A. (1972), "Strain Gage Assembly and Method of Attachment," U.S. Patent No. 3,639,875, Feb. 1, U.S. Patent Office, Washington, D.C. (Patent assigned to Brewer Engineering Laboratories, Marion, MA.)

Briaud, J. L., C. F. Raba, Jr., W. T. Johnson, Jr., R. Halvorson, and S. Ohya (1985), "Predicted and Measured Performance of a Foundation," in *Drilled Piers and Caissons II*, C. N. Baker, Jr. (Ed.), ASCE, New York, pp. 43–56.

Brooker, E. W., and D. A. Lindberg (1965), "Field Measurement of Pore Pressure in High Plasticity Soils," in *Engineer-*

ing Effects of Moisture Changes in Soils, G. D. Aitchison (Ed.), Proceedings of the International Research and Engineering Conference on Expansive Clay Soils, Texas A&M University, College Station, TX, pp. 57–68.

Brough, W. G., and W. C. Patrick (1982), "Instrumentation Rep. 1: Specification, Design, Calibration for an Experimental, High-Level, Nuclear Waste Storage Facility," Rep. No. UCRL-53248, Lawrence Livermore Laboratory, Livermore, CA.

Brown, G. L., E. D. Morgan, and J. S. Dodd (1971), "Rock Stabilization at Morrow Point Power Plant," *J. Soil Mech. Found. Div. ASCE,* Vol. 97, No. SM1, Jan., pp. 119–139.

Brown, S. F. (1977), "State-of-the-Art Report on Field Instrumentation for Pavement Experiments," Multiple Aspects of Soil Mechanics, *Trans. Res. Record,* No. 640, pp. 13–28.

Browne, R. D., and L. H. McCurrich (1967), "Measurements of Strain in Concrete Pressure Vessels," in *Proceedings of the Conference on Prestressed Concrete Pressure Vessels,* Group 1, Instrumentation and Commissioning, Institution of Civil Engineers, London, pp. 3–13.

Burke, H. H. (1957), "Garrison Dam—Tunnel Test Section Investigation—A Symposium," *J. Soil Mech. Found. Div. ASCE,* Vol. 83, No. SM4, Nov., pp. 1438-1 to 1438-50.

Burkitt, C. J. (1980), "Lightning Protection," Measurements and Control, Oct., pp. 128–132. Copy available from Measurement Technology Ltd., Power Court, Luton, Beds., LU1 3JJ, England or MTL, Inc., 7541 Gary Road, Manassas, VA 22110, U.S.A.

Burland, J. B., and J. F. A. Moore (1974), "The Measurement of Ground Displacement Around Deep Excavations," in *Proceedings of the Symposium on Field Instrumentation in Geotechnical Engineering,* British Geotechnical Society, Butterworths, London, pp. 70–84.

Burland, J. B., J. F. A. Moore, and P. D. K. Smith (1972), "A Simple and Precise Borehole Extensometer," *Géotechnique,* Vol. 22, No. 1, pp. 174–177.

Burland, J. B., T. I. Longworth, and J. F. A. Moore (1977), "A Study of Ground Movement and Progressive Failure Caused by a Deep Excavation in Oxford Clay," *Géotechnique,* Vol. 27, No. 4, pp. 557–591.

Cape, J. (1984), "When Treatment Tanks Lift and Tilt," *Public Works,* Vol. 115, No. 2, Feb., pp. 46–48.

Carlson, R. W. (1975), *Manual for the Use of Strain Meters and Other Instruments for Embedment in Concrete Structures,* 4th ed., Carlson Instruments, Campbell, CA.

Carlson, R. W. (1978), "Performance Tests on Stress Meters," *Int. Water Power & Dam Construction,* London, Vol. 30, No. 4, Apr., pp. 39–44.

Carlson, R. W. (1984), personal communication, Mar. 12.

Carpenter, L. (1984a), "New Sensors for Old Dams," *Res. News,* U.S. Dept. Interior, Bur. Reclamation, Vol. 13, No. 1, June, pp. 1–2.

Carpenter, L. (1984b), "Instrumentation Automation," *Res. News,* U.S. Dept. Interior, Bur. Reclamation, Vol. 13, No. 1, June, pp. 3–5.

Carpentier, R., and W. Verdonck (1986), "Special Pore Water Pressure Measuring System Installed in the Seabed for the Construction of the New Outer Harbour at Zeebrugge, Belgium," in *Proceedings of the International Conference on Measuring Techniques of Hydraulics Phenomena in Offshore, Coastal & Inland Waters,* London, published by BHRA, Bedford, England, pp. 121–136.

Carvalho, O. S., and K. Kovari (1983), "The Measurement of Strain Distribution in Large Diameter Steel Piles," in *Proceedings of the International Symposium on Field Measurements in Geomechanics,* Zürich, K. Kovari (Ed.), Balkema, Rotterdam, Vol. 1, pp. 361–371.

Casagrande, A. (1949), "Soil Mechanics in the Design and Construction of the Logan Airport," *J. Boston Soc. Civil Eng.,* Vol. 36, No. 2, pp. 192–221. Reprinted in *Contributions to Soil Mechanics, 1941–1953,* Boston Society of Civil Engineers, Boston, pp. 176–205.

Casagrande, A. (1958), "Piezometers for Pore Pressure Measurements in Clay," Division of Engineering and Applied Physics, Harvard University, Cambridge, MA., unpublished.

Casagrande, A., S. B. Avery, and S. J. Poulos (1967), "Measurement of Vertical Movements with an Improved Hose Level," Harvard University, Cambridge, MA, March, unpublished.

Castelli, R. J., S. K. Sarkar, and G. A. Munfakh (1983), "Ground Treatment in the Design and Construction of a Wharf Structure," in *Proceedings of the Symposium on Advances in Piling and Ground Treatment for Foundations,* Institution of Civil Engineers, London, pp. 209–215.

Catanach, R., and T. N. McDaniel (1972), "Lateral Deformation of a Dam Embankment," in *Proceedings of the ASCE, Specialty Conference on Performance of Earth and Earth Supported Structures,* Purdue University, Lafayette, IN, ASCE, New York, Vol. 1, Pt. 1, pp. 867–883.

Cerni, R. H., and L. E. Foster (1965), *Instrumentation for Engineering Measurements,* Wiley, New York.

Champa, S., and B. Mahatharadol (1982), "Construction of Srinagarind Dam," in *Transactions of the 14th ICOLD,* Rio de Janeiro, International Commission on Large Dams, Paris, Vol. 4, pp. 255–278.

Chedsey, G. L., and R. Dorey (1983), "Instrumentation of a Major Tailings Impoundment in Idaho, U.S.A.," in *Proceedings of the International Symposium on Field Measurements in Geomechanics,* Zürich, K. Kovari (Ed.), Balkema, Rotterdam, Vol. 1, pp. 661–670.

Cherry, J. A., and P. E. Johnson (1982), "A Multilevel Device for Monitoring in Fractured Rock," *Ground Water Monitoring Rev.,* Vol. 2, No. 3, Summer, pp. 41–44.

Clausen, C.-J. F., J. Graham, and D. M. Wood (1984), "Yielding in Soft Clay at Mastemyr, Norway," *Géotechnique,* Vol. 34, No. 4, pp. 581–600.

Clemente, F. M. (1979), "Downdrag—A Comparative Study of Bitumen Coated and Uncoated Prestressed Piles," Current Practices on Pile Design & Installation, Associated Pile & Fitting Corp. Piletalk Seminar, New York, March, pp. 49–71.

Clements, D. J. (1978), "Settlement in Elevation," *The Consulting Eng.,* London, Vol. 42, No. 1, Jan., pp. 46, 47.

Clements, D. J. (1982), "The Surface Installation Method," *Civil Eng.,* London, Feb., pp. 24–28.

Clements, D. J. (1984), personal communication, Sept. 26.

Clements, D. J., and A. C. Durney (1982), "Instrumentation Developments," in *Proceedings of the Autumn Conference of the British National Committee on Large Dams (BNCOLD)*, Keele University, Institution of Civil Engineers, London, pp. 45–55.

Clough, G. W., and M. W. Reed (1984), "Measured Behavior of Braced Wall in Very Soft Clay," *J. Geotech. Eng. Div. ASCE*, Vol. 110, No. 1, Jan., pp. 1–19.

Clough, G. W., P. R. Weber, and J. Lamont, Jr. (1972), "Design and Observation of a Tied-Back Wall," in *Proceedings of the ASCE, Specialty Conference on Performance of Earth and Earth-Supported Structures*, Purdue University, Lafayette, IN, ASCE, New York, Vol. I, Pt. 2, pp. 1367–1389.

Clough, G. W., W. H. Baker, D. Y. Tan, and W. M. Kuck (1977), "Effect of Chemical Solidification of Soils Around a Shallow Tunnel," ASCE Fall Convention, San Francisco, Preprint 3006.

Clough, G. W., W. H. Baker, and F. Mensah-Dwumah (1979), "Ground Control for Soft Ground Tunnels Using Chemical Stabilization—A Case History Review," in *Proceedings of the 4th Rapid Excavation & Tunneling Conference*, Atlanta, A. C. Maevis and W. A. Hustrulid (Eds.), AIME, New York, Vol. 1, pp. 395–415.

Considine, D. M. (1971), *Encyclopedia of Instrumentation and Control*, McGraw-Hill, New York.

Cook, C. W., and E. S. Ames (1979), "Borehole-Inclusion Stressmeter Measurements in Bedded Salt," in *Proceedings of the 20th Symposium on Rock Mechanics*, University of Texas, Austin, TX, pp. 481–485.

Cooke, J. B. (1986), personal communication, Jan. 24.

Cooke, R. W., and G. Price (1974), "Horizontal Inclinometers for the Measurement of Vertical Displacement in the Soil Around Experimental Foundations," in *Proceedings of the Symposium on Field Instrumentation in Geotechnical Engineering*, British Geotechnical Society, Butterworths, London, pp. 112–125.

Cooke, R. W., and G. Price (1978), "Strains and Displacements Around Friction Piles," in *Foundations and Soil Technology*, Building Research Series, Vol. 3, The Construction Press, London, pp. 197–207.

Cooke, R. W., G. Price, and K. Tarr (1979), "Jacked Piles in London Clay: A Study of Load Transfer and Settlement Under Working Conditions," *Géotechnique*, Vol. 29, No. 2, pp. 113–147.

Cooke, R. W., G. Price, and K. Tarr (1980), "Jacked Piles in London Clay: Interaction and Group Behaviour Under Working Conditions," *Géotechnique*, Vol. 30, No. 2, pp. 97–136.

Cooke, R. W., D. W. Bryden-Smith, M. N. Gooch, and D. F. Sillet (1981), "Some Observations of the Foundation Loading and Settlement of a Multi-Storey Building on a Piled Raft Foundation in London Clay," in *Proceeding of the Institution of Civil Engineers*, London, Vol. 70, Pt. 1, Aug., pp. 433–460.

Corbett, D. A., H. M. Coyle, R. E. Bartoskewitz, and L. J. Milberger (1971), "Evaluation of Pressure Cells Used for Field Measurements of Lateral Earth Pressures on Retaining Walls," Res. Rep. 169-1, Texas Transportation Institute, Texas A&M University, College Station, TX.

Cording, E. J., A. J. Hendron, W. H. Hansmire, J. W. Mahar, H. H. MacPherson, R. A. Jones, and T. D. O'Rourke (1975), "Methods for Geotechnical Observations and Instrumentation in Tunneling," Rep. No. UILU-E 75 2022, Department of Civil Engineering, University of Illinois, Urbana, IL.

Cornforth, D. H. (1974), "Performance Characteristics of the Slope Indicator Series 200-B Inclinometer," in *Proceedings of the Symposium on Field Instrumentation in Geotechnical Engineering*, British Geotechnical Society, Butterworths, London, pp. 126–135.

Corps of Engineers (1971), *Instrumentation of Earth and Rock-Fill Dams (Groundwater and Pore Pressure Observations)*, U.S. Army Corps of Engineers, Eng. Manual EM 1110-2-1908 Part 1.

Corps of Engineers (1980), *Instrumentation for Concrete Structures*, U.S. Army Corps of Engineers, Eng. Manual EM 1110-2-4300.

Corps of Engineers (1981), "Strong Motion Instruments for Recording Earthquake Motions on Dams," U.S. Army Corps of Engineers, Eng. Reg. ER 1110-2-103, Dec. 10.

Corps of Engineers (1984), "Publications Relating to Geotechnical Activities," U.S. Army Corps of Engineers, DAEN-ECE-G, Sept. 27.

Corps of Engineers (1985), "Instrumentation for Safety Evaluations of Civil Works Projects," U.S. Army Corps of Engineers, Eng. Reg. ER 1110-2-110, July 8.

Coyle, H. M., and R. E. Bartoskewitz (1976), "Earth Pressure on Precast Panel Retaining Wall," *J. Geotech. Eng. Div. ASCE*, Vol. 102, No. GT5, May, pp. 441–456.

Cragg, C. B. H. (1984a), "Instrumentation of Geotextile Reinforced Embankments," in *Proceedings of the Seminar on Recent Developments in Geotechnical Instrumentation and Monitoring*, Canadian Geotechnical Society and the Tunnelling Association of Canada, Toronto.

Cragg, C. B. H. (1984b), personal communication, Nov. 20.

Cumming, J. D. (1956), *Diamond Drill Handbook*, J. K. Smit & Son, Murray Hill, NJ.

Cunningham, J. A., and J. I. Fernandez (1972), "Performance of Two Slurry Wall Systems in Chicago," in *Proceedings of the ASCE, Specialty Conference on Performance of Earth and Earth-Supported Structures*, Purdue University, Lafayette, IN, ASCE, New York, Vol. I, Pt. 2, pp. 1425–1449.

Dames & Moore (1979), "Results of Laboratory Test Program on Cement-Bentonite Grouts for Bailly Generating Station—Nuclear 1, Northern Indiana Public Service Company," Dames & Moore, Chicago, IL.

D'Appolonia, E., R. Alperstein, and D. J. D'Appolonia (1967), "Behavior of a Colluvial Slope," *J. Soil Mech. & Found. Div. ASCE*, Vol. 93, No. SM4, July, pp. 447–473.

Dascal, O. (1987), "Postconstruction Deformations of Rockfill Dams," *J. Geotech. Eng. Div. ASCE*, Vol. 113, No. 1, Jan., pp. 46–59.

Davis, A. G., and C. S. Dunn (1974), "From Theory to Field Experience with the Non-destructive Vibration Testing of Piles," in *Proceedings of the Institution of Civil Engineers*, London, Vol. 57, Pt. 2, Dec., pp. 571–593.

Davis, A. G., and S. A. Robertson (1975), "Economic Pile Testing," *Ground Eng.*, Vol. 8, No. 3, May, pp. 40–43.

Davis, A. G., and S. A. Robertson (1976), "Vibration Testing of Piles," *Structural Eng. J.*, Vol. 54, No. 6, June, pp. A7–A10.

Davis, C. M. (1985), "Fiber Optic Sensors: An Overview," *Opt. Eng.*, Vol. 24, No. 2, Mar./Apr., pp. 347-351.

Davis, C. M., E. F. Carome, M. H. Weik, S. Ezekiel, and R. E. Einzig (1982), *Fiberoptic Sensor Technology Handbook*, Dynamic Systems, Inc., Reston, VA.

Davis, R. E., F. S. Foote, J. M. Anderson, and E. M. Mikhail (1981), *Surveying: Theory and Practice*, 6th ed., McGraw-Hill, New York.

Davisson, M. T. (1966), "Summary of Knowledge Gained from Tests on Instrumented Driven Piles," ASCE Met Section Lecture Series, New York, May.

Deadman, D. J., and J. R. Slight (1965), "A New Approach to the Ductrodding Problem—Ductmotor No. 1," *Post Office Elec. Eng. J.*, London, Vol. 58, Pt. 2, pp. 91–92.

Deardorff, G. B., A. M. Lumsden, and W. M. Hefferon (1980), "Pneumatic Piezometers: Multiple and Single Installations in Vertical and Inclined Boreholes," *Can. Geotech. J.*, Vol. 17, No. 2, May, pp. 313–320.

Debreuil, L., and R. Hamelin (1974), "Vertical Drilling of Inverse Pendulum Holes," in *Proceedings of the 2nd Canadian Symposium on Mining, Surveying and Rock Deformation Measurements*, Queen's University, Kingston.

Deere, D. U. (1982), "Cement–Bentonite Grouting for Dams," in *Proceedings of the ASCE, Conference on Grouting in Geotechnical Engineering*, New Orleans, LA, ASCE, New York, pp. 279–300.

DeLory, F. A., A. M. Crawford, and M. E. M. Gibson (1979), "Measurements on a Tunnel Lining in Very Dense Till," *Can. Geotech. J.*, Vol. 16, No. 1, Feb., pp. 190–199.

Department of Water Resources (1979), "The August 1, 1975 Oroville Earthquake Investigations," Bull. 203-78, State of California, Feb.

DiBiagio, E. (1974), Discussion, in *Proceedings of the Symposium on Field Instrumentation in Geotechnical Engineering*, British Geotechnical Society, Butterworths, London, pp. 565–566.

DiBiagio, E. (1977), "Field Instrumentation—A Geotechnical Tool," Norwegian Geotechnical Institute Pub. No. 115, pp. 29–40.

DiBiagio, E. (1979), "Use of Computers and Data Acquisition Systems in Geotechnical Instrumentation Projects," Norske Sivilingeniorers Forening, EDB I Geoteknikken, Kurs, Trondheim, Norway.

DiBiagio, E. (1983), "Instruments and Instrumentation Techniques Used to Monitor the Performance of Offshore Structures," in *Proceedings of the International Symposium on Field Measurements in Geomechanics*, Zürich, K. Kovari (Ed.), Balkema, Rotterdam, Vol. 1, pp. 405–433.

DiBiagio, E. (1986), "Comments on the Reliability and Performance of Vibrating-Wire Type Instruments," Response Paper, in *Proceedings of the Indian Geotechnical Conference*, New Delhi, published by Sarita Prakashan, New Delhi, Vol. 2, pp. 227–231.

DiBiagio, E., and B. Kjaernsli (1985), "Instrumentation of Norwegian Embankment Dams," in *Transactions of the 15th ICOLD*, Lausanne, International Commission on Large Dams, Paris, Vol. 1, pp. 1071–1101.

DiBiagio, E., and F. Myrvoll (1972), "Full Scale Field Tests of a Slurry Trench Excavation in Soft Clay," in *Proceedings of the 5th European Conference on SM & FE*, Madrid, published by Spanish Society for Soil Mechanics and Foundation Engineering, Madrid, Vol. 1, pp. 461–471. Also in Pub. No. 91, Norwegian Geotechnical Institute, 1973.

DiBiagio, E., and F. Myrvoll (1981), "Field Instrumentation for Soft Clay," in *Soft Clay Engineering*, R. P. Brenner and E. W. Brand (Eds.), Elsevier Publishing, Amsterdam, Chap. 10, pp. 699–736.

DiBiagio, E., and F. Myrvoll (1985), "Instrumentation Techniques and Equipment Used to Monitor the Perforance of Norwegian Embankment Dams," in *Transactions of the 15th ICOLD*, Lausanne, International Commission on Large Dams, Paris, Vol. 1, pp. 1169–1197.

DiBiagio, E., and J. A. Roti (1972), "Earth Pressure Measurements on a Braced Slurry-Trench Wall in Soft Clay," in *Proceedings of the 5th European Conference on SM & FE*, Madrid, published by Spanish Society for Soil Mechanics and Foundation Engineering, Madrid, Vol. 1, pp. 473–483.

DiBiagio, E., S. B. Hansen, and T. Wetlesen (1981), "Instrumentation and Data Acquisition," Norwegian Geotechnical Institute, Pub. No. 137, pp. 1–28.

DiBiagio, E., F. Myrvoll, T. Valstad, and H. Hansteen (1982), "Field Instrumentation, Observations and Performance Evaluations of the Svartevann Dam," in *Transactions of the 14th ICOLD*, Rio de Janeiro, International Commission on Large Dams, Paris, Vol. 1, pp. 789–826.

Dow (1981), *A Guide to Glycols*, Dow Chemical USA, Organic Chemicals Department, Midland, MI.

Dreyer, H. (1977), "Long-Term Measurements in Rock Mechanics by Means of Maihak Vibrating-Wire Instrumentation," in *Proceedings of the International Symposium on Field Measurements in Rock Mechanics*, Zürich, K. Kovari (Ed.), Balkema, Rotterdam, Vol. 1, pp. 109–122.

Driscoll, F. G. (1986), *Groundwater Wells*, 2nd ed., Johnson Division, St. Paul, MN.

Drnevich, V. P., and R. E. Gray (Eds.) (1981), *Acoustic Emission in Geotechnical Engineering Practice*, ASTM STP 750, ASTM, Philadelphia, PA.

Duncan Fama, M. E., and M. J. Pender (1980), "Analysis of the Hollow Inclusion Technique for Measuring the In-Situ Rock Stress," *Int. J. Rock Mech. & Min. Sci. & Geomech. Abstr.*, Vol. 17, No. 3, pp. 137–146.

Dunnicliff, J. (1968), "Instrumentation of the Plover Cove Main Dam," *Géotechnique*, Vol. 18, No. 3, pp. 283–300.

Dunnicliff, J. (1981), "Long-Term Performance of Embankment Dam Instrumentation," in *Proceedings of the ASCE, Symposium on Recent Developments in Geotechnical Engineering for Hydro Projects*, F. H. Kulhawy (Ed.), ASCE, New York, pp. 1–22.

Dunnicliff, J., D. Hampton, and E. T. Selig (1981), "Tunnel Instrumentation: Why and How?," in *Proceedings of the 5th Rapid Excavation and Tunneling Conference*, San Francisco, R. L. Bullock and H. J. Jacoby (Eds.), AIME, New York, Vol. 2, pp. 1455–1472.

Dutro, H. B. (1974), "Slope Instrumentation Using Multiple-Position Borehole Extensometers," Landslide Instrumentation, *Trans. Res. Record*, No. 482, pp. 9–17.

Dutro, H. B. (1977), "Borehole Measurements Using Portable Borehole Deflectometers," in *Proceedings of the 18th Symposium on Rock Mechanics,* Keystone, CO, Colorado School of Mines Press, Golden, CO, pp. 5C4-1 to 5C4-5.

Dutro, H. B. (1984), personal communication, Feb. 15.

Dutro, H. B. (1985), personal communication, June 15.

Dutta, P. K. (1982), "Holding Force of Irad Extensometer C-Anchor in Competent Rock," Tech. Memo. 82-9, Irad Gage, Salem, NH.

Dutta, P. K., R. W. Hatfield, and P. W. Runstadler, Jr. (1981), "Calibration Characteristics of Irad Gage Vibrating Wire Stressmeters at Normal and High Temperatures," Tech. Rep. 80-2, Irad Gage, Salem, NH.

Dyer, D. C. (1986), "The Preparation of De-aired Water Using a Method and Equipment Supplied by Geotechnical Instruments (U.K.) Ltd.," Department of Engineering, University of Warwick, England, Nov.

Easton, C. N. (1984), personal communication, Aug. 21.

Eide, O., and S. Holmberg (1972), "Test Fills to Failure on the Soft Bangkok Clay," in *Proceedings of the ASCE, Specialty Conference on Performance of Earth and Earth-Supported Structures,* Purdue University, Lafayette, IN, ASCE, New York, Vol. 1, Pt. 1, pp. 159–180.

Elson, W. K., and A. L. Reddaway (1980), "Vibrating Wire Strain Gauges in Driven Concrete Piles—Evaluation Trials," Tech. Note 99, Construction Industry Research and Information Association, London.

Enever, J. R., B. McKavanagh, and G. Carson (1977), "Some Applications of Instrumentation in the Australian Mining Industry," in *Proceedings of the International Symposium on Field Measurements in Rock Mechanics,* Zürich, K. Kovari (Ed.), Balkema, Rotterdam, Vol. 1, pp. 47–60.

England, G. L., and J. M. Illston (1965), "Methods of Computing Stress in Concrete from a History of Measured Strain," *Civil Eng. Public Works Rev.,* London, Vol. 60, No. 705, Apr., pp. 513–517, May, pp. 692–694, June, pp. 846, 847.

ENR (1984), "Damaged Bridge Returns to Health in Delicate Repair," *Eng. News Record,* Mar. 22, pp. 26, 27.

Fang, H. Y., and R. M. Koerner (1977), "An Instrument for Measuring In Situ Soil-Structure Response During Dynamic Vibration," in *Proceedings of the 14th Annual Meeting of the Society of Engineering Science,* published by Society of Engineering Science, Bethlehem, PA, pp. 1171–1180.

Farr, J. S., and R. P. Aurora (1981), "Behavior of an Instrumented Pier in Gravelly Sand," in *Drilled Piers and Caissons,* M. W. O'Neill (Ed.), ASCE, New York, pp. 53–65.

Felio, G. Y., and G. E. Bauer (1986), "Factors Affecting the Performance of a Pneumatic Earth Pressure Cell," *Geotech. Testing J. ASTM,* Vol. 9, No. 2, June, pp. 102–106.

Fellenius, B. H. (1972), "Down-drag on Piles Due to Negative Skin Friction," *Can. Geotech. J.,* Vol. 9, No. 4, Nov., pp. 323–337.

Fellenius, B. H. (1980), "The Analysis of Results from Routine Pile Load Tests," *Ground Eng.,* Vol. 13, No. 6, Sept., pp. 19–31.

Fellenius, B. H. (1984), "Ignorance Is Bliss—and That Is Why We Sleep So Well," *Geotech. News,* Vol. 2, No. 4, Dec., pp. 14, 15.

Fellenius, B. H. (1987), "Determining Load Distribution in Piles Using Strain Data and the Tangent Modulus Analysis," Lecture Notes, University of Ottawa, Ontario.

Fellenius, B. H., and B. B. Broms (1969), "Negative Skin Friction for Long Piles Driven in Clay," in *Proceedings of the 7th International Conference on SM & FE,* Mexico City, published by Mexicana de Mecanica de Suelos, Mexico, Vol. 2, pp. 93–98.

Fellenius, B. H., and T. Haagen (1969), "New Pile Force Gauge for Accurate Measurements of Pile Behaviour During and Following Driving," *Can. Geotech. J.,* Vol. 6, No. 3, Aug., pp. 356–362.

Fellenius, B. H., A. J. O'Brien, and F. W. Pita (1982), "Construction Control by Monitored Geotechnical Instrumentation for Port of Seattle New Terminal 46," The Use of Field Measurements and Observations to Design and Construct Pile Foundations, *Trans. Res. Record,* No. 884, pp. 14–22.

Fetzer, C. A. (1982), "Pumped Bentonite Mixture Used to Seal Piezometers," *J. Geotech. Eng. Div. ASCE,* Vol. 108, No. GT2, Feb., pp. 295–299.

FHWA (1980), *Proceedings of the Conference on Tunnel Instrumentation—Benefits and Implementation,* D. Hampton, S. Browne, and E. Greenfield (Eds.), U.S. Department of Transportation, Federal Highway Administration, New Orleans, Rep. No. FHWA-TS-81-201.

Figueiredo, A. F. de, and A. Negro Jr. (1981), "New Sensing System for Borehole Extensometers," *Géotechnique,* Vol. 31, No. 3, pp. 427–430.

Filho, P. R. (1976), "Laboratory Tests on a New Borehole Seal for Piezometers," *Ground Eng.,* Vol. 9, No. 1, Jan., pp. 16–18.

Fitzpatrick, M. D., T. B. Liggins, and R. H. W. Barnett (1982), "Ten Years Surveillance of Cethana Dam," in *Transactions of the 14th ICOLD,* Rio de Janeiro, International Commission on Large Dams, Paris, Vol. 1, pp. 847–866.

Forsyth, R. A., and M. McCauley (1973), "Monitoring Devices to Control Embankment Construction on Soft Foundations," Course Notes, Chap. 9, Slope Stability and Foundation Investigation, Institute of Transportation and Traffic Engineering, University of California, Berkeley, CA.

Fox, G. A. (1980), Discussion in *Proceedings of the Conference on Tunnel Instrumentation—Benefits and Implementation,* D. Hampton, S. Browne, and E. Greenfield (Eds.), U.S. Department of Transportation, Federal Highway Administration, New Orleans, Rep. No. FHWA-TS-81-201, pp. 20–23.

Franklin, J. A. (1977), "Some Practical Considerations in the Planning of Field Instrumentation," in *Proceedings of the International Symposium on Field Measurements in Rock Mechanics,* Zürich, K. Kovari (Ed.), Balkema, Rotterdam, Vol. 1., pp. 3–13.

Franklin, J. A. (1986), "Size–Strength System for Rock Characterization," in *Proceedings of the International Symposium on Application of Rock Characterization Techniques in Mine Design,* New Orleans, LA, M. Karmis (Ed.), Society of Mining Engineers, New York, pp. 11–16.

Franklin, J. A. (1988), *Rock Engineering,* McGraw-Hill, New York, in press.

Franklin, J. A., and P. E. Denton (1973), "The Monitoring of Rock Slopes," *Q. J. Eng. Geol.*, Vol. 6, No. 3/4, pp. 259–286.

Fuglevand, P. F., J. E. Zipper, and G. E. Horvitz (1984), "Instrumentation of Ground Anchors," *The Indicator*, Slope Indicator Company, Seattle, WA, Nov., pp. 4–6.

Gandahl, R., and W. Bergau (1957), "Two Methods for Measuring the Frozen Zone in Soil," in *Proceedings of the 4th International Conference on SM & FE*, London, Vol. 1, Butterworths, London, pp. 32–34.

Gemme, R. L. (1984), personal communication, Apr. 23.

General Electric (1976), *Transient Voltage Suppression Manual*, G.E. Semiconductor Products Department, Syracuse, NY.

George, P. J., and R. H. G. Parry (1974), "Field Loading Tests at Canvey Island," in *Proceedings of the Symposium on Field Instrumentation in Geotechnical Engineering*, British Geotechnical Society, Butterworths, London, pp. 152–165.

Goeke, P. M., and P. A. Hustad (1979), "Instrumented Drilled Shafts in Clay-Shale," in *Symposium on Deep Foundations*, F. M. Fuller (Ed.), ASCE, New York, pp. 149–165.

Gould, J. P. (1970), "Lateral Pressures on Rigid Permanent Structures," in *Proceedings of the ASCE, Specialty Conference on Lateral Stresses in the Ground and Design of Earth-Retaining Structures*, Cornell University, Ithaca, NY, ASCE, New York, State-of-the-Art Vol., pp. 219–269.

Gould, J. P., and J. Dunnicliff (1971), "Accuracy of Field Deformation Measurements," in *Proceedings of the 4th Panamerican Conference on SM & FE*, San Juan, ASCE, New York, Vol. 1, pp. 313–366.

Gravina, J., and G. H. Carson (1983), "CSIRO Horizontal Borehole Inclinometer—Description and Operation," Geomechanics of Coal Mining Rep. No. 53, CSIRO Division of Applied Geomechanics, Australia.

Green, G. E. (1974), "Principles and Performance of Two Inclinometers for Measuring Horizontal Ground Movements," in *Proceedings of the Symposium on Field Instrumentation in Geotechnical Engineering*, British Geotechnical Society, Butterworths, London, pp. 166–179.

Green, G. E. (1982), Discussion: "Piezometers in Earth Dam Impervious Sections," *J. Geotech. Eng. Div. ASCE*, Vol. 108, No. GT11, Nov., pp. 1526–1528.

Green, G. E. (1985), personal communication, Dec. 9.

Green, G. E. (1986), personal communication, May 22.

Green, G. E., and P. E. Mikkelsen (1986), "Measurement of Ground Movement with Inclinometers," in *Proceedings of the 4th International Geotechnical Seminar, Field Instrumentation and In Situ Measurements*, Nanyang Technical Institute, Singapore, pp. 235–246.

Green, G. E., and D. A. Roberts (1983a), "Remote Monitoring of a Coal Waste Embankment," in *Proceedings of the International Symposium on Field Measurements in Geomechanics*, Zürich, K. Kovari (Ed.), Balkema, Rotterdam, Vol. 1, pp. 671–682.

Green, G. E., and D. A. Roberts (1983b), "Remote Monitoring of a Coal Waste Impoundment in West Virginia," U.S. Department of the Interior, Bureau of Mines, Rep., Cont. No. H0282041.

Green, G. E., A. I. Feldman, and P. Thordarson (1983), "A Pile Tip Load Cell for a Driven Concrete Pile," in *Proceedings of the International Symposium on Field Measurements in Geomechanics*, Zürich, K. Kovari (Ed.), Balkema, Rotterdam, Vol. 1, pp. 463–471.

Gregory, E., T. A. Rundle, W. M. McCabe, and K. Kim (1983), "In Situ Stress Measurement in a Jointed Basalt: The Suitability of Five Overcoring Techniques," in *Proceedings of the 6th Rapid Excavation and Tunneling Conference*, Chicago, H. Sutcliffe and J. W. Wilson (Eds.), AIME, New York, Vol. 1, pp. 42–61.

Guarino, L. (1985), "Pore Pressure Changes Due to Bentonite Pellet Seals," Project Rep. for M.S. Degree, University of Massachusetts, Amherst, MA.

Guertin, J. D., Jr., and R. F. Flanagan (1979), "Construction Behavior of a Shallow Tunnel in Highly Stressed Sedimentary Rock," in *Proceedings of the 4th International Congress on Rock Mechanics*, Montreux, Vol. 2, pp. 181–188.

Guertin, J. D., Jr., and E. S. Plotkin (1979), "Observations of Construction Behavior of a Major Rock Tunnel, New York City," in *Proceedings of the 4th Rapid Excavation and Tunneling Conference*, Atlanta, A. C. Maevis and W. A. Hustrulid (Eds.), AIME, New York, Vol. 1, pp. 956–971.

Gupton, C. P., J. F. O'Brien, and T. J. Logan (1982), "Design of Drilled Shafts in South Florida Limestone," in *Proceedings of the ASCE, Conference on Engineering and Construction in Tropical and Residual Soils*, Honolulu, ASCE, New York, pp. 403–422.

Hacelas, J. E., C. A. Ramirez, and G. Regalado (1985), "Construction and Performance of Salvajina Dam," in *Proceedings of the ASCE, Symposium on Concrete Face Rockfill Dams—Design, Construction and Performance*, J. B. Cooke and J. L. Sherard (Eds.), Detroit, MI, ASCE, New York, pp. 286–315.

Hadala, P. F. (1967), "The Effect of Placement Method on the Response of Soil Stress Gages," in *Proceedings of the International Symposium on Wave Propagation and Dynamic Properties of Earth Materials*, The University of New Mexico Press, Albuquerque, NM, pp. 255–263.

Hall, C. J., and J. R. Hoskins (1972), "A Comparative Study of Selected Rock Stress and Property Measurement Instruments," U.S. Department of the Interior, Bureau of Mines, Tech. Rep. No. UI-BMR-2.

Handfelt, L. D., D. C. Koutsoftas, and R. Foott (1987), "Instrumentation for Test Fill in Hong Kong," *J. Geotech. Eng. Div. ASCE*, Vol. 113, No. 2, Feb., pp. 127–146.

Handy, R. L., B. Remmes, S. Moldt, A. J. Lutenegger, and G. Trott (1982), "In Situ Stress Determination by Iowa Stepped Blade," *J. Geotech. Eng. Div. ASCE*, Vol. 108, No. GT11, Nov., pp. 1405–1422.

Hanna, T. H. (1985), *Field Instrumentation in Geotechnical Engineering*, Trans Tech Publications, Clausthal-Zellerfeld, F.R. Germany.

Hansmire, W. H. (1978), "Collection, Processing, and Interpretation of Field Measurements," Notes for 2nd Annual Short Course on Field Instrumentation of Soil and Rock, University of Missouri, Rolla, MO.

Hansmire, W. H., and E. J. Cording (1972), "Performance of a Soft Ground Tunnel on the Washington Metro," in *Proceedings of the 1st Rapid Excavation and Tunneling Conference*, Chicago, K. S. Lane and L. A. Garfield (Eds.), AIME, New York, Vol. 1, pp. 371–389.

Hansmire, W. H., and E. J. Cording (1985), "Soil Tunnel Test Section: Case History Summary," *J. Geotech. Eng. Div. ASCE,* Vol. 111, No. 11, Nov., pp. 1301–1320.

Hansmire, W. H., H. A. Russell, and R. P. Rawnsley (1984), "Instrumentation and Evaluation of Slurry Wall Construction, Vol. 1, Interpretation of Field Measurements," U.S. Department of Transportation, Federal Highway Administration, Rep. No. FHWA/RD-84/053.

Harder, L. F. (1982), "Oroville Dam—Investigation of the Causes and Consequences of Abrupt Changes in Piezometer Readings," Mem. Rep., Department of Water Resources, Division of Design and Construction, State of California, Sacramento, CA.

Hardy, H. R., Jr., and F. W. Leighton (Eds.) (1975, 1978, 1981), *Proceedings of the 1st, 2nd, and 3rd Conferences on Acoustic Emission in Geologic Structures and Materials,* Pennsylvania State University, University Park, PA, Trans Tech Publications, Clausthal-Zellerfeld, F.R. Germany.

Harrell, A. S., and K. H. Stokoe II (1984), "Integrity Evaluation of Drilled Piers by Stress Waves," Res. Rep. 257-1F, Center for Transportation Research, University of Texas, Austin, TX.

Hawkes, I. (1978), "Measurement of Mine Roof Movement," U.S. Department of the Interior, Bureau of Mines, Final Rep., Cont. No. H0366033.

Hawkes, I., and W. V. Bailey (1973), "Design, Develop, Fabricate, Test, and Demonstrate Permissible Low Cost Cylindrical Stress Gages and Associated Components Capable of Measuring Change in Stress as a Function of Time in Underground Coal Mines," U.S. Department of the Interior, Bureau of Mines, Rep., Cont. No. H0220050.

Haws, E. T., D. C. Lippard, R. Tabb, and J. B. Burland (1974), "Foundation Instrumentation for the National Westminster Bank Tower," in *Proceedings of the Symposium on Field Instrumentation in Geotechnical Engineering,* British Geotechnical Society, Butterworths, London, pp. 180–193.

Hedley, D. G. F. (1969), "Design Criteria for Multi-Wire Borehole Extensometer Systems," Int. Rep. MR 69/68 ID, Department of Energy, Mines and Resources, Mines Branch, Mining Research Centre, Ottawa.

Herceg, E. G. (1972), *Handbook of Measurement and Control,* Schaevitz Engineering, Inc., Pennsauken, NJ.

Herget, G. (1973), "First Experiences with the C.S.I.R. Triaxial Strain Cell for Stress Determinations," *Int. J. Rock Mech. & Min. Sci. & Geomech. Abstr.*, Vol. 10, No. 6, pp. 509–522.

Higgs, J. S., and S. A. Robertson (1979), "Integrity Testing of Concrete Piles by Shock Method," *Concrete,* Oct., pp. 31–33.

Hiltscher, R., F. L. Martna, and L. Strindell (1979), "The Measurement of Triaxial Rock Stresses in Deep Boreholes and the Use of Rock Stress Measurements in the Design and Construction of Rock Openings," in *Proceedings of the 4th International Congress on Rock Mechanics,* Montreux, Balkema, Rotterdam, Vol. 2, pp. 227–234.

Hirsch, T. J., H. M. Coyle, L. L. Lowery, and C. H. Samson (1970), "Instruments, Performance and Method of Installation," in *Proceedings of the Conference on Design and Installation of Pile Foundations and Cellular Structures,* Lehigh University, Bethlehem, PA, Envo Publishing, Bethlehem, PA, pp. 173–190.

Hoar, G. J. (1982), *Satellite Surveying,* Magnavox Advanced Products and Systems Company, Torrance, CA.

Hoek, E., and J. W. Bray (1974), *Rock Slope Engineering,* Institution of Mining and Metallurgy, London.

Hoek, E., and J. W. Bray (1981), *Rock Slopes, Design, Excavation, Stabilization,* Notes for Training Course, Chap. 13, U.S. Department of Transportation, Federal Highway Administration.

Hoek, E., and P. Londe (1974), "Surface Workings in Rock," in *Advances in Rock Mechanics, Proceedings of the 3rd Congress of ISRM,* Denver, published by National Academy of Sciences, Washington, D.C., Vol. 2, Pt. A, pp. 613–654.

Holhauzen, G. R. (1985), personal communication, Sept. 30.

Holtz, R. D., and B. Broms (1972), "Long-Term Loading Tests at Skå-Edeby, Sweden," in *Proceedings of the ASCE, Specialty Conference on Performance of Earth and Earth-Supported Structures,* Purdue University, Lafayette, IN, ASCE, New York, Vol. 1, Pt. 1, pp. 435–464.

Holtz, R. D., and W. D. Kovacs (1981), *An Introduction to Geotechnical Engineering,* Prentice-Hall, Englewood Cliffs, NJ.

Hooker, V. E., and D. L. Bickel (1974), "Overcoring Equipment and Techniques Used in Rock Stress Determination," U.S. Department of the Interior, Bureau of Mines, Inf. Circ. 8618.

Hooker, V. E., J. R. Aggson, and D. L. Bickel (1974), "Improvements in the Three Component Borehole Deformation Gage and Overcoring Techniques," U.S. Department of the Interior, Bureau of Mines, Rep. Invest. 7894.

Horvath, R. G. (1985), "Instrumentation for Load-Transfer in Socketed Pier Foundations," Reliability of Geotechnical Instrumentation, *Trans. Res. Record,* No. 1004, pp. 34–46.

Hosking, A. D., and J. I. Hilton (1963), "Instrumentation of Earth Dams on the Snowy Mountain Scheme," in *Proceedings of the 4th Australian–New Zealand Conference on SM & FE,* Adelaide, published by Institution of Engineers, Barton, Australia, pp. 251–262.

Houlsby, A. C. (1982), Discussion: "Long-Term Performance of Embankment Dam Instrumentation," *J. Geotech. Eng. Div. ASCE,* Vol. 108, No. GT8, Aug., pp. 1091–1093.

Houlsby, A. C. (1983), personal communication, June 30.

Hustrulid, W. A., and A. Hustrulid (1973), "The CSM Cell—A Borehole Device for Determining the Modulus of Rigidity of Rock," in *Proceedings of the 15th Symposium on Rock Mechanics,* Custer State Park, SD, ASCE, New York, pp. 181–225.

Hvorslev, M. J. (1951), "Time Lag and Soil Permeability in Ground-water Observations," U.S. Army Corps of Engineers, Waterways Experiment Station, Vicksburg, MS, Bull. No. 36.

Hvorslev, M. J. (1976), "The Changeable Interaction Between Soils and Pressure Cells: Tests and Review at the Waterways Experiment Station," U.S. Army Corps of Engineers, Waterways Experiment Station, Vicksburg, MS, Tech. Rep. No. S-76-7.

ICOLD (1969), "General Considerations Applicable to Instrumentation for Earth and Rockfill Dams," ICOLD Bull. 21, Paris.

ICOLD (1982), "Automated Observation for the Safety Control of Dams," ICOLD Bull. 41, Paris.

IECO (1979), "Review of Geotechnical Measurement Techniques for a Nuclear Waste Repository in Bedded Salt," International Engineering Co., Inc., Rep. No. UCRL 15141, Lawrence Livermore Laboratory, Livermore, CA.

Irwin, M. J. (1964), "A Mercury-Filled Gauge for the Measurement of the Settlement of Foundations," *Civil Eng. Public Works Rev.*, London, Vol. 59, No. 692, Mar., pp. 358–360.

Irwin, M. J. (1967), "A Mercury-Filled Gauge for Measuring the Settlement of Foundations," Road Research Laboratory (U.K.), Rep. No. LR62.

ISA (1974), "American Standard for Temperature Measurement, Thermocouples, C 96.1-1964," *Standards and Practices for Instrumentation,* 4th ed., Instrument Society of America, Research Triangle Park, NC.

Ishihara, K. (1981), "Measurements of Insitu Pore Water Pressures During Earthquakes," in *Proceedings of the International Conference on Recent Advances in Geotechnical Earthquake Engineering and Soil Dynamics,* University of Missouri, Rolla, pp. 523–528.

ISRM (1981a), "Suggested Methods for Monitoring Rock Movements Using Inclinometers and Tiltmeters," *Rock Characterization Testing and Monitoring, ISRM Suggested Methods,* Pergamon Press, Oxford, pp. 187–199.

ISRM (1981b), "Suggested Methods for Monitoring Rock Movements Using Borehole Extensometers," *Rock Characterization Testing and Monitoring, ISRM Suggested Methods,* Pergamon Press, Oxford, pp. 173–183.

ISRM (1981c), "Suggested Methods for Pressure Monitoring Using Hydraulic Cells," *Rock Characterization Testing and Monitoring, ISRM Suggested Methods,* Pergamon Press, Oxford, pp. 201–211.

ISRM (1984), "Suggested Methods for Surface Monitoring of Movements Across Discontinuities," ISRM, Commission on Standardization of Laboratory and Field Tests, *Int. J. Rock Mech. & Min. Sci. & Geomech. Abstr.*, Vol. 21, No 5, pp. 265–276.

Jaggar, F. E., and J. R. Enever (1978), "An Evaluation of Different Stress Measuring Techniques Conducted in the Southern Development Headings at Pelton Colliery," Proj. Rep. 78-6, Joint Report by Australian Coal Industry Research Laboratory Ltd. and CSIRO Division of Applied Geomechanics, Australia.

Jaworski, W. E. (1973), "An Evaluation of the Performance of a Braced Excavation," Sc.D. thesis, Massachusetts Institute of Technology, Cambridge, MA.

Jefferis, S. A. (1982), "Effects of Mixing on Bentonite Slurries and Grouts," in *Proceedings of the ASCE, Conference on Grouting in Geotechnical Engineering,* New Orleans, LA, ASCE, New York, pp. 62–76.

Johannessen, I. J., and L. Bjerrum (1965), "Measurement of the Compression of a Steel Pile to Rock Due to Settlement in the Surrounding Clay," in *Proceedings of the 6th International Conference on SM & FE,* Montreal, University of Toronto Press, Toronto, Vol. 2, pp. 261–264.

Johnston, G. H. (1966), "A Compact Self-Contained Ground Temperature Recorder," *Can. Geotech. J.*, Vol. 3, No. 4, Nov., pp. 246–250.

Jones, K. (1961), "Calculations of Stress from Strain in Concrete," U.S. Department of the Interior, Bureau of Reclamation, Technical Information Branch, Eng. Monogr. No. 29, Oct.

Kallstenius, T., and W. Bergau (1961), "In Situ Determination of Horizontal Ground Movements," in *Proceedings of the 5th International Conference on SM & FE,* Paris, published by Dunod, Paris, Vol. 1, pp. 481–485.

Karlsrud, K., and F. Myrvoll (1976), "Performance of a Strutted Excavation in Quick-Clay," in *Proceedings of the 6th European Conference on SM & FE,* Vienna, published by Technische Universitat, Vienna, Vol. 1.1, pp. 157–164.

Kaufman, R. I., and F. J. Weaver (1967), "Stability of Atchafalaya Levees," *J. Soil Mech. & Found. Div. ASCE,* Vol. 93, No. SM4, July, pp. 157–176.

Keil, S., and R. Hellwig (1987), "Minimize Weighing Error Through Proper Load Cell Installation," *Instrument Technol.*, Vol. 34, No. 1, Jan., pp. 37–42.

Kennard, M. F., A. D. M. Penman, and P. R. Vaughan (1967), "Stress and Strain Measurements in the Clay Core at Balderhead Dam," in *Transactions of the 9th ICOLD,* Istanbul, International Commission on Large Dams, Paris, Vol. III, pp. 129–149.

Kennedy, B. A., K. E. Niermeyer, B. A. Fahm, and J. A. Bratt (1971), "A Case Study of Slope Stability at the Chuquicamata Mine, Chile," *Trans. Soc. Min. Eng., Am. Inst. Min., Metall. Pet. Eng.*, Vol. 250, No. 1, Mar., pp. 55–61.

Kersten, R. T., and R. Kist (Eds.) (1984), *Proceedings of the 2nd International Conference on Optical Fiber Sensors,* Liederhalle Stuttgart, F.R. Germany.

Khilnani, K. S., and J. L. Webster (1976), "Mica Dam Drainage System," *Transactions of the 12th ICOLD,* Mexico, International Commission on Large Dams, Paris, Vol. 2, pp. 129–146.

Kim, J. B., and R. J. Brungraber (1976), "Full-Scale Lateral Load Tests of Pile Groups," *J. Geotech. Eng. Div. ASCE,* Vol. 102, No. GT1, Jan., pp. 87–105.

Kinner, E. B., and J. P. Dugan, Jr. (1982), "Geotechnical Investigations for Trident Drydock," *J. Boston Soc. Civil Eng. Sec. ASCE,* Vol. 68, No. 2, pp. 237–274.

Kinner, E. B., and J. P. Dugan, Jr. (1985), Discussion: "Piezometer Installation Under Artesian Conditions," *J. Geotech. Eng. Div. ASCE,* Vol. 111, No. 11, Nov., pp. 1347–1349.

Kirschke, D. (1977), "Superficial and Underground Investigations at Sliding Rock Slopes in Greece," in *Proceedings of the International Symposium on Field Measurements in Rock Mechanics,* Zürich, K. Kovari (Ed.), Balkema, Rotterdam, Vol. 2, pp. 775–788.

Kissenpfennig, J. F., J. T. Motherwell, and L. J. LaFountain (1984), "Integrity Testing of Bored Piles Using Sonic Logging," *J. Geotech. Eng. Div. ASCE,* Vol. 110, No. 8, Aug., pp. 1079–1090.

Kjellman, W., T. Kallstenius, and Y. Liljedahl (1955), "Accurate Measurement of Settlements," *Proc. No. 10, Roy. Swed. Geot. Inst.* , Stockholm.

Kleiner, D. E. (1983), "Bath County Hydroelectric Pumped-Storage Project—Dams and Reservoirs," Third Annual USCOLD Lecture, Part IV, Staunton, VA.

Kleiner, D. E., and K. L. Logani (1982), Discussion: "Long-Term Performance of Embankment Dam Instrumentation," *J. Geotech. Eng. Div. ASCE,* Vol. 108, No. GT8, Aug., pp. 1093–1094.

Knight, D. J., D. J. Naylor, and P. D. Davis (1985), "Stress–Strain Behaviour of the Monasavu Soft Core Rockfill Dam: Prediction, Performance and Analysis," in *Transactions of the 15th ICOLD,* Lausanne, Intentional Commission on Large Dams, Paris, Vol. 1, pp. 1299–1326.

Koerner, R. M., A. E. Lord, Jr., W. M. McCabe, and J. W. Curran (1976), "Acoustic Emission Behavior of Granular Soils," *J. Geotech. Eng. Div. ASCE,* Vol. 102, No. GT7, July, pp. 761–773.

Koerner, R. M., A. E. Lord, Jr., and W. M. McCabe (1977), "Acoustic Emission Behavior of Cohesive Soils," *J. Geotech. Eng. Div. ASCE,* Vol. 103, No. GT8, Aug., pp. 837–850.

Koerner, R. M., A. E. Lord, Jr., and W. M. McCabe (1978), "Acoustic Emission Monitoring of Soil Stability," *J. Geotech. Eng. Div. ASCE,* Vol. 104, No. GT5, May, pp. 571–582.

Koerner, R. M., W. M. McCabe, and A. E. Lord, Jr. (1981), "Acoustic Emission Behavior and Monitoring of Soils," Acoustic Emissions in Geotechnical Engineering Practice, ASTM STP 750, ASTM, Philadelphia, PA, pp. 93–141.

Kohlbeck, F., and A. E. Scheidegger (1986), "Low-Cost Monitoring of Strain Changes," in *Proceedings of the International Symposium on Rock Stress and Rock Stress Measurements,* Stockholm, Centek Publishers, Luleå, Sweden, pp. 189–195.

Köppel, J., Ch. Amstad, and K. Kovari (1983), "The Measurement of Displacement Vectors with the 'Trivec' Borehole Probe," in *Proceedings of the International Symposium on Field Measurements in Geomechanics,* Zürich, K. Kovari (Ed.), Balkema, Rotterdam, Vol. 1, pp. 209–218.

Kovari, K., and Ch. Amstad (1982), "A New Method of Measuring Deformations in Diaphragm Walls and Piles," *Géotechnique,* Vol. 22, No. 4, pp. 402–406.

Kovari, K., and K. Köppel (1987), "Field Studies of Coupled Mechanical and Hydraulic Processes in the Foundation Rock of Large Dams," in *Coupled Processes Associated with Nuclear Waste Repositories,* Tsang Chin-fu (Ed.), Academic Press, Orlando, FL, pp. 739–758.

Kovari, K., Ch. Amstad, and H. Grob (1974), "Displacement Measurements of High Accuracy in Underground Openings," in *Advances in Rock Mechanics, Proceedings of the 3rd Congress of ISRM,* Denver, published by National Academy of Sciences, Washington, D.C., Vol. 2A, pp. 445–450.

Kovari, K., Ch. Amstad, and P. Fritz (1977), "Integrated Measuring Technique for Rock Pressure Determination," in *Proceedings of the International Symposium on Field Measurements in Rock Mechanics,* Zürich, K. Kovari (Ed.), Balkema, Rotterdam, Vol. 1, pp. 289–316.

Kovari, K., Ch. Amstad, and J. Köppel (1979), "New Developments in the Instrumentation of Underground Openings," in *Proceedings of the 4th Rapid Excavation & Tunneling Conference,* Atlanta, A. C. Maevis and W. A. Hustrulid (Eds.), AIME, New York, Vol. 1, pp. 817–837.

Krohn, D. A. (1983), "Fiber Optics: New Sensors for Old Problems," *Instrument Technol.,* Vol. 30, No. 5, May, pp. 57–59.

Kuesel, T. R., B. Schmidt, and D. Rafaeli (1973), "Settlements and Strengthening of Soft Clay Accelerated by Sand Drains," Soil Slopes and Embankments, *Highway Res. Record,* No. 457, pp. 18–26.

Kulhawy, F. H., and J. M. Duncan (1972), "Stresses and Movements in Oroville Dam," *J. Soil Mech. & Found. Div. ASCE,* Vol. 98, No. SM7, July, pp. 653–665.

Kulhawy, F. H., T. D. O'Rourke, J. P. Stewart. and J. F. Beech (1983), "Transmission Line Structure Foundations for Uplift-Compression Loading: Load Test Summaries," Rep. No. EL-3160, Electric Power Research Institute, Palo Alto, CA, June.

Lacy, H. S. (1979), "Load Testing of Instrumented 225-foot-long Prestressed Concrete Piles," in *Behavior of Deep Foundations,* ASTM STP 670, R. Lundgren (Ed.), ASTM, Philadelphia, PA, pp. 358–380.

Ladd, C. C. (1972), "Test Embankment on Sensitive Clay," in *Proceedings of the ASCE, Specialty Conference on Performance of Earth and Earth-Supported Structures,* Purdue University, Lafayette, IN, ASCE, New York, Vol. 1, Pt. 1, pp. 101–128.

Ladd, C. C. (1984), "Use of Precompression and Vertical Drains for Stabilization of Foundation Soils," P76-4 Rev., Department of Civil Engineering, Massachusetts Institute of Technology, Cambridge, MA.

Ladd, C. C., J. J. Rixner, and D. G. Gifford (1972), "Performance of Embankments with Sand Drains on Sensitive Clay," in *Proceedings of the ASCE, Specialty Conference on Performance of Earth and Earth-Supported Structures,* Purdue University, Lafayette, IN, ASCE, New York, Vol. 1, Pt. 1, pp. 211–242.

Lagos Marques, P., E. Maurer, and N. Buhr Toniatti (1985), "Deformation Characteristics of Foz do Areia Concrete Face Rockfill Dam, as Revealed by a Simple Instrumentation System," in *Transactions of the 15th ICOLD,* Lausanne, International Commission on Large Dams, Paris, Vol. 1, pp. 417–450.

Lambe, T. W. (1959), "Sealing the Casagrande Piezometer," *Civil Eng.* (NY), Vol. 29, No. 4, Apr., p. 66.

Lambe, T. W. (1970), "Interpretation of Field Data," in *Proceedings of the Lecture Series on Observational Methods in Soil & Rock Engineering,* Illinois Section, ASCE, Chicago, pp. 116–148.

Lambe, T. W., L. A. Wolfskill, and W. E. Jaworski (1972), "The Performance of a Subway Excavation," in *Proceedings of the ASCE, Specialty Conference on Performance of Earth and Earth-Supported Structures,* Purdue University, Lafayette, IN, ASCE, New York, Vol. I, Pt. 2, pp. 1403–1424.

Lane, K. S. (1975), *Field Test Sections Save Cost in Tunnel Support,* Underground Construction Research Council, ASCE, New York.

Lang, P. (1983), personal communication, Feb. 28.

Lau, K. C., and T. C. Kenney (1984), "Horizontal Drains to Stabilize Clay Slopes," *Can. Geotech. J.,* Vol. 21, No. 2, May, pp. 241–249.

Laurila, S. H. (1983), *Electronic Surveying in Practice,* Wiley, New York.

LBL (1982), "Recommendations for a Rock Mechanics Instrumentation Development Plan," Lawrence Berkeley Laboratories, Rep. Rockwell Hanford Operations, Richland, WA.

Leeman, E. R. (1971), "The CSIR 'Doorstopper' and Triaxial Rock Stress Measuring Instruments," *Rock Mech.*, Vol. 3, No. 1, May, pp. 25–50.

Leeuwen, J. H. van, and W. J. J. C. Volman (1976), "An Exploration of the Constructive Function of STABILENKA in an Embankment," Industrial Fibres of Enka Glazstoff, Holland.

LeFrancois, P. (1986), "Irad Gage Approach to Surveillance of Engineering Structures," Document No. 0404M, Irad Gage, Salem, NH.

Lemcoe, M. M., H. Pratt, and W. Grams (1980), "State-of-the-Art Review of Rock Mechanics Techniques for Measurement of Stress, Displacement and Strain," in *Proceedings of the International Symposium for Subsurface Space, Environmental Protection, Low Cost Storage, Energy Savings,* Rockstore '80, Stockholm, Pergamon Press, Oxford, pp. 927–942.

Lingle, R., A. Davidson, and R. Clayton (1981), "A Summary Report on Development of Stress and Displacement Measuring Techniques," Tech. Rep. No. 330, Battelle Memorial Institute, Office of Nuclear Waste Isolation, Columbus, OH.

Lion, K. S. (1959), *Instrumentation in Scientific Research, Electrical Input Transducers,* McGraw-Hill, New York.

Little, A. L., and A. J. Vail (1960), "Some Developments in the Measurement of Pore Pressure," in *Proceedings of the Conference on Pore Pressure and Suction in Soils,* Butterworths, London, pp. 75–80.

Littlejohn, G. S. (1981), "Acceptance Criteria for the Service Behaviour of Ground Anchorages," *Ground Eng.*, Vol. 14, No. 3, Apr., pp. 26–36.

Littlejohn, G. S. (1982), "Design of Cement Based Grouts," in *Proceedings of the ASCE, Conference on Grouting in Geotechnical Engineering,* New Orleans, LA, ASCE, New York, pp. 35–48.

Liu, T. K., and J. P. Dugan, Jr. (1972), "An Instrumented Tied-Back Deep Excavation," in *Proceedings of the ASCE, Specialty Conference on Performance of Earth and Earth-Supported Structures,* Purdue University, Lafayette, IN, ASCE, New York, Vol. I, Pt. 2, pp. 1323–1339.

Lo, K. Y., and C. F. Lee (1974), "An Evaluation of the Stability of Natural Slopes in Plastic Champlain Clays," *Can. Geotech. J.*, Vol. 11, No. 1, Feb., pp. 165–181.

Logani, K. L. (1983), "Piezometer Installation Under Artesian Conditions," *J. Geotech. Eng. Div. ASCE,* Vol. 109, No. 8, Aug., pp. 1121–1125.

Logani, K. L. (1985), Closure: "Piezometer Installation Under Artesian Conditions," *J. Geotech. Eng. Div. ASCE,* Vol. 111, No. 11, Nov., pp. 1351–1352.

Loh, Y. C. (1952), "Internal Stress Gauges for Cementitious Materials," *Proc. Soc. Exp. Stress Anal.*, Vol. XI, No. 2, May, pp. 13–28.

Londe, P. (1977), "Field Measurements in Tunnels," in *Proceedings of the International Symposium on Field Measurements in Rock Mechanics,* Zürich, K. Kovari (Ed.), Balkema, Rotterdam, Vol. 2, pp. 619–638.

Londe, P. (1982), "Concepts and Instruments for Improved Monitoring," *J. Geotech. Eng. Div. ASCE,* Vol. 108, No. GT6, June, pp. 820–834.

Lord, A. E., and R. M. Koerner (1977), "Modified Soil Strain Gage," *J. Geotech. Eng. Div. ASCE,* Vol. 103, No. GT7, July, pp. 789–793.

Lu, P. H. (1981), "Determination of Ground Pressure Existing in a Viscoelastic Rock Mass by Use of Hydraulic Borehole Pressure Cells," in *Proceedings of the International Symposium on Weak Rock,* Tokyo, Balkema, Rotterdam, Vol. 1, pp. 459–465.

Lu, T. D., V. G. Miller, and J. A. Fischer (1979), "Cyclic Pile Load Testing—Loading System and Instrumentation," in *Behavior of Deep Foundations,* ASTM STP 670, R. Lundgren (Ed.), ASTM, Philadelphia, PA, pp. 435–450.

Lytle, J. D. (1982), "Dam Safety Instrumentation: Automation of Data Observations, Processing and Evaluation," in *Transactions of the 14th ICOLD,* Rio de Janeiro, International Commission on Large Dams, Paris, Vol. 1, pp. 493–511.

Mahar, J. W., F. L. Gau, and E. J. Cording (1972), "Observations During Construction of Rock Tunnels for the Washington, D.C. Subway," in *Proceedings of the 1st Rapid Excavation and Tunneling Conference,* Chicago, K. S. Lane and L. A. Garfield (Eds.), AIME, New York, Vol. 1, pp. 659–681.

Mahasandana, T., and B. Mahatharadol (1985), "Monitoring Systems of Khao Laem Dam," in *Transactions of the 15th ICOLD,* Lausanne, International Commission on Large Dams, Paris, Vol. 1, pp. 1–28.

Mansur, C. I., and M. Alizadeh (1970), "Tie-backs in Clay to Support Sheeted Excavation," *J. Soil Mech. & Found. Div. ASCE,* Vol. 96, No. SM2, Mar., pp. 495–509.

Mansur, C. I., and A. H. Hunter (1970), "Pile Tests—Arkansas River Project," *J. Soil Mech. & Found. Div. ASCE,* Vol. 96, No. SM5, Sept., pp. 1545–1582.

Mansur, C. I., and J. M. Kaufman (1956), "Pile Tests, Low-Sill Structure, Old River, Louisiana," *J. Soil Mech. & Found. Div. ASCE,* Vol. 82, No. SM5, Oct., pp. 1079-1 to 1079–33.

Marchetti, S. (1980), "In Situ Tests by Flat Dilatometer," *J. Geotech. Eng. Div. ASCE,* Vol. 106, No. GT3, Mar., pp. 299–321.

Margason, E., and I. Arango (1972), "Sand Drain Performance on a San Francisco Bay Mud Site," in *Proceedings of the ASCE, Specialty Conference on Performance of Earth and Earth-Supported Structures,* Purdue University, Lafayette, IN, ASCE, New York, Vol. 1, Pt. 1, pp. 181–210.

Marsland, A. (1974a), "New Multipoint Magnetic Settlement System," in *Proceedings of the Symposium on Field Instrumentation in Geotechnical Engineering,* British Geotechnical Society, Butterworths, London, pp. 587–589.

Marsland, A. (1974b), "Instrumentation of Flood Defense Banks Along the River Thames," in *Proceedings of the Symposium on Field Instrumentation in Geotechnical Engineering,* British Geotechnical Society, Butterworths, London, pp. 287–303.

Martin, R. E., J. J. Seli, G. W. Powell, and M. Bertoulin (1987),

"Concrete Pile Design in Tidewater Virginia," *J. Geotech. Eng. Div. ASCE,* Vol. 113, No. 6, June, pp. 568 585.

Massarsch, K. R. (1975), "New Method for Measurement of Lateral Earth Pressure in Cohesive Soils," *Can. Geotech. J.*, Vol. 12, No. 1, Feb., pp. 142–146.

Massarsch, K. R., B. B. Broms, and O. Sundquist (1975), "Pore Pressure Determination with Multiple Piezometer," in *Proceedings of the ASCE, Specialty Conference on In Situ Measurement of Soil Properties,* North Carolina State University, Raleigh, NC, ASCE, New York, Vol. 1, pp. 260–265.

McCarthy, D. F. (1977), *Essentials of Soil Mechanics and Foundations,* Reston Publishing, Reston, VA.

McGuffey, V. C. (1971), "Plastic Pipe Observation Wells for Recording Ground Water Levels and Depth of Active Slide Movements," *Highway Focus,* U.S. Department of Transportation, Federal Highway Administration Vol. 3, No. 3.

Mackay, J. R. (1973), "A Frost Tube for the Determination of Freezing in the Active Layer Above Permafrost," *Can. Geotech. J.*, Vol. 10, No. 3, Aug., pp. 392–396.

McKeehan, D. S., and R. W. Griffiths (1986), "Marine Applications for a Continuous Fiber Optic Strain Monitoring System," Paper No. Eng. 031, OCEANS '86, Marine Technology Society, Sept.

McKeehan, D. S., R. W. Griffiths, and J. E. Halkyard (1986), "Marine Applications for a Continuous Fiber Optic Strain Monitoring System," in *Proceedings of the 18th Offshore Technology Conference,* Houston, Offshore Technology Conference, Richardson, TX, pp. 345–349.

McKenna, J. M., and M. Roy (1974), "Performance of the Instrumentation Used to Monitor the M5 Motorway Embankments Built on Soft Ground Between Edithmead and Huntworth," in *Proceedings of the Symposium on Field Instrumentation in Geotechnical Engineering,* British Geotechnical Society, Butterworths, London, pp. 275–286.

McRae, J. B., and J. B. Sellers (1986), personal communication, May 16.

McRostie, G. C., K. N. Burn, and R. J. Mitchell (1972), "The Performance of a Tied-Back Sheet Piling in Clay," *Can. Geotech. J.*, Vol. 9, No. 2, May, pp. 206–218.

McVey, J. R., S. R. Lewis, and E. E. Guidice (1974), "Deformation Monitoring of Underground Openings by Photographic Techniques," U.S. Department of the Interior, Bureau of Mines, Rep. Invest. 7912.

Measurements Group (1985), "Measurement of Residual Stresses by the Hole-Drilling Strain Gage Method," Tech. Note TN-503-1, Measurements Group Inc., Raleigh, NC.

Melvill, A. L. (1985), "Monitoring the Performance of Elandsjagt Dam During Construction and First Filling," in *Transactions of the 15th ICOLD,* Lausanne, International Commission on Large Dams, Paris, Vol. 1, pp. 1021–1038.

Merriam, R. (1960), "Portuguese Bend Landslide, Palos Verdes Hills, California," *J. Geol.*, Vol. 68, No. 2, Mar., pp. 140–153.

Mettier, K. (1983), "Measurements and Results of Applying the Ground Freezing Method in Tunneling," in *Proceedings of the International Symposium on Field Measurements in Geomechanics,* Zürich, K. Kovari (Ed.), Balkema, Rotterdam, Vol. 2, pp. 1103–1115.

Mikkelsen, P. E. (1982), Discussion: "Piezometers in Earth Dam Impervious Sections," *J. Geotech. Eng. Div. ASCE,* Vol. 108, No. GT8, Aug., pp. 1095–1098.

Mikkelsen, P. E. (1986a), personal communication, Feb. 8.

Mikkelsen, P. E. (1986b), "Horizontal Inclinometer Monitoring of Balsam Meadow Dam," *The Indicator,* Slope Indicator Company, Seattle, WA, Nov., p. 25.

Mikkelsen, P. E., and L. K. Bestwick (1976), "Instrumentation and Performance: Urban Arterial Embankments on Soft Foundation Soil," in *Proceedings of the 14th Annual Engineering Geology & Soils Engineering Symposium,* Boise State University, Boise, ID, published by Symposium on Engineering Geology and Soils Engineering, Boise, ID, pp. 1–18.

Mikkelsen, P. E., and S. D. Wilson (1983), "Field Instrumentation: Accuracy, Performance, Automation and Procurement," in *Proceedings of the International Symposium on Field Measurements in Geomechanics,* Zürich, K. Kovari (Ed.), Balkema, Rotterdam, Vol. 1, pp. 251–272.

Milner, R. M. (1969), "Accuracy of Measurements with Steel Tapes," Current Paper 51/69, Building Research Station, Watford, England, Dec.

Minnitti, A. (1985), personal communication, Mar. 13.

Misterek, D. L. (1986), personal communication, Jan. 14. Copy available from U.S. Department of the Interior, Bureau of Reclamation, Denver, CO.

MIT (1975), *Proceedings of the Foundation Deformation Prediction Symposium,* Constructed Facilities Division, Massachusetts Institute of Technology, U.S. Department of Transportation, Federal Highway Administration, Reps. No. FHWA-RD-75-515 and 516.

Mitchell, I. (1985), personal communication, Nov. 15.

Miyashita, K., K. Aoki, T. Hanamura, and N. Kashiwagi (1983), "An Investigation of Geomechanics and Hydraulics Around an Underground Crude Oil Storage Cavern," in *Proceedings of the International Symposium on Field Measurements in Geomechanics,* Zürich, K. Kovari (Ed.), Balkema, Rotterdam, Vol. 2, pp. 1117–1126.

Moh, Z. C., and T. F. Song (1984), "Performance of Diaphragm Walls in Deep Foundation Excavations," in *Proceedings of the International Conference on Case Histories in Geotechnical Engineering,* S. Prakash (Ed.), University of Missouri, Rolla, MO, published by University of Missouri, Rolla, MO, Vol. III, pp. 1335–1343.

Moore, J. F. A. (1973), "The Photogrammetric Measurement of Constructional Displacements of a Rockfill Dam," *Photogrammetric Rec.*, Vol. 7, No. 42, pp. 628–648. Also Current Paper 34/74, Building Research Establishment, Watford, England, Feb.

Morgenstern, N. R., and D. C. Sego (1981), "Performance of Temporary Tie-Backs Under Winter Conditions," *Can. Geotech. J.*, Vol. 18, No. 4, Nov., pp. 566–572.

Morice, P. B., and G. D. Base (1953), "The Design and Use of a Demountable Mechanical Strain Gauge for Concrete Structures," *Mag. Concrete Res.* (Cement and Concrete Assoc., London), Vol. 5, No. 13, Aug., pp. 37–42.

Müller, G., and L. Müller (1970), "Monitoring of Dams with Measuring Instruments," in *Transactions of the 10th ICOLD,* Montreal, International Commission on Large Dams, Paris, Vol. III, pp. 1033–1046.

Müller, L., and G. Spaun (1977), "Soft Ground Tunnelling under Buildings in Germany," in *Proceedings of the 9th International Conference on SM & FE,* Tokyo, Japanese Society of SM & FE, Tokyo, Vol. 1, pp. 663–668.

Müller, G., H. Voort, and M. Wohnlich (1977), "Recent Developments in Slope Monitoring Instrumentation," in *Proceedings of the International Symposium on Field Measurements in Rock Mechanics,* Zürich, K. Kovari (Ed.), Balkema, Rotterdam, Vol. 2, pp. 757–773.

Munfakh, G. A., S. K. Sarkar, and R. J. Castelli (1983), "Performance of a Test Embankment Founded on Stone Columns," in *Proceedings of the Symposium on Advances in Piling and Ground Treatment for Foundations,* Institution of Civil Engineers, London, pp. 193–199.

Muñoz, A., Jr. (1974), "The Role of Field Instrumentation in Correction of the Fountain Slide," Landslide Instrumentation, *Trans. Res. Record,* No. 482, pp. 1–8.

Murray, B. C. (1986), "Automation of Instrumentation," *USCOLD Newsletter,* Issue No. 80, July, pp. 4–15.

Neville, A. M. (1981), *Properties of Concrete,* 3rd ed., Pitman Publishing, Marshfield, MA.

New York Department of Transportation (1979), "Settlement Gages and Settlement Rods, Construction Details and Procedural Requirements," Soil Control Proc. SCP-7, Official Issuance, No. 7-41-3, Soil Mechanics Bureau, NY Department of Transportation, Albany, NY.

Newman, F. B., H. A. Salver, and R. J. Turka (1981), "1000-Ton Drilled Pier Load Test at Sammis Plant," in *Drilled Piers and Caissons,* M.W. O'Neill (Ed.), ASCE, New York, pp. 34–52.

Nicholson, D. P., and R. J. Jardine (1981), "Performance of Vertical Drains at Queenborough Bypass," *Géotechnique,* Vol. 31, No. 1, pp. 67–90.

Norton, H. N. (1969), *Handbook of Transducers for Electronic Measuring Systems,* Prentice-Hall, Englewood Cliffs, NJ.

Nowack, F., and E. Gartung (1983), "Instrumentation of Cast in Place Piles for Vertical and Horizontal Load Testing," in *Proceedings of the International Symposium on Field Measurements in Geomechanics,* Zürich, K. Kovari (Ed.), Balkema, Rotterdam, Vol. 1, pp. 513–522.

Obert, L. (1966), *Determination of Stress in Rock—A State of the Art Report,* ASTM STP 429, ASTM, Philadelphia, PA.

O'Connor, K. M. (1983), "Coal Mine Subsidence Monitoring Instrumentation," in *Proceedings of the International Symposium on Field Measurements in Geomechanics,* Zürich, K. Kovari (Ed.), Balkema, Rotterdam, Vol. 2, pp. 1137–1151.

O'Connor, K. M., and C. H. Dowding (1984), "Application of Time Domain Reflectometry to Mining," in *Proceedings of the 25th Symposium on Rock Mechanics,* Northwestern University, Evanston, IL, pp. 737–746.

Ohdedar, D. N., and B. J. Dawes (1983), "Instrumenting Thailand's Khao Laem Dam," *Water Power & Dam Construction,* London, Vol. 35, No. 12, Dec., pp. 31–34.

Oil & Gas Journal (1985), "Electro-Optic Monitoring Studied for Arctic Lines," *Oil & Gas Journal,* PennWell Publishing Co., Tulsa, OK, May 6.

Olson, L. D., and R. W. Thompson (1985), "Case Histories Evaluation of Drilled Pier Integrity by the Stress Wave Propagation Method," in *Drilled Piers and Caissons II,* C. N. Baker, Jr. (Ed.), ASCE, New York, pp. 28–41.

O'Neill, M. W., R. A. Hawkins, and J. M. E. Audibert (1982a), "Installation of Pile Group in Overconsolidated Clay," *J. Geotech. Eng. Div. ASCE,* Vol. 108, No. GT11, Nov., pp. 1369–1386.

O'Neill, M. W., R. A. Hawkins, and L. J. Mahar (1982b), "Load Transfer Mechanisms in Piles and Pile Groups," *J. Geotech. Eng. Div. ASCE,* Vol. 108, No. GT12, Dec., pp. 1605–1623.

O'Rourke, J. E. (1974), "Performance Instrumentation Installed in Oroville Dam," *J. Geotech. Eng. Div. ASCE,* Vol. 100, No. GT2, Feb., pp. 157–174.

O'Rourke, J. E. (1978), "Soil Stress Measurement Experiences," *J. Geotech. Eng. Div. ASCE,* Vol. 104, No. GT12, Dec., pp. 1501–1514.

O'Rourke, J. E., and B. B. Ranson (1979), "Instruments for Subsurface Monitoring of Geothermal Subsidence," Rep. No. LBL-8616, Lawrence Berkeley Laboratory, Berkeley, CA.

O'Rourke, T. D., and E. J. Cording (1975), "Measurement of Strut Loads by Means of Vibrating Wire Strain Gages," in *Performance Monitoring for Geotechnical Construction,* ASTM STP 584, ASTM, Philadelphia, PA, pp. 58–77.

O'Rourke, T. D., and A. S. Kumbhojkar (1984), "Field Testing of Cast Iron Pipeline Response to Shallow Trench Construction," Rep. No. 84-3, School of Civil and Environmental Engineering, Cornell University, Ithaca, NY.

Osterberg, J. O. (1984), "A New Simplified Method for Load Testing Drilled Shafts," *Foundation Drilling,* Vol. 20, No. 17, Aug., pp. 9–11.

Palladino, D. J., and R. B. Peck (1972), "Slope Failures in an Overconsolidated Clay, Seattle, Washington," *Géotechnique,* Vol. 22, No. 4, pp. 563–595.

Panek, L. A., and J. A. Stock (1964), "Development of a Rock Stress Monitoring Station Based on the Flat Slot Method of Measuring Existing Rock Stresses," U.S. Department of the Interior, Bureau of Mines, Rep. Invest. 6537.

Panek, L. A., E. E. Hornsey, and R. L. Lappi (1964), "Determination of the Modulus of Rigidity of Rock by Expanding a Cylindrical Pressure Cell in a Drill Hole," in *Proceedings of the 6th Symposium on Rock Mechanics,* University of Missouri, Rolla, MO, published by University of Missouri, Rolla, MO, pp. 427–449.

Pariseau, W. G. (1985), "Linearization of In-Situ Stress Change Formulae for Gauges at Arbitrary Down-Hole Orientation," *Int. J. Num. Anal. Meth. Geomech.,* Vol. 9, No. 3, May-June, pp. 277–283.

Parry, R. H. G. (1971), "A Simple Driven Piezometer," *Géotechnique,* Vol. 21, No. 2, pp. 163–167.

Partially Integrated (1962), "Pore Portents," *Water and Water Eng.,* London, Apr., p. 167.

Patton, F. D. (1979), "Groundwater Instrumentation for Mining Projects," pp. 123–153 in *Mine Drainage,* G. O. Argall, Jr. and C. O. Brawner (Eds.), Proceedings of the First International Mine Drainage Symposium, Denver, CO, Miller Freeman Publications, San Francisco, CA.

Patton, F. D. (1983), "The Role of Instrumentation in the Analysis of the Stability of Rock Slopes," in *Proceedings of the*

International Symposium on Field Measurements in Geomechanics, Zürich, K. Kovari (Ed.), Balkema, Rotterdam, Vol. 1, pp. 719–748.

Peattie, K. R., and R. W. Sparrow (1954), "The Fundamental Action of Earth Pressure Cells," *J. Mech. Phys. Solids*, Vol. 2, pp. 141–155.

Peck, R. B. (1969a), "Advantages and Limitations of the Observational Method in Applied Soil Mechanics," *Géotechnique*, Vol. 19, No. 2, pp. 171–187. Reprinted in *Judgment in Geotechnical Engineering: The Professional Legacy of Ralph B. Peck* (1984), J. Dunnicliff and D. U. Deere (Eds.), Wiley, New York, pp. 122–127.

Peck, R. B. (1969b), "Deep Excavations and Tunneling in Soft Ground; State-of-the-Art Report," in *Proceedings of the 7th International Conference on SM & FE*, Mexico, published by Mexicana de Mecanica de Suelos AC, Mexico, State-of-the-Art Vol., pp. 225–290.

Peck, R. B. (1972), "Observation and Instrumentation: Some Elementary Considerations," *Highway Focus*, U.S. Department of Transportation, Federal Highway Administration, Vol. 4, No. 2, pp. 1–5. Reprinted in *Judgment in Geotechnical Engineering: The Professional Legacy of Ralph B. Peck* (1984), J. Dunnicliff and D. U. Deere (Eds.), Wiley, New York, pp. 128–130.

Peck, R. B. (1973), "Influence of Nontechnical Factors on the Quality of Embankment Dams," *Embankment-Dam Engineering*, Casagrande Vol., Wiley, New York, pp. 201–208. Reprinted in *Judgment in Geotechnical Engineering: The Professional Legacy of Ralph B. Peck* (1984), J. Dunnicliff and D. U. Deere (Eds.), Wiley, New York, pp. 137–144.

Peck, R. B. (1984), "Observation and Instrumentation, Some Elementary Considerations, 1983 Postscript," in *Judgment in Geotechnical Engineering: The Professional Legacy of Ralph B. Peck*, J. Dunnicliff and D. U. Deere (Eds.), Wiley, New York, pp. 128–130.

Peck, R. B. (1985), "The Last Sixty Years," in *Proceedings of the 11th International Conference on SM & FE*, San Francisco, CA, Balkema, Rotterdam, Golden Jubilee Vol., pp. 123–133.

Peng, S. S., W. H. Su, and S. Okubo (1982), "A Low Cost Stressmeter for Measuring Complete Stress Changes in Underground Mining," *Geotech. Testing J. ASTM*, Vol. 5, No. 1/2, Mar./June, pp. 50–53.

Penman, A. D. M. (1960), "A Study of the Response Time of Various Types of Piezometers," in *Proceedings of the Conference on Pore Pressure and Suction in Soils*, Butterworths, London, pp. 53–58.

Penman, A. D. M. (1971), "Instrumentation for Embankment Dams Subjected to Rapid Drawdown," in *Proceedings of the International Conference on Pumped Storage Development and Its Environmental Effects*, University of Wisconsin, Milwaukee, pp. 213–229.

Penman, A. D. M. (1975), "Earth Pressures Measured with Hydraulic Piezometers," in *Proceedings of the ASCE, Specialty Conference on In Situ Measurement of Soil Properties*, North Carolina State University, Raleigh, NC, ASCE, New York, Vol. II, pp. 361–381.

Penman, A. D. M. (1978), "Pore Pressure and Movement in Embankment Dams," *Int. Water Power & Dam Construction*, London, Vol. 30, No. 4, Apr., pp. 32–39.

Penman, A. D. M. (1982), "Instrumentation Requirements for Earth and Rockfill Dams," in *Proceedings of the Symposium on Geotechnical Problems and Practice of Dam Engineering*, Bangkok, A. S. Balasubramaniam, Yudhbir, A. Tomiolo, and J. S. Younger (Eds.), Balkema, Rotterdam, pp. 183–209.

Penman, A. D. M. (1986), personal communication, June 6.

Penman, A. D. M., and J. A. Charles (1973), "Constructional Deformations in Rockfill Dam," *J. Soil Mech. & Found. Div. ASCE*, Vol. 99, No. SM2, Feb., pp. 139–163.

Penman, A. D. M., and J. A. Charles (1982), "An Improved Horizontal Plate Gauge," *Géotechnique*, Vol. 32, No. 3, pp. 278–282.

Penman, A. D. M., and J. A. Charles (1985), "A Comparison Between Observed and Predicted Deformations of an Embankment Dam with a Central Asphaltic Core," in *Transactions of the 15th ICOLD*, Lausanne, International Commission on Large Dams, Paris, Vol. 1, pp. 1373–1389.

Penman, A. D. M., and A. Hussain (1984), "Deflection Measurements of the Upstream Asphaltic Membrane of Marchlyn Dam," *Water Power & Dam Construction*, Vol. 36, No. 9, Sept., pp. 33–37.

Penman, A. D. M., and M. F. Kennard (1981), "Long-Term Monitoring of Embankment Dams in Britain," in *Proceedings of the ASCE, Symposium on Recent Developments in Geotechnical Engineering for Hydro Projects*, F. H. Kulhawy (Ed.), ASCE, New York, pp. 46–67.

Penman, A. D. M., and P. B. Mitchell (1970), "Initial Behaviour of Scammonden Dam," in *Transactions of the 10th ICOLD*, Montreal, International Commission on Large Dams, Paris, Vol. 1, pp. 723–747.

Penman, A. D. M., J. A. Charles, J. K. T. L. Nash, and J. D. Humphreys (1975), "Performance of Culvert Under Winscar Dam," *Géotechnique*, Vol 25, No. 4, pp. 713–730.

Perlman, R. S. (1985), "Sole Source Contracts, Basic Principles and Guidelines," Briefing Papers, No. 83-7, Vol. 6, *Briefing Papers Collection 187*, Federal Publications, Inc., Washington, DC, July, pp. 187–222.

Perry, C. C., and H. R. Lissner (1962), *The Strain Gage Primer*, McGraw-Hill, New York.

Peters, N., and W. C. Long (1981), "Performance Monitoring of Dams in Western Canada," in *Proceedings of the ASCE, Symposium on Recent Developments in Geotechnical Engineering for Hydro Projects*, F. H. Kulhawy (Ed.), ASCE, New York, pp. 23–45.

Piezometer R&D (1968), "Procedure for Installation of Positive Operating Piezometer, Inspector's Booklet," Piezometer Research & Development, Stamford, CT, unpublished.

Piezometer R&D (1983), "The Subsurface Problems of the Porous Tube Piezometer and Their Solutions," Piezometer Research and Development, Stamford, CT, unpublished.

Pirtz, D., and R. W. Carlson (1963), "Tests of Strain Meters and Stress Meters under Simulated Field Conditions," Rep. No. SP-6, in *Proceedings of the Symposium on Mass Concrete*, American Concrete Institute, Detroit, pp. 287–308.

Piteau, D. R., F. H. Mylrea, and J. G. Blown (1978), "The Downie Slide, Columbia River, British Columbia," in *Rock-*

slides and Avalanches, 1, Natural Phenomena, B. Voight (Ed.), Elsevier Publishing, Amsterdam, Vol. 1, pp. 365–392.

Pope, R., R. C. Weeks, and P. N. Chipp (1982), "Automatic Recording of Standpipe Piezometers," in *Proceedings of the 7th Southeast Asian Geotechnical Conference*, Hong Kong, published by The Southeast Asian Geotechnical Society and Hong Kong Institution of Engineers, Hong Kong, Vol. 1, pp. 77–89.

Post, G. (1985), "General Report Q56, Dams and Foundation Monitoring," in *Transactions of the 15th ICOLD*, Lausanne, International Commission on Large Dams, Paris, Vol. 1, pp. 1621–1727.

Potts, E. L. (1957), "Underground Instrumentation," *Q. Colo. Sch. Mines*, Vol. 52, No. 3, pp. 135–182.

Potts, E. L., and N. Tomlin (1960), "Investigations into the Measurement of Rock Pressures in the Mines and in the Laboratory," in *3rd International Strata Control and Rock Mechanics Congress*, Paris, published by Revue de l'Industrie Minerale, Saint-Etienne, pp. 281–299.

Preiss, K., and A. Caiserman (1975), "Non-destructive Integrity Testing of Bored Piles by Gamma Ray Scattering," *Ground Eng.*, Vol. 8, No. 3, May, pp. 44–46.

Premchitt, J., and E. W. Brand (1981), "Pore Pressure Equalization of Piezometers in Compressible Soils," *Géotechnique*, Vol. 31, No. 1, pp. 105–123.

Prensky, S. D. (1963), *Electronic Instrumentation*, Prentice-Hall, Englewood Cliffs, NJ.

Price, G. (1979), "Field Tests on Vertical Piles Under Static and Cyclic Horizontal Loading in Overconsolidated Clay," in *Behavior of Deep Foundations*, ASTM STP 670, R. Lundgren (Ed.), ASTM, Philadelphia, PA, pp. 464–483.

Price, G., and I. F. Wardle (1983), "Recent Developments in Pile/Soil Instrumentation Systems," in *Proceedings of the International Symposium on Field Measurements in Geomechanics*, Zürich, K. Kovari (Ed.), Balkema, Rotterdam, Vol. 1, pp. 533–542.

Rabcewicz, L., and J. Golser (1973), "Principles of Dimensioning the Supporting System for the New Austrian Tunneling Method," *Water Power*, Vol. 25, No. 3, Mar., pp. 88–93.

Ramalho-Ortigão, J. A., W. A. Lacerda, and M. L. G. Werneck (1983), "The Behaviour of the Instrumentation of an Embankment on Soft Clay," in *Proceedings of the International Symposium on Field Measurements in Geomechanics*, Zürich, K. Kovari (Ed.), Balkema, Rotterdam, Vol. 1, pp. 703–717.

Rausche, F., G. G. Goble, and G. E. Litkins, Jr. (1985), "Dynamic Determination of Pile Capacity," *J. Geotech. Eng. Div. ASCE*, Vol. 111, No. 3, Mar., pp. 367–383.

Rawnsley, R. P., H. A. Russell, and W. H. Hansmire (1985), "Instrumentation Reliability at Harvard Square Station," Reliability of Geotechnical Instrumentation, *Trans. Res. Record*, No. 1004, pp. 27–34.

Rebull, P. M. (1972), "Earth Responses in Soft Ground Tunneling," in *Proceedings of the ASCE, Specialty Conference on Performance of Earth and Earth-Supported Structures*, Purdue University, Lafayette, IN, ASCE, New York, Vol. 1, Pt. 2, pp. 1517–1535.

Reese, L. C. (1978), "Design and Construction of Drilled Shafts," *J. Geotech. Eng. Div. ASCE*, Vol. 104, No. GT1, Jan., pp. 95–116.

Reese, L. C., and R. C. Welch (1975), "Lateral Loading on Deep Foundations in Stiff Clay," *J. Geotech. Eng. Div. ASCE*, Vol. 101, No. GT7, July, pp. 633–649.

Reese, L. D., J. C. Brown, and H. H. Dalrymple (1968), "Instrumentation for Measurements of Lateral Earth Pressure in Drilled Shafts," Res. Rep. 89-2, Center for Highway Research, University of Texas, Austin, TX.

Reese, L. C., F. T. Touma, and M. W. O'Neill (1976), "Behavior of Drilled Piers Under Axial Loading," *J. Geotech. Eng. Div. ASCE*, Vol. 102, No. GT5, May, pp. 493–510.

Rehtlane, E. A., and F. D. Patton (1982), "Multiple Port Piezometers vs. Standpipe Piezometers: An Economic Comparison," in *Proceedings of the 2nd National Symposium on Aquifer Restoration and Ground-Water Monitoring*, Columbus, OH, published by National Water Well Association, Worthington, OH, pp. 287–295.

Reyes, M. S. (1985), Discussion: "Piezometer Installation Under Artesian Conditions," *J. Geotech. Eng. Div. ASCE*, Vol. 111, No. 11, Nov., pp. 1349–1351.

Rieke, R. D., and J. C. Crowser (1986), "Instrumentation of Driven Piles," *The Indicator*, Slope Indicator Company, Seattle, WA, Nov., pp. 2–5.

Rieke, R. D., and J. C. Crowser (1987), "Interpretation of Pile Load Test Considering Residual Stresses," *J. Geotech. Eng. Div. ASCE*, Vol. 113, No. 4, Apr., pp. 320–334.

RKE/PB (1984), "In Situ Instrumentation," Raymond Kaiser Engineers and Parsons Brinckerhoff Quade & Douglas, Inc., Rep., Task V, Eng. Study No. 8, Proj. B-301, Basalt Waste Isolation Project, U.S. Department of Energy.

Roberts, A., and I. Hawkes (1979), "Photoelastic Instrumentation, Principles and Techniques," U.S. Army Corps of Engineers, Cold Regions Research and Enginering Laboratory, Hanover, NH, Spec. Rep. 79-13.

Robertson, E. C., R. Raspet, J. H. Swartz, and M. E. Lillard (1966), "Properties of Thermistors Used in Geothermal Investigations; Preparation of Thermistor Cables Used in Geothermal Investigations," *U.S. Geol. Surv. Bull.*, 1203-B,C.

Robertson, K. (1977), "Development of a High Precision Capability for Monitoring Structural Movements of Locks and Dams," U.S. Army Corps of Engineers, Engineer Topographic Laboratories, Fort Belvoir, VA, Rep. ETL-0121, Sept.

Robertson, K. (1979), "The Use and Calibration of Distance Measuring Equipment for Precise Mensuration of Dams (Revised)," U.S. Army Corps of Engineers, Engineer Topographic Laboratories, Fort Belvoir, VA, Rep. ETL-0190, June.

Robertson, S. A. (1979), "Horizontal Pile Testing," *Civil Eng.*, London, Jan., p. 17.

Robertson, S. A. (1982), "Integrity and Dynamic Testing of Deep Foundations in S.E. Asia, 1979–82," in *Proceedings of the 7th Southeast Asian Geotechnical Conference*, Hong Kong, published by The Southeast Asian Geotechnical Society and the Hong Kong Institution of Engineers, Hong Kong, pp. 403–421.

Robinson, R. A. (1985), personal communication, Mar. 26.

Robinson, R. A., E. J. Cording, D. A. Roberts, N. Phienweja, J. W. Mahar, and H. W. Parker (1985), "Ground Deformations Ahead of and Adjacent to a TBM in Sheared Shales at Stillwater Tunnel, Utah," in *Proceedings of the 7th Rapid Excavation & Tunneling Conference,* New York, C. D. Mann and M. N. Kelley (Eds.), AIME, New York, Vol. 1, pp. 34–53.

Rocha, M., and A. Silverio (1969), "A New Method for the Complete Determination of the State of Stress in Rock Masses," *Géotechnique,* Vol 19, No. 1, pp. 116–132.

Rocha, M., A. Silverio, J. O. Pedro, and J. S. Delgado (1974), "A New Development of the LNEC Stress Tensor Gauge," in *Advances in Rock Mechanics,* Geotecnia No. 12, *Proceedings of the 3rd Congress of ISRM,* Denver, CO, Vol. 2A, pp. 464–467.

Rosati, E., and R. F. Esquivel (1981), "Instrumentation Performance for El Infiernillo Dam After 18 Years of Observation," in *Proceedings of the ASCE, Symposium on Recent Developments in Geotechnical Engineering for Hydro Projects,* F. H. Kulhawy (Ed.), ASCE, New York, pp. 104–124.

Rosenberg, P., G. St. Arnaud, N. L. Journeaux, and H. Vallée (1977), "Design, Construction, and Performance of a Slurry Trench Wall Next to Foundations," *Can. Geotech. J.,* Vol. 14, No. 3, Aug., pp. 324–339.

Rowe, R. K., M. D. MacLean, and A. K. Barsvary (1984a), "The Observed Behaviour of a Geotextile-Reinforced Embankment Constructed on Peat," *Can. Geotech. J.,*Vol. 21, No. 2, May, pp. 289–304.

Rowe, R. K., M. D. MacLean, and K. L. Soderman (1984b), "Analysis of a Geotextile-Reinforced Embankment Constructed on Peat," *Can. Geotech. J.,* Vol. 21, No. 3, Aug., pp. 563–576.

Rózsa, L., and L. Vidacs (1983), "New and Easy Method for Measuring Settlement of Embankments," in *Proceedings of the International Symposium on Field Measurements in Geomechanics,* Zürich, K. Kovari (Ed.), Balkema, Rotterdam, Vol. 1, pp. 765–771.

Saito, M. (1965), "Forecasting the Time of Occurrence of a Slope Failure," in *Proceedings of the 6th International Conference on SM & FE,* Montreal, University of Toronto Press, Toronto, Vol. 2, pp. 537–541.

Sandroni, S. S. (1980), "A Simple, Accurate and Fault-Proof Water Level Indicator," *Géotechnique,* Vol. 30, No. 3, pp. 319–320.

Sauer, G., and B. Sharma (1977), "A System for Stress Measurement in Constructions in Rock," in *Proceedings of the International Symposium on Field Measurements in Rock Mechanics,* Zürich, K. Kovari (Ed.), Balkema, Rotterdam, Vol. 1, pp. 317–329.

Saxena, S. K. (1974), "Measured Performance of a Rigid Concrete Wall at the World Trade Center," in *Proceedings of the International Conference on Diaphragm Walls and Anchorages,* Institution of Civil Engineers, London, pp. 107–112.

Schmertmann, J. H. (1982), "A Method for Determining the Friction Angle in Sands from the Marchetti Dilatometer Test," in *Proceedings of the 2nd European Symposium on Penetration Testing,* Amsterdam, Balkema, Rotterdam, Vol. 2, pp. 853–861.

Schmidt, B. (1976), "Monitoring Soft Ground Tunnel Construction: A Handbook of Rational Practices for Planners and Designers," U.S. Department of Transportation, Urban Mass Transportation Administration, Rep. No. UMTA-MA-06-0025-76-6.

Schrauf, T. W., and H. R. Pratt (1979), "Review of Current Capabilities for the Measurement of Stress, Displacement, and In Situ Determination of Modulus," Tech. Rep. No. 95, Battelle Memorial Institute, Office of Nuclear Waste Isolation, Columbus, OH.

Scott, J. P., N. E. Wilson, and G. Bauer (1972), "Analysis and Performance of a Braced Cut in Sand with Large Deformations," *Can. Geotech. J.,* Vol. 9, No. 4, Nov., pp. 384–406.

Selig, E. T. (1964), "A Review of Stress and Strain Measurement in Soil," in *Proceedings of the Symposium on Soil–Structure Interaction,* University of Arizona, Tucson, AZ, published by University of Arizona, Tucson, AZ, pp. 172–186.

Selig, E. T. (1975a), "Soil Strain Measurement Using Inductance Coil Method," in *Performance Monitoring for Geotechnical Construction,* ASTM STP 584, ASTM, Philadelphia, PA, pp. 141–158.

Selig, E. T. (1975b), "Instrumentation of Large Buried Culverts," in *Performance Monitoring for Geotechnical Construction,* ASTM STP 584, ASTM, Philadelphia, PA, pp. 159–181.

Selig, E. T. (1980), "Soil Stress Gage Calibration," *Geotech. Testing J. ASTM,* Vol. 3, No. 4, Dec., pp. 153–158.

Sellers, J. B. (1970), "The Measurement of Rock Stress Changes Using Hydraulic Borehole Gages," *Int. J. Rock Mech. & Min. Sci.,* Vol. 7, No. 4, July, pp. 423–435.

Sellers, J. B. (1977), "The Measurement of Stress Changes in Rock Using the Vibrating Wire Stressmeter," in *Proceedings of the International Symposium on Field Measurements in Rock Mechanics,* Zürich, K. Kovari (Ed.), Balkema, Rotterdam, Vol. 1, pp. 275–288.

Sellers, J. B., G. R. Haworth, and P. G. Zambas (1972), "Rock Mechanics Research on Oil Shale Mining," *Trans. Soc. Min. Eng.* AIME, Vol. 252, No. 2, June, pp. 222–232.

Senger, J. A., and W. M. Perpich (1983), "An Alternative Well Seal in Highly Mineralized Ground Water," in *Proceedings of the 3rd National Symposium on Aquifer Restoration and Ground-Water Monitoring,* Columbus, OH, D. M. Nielsen (Ed.), published by National Water Well Association, Worthington, OH, May, pp. 230–234.

Senne, J. H. (1980), "Optical Tooling, Surveying, and Photogrammetry," Notes for 4th Annual Short Course on Field Instrumentation of Soil and Rock, University of Missouri, Rolla, MO.

Sherard, J. L. (1980), personal communication, Dec. 17.

Sherard, J. L. (1981), "Piezometers in Earth Dam Impervious Sections," in *Proceedings of the ASCE, Symposium on Recent Developments in Geotechnical Engineering for Hydro Projects,* F. H. Kulhawy (Ed.), ASCE, New York, pp. 125–165.

Sherard, J. L. (1982), Closure: "Piezometers in Earth Dam Impervious Sections," *J. Geotech. Eng. Div. ASCE,* Vol. 108, No. GT8, Aug., pp. 1098–1099.

Sherard, J. L. (1985), "Filters and Leakage Control in Embankment Dams," in *Proceedings of the ASCE, Symposium on Seepage and Leakage from Dams and Reservoirs,* Denver, CO, ASCE, New York, pp. 1–30.

Sherard, J. L. (1986), "Hydraulic Fracturing in Embankment Dams," *J. Geotech. Eng. Div. ASCE*, Vol. 112, No. 10, Oct., pp. 905–927.

Sherard, J. L., L. P. Dunnigan, and J. R. Talbot (1984), "Filters for Silts and Clays," *J. Geotech. Eng. Div. ASCE*, Vol. 110, No. 6, June, pp. 701–718.

Shoup, D., and H. B. Dutro (1985), "Battery Power for Instrumentation," *The Indicator*, Slope Indicator Company, Seattle, WA, Dec., pp. 19–22.

Shuri, F. S., and G. E. Green (1987), "Instrumentation for Nuclear Waste Studies in Salt," submitted to 2nd International Symposium on Field Measurements in Geomechanics, Kobe, Japan.

Slope Indicator Company (1987), *Manual for Digitilt 50309E Inclinometer*, Slope Indicator Company, Seattle, WA.

Smith, E. A. L. (1960), "Pile Driving Analysis by the Wave Equation," *J. Soil Mech. & Found. Div. ASCE*, Vol. 86, No. SM4, Aug., pp. 35–61.

Smith, R. L. (1972), "Mine Installation of Two Bureau of Mines Hydraulic Pressure Cells and a Borehole Deformation Gage," in *Proceedings of the Seminar on Rock Mechanics Instrumentation for Mine Design*, U.S. Department of the Interior, Bureau of Mines, Inf. Circ. 8585, p. 49–56.

Smith, T., and R. Forsyth (1971), "Potrero Hill Slide and Correction," *J. Soil Mech. & Found. Div. ASCE*, Vol. 97, No. SM3, Mar., pp. 541–564.

Soares, M. M. (1983), "The Instrumentation of a Diaphragm Wall for the Excavation for the Rio de Janeiro Underground," in *Proceedings of the International Symposium on Field Measurements in Geomechanics*, Zürich, K. Kovari (Ed.), Balkema, Rotterdam, Vol. 1, pp. 553–563.

Soderman, L. G. (1961), Discussion: "Field Studies on the Consolidation Properties of Leda Clay," in *Proceedings of the 14th Canadian Soil Mechanics Conference*, Ottawa, published by National Research Council, Canada, Ottawa, pp. 119–120.

Soil Instruments Ltd. (1983), *Hydraulic Piezometer Users' Manual*, Soil Instruments Ltd., Uckfield, England.

Sokolov, I. B., A. N. Marchuk, V. S. Kuznetsov, E. A. Aleksandrovskaya, K. K. Kuzmin, V. L. Pavlov, A. I. Tsaryov, and V. V. Alipov (1985), "Analysis and Interpretation of Measurement Data Illustrated by the Construction and Staged Commissioning of the Sayano-Shushenskaya and Nurek Hydro Power Plants," in *Transactions of the 15th ICOLD*, Lausanne, International Commission on Large Dams, Paris, Vol. 1, pp. 1471–1482.

Sparrow, E. C. (1967), "Stress Cell Systems for Monitoring Static and Dynamic Stresses in Earth and Rock Fill Dams," *World Dams Today, Japan Dam Assoc.*, Nov., pp. 255–259.

Spathis, A. T., and D. Truong (1987), "Analysis of a Biaxial Elastic Inclusion Stressmeter," *Int. J. Rock Mech. & Min. Sci. & Geomech. Abstr.*, Vol. 24, No. 1, Feb., pp. 31–39.

Spitzer, F., and B. Howarth (1972), *Principles of Modern Instrumentation*, Holt, Rinehart & Winston, New York.

Stacey, T. R., and B. P. Wrench (1985), "The Convergence Meter," *Can. Geotech. J.*, Vol. 22, No. 4, Nov., pp. 604–607.

Stain, R. T. (1982), "Integrity Testing," *Civil Eng.*, London, Apr., pp. 53–59, May, pp. 71–73.

State of California (1968), "Evaluation of Commercial Soil Pressure Cells," Res. Rep. No. M&R 636342, Materials and Research Department, State of California, Sacramento, CA.

State of California (1971), "Field Performance Evaluation of Soil Pressure Cells," Res. Rep. No. M&R 632954-1, Materials and Research Department, State of California, Sacramento, CA.

Stateham, R. M., and J. S. Vanderpool (1971), "Microseismic and Displacement Investigations in an Unstable Slope," U.S. Department of the Interior, Bureau of Mines, Rep. of Invest. 7470.

Steenfelt, J. S. (1983), "Automated Alarm and Data Acquisition System for Reinforcement Works at Coal Terminal Quay Wall," in *Proceedings of the International Symposium on Field Measurements in Geomechanics*, Zürich, K. Kovari (Ed.), Balkema, Rotterdam, Vol. 1, pp. 565–574.

Stein, P. K. (1964), *Measurement Engineering, Vol. 1, Basic Principles*, Stein Engineering Services, Phoenix, AZ.

Stepanov, V. Y. (1983), "Monitoring of Rock Slope Stability with Quartz Extensometers and Hydrostatic Levels," in *Proceedings of the International Symposium on Field Measurements in Geomechanics*, Zürich, K. Kovari (Ed.), Balkema, Rotterdam, Vol. 1, pp. 773–780.

Steward, J. E., R. Williamson, and J. Mohney (1977), "Guidelines for Use of Fabrics in Construction and Maintenance of Low-Volume Roads," U.S. Department of Agriculture, Forest Service, Portland, OR.

Stickney, R. G., P. E. Senseny, and E. C. Gregory (1984), "Performance Testing of the Doorstopper Biaxial Strain Cell," Rep. No. RHO-BW-SA-334P, Rockwell Hanford Operations, Richland, WA.

Stillborg, B., and B. A. Leijon (1982), "A Comparative Study of Rock Stress Measurements at the Luossavaara Mine," SB8270, Swedish Mining Research Foundation, Kiruna, Sweden.

Stillborg, B., S. Pekkari, and R. Pekkari (1983), "An Advanced Rock Mechanics Monitoring System," in *Proceedings of the International Symposium on Field Measurements in Geomechanics*, Zürich, K. Kovari (Ed.), Balkema, Rotterdam, Vol. 2, pp. 1215–1227.

Swiger, W. F. (1972), "Subsurface Investigations for Design and Construction of Foundations of Buildings, Parts III and IV," *J. Soil Mech. & Found. Div. ASCE*, Vol. 98, No. SM7, July, pp. 749–764.

Szalay, K., and M. Marino (1981), "Instrumentation of Tarbela Dam," in *Proceedings of the ASCE, Symposium on Recent Developments in Geotechnical Engineering for Hydro Projects*, F. H. Kulhawy (Ed.), ASCE, New York, pp. 68–103.

Tao, S. S. (1979), "Fluid Settlement Profiler: Error Analysis for Installation at Blackstrap, Saskatchewan," *Can. Geotech. J.*, Vol. 16, No. 2, May, pp. 401–405.

Tao, S. S., D. M. Guenter, and L. J. Snodgrass (1980), "Design, Construction, and Performance of a Pneumatic Seal for Closed System Piezometers," *Can. Geotech. J.*, Vol. 17, No. 3, Aug., pp. 436–439.

Tattersall, F., T. R. M. Wakeling, and W. H. Ward (1955), "Investigations into the Design of Pressure Tunnels in London Clay," in *Proceedings of the Institution of Civil Engineers*, London, Vol. 4, Pt. 1, pp. 400–471.

Tavenas, F. A., C. Chapeau, P. LaRochelle, and M. Roy (1974), "Immediate Settlements of Three Test Embankments on Champlain Clay," *Can. Geotech. J.*, Vol. 11, No. 1, Feb., pp. 109–141.

Taylor, H., and Y. M. Chow (1976), "Design, Monitoring and Maintaining Drainage System of a High Earthfill Dam," in *Transactions of the 12th ICOLD*, Mexico, International Commission on Large Dams, Paris, Vol. 2, pp. 147–167.

Taylor, H., V. S. Pillai, and A. Kumar (1985), "Embankment and Foundation Monitoring and Evaluation of Performance of a High Earthfill Dam," in *Transactions of the 15th ICOLD*, Lausanne, International Commission on Large Dams, Paris, Vol. 1, pp. 173–198.

Taylor, K. V. (1968), "New Don Pedro Dam," in *Proceedings of the Western Water & Power Symposium*, Western Periodicals Co., North Hollywood, CA, pp. A-111 to A-122.

Teal, D. L. E. (1986), "Cordless Geo-Data Link," *Water Power & Dam Construction*, Vol. 38, No. 9, Sept., pp. 13–15.

Terzaghi, K. (1938), "Settlement of Structures in Europe and Methods of Observation," *Trans. ASCE*, Vol. 103, p. 1432.

Terzaghi, K., and R. B. Peck (1967), *Soil Mechanics in Engineering Practice*, 2nd ed., Wiley, New York.

Thamm, B. R. (1984), "Field Performance of Embankment Over Soft Soil," *J. Geotech. Eng. Div. ASCE*, Vol. 110, No. 8, Aug., pp. 1126–1146.

Tharp, D. (1986), personal communication, Oct. 2.

Thomas, H. S. H. (1966), "The Measurement of Strain in Tunnel Linings Using the Vibrating-Wire Technique," *Strain*, Vol. 2, July, pp. 16–21.

Thomas, H. S. H., and W. H. Ward (1969), "The Design, Construction and Performance of a Vibrating Wire Earth Pressure Cell," *Géotechnique*, Vol. 19, No. 1, pp. 39–51.

Thomas, R. (1985), "Time & Cost Comparisons Between Slope Indicator Company Inclinometer Indicators," *The Indicator*, Slope Indicator Company, Seattle, WA, Dec., pp. 4–7.

Thut, A. (1987), personal communication, Mar. 4.

Tice, J. A., and C. E. Sams (1974), "Experiences with Landslide Instrumentation in the Southeast," Landslide Instrumentation, *Trans. Res. Record*, No. 482, pp. 18–29.

Torstensson, B.-A. (1975), "Pore Pressure Sounding Instrument," in *Proceedings of the ASCE, Specialty Conference on In Situ Measurement of Soil Properties*, North Carolina State University, Raleigh, NC, ASCE, New York, Vol. II, pp. 48–54.

Torstensson, B.-A. (1984a), "A New System for Ground Water Monitoring," *Ground Water Monitoring Rev.*, Vol. 4, No. 4, Fall, pp. 131–138.

Torstensson, B.-A. (1984b), personal communication, Nov. 29.

Trow, W. A. (1974), "Temporary and Permanent Earth Anchors: Three Monitored Installations," *Can. Geotech. J.*, Vol. 11, No. 2, May, pp. 257–268.

Truong, D. (1977), "Investigation of a Biaxial Rigid Inclusion Stressmeter," B.S. Thesis, University of Queensland, Australia.

Uff, J. F. (1970), "In Situ Measurements of Earth Pressure for a Quay Wall at Seaforth, Liverpool," in *Proceedings of the Conference on In Situ Investigations in Soil & Rock*, British Geotechnical Society, London, pp. 229–239.

USBM (1984), "Ultrasonic Instrument for Closure Rate Measurement," Technology News, U.S. Department of the Interior, Bureau of Mines, Technology Transfer Group, No. 211, Nov.

USBR (1974), *Earth Manual*, 2nd ed., U.S. Department of the Interior, Bureau of Reclamation.

USCOLD (1979), "General Considerations on Reservoir Instrumentation," Committee on Measurements, USCOLD.

USCOLD (1985), "Compilation of United States Dams with Strong Motion Instruments and Reservoir Seismicity Networks," Committee on Earthquakes, USCOLD, Oct.

USCOLD (1986), "General Considerations Applicable to Performance Monitoring of Dams," Committee on Measurements, USCOLD, Dec.

USCOLD (1988), "General Considerations and Current U.S. Practice in Automated Performance Monitoring of Dams," Committee on Measurements, USCOLD, in preparation.

Vargas, M., and S. J. C. Hsu (1975), "Design and Performance of Ilha Solteira Embankments," in *5th Panamerican Conference on SM & FE*, Buenos Aires, published by Argentina Society for Soil Mechanics and Foundation Engineering, Buenos Aires, Vol. 2, pp. 199–218.

Vaughan, P. R. (1969), "A Note on Sealing Piezometers in Bore Holes," *Géotechnique*, Vol. 19, No. 3, pp. 405–413.

Vaughan, P. R. (1974a), "The Measurement of Pore Pressures with Piezometers," in *Proceedings of the Symposium on Field Instrumentation in Geotechnical Engineering*, British Geotechnical Society, Butterworths, London, pp. 411–422.

Vaughan, P. R. (1974b), Discussion: "Pore Pressure Reading Equipment for Hydraulic Piezometers," in *Proceedings of the Symposium on Field Instrumentation in Geotechnical Engineering*, British Geotechnical Society, Butterworths, London, pp. 559–560.

Vaughan, P. R., and M. F. Kennard (1972), "Earth Pressures at a Junction Between an Embankment and a Concrete Dam," in *Proceedings of the 5th European Conference on SM & FE*, Madrid, published by Spanish Society for Soil Mechanics and Foundation Engineering, Madrid, Vol. 1, pp. 215–221.

Vischer, W. (1975), "Use of Synthetic Fabrics on Muskeg Subgrades in Road Construction," U.S. Department of Agriculture, Forest Service, AK.

Von Thun, J. L. (1984), "Grand Coulee Riverbank Stabilization—Case History of the Design of Remedial Measures," in *Proceedings of the International Conference on Case Histories in Geotechnical Engineering*, S. Prakash (Ed.), University of Missouri, Rolla, MO, published by University of Missouri, Rolla, MO, Vol. II. pp. 553–559.

Wade, L. V., and P. J. Conroy (1980), "Rock Mechanics Study of a Longwall Panel," *Min. Eng.*, Vol. 32, No. 12, Dec., pp. 1728–1735.

Walker, L. K., and P. L. Darvall (1973), "Dragdown on Coated and Uncoated Piles," in *Proceedings of the 8th International Conference on SM & FE*, Moscow, USSR National Society for SM & FE, Moscow, Vol. 2.1, pp. 257–262.

Walton, R. J., and P. G. Fuller (1980), "An Investigation of a Bolted Coal Mine Roof During Mining at Nattai North Colliery," Geomechanics of Coal Mining Rep. No. 24, CSIRO Division of Applied Geomechanics, Australia.

Walton, R. J., and S. M. Matthews (1978), "Measurement of Rock Deformation, Stress Changes and Rock Competency in the 18CC/12CZ2 Crown Pillar at the CSA Mine Cobar—Part One," Support and Stabilization of Stopes, Proj. Rep. No. 33, CSIRO Division of Applied Geomechanics, Australia.

Walton, R. J., and G. Worotnicki (1978), "Rock Stress Measurements in the 18CC/12CZ2 Crown Pillar Area of the CSA Mine, Cobar, N.S.W.," Proj. Rep. No. 38, CSIRO Division of Applied Geomechanics, Australia.

Walton, R. J., and G. Worotnicki (1986), "A Comparison of Three Borehole Instruments for Monitoring the Change of Rock Stress with Time," in *Proceedings of the International Symposium on Rock Stress and Rock Stress Measurements,* Stockholm, Centek Publishers, Luleå, Sweden, pp. 479–488.

Ward, W. H., J. B. Burland, and R. W. Gallois (1968), "Geotechnical Assessment of a Site at Mundford, Norfolk, for a Large Proton Accelerator," *Géotechnique,* Vol. 18, No. 4, pp. 399–431.

Ward, W. H., D. J. Coats, and P. Tedd (1976), "Performance of Tunnel Support Systems in the Four Fathom Mudstone," Current Paper 25/76, Building Research Establishment, Watford, England, Mar.

Ward, W. H., P. Tedd, and N. S. M. Berry (1983), "The Kielder Experimental Test Tunnel: Final Results," *Géotechnique,* Vol. 33, No. 3, pp. 275–291.

Wardlaw, E. G., L. A. Cooley, A. E. Templeton, and R. L. Fleming, Jr. (1984), "Slide Stabilization with Gravel Trenches," in *Proceedings of the International Conference on Case Histories in Geotechnical Engineering,* S. Prakash (Ed.), University of Missouri, Rolla, MO, published by University of Missouri, Rolla, MO, Vol. II, pp. 743–748.

Warlam, A. A., and E. W. Thomas (1965), "Measurement of Hydrostatic Uplift Pressure on Spillway Weir with Air Piezometers," in *Instruments and Apparatus for Soil and Rock Mechanics,* ASTM STP 392, ASTM, Philadelphia, PA, pp. 143–151.

Warner, J. (1978), "Compaction Grouting—A Significant Case History," *J. Geotech. Eng. Div. ASCE,* Vol. 104, No. GT7, July, pp. 837–847.

Water Power & Dam Construction (1985), "15th International Congress on Large Dams, Part One," *Water Power & Dam Construction,* London, Vol. 37, No. 9, Sept., pp. 40–45.

Watson, D. A. (1964), *Specifications Writing for Architects and Engineers,* McGraw-Hill, New York.

Webster, J. L. (1970), "Mica Dam Designed with Special Attention to Control of Cracking," in *Transactions of the 10th ICOLD,* Montreal, International Commission on Large Dams, Paris, Vol. 1, pp. 487–510.

Weeks, R. C., and P. Starzewski (1986), "Automatic Monitoring and Data Processing," in *Proceedings of the U.K. Geotechnical Conference,* Birmingham, England, published by The Midland Geotechnical Society, Birmingham, England, pp. 117–126.

Weiler, W. A., and F. H. Kulhawy (1978), "Behavior of Stress Cells in Soil," Rep. 78-2, School of Civil and Environmental Engineering, Cornell University, Ithaca, NY.

Weiler, W. A., and F. H. Kulhawy (1982), "Factors Affecting Stress Cell Measurements in Soil," *J. Geotech. Eng. Div. ASCE,* Vol. 108, No. GT12, Dec., pp. 1529–1548.

Weir-Jones, I., and T. G. Bumala (1975), "The Design of Slope Stability Monitoring Systems," Soc. of Mining Engineers, AIME, Fall Meeting, Salt Lake City, UT, Sept., Preprint No. 75-AM-338.

Weltman, A. J. (1977), "Integrity Testing of Piles: A Review," Rep. PG4, Construction Industry Research and Information Association (C.I.R.I.A.), London.

Whitaker, T. (1964), "Load Cells for Measuring the Base Loads in Bored Piles and Cylinder Foundations," Current Paper, Eng. Ser. No. 11, Building Research Station, Watford, England.

Whitaker, T., and R. W. Cooke (1966), "An Investigation of the Shaft and Base Resistances of Large Bored Piles in London Clay," in *Proceedings of the Symposium on Large Bored Piles,* Institution of Civil Engineers, London, pp. 7–49.

Whittaker, B. N., and G. J. M. Woodrow (1977), "The Constant-Tension Strain Wire Borehole Extensometer and Its Application to Instrumentation of Underground Openings," in *Proceedings of the International Symposium on Field Measurements in Rock Mechanics,* Zürich, K. Kovari (Ed.), Balkema, Rotterdam, Vol. 1, pp. 437–448.

Wightman, W. D., S. J. Calabrese, and E. L. Foster (1980), "Special Study of Precast Concrete Tunnel Liner Demonstration, Lexington Market Tunnels, Baltimore, Maryland," U.S. Department of Transportation, Urban Mass Transportation Administration, Rep. No. UMTA-MD-06-0039-81-1.

Wilen, B. O., and W. P. MacConnell (1973), "The Modified Fernow Frost Gauge," *Soil Sci.,* Vol. 115, No. 4, Apr., pp. 326–328.

Wilkes, P. F. (1972), "An Induced Failure at a Trial Embankment at King's Lynn, Norfolk, England," in *Proceedings of the ASCE, Specialty Conference on Performance of Earth and Earth-Supported Structures,* Purdue University, Lafayette, IN, ASCE, New York, Vol. 1, Pt. 1, pp. 29–63.

Wilson, A. H. (1961), "A Laboratory Investigation of a High Modulus Borehole Plug Gage for Measurement of Rock Stress," in *Proceedings of the 4th Symposium on Rock Mechanics,* Pennsylvania State University, University Park, PA, published by the Pennsylvania State University, University Park, PA, pp. 185–195.

Wilson, S. D. (1967), "Investigation of Embankment Performance," *J. Soil Mech. & Found. Div. ASCE,* Vol. 93, No. SM4, July, pp. 135–156.

Wilson, S. D. (1970), "Observational Data on Ground Movements Related to Slope Stability," *J. Soil Mech. & Found. Div. ASCE,* Vol. 96, No. SM5, Sept., pp. 1519–1544.

Wilson, S. D. (1973), "Deformation of Earth and Rockfill Dams," in *Embankment-Dam Engineering,* Casagrande Vol., Wiley, New York, pp. 365–417.

Wilson, S. D. (1974), "Landslide Instrumentation for the Minneapolis Freeway," Landslide Instrumentation, *Trans. Res. Record,* No. 482, pp. 30–42.

Wilson, S. D. (1982), "Instrumentation and Performance of Embankment Dams," Lecture Notes, Nigerian Geotechnical Association, The Nigerian Society of Engineers, Lagos, Nigeria, Oct. 2.

Wilson, S. D. (1984), personal communication, Aug. 29.

Wilson, S. D., and P. E. Mikkelsen (1977), "Foundation Instrumentation—Inclinometers" (color video tape, film and reference manual), U.S. Department of Transportation, Federal Highway Administration, Implementation Division, Office of Research and Development.

Wilson, S. D., and P. E. Mikkelsen (1978), "Field Instrumentation," in *Landslides, Analysis and Control,* Transportation Research Board, Special Rep. 176, Chap. 5.

Windsor, C. R., and G. Worotnicki (1986), "Monitoring Reinforced Rock Mass Performance," submitted to International Symposium on Large Rock Caverns, Helsinki.

Wissa, A. E. Z., R. T. Martin, and J. E. Garlanger (1975), "The Piezometer Probe," in *Proceedings of the ASCE, Specialty Conference on In Situ Measurement of Soil Properties,* North Carolina State University, Raleigh, NC, ASCE, New York, Vol. I, pp. 536–545.

Wnuk, S. P., Jr. (1981), "On the Use of Bonded Weldable Strain Gages for Field Measurements in Hostile Environments," in *Proceedings of the International Conference on Measurements in Hostile Environments,* British Society for Strain Measurements, Edinburgh.

Wolf, S. (1973), *Guide to Electronic Measurements and Laboratory Practice,* Prentice-Hall, Englewood Cliffs, NJ.

Wolfskill, L. A., and C. Soydemir (1971), "Soil Instrumentation for the I-95 MIT-MDPW Test Embankment," *J. Boston Soc. Civil Eng.,* Vol. 58, No. 4, pp. 193–229.

Wolosick, J. R., and A. I. Feldman (1987), "Reliability of Bending Moments Backcalculated From Inclinometer Measurements," *Geotech. News,* Vol. 5, No. 3, Sept.

Wood, L. A., and A. J. Perrin (1984), "Observations of a Strutted Diaphragm Wall in London Clay: A Preliminary Assessment," *Géotechnique,* Vol. 34, No. 4, pp. 563–579.

Worotnicki, G., and R. J. Walton (1976), "Triaxial Hollow Inclusion Gauges for Determination of Rock Stresses in Situ," in *Proceedings of the Symposium on Investigation of Stresses in Rock—Advances in Stress Measurement,* Institution of Engineers, Australia, National Conference, Pub. No. 76/4, pp. 1–8.

Wroth, C. P. (1975), "In Situ Measurement of Initial Stresses and Deformation Characteristics," in *Proceedings of the ASCE, Specialty Conference on In Situ Measurement of Soil Properties,* North Carolina State University, Raleigh, NC, ASCE, New York, Vol. II, pp. 181–230.

Wroth, C. P., and J. M. O. Hughes (1973), "An Instrument for the In Situ Measurement of the Properties of Soft Clays," in *Proceedings of the 8th International Conference on SM & FE,* Moscow, published by USSR National Society for SM & FE, Moscow, Vol. 1.2, pp. 487–494.

Yu, T. R. (1983), "Rock Mechanics to Keep a Mine Productive," *Can. Min. J.,* Vol. 104, No. 4, Apr., pp. 61–66.

Zeigler, E. J., J. L. Wirth, and J. T. Miller (1984), "Slurry Trench Wall Replaces Structure Underpinning," in *Proceedings of the International Conference on Case Histories in Geotechnical Engineering,* S. Prakash (Ed.), University of Missouri, Rolla, MO, published by University of Missouri, Rolla, MO, Vol. III, pp. 1287–1296.

INDEX

Bold face numbers indicate primary discussion(s) of a topic.

Accelerograph, 425
Accelerometer, 106
Acceptance tests:
  after installation, 364
  before installation, 343
Accuracy, 75
Acoustic emission monitoring, **295**, 431, 445, 447, 457, 459, 461
Ailtech gages, *see* Eaton Corporation
Air bubbles, 81
Air entry value, 141
Alignment stakes, 208
American Consulting Engineers Council, 64
Apparatus Specialties Company:
  instruments manufactured by, 511
  pneumatic transducer, 90
Applied Geomechanics, Inc. tiltmeter, **218**, 511
Approval of proposed instruments, 50
Aquaducer, 287
Aquifer:
  artesian, 19
  definition, 19
Assigned subcontract, **58**, 60
Association of Soil and Foundation Engineers, 64
Atkins Technical, Inc. thermistors, **333**, 511
Auger:
  dimensions, 527
  drilling, 350
Automatic data acquisition systems:
  advantages and limitations, 368
  for electrical instruments, 108
  for hydraulic instruments, 87
  for pneumatic instruments, 91
Avongard calibrated crack monitor, **212**, 511

BAT piezometer, **134**, 511
Batteries:
  charging, 116
  disposable, 112
  maintenance, 116
  rechargeable, 112
Bellow-hose gage, 232
Benchmark, **206**, 207
Benefits of instrumentation, 3, **33**, 38
  braced excavations, 389
  drilled shafts, 483
  driven piles, 467
  embankment dams, 417
  embankments on soft ground, 407
  slopes, 443
  underground excavations, 453
Bentonite:
  cutoffs, 434
  pellets, **158**, 357
Biaxial strain cells, *see* Soft inclusion gages
Bidding:
  for field services, 61
  for instrument procurement, 47
Bimetal thermometer, 332
Biodegradable drilling mud, 154
Bishop piezometer tip, 124
Bison Instruments, Inc. soil strain gage, **236**, 511
Blind hole drilling method, 326
Bored piles, *see* Drilled shafts
Borehole deflectometers, *see* Transverse deformation gages
Borehole deformation gages, *see* Soft inclusion gages
Borehole directional survey instruments, **274**, 511
Borehole extensometers, *see* Fixed borehole extensometers; Probe extensometers
Borehole pressure cells, *see* Soft inclusion gages
Borros AB:
  anchor, 234, **249**
  instruments manufactured by, 511
Bottom heave, 21, **398**
Bourdon tube pressure gage, 80
Braced excavations, *see* Excavations, braced
Bracing, *see* Excavations, braced

Brewer Engineering Laboratories, 511
Bubbling pressure, 141
Budget, 43, 44
Buried plate, 235

Cable, *see* Electrical cable
Cable tension meters, **303**, 471, 472, 486
Caissons, *see* Drilled shafts
Calibrated crack monitor, 211
Calibrated telltale, 211
Calibration, 343
  acceptance tests, 343
  contractual arrangements, **51**, **61**, 64, 68
  factory, 343
  function checks, 344
  planning, 44
  during service life, 344
Cambridge drive-in piezometer, 131
Cambridge Insitu:
  demountable extensometer, 314
  instruments manufactured by, 511
  pressuremeter, 4
Capacitance discharge spot welder, 311
Capillary reader, 120
Carlson Instruments:
  elastic wire strainmeter, 93, **321**
  instruments manufactured by, 511
  pore pressure cell, 129
  stress meter, 327
Casagrande piezometer, 118
Casing, drill:
  backfilling, 156
  cleaning, 156
  dimensions, 527
  drilling with, 349
  pulling, 156
Cast-in-place piles, *see* Drilled shafts
Cause and effect, **38**, 43
Chain deflectometers, 272
Checklist:
  field service specifications, 507
  planning steps, 501
  procurement specifications, 505
Clinometers, *see* Tiltmeters
Closed hydraulic:
  piezometer, *see* Piezometer, twin-tube hydraulic
  systems, 86
Cohesionless soil, 13
Cohesive soil, 13
Cold conditions, 116
Collection of data, *see* Data collection
Commercially available instruments, 511
Communication systems, 113
Compression fittings:
  for hydraulic instruments, 85
  for pneumatic instruments, 92
Computer, 108, 374
Conax Corporation thermocouple, **334**, 511
Concrete:
  causes of strain in, 316
  conversion of strain to stress, 317
Concrete stress cells, 327
  for underground excavations, 457, 458
Conformance, 75
Connectors for electrical cable, 115
Consolidation, 15
Construction control, 35
Consulting engineer, selection of, 59
Contingency allowance, 58
Contractual arrangements:
  field services, 57
    bidding, 61
    content, 63, 507
    goals, 57
    negotiation, 61
    planning, 44
    recommendations, 61
    sole source, 60
  procurement of instruments, *see* Procurement specifications
Control stakes, 208
Convergence gages, *see* also Surface extensometers
  for slurry trenches, **293**, 392
Conversion factors, 539
Cooperation during field work, 65
Coring bit dimensions, 527
Correctness of readings, **43**, 75
Crack gages, *see* Surface extensometers
Crisis situations, 34
Cross-lot bracing, *see* Excavations, braced
CSIR:
  doorstopper cell, 189
  instruments manufactured by, 511
  triaxial strain cell, 189
CSIRO:
  hollow inclusion cell, 189
  instruments manufactured by, 511
  yoke gage, 188
Curvometer, 307
Cylindrical pressure cell, *see* Soft inclusion gages

Damage to instrumentation:
  contractual arrangements, 70
  protection, 363
Dams, *see* Embankment dams
Data acquisition systems, 79
  automatic, *see* Automatic data acquisition systems
  for electrical instruments, 108
  handling and transporting, 116
  for hydraulic instruments, 86
  for pneumatic instruments, 91
Data availability, 69
Data collection, 367
  automatic systems, 368
  contractual arrangements, **61**, 64, 68
  field books, 370
  field data sheets, 370
  field records, 370
  frequency, 373
  initial readings, 373
  personnel qualifications, 367
  personnel responsibilities, 367
  planning, 44
  written procedure, 368

Data controller, 109
Data implementation:
  contractual arrangements, 69
  planning, 44
Data interpretation, 382
  contractual arrangements, **61**, 64
  guidelines, 383
  personnel qualifications, 382
  planning, 44
Data logger, 109
Data processing and presentation, 374
  automatic systems:
    examples of plots, 375
    role, 374
  calculation sheets, 377
  contractual arrangements, **61**, 64, 68
  personnel qualifications, 374
  planning, 44
  plots:
    guidelines on plotting, 381
    types, 379
  screening of data, 377
  written procedure, 375
Data reporting, 384
  contractual arrangements, **61**, 64
  final report, 384
  interim report, 384
  planning, 44
  technical publication, 385
DCDT, 101
De-aired liquid, 82
Deep foundations, *see* Drilled shafts; Driven piles
Deep settlement point, *see* Fixed borehole extensometers
Deflectometers:
  extenso-, 273
  multiple, 272
  portable borehole, 273
Deformation, 199
  fixed borehole extensometers, *see* Fixed borehole extensometers
  fixed embankment extensometers, *see* Fixed embankment extensometers
  inclinometers, *see* Inclinometers
  instrument categories, 199
  liquid level gages, *see* Liquid level gages
  probe extensometers, *see* Probe extensometers
  surface extensometers, *see* Surface extensometers
  surveying methods, *see* Surveying methods
  tiltmeters, *see* Tiltmeters
  transverse deformation gages, *see* Transverse deformation gages
Delay to construction, 69
Delivery of instruments, 53
Demec strain gage, 306
Demountable extensometer, 314
Dial gage, **79**, 212
Dial indicator, **79**, 212
Dilatometer, 167
Dipmeter, 120
Direct current differential transformer (DCDT), 101
Directional survey instruments, 274
Displacement, *see* Deformation
Displacement stakes, 208
Disposition of instruments, 70
Distofor extensometer, 246
Distomatic convergence gage, 214
Distometer convergence gage, 214
Doorstopper, 189
Double fluid settlement gages, 289
Dowell chemical sealant, 160
Downdrag, *see* Drilled shafts; Driven piles
Drilled caissons, *see* Drilled shafts
Drilled piers, *see* Drilled shafts
Drilled piles, *see* Drilled shafts
Drilled shafts, 483
  applications, routine and special, 489
  behavior mechanisms, **19**, 483
  benefits of instrumenting, 483
  case histories, 489
  downdrag on, 19, 488
  echo test, 488
  falling weight test, 487
  geotechnical questions, 483
  integrity tests, 488
  load-movement relationships:
    cyclic testing, 486
    dynamic testing, 487
    miscellaneous static tests, 486
    static axial testing, 484
    static lateral testing, 486
  load tests, **484**, **486**, 487
  nuclear density probe, 488
  sonic coring, 488
  steady-state vibration testing, 487
  transient dynamic response testing, 487
Drilling methods:
  alignment control, 353
  auger, 350
  core:
    conventional, 352
    wireline, 352
  double tube reverse air circulation, 351
  guidelines for, 353
  hydraulic rotary, 350
  mud rotary, 350
  rotary, 350
  rotary percussion, 352
  straight rotary, 350
  wash boring, 349
Drill rod dimensions, 527
Driven piles, 467
  applications, routine and special, 479
  behavior mechanisms, **19**, 467
  benefits of instrumenting, 467
  case histories, 479
  Case Method, **473**, 476
  defects during driving, 476
  deformation of ground, 478
  downdrag on, 19, 475
  echo test, 476
  geotechnical questions, 467
  integrity tests, 476
  load-movement relationships:
    cyclic testing, 472

Driven piles (Continued)
dynamic testing, 472
miscellaneous static tests, 472
static axial testing, 467, 469
static lateral testing, 471
load tests, **467**, **469**, **471**, 472
nuclear density probe, 477
Pile Driving Analyzer, 473
sonic coring, 477
steady-state vibration testing, **473**, 474, 477
transient dynamic response testing, **473**, 474, 477
Dudgeon hydraulic jack, **302**, 511
Dynamometers, *see* Load cells

Earth pressure:
active, 21
passive, 21
Earth pressure cells:
applications, 165
bedding, nonuniform, 172
calibration chambers, 170
contact cells, 177
accuracy, 177
for braced excavations, 397, 398
for embankment dams, 423
factors affecting measurements, 180
installation, 181
irregularity of structure surface, 181
for load at tips of drilled shafts, **178**, 183, 484
for load at tips of driven piles, **178**, 183, 468, 470, 475
number of cells, 180
standard types, 177
for stress on sides of drilled shafts, 486
for stress on sides of driven piles, **180**, 183, 476
for underground excavations, 458
embedment cells, 165
accuracy, 176
aspect ratio, 169
diaphragm cells, 166
factors affecting measurements, 167
hydraulic cells, **166**, 175
installation in boreholes, 167
installation in fill, 172, **176**
limitations imposed by environment, 167
orientation, 175
field placement effects, 171
grooves in active face, **174**, 175
laboratory calibrations, 170, 180
rim design, 175
size, 169
stiffness, 169, 181
temperature effects, 168, 181
Eastman Christensen Company borehole directional survey instruments, **275**, 511
Eaton Corporation:
embedment strain gage, 324
instruments manufactured by, 511
spot welder, 311
surface mounted weldable strain gage, 94, 312
Effective stress:
definition, 14
principle of, 15

Electrical cable:
connectors, 115
lightning protection, 116
selection, 114
splices, 115
water-blocked, 116
Electrical instruments:
communication systems, 113
data acquisition systems, 108
power supplies, 112
transducers, 92
Electrical resistance strain gages:
basic types, 92
circuit tests, **98**, 116
embedment types, 320
accuracy, 322
advantages and limitations, 322
conversion to stress, **316**, 325
Eaton Corporation gage, 324
inclusion effect, 325
initial readings, 325
installation, 325
length, 322
locations, 325
Mustran cell, 324
plastic encased gages, 324
range, 322
sensitivity, 322
sister bars, 320
unbonded gages, 323
longevity, 94
manual data acquisition systems, 100
surface mounted types, 311
accuracy, **307**, 312
advantages and limitations, 307
bonded foil, **93**, 312
bonded wire, 93
conversion to stress, 316
cost, 312
environmental conditions, 312
for geotextiles, 314
initial readings, 316
installation, 313, 314, 316
length, **307**, 312, 315
locations, 315
range, 307
recommendations for choice, 312, 313, 314
semiconductor, 93
sensitivity, 307
for timber piles, 314
unbonded wire, 93
weldable, 94
Wheatstone bridge circuits, 94
zero drift, 319
Electrolytic level, 108
Electronic distance measurement, 203
ELE International Ltd., 511
Embankment dams, 417
acoustic emission monitoring, 431
automatic monitoring, 431
behavior mechanisms, 21
benefits of instrumenting, 417

case histories, 435
concrete face rockfill, 422
construction phase, 419
core settlement, 435
drawdown, 422
first filling, 420
frequency of readings, 435
geotechnical questions, 418
initial conditions, 418
installation of instruments, 347, **432**
lateral deformation, long-term monitoring of, 430
leakage, **420**, 423, 424, 425
long-term performance, 422, **423**
operating and maintenance, 431
pore water pressure, long-term monitoring of, 426
problems caused by instrumentation, 418
rogue instruments, 418, 430
seismic events, 425
settlement, long-term monitoring of, 430
state of the art improvement, 422
streaming potential, 431
terminal enclosure, 434
thermal monitoring, 431
visual observations, 424
Embankments:
on hard ground, 21
on soft ground, *see* Embankments on soft ground
Embankments on soft ground, 407
applications, routine and special, 410
behavior mechanisms, **21**, 407
benefits of instrumenting, 407
case histories, 410
consolidation progress, 408
fill quantities, 410
geotechnical questions, 407
initial conditions, 408
stability, 409
staged construction, 409
surcharging, 409
test embankment, 408
vertical drains, 409
Environment for instruments, **41**, 50
Error, 77
Excavations, braced, 389
applications, routine and special, 399
behavior mechanisms, **21**, 389
benefits of instrumenting, 389
bottom heave, 21, **398**
case histories, 400
cross-lot bracing, **21**, 389, **393**, **396**
geotechnical questions, 390
groundwater lowering, 398
initial conditions, 390
long-term performance, 398
rakers, **21**, 389
sheeting and walers, 397
slurry trench, 392, 393
slurry wall, 389, 397
stability, 392, **394**
structures nearby, 394
test section, **391**, 392
tieback anchors, **21**, 389, **391**, **393**, **396**
performance testing, 389
proof, 389
test, 389, **391**
Existing stress, 326
Experience requirements:
for field personnel, 64
for instrument manufacturers, 50
Extenso-deflectometer, 273
Extensofor extensometer, 228
Extensometers:
fixed borehole, *see* Fixed borehole extensometers
fixed embankment, *see* Fixed embankment extensometers
probe, *see* Probe extensometers
surface, *see* Surface extensometers

Fiber-optic sensors, 274, **294**
Filters for piezometer tips:
requirements, 144
saturation, 146
types, 141
Fixed borehole extensometers, 237
advantages and limitations, 242
anchors, 239
applications, 237
bayonet disconnect on rod, 238, **241**
for braced excavations, 394
definition, 237
for embankments on soft ground, 410
heads, 241
as heave gage, 246
installation, 247
multipoint (MPBX), 240
operating principle, 237
precision, 243
rods, 238
rod settlement gage, 250
single-point (SPBX), 240
for slopes, 445, 447, 448
subsurface settlement points, 248
Borros anchor, 249
deep settlement point, 249
spiral-foot anchor, 250
transducers for, 241
DCDT, 245
dial indicator, 243
electrical resistance strain gage, 244
induction coil, 246
LVDT, 245
magnetostrictive, 247
micrometer, 243
potentiometer, 245
slack-wire indicator, 244
sonic probe, 247
suspended weights, 243
vibrating wire, 246
for underground excavations, 456, 457, 459, 460, 461
wires, 238
Fixed embankment extensometers, 233
applications, 233
buried plate, 235
definition, 233
for driven piles, 476

Fixed embankment extensometers (Continued)
  with electrical transducers, 235
  for embankment dams, 419, 431
  for embankments on soft ground, 410
  settlement platform, 233
  soil strain gage, 236
  tensioned wire gages, 235
Flat plate dilatometer, 167
Force account, 58
Force balance accelerometer, 106
Frost gages, 335
Full profile gages, 287
Fulmer tension meter, **304**, 511
Future trends, 10

Gage Technique Ltd., 511
Gas bubbles, 81
Gas for pneumatic instruments, 91
Generators:
  diesel, 112
  gasoline, 112
  thermoelectric, 113
Geokon, Inc.:
  borehole deformation gage, 4
  borehole pressure cell, 187
  groutable anchor, 241
  instruments manufactured by, 511
  load cells, 300
  multipoint liquid level gage, 286
  piezometer with duplicate transducers, 136
  push-in vibrating wire piezometer, 128
  vibrating wire strain gages, 310
  vibrating wire stressmeters, 6, **194**
Geomation, Inc. automatic data acquisition systems, 511
Geonor A/S:
  earth pressure cell, 177
  field piezometer, Model M-206, 130
  instruments manufactured by, 511
  remote reading weir, 421
  settlement probe, 249
  vane shear equipment, 4
  vibrating wire piezometer, 133, 136
Geosistemas:
  instruments manufactured by, 511
  pneumatic transducer, 90
Geotech AB vibrating wire piezometer, **133**, 511
Geotechnical construction, 3
Geotechnical Engineering and Mining Services (U.S.A.), Inc.
    automatic data acquisition systems, 511
Geotechnical Instruments (U.K.) Ltd.:
  boilers for de-aired water, 82
  electrical dipmeter, 120
  flushing equipment, 521
  inclinometer casing, 258
  inclinometer data logger, 254
  inclinometer test stands, 263
  instruments manufactured by, 511
  magnetic extensometer, 229
  pneumatic transducer, 90
  purge bubble system, 123
  slip indicator, 269
  twin-tube hydraulic piezometer, 6
Geotechnical questions, 38
  braced excavations, 390
  drilled shafts, 483
  driven piles, 467
  embankment dams, 418
  embankments on soft ground, 407
  slopes, 443
  underground excavations, 455
Geotechniques International, Inc.:
  instruments manufactured by, 511
  piezocone, 4
Geotextiles, strain gages for, 314
G+G Technics AG tiltmeter, **218**, 511
Glass plates, 210
Global positioning system, 206
Glötzl GmbH:
  concrete stress cell, 329
  earth pressure cell, 167
  hydraulic load cell, 299
  instruments manufactured by, 511
  liquid level gage, 281
  pneumatic transducer, 90
Grid crack monitor, 211
Groundwater level:
  definition, 16
  measurement, *see* Observation wells; Piezometers
  perched, 19
Groundwater pressure measurement, 117
Grout, backfilling boreholes with, 355

Halcrow bucket, 122
Hall, Earl B., Inc.:
  instruments manufactured by, 511
  pneumatic transducer, 90
Hanging pendulum, 270
Hard inclusion gages, *see* Rigid inclusion gages
Heave gage:
  in braced excavations, 398
  fixed borehole extensometer type, 246
  mechanical type, 219
  magnet/reed switch type, 231
Heavy liquid piezometer, 131
High air entry filter, 141
Historical perspective, 5
HITEC Products, Inc. weldable strain gage, **94**, **312**, 511
Hole-drilling strain gage method, 326
Horizontal control station, 207
Horizontal plate gage, 224
Horizontal strainmeter, 236
Hose level, 276
Huggenberger AG, 511
Human eye, 199
Hydraulic fracturing:
  of embankment dams, 428
  for stress measurement, 167
Hydraulic instruments:
  data acquisition systems, 86
  transducers, 80
Hydraulic jacks, **301**, 391
Hydraulic leveling devices, 280
Hydraulic strain gage, 314
Hydrodynamic time lag, 139

- Hydrostatic profile gaûge, 288
- Hysteresis, 76

- Idel sonde, **224**, 511
- Igneous rocks, 23
- Implementation of data:
  - contractual arrangements, 69
  - planning, 44
- Inclinometers, 250
  - advantages and limitations, 253
  - applications, 251
  - automatic readout units, **252**, 265, 266
  - for bending moments, 267
  - bias, 255
  - for braced excavations, 392, 393, 394
  - calibration, 263
    - field casing, 263
    - horizontal casing, 264
    - inclined casing, 264
    - test casing, 263
    - test stand, 263
  - casing:
    - alignment, 256
    - aluminum alloy, 259
    - diameter, 256
    - fiberglass, 258
    - plastic, 258
    - selection of, 259
    - spiraling of, **256**, 266
    - steel, 259
    - types, 258
  - check-sums, 264, 265, **266**
  - combined with probe extensometer, 232
  - components, 251
  - data:
    - collection, 257, 258, **264**
    - interpretation, 266
    - processing, 265
  - definition, 251
  - for drilled shafts, 486
  - for driven piles, 472, 476, 479
  - for embankment dams, 420, 423, 430, 431
  - for embankments on soft ground, 408, 410
  - error:
    - azimuth rotation, 255
    - scaling, 255
    - zero offset, 255
  - in-place, 271
  - installation, 260
    - borehole backfilling, 256
    - in boreholes, 261
    - buoyancy, 262
    - check valve, 261
    - coupling requirements, 260
    - in fill, 261
    - large movements expected, 259, **262**
    - on piles, 262
    - spiraling, 257
    - spiral survey, 257
  - maintenance, 264
  - operating principle, 251
  - plots, 266
  - precision:
    - factors affecting, 255
    - magnitude, 253
  - range, 253
  - reading, 264
    - automatic, 265
    - depth interval, 257
    - field checks, 264
    - handling of probe, 258
    - initial, 265
    - method, 264
    - repeatability of position, 257
    - telescoping casing, 265
    - temperature effects, 258
  - Recorder–Processor–Printer (RPP), **252**, 254, 266
  - for slopes, 445, 446, 447
  - for slurry trenches, 392, 394
  - transducers for:
    - accelerometer, 252
    - electrical resistance strain gage, 254
    - electrolytic level, 255
    - precision of, 255
    - Slope Indicator 200B, 254
    - vibrating wire, 255
  - for underground excavations, 456, 459, 460, 461
  - wheel assembly condition, 255
- Inclusion effect of strain gages, 325
- Induction coil:
  - gages, 223
  - transducers, 107
- In-place inclinometers, 271
- In situ properties, 3
- In situ stress:
  - definition, 26
  - measurement, 186, 326
- Installation, 347
  - acceptance tests, 364
  - in boreholes, 348
    - backfill material, 355
    - bentonite pellets, 357
    - borehole requirements, 349
    - downhole components, 354
    - drilling methods, 349
    - grout, 355
    - pea gravel, 357
    - sand, 357
  - contractual arrangements, **59**, 64, 66, 347
  - detailed procedures, **348**, 533
  - in fill, 358
    - exit points from conduit or boreholes, 362
    - horizontal pipes, 359
    - horizontal tubes and cables, 360
    - surface installation method, 360
    - vertical pipes, 360
    - vertical tubes and cables, 363
  - golden rules for field work, 365
  - at ground surface, 348
  - locations of instruments, 347
  - planning, **44**, 365
  - protection from damage, 363
  - records, 364
  - report, 366

Installation (Continued)
schedule, 364
in underground excavations, 363
Instruction manual, 52
Instrumented sections:
primary, 42
secondary, 42
Intact rock, 25
Interfels GmbH:
extensometer head, 245
instruments manufactured by, 511
Inverted pendulum, 270
Irad Gage:
electrical crack gage, 213
instruments manufactured by, 511
rockbolt expansion shell anchor, 239
sonic probe, 232, 248
vibrating wire strain gage, 6, 320
vibrating wire stressmeter, 193
ISETH:
Distometer, 214
sliding micrometer, 221

Jamin effect, 81
Jointmeters, *see* Surface extensometers
Joint water pressure:
definition, 26
measurement of, *see* Piezometers

Kern Instruments, Inc.:
automatic level, 202
instruments manufactured by, 511
ISETH Distometer, 214
stereoplotter, 205
Kyowa Electronic Instruments Company, 511

Laser beam surveying, 203
Legal protection, 35
Lightning protection, 116
Linearity, 76
Linear potentiometer, 101
Linear variable differential transformer (LVDT), 100
Liquid in hydraulic instruments, 81
density changes, 83
discontinuity, 81
recommendations, 85
surface tension effects, 83
temperature variation, 83
Liquid level gages, 275
advantages and limitations, 276
applications, 275
definition, 275
density changes in liquid, 83
discontinuity of liquid, 81
double fluid settlement gage, 289
for embankment dams, 418, 420, 422, 430
for embankments on soft ground, 408, 410
full profile gages, 287
hose level, 276
installation, 292
liquid for, 85
mercury in, 275
multipoint gages, 285
overflow gages:
with backpressure, 284
full profile type, 287
simple type, 280
with suction, 284
precision, 276
with pressure transducer in cell, 282
with pressure transducer in readout unit, 283
recommendations for choice, 292
for slopes, 445
surface tension effects, 83
temperature variation, 83
tubing for:
diameter, 84
filling, 291
fittings, 85
flushing, 291
material, 84
routing, 86
LNEC triaxial strain cell, **189**, 511
Load cells, 297
accuracy, 298
advantages and limitations, 298
applications, 297
for braced excavations, 391, 393, 395, 396, 398
cable tension meters, 303
calibration, 305
capacity, 304
configurations, 304
definition, 297
disk type, 297
for drilled shafts:
at butt, 484, 486
at tip, *see* Earth pressure cells
for driven piles:
at butt, 467, 471, 473
within pile, 470, 475
at tip, *see* Earth pressure cells
electrical resistance, 299
height/diameter ratio, 304
hydraulic, 298
hydraulic jacks, 301
installation, 305
mechanical, 297
photoelastic, 301
pretreatment, 305
for slopes, 448
telltale, 298, **308**
for underground excavations, 456, 457, 458, 460
vibrating wire, 300
Load in structural members, 297
Load transfer assembly, **468**, 485
Locations of instruments, **42**, 66
Low air entry filter, 141
Luleå triaxial strain cell, **189**, 511
LVDT, 100

Magnetic extensometer, 229
Magnetic probe extensometer, 229
Magnetostrictive:
gage, 232
transducer, 107

Magnet/reed switch:
gage, 229
transducer, 106
Maihak AG:
instruments manufactured by, 511
load cell, 300
Mains power, 112
Maintenance, 345
contractual arrangements, **61**, 64, 68
embedded components, 345
field terminals, 345
planning, 44
readout units, 345
Manometer, 80
Manufacturers of instruments, 8, **511**
Material requirements, 50
Mayes Demec strain gage, **308**, 511
Measurement and payment:
for field services, 70
for instrument procurement, 54
Measurements Group, Inc.:
bonded foil gage, 312
hole-drilling strain gage method, 326
instruments manufactured by, 511
Measuring points for surveying methods, 207
Mechanical extensometer, 306
Mechanical instruments, 79
Mechanical strain gages:
embedment types, 320
accuracy, 322
advantages and limitations, 322
conversion to stress, **316**, 325
inclusion effect, 325
initial readings, 325
installation, 325
length, 322
locations, 325
range, 322
sensitivity, 322
surface mounted types, 306
accuracy, 307
advantages and limitations, 307
conversion to stress, 316
initial readings, 316
installation, 316
length, **307**, 315
locations, 315
portable gages, 306
range, 307
scratch gages, 308
sensitivity, 307
Mercury thermometer, 332
Metamorphic rocks, 23
Micrometer, 79
Microseismic detection, 295
Microseismograph, 426
Mineral skeleton, 13
Mines, *see* Underground excavations
Monitoring well, 118
MPBX, *see* Fixed borehole extensometers
Multiple deflectometers, 272
Multiple telltales, **292**, **309**, 320, 392, 469, 471, 476, 484
Multipoint settlement gages, *see* Liquid level gages
Mustran cell, 324

Negative skin friction, *see* Drilled shafts; Driven piles
Negotiation:
for field services, 61
for instrument procurement, 47
Noise, 77
Nold, Walter Company, Inc.:
Aquaducer, 287
DeAerator, 82
instruments manufactured by, 511
pressurized de-aired water containment system, 83, **521**
Normally consolidated soil, 16

Observational method, 34
Observation point, 207
Observation wells, 118
advantages and limitations, 142
applications, **117**, 143
definition, **17**, 117
description, 118
Offsets from a baseline, 202
Omega Engineering, Inc.:
bimetal thermometer, 333
instruments manufactured by, 511
resistance temperature device, 334
Open cut excavations, *see* Excavations, braced
Optical leveling, 202
Overconsolidated soil, 16
Overflow gages, 280, 284, 287
Overflow weirs, 280
Overvoltage protection, 116
Oyo Corporation, 511

Pajari Instruments Ltd., **274**, 511
Partially saturated soil, 13
Payment:
for field services, 70
for instrument procurement, 54
Pendulum, 270
Perard Torque Tension Ltd. load cell, **302**, 511
Permeability, 16
Petur Instruments, *see* Thor International, Inc.
Photoelastic:
load cell, 301
strain gage, 314
Photogrammetric surveying methods, 204
Photographic borehole directional survey instruments, 274
Piezocone, 4
Piezodex piezometer, 139
Piezofor piezometer, 138
Piezometer Research and Development:
bentonite pellets, 158
heavy liquid piezometer, 132
instruments manufactured by, 511
open standpipe piezometer, 119
piezometer installation method, 153
Piezometers, 117
accuracy, 143
applications, 117, 144
bentonite balls, 159

Piezometers (Continued)
  bentonite gravel, 160
  bentonite pellets, 158
    placement in boreholes, 156
    pore water pressure changes, 161
    properties, 158
    retarding swell, 158
    in saline groundwater, 159
    tests for swell, 159
  bentonite rings, 160
  for braced excavations, 390, 392, 393, 395, 398
  chemical sealant, 160
  in compacted fills, 145
  definition, **17**, 117
  for drilled shafts, 486, 488
  for driven piles, 469, 472, 475, 476, 479
  duplicate transducers, 135
  electrical resistance, 128
    advantages and limitation, 142
    applications, 144
    bonded type, 128
    push-in type, 129, 148
    unbonded type, 128
  for embankment dams, **145**, 418, 422, 425, 426
  for embankments on soft ground, 408, 410
  filters for:
    requirements, 144
    saturation of, 146
    types, 141
  gas generation, 149
  grout in boreholes, 157
  installation:
    in boreholes with artesian conditions, 161, 164
    in boreholes in rock, single piezometer, 163
    in boreholes in rock, two or more piezometers, 164
    in boreholes in soil, single piezometer, 150
    in boreholes in soil, two or more piezometers, 163
    in fill, 148
    push-in method, 148
  multipoint, 136
    advantages and limitations, 143
    applications, 144
    with grout, 136
    with movable probes, 137
    with packers, 136
    push-in type, 137
  open standpipe, 118
    advantages and limitations, 119, **142**
    applications, 144
    checking seal integrity, 123
    combined with probe extensometer, 233
    in consolidating soils, 123
    conversion to twin-tube hydraulic, 124
    description, 118
    freezing of water in, 123
    push-in types, **130**, 148
    reading methods, 120
    in unsaturated or gaseous soils, 122
  in organic soil deposits, 145
  pneumatic, 126
    advantages and limitations, 142
    applications, 144
    description, 126
    push-in type, 134, 148
  recommendations for choice:
    saturated soil and rock, 141
    unsaturated soil, 144
  sand in boreholes, 156
  sealing:
    in boreholes, 150
    of push-in type, 148
  for slopes, 444, 445, 446, 447
  smearing and clogging, 149
  twin-tube hydraulic, 123
    advantages and limitations, 125, **142**
    applications, 124, **144**
    Bishop tip, 124
    description, 123
    details of system, 519
    USBR tip, 124
  for underground excavations, 455, 458, 459, 460, 461
  vibrating wire, 127
    advantages and limitations, 142
    applications, 144
    description, 127
    for embankment dams, 135
    in-place check feature, 132
    push-in type, 128, 148
Piezometric elevation, 19
Piezometric level, 19
Pile Dynamics, Inc. Pile Driving Analyzer, 473
Piles, *see* Drilled Shafts; Driven piles
Pipe dimensions, 527
Piping, 22
Planning, **37,** 501
Plaster patches, 210
Plastic encased strain gages, 324
Plumb line, 270
Pneumatic instruments:
  readout units, 91
  recommendations for use, 91
  transducers for, 87
Poor boy, 268
Poor man's inclinometer, 255, **268**
Pore air pressure, 18
Pore gas pressure, 18
Pore pressure cells, *see* Piezometers
Pores, 13
Pore spaces, 13
Pore water pressure:
  definition, 14
  dissipation, 15
  excess, 15
  measurement of, *see* Piezometers
  negative, 17
  positive, 17
Portable borehole deflectometers, 273
Portable clinometer, 217
Potentiometer, 101
Power supplies, 112
Precipitation gage, 422
Precision, 75
Precision settlement gage, 250
Preconstruction conditions survey, 390, 455

Predictions, 38
Presentation of data, *see* Data processing and presentation
Pressure, 13
Pressure gage, 80
Pressuremeter, 4, 167
Prewitt scratch gage, 308
Price of instruments, 47
Probe extensometers, 219
  accuracy, 220
  advantages and limitations, 220
  applications, 219
  bellow-hose gage, 232
  for braced excavations, 392, 394, 397
  combined with inclinometer casing, 232
  combined with open standpipe piezometer, 233
  crossarm gage, 219
  definition, 219
  for drilled shafts, 484, 486
  for driven piles, 476
  for embankment dams, 419, 420, 430, 431
  for embankments on soft ground, 408, 410
  induction coil gages:
    current-displacement type, 223
    frequency-displacement type, 228
    installation in boreholes, 225
    installation in fill, 228
  magnetostrictive gage, 232
  magnet/reed switch gage:
    description, 229
    installation in boreholes, 230
    installation in fill, 231
    installation as heave gage, 231
  mechanical heave gage, 219
  mechanical probe within inclinometer casing, 221
  precision, 224, 230
  recommendations for choice, 233
  sliding micrometer, 221
  for underground excavations, 456, 459, 460, 461
Proceq SA load cell, 6, **299**, 511
Processing of data, *see* Data processing and presentation
Procurement specifications, 45
  bidding, 47
  content, 49, 505
  descriptive, 46
  or equal provision, 47, 49, **50**
  negotiation, 47
  performance, 46
  planning, 44
  recommendations, 47
  sole source, 47
Profiling piezometer, 129, 130
Proof testing, 33
Public relations, 35
Purge bubble principle, 121
Purpose of instrumentation, *see* Benefits of instrumentation

Qualifications:
  of field personnel, 64
  of instrument manufacturers, 50
Quality assurance, 51

Radiofor extensometer, 247
Radio transmission, 113
Reading instruments, *see* Data collection
Reasons for instrumentation, *see* Benefits of instrumentation
Reference datums for surveying methods, 206
Reference monument, 206
Reference stakes, 208
Reliability, 9, 41, **341**
Remedial action, 39
Remote settlement gages, *see* Liquid level gages
Repeatability, 75
Reporting of data, *see* Data reporting
Reproducibility, 75
Residual stress in piles, 469
Resistance strain gages, *see* Electrical resistance strain gages
Resistance temperature device (RTD), 334
Resolution, 76
Response time, 140
Retaining wall, behavior mechanisms, 20
Reverse circulation method, 156
Revert, 154
Review of proposed instruments, 50
Rigid inclusion gages, 191
  accuracy, 196
  advantages and limitations, 195
  applications, **185**, 186
  definition, **186**, 191
  photoelastic stressmeters, 194
  recommendations for choice, 195
  tapered plugs, 194
  for underground excavations, 457, 460
  vibrating wire stressmeters, 191
    biaxial, 193
    uniaxial, 191
Rock behavior, 23
  discontinuities, 25
  elastic, 24
  engineers' view, 24
  geologists' view, 23
  mechanisms, **27**, 38
  viscoelastic, 24
Rockfill dams, *see* Embankment dams
Rock mass, 25
Rock stressmeters, *see* Rigid inclusion gages
Roctest Ltd.:
  cable tensiometer, 304
  convergence gage, 215
  fixed borehole extensometer, 246
  instruments manufactured by, 511
Rod settlement gage, 250
Rogers Arms & Machine Company borehole deformation gage, **188**, 511
RS Technical Instruments Ltd., 511
Rupture stakes, 268

Safety, 34
Sand, backfilling boreholes with, 357
Satellites, 113, 206
Saturated soil, 13
Scratch gages, 308
Sedimentary rocks, 23
Seismographs, 425
Selection of instruments, 40
Self-potential, 431

Sensitivity, 76
Serata Geomechanics, Inc., 511
Settlement, 16
Settlement platform, 233
Settlement probe, 249
Shallow foundations, behavior mechanisms, 19, 27
Shear plane indicators, 268
Shear probe, 255, 262, **268**
Shear strength, 16
Shear strip, 269
Sheet piles, *see* Excavations, braced
Shipment of instruments, 53
Sister bar:
  description, 320
  inclusion effect, 326
Sliding micrometer, **221**, 325, 397, 484, 486
Slip indicator, 268
Slope extensometer, 262, **269**
Slope Indicator Company:
  deflectometer, 273
  electrical resistance strain gage piezometers, 130
  extensometer head, 244
  flat wedge spring anchor, 239
  groutable anchor, 241
  hydraulic anchors, 241
  IDA system, 110
  inclinometer, 6, 251
    casing, 258
    casing check valve system, 261
    in-place, 272
    PC-SLIN program, 266
    plots, 267
    Recorder-Processor-Printer, **252**, 254, 266
    Series 200B, 254
    spiral sensors, 257
  instruments manufactured by, 511
  liquid level gage, 282
  pneumatic transducers, 90
  push-in pneumatic piezometers, 135
  settlement probes, 222
  Sondex, 224
  tape extensometer, 214
  tiltmeter, 217
Slope monitoring stakes, 208
Slopes, excavated and natural, 443
  applications, routine and special, 448
  behavior mechanisms, **20**, **27**, 443
  benefits of instrumenting, 443
  case histories, 448
  geotechnical questions, 443
  initial conditions, 444
  long-term performance, 447
  movement zone, 446
  stability, 445, 447
Slurry trench, *see* Excavations, braced
Slurry wall, *see* Excavations, braced
Soft inclusion gages, 186
  accuracy, 196
  advantages and limitations, 191
  applications, **185**, 186
  biaxial strain cell, 189
  borehole deformation gages, 188
    CSIRO yoke gage, 188
    U.S. Bureau of Mines gage, 188
  borehole pressure cell, 186
  definition, 186
  doorstopper, 189
  recommendations for choice, 195
  solid cells, 191
  triaxial strain cells, 189
    CSIR, 189
    CSIRO, 189
    LNEC, 189
    Luleå, 189
  for underground excavations, 457, 460
Soil behavior, 13
  mechanisms, **19**, 38
Soil Instruments Ltd.:
  Bishop piezometer tip, 126
  borehole extensometer, 6
  Cambridge drive-in piezometer, 132
  double fluid settlement gage, 291
  earth pressure cells, 166, 167, 176
  hydrostatic profile gauge, 288
  instruments manufactured by, 511
  inverted pendulum, 271
  liquid level gage, 282
  plumb line, 270
  pneumatic transducer, 90
  portable clinometer, 217
Soil pressure cells, *see* Earth pressure cells
Soil strain gage, 107, **236**, 293, 314, 324
Soil stress cells, *see* Earth pressure cells
Soiltest, Inc.:
  instruments manufactured by, 511
  invar tube convergence gage, 215
  Terzaghi water level meter, 280
Solar power, 112
Soldier piles and lagging, *see* Excavations, braced
Solexperts AG:
  deflectometer, 273
  instruments manufactured by, 511
  Piezodex piezometer, 139
  tiltmeter, 217
  Trivec, 223
Solinst Canada Ltd.:
  instruments manufactured by, 511
  Waterloo Multilevel System, 137
Sondex settlement system, 223
Sonic probe, 107
Sounding hammer, 157
Spare parts, **54**, 66, 116
Specialist work, 58
Specifications:
  for field services, *see* Contractual arrangements
  for procurement of instruments, *see* Procurement specifications
Spectra-Physics electronic level, 203
Sperry tilt sensing system, **218**, 511
Spiral-foot anchor, 250
Splices for electrical cable, 115
Spot welder, 311
State of the art advance, 35
Stepped blade, 167

Storage of instruments, 66
Strain gages, 306, 322
  applications, 297
  for braced excavations, 392, 395, 396, 397
  conversion to stress, **316**, 325
  definition, 297
  for drilled shafts, 484, 486, 487
  for driven piles, 468, 469, 471, 472, 473, 475
  electrical resistance, *see* Electrical resistance strain gages
  for embankment dams, 423
  initial readings, 316
  installation, 316, 325
  locations, 315
  mechanical, *see* Mechanical strain gages
  for underground excavations, 456, 457, 458, 460
  vibrating wire, *see* Vibrating wire strain gages
Strainsert Company cable tension meter, **303**, 511
Strain/stress relationships, **316**, 325
Strain in structural members, 297
Streaming potential, 431
Stress:
  in rock, 26
  in soil, 13
Stress change in rock, 185
  applications, 185
  measurement of, *see* Rigid inclusion gages; Soft inclusion gages
Stressmeters, *see* Rigid inclusion gages
Stress/strain relationships, **316**, 325
Strong motion accelerographs, 425
Structural Behavior Engineering Laboratories, 511
Subaudible rock noise, 295
Submittals:
  for field service specifications, 65
  for procurement specifications, 50
Subsurface settlement points, *see* Fixed borehole extensometers
Support work:
  biddable, 57
  incidental, 66
  nonbiddable, 58
  typical items, 67
Surface extensometers, 209
  advantages and limitations, 210
  applications, 209
  for braced excavations, 391, 394
  convergence gages, 213
  definition, 209
  electrical crack gages, **212**, 324
  for embankment dams, 422
  mechanical crack gages, 210
  precision, 210
  for slopes, 445, 446, 447
  for underground excavations, 455, 456, 457, 458, 459, 460, 461
Surface installation method, **360**, 434
Surveying methods, 199
  accuracy, 201
  advantages and limitations, 201
  applications, 199
  for braced excavations, 391, 392, 394, 398
  for drilled shafts, 484, 485, 486, 488
  for driven piles, 469, 470, 471, 472, 475, 479
  electronic distance measurement, 203
  for embankment dams, 420, 423, 430
  for embankments on soft ground, 408, 410
  global positioning system, 206
  laser beam, 203
  measuring points on surfaces, 207
  offsets from baseline, 202
  optical leveling, 202
  photogrammetric methods, 204
  reference datums, 206
    benchmarks, 206
    horizontal control stations, 207
  for slopes, 445, 446, 447
  taping, 202
  traverse lines, 202
  triangulation, 202
  trigonometric leveling, 204
  for underground excavations, 455, 456, 457, 458, 460, 461

Tamping hammer, 157
TAMS double fluid settlement device, 290
Tape extensometer, 213
Taping, 202
Task assignment, **39**, 45
Tektronix, Inc. TDR cable tester, **294**, 511
Telemac:
  Distofor extensometer, 247
  Distomatic convergence gage, 214
  Extensofor extensometer, 229
  instruments manufactured by, 511
  Piezofor piezometer, 138
  Radiofor extensometer, 247
  tiltmeter, 218
  vibrating wire piezometer, 6, **128**
Telephone lines, 113
Telltale:
  load cell, 298, **308**
  multiple, *see* Multiple telltales
  single, **292**, 308
Temperature measurement, 331
  applications, 331
  bimetal thermometer, 332
  frost gages, 335
  installation, 338
  mercury thermometer, 332
  recommendations for choice, 336
  resistance temperature device (RTD), 334
  for slopes, 448
  thermistor, 333
  thermocouple, 333
  vibrating wire gage, 336
Tensioned wire gages, 235
Terrascience Systems Ltd.:
  in-borehole data logger, 110
  instruments manufactured by, 511
  slack-wire extensometer, 244
Terra Technology Corporation:
  instruments manufactured by, 511
  pneumatic piezometer, 127
  pneumatic transducer, 90
Terzaghi water level meter, 276

Test section, 391, 392, 408, 456, 475, 488
Texas Measurements, Inc. strain gage, **324**, 511
Thermal coefficients of expansion, 332
Thermal monitoring of dams, 431
Thermistor, 333
Thermocouple, 333
Thermotic surveys, 431
Thor International, Inc.:
  earth pressure cell, 6
  instruments manufactured by, 511
  liquid level gage, 285
  multipoint pneumatic piezometer, 137
  piezometer seal system, 159
  pneumatic piezometer, 127
  pneumatic transducer, 90
Tieback anchors, *see* Excavations, braced
Tied back excavation, *see* Excavations, braced
Tiltmeters, 216
  with accelerometer transducer, 216
  applications, 216
  definition, 216
  for drilled shafts, 486
  for driven piles, 472
  earthtide, 218
  with electrolytic level transducer, 108, **218**
  with geodetic sensitivity, 218
  mechanical, 216
  for slopes, 445, 447
  with vibrating wire transducer, 217
Time domain reflectometry, 294
Toe stakes, 208
Topcon EDM equipment, 204
Total stress in soil, 165
  definition, 14
  measurement, *see* Earth pressure cells
Transducers, 79
  electrical, 92
  handling and transporting, 116
  hydraulic, 80
  mechanical, 79
  pneumatic, 87
Transient protection devices, 116
Transverse deformation gages, 268
  applications, 268
  borehole directional survey instruments, 274
  manufacturers, 511
    Pajari method, 274
    photographic method, 274
  for braced excavations, 394
  definition, 268
  deflectometers:
    extenso-, 273
    multiple, 272
    portable borehole, 273
  for driven piles, 476, 479
  for embankments on soft ground, 408
  fiber-optic sensors, 274
  inclinometers, *see* Inclinometers
  in-place inclinometer, 271
  inverted pendulum, 270
  plumb line, 270
  shear plane indicators, 268
    rupture stakes, 268
    shear probe, 268
    shear strip, 269
    slope extensometer, 269
  for slopes, 445
Traverse lines, 202
Triangulation, 202
Triaxial strain cells, *see* Soft inclusion gages
Trigonometric leveling, 204
Trilateration, 203
Trivec, 223
Tubing:
  diameter for hydraulic instruments, 84
  diameter for pneumatic instruments, 92
  material for hydraulic instruments, 84
  material for pneumatic instruments, 92
  routing for liquid-filled tubes, 86
Tubing fittings:
  for hydraulic instruments, 85
  for pneumatic instruments, 92
Tunnels, *see* Underground excavations

Uncertainty, 75
Underground excavations, 453
  applications, routine and special, 461
  behavior mechanisms, **22**, **27**, 453
  benefits of instrumenting, 453
  case histories, 461
  exploratory excavations, 455
  geotechnical questions, 455
  initial conditions, 455
  movement zone, 460
  New Austrian Tunneling Method, 459
  stability, 458
  structures nearby, 459
  test section, 456
Unsaturated soil, 13
USBM:
  borehole deformation gage, 188
  ultransonic convergence gage, 215
USBR:
  crossarm gage, 219
  piezometer tip, 124
  settlement measuring point, 209

Vane shear equipment, 4
Variable reluctance transducer, 102
Vibrating wire strain gages:
  embedment types, 320
    accuracy, 322
    advantages and limitations, 322
    conversion to stress, **316**, 325
    inclusion effect, 325
    initial readings, 325
    installation, 325
    length, 322
    locations, 325
    range, 322
    sensitivity, 322
    sister bar, 320
  surface mounted types, 310
    accuracy, 307

advantages and limitations, 307
arc welded type, 310
conversion to stress, 316
initial readings, 316
installation, 310, 311, **316**
length, **307**, 315
locations, 315
range, 307
sensitivity, 307
spot welded type, 311
zero drift, **104**, 319
Vibrating wire transducer, 102
advantages of frequency signal, 103
error, sources of, 103
external vibration, effect of, 105
frequency/strain relationships, 103
operating principle, 102
reading methods, 102
zero drift, 104
Virgin stress, 26
Voids, 13
Volclay sausages, 160
Volt–ohm–milliammeter, 116
Warranty, 52
Waterloo Multilevel System, 137
Water overflow pots, 280
Water power, 113
Weir gage, 421
Weldable strain gages, 94, 311, 312
Wellpoint piezometer, 130
Westbay Instruments Ltd.:
combined piezometer-inclinometer, **139**, 446
inclinometer casing coupling, 258
instruments manufactured by, 511
multiple piezometer, 138
Wheatstone bridge circuits, 94
Whittemore gage, 307
Wild Heerbrugg Ltd.:
instruments manufactured by, 511
theodolite, 203
Wind power, 113
Wire extensometer, 213
Wooden wedges, 210

Yoke gage, 188

Zeiss, Carl, 511

9 780471 005469